Lecture Notes on Data Engineering and Communications Technologies

Volume 81

Series Editor

Fatos Xhafa, Technical University of Catalonia, Barcelona, Spain

The aim of the book series is to present cutting edge engineering approaches to data technologies and communications. It will publish latest advances on the engineering task of building and deploying distributed, scalable and reliable data infrastructures and communication systems.

The series will have a prominent applied focus on data technologies and communications with aim to promote the bridging from fundamental research on data science and networking to data engineering and communications that lead to industry products, business knowledge and standardisation.

Indexed by SCOPUS, INSPEC, EI Compendex.

All books published in the series are submitted for consideration in Web of Science.

More information about this series at http://www.springer.com/series/15362

Jemal Abawajy · Zheng Xu ·
Mohammed Atiquzzaman ·
Xiaolu Zhang
Editors

2021 International Conference on Applications and Techniques in Cyber Intelligence

Applications and Techniques in Cyber Intelligence (ATCI 2021) Volume 2

 Springer

Editors
Jemal Abawajy
Faculty of Science, Engineering
and Built Environment
Deakin University
Geelong, VIC, Australia

Zheng Xu
School of Computer Engineering
and Sciences
Shanghai University
Shanghai, China

Mohammed Atiquzzaman
School of Computer Science
University of Oklahoma
Norman, OK, USA

Xiaolu Zhang
Department of Information Systems
and Cyber Security
University of Texas at San Antonio
San Antonio, TX, USA

ISSN 2367-4512   ISSN 2367-4520   (electronic)
Lecture Notes on Data Engineering and Communications Technologies
ISBN 978-3-030-79196-4   ISBN 978-3-030-79197-1   (eBook)
https://doi.org/10.1007/978-3-030-79197-1

This Springer imprint is published by the registered company Springer Nature Switzerland AG
The registered company address is: Gewerbestrasse 11, 6330 Cham, Switzerland

Foreword

The 2021 International Conference on Applications and Techniques in Cyber Intelligence (ATCI 2021), building on the previous successes in Fuyang China (2020), Huainan, China (2019), Shanghai, China (2018), Ningbo, China (2017), Guangzhou, China (2016), Dallas, USA (2015), Beijing, China (2014), and Sydney, Australia (2013), is proud to be in the 8th consecutive conference year. ATCI 2021 has moved online due to COVID-19.

The purpose of ATCI 2021 is to provide a forum for presentation and discussion of innovative theory, methodology and applied ideas, cutting-edge research results, and novel techniques, methods, and applications on all aspects of cyber and electronics security and intelligence. The conference establishes an international forum and aims to bring recent advances in the ever-expanding cybersecurity area including its fundamentals, algorithmic developments, and applications.

Each paper was reviewed by at least two independent experts. The conference would not have been a reality without the contributions of the authors. We sincerely thank all the authors for their valuable contributions. We would like to express our appreciation to all members of the program committee for their valuable efforts in the review process that helped us to guarantee the highest quality of the selected papers for the conference.

We would like to express our thanks to the strong support of the general chairs, publication chairs, organizing chairs, program committee members, and all volunteers.

Our special thanks are due also to the editors of Springer book series "Advances in Intelligent Systems and Computing," Thomas Ditzinger, Holger Schaepe, Beate Siek, and Gowrishankar Ayyasamy for their assistance throughout the publication process.

Jemal Abawajy
Zheng Xu
Mohammed Atiquzzaman
Xiaolu Zhang

Welcome Message

The 2021 International Conference on Applications and Techniques in Cyber Intelligence (ATCI 2021), building on the previous successes in Fuyang China (2020), Huainan, China (2019), Shanghai, China (2018), Ningbo, China (2017), Guangzhou, China (2016), Dallas, USA (2015), Beijing, China (2014), and Sydney, Australia (2013), is proud to be in the 8th consecutive conference year. ATCI 2021 has moved online due to COVID-19.

The purpose of ATCI 2021 is to provide a forum for presentation and discussion of innovative theory, methodology and applied ideas, cutting-edge research results, and novel techniques, methods, and applications on all aspects of cyber and electronics security and intelligence. The conference establishes an international forum and aims to bring recent advances in the ever-expanding cybersecurity area including its fundamentals, algorithmic developments, and applications.

Each paper was reviewed by at least two independent experts. The conference would not have been a reality without the contributions of the authors. We sincerely thank all the authors for their valuable contributions. We would like to express our appreciation to all members of the program committee for their valuable efforts in the review process that helped us to guarantee the highest quality of the selected papers for the conference.

We would like to express our thanks to the strong support of the general chairs, publication chairs, organizing chairs, program committee members, and all volunteers.

Our special thanks are due also to the editors of Springer book series "Advances in Intelligent Systems and Computing," Thomas Ditzinger, Holger Schaepe, Beate Siek, and Gowrishankar Ayyasamy for their assistance throughout the publication process.

Jemal Abawajy
Zheng Xu
Mohammed Atiquzzaman
Xiaolu Zhang

Organization

General Chairs

Hui Zhang   Tsinghua University, China
Liang Wang   Chinese Academy of Sciences, China

Online Conference Organizing Chairs

Bingkai Zhang   Fuyang Normal University, Director of Scientific
                              Research, China
Shibing Wang   Fuyang Normal University, Dean of School
                              of Computer Information Engineering, China

Program Chairs

Jemal Abawajy   Deakin University, Australia
Zheng Xu   Shanghai Dianji University, China
Mohammed Atiquzzaman   University of Oklahoma, USA
Xiaolu Zhang   The University of Texas at San Antonio, USA

Publication Chairs

Mazin Yousif   T-Systems International, USA
Vijayan Sugumaran   Oakland University, USA

Publicity Chairs

Kewei Sha   University of Houston, USA
Neil. Y. Yen   University of Aizu, Japan
Shunxiang Zhang   Anhui University of Science and Technology,
                              China

Website and Local Service Chairs

Xianchao Wang	Fuyang Normal University, China
Jia Zhao	Fuyang Normal University, China

Program Committee Members

William Bradley Glisson	Sam Houston State University, USA
George Grispos	University of Nebraska at Omaha, USA
V. Vijayakumar	VIT Chennai, India
Aniello Castiglione	Universit di Salerno, Italy
Florin Pop	University Politehnica of Bucharest, Romania
Neil Yen	University of Aizu, Japan
Xianchao Wang	Fuyang Normal University & Technology, China
Feng Wang	Fuyang Normal University & Technology, China
Jia Zhao	Fuyang Normal University & Technology, China
Xiuyou Wang	Fuyang Normal University & Technology, China
Gang Sun	Fuyang Normal University & Technology, China
Ya Wang	Fuyang Normal University & Technology, China
Bo Han	Fuyang Normal University& Technology, China
Xiuming Chen	Fuyang Normal University& Technology, China
Xiangfeng Luo	Shanghai University, China
Xiao Wei	Shainghai University, China
Huan Du	Shanghai University, China
Zhiguo Yan	Fudan University, China
Abdulbasit Darem	Northern Boarder University, Saudi Arabia
Hairulnizam Mahdin	Universiti Tun Hussein Onn, Malaysia
Anil Kumar K. M.	JSS Science & Technology University, Mysore, Karnataka, India
Haruna Chiroma	Abubakar Tafawa Balewa University Bauchi, Nigeria
Yong Ge	University of North Carolina at Charlotte, USA
Yi Liu	Tsinghua University, China
Foluso Ladeinde	SUNU, Korea
Kuien Liu	Pivotal Inc, USA
Feng Lu	Institute of Geographic Science and Natural Resources Research, Chinese Academy of Sciences, China
Ricardo J. Soares Magalhaes	University of Queensland, Australia
Alan Murray	Drexel University, USA
Yasuhide Okuyama	University of Kitakyushu, Japan
Wei Xu	Renmin University of China, China

ATCI 2021 Keynotes

Vijayan Sugumaran

Vijayan Sugumaran is Professor of Management Information Systems and Chair of the Department of Decision and Information Sciences at Oakland University, Rochester, Michigan, USA. He is also Co-Director of the Center for Data Science and Big Data Analytics at Oakland University. He received his Ph.D. in Information Technology from George Mason University, Fairfax, Virginia, USA. His research interests are in the areas of big data management and analytics, ontologies and semantic web, and intelligent agent and multi-agent systems. He has published over 200 peer-reviewed articles in journals, conferences, and books. He has edited twelve books and serves on the editorial board of eight journals. He has published in top-tier journals such as Information Systems Research, ACM Transactions on Database Systems, Communications of the ACM, IEEE Transactions on Big Data, IEEE Transactions on Engineering Management, IEEE Transactions on Education, and IEEE Software. He is Editor-in-Chief of the International Journal of Intelligent Information Technologies. He is Chair of the Intelligent Agent and Multi-Agent Systems Mini-track for Americas Conference on Information Systems (AMCIS 1999–2019). He has served as Program Chair for the 14th Workshop on E-Business (WeB2015), the International Conference on Applications of Natural Language to Information Systems (NLDB 2008,

NLDB 2013, NLDB 2016, and NLDB 2019), the 29th Australasian Conference on Information Systems (ACIS 2018), the 14th Annual Conference of Midwest Association for Information Systems, 2019, and the 5th IEEE International Conference on Big Data Service and Applications, 2019. He also regularly serves as Program Committee member for numerous national and international conferences.

Jemal Abawajy

Jemal Abawajy is Faculty Member at Deakin University and has published more than 100 articles in refereed journals and conferences as well as a number of technical reports. He is on the editorial board of several international journals and edited several international journals and conference proceedings. He has also been Member of the organizing committee for over 60 international conferences and workshops serving in various capacities including best paper award chair, general co-chair, publication chair, vice-chair, and program committee. He is actively involved in funded research in building secure, efficient, and reliable infrastructures for large-scale distributed systems. Toward this vision, he is working in several areas including pervasive and networked systems (mobile, wireless network, sensor networks, grid, cluster, and P2P), e-science and e-business technologies and applications, and performance analysis and evaluation.

Contents

Cyber Intelligence for CV Process and Data Mining

Cyber Intelligence for Health and Education Informatics

Short Paper Session

Cyber Intelligence for Industrial, IoT, and Smart City

Construction of User Evaluation System of Distribution Network Based on Big Data Technology

Yaheng Su[✉], Yan Li, Xiaopeng Li, and Chenxu Zhao

Inner Mongolia Electric Power Economic and Technological Research Institute, Baotou City 010090, Inner Mongolia Autonomous Region, China

Abstract. A variety of information systems have been widely used in the distribution network, and the intelligent and information level of distribution network has also been improved. This paper constructs a comprehensive index system. The highest layer is the total target layer. When calculating the weight of sub objective and criterion layer, the relative importance value of each element is evaluated according to the scale table, and the fuzzy evaluation matrix is formed. Then the weights are obtained according to the above method. In the evaluation, the damage of each element to the system is considered. The data shows that the average response time is 460 ms when the concurrency number is 10 and 1012 ms when the concurrency number is 100.

Keywords: Big data technology · Distribution network · User evaluation system · Parallel computing

1 Introduction

In order to change the status quo of uncoordinated development of power system construction and lag of distribution network construction, and narrow the gap between China and developed countries in terms of power quality, power consumption and reliability, the state has increased the construction and transformation of distribution network.

It is also necessary to investigate the distribution network structure, equipment status and equipment load, and combine the power grid data with the overall industrial layout, population distribution and development hotspots of the city [1, 2]. With the continuous development of power grid research, new indicators will be proposed one after another, which requires that the index system has certain expansibility [3, 4]. The application of electric energy has been integrated with our life, and is closely related to our life [5, 6]. Once the power grid power supply problems occur, it will bring inconvenience to our lives, cause adverse impact on China's economic development, and even pose a certain threat to people's life safety [7, 8]. Accurate grid load forecasting can improve the efficiency of investment, energy utilization and environment, realize the optimal configuration of grid variable capacitance, which is conducive to the management of medium voltage distribution network target grid [9, 10].

J. Abawajy et al. (Eds.): ATCI 2021, LNDECT 81, pp. 3–9, 2021.
https://doi.org/10.1007/978-3-030-79197-1_1

With the establishment of operation and the opening of business distribution data, the electricity consumption data of power customers can be associated with customer files, payment and other data.

2 Big Data and Distribution Network

2.1 Big Data Technology

Unstructured data can be distributed file system, and semi-structured data can be stored in distributed database. In the process of MPI programming, users need to consider many problems, including data storage, partition, distribution, result collection and fault-tolerant processing, which makes the parallel programming calculation less automatic and programming too complex. Compared with the traditional electricity meter, it is more accurate and diversified. However, with the diversity of smart meters in the market, the data structure is not unified. With the promotion of lean management and control of power grid technical renovation and overhaul, lean management of distribution network automation, power quality monitoring and analysis and other business lean management, the massive data involved in power quality monitoring and analysis cannot be guaranteed to complete the power quality analysis within the specified time by the existing computing resources.

2.2 Distribution Network

Power distribution equipment is mainly for power customers, which needs to change frequently according to their electricity demand. The new power consumption of customers needs to increase the distribution equipment, and the power distribution equipment needs to be reduced or cancelled if the power consumption demand of power customers is reduced. At the same time, due to equipment technical transformation and overhaul and other reasons, the distribution equipment information is changing at any time, so compared with the basic stable transmission and transformation equipment, the distribution equipment changes quite frequently. Its expression is as follows:

$$R = \sum_{i=1}^{n} \omega_i Val_i \tag{1}$$

In the formula, R is the evaluation result. It is expressed as a percentage of the actual power supply time and the total time during a certain statistical period. The data used in the index calculation is the annual statistical value, which is calculated by:

$$Ind1 = \frac{\sum_{i=1}^{N_{Ld}} N_{usi} \times 8760 - \sum_{i=1}^{N_{Ld}} t_{tdi} N_{usi}}{\sum_{i=1}^{N_{Ld}} N_{usi} \times 8760} \times 100\% \tag{2}$$

According to the feature quantity A, the sample D is divided into v different categories $D_j(j = 1, 2, ..., v)$, and the corresponding expected information is calculated by the following formula:

$$Info_A(D) = \sum_{j=1}^{v} \frac{|D_j|}{|D|} \cdot Info(D_j) \tag{3}$$

In the formula, $|D|$ is the total number of samples.

3 Simulation Experiment of User Evaluation System of Distribution Network

3.1 Experimental Environment

This paper builds a Linux Cluster with three nodes on VM ware work station. The configuration of host and virtual machine, software version number and cluster network configuration are shown in Table 1.

Table 1. Environment configuration

Configuration	Description
Operating system	Windows10
CPU	Inter i7-8550U
RAM	16 GB
Spark	2.2.0
Hadoop	2.6.0-cdh5.7.0
JDK	1.8.0_144

3.2 Construction of the Evaluation System

This paper constructs a comprehensive index system, and puts forward a reasonable and feasible comprehensive index system. The highest layer is the total target layer, which is used to represent the ultimate goal to be achieved in the evaluation process of distribution network. For each index in the index system, the method of expert survey is adopted, and the evaluation of the index is made by several experts.

4 Discussion

4.1 Model Simulation Results

Power grid construction projects are not independent of each other, and the construction effect of each project is not unique. Some projects improve the service quality of the power grid, thus contributing to a number of basic indicators of customer satisfaction evaluation system. We randomly allocate the proportion of the number of feeder nodes in three groups, and compare the computing time of map function in serial mode, Hadoop in load balanced parallel mode and Hadoop in load unbalanced parallel mode. The calculation time comparison of map function is shown in Table 2. We find that the optimal power flow calculation time of large-scale distribution network in Hadoop environment is obviously better than that of serial calculation. In the case of load balancing, the computing time of Hadoop cluster is greatly shortened, which is obviously better than that of unbalanced load. The experiment verifies the effectiveness of the performance model and load balancing algorithm, and achieves a certain speedup ratio. The city's distribution network still needs to be further improved in terms of the average outage time, so as to effectively guarantee the reliability and quality of power supply. In terms of economy, the loss rate of medium and high voltage lines are significantly reduced, and the power supply per unit asset is increased, which indicates the feasibility of the optimization plan and the investment effect of the planning scheme.

Table 2. Comparison of map function calculation time

Mode	Serial	Unbalanced load	Load balancing
1	1160 s	360 s	210 s
2	1013 s	318 s	179 s
3	1044 s	339 s	183 s

4.2 Algorithm Performance Analysis

The number of iterations and prediction errors vary with the number of neurons in the hidden layer, as shown in Fig. 1. In short-term load forecasting, Elman neural network is better than BP neural network algorithm in mrmse and mmape, which makes the two indexes decrease. The mrmse index is 16.9% lower than BP neural network, and mmape index is 19.0% lower than BP neural network. Because Web service and distributed cluster are separated in the design of the system, the response time is 460 ms when the number of concurrency is 10 and 1012 ms when the number of concurrency is 100. At the same time, in 100 independent repeated experiments, the number of times that cloud algorithm obtains the optimal solution is 97 times, while that of traditional algorithm is 95 times, which shows that the accuracy of cloud algorithm is higher. From this analysis, it can be concluded that the cloud algorithm has a high computational efficiency and accuracy. To sum up, the free entropy index can better characterize the scene characteristics, and combine with clustering analysis to achieve scene matching quickly and effectively. More importantly, the reactive power optimization method based on big data free entropy and scene matching is no longer dependent on the model and parameters of the system. It can quickly get the optimization strategy through online decision-making, which can effectively reduce the active power loss and node voltage offset of the distribution network. According to the different grid characteristics, the load release curve can be checked according to different grid types.

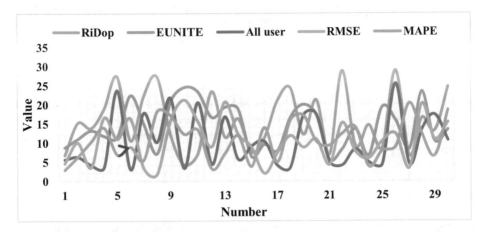

Fig. 1. The number of iterations and prediction errors vary with the number of hidden layer neurons

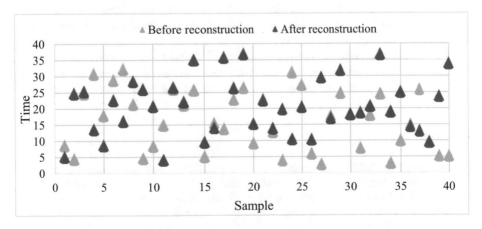

Fig. 2. Node voltage comparison before and after reconstruction

The comparison of node voltage before and after reconstruction is shown in Fig. 2. The node voltage has generally been improved to a certain extent, especially for 10 and 11 nodes, which fully illustrates the effectiveness of the results. The main reason is that this paper uses a fast radial power flow calculation method in power flow calculation, which can converge quickly and has less iterations. On average, it can converge about 9 times, and there will be no non convergence, which greatly shortens the execution time of the algorithm. It can be seen from the figure that the node voltage has generally been improved to a certain extent, especially for 10 and 11 nodes. After the reconstruction, the minimum voltage still appears at node 12, and the minimum voltage is 0.9715p.u, which is 0.23% higher than that before the structure, which fully illustrates the effectiveness of the results. The algorithm process does not need iteration, and the running time is short, and only one optimal solution is finally generated. The mechanism of Elman model is that the threshold and weight of each layer are updated by error back-propagation algorithm, so the training time is long and local minimum value is easy to be generated. It can be concluded from the graph that compared with the traditional PSO, HS can solve the objective function more accurately, and its convergence speed is faster. It can search the optimal value of operating cost in less evolution times.

5 Conclusions

Big data is the trend of the current information society, but also the inevitable development trend. This paper provides data storage means for distribution network diagnosis by integrating time series database.

This paper constructs a big data platform based on Hadoop, uses zookeeper to ensure the coordination and consistency of data resources, uses HBase database to store massive distribution network data, and finally constructs an efficient, highly reliable and fault-tolerant big data platform.

This paper analyzes the current situation of distribution network operation and maintenance data. For residential areas with small installed capacity, the remaining intervals can be redistributed to achieve reasonable utilization of power grid resources.

References

1. Potyralla, M.: Geostatistical methods in water distribution network design - a case study. Ecol. Chem. Eng. S = Chemia i Inżynieria Ekologiczna. S 26(1), 101–118 (2019)
2. Ghasemi, S.: Balanced and unbalanced distribution networks reconfiguration considering reliability indices. Ain Shams Eng. J. 9(4), 1567–1579 (2018)
3. Pimm, A.J., Cockerill, T.T., Taylor, P.G.: The potential for peak shaving on low voltage distribution networks using electricity storage. J. Energy Storage 16(4), 231–242 (2018)
4. Zhang, L., Tang, W., Liang, J., et al.: Coordinated day-ahead reactive power dispatch in distribution network based on real power forecast errors. IEEE Trans. Pow. Syst. 31(3), 2472–2480 (2019)
5. Yang, C., Zhang, H., Su, J.: Quantum key distribution network: optimal secret-key-aware routing method for trust relaying. China Commun. 15(2), 33–45 (2018)

6. Hu, S., Zhou, H., Cong, L., et al.: Reliability analysis of distribution network with power electronic substation based on fault tree. Dianli Xitong Baohu yu Kongzhi/Pow. Syst. Protect. Control **46**(21), 25–31 (2018)
7. Kong, X., Lü, X., Dong, X., et al. Reliability analysis method of distribution network based on load classification evaluation. Hunan Daxue Xuebao/J. Hunan Univ. Nat. Sci. **45**(4), 104–111 (2018)
8. Weng, J., Liu, D., Wang, Y., et al.: Research on fault simulation of active distribution network based on cyber physical combination. Zhongguo Dianji Gongcheng Xuebao/Proc. Chin. Soc. Electr. Eng. **38**(2), 497–504 (2018)
9. Wang, X., Wang, C., Xu, T., et al.: Optimal voltage regulation for distribution networks with multi-microgrids. Appl. Energy **210**(1), 1027–1036 (2018)
10. Wang, L., Sharkh, S., Chipperfield, A.: Optimal decentralized coordination of electric vehicles and renewable generators in a distribution network using a search. Int. J. Electr. Pow. Energy Syst. **98**(6), 474–487 (2018)

Distribution Network Load Forecasting Model Based on Big Data

Xiaopeng Li$^{(\boxtimes)}$, Xiaolei Cheng, Peng Wang, and Yuan Wang

Inner Mongolia Electric Power Economic and Technological Research Institute,
Baotou City 010090, Inner Mongolia Autonomous Region, China

Abstract. Power load forecasting is an important work of power sector, which is very important for making scientific and reasonable power grid planning scheme. The concept of grid planning of distribution network puts forward new requirements for power load forecasting. In this paper, under the background of grid planning of distribution network, according to the characteristics of power load in the existing power grid, the spatial load forecasting technology in the area with regulatory planning and without planning is studied respectively, and the load forecasting methods of three levels of equipment capable of providing electricity, power supply grid and area where the electricity is supplied are given from bottom to top. The results show that the error between users and residents is better than 5% in the experiment, and the error between users and residents is not more than 5%.

Keywords: Distribution network load · Load forecasting · Model design · Forecasting method

1 Introduction

The traditional distribution network planning takes the whole planning area as the research object, and the construction projects of lower level power grid need to meet the planning and construction of higher level power grid. This traditional distribution network planning method gradually exposed its disadvantages in the process of power grid construction: the power supply scope of high-voltage substation is not clear, the load rate distribution of main transformer is uneven; the medium voltage distribution line is circuitous and complex, and there is cross power supply phenomenon; the low-voltage outgoing lines are scattered and disordered, and the power supply reliability is low, and the operation and maintenance management of distribution network is also more difficult. In recent years, the "grid" planning concept of distribution network has gradually taken shape, and has been successfully promoted in many areas of the country, which has improved the precision of distribution network planning, and laid the foundation for building a strong and reliable target grid [1].

After the initial establishment of the electricity sales market, the electricity market transactions mainly include three ways: the first is to sign a contract between the power sales company and the power plant to agree on the power consumption time and demand amplitude; the second is that both the power selling company and the power plant quote on their own in the power market, and the power market manager decides

J. Abawajy et al. (Eds.): ATCI 2021, LNDECT 81, pp. 10–18, 2021.
https://doi.org/10.1007/978-3-030-79197-1_2

the buyer and the seller according to the relationship between supply and demand; the third is to comprehensively analyze the real-time power sales situation of the power sales company The power market manager issues an order to the power plant to increase or reduce the issue at any time. Power load forecasting is the premise and foundation of making scientific and reasonable distribution network planning. The prediction results can be used to guide the construction scale and time sequence of power equipment such as high-voltage substation, medium voltage line and switching station in the future. Its accuracy will determine the operability and applicability. In the distribution network planning, load forecasting takes the whole planning area as the object, and forecasts the total load of the whole planning area [2]. Under the background of grid, the process of load forecasting is changed from the load forecasting of the whole planning area to the load forecasting of three levels of equipment capable of providing electricity, power supply grid and area where the electricity is supplied from bottom to top. In the process of transformation, some new problems appear, such as how to determine the load density index and the simultaneity rate between equipment capable of providing electricitys in the transition year, and how to determine each power supply How to calculate the land area of different land use types in the unit conveniently and quickly, and how to forecast the load of rural areas more accurately. How to carry out accurate spatial load forecasting under the background of grid planning of distribution network needs further analysis and research [3].

According to the requirements of grid planning of distribution network, based on the division of area where the electricity is supplied, power grid and equipment capable of providing electricity, according to the characteristics of power load in the existing power grid, this paper studies the spatial load forecasting technology in the areas with and without regulatory planning. In the area with regulatory planning, this paper mainly puts forward the determination method of load density index and simultaneity rate, as well as the realization method of computer program for calculating the nature area of each land use of equipment capable of providing electricity; in the area without regulatory planning, the construction of grey prediction model and the optimization of prediction accuracy are mainly studied. These studies can not only enrich the means of power load forecasting in the grid planning of distribution network, but also improve the quality and effect of load forecasting, and lay a solid foundation for determining the planning stage of each voltage level construction scheme in the grid planning of distribution network.

2 Related Concepts and Definitions

2.1 Three Level Network

The traditional distribution network planning mode is to take the high voltage substation as the center, and the medium voltage line extends to all sides. This "radial" distribution network structure has the following disadvantages [4]:

(1) The power supply scope of high voltage substation is not clear, the distribution of capacity load ratio of main transformer is uneven, there are heavy overload of main transformer, and there are some cases of light load of main transformer.

(2) The medium voltage line is too long and there is cross power supply.
(3) Low voltage outgoing lines are scattered and disordered, and the line loss rate is high. The power supply scope of substation is clearer, the distribution network planning is more refined, and the operation and maintenance management is more convenient.

The area where the electricity is supplied is divided into power supply levels according to the administrative level of the region or the load density of the planning year, considering the regional power consumption level, economic development and other factors. The area where the electricity is supplied is generally divided into five categories: a+, a, B, C and D. For a certain area, the type of area where the electricity is supplied with regulatory planning is one level higher than that of the area without regulatory planning. For example, the type of area where the electricity is supplied in the area with regulatory planning is B and the type of area where the electricity is supplied without regulatory planning is C [5].

The equipment capable of providing electricity is a number of relatively independent units which are divided into several relatively independent units on the basis of power supply grid division and combined with land use nature, function orientation, load density and load characteristics. For rural areas, in order to facilitate the management of the power system, equipment capable of providing electricitys are generally divided into towns, and a township is an equipment capable of providing electricity, which is managed by the township power supply station [6].

In the grid planning of distribution network, three levels of area where the electricity is supplied, power supply grid and equipment capable of providing electricity correspond to different planning levels. In the area where the electricity is supplied level, the superior power supply planning is considered, and the 110 kV and 35 kV high-voltage grid structure planning is focused to build a strong high-voltage grid; in the power supply grid level, the high-voltage substation layout planning and medium voltage target grid planning are carried out. According to the load of the power supply grid, the required substation capacity and number are determined, and the power supply grid is determined from the global optimization In principle, the substation in saturation year shall not cross the grid, that is, the same substation shall not conduct medium voltage outgoing lines to two or more power supply grids at the same time in saturation year; at the level of equipment capable of providing electricity, the 10 kV target grid structure shall be focused on, and the specific medium voltage lines, distribution transformers, distribution transformers, etc. shall be determined According to the planning and construction scheme of distribution facilities such as ring main unit, that is, the same medium voltage line in saturation year shall not supply power to two or more equipment capable of providing electricitys at the same time.

2.2 Distribution Network Load Forecasting

The main purpose of spatial load forecasting is to obtain the spatial distribution information of power load in the planning area, so as to put forward the distribution transformer demand of each plot supply unit and the scale demand of medium voltage feeder. Spatial load forecasting in the grid planning of distribution network is to forecast the load of equipment capable of providing electricity from bottom to top under the background of three-level network of "area where the electricity is supplied, power supply grid and equipment capable of providing electricity". Then, according to the load forecast results of equipment capable of providing electricity, the load scale of power supply grid level and area where the electricity is supplied level will be collected in turn. For a certain area, the area where the electricity is supplied is generally divided according to the regulatory planning and non-regulatory planning. According to the regional characteristics of the two regions, different spatial load forecasting methods are used to forecast. At present, in the practical engineering application, the load density method is generally used in the saturated year load forecasting of the area with regulatory planning, the large user + natural growth rate method is generally used in the transition year, the average household distribution variable capacity method is generally used in the saturated year load forecasting of the uncontrolled area, and the large user + natural growth rate method is generally used in the transition year. The spatial load forecasting process in the grid planning of distribution network. The load power supply grid and area where the electricity is supplied are forecasted from bottom to top in the area with and without regulatory planning, and the spatial load forecasting results of the whole planning area are finally collected [7].

2.3 Load Forecasting Method

Load density method is a method of planning and forecasting load according to the nature, which is widely used in the load forecasting in the area with regulatory planning. Under the background of grid planning of distribution network, the load forecasting process of load density method from bottom to top is as follows: firstly, the load scale of the plot is calculated by the land area and load density index, and then the load scale of all levels of spatial partition is collected through the simultaneity rate. The load forecast of area where the electricity is supplied is directly accumulated by the load forecast of power supply grid, because the area of power supply grid is large. The simultaneity rate is very close to 1, and it is not considered when accumulating [8].

(1) Load forecasting of equipment capable of providing electricity
   The load of the equipment capable of providing electricity is obtained by the product of the land use area and the corresponding load density index, and considering the simultaneity rate in the equipment capable of providing electricity. Its load is

$$P = S * R * \rho \tag{1}$$

For the equipment capable of providing electricity, its load is

$$P_{DY} = t_1 * \left(\sum_{i=1}^{m} (S_i * R_i * \rho_i) + \sum_{j=1}^{n} (S_j * \sigma_j) / 10000 \right) \tag{2}$$

(2) Load forecasting of power supply network
   The load of power supply grid is the result of load forecasting of equipment capable of providing electricity, and the superposition of simultaneity rate is considered

$$P_{WG} = t_2 * \sum_{i=1}^{n} P_{DYi} \tag{3}$$

(3) Load forecasting of area where the electricity is supplied

The load in the area where the electricity is supplied is the accumulation of the load forecast results of the power supply grid. Since the power supply grid is a comprehensive area with a large area, the simultaneity rate between them is very close to 1. Generally, the simultaneity rate is not considered in the accumulation.

$$P_{QY} = \sum_{i=1}^{k} P_{WGi} \tag{4}$$

2.4 Limitations of Traditional Load Forecasting

The area with regulatory planning generally refers to the urban area and the industrial agglomeration area located in the rural area. For this kind of useful land planning area, the traditional load forecasting method generally adopts the load density method to forecast the saturated year, and the large user + natural growth rate method to forecast the load in the transition year. Under the background of grid planning of distribution network, spatial load forecasting adopts the "bottom-up" method to forecast the load, power supply grid and area where the electricity is supplied in turn [9, 10]. Compared with the traditional load forecasting method, this bottom-up spatial load forecasting method has some new problems.

(1) Since the equipment capable of providing electricity is a man-made area, which is inconsistent with the management scope of power system such as power supply station and substation, the historical data and current data of equipment capable of providing electricity need to be re summarized. The status data of equipment capable of providing electricity can be determined by distribution transformer account and single line diagram, and then the status data can be summarized. The process of statistical summary is more complex, but the historical information of distribution and transformation in equipment capable of providing electricity is more difficult to be summarized. Therefore, it is not suitable to use the method of large user + natural growth rate to predict the load of equipment capable of providing electricity in transition year;

(2) If the load density method is used for spatial load forecasting, it is difficult to determine the load density index, especially the transition year load density index in the process of equipment capable of providing electricity load forecasting, and the selection of load density index will directly determine the accuracy of load forecasting;

(3) For the load density method under the background of grid planning of distribution network, the area of different land use nature in each equipment capable of providing electricity needs to be counted for load forecasting of equipment capable of providing electricity. If manual measurement is adopted, it will consume a lot of time and energy;

(4) In the past, in the process of load forecasting, the simultaneity rate of the whole region only needs to be considered. However, under the background of grid planning of distribution network, the whole area is artificially divided into three levels: power supply grid and equipment capable of providing electricity the simultaneity rate between the equipment capable of providing electricitys in the grid is used to calculate the load value of the power grid.

3 Design of Distribution Network Load Forecasting Model Based on Big Data.

3.1 Incremental Network

How to determine whether to delete hidden nodes, this paper studies and analyzes three nodes: the weight of hidden node, the contribution of output of hidden node to output layer, and error reduction rate. In addition, considering the time series data, the closer the segment to the current time, the greater the impact, and the more reflect the latest time series information, the greater the value. Therefore, we can add time impact factor for the weight update.

3.2 Based on Power Data

(1) The load data and meteorological data are fused according to the unified format, and the null, zero data are deleted.
(2) Normalizing the input load data, It can avoid numerical problems and make the neural network converge quickly.
(3) According to the input value and output result after data processing, network is trained in batch.
(4) Input the new data and use the method network.

4 Example Analysis

4.1 Commercial User Experiment

Table 1. Business user experiment group information

Commercial users	Training methods	Time slot
A	Batch learning	1 January–31 January
B	Batch learning	1 January–14 February
C	Incremental learning	1 January–14 February

Table 1 shows the training mode and data set information of commercial user experimental groups.

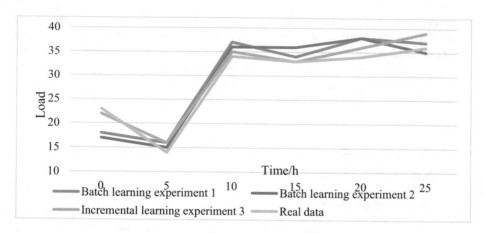

Fig. 1. Chart of forecast results for commercial users

Figure 1 shows the actual and predicted load values for business users as of February 15. It can be seen from the figure that the network prediction effect of incremental learning method training is better than that of batch learning mode.

4.2   Residential User Experiment

For residents: Resident batch learning experiment 1, using load data from January 1 to January 31 for training; batch learning experiment 2, using load data from January 1 to February 14; incremental learning experiment 3, Use load forecasting model based on incremental network, read data in real time for training. Table 2 shows the training method and data set information of the resident experimental group.

Table 2. Resident user experimental grouping information

Resident users	Training methods	Time slot
A	Batch learning	1 January–31 January
B	Batch learning	1 January–14 February
C	Incremental learning	1 January–14 February

Figure 2 shows the actual load value and predicted value curve of residential users on February 15. It can be seen from the figure that due to the low load of residents during the period 7–13, the prediction effects of the two batch learning methods are not ideal, but they are greatly affected by other periods. In contrast, incremental learning can reflect the latest time series, and its prediction effect is better than batch learning.

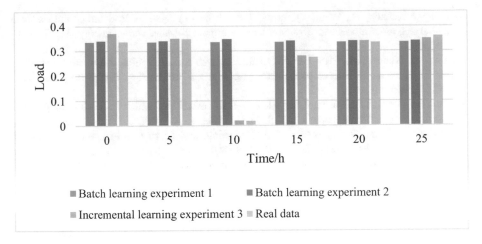

Fig. 2. Resident user forecast

5 Conclusion

Power load forecasting is an important task of the power sector, and it is of great significance for formulating scientific and reasonable power grid planning schemes. In this context, the original load forecasting method can no longer meet the needs of distribution network planning, and it needs to be enriched and improved. In order to improve the accuracy of distribution network spatial load forecasting and the adaptability of spatial load forecasting to power grid planning, this paper studies spatial load forecasting in controlled planning areas and unregulated planning areas.

References

1. Dong, J.-R., Zheng, C.-Y., Kan, G.-Y., Zhao, M., Wen, J., Yu, J.: Applying the ensemble artificial neural network-based hybrid data-driven model to daily total load forecasting. Neural Comput. Appl. **26**(3), 603–611 (2014). https://doi.org/10.1007/s00521-014-1727-5
2. Xiao, A.L., Shao, A.W., Wang, B.C., et al.: Research and application of a hybrid model based on multi-objective optimization for electrical load forecasting - ScienceDirect. App. Energy 180(C), 213–233 (2016)
3. Dongxiao, N., Shuyu, D.: A short-term load forecasting model with a modified particle swarm optimization algorithm and least squares support vector machine based on the denoising method of empirical mode decomposition and grey relational analysis. Energies **10**(3), 408–409 (2017)
4. Fan, G.F., Peng, L.L., Hong, W.C.: Short term load forecasting based on phase space reconstruction algorithm and bi-square kernel regression model. Appl. Energy **224**(Aug.15), 13–33 (2018)
5. Gao, X., Li, X., Zhao, B., et al.: Short-term electricity load forecasting model based on EMD-GRU with feature selection. Energies **12**(6), 1140–1141 (2019)

6. Gong, M., Zhou, H., Wang, Q., et al.: District heating systems load forecasting: a deep neural networks model based on similar day approach. Adv. Build. Energy Res. **2**, 1–17 (2019)
7. Yuqi, D., Xuejiao, M., Chenchen, M., et al.: Research and application of a hybrid forecasting model based on data decomposition for electrical load forecasting. Energies **9**(12), 1050–1052 (2016)
8. Cui, H., Peng, X.: Short-term city electric load forecasting with considering temperature effects: an improved ARIMAX model. Math. Probl. Eng. **2015**(PT.12), 1–10 (2015)
9. Zhang, Z., Gong, W.: Short-term load forecasting model based on quantum elman neural networks. Math. Probl. Eng. **2016**(pt.9), 1–8 (2016)
10. Tarsitano, A., Amerise, I.L.: Short-term load forecasting using a two-stage sarimax model. Energy **133**(Aug.15), 108–114 (2017)

An Exploration on Sanya City Image Communication in the Era of Big Data

Yongfei Xie[✉]

School of Humanities and Communication, Sanya University,
Sanya, Hainan, China
yongfeixie@sanyau.edu.cn

Abstract. The advent of globalization has intensified the competition between cities, and the spread of city images has also received people's attention. With the development of technologies such as the Internet, the Internet of Things, and the Internet of Vehicles, as well as the improvement of data mining and analysis technologies, mankind has entered the era of big data. Big data has played an important role in accurately insight into the needs of the audience, efficient integration of communication methods, and real-time guidance of public opinion. It can be said that big data has reversed the traditional urban image communication relying on administrative power and made up for the limitations of urban image communication. This article uses literature method, logical analysis method and other methods to explore the importance of Sanya city image communication under big data, analyze the problems of Sanya city image communication under big data, and finally analyze how to use big data to effectively spread Sanya city image.

Keywords: Big data Era · Sanya city image · Communication

1 Introduction

With the advent of economic globalization, various countries and regions have to show themselves through the dissemination of city images in order to gain the recognition of audiences from all over the world and thus grasp the initiative of urban development. With the popularization of big data, a series of changes have taken place in the urban image communication ecology [1]. As an important city in the construction of Hainan Free Trade Port, using big data to spread the city image of Sanya can create a new city landmark in Sanya, establish a good city brand image, promote the development of the city's economy, and scientifically and effectively solve the crisis of online public opinion [2].

At present, many scholars at home and abroad have conducted research on the impact of big data on the development of smart cities and the role of social media on the development of urban image, and discussed the impact of big data on urban development. However, most domestic scholars currently focus on Research is carried out from the perspective of urban intelligence brought about by big data, while the research on city image communication is carried out from other media forms and media

J. Abawajy et al. (Eds.): ATCI 2021, LNDECT 81, pp. 19–25, 2021.
https://doi.org/10.1007/978-3-030-79197-1_3

symbols. There are still few literatures on the impact of big data on city image communication.

This study combines domestic and foreign research results to explore how Sanya, a famous tourist city in China, can better spread the city's image in the new era with the background of big data. Through research, it can provide reference for the image dissemination of similar cities, and at the same time provide suggestions for the construction of Hainan Free Trade Port.

2   The Dilemma of Sanya's Urban Image Communication in the Era of Big Data

2.1   Lack of the Idea of Using Big Data to Spread the Image of the City

Since McKinsey first formally put forward the view that the era of big data has arrived in 2011, big data has become the key to improving competitiveness and productivity. Big data has not only changed people's daily life and working methods, but also caused a series of new changes in the communication ecology of the city image [3]. The advanced communication technologies and means brought by big data not only enrich the content and form of urban image communication, but also open up a new way to improve the communication effect through the integrated use of cross-media, interactive, experiential and intelligent. Although the Sanya Big Data Center was put into use in Sanya Yungang Park in 2017, it provides important support for Sanya to use big data to spread the city's image. However, the main body of city image communication at all levels has weak awareness of big data and lack of knowledge about big data. Therefore, traditional image communication methods are still used, and big data is not fully used to spread the city image, which seriously restricts cities in the information age. The pace of image spread [4].

2.2   The Brand Awareness is Not Strong, and the Urban Image Communication Lacks Personality

The dissemination of the city's image must be based on precise positioning, focusing on themes, and consistent communication inside and outside, so as to strengthen the city's brand image in the eyes of the public. But at present, in the long-term external publicity of Sanya's city image, due to the needs of big data dissemination, the principle of audience orientation has been complied with, too much attention has been paid to catering to the needs and psychological expectations of consumption and individualization, so "Oriental Hawaii" is played. Many labels, such as "International Coastal Tourist City", have not been able to form a brand image with a concentrated theme in the eyes of the public, thus greatly weakening the impression in the eyes of consumers [5]. Only under the premise of taking into account the needs of the audience and creating a city image that can give consumers a visual impact and soul shock based on their own historical background, resource conditions and characteristics, can a city brand be truly recognized by consumers.

2.3   Weak Foundation for Big Data Application

Although big data has become an important strategic resource for city image dissemination, at present, big data in Sanya is in the initial stage of development. There are still very few real big data companies, there is a lack of leading companies, and no industry standards have been established. The formation of a big data industry chain and the formation of an effective and reasonable development model have caused certain obstacles to the spread of big data in Sanya's urban image. In addition, Internet tools for collecting big data information need to be built and improved. Take tourism companies as an example. Although many tourism companies have begun to pay attention to the construction of WIFI networks, client development, WeChat, Weibo and other network tools with the help of the government. However, in many scenic spots, hotels or travel agencies, the use and mining of big data has not yet started. The weak application basis of big data will lead to the comprehensiveness and scientific nature of the information collected, and ultimately affect the effect of city image dissemination [6].

2.4   There is a Shortage of Relevant Talents for Big Data to Spread the City Image

In the process of using big data to spread the city image in Sanya, there are often people who understand city image communication but do not understand professional analysis of data, or people who understand data but do not understand city image communication, who understand both city image communication and data analysis Comprehensive talents are very scarce. This has caused a disconnect between city image communication and data analysis, and hindered the effectiveness of Sanya city image communication. Due to the lack of relevant talents, it is difficult for Sanya to have a comprehensive understanding of big data when using big data to disseminate the image of the city, which will eventually lead to blind dissemination of the city's image and ultimately cause a certain waste of resources.

3   The Shaping Strategy of Sanya City Image Communication in the Era of Big Data

3.1   Establish a Big Data Concept for City Image Communication

The idea is the forerunner of action. In the era of big data, if you want to better spread the image of the city, you must first change your thinking. Only when you have the communication thinking of big data and build a new cognitive system for city image communication, can you better play. The role of big data in the spread of city image. Today, when big data is all-pervasive, some people still hold prejudices against big data, either they have vague and ambiguity about big data, or they think big data is fudge and formalism, which greatly hinders the pace of big data spreading the image of cities. Therefore, the main body of urban communication should establish the concept of big data dissemination of city image, and pay attention to the use of big data in all aspects of city image dissemination. The main bodies must coordinate and cooperate to form a joint force to give full play to the enthusiasm of each subject. Come out and focus on spreading the image of a data city.

3.2 Reshaping a New Sanya City Image Communication System

3.2.1 Strategy of the Main Body of Communication: Give Full Play to the Advantages of Big Data and Optimize the Main Structure of Big Data Dissemination

In the dissemination of Sanya city image, government departments, media agencies, non-governmental organizations and individuals are the main body of city image dissemination. In the era of big data, each subject has its own dissemination status and manifestation. Big data must be used to optimize each dissemination body, strengthen the methods and channels of each subject in Sanya city image communication, and effectively enhance the effect of Sanya city image communication. Through the integration of the responsibilities of various government departments, the formation of a joint force for image dissemination, optimizing the structure of various media, and guiding non-official organizations and individuals to become conscious people in the dissemination of Sanya's city image, so that they can become rational disseminators of Sanya's city image.

3.2.2 Communication Content Strategy: Optimize the Content of City Image Communication

In the era of big data, high-quality content is the core of city image communication, the premise and guarantee for improving the quality of city image communication, and the key to attracting audiences to city communication. Therefore, in the era of big data, it is extremely urgent to optimize the content of city image dissemination. As for the communication of Sanya's urban image, the content of communication should not just stop at the sun, beach, and air quality. Big data should be used to integrate historical culture, ocean culture, island culture, folk culture, and red culture, and tap the citizens from it. Stories and humanistic allusions have won the favor of the public with more abundant content [7].

3.2.3 Audience Dissemination Strategy: Insight into the Needs of the Audience and Accurate Dissemination of the City Image

In the context of big data, the analysis of big data can more accurately find the target objects of the city image dissemination, and real-time tracking and measurement of the target objects, insight into their needs, mining their attitude preferences, and finally achieving precise and personalized Spread to improve the effect of city image spread. Specifically, according to the SICAS model, according to the behavior trajectory of the audience at different stages, and the targeted communication goals at each stage of the system, we can reconnect the image of the city in the minds of the audience, ultimately improve the communication of the city image Effect [8] (Fig. 1).

Fig. 1. The original image

3.2.4 Communication Channel Strategy: Efficiently Integrate the Channels and Means of City Image Communication

Only by choosing the proper channels and means of city image dissemination can a good dissemination effect be achieved. Through the analysis of big data, the main body of city image communication should target the target audience and choose the most effective communication channels and means. On the one hand, for tourism promotional films and documentaries, traditional media with depth and precision can be selected for dissemination; on the other hand, for other dissemination content, in the context of big data, the new media dissemination matrix should be emphasized. By building platforms such as WeChat, Weibo, and thematic websites to spread through small videos, pictures, text, animation, H5 pages, FLASH, rap, etc., so that the audience can feel more vivid, life-oriented, and more intimate content. Through the efficient integration of communication channels and means, it can increase the audience's contact points and ultimately improve the effect of Sanya city image communication [9].

3.3 Improve the Foundation of Big Data and Cultivate a Team of Talents with Multiple Skills

On the one hand, we must strive to improve the infrastructure of big data. Improve the big data sharing and exchange platform and the Internet data collection platform, strengthen the coordination of policies and measures of big data related departments, and realize the data sharing at all levels and regions within the government [10]. On the other hand, increase the training of professional communication personnel. Professional communication talents must not only have the core technology of big data, but also understand the industry or domain knowledge involved in the communication, so as to lay a good foundation for the communication of Sanya's city image.

4 The Orthopedic Strategy of Sanya City Image Communication in the Era of Big Data

4.1 Use Big Data for Agenda Setting and Guide Public Opinion

When an urban emergency breaks out, the audience can understand the truth of the matter through the guidance of public opinion, and improve the efficiency of emergency resolution, so as to prevent the emergency from affecting the image of the city. In this context, city managers in Sanya should use big data to collect information about emergencies, realize multi-party interactions such as media and opinion leaders, and give full play to the function of guiding public opinion through agenda setting.

4.2 Use Effective Monitoring of Public Opinion to Actively Resolve the Crisis

During the crisis, the main body of city image dissemination uses big data to capture massive amounts of Internet information, conduct tendency research through automatic classification, effectively track the public's response and formulate targeted counter-measures to resolve the crisis, and finally effectively resolve the crisis.

4.3 Use Big Data to Restore the Image of the City

After a crisis event occurs, no matter how the crisis is resolved, it will have varying degrees of impact on the city's image. At this time, with the help of big data to gain insight into the psychological status of the audience, accurately identify the various stages of public opinion development and timely release various authoritative information, guide the trend of public opinion, and ultimately eliminate the impact of crisis events on the image of the city.

5 Conclusion

Big data is no longer a new thing, it has penetrated into all areas of people's life and work, and has penetrated into all stages of urban image communication. As an important city of Hainan Free Trade Port, Sanya uses the predictability and accuracy of big data, analyzes and mines the data, and accurately portrays the audience, allowing the main body of city image communication to pass the city image through the target audience more accurately To reach the target audience, establish a public opinion monitoring mechanism and better resolve crises, ultimately reduce the cost of city image communication, and improve the efficiency of city image communication.

References

1. Huang, J., Obracht-Prondzynska, H., Kamrowska-Zaluska, D., et al.: The image of the city on social media: a comparative study using "Big Data" and "Small Data" methods in the Tri-City Region in Poland. **206**, 103977 (2021)
2. Hussein, A.S.: City branding and urban tourist revisit intention: the mediation role of city image and visitor satisfaction. **10**(3) (2020)
3. McKinsey Global Institute (MGI): Big Data the Next Frontier for Innovation, Competition, and Productivity. McKinsey, United States (2011)
4. Duan, Z.: Analysis on the application of PPP mode in Hangzhou city management—take the infrastructure construction as an example. **5**(12), 14–26 (2017)
5. Identifying unique features of the image of selected cities based on reviews by TripAdvisor portal users. **20**(5), 503–519(2020)
6. Gao, H.: Big data development of tourism resources based on 5G network and internet of things system. **80** (2021)
7. Viriya, T.: City-district divergence grid: a multi-level city brand positioning tool. **14**(2), 101–114 (2018)

8. Alcocer, N.H., Ruiz, V.R.L.: The role of destination image in tourist satisfaction: the case of a heritage site. **33**(1), 2444–2461 (2019)
9. Wu, J., Zhao, M., Chen, Z.: Small data: effective data based on big communication research in social networks. **99**(3), 1391–1404 (2018)
10. Gunn, L.D.: Can a liveable city be a healthy city, now and into the future? **50**(11), 1405–1408 (2020)

IOT Gateway Based on Microservices

Chunling Li[1(✉)] and Ben Niu[2]

[1] School of Electronics and Internet of Things, Chongqing College
of Electronic Engineering, Chongqing 400000, China
[2] Chongqing Research Institute, ZTE Corporation, Chongqing 400000, China

Abstract. Microservice architecture is an architectural pattern that has emerged in the field of software architecture patterns in recent years. The microservice architecture can decouple the complex functions of the IoT gateway, and it can well solve the service deployment problem that the microservice architecture requires independent deployment and independent operation of applications. Therefore, it is of great significance to study the IoT gateway based on microservices. The purpose of this article is to research the IoT gateway based on microservices. Aiming at the requirements of the Internet of Things safety supervision system, this paper designs an Internet of Things gateway suitable for the safety supervision microservice architecture system based on the basic principle of information transmission safety. The Internet of Things gateway based on microservices can concurrently collect and process data in the underlying nodes and upper-layer services, and at the same time has the ability of node offline and node parameter management. This article is a highly available IoT platform based on microservice architecture. Study the service communication, service deployment and service discovery solutions of the microservice architecture, through the service splitting of the components of the Internet of Things platform, formulate the service communication method that adapts to the service characteristics, and use the container technology to deploy services, and build a high Available IoT platform. This article is based on the existing AES algorithm in ECB mode, which has a small resource occupancy rate and is more efficient. Aiming at the defects of fixed key and the same ciphertext key, it compares it with the non-periodic and initial value sensitive. The chaos mapping is merged, and the chaos-AES encryption algorithm that can be used on the IoT gateway is designed and implemented. Experimental research shows that although the efficiency of the algorithm proposed in this paper is reduced by 0.36 points, compared to other chaotic encryption algorithms, the efficiency of the algorithm is improved. However, this algorithm can greatly improve the security. Considering comprehensively, this algorithm is feasible.

Keywords: IoT platform · Micro server architecture · IoT gateway · Container deployment

J. Abawajy et al. (Eds.): ATCI 2021, LNDECT 81, pp. 26–33, 2021.
https://doi.org/10.1007/978-3-030-79197-1_4

1 Introduction

The security problem of the Internet of Things system has become one of the factors hindering the development of the Internet of Things [1, 2], and it is also an urgent problem for Internet of Things technicians at this stage. Therefore, the ability of the security of the Internet of Things network equipment to meet the requirements of the system has become one of the contents of this paper [3, 4]. The combination of the microservice architecture and the Internet of Things can achieve the Internet of Things efficiently while making full use of resources. Redundant backup and horizontal expansion of the platform to solve the high availability problem of the Internet of Things platform [5, 6].

In the research of micro-service-based IoT gateways, many researchers have studied them and achieved good results [7, 8]. For example, Taneja M researched a general IoT framework based on micro-service architecture. Compared with the traditional Internet of Things framework, it has higher scalability and maintainability. By constructing a core service to schedule and manage various types of Internet of Things application services, it can improve the relationship between big data services and geographic services. Jarwar M proposed a construction method of power cloud platform based on microservice architecture. By abstracting business logic into fine-grained reusable services, the continuous delivery component is designed for service deployment, verification and registration, and the service gateway component is responsible for intercepting and positioning service access requests.

This article is a highly available IoT platform based on microservice architecture. Study the service communication, service deployment and service discovery solutions of the microservice architecture, through the service splitting of the components of the Internet of Things platform, formulate the service communication method that adapts to the service characteristics, and use the container technology to deploy services, and build a high Available IoT gateway platform [9, 10].

2 IoT Gateway Based on Microservice Architecture

2.1 Construction Scheme of a Microservice-Based IoT Gateway Platform

(1) Microservice architecture
   The microservice architecture analyzes and models the function and structure of the application, and splits a complex application into a set of fine-grained, single-business logic, low-coupling services that can be deployed and run independently.
   1) Communication between services
      Through the lightweight communication mechanism, the interconnection and collaboration between services are realized to provide users with the ultimate application value. The lightweight communication mechanism is reflected in the communication mechanism is a language-independent and platform-independent interactive mode.

2) Service deployment

Using the Docker container technology, the service deployment package can be made into a container image, which can run in any environment where Docker is installed to realize service deployment. At the same time, because the Docker container itself is based on virtualization technology, it is possible to build multiple containers on one machine, which helps to improve the utilization of machine resources.

3) Service discovery

In the distribution layer service discovery scenario, the client makes a service access request to the distribution layer. The distribution layer first obtains the list of available services from the service registration and discovery module during the distribution request process, and then forwards the access request to the service instance according to the distribution routing strategy. The server-side framework is the solution for this scenario.

(2) Container orchestration and deployment technology

The container orchestration and deployment technology mainly solves the dependencies between the various services of the application, the basic environment for the interaction between the services, and performs cross-host orchestration deployment in a cloud environment, scheduling services to respond to business requests, and service nodes according to business requests. A series of complex management tasks such as dynamic scaling.

2.2   Implementation of the IoT Gateway Hardware Platform

(1) Realization of ARM9 core board

The ARM9 core board module is the central control module of the IoT gateway in this article, responsible for the normal operation of the entire system data modules and reliable data transmission. This article uses the S3C2440 core module mainly to consider:

1) In the Internet of Things network, the data that needs to be processed is very large, and the processing speed of the processor is very high. In the Internet of Things environment, TCP/IP can process the data of 10 users, and a ZigBee coordinator can access 64 at the same time Sensing nodes, in the safety supervision system, there may be thousands of ZigBee nodes in the environment; the IoT gateway needs to process the above data concurrently, and the S3C2440 adopts the ARM920 architecture, the main frequency can reach 400 MB, and the S3C2440 can well integrate these data Perform scheduling and processing.

2) The storage part of S3C2440 supports 128M memory, and Nandflash storage space up to 4097 bytes/page. The IoT gateway needs an operating system to manage the data in the IoT environment. This part of the memory is fully capable of running the operating system and can realize the processing and storage of large amounts of data.

(2) Implementation of serial port transparent transmission module

The serial port transparent transmission module can convert the TCP/IP protocol and the serial port protocol in the entire physical network gateway. Because the S3C2440 itself does not have a wireless network card, it cannot transmit data through the TCP/IP protocol family. The serial port transparent transmission module can provide this function conveniently and quickly. In summary, the MCU selection of the serial port transparent transmission module must be able to support the operation of the TCP/IP protocol stack. The MCU used in this article is HLK-M30, powered by 3.3 V, and UART2 is the interface for communicating with the central control chip.

2.3 Design and Implementation of IoT Gateway Software

(1) Transplantation of embedded operating system

The work to be done in this article includes: porting serial port driver, porting USB module and porting test module driver. The file system provides users with an environment for storing and running applications, and provides a kernel function interface that is convenient for users to program, so that user data and kernel data can interact. The file system is a bridge between Linux systems and applications. The construction of the file system mainly includes the following tasks: transplantation of Nandflash; construction of the basic environment of the file system; construction of the file system.

(2) Communication module programming

After Linux is successfully transplanted, it provides users with services such as process, thread, and inter-process thread communication. This article combines modular hierarchical design ideas with Linux multi-threaded multi-process services to design communication function modules. Each protocol runs in a process, implemented by the function fork, and uses memory sharing services between processes to achieve data sharing Sharing to realize the conversion between various protocols.

2.4 Data Security Communication Algorithm of IoT Gateway

In the field of the Internet of Things, in order to ensure the security of the Internet of Things system, the terminal devices in the system need to be further improved in terms of data encryption. Based on the existing algorithm, this paper focuses on whether it is suitable for embedded systems and high efficiency as the core point of the research, researches a secure encryption algorithm that can be applied to the Internet of Things network environment, and tests the algorithm.

(1) S-box byte transformation, including multiplication inverse and radiation process:

In order to improve the efficiency of the algorithm, the creator of the AES algorithm created the S-box through a non-linear algorithm. The user can quickly obtain the corresponding replacement data by using the table look-up method to improve the calculation speed.

(2) Row shift

Specific implementation process: Forward row shift This process is to cyclically shift each element in the matrix to the left based on where it is. The reverse row shift is to move each element in the matrix to the right based on the row.

(3) Column transformation

Multiply each column of the state matrix by a fixed polynomial f (x). Because of the AES algorithm, mod (x^4 + 1) is performed. The definition of f (x) is shown in (1).

$$f(x) = 03x^3 + 01x^2 + 01x + 02 \tag{1}$$

The inverse column transformation multiplies each column of the state matrix by the polynomial g(x) to perform mod(x^4 + 1) modular transformation. The definition of g (x) is shown in (2).

$$g(x) = 0Bx^3 + 0Dx^2 + 09x\_0E \tag{2}$$

The column transformation has the characteristics of formula (3).

$$g(x) = f^{-1}(x)mod(x^4 + 1) \tag{3}$$

AES encryption first performs matrix processing on the plaintext, and then each round of transformation is performed in the order of exclusive OR processing, S-box processing, row shifting, and column transformation exclusive. Carry out $Nr - 1$ round of transformation, the last round of transformation does not carry out column transformation processing, and finally form the corresponding bit ciphertext; decryption is the inverse process of encryption, and will not be described here.

3   Experimental Research on IoT Gateway Based on Microservice Architecture

3.1   Software Portability and Efficiency Test

Software portability means that the algorithm can be well adapted on different platforms without affecting its function and efficiency. The implementation language of this algorithm is C language, which has compatibility on both PC and embedded platforms, so this algorithm first has language compatibility. First, run this algorithm and the traditional AES algorithm on the PC for efficiency comparison tests, and keep the results to compare with the embedded runtime.

3.2   Experimental Algorithm Design

In this paper, through the extended algorithm in the traditional AES algorithm, the extended key for each round of encryption can be obtained. The end processing adopts the XOR method of the chaotic result and the plaintext end. In the encryption/decryption process, chaotic data is used to replace the public S-box. This algorithm realizes "one block and one encryption" without greatly affecting the efficiency of traditional AES, which improves the security of the algorithm.

4 Experimental Research and Analysis of IoT Gateway Based on Microservice Architecture

4.1 Software Portability and Efficiency Test Analysis

In the PC compilation environment, this algorithm is tested with 20000 bytes to 120000 bytes as input, and the encryption time is recorded using the time() function. The results are shown in Table 1.

Table 1. PC platform speed test (unit: ms)

Plaintext size (byte)	The algorithm used in this article	Standard AES-256
20000	43	37
40000	89	75
60000	127	115
80000	169	152
100000	211	189
120000	253	231

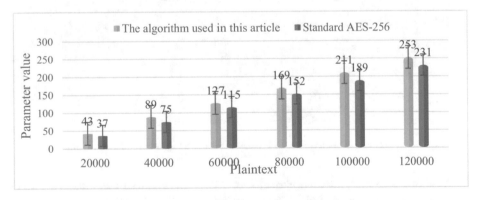

Fig. 1. PC platform speed test (unit: ms)

As shown in Fig. 1, although the efficiency of this algorithm is reduced by 0.36 points, compared to other chaotic encryption algorithms, the efficiency of the algorithm is improved. However, this algorithm can greatly improve the security. Considering comprehensively, this algorithm is feasible.

4.2   Speed Test of Embedded Platform

The rate test of the embedded platform is now carried out, and the algorithm is transplanted to the embedded IoT gateway running the Linux system. The size of 2000 bytes, 4000 bytes 12000 bytes are respectively encrypted, and the traditional AES the algorithm is applied in the IoT gateway, and the two results are compared. The results are shown in Table 2.

Table 2. Encryption and decryption speed test of IoT gateway platform (unit: ms)

Plaintext size (byte)	The algorithm used in this article	Standard AES-128
2000	38	34
4000	74	66
6000	126	113
8000	163	142
10000	186	167
12000	211	194

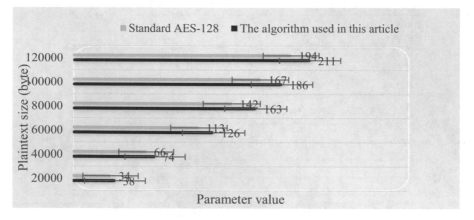

Fig. 2. Speed test of encryption and decryption of IoT gateway platform (unit: ms)

As shown in Fig. 2, this algorithm affects 0.72 points compared with the traditional AES algorithm, which is similar to the traditional AES algorithm. Therefore, this algorithm can be used in embedded systems, and it is feasible to use this algorithm to encrypt data in the IoT environment.

5   Conclusions

This paper proposes an AES encryption algorithm based on Hénon and Tent double chaotic mapping. This encryption algorithm first creates a chaotic system, fully discretizes the two chaos and makes them enter the chaotic state, and then iterates the chaotic state again, and performs the specified processing on the result of the iteration to obtain the key of each block of plaintext. Through the extended algorithm in the traditional AES algorithm, the extended key for each round of encryption can be obtained.

Acknowledgements. This work was supported by research projects of Chongqing College of Electronic Engineering of China (XJZK201901).

References

1. Xu, R., Jin, W., Kim, D.: Microservice security agent based on API gateway in edge computing. Sensors **19**(22), 4905 (2019)
2. Morabito, R., Petrolo, R., Valeria, L., et al.: LEGIoT: a lightweight edge gateway for the internet of things. Future Gener. Comput. Syst. **81**(Mar.), 1157–1171 (2017)
3. Herrera-Quintero, L.F., Vega-Alfonso, J.C., Banse, K.B.A., et al.: Smart ITS sensor for the transportation planning based on IoT approaches using serverless and microservices architecture. IEEE Intell. Transp. Syst. Mag. **10**(2), 17–27 (2018)
4. Filev, R., Lurbe, C.B., Baniya, A.A., et al.: IRRISENS: an IoT platform based on microservices applied in commercial-scale crops working in a multi-cloud environment. Sensors **20**(24), 7163 (2020)
5. Ortin, F., O'Shea, D.: Towards an easily programmable IoT framework based on microservices. J. Softw. **13**(1), 90–102 (2018)
6. Gaur, A.S., Budakoti, J., Lung, C.H.: Vertical handover decision for mobile IoT edge gateway using multi-criteria and fuzzy logic techniques. Adv. Internet Things **10**(4), 57–93 (2020)
7. Muhammad, A., Joao, R., Joaquim, F., et al.: Orchestration of microservices for IoT using docker and edge computing. IEEE Commun. Mag. **56**(9), 118–123 (2018)
8. Díaz-Sánchez, D., Marín-Lopez, A., Almenárez Mendoza, F., et al.: DNS/DANE collision-based distributed and dynamic authentication for microservices in IoT. Sensors **19**(15), 3292 (2019)
9. Taneja, M., Jalodia, N., Byabazaire, J., et al.: SmartHerd management: a microservices-based fog computing-assisted IoT platform towards data-driven smart dairy farming. Software **49**(7), 1055–1078 (2019)
10. Jarwar, M., Kibria, M., Ali, S., et al.: Microservices in web objects enabled IoT environment for enhancing reusability. Sensors **18**(2), 352 (2018)

Comparative Research on Blockchain Consensus Algorithms Applied in the Internet of Things

Xinyan Wang[✉], Zheng Jia, and Jing Zhang

State Grid Henan Information and Telecommunication Company (Data Center),
Zhengzhou 450052, Henan, China

Abstract. The current Internet of Things system has security and privacy issues, insufficient resources and network transmission delays. The advantages of introducing blockchain technology are analyzed, and the blockchain architecture of the Internet of Things and the traditional Internet of Things are compared. This article mainly introduces the DPoS algorithm and the workload proof mechanism method. This paper uses the blockchain consensus algorithm of the Internet of Things to detect the Internet of Things, and establishes a potential DPoS algorithm model. The model is solved using the DPoS algorithm, and the blockchain consensus algorithm is evaluated, and the model is revised using historical data to improve the accuracy of the blockchain consensus algorithm evaluation. The experimental results of this paper show that the blockchain consensus algorithm improves the efficiency of the application of the Internet of Things by 53%, and reduces the false alarm rate and the false alarm rate. Finally, by comparing the blockchain consensus algorithm, the influence of the blockchain consensus algorithm on improving the application of the blockchain consensus algorithm in the Internet of Things is analyzed.

Keywords: Internet of Things · Blockchain consensus algorithm · DPoS algorithm · Proof of work mechanism

1 Introduction

1.1 Background and Significance

The development of the Internet of Things has brought great convenience to everyone. At the same time, the development of the Internet of Things industry has also encountered many problems, especially data security has become the most urgent problem today [1]. The latest technology of blockchain solves many information technology and security, network security issues, and solves technical information security issues through the Internet of Things [2]. On the Internet and blockchain, art in heaven is even [3]. Therefore, the understanding of the blockchain fills the application of the Internet of Things loopholes, and fully reflects the characteristics of the blockchain of the Internet of Things. After long-term research and development, the application of this technology can be realized [4]. It has at least three roles, namely the security and transparency of big data management, the convenience of micro-operation

J. Abawajy et al. (Eds.): ATCI 2021, LNDECT 81, pp. 34–41, 2021.
https://doi.org/10.1007/978-3-030-79197-1_5

and intelligence, efficiency and equipment. In the blockchain system, the method of establishing consensus plays a key role. It not only helps data to maintain data consistency, but also has the problem of tokens [5, 6]. The blockchain consensus algorithm is also constantly developing and evolving. In recent years, branches have been continuously developing, and attack prevention has certain functions [7].

1.2 Related Work

Blockchain technology has completely changed the digital currency space through a pioneering cryptocurrency platform called Bitcoin [8]. From an extraordinary point of view, the discontinuation is a condensed book that can continue to record activities available online [9]. In recent years, this technology has aroused major scientific interest in research outside the financial world, one of which is the Internet of Things [10]. In this case, the blockchain is seen as the missing link in building a truly decentralized, untrusted and secure environment for the Internet of Things. In this survey, it aims to shape a coherent and comprehensive picture of the current state. Efforts in this direction start from the basic working principle of the blockchain and how the blockchain-based system realizes the characteristics of decentralization, security and auditability [11]. Here, based on the challenges brought by the current centralized IoT model, we will make the latest progress in the industry and research fields to effectively solve these challenges. Ali MS pointed out the latest research directions of these technologies, summarized the technical and strategic challenges faced, and finally analyzed the nature of the technology, providing important guidance for the construction of the innovation and entrepreneurship education curriculum system of local universities under the background of informationization [12]. But because the message collection process is too complicated, the data result is not very accurate.

1.3 Main Content

The innovation of this paper is to propose DPoS algorithm and workload proof mechanism method. Based on the comparative research of blockchain consensus algorithms applied to the Internet of Things, the blockchain consensus algorithm is evaluated through the Internet of Things. The establishment of the DPoS algorithm combined with the proof of work mechanism method provides research guidance for the application of the blockchain consensus algorithm to the Internet of Things.

2 Comparison Methods of Blockchain Consensus Algorithms Applied to the Internet of Things

2.1 Dpos Algorithm

The DPoS algorithm uses the witness mechanism to solve the central problem. DPoS selects multiple middle boards by choosing between books. By using polling points in the proxy submission committee, the need to reduce the participation of consensus nodes can be met. Generally, consensus voting among dozens of nodes among hundreds of nodes can be completed at most, and consensus can be reached within a few seconds. Therefore, the DPOS system can raise the outpost to another level by reducing the number of polling nodes or using the token ring system, which can be reduced to the millisecond level, and the use of computer and storage resources is much less than that of PW.

If the Internet of Things blockchain adopts DPoS, not all sensor nodes and microcontrollers perform account book recording functions, but a special system selects strange nodes as proxy accounts. These nodes are first used in the blockchain network and can be designed by several different IoT service providers to better solve efficiency and performance security issues. However, DPoS performance cannot be improved infinitely.

2.2   Proof of Work Mechanism

The mechanism to prove workload, as a decentralized system of virtual currency, and how to ensure the fairness and correctness of records without the authorization of a third party. The so-called task is that the nest must spend its own computer resources to "mine". The node that has completed the "mining" operation can first generate the signal and charging information of the operation to other nodes. Once other nodes receive this message, they will first verify the authenticity of the work. If it is valid and the accounting information has been verified, the node will add the block to the locally stored network and start the next round of competition for computing power. If the verification fails, the data will be discarded and the current "mining" will continue. Therefore, PW uses a hash calculation method: for a specific input, the hash result is the same every time, anyone can copy and verify it; on the contrary, the input can only be continuously changed to generate a certain hash value, Instead of direct selection, the specific formula is as follows:

The first is defined as the possibility of public nodes in the logistics network passing through the logistics blockchain, and the damaged node can control the nodes in the entire network:

$$\phi_a = \left(\frac{b}{\phi}\right)^2, b \leq \phi$$
$$\phi_a = 1, \phi \leq b \tag{1}$$

According to the law of the number of blocks of negative nodes and positive nodes, the probability of Poisson expansion of set b is satisfied, and the probability of a fake block node being attacked is Q:

$$Q = \lim_{\lambda \to \infty} \sum_{\alpha \leq a \leq \theta} \frac{\lambda^a e^{-\lambda}}{\kappa} \tag{2}$$

Among them, α and θ are set intervals.

3   Comparative Experiment of Blockchain Consensus Algorithms Applied to the Internet of Things

3.1   Throughput Data Collection of Experimental Verification Algorithm

When the number of sites is less than 17, the output of these three algorithms represents the maximum increase limit, and the output of CloudPBFT is very large, up to 45 times. CloudPBFT can be completed flexibly and incrementally, up to 54 times. The MinBFT algorithm shows a slight skew of 20–50 nodes, and then continues to increase the number of nodes, and the output increases accordingly, indicating a higher trend, but it is also filled 48 times. There is no change in PBFT input between 10–11 nodes, and there is a slight downward trend, which shows that under the same circumstances, the two CloudPBFT and MinBFT optimization algorithms are better than PBFT. CloudPBFT has a dynamic fault tolerance mechanism and the ability to distribute copies in real time, so it can process defective nodes faster in the information processing process to ensure continuous improvement in productivity. The experimental data results are shown in Table 1:

Table 1. Algorithm throughput data table

Total delay point	10	20	30	40
CloudPBFT	45	54	67	87
MinBFT	48	56	68	88
PBFT	49	57	69	89

It can be seen from the Table that under the same environment, by simulating the main nodes of the transmission service, these three algorithms show a growing trend in the network output of 0–40 ms, with little fluctuation. Among them, the original PBFT algorithm is not optimized. The Table shows that in the same environment, simulating the main nodes of the transmission service, these three algorithms show that the network output shows an upward trend in the range of 0–40 ms, and the fluctuation is small. In terms of total bandwidth, CloudPBFT is the lowest bandwidth in the algorithm, only 12 min. Experiments also show that the larger the node, the slower the production speed. It can be concluded that the comparison of PBFT > MinBFT > CloudPBFT with the algorithm delay shows that the algorithm has relatively good advantages in network communication, and the node fault tolerance makes the distributed node performance of the algorithm has sufficient advantages. Hadoop cloud computing has improved the speed of blocking chain technology in the actual consensus authentication process. The analysis results are shown in Fig. 1:

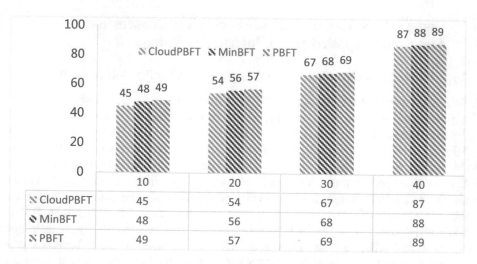

Fig. 1. Throughput data analysis diagram of the algorithm

From the above Figure, the continuous increase in the number of nodes in the cluster will lead to a linear increase in the speedup of the algorithm. Use data to compare the execution performance of PBFT, MinBFT, and CloudPBFT. Set the support degree to 0.13, and use PBFT, MinBFT and CloudPBFT to mine frequent itemsets respectively. Since the test set is smaller than the usual normal value, this article defines the data block of the imported function as 5, so the test set is divided into 4 blocks evenly and sent to each node of the cluster for execution.

3.2 Blockchain Consensus Algorithm Design Applied to the Internet of Things

At the edge, there are many raw materials, including more personal information. For example, videos or photos taken on smart phones, more GPS, health information that can be provided through porTable devices, smart home functions, and device sensors in precious homes. IoT devices are used to monitor and collect various social infrastructure environments, and send local filtering data to the fog layer to obtain requested services. Each fog in the fog layer contains a high-performance distributed SDN controller. Each fog node covers a small associated area, is responsible for data analysis, provides real-time services and reports the processed output results to the cloud and edge layer. The fog layer provides positioning, while the cloud layer provides monitoring and control of the entire area. The flowchart topology generator generates parsed data sets to construct and modify flowcharts related to the flow network.

At the cloud level, distributed clouds based on blockchain technology can provide secure, cost-effective and on-demand access to the most competitive IT infrastructure. The client can search, search, provide, use and automatically release all necessary computing resources, such as servers, data, and applications. Unlike traditional blockchains that use PoW consensus protocols, workloads are activities that occur outside the blockchain.

4   Comparative Analysis of Blockchain Consensus Algorithms Applied to the Internet of Things

4.1   Analysis Standard of Consensus Algorithm

The advantages and disadvantages of contractual agreements have a significant impact on the implementation of distributed systems. First of all, security is to prevent potential attacks. Scale, attack performance and network response speed, the possibility of general attacks on algorithm machines. Second, the fault tolerance of the algorithm must be considered. Generally, if you follow the distributed modular hierarchical application design, you can greatly improve fault tolerance. Therefore, the algorithm should be simplified as needed. One of the characteristics of blockchain technology is decentralization. When the merge solution project encounters this concept, this is a problem that needs to be studied in depth.

4.2   Consensus Algorithm Mechanism Data Analysis

The main problem of distributed networks is how to reach an effective consensus. The degree of centralization is low, and it is more difficult for a society with decentralized decision-making power to reach consensus. The purpose of balancing consistency and applicability and ensuring relatively reliable consistency without affecting actual use experience is the purpose of studying the consensus mechanism. The operation of Internet-based networks is the main trend in the use of the Internet of Things. With the integration of the blockchain and the Internet of Things, the technical advantages of the blockchain can provide great convenience for the online transaction system of Internet objects, that is, by creating a complete data transaction system, rational use of the Internet of Things based on the blockchain Technology establishment. The consensus algorithm mechanism data is shown in Table 2:

Table 2. Data Table of consensus algorithm mechanism

Algorithm type	Algorithm block height	Algorithm delay time	Trend of success probability of forged blocks	Algorithm throughput comparison
CloudPBFT	7000	150	99	0.6
MinBFT	6780	142	87	0.49
PBFT	5790	138	67	0.31

Through the analysis of Table 2, the data mining in this paper forms a new data table, and the data is based on the development of teaching reform, Leaders' emphasis and hardware facilities are used for further data collection, so that the data in the original database is reorganized according to the target requirements, which is more conducive to decision-making analysis, and can more clearly see the internal relationship between the data. The analysis result is shown in Fig. 2:

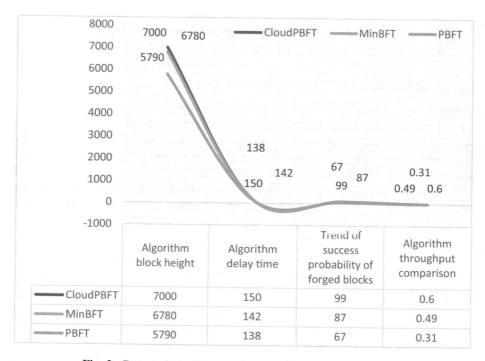

Fig. 2. Data analysis diagram of consensus algorithm mechanism

Simulation experiments show that the graph has high security, stability and transferability. CloudPBFT, MinBFT and PBFT used in the logistics blockchain of the Internet of Things provide decentralization and non-deceptive requirements, and use the powerful functions of cloud computing and distributed storage functions to solve large-scale consensus computing problems. The problem of computing power provides blockchain solutions to many problems, such as the current opaque practices in the logistics industry.

5   Conclusions

Although this paper has made some research results on DPoS algorithm and workload proof mechanism, it still has many shortcomings. There are still many in-depth content to be studied in the comparative research methods of blockchain consensus algorithms applied to the Internet of Things. There are many steps in the decision-making process that have not been involved due to reasons such as space and personal ability. In addition, the actual application effect of the improved algorithm can only be compared with the traditional model from the level of theory and simulation.

References

1. Yang, Y., Jia, Z.: Application and challenge of blockchain technology in the field of agricultural Internet of Things% application and challenge of blockchain technology in the field of agricultural Internet of Things. Agric. Netw. Inf. (12), 24–26 (2017)
2. Chen, S., Wang, H., Zhang, L.J.: Using Ethereum Blockchain in Internet of Things: A Solution for Electric Vehicle Battery Refueling. In: Blockchain – ICBC 2018. Lecture Notes in Computer Science, vol. 10974, pp. 3–17 (2018). Chapter 1. https://doi.org/10.1007/978-3-319-94478-4
3. Ali, M.S., Vecchio, M., Pincheira, M., et al.: Applications of blockchains in the Internet of Things: a comprehensive survey. IEEE Commun. Surv. Tutor. PP(99), 1 (2018)
4. Yu, Y., Li, Y., Tian, J., et al.: Blockchain-based solutions to security and privacy issues in the Internet of Things. IEEE Wirel. Commun. 25(6), 12–18 (2019)
5. Yatskiv, N.G., Yatskiv, S.V.: Perspectives of the Usage of Blockchain Technology in the Internet Of Things. Sci. Bull. UNFU 26(8), 381–387 (2016)
6. Cekerevac, Z., Prigoda, L., Maletic, J.: Blockchain technology and industrial Internet of Things in the supply chains. MEST J. 6(2), 39–47 (2018)
7. Li, F., Wang, D., Wang, Y., et al.: Wireless communications and mobile computing blockchain-based trust management in distributed Internet of Things. Wirel. Commun. Mob. Comput. 2020(5), 1–12 (2020)
8. Wang, P., Susilo, W.: Data security storage model of the Internet of Things based on blockchain. Comput. Syst. Sci. Eng. 36(1), 213–224 (2021)
9. Alghamdi, N.S., Khan, M.A.: Energy-efficient and blockchain-enabled model for Internet of Things (IoT) in smart cities. Comput. Mater. Continua 66(3), 2509–2524 (2021)
10. Hattab, S., Alyaseen, I.F.T.: Consensus algorithms blockchain: a comparative study. Int. J. Perceptive and Cogn. Comput. 5(2), 66–71 (2019)
11. Wang, P., Qiao, S.: Emerging applications of blockchain technology on a virtual platform for English teaching and learning. Wirel. Commun. Mob. Comput. 2020(2), 1–10 (2020)
12. Ali, M.S., Vecchio, M., Pincheira, M., et al.: Applications of blockchains in the Internet of Things: a comprehensive survey. IEEE Commun. Surv. Tutor. 21(2), 1676–1717 (2019)

Public Attention and Housing Prices: City Panel Data Based on Search Big Data

Linyan Wang[1], Haiqing Hu[1], and Xianzhu Wang[2(✉)]

[1] School of Economics and Management, Xi'an University of Technology,
Xi'an, Shaanxi, China
[2] School of Business, Anhui University of Technology,
Ma'anshan, Anhui, China

Abstract. From the perspective of big data, this paper uses panel vector autoregressive model (PVAR) and Granger causality test model to analyze the dynamic relationship between public attention and housing prices, and further uses variance house to verify the level of economic development and macroeconomic development of external factors. Control policies and the degree of supply in the real estate market affect the difference between public attention and housing prices. Granger causality shows that in the case of a lagging period, public attention affects housing price fluctuations in one direction; variance decomposition shows that long-term quantitative monetary policy can regulate the healthy development of the real estate market.

Keywords: Public attention · House price · Big data

1 Introduction

The real estate industry is one of the pillar industries of China's national economy. The stability of the real estate market is related to the steady development of China's economy. In response to this, the Chinese government has successively introduced different policies to guide the healthy development of the real estate market. For example, in order to get rid of the global financial crisis, it has repeatedly lowered the deposit reserve ratio and the benchmark interest rate to stimulate the economy to stabilize and rebound; in order to prevent the housing price from rising again, implement a tightening currency Policies curb speculative demand in real estate, and regulation of the real estate market often leads to a vicious circle of "increasing adjustments." The public's attention to housing prices and their expectations of rising prices are important factors that increase the volatility of housing prices. Paying attention to the public's expectations of the real estate market has become one of the key factors in studying the stable development of the real estate market.

Big data has initiated the revolution of the times, and it has been integrated into all levels of society. China formally proposed a national big data strategy in the "13th Five-Year Plan" to comprehensively promote the high-quality collection and integration of big data in key areas and industries. According to the "China Internet Development Report 2020", as of the end of 2019, the number of mobile Internet users in

China reached 1.319 billion. The public's interest in the real estate market and housing prices was searched through the Internet to adjust their expectations and decisions.

In the existing research, the influencing factors of housing prices are mostly concentrated on development costs, market supply and demand, residents and market expectations, macroeconomic environment, population migration, etc., while research on residents and market expectations is mostly concentrated on the micro level, and there are few Analyze through the big data level [1, 2]. Therefore, this article takes the housing prices of China's first-tier and new-tier cities as the research object, cuts in from the perspective of big data, and empirically analyzes the impact of public attention on urban housing prices through the annual average daily search volume of "city + housing prices" in the Baidu index. The conclusion of the article. The innovation of this paper is to study the impact of public attention on housing price changes from the perspective of big data, and to provide a valuable reference for the regulation of the real estate market from the level of social psychological expectations.

2   Data and Methods

2.1   Data Sources

The research samples in this article are first-tier and new-tier cities from 2011 to 2019. The public attention index comes from the daily search volume of house prices in the Baidu Index. The house price index comes from Anju Guest House Property Information Service Platform. The macro-control policies come from the People's Bank of China. The indicators of economic development level and real estate market supply situation are derived from the corresponding city statistics bureau, and finally a total of 171 cities-annual samples from 19 cities were obtained. Except for the monetary policy to deal with dummy variables, in view of the different units among the remaining variables, the logarithmic processing method is used for dimensionless processing.

(1) Public attention. The Baidu Index is one of the most important statistical analysis platforms for China's current Internet and even the entire data era. It conducts data analysis based on the search data of Baidu users. Therefore, the Baidu index is used as a proxy variable of public attention, and the search keywords of "city name + house price" are used for trend analysis, such as "Beijing house price", "Shanghai house price", etc., and the annual average is taken as the current year's attention Degree data, the time range is from January 1, 2011 to December 31, 2019.

(2) House price. Anjuke is China's leading real estate information service platform, and its real estate research institute produces corresponding reports based on previous real estate information. The proxy variable of house price in this article comes from the house price report of Anjuke platform. The average sales price of 19 cities in the first-tier and new-tier cities are used as the statistical data of the current year's house prices.

(3) Economic development level There are many indicators to measure the level of economic development of a country and region, and GDP is recognized as the best indicator to measure economic conditions. This article draws on the research of scholars of Alvarado, Fernández, and Liu, and uses GDP as the proxy variable of the economic development level of this article [3–5].

(4) Monetary policy. Monetary policy includes quantitative and price-based monetary policies. China has long adopted a combination of quantitative and price-based monetary policies. From the perspective of research on controlling macroeconomic operations, expectation management, and stabilizing housing prices, the implementation of quantitative monetary policy The effect is more prominent [6, 7]. Therefore, this article draws on the research of Zhang and other scholars, and selects the actual growth rate of M2 minus the difference between the actual GDP growth rate and the CPI growth rate as the proxy variable of quantitative monetary policy [8]. If the difference is greater than 0, it is a loose monetary policy and recorded as 1, otherwise it is 0.

(5) Real Estate Market Supply. The real estate market supply includes variables such as the amount of real estate development investment, the newly started area of houses, and the completed area of houses. This article draws on the research of Wang, and then considers the availability of data, and uses real estate development investment as the proxy variable of the real estate market supply [9].

2.2   Methods

This paper uses Granger causality test model and panel vector autoregressive model (PVAR) to analyze the relationship between public attention and housing prices. The Granger causality test model is:

$$Y_t = \sum_{i=1}^{s} \alpha_i Y_{t-i} + \sum_{i=1}^{m} \beta_i X_{t-i} + u_t \tag{1}$$

In formula (1), X and Y represent two sets of variables, X_{t-1} is X lagging i period, Y_{t-1} is Y lagging i period, α_i is a constant, β_i is the regression coefficient, and μ_i is a random error term. Null hypothesis H_0: "X is not the cause of Y's development." If at least one of the coefficients $\beta_1, \beta_2, \beta_3 \ldots \beta_i$ is not zero, reject the null hypothesis and accept that X is the cause of Y's development. Combined with the research of this article, it analyzes the causal relationship between citizen attention and housing prices.

The PVAR model is:

$$U_{it} = \gamma_0 + \sum_{j=1}^{k} \gamma_j U_{it-j} + \alpha_i + \beta_t + \varepsilon_{it} \tag{2}$$

In the formula (2), i = 1, 2, ..., N, representing different cities; t = 1, 2, ..., T, representing the year; U_{it} means that the public attention (U_{1it}) and the housing price (U_{2it}) are sufficient Two-dimensional column vector; γ_0 is the intercept term; k is the lag order; γ_j is the parameter matrix of U_{it} lag j order; α_i and β_t are the individual effect and time effect vectors, respectively representing the difference between cities and the time change its impact; ε_{it} is a random disturbance term.

2.3   Descriptive Statistics

Before the empirical analysis, a descriptive statistical analysis of the main variables involved in this article (Table 1). It can be seen from the table that during the period from 2011 to 2019, the lowest house price was RMB 6,289 per square meter and the highest was RMB 59,696 per square meter, corresponding to the average house price in Chongqing in 2015 and the average house price in Beijing in 2019. 17,234, accounting for 33.92% of all housing price data; the lowest public attention is 351 and the highest is 3739, corresponding to the average daily search volume of Dongguan in 2012 and 2016, respectively, with an average of 1536, accounting for all public attention data The average value of quantitative monetary policy is 0.778, indicating that generally loose monetary policy has been implemented for more years.

Table 1. Descriptive statistics

Variable	Variable meaning	N	Mean	sd	min	max
hp	House price	171	17234	12056	6289	59696
pa	Public attention	171	1536	803.60	351	3739
lnhp	Logarithm of house price	171	9.568	0.584	8.747	11.00
lnpa	Logarithm of public attention	171	7.198	0.542	5.861	8.227
gdp	Logarithm of GDP	171	6.988	0.542	5.533	8.247
mp	Virtual variable(0, 1)	171	0.778	0.417	0	1
invest	Logarithm of real estate development investment	171	12.070	0.531	10.53	13.00

3   Empirical Analysis

3.1   Stationary Analysis

The premise of Granger causality test and PVAR model analysis is that the data must be stable. If it is stable, co-integration and causality tests can be performed, indicating that there is a long-term stable relationship between variables; if it is non-stationary, the variables should be differentiated, Until stable. And use IPS inspection method to carry out unit root inspection on public attention and housing price. The results are shown in Table 2. The public attention and housing prices become stable after the first-order first-order difference processing. This article draws on Zhang's research and determines the minimum value of the test results of AIC, BIC, and HQIC as the lag order is lag level one [10].

Table 2. Unit root test result

Variable	P value	Is it stable
lnhp	0.6152	No
lnpa	0.3085	No
Δlnhp	0.0000	Yes
Δlnpa	0.0000	Yes

3.2   Analysis of Autoregressive

This paper uses the GMM estimation method in STATA16.0 to fit the panel data of the dependent variable, its lagging variable and the independent variable. The results are shown in Table 3. Explanatory variables, the autoregressive results of housing price lagging one period and public attention lagging one period are both significant, and when public attention is used as the explained variable, only the autoregressive results of housing price lagging one period and public attention lagging one period are explained. The variable is significantly related to its own lag. Based on the results of the Granger causality test in Table 4, the original hypothesis "*lnpa* is not the Granger cause of *lnhp*" passes the 1% significance test, and the null hypothesis is rejected, and "*lnpa* is the Granger cause of *lnhp*", that is, public attention has caused changes in housing prices; and the original hypothesis "*lnhp* is not the Granger reason for *lnpa*" does not pass the 10% significance test, accept the original hypothesis. Generally speaking, when public attention and housing prices lag by one order, public attention unilaterally causes price changes, that is, public attention has an impact on housing price fluctuations.

Table 3. Autoregressive estimation result

Y	X	Coef	P value
lnhp	L.*lnhp*	0.569***	0.000
	L.*lnpa*	0.203***	0.000
lnpa	L.*lnhp*	−0.279	0.148
	L.*lnpa*	0.956***	0.000

Note: ***p < 0.01.

Table 4. Granger causality test results

H_0	chi2	p	Accept the H_0	Causality conclusion
lnpa is not the Granger reason for *lnhp* changes	15.109	0.000	No	*lnpa* has an impact on *lnhp* changes
lnhp changes are not the Granger cause of *lnpa*	2.094	0.148	Yes	*lnhp* changes are not the cause of the *lnpa*

3.3 Impulse Response

This article uses impulse response to re-verify the interactive response mechanism between public attention and housing prices. In Fig. 1, the horizontal axis represents the number of prediction periods of the response generated by the impact, the vertical axis represents the response degree of the impact, the red curve represents the impulse response function value, and the blue and green curves represent the confidence interval band of the impulse response function value. It can be seen from Fig. 1 that both *lnhp* and *lnpa* respond to their own shocks in the current period; *lnhp* will make corresponding changes when facing the shock of *lnpa*; while *lnpa* does not change in the current period when facing the shock of *lnhp*. To further verify the results of the above PVAR model and Granger causality test.

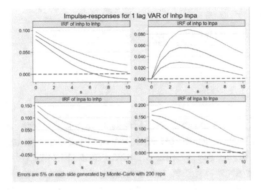

Fig. 1. Impulse response graph

3.4 Variance Analysis

Variance decomposition can decompose the standard deviation of the prediction residuals into the parts related to the endogenous variables of the system according to the cause, so as to evaluate the contribution degree of the changes according to the importance of the impact of each endogenous variable. This article will introduce the external factors affecting public attention and housing prices, the level of economic development, macro-control policies, and the degree of real estate market supply to analyze its contribution to the relationship between public attention and housing prices. This paper selects 8 periods as the lag period of variance decomposition, and the results are shown in Table 5.

In the first period, house price changes were only affected by its own fluctuations, and gradually stabilized in the sixth period. In the eighth period, the lagging items of house prices accounted for 64.3%, indicating that the most important factor for house price fluctuations is its own Secondly, the degree of public attention has a greater impact on the changes in housing prices; the level of economic development, macro-control policies, and the degree of supply in the real estate market have three external factors that affect housing prices as the number of lag periods increases. Moreover, the proxy variables of monetary policy account for a large proportion, indicating that the implementation of quantitative monetary policy in long-term regulation can have a certain impact on the fluctuation of house prices, which is consistent with Eric's research conclusion.

Table 5. Variance decomposition result

	lnhp	lnpa	gdp	mp	invest
1	1.000	0.000	0.000	0.000	0.000
2	0.911	0.052	0.000	0.036	0.000
3	0.821	0.114	0.002	0.060	0.003
4	0.753	0.164	0.003	0.072	0.008
5	0.706	0.199	0.004	0.075	0.016
6	0.674	0.220	0.005	0.074	0.026
7	0.655	0.231	0.006	0.073	0.035
8	0.643	0.236	0.007	0.071	0.043

4   Conclusion

This paper uses the PVAR model and Granger causality test model to analyze the relationship between public attention and house price fluctuations. The results show that when the period is lagging, public attention affects house price fluctuations in one direction, and further uses variance decomposition to analyze the level of economic development, Macro-control policies and the degree of supply in the real estate market analyze their contribution to the relationship between public attention and housing prices. The results show that quantitative monetary policy plays an important role in the long-term regulation of the real estate market.

The enlightenment of this article: First, we must pay attention to the public's expectations of housing prices, which can provide a certain reference for the steady development of the real estate market at the demand level; second, in the long-term regulation of the real estate market, monetary policy plays a positive role. Therefore, in terms of macro supply, "stabilizing land prices, stabilizing housing prices, and stabilizing expectations" can serve as the goal of stable development of the real estate market.

Acknowledgments. This work was financially supported by National Natural Science Foundation of China (71974003).

References

1. Ting, L., Tao, L., Jiajie, L., Tianyi, C.: Spatial analyses of stem families in China: based on 2015 one-percent population sample survey. Popul. Res. **44**(06), 3–19 (2020). (in Chinese)

2. Yong, H.: The measurement of regional housing price linkage and its influencing factors in China—based on the research of panel data of 35 large and medium cities. Manag. Rev. **32** (06), 62–71 (2020). (in Chinese)
3. Alvarado, R., Deng, Q., Tillaguango, B., et al.: Do economic development and human capital decrease non-renewable energy consumption? Evidence for OECD countries. Energy **215**, 119–147 (2021)
4. Fernández-Rodríguez, E., García-Fernández, R., Martínez-Arias, A.: Business and institutional determinants of effective tax rate in emerging economies. Econ. Model. **94**, 692–702 (2021)
5. Xue, L.Y., Shi, T.C., Yan, C.L.: Measurement and comparison of high-quality development of world economy. Economist **05**, 69–78 (2020). (in Chinese)
6. Hjalmarssona, E., Österholm, P.: Heterogeneity in households' expectations of housing prices - evidence from micro data. J. Hous. Econ. **50**, (2020). Sciencedirect
7. Guo, Y.M., Chen, W.Z., Chen, Y.B.: Research on the decline of China's monetary policy effectiveness and expectation management. Econ. Res. J. **1**, 28–41 (2016). (in Chinese)
8. Zhang, L., Yin, H., Wang, Q.: Quantity type or price type—empirical evidence from the "non-linear" effectiveness of monetary policy. China Ind. Econ. (07), 61–79 (2020). (in Chinese)
9. Wang, X.Z., Yang, Y.W.: Differentiated expectations, policy regulation and housing price fluctuations: an empirical study of 35 large and medium-sized cities in China. J. Finance Econ. **41**(12), 51–61 + 71 (2015). (in Chinese)
10. Zhang, G.H., Zhao, W.S.: The dynamic relationship, mechanism and regional differences of urbanization and touristization in China—based on provincial panal data by PVAR model. Bus. Manag. J. **39**(11), 116–133 (2017). (in Chinese)

Transmission Delay in Optical Fiber Communication System of Power Transmission and Transformation

Hongzhen Yang[1(✉)], Yanbo Wang[1], Hao Chi[2], and Daochun Yuan[3]

[1] Information and Telecommunication Branch, State Grid Zhejiang Electric Power Co., Ltd., Hangzhou, Zhejiang, China
tlnwu@nwu.edu.cn
[2] School of Communication Engineering, Hangzhou Dianzi University, Hangzhou, Zhejiang, China
[3] Enterprise Transmission and Access Product Line, Huawei Technologies Co., Ltd., Chengdu, Sichuan, China

Abstract. An OTN optical service unit (OSU) solution uses dedicated DM bytes for delay information transmission. The NMS can visualize network delay in real time, which is better than the manual delay evaluation method of SDH. The OSU defines the timestamp byte for the transmission clock at the physical layer. The physical-layer pre-processing is not required when the IP-based IEEE 1588v2 protocol is used, and the natural time transmission protocol is formed. The precision of the OSU is higher than the SDH delay estimation mode. In addition, the clock transmission protocols of different vendors are consistent and interconnected. Compared with SDH and IP networks, the OTN OSU networks are more suitable for WAN interconnection of the power production services and the time-sensitive network services.

Keywords: Tele protection · Power transmission · Electric power network · Time-sensitive network

1 Introduction

With the development of communication technologies and the line tele protection technologies, the transmission line tele protection channels have evolved from carrier waves to optical fiber channels [1]. Due to its advantages in high stability, high reliability, strong anti-interference capability, large amount of information, and easy setting and maintenance, the optical fiber communication technology gradually becomes the preferred protection channel mode. There are two optical fiber protection methods [2, 3]. One is the dedicated optical fiber, in which the signals are transmitted directly without being converted. However, fiber core resources are prone to waste and have short transmission distances. The other is multiplexing, which uses the 2 Mbit/s or 64 Kbit/s digital channel of an SDH optical fiber communication circuit to transmit protection signals. This method has the advantages of long transmission distance and saving optical fiber resources. Based on the SDH optical communication, the WDM OTN technology is developed to transmit different wavelengths in one optical fiber, increasing the number of physical channels and bandwidth of optical fibers.

J. Abawajy et al. (Eds.): ATCI 2021, LNDECT 81, pp. 50–57, 2021.
https://doi.org/10.1007/978-3-030-79197-1_7

SDH services include Ethernet services, 2M optical services, and legacy-network PCM tele protection services [4, 5]. These services provide a stable WAN with fixed low delay. The application value of SDH in the network structure decreases every year. However, SDH still has obvious technical advantages in the industrial real-time control field and some premium private line services. OSU is a small-granularity technology developed under the WDM OTN system. OSU applies to similar scenarios as SDH, but has the potential to replace the latter due to channelized transmission of any bandwidth, real-time delay visibility, and high-bandwidth transmission compatibility [6].

2 Electric Power Production Service Networking

2.1 OTN/SDH-Based Electric Power Service Networking

Electric power optical transmission technologies and devices have gone through PDH and SDH to the OTN era [7, 8]. The IEC2015 technical reported that SDH/SONET transmission technology is widely used to carry mission-critical services on production control networks, such as tele protection. In 2015, the OTN technology, as a new technology, attracted attention from the electric power industry worldwide. It is widely recognized in the industry that the OTN technology is an extension of the traditional SDH/SONET technology [9].

OTN is gradually taking over optical transmission networks in the electric power industry from SDH/MSTP. SDH and WDM are the most widely used technologies in transmission networks. OTN integrates the advantages of SDH and WDM and has become the mainstream technology for large-granularity service transmission [10, 11]. Currently, OTN devices can satisfy the requirements of electrical power access from small-granularity SDH services to large-granularity unified switching of PKT and ODU services. The unified structure is showed in Fig. 1.

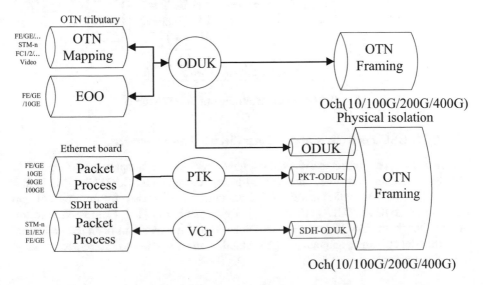

Fig. 1. Unified switching structure

From the Fig. 1, the mainstream SDH-compatible OTN architecture of State Grid Corporation of China (SGCC) is used to carry three types of services in transmission networking.

2.2   OTN OSU-Based Electric Power Service Networking

Currently, small-granularity services such as tele protection and security- and stability-sensitive services are encapsulated using SDH/VC-12 and then mapped into ODU1/OTN. There is no OTN device that can directly add or drop these services [12, 13].

To efficiently carry production service signals at a rate lower than 1 Gbit/s, the flexible small-container optical service units based on the production services need to be introduced to the existing OTN architecture [14]. The generated OTN OSU structure efficiently carries the existing WDM and MSTP optical transmission networks, making it the best choice for the dedicated WAN technology of the electric power industry. The layered OTN architecture in the OSU technology is shown in Fig. 2.

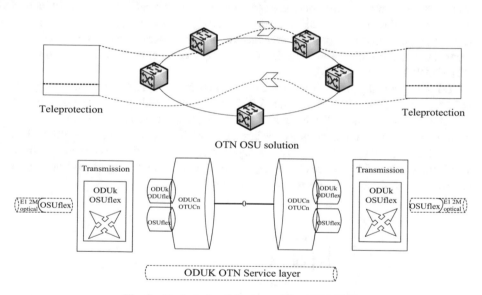

Fig. 2. Technical architecture of the OTN OSU

2.3   OTN OSU Perfectly Carries Core Electric Power Services

Low-rate services, packet services, and new services are adapted and mapped to OSUflex containers, and then multiplexed to higher-order ODUk/OTUk/OTUCn signals for transmission. Alternatively, OSU containers can be directly multiplexed into higher-order ODUk/OTUk/OTUCn signals for transmission [15]. Compared with traditional ODUk, the OSU is more refined, flexible, and independent from the traditional OTN timeslot structure, achieving high scalability and adjusting to future packet-based service evolution.

3 Channel Transmission Delay in the OTN+SDH Architecture

In OTN and SDH optical fiber communication systems, the transmission delay includes the delay of optical signals in optical fibers and that generated during the processing of optical signals on OTN and SDH devices. In the engineering design process, the manual delay estimation method is mainly used.

3.1 Optical Fiber Transmission Delay

The propagation speed (v) of light wave signals depends on the refractive index (n) of the transmission medium, or the transmission capability of the medium. Specifically, n = c (speed of light in vacuum)/v. Because time (t) = distance (l)/speed (v), when n = c/v, the time T_0 (transmission delay) can be calculated as:

$$T_0 = \frac{l}{v} = \frac{l \cdot n}{c} \tag{1}$$

Where c is the speed of light in the vacuum ($3*10^5$ km/s). l indicates the transmission distance (km). n indicates the refractive index of an optical fiber. The refractive index of G.652 fibers that are widely used is about 1.47. Therefore, the transmission delay of optical signals in optical fibers is about 4.9 μs, which is rounded to 5 μs in engineering applications.

3.2 Calculation of Transmission Delay

During signal transmission in a communications network, the network node devices (a digital switch and a digital cross-connect device) may include a buffer, a timeslot switching unit, and other digital processing devices. These devices all generate certain transmission delay. Additionally, the PCM devices, multiplexers, and multiplexing converters also produce varying degrees of delay. This paper mainly discusses the device delay of the transmission line tele protection information transmitted through the 2 Mbit/s channel in the SDH optical fiber communication system. Currently, China's power grid communication network has been gradually transformed from SDH devices to MS-OTN devices, and a 2 Mbit/s channel will be added to an SDH channel and then mapped to an OTN channel.

An SDH device consists of different functional modules. When signals pass through these modules, the delay occurs. The total delay of signals passing through an SDH device is related to factors such as fixed padding, connection processing, positioning, mapping, and damping.

Device delay is an important performance parameter in the OTN and SDH devices. The delay varies according to vendors and device models. The related standard organizations have also defined the transmission delay specifications of OTN and SDH devices. According to the ETSI standard, the line-to-line delay of OTN and SDH devices should be lower than or equal to 60 μs, and the tributary-to-line or line-to-tributary delay of OTN and SDH devices should be lower than or equal to 110 μs.

According to China's telecom industry standards, the maximum conversion delay of OTN and SDH digital cross-connect devices should be lower than or equal to 125 μs.

In an OTN+SDH optical fiber communication system, the 2 Mbit/s channel delay T is the sum of the transmission delays of OTN and SDH devices and the optical cables that form the transmission channel. The following formula can be used to calculate the 2 Mbit/s channel delay T:

$$T = T_{OTN-SDH} + N * T_N + T_0 \tag{2}$$

In Eq. (2), $T_{OTN-SDH}$ is the multiplexing/demultiplexing delay from the tributary to the line and from the line to the tributary. N indicates the total number of intermediate nodes. T_N indicates the device delay of the intermediate node. T_0 is the fiber delay.

4   Calculation of Transmission Delay in the OTN OSU Architecture

In the OTN OSU structure, dedicated DM bytes are defined for delay calculation, and the port uses the IEEE 1588v2 protocol to calculate the delay of source and sink nodes in real time. The NMS can automatically display the real-time network delay.

4.1   Functional Model of the OTN OSU Device

An OTN OSU device consists of the transmission plane, management plane and the control plane. The transmission plane includes the service adaptation, cross-connection grooming, OAM, protection, and the synchronization modules. The management plane is connected to devices through the management interface to perform the management functions of the transmission plane and the entire system and coordinate the planes. It provides performance management, fault management, configuration management, and security management functions. This standard does not involve the control plane.

4.2   OTN OSU Access Devices

Electrical-layer OTN OSU devices are generally deployed at the start and end nodes of a customer network. Based on specific requirements, a fixed number of E1, FE, GE, 10GE LAN, STM-1, and STM-4 services can be accessed and carried on OTN networks. These devices support OSU cross-connections and optionally support ODUk cross-connections. The devices should be easy to manage and support plug and play. The management information of these devices should be transmitted to the OTN devices deployed at central offices (COs) through GCC channels. The functional model of an OTN OSU access device is showed in Fig. 3.

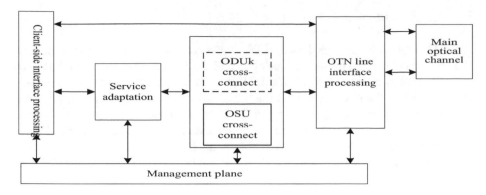

Fig. 3. Functional model of the OTN OSU access device

Generally, two main optical channels are used in active/standby mode. Based on the OSU cross-connections, OTN OSU access devices need to collaborate with optical-layer devices at backbone nodes. The logical function model of optical-layer and electrical-layer centralized cross-connections of OTN devices is showed in Fig. 4.

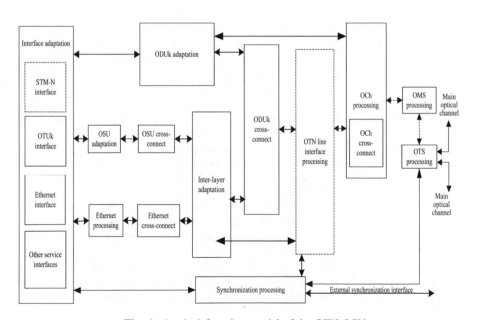

Fig. 4. Logical function model of the OTN OSU

5 Conclusion

The OTN OSU solution uses dedicated DM bytes for delay information transmission. The NMS can visualize the network delay in real time, which is better than the manual delay evaluation method of SDH. The OSU defines the timestamp byte for the transmission clock at the physical layer of the transmission protocol. In this way, physical-layer pre-processing is not required when the IP-based IEEE 1588v2 protocol is used, and the natural time transmission protocol is formed. The precision of the OSU is higher than that of the SDH delay estimation mode. In addition, the clock transmission protocols of different vendors are consistent and can be interconnected. Compared with SDH and IP networks, the OTN OSU networks are more suitable for WAN inter-connection of time-sensitive network services.

Acknowledgments. This work was financially supported by the Science and Technology Project (5211XT18008R) of China State Grid Zhejiang Electric Power Co., Ltd.

References

1. Mendiola, A., Astorga, J., Jacob, E., Stamos, K.: Enhancing network resources utilization and resiliency in multi-domain bandwidth on demand service provisioning using SDN. Telecommun. Syst. **71**(3), 505–515 (2018). https://doi.org/10.1007/s11235-018-0523-4
2. Francisco, C., Martins, L., Medhi, D.: Dynamic multicriteria alternative routing for single- and multi-service reservation-oriented networks and its performance. Ann. Telecommun. **74**(6), 697–715 (2019). https://doi.org/10.1007/s12243-019-00715-9
3. Costa, L.R., Lima, F.R.M., Silva, Y.C.B., et al.: Radio resource allocation in multi-cell and multi-service mobile network based on QoS requirements. Comput. Commun. **135**(02), 40–52 (2019)
4. Nguyen, N.T., Liu, B.H., Chu, S.I., et al.: Challenges, designs, and performances of a distributed algorithm for minimum-latency of data-aggregation in multi-channel WSNs. IEEE Trans. Netw. Serv. Manage. **16**(1), 192–205 (2019)
5. Desogus, C., Anedda, M., Murroni, M., et al.: A traffic type-based differentiated reputation algorithm for radio resource allocation during multi-service content delivery in 5G heterogeneous scenarios. IEEE Access **7**, 27720–27735 (2019)
6. Popkov, G.: Application of Siem solutions on multi-service communications networks. Interexpo GEO Siberia **9**, 61–65 (2019)
7. Rowe, B.M., Feijoo, R., Kludy, T.M., et al.: U.S. Patent 15, 714460 (2019)
8. Globa, L., Sulima, S., Skulysh, M., et al.: An Approach for virtualized network slices planning in multiservice communication environment. Inf. Telecommun. Sci. **10**(1), 37–44 (2019)
9. Maharramov, Z., Abdullayev, V., Mammadova, T.: Modelling self-similar traffic of multiservice networks. EUREKA Phys. Eng. **1**, 46–54 (2019)
10. Anitha, N.R., Saran, A., Vinoth, R.: Adaptive resource allocation and provisioning in multi-service cloud environments. Int. J. Sci. Res. Comput. Sci. Eng. Inf. Technol. **5**, 372–381 (2019)
11. Kist, M., Santos, J.F., Collins, D., et al.: Airtime: end-to-end virtualization layer for RAN-as-a-service in future multi-service mobile networks. IEEE Trans. Mob. Comput. 14–18 (2020)

12. Ageev, S., Karetnikov, V., Ol'Khovik, E., et al.: Adaptive method of detecting traffic anomalies in high-speed multi-service communication networks. E3S Web Conf. **157**(9), 04027 (2020)
13. Li, Y., Li, J., Zhao, Y., et al.: End-to-end URLLC slicing based on packet duplication in 5G optical transport networks. IEEE/OSA J. Opt. Commun. Networking **12**(7), 192–199 (2020)
14. Zhao, Y., Yan, B., Li, Z., et al.: Coordination between control layer AI and on-board AI in optical transport networks. J. Opt. Commun. Networking **12**(1), A49–A57 (2020)
15. Fichera, S., Martinez, R., Martini, B., et al.: Latency-aware resource orchestration in SDN-based packet over optical flexi-grid transport networks. IEEE/OSA J Opt. Commun. Networking **11**(4), 83–96 (2019)

Terminal Security Protection System of Power Internet of Things Based on Machine Learning

Yongjun Qi[1] and Haiyan Wu[2(✉)]

[1] Faculty of Megadate and Computing, Guangdong Baiyun University,
Guangzhou, Guangdong, China
[2] Management School South China Business College, Guangdong University
of Foreign Studies, Guangzhou, Guangdong, China

Abstract. The power IoT terminal security system is a key system to understand the operation and control of low-voltage and medium-voltage distribution networks. The distribution network has the function of monitoring and controlling information. The technical characteristics of the Internet of Things are embodied in both error handling and implementation of the structure of the application. The development of power Internet of Things terminal security system technology is an inevitable trend of power system modernization, and power Internet of Things terminal security protection is the main means to ensure reliability. However, with our country's network security situation becoming tense year after year, and the second round of overhaul of the national power system, the network security of distributed network automation systems has attracted more and more attention. This article starts with the Internet of Things network security protection system, focusing on the Internet of Things power terminal to discuss the analysis and research of the security protection system. This article mainly introduces the decision tree method and the reinforcement learning method. This paper uses decision tree method and reinforcement learning method to analyze and research the terminal security protection system of power Internet of Things, and establish a potential mathematical model. The model is solved by the reinforcement learning method, and the analysis status is evaluated, and the model is revised using historical data to improve the accuracy of analysis and detection. The experimental results of this paper show that the reinforcement learning method improves the analysis efficiency of the power IoT terminal security protection system by 13%. Finally, by comparing the security protection system and the efficiency before and after protection, the influence of machine learning on the analysis of the terminal security protection system of the power Internet of Things is analyzed.

Keywords: Machine learning · Power Internet of Things · Terminal security protection · Reinforcement learning method

1 Introduction

1.1 Background and Significance

In the era of machine learning, the Internet of Things plays an increasingly important role in social life and production, and network security is closely related to social

J. Abawajy et al. (Eds.): ATCI 2021, LNDECT 81, pp. 58–65, 2021.
https://doi.org/10.1007/978-3-030-79197-1_8

stability and national security [1]. The combination of network security technology and machine learning technology provides new ideas and guidelines for the development of network security technology, which is worthy of in-depth discussion and research [2]. The Internet of Things mainly uses sensor technology, computer technology and communication technology to detect the docking of entities and networks. It is a new type of network developed in the new era of information and communication. Perceived physical characteristics, wireless information transmission and intelligent information processing are the main features of the Internet of Things, because it can realize rapid communication and information processing [3, 4]. But at the same time, the uniqueness of this method of sensing and transmitting information also promotes the attack of external factors in information transmission, leading to the leakage and reproduction of information, thus threatening the basic security of the entire system [5, 6]. If the security of the Internet of Things is unavailable, its basic security will be compromised. Therefore, strengthening the security protection of the power Internet of Things is a key issue for the development of power companies, and it is also an important support for power companies to achieve innovation and upgrade in the new era [7].

1.2   Related Work

LI Ni provides a method that can evaluate participatory stakeholder innovation in a complex stakeholder environment to solve essential problems [10]. Based on the principle of common value creation, he proposed an analytical framework that illustrates the security protection process, during which stakeholders integrate their resources and capabilities to develop innovative security protection system construction [8, 9]. In order to evaluate this evaluation framework, a number of data were collected in the study. This case represents the significance of machine learning to the security protection system of the power Internet of Things terminal [11, 12]. But because the message collection process is too complicated, the data result is not very accurate.

1.3   Main Content

The innovation of this article is to propose a decision tree method and a reinforcement learning method. Based on the research of the terminal security protection system of the power Internet of Things under the background of machine learning, through the reinforcement learning method, the terminal security protection system of the power Internet of Things is analyzed. Establish calculation methods of decision tree method and reinforcement learning method to provide research guidance for analyzing the terminal security protection system of power Internet of Things in the context of machine learning.

2 Power Internet of Things Terminal Security Protection System Based on Machine Learning

2.1 Decision Tree Method

As the name implies, the so-called decision tree is a tree structure constructed for strategic decision-making. The obvious advantages of some decision trees over other algorithms are: strong interpretability and comprehensibility; can intuitively reflect the characteristics of the data itself, users do not need to know their knowledge too much, they can understand what is expressed through explanation the meaning of; the requirements for data are not high; other technologies often require that the processed data be of a uniform type, but the decision tree can handle data types and common attributes.

Therefore, the data preparation of the decision tree is simpler than other algorithms, and good results can be quickly obtained when dealing with more types of data. The accuracy of the determination model of the decision tree can be measured, and the decision tree is a white box model. As long as any model can generate a decision tree, it is easy to infer the corresponding logical process. The calculation amount of the decision tree is small and can be easily converted into classification rules. Along the root node of the decision tree all the way down to the leaf nodes, a unique classification result can be determined. Because of its simplicity, intuitiveness, ease of grasping, strong applicability, and ability to combine qualitative and quantitative analysis, the decision tree method has gradually attracted the attention and application of power Internet of Things security protection system builders.

2.2 Reinforcement Learning Method

Reinforcement learning is a learning method that establishes a communication agent between the agent and the environment, makes the two interact, and learns from it. The agent can obtain the current state of the environment and perform operations based on the current state. The action of the agent will affect the state of the environment at the next moment, and the agent will receive feedback as the state of the environment changes. If the environment changes to the desired situation, the agent will receive a positive response, otherwise it will receive a negative response. Through this repeated interaction with the environment, it can be expected that through this repeated interaction with the environment, it can be expected that the agent will eventually be able to understand the best solution to achieve the best return according to the environment, thereby obtaining the greatest return.

The agent can learn the strategy function a to obtain the function distribution of the specified action in a given state m.

$$\alpha(b|m) = Q[B_t = b|M_t = m] \tag{1}$$

Alternatively, the value function U can be studied according to a given state M and action f to obtain an estimate of the quality of the future state, thereby indirectly assisting the selection of action m, and thus the development trend of the selected m.

$$U_{*(m)} = \max_a \sum_{m'f} q(m',f|m,a)\,[m + \beta U_*(m')] \qquad (2)$$

Reinforcement learning value function or strategy function can be realized in different ways according to the actual situation.

In the field of cyber security, the environment itself is digital. The network throughput, service process operation status, resource reservation, etc. can be easily assigned to the neural network for learning, and the learning time is almost the same as other learning methods. Reinforcement learning can use verified learning to detect network security and eliminate server vulnerabilities. Use the deep neural network model to learn, regard the state of the server as the state of the environment, and attack the server to gain access rights or interfere with its normal operation as a trained object. The benefit of using reinforcement learning is also reflected in the ability to change its attack strategy at any time and continue to function as the server software version is upgraded.

3 Terminal Security Protection System of Power Internet of Things Based on Machine Learning

3.1 Experimental Design of Safety Protection System

The use of the Internet of Things in various industries provides a new impetus for the transformation of related industries. As a major part of society, power companies have also shown a rapid development trend in the use of the Internet of Things, but it should be noted that while making full use of the Internet of Things technology, they should also be widely used to achieve their security. It is also organizing targeted safeguard measures to gradually improve its own safety level and create a strong impetus for the modern development of new energy power generation.

The first is to use encryption technology, mainly electronic tag RFID technology, which is essentially a wireless communication technology. The current academic and practical circles are rich in research on RFID. The second is that the authentication technology is based on the actual work of the Internet of Things, the process of authenticating the true identity of the document applicant, and eliminating potential security threats through authentication, thereby effectively preventing resource leakage. It is also a recognized typical method for achieving hierarchical management.

Finally, the mechanism of intrusion detection and protection: This technology is an "after the fact" protection mechanism, that is, after the administrator determines that they are threatened by the Internet of Things, they will organize many security operations. Intrusion detection includes network intrusion detection, host-based intrusion detection and component-based intrusion detection. It is found that only through the cooperative work of detection and protection can the ideal protection effect be achieved.

3.2   Experimental Data Collection of Safety Protection System

The focus of security management is to protect the power company's data, infrastructure, strategies, etc. from external factors, and to institutionalize management through a series of rules and systems. In fact, given the complexity of the IoT security environment, it is difficult to achieve comprehensive security protection only by relying on this technology, so it is necessary to pay close attention to guidance. Network communication security aims to protect the data in the network and the communication software and hardware supporting the operation of the Internet of Things, and prevent network communication equipment from being interrupted or even paralyzed. Media security has a fundamental impact on the security of the power Internet of Things. Its core goal is to ensure that hardware media such as infrastructure and communication equipment in the Internet of Things are not disturbed due to diversified reasons such as man-made damage, negligent operation, and natural disasters. Based on the current basic situation, this article chooses to adopt specific technologies such as encryption technology, authentication technology, and intrusion detection and protection mechanism to detect the security level of the security protection system. Therefore, this paper conducts security protection tests on encryption technology, authentication technology, intrusion detection, and protection mechanism. The data results are shown in Table 1:

Table 1. Test data table of safety protection system

Working methods	Encryption technology	Authentication technology	Intrusion detection	protection mechanism
Traffic safety	78	77	61	66
Information confidentiality	69	78	56	71
Virus detection	59	61	49	63

It can be seen from Table 1 that when security management cannot be managed systematically in the four aspects of encryption technology, authentication technology, intrusion detection and protection mechanism, security management has the effect of data, infrastructure, and policy flow security and information confidentiality. The average value of the data obtained from the three aspects of virus effectiveness is 68.6, 72, 53.3 and 66.6 respectively. From the average point of view, the overall level is low, which also indicates that the security system is relatively low.

4   Data Collection and Analysis of Safety Protection System Experiment

4.1   Safety Protection Reliability Analysis

The reliability of security protection is the core part of the experimental analysis. The experiment will start from the perspectives of encryption technology, authentication technology, intrusion detection and protection mechanism, use the application protection system implemented in this article to process the test code, and then use the disassembly tool to test Compare with the unprotected. The analysis results are shown in Fig. 1.

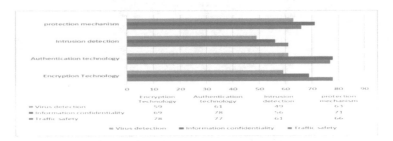

Fig. 1. Experimental data diagram of safety protection system

It can be seen from Fig. 1 that when the security management cannot be systematically managed in the four aspects of encryption technology, authentication technology, intrusion detection and protection mechanism, the security management is for data, infrastructure, policy traffic security and information confidentiality. The three aspects of virus detection degree are not very effective, especially the virus detection degree data in Fig. 1 shows that the security system security degree is relatively low.

4.2 Efficiency Data Before and After Protection

The application security protection scheme proposed in this paper is mainly aimed at improving the operation mode of the ART virtual machine adopted by the power IoT terminal security protection system based on machine learning. The prerequisite for the security protection scheme proposed in this paper to ensure the security protection performance must not affect the efficiency of the ART virtual machine when the application is started. It is not worth the gain to improve the security but lose the efficiency.

In the experimental analysis at this stage, the main purpose is to analyze the impact of the program on the startup of the power Internet of Things terminal security protection system based on machine learning. It is very time-consuming for the ART virtual machine to optimize and transcode the DEX bytecode into machine code. The time it takes is much longer than the optimization time under the Dalvik virtual machine. The application packer method is used to put the unpacking step in operation. It will greatly affect the efficiency of application startup, so this article proposes to strip the core code for packer processing, and only a small part of the code is optimized at startup. This processing method is a good balance between safety and efficiency. In this paper, the following experiments are done in response to this problem. The efficiency data results before and after protections are shown in Table 2:

Table 2. Efficiency data before and after protection

Before and after protection	Start time before safety protection (ms)	Start-up time after safety protection (ms)
ART 5.1	110	139
ART 5.0	120	138
Dalvik 4.3	105	140
Dalvik4.0	112	128

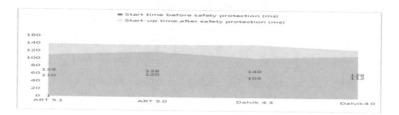

Fig. 2. Efficiency data graph before and after protection

In Table 2, the average start-up time of ART 5.1, ART 5.0, Dalvik 4.3, and Dalvik 4.0 before and after the security protection is 111.75 and 136.25, respectively. There is no big difference between the two, and they are at a normal level. This means that the protection has little effect on the ART virtual machine and the Dalvik virtual machine. After reorganizing according to the target requirements, it is more conducive to decision-making analysis, and the inner relationship between the data can be seen more clearly. The analysis results are shown in Fig. 2.

5 Conclusions

Although this article has made certain research results on decision tree method and reinforcement learning method, there are still many shortcomings. In the context of machine learning, there are many in-depth research methods for the terminal security protection system of electric power Internet of Things. This article conducts experiments on the startup efficiency of the security protection system and the protection of the virtual machine. The test results show that the security protection system needs to be continuously improved. The protection does not have a major impact on the startup of the virtual machine. However, there are still many steps in the analysis process that have not been involved due to reasons such as length and personal ability. In addition, the actual application effect of the improved algorithm can only be compared with the traditional model from the level of theory and simulation.

Acknowledgements. Supported by ① the Characteristic Innovation Project in Natural Science of Department of Education of Guangdong Province.(No. 2018KTSCX256) ② the Project in Key Fields of New Generation Technology of Department of Education of Guangdong Province (No. 2020ZDZX3001).

References

1. Liu, Z., Xu, G., Tong, H.: Magnetic smart and intelligent security systems based on Internet of Things (2017)
2. Wei, Y.: Research on the application and security protection of HTTPS in power business system% HTTPS encryption protocol in the power business system and security protection research. Electr. Power Inf. Technol. **016**(001), 39–43 (2018)
3. Zhu, W., Yin, N., Zhu, W., et al.: Research on the mass sports fitness based on internet of things and intelligent system. Revista De La Facultad De Ingenieria **32**(3), 376–382 (2017)
4. Zhang, Y., Zhang, Y., Zhao, X., et al.: Design and data analysis of sports information acquisition system based on internet of medical things. IEEE Access **8**, 84792–84805 (2020)
5. Song, Z., Lee, I., Zhou, Z.: Intelligent power management system of apartment based on internet of things. Bol. Tec./Tech. Bull. **55**(20), 631–636 (2017)
6. Wang, Z., Zhuang, Y., Yan, Z.: TZ-MRAS: a remote attestation scheme for the mobile terminal based on ARM TrustZone. Secur. Commun. Netw. **2020**(7), 1–16 (2020)
7. Chen, S., Bai, R., Wang, W.: Research and analysis of wireless smart home security system based on internet of things. J. Jiamusi Educ. Coll. (010), 374–375 (2017)
8. Meng, Q., Yu, R.: Research on safety protection of industrial control system in thermal power plant based on risk analysis% research on safety protection of industrial control system in thermal power plant based on risk analysis. New Comput. **001**(001), 35–39 (2018)
9. Li, R.: Research on the reliability of wide area protection communication system for power grid and simulation analysis. Bol. Tec./Tech. Bull. **55**(14), 518–524 (2017)
10. Ni, L., Pu, C., Menglin, L., et al.: Research on new security protection system and technology of power IMS administrative switching netword. Comput. Netw. **045**(011), 62–65 (2019)
11. Wang, Z.: Research the analysis of information security based on WEB access behavior. Sci. Technol. Vis. (012), 247–248 (2016)
12. Singh, R., Ranga, V.: Performance evaluation of machine learning classifiers on internet of things security dataset. Int. J. Control Autom. **11**(5), 11–24 (2018)

Analysis of Output and Load Characteristics of VPP in Consideration of Uncertainty

Caixia Tan[1(✉)], Zhongfu Tan[1], Jianbin Wu[2], Huiwen Qi[2], Xiangyu Zhang[2], and Zhenbo Xu[2]

[1] School of Economics and Management, North China Electric Power University, Beijing 102206, China
[2] Economic and Technical Research Institute of Shanxi Electric Power Company, State Grid, Taiyuan 030000, Shanxi Province, China

Abstract. With the development of virtual power plants (VPP), its output and load characteristics are the key factors affecting the capacity configuration of virtual power plants. Therefore, in order to reduce the cost of VPP capacity allocation, it is necessary to study the output and load characteristics of VPP under different scenarios. This paper first considers the uncertainty of the VPP, and builds the VPP scheduling optimization model to maximize the expected profit. Secondly, discover the key factors that affect the output and load characteristics of the VPP. Finally, different scenarios are set, and a VPP is used for example analysis. The results of the calculation example show that the degree of compensation of flexible load and the reliability requirements of energy use are the key factors affecting the load of the VPP. The wind and solar load forecast accuracy and the efficiency of each unit are the key factors affecting the output of the VPP.

Keywords: Uncertainty · Virtual power plant · Output characteristics · Load characteristics

1 Introduction

As the problem of environmental pollution is becoming more and more prominent, the agglomeration of distributed energy in VPP has become an effective way to solve environmental problems. However, in the planning and capacity allocation stage of VPP, wind power and photovoltaic power generation are uncertain, which leads to great uncertainty in the capacity allocation of VPP. The output and load characteristics of the VPP are the key factors that affect the capacity configuration of the VPP. In order to reduce the capacity configuration cost of the VPP and improve the configuration efficiency of the VPP, it is extremely important to study the output and load characteristics of the VPP under different scenarios.

Regarding the research on the output and load characteristics of the VPP, literature [1] considers the uncertainties faced by the VPP and its internal profit distribution problems on this basis, and establishes a day-ahead scheduling model of the VPP that takes into account the risk preference. Literature [2, 3] considers the operating characteristics of different types of user flexible loads, analyzes and models the characteristics of industrial loads, commercial loads and household loads, and establishes a VPP purchase and sale of electricity considering the deep interaction of diversified

© The Author(s), under exclusive license to Springer Nature Switzerland AG 2021
J. Abawajy et al. (Eds.): ATCI 2021, LNDECT 81, pp. 66–78, 2021.
https://doi.org/10.1007/978-3-030-79197-1_9

flexible loads. Optimization model. Literature [4, 5] analyzed the energy consumption characteristics of temperature control load, and proposed a VPP collaborative optimization control strategy based on temperature control load. Literature [6, 7] studied the potential of air conditioning cluster control, and proposed a load group scheduling control strategy based on diversity maintenance. Existing research can propose collaborative optimization control strategies based on load characteristics, but the key influencing factors of VPP output and load are rarely involved.

Based on the above research, this paper firstly constructs an optimal scheduling model of VPP that takes into account the uncertainty of the VPP's expected return. Second, the key factors that affect the output and load of the VPP are identified. Finally, take VPP in different scenarios as examples to analyze the calculation examples to verify the effectiveness of the model.

2 VPP Scheduling Optimization Model Considering Uncertainty

Since wind power and photovoltaic output are directly related to wind speed and solar radiation, they are significantly random and intermittent. If we analyze wind speed and solar radiation data on longer time scales such as monthly, quarterly, or annual, we can see that they show very obvious seasonal periodicity. Therefore, scenario analysis methods will be used to generate multiple scenario sets based on historical weather data to solve the uncertainty of distributed wind power and distributed photovoltaic power output. The details are shown in Fig. 1:

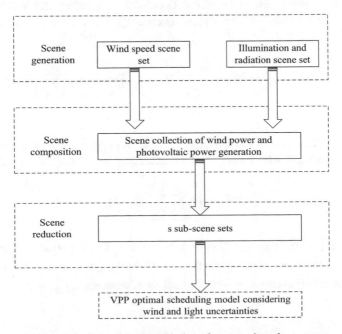

Fig. 1. Process of dealing with uncertainties of scenery based on scene analysis

2.1   Objective Function

After considering the uncertainty of wind and light, the objective function of the optimal scheduling model of the VPP is to maximize the expected return. Specifically as shown in Eq. (1):

$$maxTR^{all} = I_{sell}^{PB} - \sum_{i=1}^{N_{IB}} C_i^{IB} - \sum_{i=1}^{N_W} C_i^W - \sum_{i=1}^{N_{PV}} C_i^{PV} - \sum_{i=1}^{N_{FC}} C_i^{FC} - \sum_{i=1}^{N_{BT}} C_i^{BT} - C^{in} \quad (1)$$

Where I_{sell}^{PB} is the revenue of the VPP from selling electricity to end users in a specific area; $\sum_{i=1}^{N_{IB}} C_i^{IB}$ is the economic compensation to the user after the system calls the interruptible load; $\sum_{i=1}^{N_W} C_i^W$ is the power generation cost of distributed wind power; $\sum_{i=1}^{N_{PV}} C_i^{PV}$ is the power generation cost of distributed photovoltaic; $\sum_{i=1}^{N_{FC}} C_i^{FC}$ is the cost of gas distributed power generation; $\sum_{i=1}^{N_{BT}} C_i^{BT}$ is the charging cost of the battery; C^{CW} is the pumping cost of the pumped storage system; C^{in} is the cost of purchased electricity; N_{IB} is the number of users participating in PBDR and IBDR; N_W is the number of distributed wind power in the VPP; N_{PV} is the number of distributed photovoltaics; N_{FC} is the number of distributed gas generating sets; N_{BT} is the number of batteries. These eight parts can be expressed in the following form:

Based on the peak-valley time-of-use electricity price policy and load transfer, the VPP operator will obtain electricity sales revenue I_{sell}^{PB} from the user side, as shown in Eq. (2):

$$I_{sell}^{PB} = \rho_p \sum_{i=1}^{N_{IB}} \sum_{t=1}^{T_p} (L_{i,t}^{PB} - \mu_{i,t}^{IB} \Delta L_{i,t}^{IB}) + \rho_f \sum_{i=1}^{N_{IB}} \sum_{t=1}^{T_f} (L_{i,t}^{PB} - \mu_{i,t}^{IB} \Delta L_{i,t}^{IB}) + \rho_v \sum_{i=1}^{N_{IB}} \sum_{t=1}^{T_p} (L_{i,t}^{PB} - \mu_{i,t}^{IB} \Delta L_{i,t}^{IB})$$

$$(2)$$

When the VPP invokes the interruptible load, it needs to provide economic compensation to the users who reduce the load according to the plan at the price ρ_{IB}:

$$C_{i,t}^{IB} = \rho^{IB} \sum_{t=1}^{T} \sum_{i=1}^{NB} \mu_{i,t}^{IB} \Delta L_{i,t}^{IB} \quad (3)$$

Where $\Delta L_{i,t}^{IB}$ is the load reduction of calling user i in time period t according to the contract; $\mu_{i,t}^{IB}$ is the load reduction status of user i during t, if user i responds to the load reduction command during t period, then $\mu_{i,t}^{IB} = 1$, otherwise, $\mu_{i,t}^{IB} = 0$.

The power generation costs C_i^W and C_i^{PV} of distributed wind power and distributed photovoltaic can be expressed as the following quadratic function:

$$C_i^W = \sum_{t=1}^{T} [a_i^W (G_{i,t}^W)^2 + b_i^W G_{i,t}^W + c_i^W] \tag{4}$$

$$C_i^{PV} = \sum_{t=1}^{T} [a_i^{PV} (G_{i,t}^{PV})^2 + b_i^{PV} G_{i,t}^{PV} + c_i^{PV}] \tag{5}$$

Where a_i^W, b_i^W and c_i^W are the operating cost coefficients of wind turbine i; a_i^{PV}, b_i^{PV} and c_i^{PV} is the operating cost coefficient of photovoltaic generator set i.

The VPP will purchase electricity from the external grid when the power supply is insufficient, and its electricity purchase cost C^{in} is shown in Eq. (6):

$$C^{in} = \sum_{t=1}^{T} (\rho_{in} + \rho_{sp}) \times \mu_t^{in} L_t^{in} \tag{6}$$

Where ρ_{in} is the price at which the VPP purchases electricity from the external large grid; ρ_{sp} is the transmission and distribution price; μ_t^{in} is the state of the power grid supplying power to the VPP, $\mu_t^{in} = 1$ means that the power grid supplies power to the VPP, otherwise $\mu_t^{in} = 0$.

2.2 Constraints

The optimal scheduling of the VPP should meet conditions such as power balance constraints, distributed power output constraints, interruptible load constraints, and battery operation constraints.

(1) Power balance constraints

$$\sum_{i=1}^{N_W} \mu_{i,t}^W G_{i,t}^W (1 - \varphi_W) + \sum_{i=1}^{N_{PV}} \mu_{i,t}^{PV} G_{i,t}^{PV} (1 - \varphi_{PV})$$

$$+ \sum_{i=1}^{N_{FC}} G_{i,t}^{FC} (1 - \varphi_{FC}) + \sum_{i=1}^{N_{BT}} \mu_{i,t}^{dis} G_{i,t}^{BT,dis} + \mu_i^{CWg} G_i^{CWg} + \mu_t^{in} L_t^{in}$$

$$= \sum_{i=1}^{IB} L_{i,t}^{PB} - \sum_{i=1}^{IB} \mu_{i,t}^{IB} \Delta L_{i,t}^{IB} + \sum_{i=1}^{N_{BT}} \mu_{i,t}^{chr} G_{i,t}^{BT,chr} \tag{7}$$

Where φ_W, φ_{PV} and φ_{FC} are the plant power consumption rate of wind power, photovoltaic power generation and gas power generation.

(2) Distributed power output constraints.

Distributed wind power, distributed photovoltaic and gas distributed power generation should meet the upper and lower limits:

$$
\begin{cases}
0 \leq G_{i,t}^{W} \leq \mu_{i,t}^{W} \bar{G}_t^{W} \\
0 \leq G_{i,t}^{PV} \leq \mu_{i,t}^{PV} \bar{G}_t^{PV} \\
\mu_{i,t}^{FC} G_{i,min}^{FC} \leq G_{i,t}^{FC} \leq \mu_{i,t}^{FC} G_{i,max}^{FC}
\end{cases}
\tag{8}
$$

Where $\bar{G}_t^{W}$ and $\bar{G}_t^{PV}$ are the maximum output power of distributed wind power and distributed photovoltaic i; $G_{i,min}^{FC}$ and $G_{i,max}^{FC}$ are the maximum and minimum power generation output of gas-fired distributed unit i, respectively.

Gas-fired distributed power generation should also meet the ramp rate constraints and minimum start-up and stop time constraints, as shown in Eq. (9):

$$
\begin{cases}
\mu_{i,t}^{FC} \Delta G_{min,t}^{FC} \leq G_{i,t}^{FC} - G_{i,t-1}^{FC} \leq \mu_{i,t}^{FC} \Delta G_{max,t}^{FC} \\
(T_{t-1}^{FC,on} - T_{min}^{FC,on})(\mu_{i,t-1}^{FC} - \mu_{i,t}^{FC}) \geq 0 \\
T_{t-1}^{FC,off} - T_{min}^{FC,off})(\mu_{i,t-1}^{FC} - \mu_{i,t}^{FC}) \geq 0
\end{cases}
\tag{9}
$$

Where $\Delta G_{min,t}^{FC}$ and $\Delta G_{max,t}^{FC}$ are the upper and lower limits of the climbing power of gas distributed unit i; $T_{t-1}^{FC,on}$ and $T_{t-1}^{FC,off}$ are the continuous operation time and continuous shutdown time of the unit at t-1; $T_{min}^{FC,on}$ and $T_{min}^{FC,off}$ are the shortest starting time and shortest down time of the unit.

(3) Constraint on callable capacity of interruptible load

$$
\Delta L_{min,t}^{IB} \leq \sum_{i=1}^{N_{IB}} \mu_{i,t}^{IB} \Delta L_{i,t}^{IB} \leq \Delta L_{max,t}^{IB}
\tag{10}
$$

Where $\Delta L_{min,t}^{IB}$ and $\Delta L_{max,t}^{IB}$ are the upper and lower limits of the interruptible load of the VPP dispatching at time t.

(4) Battery charge and discharge constraints.

In order to ensure the service life of the battery, it cannot run the two working modes of charging and discharging at the same time during the scheduling period, as shown in Eq. (11):

$$
0 \leq \mu_{i,t}^{dis} + \mu_{i,t}^{chr} \leq 1
\tag{11}
$$

The charge and discharge power of the battery should meet the upper and lower limits as shown in Eq. (12):

$$
\begin{cases}
0 \leq G_{i,t}^{BT,dis} \leq \mu_{i,t}^{BT,dis} \bar{G}_i^{BT,dis} \\
0 \leq G_{i,t}^{BT,chr} \leq \mu_{i,t}^{BT,chr} \bar{G}_i^{BT,chr} \\
SOC_t^{min} \leq SOC_{i,t} \leq SOC_t^{max}
\end{cases}
\tag{12}
$$

Where $\bar{G}_i^{BT,dis}$ and $\bar{G}_i^{BT,chr}$ are the maximum discharge power and maximum charging power of battery i, respectively; SOC_t^{min} and SOC_t^{max} are respectively the upper and lower limits of the state of charge of battery i.

3 Identification of Key Influencing Factors of Output and Load Characteristics of VPP

Based on the scheduling optimization objective function and constraint conditions of the VPP, the factors that affect the load characteristic curve of the VPP can be determined, including the degree of compensation of the VPP for flexible loads such as interruptible loads and the degree of reliability required by various users. In addition to the factors affecting the load curve, the factors that affect the output characteristic curve of the VPP also include the forecast deviation degree of the wind and solar and the operating efficiency of each unit in the VPP.

3.1 Influence Factors of VPP Load

(1) Compensation degree of flexible load.
    On the premise of meeting the basic power demand, the higher the compensation that the VPP provides to the user, the greater the user's participation in demand response, that is, the greater the possibility of the load curve being adjusted, which is an important factor affecting the VPP load curve factor.
(2) Reliability requirements for energy use.
    The reliability requirements of energy use will also have an important impact on the load curve of the VPP. The users in the VPP, such as hospitals, have high requirements on the reliability of electric power, and the high degree of compensation provided has little influence on their demand response, that is, the flexibility of electric power adjustment is small. For ordinary users, after meeting the basic energy demand, demand elasticity such as air conditioning is relatively large, so the load curve is easily changed.

3.2 Influencing Factors of VPP Output

(1) Wind solar and load forecast accuracy.
    The dispatching of each unit of the VPP arranges the output of each unit according to the forecast value of the wind and load side. Therefore, the output of the VPP depends on the prediction accuracy of the source side such as wind and wind; on the other hand, it depends on the prediction accuracy of the load side. When the forecast accuracy of the wind and load side is higher, the output curve is less likely to be adjusted.

(2) Operating efficiency of each unit.
    The operating efficiency of each unit of the VPP refers to the operating efficiency of gas turbines, boilers and other units on the one hand, and also includes the conversion efficiency between various energies. The operating efficiency and energy conversion efficiency of each unit directly affect the output of each unit. The higher the operating efficiency and conversion efficiency of the unit, the smaller the output value of the unit; otherwise, the greater the output value of the unit.

4 Case Analysis

4.1 Basic Data

In order to analyze the output and load curves of the VPP under different scenarios, four 100 MW wind turbines, two 10 MW photovoltaics unit, ten 40 MW gas generators, five 10 MW energy storage systems, and a 100 MW pumped storage are selected. The output power of the wind turbine at a rated wind speed of 7 m/s is 220 MW, the cut-in wind speed is 1 m/s, and the cut-out wind speed is 10 m/s. The first-order coefficient of the wind power generation cost function is 42 and the second-order coefficient is 15. The constant term is 90. Each photovoltaic generator set covers an area of about 60,000 square meters, the photoelectric conversion efficiency is 15%, and the maximum output power is 10 MW when the light intensity is 1200 W/m^2. The first-order coefficient of photovoltaic power generation cost is 46, and the second-order coefficient is 18. The coefficient of the constant term is 50. The upper and lower limits of the ramp rate of the gas generator set are 30 kW/min to 100 kW/min, the minimum start-stop time is 1 h, the output upper limit is 10 MW, and the lower limit is 3 MW. The rated capacity of the battery energy storage system is 10 MWh, the initial capacity is 1 MWh, the minimum capacity is 1 kWh, the upper limit of charging power is 4 MW, the upper limit of discharging power is 3 MW, and the charging and discharging efficiency is 95%. Pumped storage bears reservoir water volume of 200 m^3, initial reservoir volume is 50, electric energy conversion coefficient during pumping period is 0.9 m^3/MW, working rated power is 100 MW; electric energy conversion coefficient during generation period is 1.2 m^3/MW, generating power The upper and lower limits of power are 50 MW and 150 MW respectively [8–10]. Assume that the electricity price of the external power grid is 0.55 yuan/kWh, the transmission and distribution price is 0.2 yuan/kWh, and the on-grid power price of the VPP is 0.6 yuan/kWh.

4.2 Result Analysis

(1) Load curve analysis of VPP.
    (1) Influence analysis of flexible load compensation degree.
        In order to analyze the influence of different flexible load compensation degree on the load curve of VPP, it is assumed that the reliability requirements of users are moderate and other factors are not changed. Set up three scenarios as follows:

Scenario 1: Set other factors the same, the VPP's compensation for the user's interruptible load is 0.09 yuan/kWh.

Scenario 2: Set other factors to be the same, and the VPP's compensation for the user's interruptible load is 0.05 yuan/kWh.

Scenario 3: Set other factors the same, the VPP's compensation for the user's interruptible load is 0.03 yuan/kWh.

Based on different scenarios, the interruptible load and load demand of users in different scenarios are shown in Fig. 2:

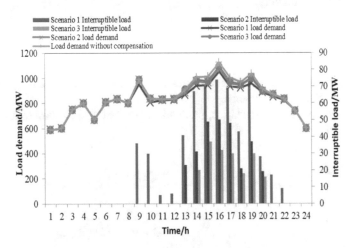

Fig. 2. Interruptible load and load demand of users in different scenarios

It can be seen from Fig. 2 that the interruptible load time under the three scenarios is concentrated in the time period of 13:00–20:00. Among them, the interruptible time distribution of scenario 1 is wider, with interruptible loads distributed from 9:00 am to 22:00 pm. At the same time, in each time period, the participating interruptible loads are ranked as scenario 1 > scenario 2 > scenario 3, indicating that the higher the interruptible load compensation, the higher the enthusiasm of users to participate in the interruptible load.

At the same time, it can be seen from Fig. 2 that the peak load period is 8:00 am and 14:00–19:00 pm. Compared with uncompensated load demand, the three scenarios set in this paper can effectively reduce the peak load, and the load reduction degree of scenario 1 > the load reduction degree of scenario 2 > the load reduction degree of scenario 3. Taking 16:00 in the afternoon as an example, the load demand of each scenario is 1049 MW in scenario 1 > 1073 MW in scenario 2 > 1091 MW in scenario 3 > 1123 MW without compensation.

(2) Impact analysis of user reliability requirements.

In order to analyze the influence of the reliability requirements of different users on the load curve of the VPP, suppose that the interruptible load compensation is 0.09 yuan/kW·h, and other factors are not changed. The reliability of the user is required to be characterized by the user's power failure rate. Set up three scenarios as follows:

Scenario 1: Set other factors to be the same, and the user's power failure rate does not exceed 1%, which means that the user has a high demand for power reliability.

Scenario 2: Set other factors to be the same, and the user's power outage failure rate does not exceed 3%, which means that the user has moderate power reliability requirements.

Scenario 3: Set other factors to be the same, and the user's power failure rate does not exceed 5%, which means that the user has relatively low power reliability requirements.

Based on different scenarios, the load demand of users in different scenarios is shown in Fig. 3:

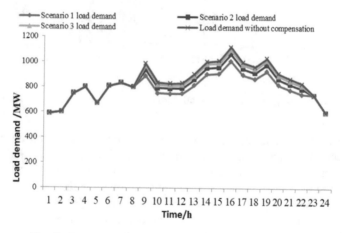

Fig. 3. User load demand curve under different scenarios

It can be seen from Fig. 3 that under different user failure rates, there are differences in user load demand curves. The lower the user outage failure rate, the higher the reduction in user peaks. Taking 9:00 am as an example, the load demand of scenario 1 is 889MW < 938MW in scenario 2 < 968MW in scenario 3 < 988MW without compensation.

(2) Analysis of output curve of VPP.
   (1) Analysis of the impact of wind forecast accuracy.
       In order to analyze the impact of different wind and solar forecasting accuracy on the output curve of the VPP, assuming that other factors are also unchanged, the wind and solar forecasting accuracy is characterized by the wind and solar forecast deviation, and three scenarios are set as follows:

   Scenario 1: Set other factors to be the same, and the wind-solar forecast deviation is 2%, which means that the wind-solar forecast accuracy is higher.
   Scenario 2: Set other factors to be the same, and the wind-solar forecast deviation is 5%, which means that the wind-solar forecast accuracy is moderate.
   Scenario 3: Set other factors to be the same, and the wind-solar forecast deviation is 8%, which means that the wind-solar forecast accuracy is low.
   Based on different scenarios, the output curves of each unit of the VPP under different scenarios are shown in Figs. 4, 5, and 6:

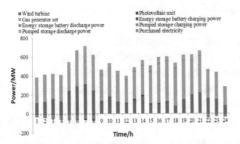
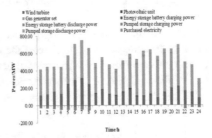

Fig. 4. The output of the VPP in scenario 1   **Fig. 5.** The output of the VPP in scenario 2

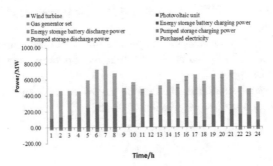

Fig. 6. The output of the VPP in scenario 3

From Figs. 4, 5, and 6, we can see that, on the one hand, the output of the VPP is mainly provided by wind turbines and gas generators. During the low load period from 2:00 am to 8:00 am, energy storage batteries and pumped storage power stations charging; during the peak load period from 9:00 a.m. to 21:00 p. m., the energy storage battery and the pumped storage power station are discharged. At the same time, the VPP will purchase electricity from the Internet during the peak load period.

On the other hand, as the forecast deviation of wind power and photovoltaics increases, the output power of gas turbine units increases accordingly. Taking 7:00 in the morning as an example, the output power of the gas turbine unit in scenario 1 is 401 MW > the output power in scenario 2 is 441 MW > the output power in scenario 3 is 461 MW. This is because the forecast deviation of wind power and photovoltaics has increased. In order to maintain the stable operation of the system, it is necessary to increase the reserve capacity of the system. In the VPP, the unit that provides the reserve capacity is the gas turbine unit. Therefore, as the forecast deviation increases, the gas turbine unit The output power increases. On the other hand, with the increase of gas turbine units, in order to maintain the power supply and demand balance of the system, the charging and discharging power of pumped storage power stations and energy storage batteries decrease with the increase of forecast deviation.

(2) Analysis of the impact of the operating efficiency of each unit.

In order to analyze the influence of the operating efficiency of different units on the output curve of the VPP, assuming that other factors are also constant, the output of the VPP mainly comes from wind turbines and gas generators, so this article only analyzes the impact of gas turbines on the output of the VPP. On this basis, three scenarios are set up as follows:

Scenario 1: Set other factors the same, the operating efficiency of the gas generator set is 95%, which means that the operating efficiency of the gas generator set is higher.

Scenario 2: Set other factors to be the same, and the operating efficiency of the gas generator set is 90%, which means that the operating efficiency of the gas generator set is moderate.

Scenario 3: Set other factors the same, the operating efficiency of the gas generator set is 85%, which means that the operating efficiency of the gas generator set is poor.

Based on different scenarios, the output curves of each unit of the VPP under different scenarios are shown in Figs. 7, 8, and 9:

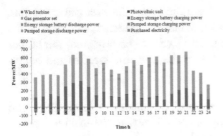

Fig. 7. The output of the VPP in scenario 1   **Fig. 8.** The output of the VPP in scenario 2

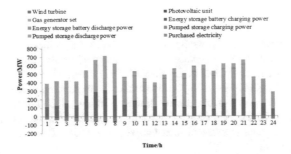

Fig. 9. The output of the VPP in scenario 3

As can be seen from Figs. 7, 8, and 9, as the operating efficiency of the gas turbine unit decreases, the internal output of the VPP is still provided by the gas turbine unit and the wind turbine unit, but the output value of the gas turbine unit decreases. Taking 9:00 in the morning as an example, the output values of gas generator sets under the three scenarios are 267 MW, 252 MW, and 239 MW, respectively. The discharge powers of pumped storage and energy storage batteries have increased. The discharge powers of pumped storage are 43 MW, 50 MW, and 57 MW, respectively; the discharge powers of energy storage batteries are 7 MW, 14 MW, and 21 MW. This is because the lower the operating efficiency of the gas turbine unit, the more the gas turbine unit loses and the lower the output value under the same natural gas scenario. In order to maintain the system's supply and demand balance, pumped storage power stations and energy storage batteries need to increase their output.

5 Conclusions

In this paper, considering the uncertainty, a VPP scheduling optimization model is constructed, and the key influencing factors of the VPP load and output are mined. A case study of VPP in different scenarios is carried out, and the following conclusions are obtained: the higher the interruptible load compensation, the higher the user's enthusiasm for participating in the interruptible load; the lower the user outage failure rate, the higher the reduction in peak user peaks; When the forecast accuracy of the wind and load side is higher, the output curve is less likely to be adjusted; the higher the operating efficiency and conversion efficiency of the unit, the smaller the output value of the unit.

Acknowledgements. This work is supported by Science and Technology Project of State Grid Corporation of China. (Research on market mechanism and business model of virtual power plant under the background of energy Internet; No.: 1400-202057442A-0-0-00).

References

1. Xin, L, et al.: Economic dispatch of virtual power plant considering demand response in market environment. Elect. Power **53**, 172–180 (2020). (in Chinese)
2. Shengyong, Y., et al.: Power purchase and sale strategy of virtual power plant based on deep interaction of diversified flexible loads. Elect. Power Constr. **42**, 59–66 (2021). (in Chinese)
3. LI, J., Lu, B., Wang, Z., Mengshu, Z.. Di-level optimal planning model for energy storage systems in a virtual power plant. Renew. Energy **188–197**, 165 (2021)
4. Fumin, Q., Jian, Z., Zhi, C., et al.: Cooperative optimization control strategy of electric vehicle and temperature-controlled load virtual power plant. Proc. CSU-EPSA **33**, 48–56 (2021). (in Chinese)
5. Guo, W., Liu, P., Shu, X.: Optimal dispatching of electric-thermal interconnected virtual power plant considering market trading mechanism. J. Clean. Prod. **266–278**, 279 (2021)
6. Yang, D., Lin, H., Lai, J., et al.: Air-conditioning load control strategy under the background of virtual power plant. Power Demand Side Manage. **22**, 48–51 (2020). (in Chinese)
7. Zhu, Y., Wang, J., Cao, X., et al.: Central air-conditioning load direct control strategy and its dispatchable potential evaluation. Elect. Power Autom. Equip. **38**, 234–241 (2018). (in Chinese)
8. Pu, L., Wang, X., Tan, Z., et al.: Is China's electricity price cross-subsidy policy reasonable? Comparative analysis of eastern central and western regions. Energy Policy **138**, 111250 (2020)
9. Pu, L., Wang, X., Tan, Z., et al.: Feasible electricity price calculation and environmental benefits analysis of the regional nighttime wind power utilization in electric heating in Beijing. J. Clean. Prod. **212**, 1434–1445 (2019)
10. Löschenbrand, M.: Modeling competition of virtual power plants via deep learning. Energy **214**, 2277–2291 (2021)

Path Analysis of Implementing Whole Process Engineering Consulting Mode in Power Grid Project

Xiaohu Zhu[1], Si Shen[1], and Cuiliu Liu[2(✉)]

[1] State Grid Anhui Electric Power Co., Ltd., Hefei, Anhui, China
[2] Electric Power Development Research Institute CEC, Beijing, China

Abstract. This paper explores the transformation from the traditional consulting mode to the whole process engineering consulting mode based on the current consulting mode of power grid project and the understanding of the Whole Process Engineering Consulting, and analyzes the path of implementing the whole process engineering consulting mode in power grid projects of Provincial Power Grid Corporation. Through contrastive analysis, three aspects will change caused by mode, then discuss how to change used by deductive reasoning. Based on the realistic, proposing Province electric power company and consulting companies should do some changes in short-term/medium- term/ long -term, conjecturing Whole Process Engineering Consulting Mode will be used in Power Grid Project successfully, which has a positive role in promoting the full process engineering consultation of power grid projects.

Keywords: Transmission and transformation project · Life cycle engineering consulting · Path analysis

1 Introduction

At present, the management mode of power grid projects is DBB (design bidding construction) in the construction process. The construction unit carries out the construction management work, and the construction unit and other participating units sign special technical consulting contracts such as survey and design, engineering construction, engineering supervision, cost consulting, etc. While the consulting work of project is undertaken by the whole process consulting unit under the whole process engineering consulting mode, such as investment consultation, survey and design, bidding agency, project supervision, cost consultation, project management, etc. It has changed the status of various subject responsibilities, the communication becomes from external to internal, greatly reduced the project coordination work, improved the work efficiency, and ensured the project quality [1–3].

1.1 The Management Structure of Whole Process Consulting Mode

The management structure of Whole Process Consulting Mode can be divided into three consulting modes: agent mode, collaborative mode, consultant mode. The

J. Abawajy et al. (Eds.): ATCI 2021, LNDECT 81, pp. 79–85, 2021.
https://doi.org/10.1007/978-3-030-79197-1_10

collaborative mode can be divided into collaborative mode (mainly consulting unit) and collaborative mode (mainly construction unit) according to the main unit [4–6]. The project management structure under the four modes is different, as follows.

(1) Agent mode (Fig. 1)

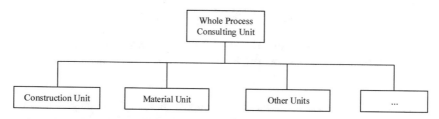

Fig. 1. The department setting of agent mode

(2) Collaborative Mode (mainly consulting unit) (Fig. 2)

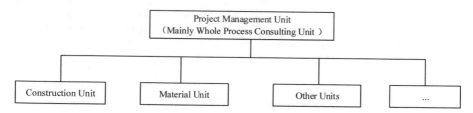

Fig. 2. The department setting of collaborative mode (mainly consulting unit)

(3) Collaborative Mode (mainly investment unit) (Fig. 3)

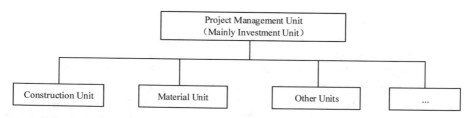

Fig. 3. The department setting of collaborative mode (mainly investment unit)

(4) Consultant Mode (Fig. 4)

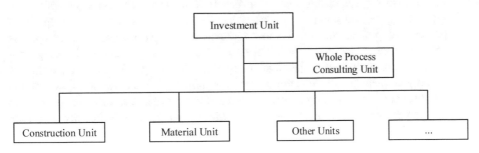

Fig. 4. The department setting of consultant mode

2 Analysis the Change Caused by Mode

In order to actively respond to the national engineering construction management policy and adapt to the change trend of the national engineering construction mode from the traditional parallel contracting mode to the whole process engineering consulting mode [7]. The construction management mode of power grid project needs to carry out relevant changes, but the change of the new mode is not achieved overnight, but gradually [8].

2.1 The Analysis of Change Content Analysis

According to the power grid project comparative analysis of the current management mode and the whole process engineering consulting mode, three dimensions of construction management will be changed, like management unit, professional consultant and project service cycle (Table 1).

Table 1. The transformation caused by project construction mode

Dimension	DBB mode	Whole process consulting
Management unit	Investment unit	Whole process consulting unit
Professional consultant	Many professional consulting units	Whole process consulting unit
Project service cycle	Stage consulting	Whole process consulting

2.2 The Analysis of Transformation Route

In order to adapt to the change of the three dimensions, the management mode of power grid project can be gradually changed from the three dimensions, and the specific transformation path of each dimension is as follows.

(1) The management unit of power grid project is gradually transformed into the whole process engineering consultation unit.

Power grid project management is a professional and technical work with long working cycle and complexity. After decades of construction and management, power grid project construction management units have formed a set of feasible and effective management experience, while the whole process engineering consulting units are lack of project management experience [9]. Based on the difficulty of project management transformation, combined with the service types of the whole process engineering consulting project management, in order to inheritance the power grid project management experience and ensure the quality of power grid project management, it is suggested that the project management mode should be gradually changed from collaborative type (mainly construction unit), collaborative type (mainly consulting unit) to agent type (Fig. 5).

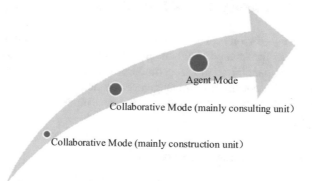

Agent Mode

Collaborative Mode (mainly consulting unit)

Collaborative Mode (mainly construction unit)

Fig. 5. The route of management

(2) The professional consultation of power grid project is gradually transformed into the whole process engineering consultation unit.

The guiding opinions on promoting the development of whole process engineering consulting services (fgzyg [2019] No. 515) proposes to "encourage construction to entrust relevant units to provide whole process consulting services such as investment consulting, bidding agency, survey, design, supervision, cost and project management", which provides a direction for the integration of professional consulting business. The survey and design work is mainly in the charge of the Electric Power Design Institute; the bidding agency, supervision and cost are all in the charge of the professional consulting company [10]. From the perspective of the difficulty of the expansion of the professional consulting business and the integration with the international "architect responsibility system", it is suggested to give priority to the design unit as the whole process engineering consulting unit (Fig. 6).

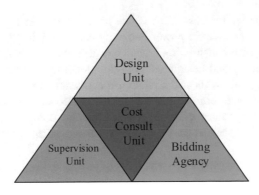

Fig. 6. The selection of professional consultation

(3) The project service cycle is gradually transformed into the whole life cycle.
The service cycle of the project includes the whole life cycle of investment decision, survey and design, bidding, project construction, project completion, project
operation. [11, 12]. According to the current situation that the investment decision,
project construction and project operation and maintenance of power grid project
are led by different management departments, combined with the law of power grid
project construction, it is suggested that the whole process of engineering consulting service cycle should be divided into survey and design, project completion
and investment The three stages of decision-making project completion and
investment decision-making project operation are gradually extended to the whole
life cycle of the project (Fig. 7).

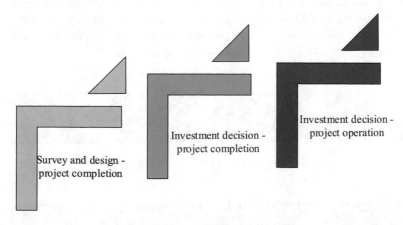

Fig. 7. The route of service cycle

2.3 The Analysis of Transformation Route

Based on the above actual demand of power grid project construction and the
assumption that the whole process engineering consulting is implemented in power grid

project, the development forecast of the whole process engineering consulting in power grid project of province electric power company can be obtained.

Short-term: province construction branch can absorb the relevant units, departments and personnel of survey and design, cost consultation and bidding agency in Province to form a consulting unit to meet the whole process engineering consulting service of survey and design project completion stage, and carried out the whole process engineering consulting of power grid projects within the company's bearing capacity. Through practice, it has formed the whole process engineering consulting experience of the project.

Medium-term: the ministry of construction of province electric power company will gradually increase the number of whole process engineering consulting projects in combination with the effect of whole process engineering consulting projects. The whole process engineering consulting units formed by construction branches can gradually absorb consulting units in the market to form a consortium for whole process engineering consulting, which can not only improve the ability of consulting services, expand the scope of consulting services, but also improve the quality of consulting services. It will form a demonstration effect for the relevant grid consulting service units in the market, gradually transform to the whole process engineering consulting units, and form a multi- market competition subject (Fig. 8).

Long-term: with the emergence of the whole process engineering consulting service effect of power grid project, under the background of multi market competition, the whole process engineering consulting service stage of power grid project will expand to the investment decision-making and project operation stage, and the service scope will increase the project investment consulting and project operation and maintenance services. With the continuous extension of the depth and breadth of the whole process engineering consulting, the power grid will gradually form Project life cycle engineering consulting. In the real sense of the whole process engineering consulting mode, province electric power company will put more energy into the investment control of power grid projects or other more important things (Fig. 9).

Fig. 8. Development path of construction branch.

Fig. 9. Development path of professional consulting company.

3 Conclusions

This paper analyzes the management structure of the whole process consulting mode of power grid project; Based on the current situation of power grid project construction, it predicts the short-term, medium-term and long-term transformation path of construction branch and professional consulting company under the condition of implementing the whole process consulting mode of power grid project, and provides feasible suggestions, which will play an important role in realizing the whole process consulting of power transmission and transformation project in an all-round way To promote the positive role.

References

1. Zhang, S., Wu, J.: Analysis of the strategy of the whole process of the development of engineering cost consulting enterprises. Build. Technol. Res. **3**(2) (2020)
2. Fang, L.: Analysis of the whole process cost consulting management of engineering project construction. Int. J. Sci. **7**(5) (2020)
3. Zhu, D.: The investment consulting and control under the whole process engineering consulting service mode. Build. Technol. Res. **3**(1) (2020)
4. Heigermoser, D., de Soto, B.G., Leslie, E., Abbott, S., et al.: BIM-based Last Planner System tool for improving construction project management. Autom. Constr. **104** (2019)
5. Song, H., Yin, Y.: Research on construction of integrated management platform for whole process engineering consulting based on BIM technology. Taking Urban Rail Transit Project as an Example (2019)
6. Li, S.: Whole process engineering consulting. In: Proceedings of the 2018 International Symposium on Humanities and Social Sciences, Management and Education Engineering (HSSMEE 2018), pp. 140–143. Ed. Atlantis Press (2018)
7. Wang, Q., Zhang, J., Fan, L., Cao, X.: Research on the whole process engineering consulting mode of smart hospital based on BIM project management mode. J. Phys. Conf. Ser. **1744**(3) (2021)
8. Guo, X.: Research on cost management in the whole process of construction project based on computer BIM. J. Phys. Conf. Ser. **1744**(2) (2021)
9. Yu, H., Zhang, T.: Research on application of cloud platform monitoring technology in the whole process of asphalt pavement construction. **10**(01) (2021)
10. Xu, N., Nan, X., Yang, H., Song, Y., Liu, Z.: Research on general cost and engineering investment analysis method of china national network using computer information technology. J. Phys. Conf. Ser. **1578**(1) (2020)
11. Jiao, L.: Analysis of the application of refined management in engineering construction management. In: Proceedings of 2018 1st International Conference on Education, Art, Management and Social Sciences(EAMSS 2018), pp. 122–126. Ed. Clausius Scientific Press, Canada (2018)
12. Wei, G.: Study of engineering construction management based on refined management. Acad. J. Eng. Technol. Sci. **3**(3) (2020)

Application of Computer Simulation Technology in Safety Management of Civil Engineering Projects

Xiongfei Xue[⊠]

Lanzhou Resources and Environment Voc-Tech College, Lanzhou 730021, Gansu, China

Abstract. Since the reform and opening up, they have made great progress in digital technology, information age, science and technology, and information processing technology. With the development and improvement of professional related research in related fields, computer simulation technology has gradually entered people's lives. In the eyes, it continues to develop and grow, and has been applied to many industries and fields. Through specific research on system operation status, computer simulation technology can clearly explain to the public the law of dynamic operation and system status. This article mainly introduces the general programming language of simulation language and Born–Oppenheimer approximation algorithm. This paper uses computer simulation technology to apply the safety management of civil engineering projects, and establishes a potential Born–Oppenheimer approximate algorithm model. The model is solved by the Born–Oppenheimer approximation algorithm, and the application of computer simulation technology in the safety management of civil engineering projects is evaluated, and historical data is used to revise the model to improve the safety management of computer simulation technology in civil engineering projects. The accuracy of the application status assessment. The experimental results in this paper show that the Born–Oppenheimer approximation algorithm improves the safety management efficiency of civil engineering projects by 13%, and reduces the false alarm rate and false alarm rate. Finally, through comparison, it is found that the traditional technology is better than the computer simulation technology in the safety management of civil engineering projects.

Keywords: Computer simulation technology · Safety management of civil engineering projects · Born–Oppenheimer approximation algorithm · General programming language for simulation language

1 Introduction

1.1 Background and Significance

With the advancement of science and technology and the advancement of society, computer application technology provides the possibility for scientific, accurate, multifaceted and effective management of engineering projects [1]. The use of project management means scientific and efficient management of engineering projects.

© The Author(s), under exclusive license to Springer Nature Switzerland AG 2021
J. Abawajy et al. (Eds.): ATCI 2021, LNDECT 81, pp. 86–93, 2021.
https://doi.org/10.1007/978-3-030-79197-1_11

Multi-department management includes engineering projects, multiple data sets, multiple connections, complex management and management of advanced management requirements [2]. Demonstrated the use of computer technology, application management, the use of computer technology and existing demonstration goals. The use of project management refers to the scientific and efficient management of engineering projects [3]. Multi-department management involves project management, multiple data connections, difficult management and high-level management requirements [4]. Traditional project management is based on paper materials and uses manual techniques. Management efficiency and management quality are not high. The use of computer technology can solve major problems, such as data collection and processing, sharing information and design results with the department, exhibition, negotiation and coordination, project schedule management, quality control and monitoring, and assistance to achieve efficient project management [5].

1.2 Related Work

Zhao Lei provides a method to evaluate participatory stakeholder innovation in a complex multi-stakeholder environment to solve essential problems [6, 7]. Based on the principle of common value creation, he proposed an evaluation framework that illustrates the process of social interaction, during which stakeholders integrate their resources and capabilities to develop innovative products and services [8, 9]. In order to evaluate this evaluation framework, a number of data were collected in the study. This case represents the multi-stakeholder environment related to the construction of the innovation and entrepreneurship education curriculum system in local universities [10, 11]. But because the message collection process is too complicated, the data result is not very accurate [11].

1.3 Main Content

The innovation of this paper is to propose a general programming language for simulation language and Born–Oppenheimer approximation algorithm. Based on the application of computer simulation technology in the safety management of civil engineering projects, the Born–Oppenheimer approximate algorithm is used to evaluate the application of computer simulation technology in the safety management of civil engineering projects. The Born–Oppenheimer approximation algorithm combined with a general programming language for simulation language is established to provide research guidance for the application of computer simulation technology in the safety management of civil engineering projects.

2 Application Method of Computer Simulation Technology in Safety Management of Civil Engineering Projects

2.1 Simulation Language Universal Programming Language

It is possible to use computer simulations such as FORTRAN, CII, etc., but the more complex the system, the more difficult it is to solve, which will greatly affect the use of simulation technology. Although the execution efficiency of the simulation language program is very low. However, programming efficiency, consistency and portability are much higher than other high-level languages. Because it contains a large number of functions and software packages, it can intuitively simulate the system without too much work, which can significantly improve the reliability and quality of programming.

Water-saving project is a huge and complex system. In order to simplify the process and simplify computer simulation, the simulation system first needs to create a water-saving project virtual environment, such as land, geography, construction site, navigation building, power station building and water storage area. Modeling objects and 3D simulation environments, such as lighting and sound effects.

Secondly, it is necessary to establish a mathematical model of the movement of the main equipment of the water-saving project (such as ship locks, flume gates, hydraulic turbines, etc.) and simulate the operation mechanism of the water-saving project. Through the above simulation, the virtual roaming system for the water conservancy project is realized in a three-dimensional simulation environment, which dynamically displays information about the running status, visualization and process intelligence of the water conservancy project, and at the same time simplifies the data request for the design and construction of the project attributes and drawing.

2.2 Born–Oppenheimer Approximation Algorithm

The most commonly used method of Born–Oppenheimer approximation algorithm is to deal with the movement of the nucleus and the movement of electrons separately. Although the motion of the nucleus is related to the electron, the mass of the nucleus is a thousand times the mass of the electron. In proportion to electrons, the movement of the nucleus is very slow, so the movement speed of the electron is usually at least one order of magnitude faster than the speed of the nucleus. Therefore, it can be assumed that electrons move freely from the nucleus, thus dividing the system equation into two parts. Under a specific structural configuration, the equation of motion of all electrons is:

$$\left[-\frac{1}{3} \sum_j \Delta_j^3 + \sum_E \sum_{F \geq E} \frac{B_E B_F}{R_{EF}} \right] = \phi = F(R)\phi \tag{1}$$

Among them, REF represents the distance between the nucleus and the nucleus, and Φ is a constant; it is a wave function describing the movement of electrons in a specific nuclear configuration; F(R) represents the total energy of electrons when the molecule is in this configuration. Another equation of nuclear motion, namely:

$$\left[-\frac{1}{4} \sum_B \Delta_B^4 + F(R) \right] \phi = F\phi \tag{2}$$

Among them, F(R) is the wave function of extranuclear motion, and F is the total energy of the molecular system.

3 Application Experiment of Computer Simulation Technology in Safety Management of Civil Engineering Projects

3.1 3D Modeling Design

Taking water conservancy engineering as the research goal, using visual modeling software and real-time driving software, combined with high-performance graphics workstations, accurate and realistic 3D models. The creation of a 3D geometric model is the basis for creating a virtual scene of the entire project, and the creation of the model is mainly divided into 3D terrain modeling and building 3D model. First, optimize the 3D contour model of the earth, use appropriate programs to create a terrain model, and paste textures, such as a map aerial project. By generating a terrain model with accurate "size" and "location" from the perspective of "from above", you can establish a virtual stage layout at a ratio of 1:1, and divide the entire dam into several blocks. After dividing the blocks, the entire modeling task can be shared. The entire large-scale modeling task can be decomposed into block modeling and environment modeling. You also need to determine the location, top shape (or bottom shape) and height of each building, and then you can generate basic 3D shapes from this information and apply texture mapping to the sides and top of the object.

3.2 Interactive Virtual Tour and Visual Analysis Environment Data Collection

In the virtual environment, it is required to provide spatial information and project feature information of the geographical environment and related buildings. Unit attributes include coordinates, location, unit location and other attribute information. Database management technology can be used to perform various operations, such as search and analysis. Fluid simulation is a state-of-the-art mixing technology that uses a particle system to simulate floods with water changes and different changes. Hydrodynamic excitation is a technique that imitates a dynamic environment and uses a particle system to trigger changes in flood height and gates. The water level and the opening angle of the gate, the reaction speed of the obtained particles, the reaction angle, the diffusion angle, the life cycle, the initial color, the final color and other simulation data results are shown in Table 1:

Table 1 Interactive virtual tour and visual analysis environment simulation data table

Simulation data	Dynamic viscosity	Temperature °C	Stability index	Proportion of special phenomena
Geographical environment	99	78	56	5%
Spatial information	87	69	57	6%
Three-dimensional solid	90	88	57	5%

The data in Table 1 is further collected according to Geographical environment, Spatial information, and Three-dimensional solid. In this way, the data in the original database is reorganized according to the target requirements, which is more conducive to decision-making analysis and can be clearer. Seeing the inherent relationship between the data, the analysis results are shown in Fig. 1:

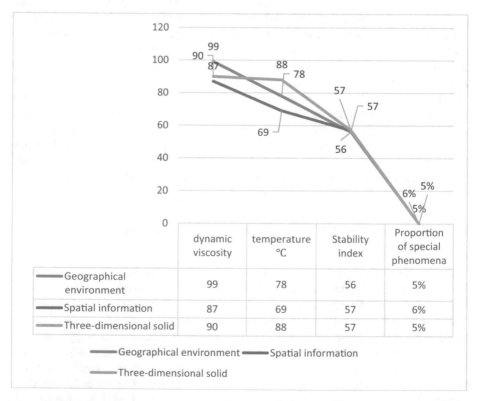

Fig. 1. Interactive virtual tour and visual analysis environment simulation data analysis diagram

From the above figure, the continuous increase in the number of nodes in the cluster will lead to a linear increase in the speedup of the algorithm. Use data to compare the execution performance of Geographical environment, Spatial information, and Three-dimensional solid. Set the support degree to 0.13, and use Geographical environment, Spatial information, and Three-dimensional solid to mine frequent itemsets respectively. Since the test set is smaller than the usual normal value, this article defines the data block of the imported function as 7, so the test set is divided into 4 blocks and sent to each node of the cluster for execution. Temperature and viscosity are the most important factors. The relationship between viscosity and temperature is very close. The lower the temperature, the lower the viscosity and temperature, and the lower the viscosity of the liquid as the temperature increases. Since moisture comes from molecular forces, the temperature increases, the greater the distance between molecules, the smaller the molecular strength and internal turbulence, and therefore the viscosity decreases.

4 Application Analysis of Computer Simulation Technology in Safety Management of Civil Engineering Projects

4.1 Status Quo of Safety Management of Civil Engineering Projects

Civil engineering management personnel did not really realize the responsibility of civil engineering safety in production, and the supervision of safety and civilized construction was weak, and there was no inspection and no fines. In addition, some civil construction managers are busy with daily work, only talking about safe production and civilized construction verbally, or undergoing field inspections that affect civil construction safety.

At present, many engineering construction companies are not strong in safety inspections. They not only have not formed a standardized safety inspection system, but also have not fully implemented the existing safety inspection system, and lack effective inspections for certain items or links that are prone to safety. Some safety inspections are only formal and may not play a real role in supervision and inspection.

The construction site and location of the project are complex, the construction site has a lot of machinery and equipment, the workload is large, the construction time is relatively short, and there are many external objects, buildings and geological sites. Closely related, there are more uncertainties and life-threatening issues. To make matters worse, any problems during the construction process will cause major losses and other potential safety hazards.

4.2 Application Analysis of Computer Simulation Technology

Engineering construction and management technology is a very complex and huge system engineering. Computer simulation technology is built on the theoretical basis of control theory, comparison theory, information processing technology and computer technology. Use computers and other special equipment as tools, use system models to test real or virtual systems, and use experience and expertise to analyze and study statistical data, information and test results, and then develop complete experimental decision-making techniques. In order to visualize the simulation technology, this paper analyzes the terrain treatment of the water-retaining structure, the three-dimensional model of the building, and the flood discharge of the surface hole. The data is shown in Table 2:

Table 2 Application data sheet of computer simulation technology

Simulation data	Efficient	Stability index	Proportion of special phenomena
Geoprocessing	90%	79	83%
Basic three-dimensional building	89%	88	82%
Flood discharge from surface holes of retaining structures	88%	91	83%

In this paper, the data is further collected according to Geoprocessing, Basic three-dimensional building and Flood discharge from surface holes of retaining structures, etc., so that the data in the original database is reorganized according to the target requirements, which is more conducive to decision analysis. You can see the inner relationship between the data more clearly, and the analysis results are shown in Fig. 2:

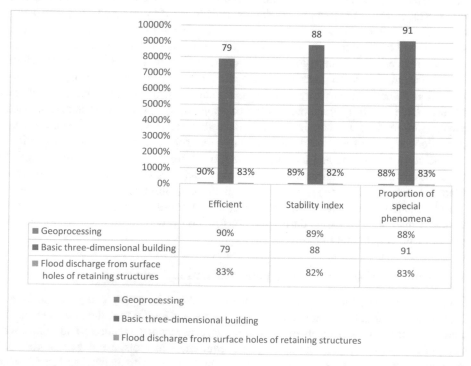

Fig. 2. Data analysis diagram of computer simulation technology application

The Figure shows that the Knowledge representation, Search technology and Compressed image used by computer simulation technology provide the most suitable data results. This is a digital manifestation of civil engineering safety management and an improvement in safety.

5 Conclusions

Although this dissertation has made some research results on the general programming language of simulation language and Born–Oppenheimer approximation algorithm, there are still many shortcomings. Computer simulation technology has a lot of in-depth content worth studying for the safety management of civil engineering projects. There are still many steps in the application process that have not been involved because of space and personal ability. In addition, the actual application effect of the improved algorithm can only be compared with the traditional model from the level of theory and simulation.

References

1. Wang, C., Guo, M., Du, H.: Application of simulation technology in safety production management of petrochemical enterprises. Chem. Eng. Des. Commun. **043**(004), 53 (2017)
2. Xu, H., Zhou, G., Jiang, F., et al.: Application of BIM technology in safety management of engineering construction. Jiangsu Sci. Technol. Inf. **035**(007), 49–51 (2018)
3. Baolin, M.: Analysis of the application of BIM technology in safety management of expressway bridge construction. Eng. Constr. Des. **000**(004), 226–227 (2019)
4. Krejsa, M., Janas, P., Krejsa, V.: Application of the DOProC method in solving reliability problems. Appl. Mech. Mater. **821**, 717–724 (2016)
5. Veblen, M., Ferdinand, B.: Application of computer simulation in the optimization of plastics processing technology. Appl. Comput. Simul. **001**(002), 37–42 (2018)
6. Tong, L.: Application of BIM technology in construction safety management. Value Eng. **037**(003), 27–28 (2018)
7. Zhao, H.B., Long, Y., Li, X.H., Liang, L.: Experimental and numerical investigation of the effect of blast-induced vibration from adjacent tunnel on existing tunnel. KSCE J. Civ. Eng. **20**(1), 431–439 (2016). https://doi.org/10.1007/s12205-015-0130-9
8. García-Macías, E., Castro-Triguero, R., Flores, E.I.S., et al.: An interactive computational strategy for teaching the analysis of silo structures in civil engineering. Comput. Appl. Eng. Educ. **27**(4), 1–15 (2019)
9. Wu, W., Cheng, S., Liu, L., et al.: Source apportionment of PM2.5 in Tangshan, China—hybrid approaches for primary and secondary species apportionment. Front. Environ. Sci. Eng. **10**(5), 1–14 (2016)
10. Santhiya, S., Muthu, D.: Safety management and its application with case studies. Int. J. Civ. Eng. Technol. **8**(6), 196–202 (2017)
11. Zhao, L., Qin, Y., Ma, X.: Research on the application of BIM technology in engineering quality and safety management. Value Eng. **036**(018), 102–105 (2017)

Car Navigation System Based on Cloud Computing Technology

Zhou Fan[✉]

Lanzhou Resources and Environment Voc-Tech College, Lanzhou 730000, Gansu, China

Abstract. Navigation and positioning applications are playing an increasingly important role in the development of the national economy and the improvement of people's lives. Most of the existing navigation and positioning applications show the characteristics of singleness and specificity, and cannot provide comprehensive services for multiple enterprises or the public. This requires research on new technologies and development of integrated navigation and positioning integrated systems that serve the public. In this regard, the purpose of this article is to study car navigation systems based on cloud computing technology. This article first discusses the current situation of my country's car navigation and the main problems. Then, through the method of questionnaire survey, to understand the car owner's like and cognition of the existing car navigation system, so as to more truly understand the car owner's view of the car navigation system. After that, this paper analyzes the related theories and technologies of cloud computing, and demonstrates the theory and technical foundation of cloud computing technology. This article focuses on cloud computing technology and combines its operational characteristics in navigation and positioning applications to design and implement a "cloud computing technology navigation system". This system has multiple functions such as business strategy support, resource monitoring, service migration, and rapid business deployment, and upgrades local navigation and positioning applications for specific groups of people to comprehensive navigation and positioning applications for multiple organizations. In addition, this article uses the communication module to transfer the collected data to the cloud computing platform for processing and analysis. The terminal navigation data analysis and processing module can complete the map display of the path with less fuel consumption. Finally, this article compares the cloud computing navigation system with the traditional navigation system, and analyzes the actual application effect of the cloud computing navigation system. The experimental results show that the car navigation system using cloud computing technology is more recognized and loved by car owners, and the accuracy of the cloud computing navigation system is increased by about 20%, the navigation effect and safety factor are better, and it will be more effective for car navigation products. R&D has certain practical reference value.

Keywords: Cloud computing technology · Car navigation system · Comprehensive navigation · Comparative analysis experiment

© The Author(s), under exclusive license to Springer Nature Switzerland AG 2021
J. Abawajy et al. (Eds.): ATCI 2021, LNDECT 81, pp. 94–102, 2021.
https://doi.org/10.1007/978-3-030-79197-1_12

1 Introduction

Although the application of navigation and positioning application systems is becoming more and more extensive, there are still problems in the market: lack of standardization and scale [1, 2]. On the one hand, many navigation and positioning manufacturers are committed to the development and production of navigation terminals and navigation monitoring, but navigation and positioning application service providers are scarce, and most of the services provided are for specific organizations [3, 4], and specific groups of people. Special services lack popular comprehensive services [5]; on the other hand, due to the lack of infrastructure supporting facilities industry, it is impossible to provide users with complete supporting services [6]. In the market, a certain scale cannot be formed, and there is no navigation and positioning service industry chain that benefits both parties. Therefore, the existing navigation and positioning services can no longer meet the needs of existing navigation and positioning applications [7].

At present, with the continuous development of cloud computing technology, many research results of car navigation systems have emerged at home and abroad [8]. In foreign countries, Petrera has studied an architecture design based on a single service model. A database, a set of navigation and positioning application software and a supporting monitoring system are installed on the operating system of a physical server. Users can navigate and locate smart terminals. To use navigation and location services [9]. In China, Zhang W pointed out the application of cloud computing technology in navigation and positioning application systems and location service systems, and integrated similar applications on the Internet through cloud computing to provide external interfaces for users to use [10].

This article is based on cloud computing technology to study the car navigation system. This article first introduces the status quo and main problems of car navigation systems. Next, through the questionnaire survey method, to understand the car owner's understanding and views on the existing car navigation system. Then, this article analyzes the theories and technologies related to cloud computing, and combines the operational characteristics of navigation and positioning applications to design and implement a "cloud computing technology navigation system". In addition, this article uses the communication module to transfer the collected data to the cloud computing platform for processing and analysis. Finally, this article compares the cloud computing navigation system with the traditional navigation system, and analyzes the practical application effectiveness of the cloud computing navigation system. The car navigation system using cloud computing technology has greatly improved the accuracy and convenience of navigation, and has important reference significance for the further development of car navigation systems.

2 Technical Research of Car Navigation System Based on Cloud Computing Technology

2.1 Cloud Computing Technology

Cloud computing is a new type of data-centric data-intensive supercomputing method. It has its own unique technology in many aspects such as virtualization, data storage, distributed parallel architecture, programming mode, and data management, and it also involves many Other technologies. As a brand-new business operation model, cloud computing virtualizes hardware resources such as CPU, memory, and disk into a "resource pool" through virtualization technology, and disperses the computing tasks of various applications in the "resource pool" and shares them. Provide various computing, storage, service and other resources without any physical limitations. It can not only use large-scale data centers or supercomputer clusters, but also provide users with large-scale, low-cost IT services of computing units through the IP network connected to the five, but also can use the calculations and transferring storage to supercomputers Among cloud computing technologies, programming models, data storage technologies, virtualization technologies, and cloud computing platform management technologies are the most critical.

(1) Programming model

It is a simplified distributed programming model and an efficient task scheduling model for parallel operations of large-scale data sets. Strict programming model can make programming in cloud computing environment simple and fast. The idea of the programming model is to decompose the problem to be executed into a way of mapping and simplifying. First, the data is cut into irrelevant blocks through the mapping program, and the scheduling is assigned to a large number of computer processing to achieve the effect of distributed computing, and then through the simplified program reduces the result to output. The method of principal component estimation is to transform the original regression independent variable to another set of variables, namely the principal component, and select some of the important principal components as the new independent variable (some independent variables with little influence are discarded at this time. Actually achieved the purpose of dimensionality reduction), then use the least square method to estimate the model parameters after selecting the principal components, and finally transform back to the original model to obtain the parameter estimation. Consider the programming model:

$$Y = X\beta + \varepsilon \tag{1}$$

$$E(\varepsilon) = 0, \mathrm{cov}(\varepsilon) = \sigma^2 \tag{2}$$

First standardize the columns of X, then $X'X$ is the correlation matrix, its eigenvalue is $\lambda_1 > \lambda_2 > ... > \lambda_p$, and the corresponding normalized orthogonal eigenvector is $\eta_1, \eta_2...\eta_p$, denoted as

$$Q = \left(\eta_1|\eta_2|...|\eta_p\right)_{p*p} \tag{3}$$

Among them, Q is a standardized orthogonal array, and a new parameter $a = Q'\beta$ is introduced, then the model is transformed into:

$$Y = Za + \varepsilon \tag{4}$$

$$E\varepsilon = 0, \text{cov}(\varepsilon) = \sigma^2 I_n \tag{5}$$

Then the estimate of β can be determined $\tilde{\beta} = Q\tilde{a}$ by the relation $\beta = Qa$, which is called the principal component estimate of β.

(2) Mass data distributed storage technology

Mass data distributed storage technology includes unstructured data storage and structured data storage. Among them, unstructured data storage mainly uses distributed file storage and distributed object storage technology, while structured data storage mainly uses distributed database technology. Different from traditional file systems, this technology is designed for large-scale distributed file storage and application characteristics. It can run on cheap and unreliable nodes to provide users with highly reliable cloud services. In the design, multiple copies can be automatically copied, metadata and user data can be stored separately, and storage and calculation can be combined to use data location correlation for efficient parallel calculation.

(3) Virtualization technology

Virtualization technology can realize the isolation of software applications from the underlying hardware. It includes a split mode that divides a single resource into multiple virtual resources and an aggregation mode that integrates multiple resources into one virtual resource.

(4) Cloud computing platform management technology

The scale of cloud computing resources is huge, the server nodes are numerous and distributed in different locations, and hundreds of applications are running at the same time. How to effectively manage these servers to ensure that the entire system provides uninterrupted services is a huge challenge. The platform management technology of the cloud computing system can enable a large number of servers to work together, facilitate business deployment and activation, quickly discover and restore system failures, and achieve reliable operation of large-scale systems through automated and intelligent means.

2.2 Using Cloud Computing Technology to Innovate Car Navigation Systems

(1) Network communication

At present, some research institutions, scholars and manufacturers integrate GSM/GPRS, WIFI, CAN, Bluetooth, infrared and FM radio frequency and other communication methods into the design of car navigation terminals to provide comprehensive, multi-level and dynamic navigation information and services. At present, it has been widely used in the navigation, positioning and monitoring management of vehicles such as logistics transportation, taxis, buses, traffic police, and bank money transportation. Taking FM radio frequency as an example, the GPS navigation terminal can transmit FM signals, and the car audio system is responsible for receiving FM signals, so as to realize the playback of GPS navigation sounds through the car audio system.

(2) Multimedia support

With the increasing demand for the diversified functions of automotive electronic products, the current car navigation terminal is becoming more and more diversified. With a better visual interface and user operability, it has become an important carrier of car multimedia functions. In addition to maintaining the original navigation capabilities, various multimedia functions have been gradually developed, such as music, movies, mobile TV, web browsing, and Internet chat. Taking mobile TV as an example, the wireless TV signal is received through the CMMB receiver module built into the car GPS terminal, and the user can enjoy TV programs while driving.

(3) Environmental measurement and control

After the joint control, the temperature can be automatically adjusted. The radar ranging can assist the user to reverse or park the car. The electronic dog can receive radar waves. If a radar speedometer is detected at a certain distance, the mobile test instrument used by the police can be warned in advance, Remind the driver to slow down and avoid the ticket. The camera can realize visual reversing and anti-theft video functions through the video detection of the environment inside and outside the car.

(4) Navigation optimization

Navigation optimization mainly includes two aspects: route optimization and map optimization. Some experts and scholars have devoted themselves to the optimization research of the shortest path algorithm, and applied the neural network, bootstrap particle filter and other algorithms to the car navigation terminal. Practice has proved that the application effect is good. At the same time, some electronic map design manufacturers attach great importance to users' needs for visual effects, and gradually improve the functions of electronic maps such as three-dimensional, real map, geographic information and real-time information.

(5) Service operation management

Users can access the required navigation and positioning services through the unified access interface provided by the platform. In order to provide users with convenient and efficient access to services, this platform is designed with a service operation management system for unified management of various navigation and positioning services, including service registration, release, billing, discovery, and use.

3 Experimental Research on Car Navigation System Based on Cloud Computing Technology

3.1 Experimental Data

In this paper, 100 car owners were randomly selected as the survey subjects. In order to ensure the validity and representativeness of the research, we conducted an anonymous questionnaire survey on thcm. This article distributed 100 questionnaires to him, and returned 100 questionnaires, including 100 valid questionnaires.

3.2 Experimental Process

This article issued a questionnaire survey to these 100 residents to understand their views and suggestions on car navigation systems using cloud computing technology and traditional navigation systems. Then, based on the results of the questionnaire survey, this article randomly divided the 100 car owners into two groups A and B, each with 50 people. Among them, the car owners in Group A use the car navigation system using cloud computing technology, while the owners in Group B use the traditional car navigation system. The car navigation system is used for 1 month. This article conducts data statistics during the experiment and compares them to analyze the effect of the car navigation system using cloud computing technology.

4 Experimental Analysis of Car Navigation System Based on Cloud Computing Technology

4.1 Car Owners' Views on Two Car Navigation Systems

In order to study the status quo of my country's car navigation, the main problems that exist, and the car owners' liking and cognition of the existing car navigation system, this paper uses a questionnaire survey and classifies and counts 100 selected car owners. As shown in Table 1 and Fig. 1.

Table 1. Performance comparison of two car navigation systems

	Good intelligence	High accuracy	Highly entertaining	Good convenience	Multiple functions
Cloud computing navigation system	96%	94%	92%	95%	96%
Traditional navigation system	84%	79%	88%	82%	80%

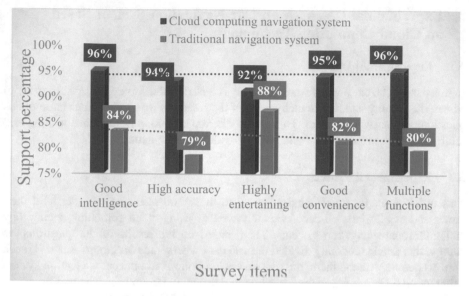

Fig. 1. Relevant views of community residents

It can be seen from the survey data that car owners are satisfied with all aspects of the car navigation system using cloud computing technology, and the support for each survey item has reached more than 90%, which shows that car navigation integrated with cloud computing technology. The system can meet the needs of most car owners for various functions of the car navigation system, and can get their support. However, the traditional navigation system has poor performance and fewer functions, which leads to and cannot meet the actual needs of car owners. The car navigation system designed in this paper based on cloud computing technology greatly expands the functions of the navigation system. It not only has a variety of information processing functions and human-computer interaction functions, but more importantly, it can also transmit and share online navigation information resources to meet people's needs.

4.2 Actual Effect Analysis of Car Navigation System Using Cloud Computing Technology

In order to further verify the actual effect of the car navigation system using cloud computing technology, this article allows the owners of Group A to use the car navigation system incorporating cloud computing technology, while the owners of Group B use the traditional car navigation system for one month. Then the changes in the use effect of these residents are visually expressed, and curve fitting is performed according to the mean value. As shown in Table 2 and Fig. 2.

Table 2. Changes in the effects of the two groups of experiments

	5 days	10 days	15 days	20 days	25 days	30 days
Group A	50%	58%	65%	77%	85%	90%
Group B	50%	53%	55%	60%	64%	70%

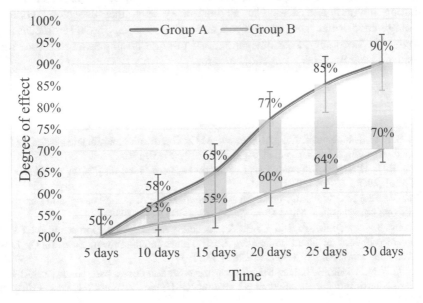

Fig. 2. Changes in the effects of the two groups of experiments

From the experimental data, it can be seen that the actual effect of the car navigation system incorporating cloud computing technology has reached a theoretical level, and the satisfaction of the group A owners who use the car navigation system and the degree of effect have improved over time The speed is very fast, far higher than that of Group B owners who use traditional car navigation systems. This shows that car navigation systems incorporating cloud computing technology can effectively improve people's use effects and truly realize intelligent navigation. Moreover, the car navigation system integrated with cloud computing technology has formed an interactive atmosphere. The owner of the car fully reflects the versatility of the navigation system in the process of using the navigation system, so that its functions can be fully implemented, and the user feels better.

5   Conclusions

This article is based on cloud computing technology to study the car navigation system. This article first introduces the current status of car navigation and its main problems. In addition, understand the car owner's view of the existing car navigation system. After that, this article analyzes the theories and technologies related to cloud computing, and puts forward the theory and technical basis of cloud computing technology. This article focuses on the introduction of cloud computing technology, and combines the operational characteristics of navigation and positioning applications to design and implement a "cloud computing technology navigation system". Finally, this article compares the cloud computing navigation system with the traditional navigation system, and analyzes the practical application effectiveness of the cloud computing navigation system. Car navigation systems that use cloud computing technology have better navigation effects and safety factors, play a significant role in improving the accuracy of car navigation, and provide an important reference for the innovative research of car navigation systems.

References

1. Al-Balushi, I.A., Yousif, J.H., Al-Shezawi, M.O.: Car accident notification based on mobile cloud computing. Int. J. Comput. Appl. Sci. **2**(2), 46–50 (2017)
2. Wu, B.H., Huang, S.J.: A new in-car navigation system based on V2C2V and data science. IT Prof. **20**(5), 68–73 (2018)
3. Jin, W., Fang, L., Wang, L.: Research on the application of mobile navigation system based on cloud computing. J. Phys. Conf. Ser. **1648**(3), 032086 (2020)
4. Huang, S.-C., Jiau, M.-K., Lin, C.-H.: A genetic-algorithm-based approach to solve carpool service problems in cloud computing. IEEE Trans. Intell. Transp. Syst. **16**(1), 352–364 (2015). https://doi.org/10.1109/TITS.2014.2334597
5. Jeong, Y.N., Jeong, E.H., Lee, B.K.: An app visualization design based on IoT self-diagnosis micro control unit for car accident prevention. KSII Trans. Internet Inf. Syst. **11**(2), 1005–1018 (2017)
6. Haimes, Y.Y., Horowitz, B.M., Guo, Z., et al.: Assessing systemic risk to cloud-computing technology as complex interconnected systems of systems. Syst. Eng. **18**(3), 284–299 (2015)
7. Hairu, Z.: Application of cloud computing technology in the university's information construction and development. Softw. Eng. Appl. **08**(2), 32–37 (2019)
8. Yang, D., Jiang, L.: Simulation study on the natural ventilation of college student' dormitory. Procedia Eng. **205**(4), 1279–1285 (2017)
9. Petrera, M., Pfadler, A., Suris, Y.B., Fedorov, Y.N.: On the construction of elliptic solutions of integrable birational maps. Exp. Math. **26**(3), 324–341 (2017). https://doi.org/10.1080/10586458.2016.1166354
10. Zhang, W., Xu, L., Duan, P., Gong, W., Lu, Q., Yang, S.: A video cloud platform combing online and offline cloud computing technologies. Pers. Ubiquit. Comput. **19**(7), 1099–1110 (2015). https://doi.org/10.1007/s00779-015-0879-3

Indoor Air Quality Detector Design Based on ATMEGA32 Single Chip Computer

Wenkang Zhang and Shiying Wang[✉]

College of Information Science and Engineering, Linyi University,
Linyi, Shandong, China

Abstract. In recent years, with the rapid development of the modern science and technology in our country, smart home has gradually changed the life of every family. This design is an air quality test system, which is based on ATMEGA32 microcontroller as the control center. The system can monitor and adjust the concentration of harmful substances in the home environment, such as benzene, formaldehyde, carbon monoxide and solid particles PM2.5. The result of detection can be displayed on LCD screen, and also can be sent to the subscribers' mobile phone by the way of SMS through GSM. The main controller further analyzes collected data, and controls the indoor electrical appliances to adjust the temperature and humidity. In order to save power as much as possible, the system takes the obtained indoor environment parameters into the corresponding algorithm to calculate, so it can be adaptive to adjust itself effectively to keep the indoor air environment good.

Keywords: ATMEGA32 · PM2.5 · Smart home · Sensor · LCD screen

1 Introduction

In today's society, people have long lived in an environment with unsuitable temperature and humidity and harmful gases, which will cause continuous damage to the human body, mainly formaldehyde, benzene, ammonia, radon, volatile organic compounds (TVOC) and the like. Especially harmful gases such as formaldehyde and benzene have been identified as strong carcinogens by the World Health Organization. Since formaldehyde is a major indoor environmental pollutant and highly toxic, it has been identified as a carcinogen and teratogenic substance on the World Health Organization list. It is a recognized source of allergic reactions and is potentially induced by mutagens [1]. One of the strengths. Formaldehyde is a highly irritating volatile organic compound that can irritate human eyes, nose and respiratory tract. Its diameter is smaller than or PM2.5 particles equal to 2.5 microns. It is also called lung particulate matter, although it has little impact on the human body in a short time., But long-term exposure to an environment where PM2.5 exceeds the standard is also very harmful; CO-based gas and alkanes-based natural gas "rampant" in the kitchen, becoming a health "killer" that cannot be ignored: its leakage Human body poisoning and gas explosion caused by diffusion are also shocking everyone's heart; therefore, in order to improve our home environment, a practical smart home environment detection system

J. Abawajy et al. (Eds.): ATCI 2021, LNDECT 81, pp. 103–108, 2021.
https://doi.org/10.1007/978-3-030-79197-1_13

will be an indispensable equipment for housing [2]. The research and development idea of this system came into being based on this environment background.

2 Hardware Design of the System

2.1 The Overall Design of the System

The system is based on ATMEGA32 single-chip microcomputer. The detection of external CO, alkanes and other gases uses MQ-2 type sensors; the PM2.5 detection module selects GP2Y1010AU0F optical dust sensor module to detect the concentration of PM2.5, which is reflected in the module this sensor module can effectively detect solid particles with a very small diameter and ultra-low power consumption with a maximum current of 20mA. The HS1101 is selected as the humidity sensor. The original solid polymer structure of the humidity sensor can make the HS1101 possess Fast response speed, high reliability, etc.; temperature detection uses NTC thermistor, this sensor has the advantages of low price and wide linear working range, easy to use; comprehensive detection modules such as formaldehyde, C6H6, etc., choose to have a thick chip ZP01 air quality module with membrane semiconductor gas sensor. It has good sensitivity to C6H6, formaldehyde and other organic volatile gas sensors [3].

The output display selects LCD12864 character liquid crystal display, which displays a large amount of information, comes with a Chinese character library, which is convenient for program writing to control the display content, and has two communication modes: parallel and serial; at the same time, the system also has GSM communication module, temperature and humidity The control module, motor control module [4], and the system block diagram are shown in Fig. 1.

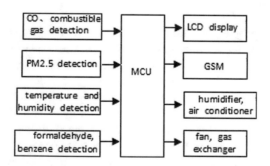

Fig. 1. The structure diagram of the system

2.2 System Power Design

The power supply of the air quality detector adopts a ±5V power supply module composed of LM7805 and LM7905 integrated voltage stabilizers. The initial voltage input is +12V, and the normal operation outputs ±5V DC voltage. D1 is used to determine the direction of current, and C7 is LM7805 voltage stabilizer. The filter capacitor and C4 at the input of the module are the filter capacitors at the output, and

C9 and C8 provide filtering for the LM7905. The power conversion chip TLV1117–3.3 also uses capacitor filtering, which outputs a voltage of 3.3V to supply power to the corresponding modules of the system. A 5.1V Zener diode is added to the input port of the TLV1117 power supply, the purpose is to limit the voltage to the maximum input voltage of the TLV1117 to increase the anti-interference ability of the circuit [5]. The DC stabilized power supply is shown in Fig. 2.

2.3 GSM Communication

The GSM communication module realizes that when an abnormality occurs, the main control sends a short message alarm to the user. The SMS sending and receiving in the gas detector is realized by SIM900A. The GSM module based on the SIM900A chip has two different working frequency bands, the frequency bands are: DCS1800 MHz and EGSM 900 MHz the module has two types of serial ports: ordinary serial port and development and debugging serial port. Users can use serial communication to develop and use conveniently based on this module. When using an external SIM card, you can send and receive short messages only by using AT commands. The SIM900A module adopts the widely popular embedded TCP/IP protocol. The extended TCP/IP AT command allows developers to conveniently call the TCP/IP protocol, which greatly facilitates the user's data transmission. The normal working voltage of the GSM communication module is 3.1 $\sim$ 6.2V, which requires low external power supply [6].

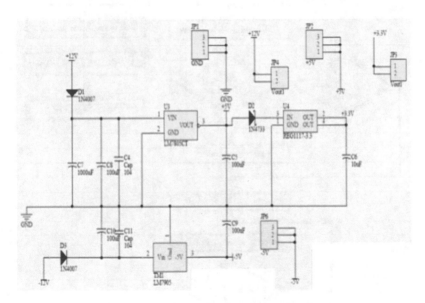

Fig. 2. Circuit diagram of power supply circuit

3 System Software Design

3.1 Overall Software Design of the System

Based on the design requirements of the entire air quality detector, we have planned the overall software design flow chart of the system. The specific process is as follows: When the microcontroller is powered on, first perform the initialization of each function: A/D acquisition initialization, I/O port initialization, serial port initialization, GSM, initialization, LCD12864 initialization, etc. [7]. In the real-time data collection, the temperature and humidity collected by each module and the concentration of each gas are collected into the single-chip microcomputer, and the collected data is converted and processed, and compared with the set threshold value to determine whether an alarm is required or corresponding processing is required [8]. The system flow chart is shown in Fig. 3:

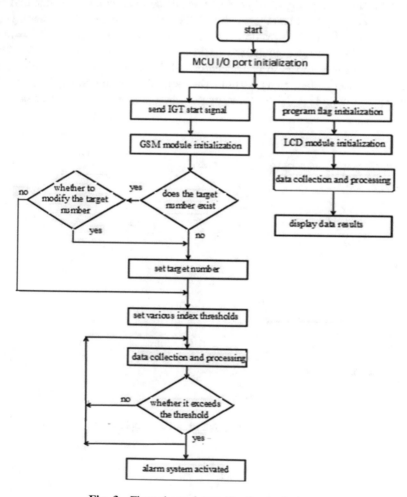

Fig. 3. Flow chart of overall software design

3.2   GSM Short Message Module Software Design

First, at the beginning of the program, perform the initial configuration of the GSM module, using the standard AT command in the module reference manual, which is sent by the single-chip microcomputer via the serial port to the GSM module for initialization, and automatically set the SMS address; in the main program, real-time loop detection of air quality information, Once it is greater than the alarm value, an alarm message is sent to the user [9]. The content of the message is the type and specific parameters of the alarm. After sending, continue the above process. If no user return information is obtained, an alarm message will be sent every 30 min, knowing that user feedback information has been received or the corresponding parameter has fallen below the dangerous value [10]. The GSM program flow chart is shown in Fig. 4.

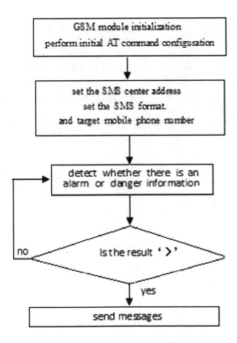

Fig. 4. GSM communication flow chart

4   Conclusion

The air quality detector can detect the hidden dangers of production and life in time, reduce the damage to the human body and family property due to environmental problems, and can meet the increasing requirements of consumers for the home environment. The air quality detection system is easy to operate and intelligent High, convenient installation, friendly man-machine interface, and high reliability. Can meet the needs of different users. The air quality detector has high application value and promotion value.

References

1. Wei, L.: Research on indoor air environmental quality detection and pollution control technology. Constr. Eng. Technol. Des. **29** (2019). (in Chinese)
2. Wang, F., Song, C.: Discussion on indoor air quality detection and control strategies of green residential buildings. Decoration World. **22** (2019). (in Chinese)
3. Li, Z. Zhang, Q., Wang X.: Based on the air quality PM2.5 detector system design. Digital World. **7** (2018). (in Chinese)
4. Chen, W., Xue, H.: Development of a smart home indoor air quality detection device. Electron. Testing. **15** (2019). (in Chinese)
5. Wang, S.,, Wang, Y.: Design of power supply system of ambient air quality detector. Commun. Power Supply Technol. **36**(7) (2019). (in Chinese)
6. Shao, M., Liu, S., Hong, H.: Design of intelligent air quality detection system based on Internet of Things technology. Comput. Sci. Appl. **4**(12) (2014)
7. Kenderian, S., Case, J.T.: Orion heat shield bond quality inspection: complete inspection system. Mater. Eval. **77**(1), 83–93 (2019)
8. Rautiainen, P., Hyttinen, M., Ruokolainen, J., et al.: Indoor air-related symptoms and volatile organic compounds in materials and air in the hospital environment. Int. J. Environ. Health Res. **29**(5), 479–488 (2019)
9. Chen, X., Peng, D.: Design of indoor air quality detection system based on WIFI. Instrum. Users. **25**(3) (2018). (in Chinese)
10. Liu, R., Tong, X., Zhou, T.: Design of portable air quality detection system. Internet of Things Technol. **7**(12) (2017). (in Chinese)

A Traffic Safety Early Warning Method Based on Internet of Vehicles

Rongxia Wang[✉] and Weihuang Yang

Guangzhou Nanyang Polytechnic College, Guangzhou, Guangdong, China

Abstract. Traffic accidents cause huge casualties and economic losses every year. The main causes of traffic accidents are the untimely response and improper disposal of drivers. The Internet of vehicles (IOT) system is a large system network based on the intranet, Internet and mobile Internet. According to the agreed communication protocol and data exchange standard, it carries out wireless communication and information exchange between vehicles and roads. It provides technical support for the realization of vehicle danger identification and creates new opportunities for safety early warning. Therefore, from the perspective of assistant driving, this paper discusses the dangerous state identification and safety warning methods of vehicles in different traffic scenes in the environment of Internet of vehicles. In this paper, low order car following model is studied for road traffic scene, and response time and following distance characteristic parameters are obtained by using different calculation methods. Through the above parameters, the curve of vehicle speed and following distance can be obtained. This paper analyzes the speed and following distance curves of all vehicles in the fleet, discusses the representativeness of different mean values to the data as a whole, and determines the safety threshold of following distance in normal driving. When the following distance of vehicles is less than the safety threshold, it is judged as a dangerous state; and according to the relationship between the actual value and the safety threshold, it uses the piecewise function to calculate the probability of vehicle collision accident. The algorithm is verified by the traffic data provided by ngsim project, and the results show that the method can accurately describe the dangerous state of vehicles and the possibility of collision accidents. The experimental results show that in the two traffic scenarios of road section and intersection, this paper uses the low-order car following model and the dangerous state recognition method combining time conflict zone with space conflict zone to realize the dangerous state recognition of vehicles, and expresses it by probability. The dispersion coefficient of arithmetic mean is 0.1840. The utility based early warning method can reduce vehicle collision accidents.

Keywords: Internet of Vehicles · Low order car following model · Speed and distance curve · Traffic safety

© The Author(s), under exclusive license to Springer Nature Switzerland AG 2021
J. Abawajy et al. (Eds.): ATCI 2021, LNDECT 81, pp. 109–116, 2021.
https://doi.org/10.1007/978-3-030-79197-1_14

1 Introduction

The purpose of the Internet of vehicles is to establish communication networks between vehicles and roadside devices, so that all traffic participants in the network can realize information sharing, so as to improve traffic efficiency and reduce traffic accidents [1]. Vehicles driving on the road collect vehicle information through on-board equipment, and then establish link channels between vehicles and vehicles, between vehicles and roadside equipment through wireless network system, so that vehicles can obtain information of themselves and other vehicles around in real time, and can transmit these information to roadside equipment [2]. Finally, through the connection between the roadside equipment and the control center, the mutual information transmission is realized. The control center stores and processes the information of all vehicles driving on the road, provides service information to vehicles, strengthens traffic management, improves road operation efficiency, and improves road safety [3].

This paper first introduces the background and current situation of the traffic safety early warning of the Internet of vehicles. In order to improve the research methods of traffic safety, different domestic scholars have their own views. Some scholar Lu Xin proposed a prediction based analysis method, which does not fully use the historical data of accidents, but uses other indicators to measure the possible risks of vehicles, such as safety signs or prediction models. This kind of method provides another idea for the study of traffic safety, which reflects the proximity between traffic participants through time or space metrics, so as to map the possible accident points of collision [4]. At the same time, it can also meet the needs of assistant driving in intelligent transportation system for dangerous state identification and safety warning. Another scholar Li Zhihan's realization of the Internet of vehicles technology provides technical support for the selection and calculation of safety signs. Only in the Internet of vehicles environment, can vehicles accurately obtain the information of other vehicles, and on this basis, use these data to judge the dangerous state of vehicles [5].

In this paper, a large number of experiments are carried out to verify the proposed method. The influence of simulation parameters on the experimental results is analyzed. In order to accurately express the vehicle braking distance considering the driver's reaction time, the simulation step is determined as 0.1 s. The experimental results of early warning efficiency show that the proposed method can effectively reduce the occurrence of traffic accidents; compared with the collision time early warning method, the experimental results show that the proposed method is effective in reducing the false alarm rate and false alarm rate; the impact of real-time early warning system on reducing the severity of collision accidents is analyzed, by comparing the average speed of vehicles in normal driving and the presence of early warning system, the simulation results show that this method can reduce the severity of the accident.

2 Traffic Safety Early Warning Technology Based on Internet of Vehicles

2.1 Vehicle Dangerous State Recognition

(1) Mazda Algorithm

Under the assumption of initial conditions, two vehicles drive at a constant speed, and the speed is V_L and V_F. The car in front begins to decelerate. The deceleration is a_L. The deceleration time is τ_2. After the rear car finds that the front car decelerates, it passes through τ_1 time at deceleration a_F starts to slow down until both cars stop completely [6, 7].

The formula of safe distance is as follows

$$R_{warning} = f(V_L, V_F, V_{rel}) = 0.5\left[\left(\frac{V_F^2}{a_F}\right) - \left(\frac{V_L^2}{a_L}\right)\right] + V_F\tau_1 + V_{rel}\tau_2 + r_{mn} \qquad (1)$$

Among them, R_{min} is the minimum distance?

(2) Path Algorithm

If the distance between two vehicles is less than the safe distance, the safety warning system will alarm the driver. The distance calculation formula is as follows:

$$R_{warning} = f(V_L, V_F) = 0.5\left[\left(\frac{V_F^2}{a}\right) - \left(\frac{V_L^2}{a}\right)\right] + V_F\tau + r_{mn} \qquad (2)$$

(3) Honda Algorithm

According to Honda algorithm, when the collision time is equal to 2.2 s, the early warning system will alarm the driver. The calculation formula of safety distance is as follows:

$$R_{warning} = f(V_{rel}) = 2.2V_{vel} + 6.2 \qquad (3)$$

2.2 Honda Anti Collision Braking System

Most of the researches on collision time set this index as a constant to judge whether it is safe or not. Based on Unscented Kalman filter, a probabilistic motion prediction algorithm is proposed to evaluate the dangerous state of vehicles based on vector calculation, which is used in the front collision prevention system [8, 9]. It is proposed that the warning threshold of collision time should increase with the increase of speed penalty, so as to realize the early warning at an appropriate time. In the algorithm, the threshold of collision time is set to 3 s, which is adjusted according to the speed penalty of the vehicle. The speed penalty is equivalent to 0.4905 m/km//h for the rear vehicle. The calculation formula of safety distance is as follows:

$$R_{warning} = f(v_{rel}, v_F) = \frac{3dR}{dt} + 0.4905 V_F \qquad (4)$$

Where, $dR/d_t = \Delta V$ is the relative speed between two vehicles, V_F is the rear vehicle speed:

$$u = k_p \left[e + \frac{1}{T_1} \int_o^t edt + \frac{T_D de}{dt} \right] \qquad (5)$$

Through image processing technology, using light matching technology to realize the night recognition of the car in front. Due to the large error of the workshop distance obtained by image processing technology, the method of calculating the workshop distance by using the camera installed in the car is proposed. A vehicle anti-collision system based on radar distance detection is also proposed. The system includes distance detection equipment and early warning equipment. The distance detection equipment determines the position of the object by transmitting signals to the surrounding environment and by echo, and the early warning equipment judges the distance and provides an alarm. A vehicle cooperative anti-collision system based on future trajectory prediction is proposed [10]. It is assumed that vehicles are equipped with GPS equipment, and vehicles and roadside equipment can communicate with each other to obtain each other's information. On this basis, it is proposed that when the collision time TTC meets the following two conditions, the vehicle has potential collision risk.

1) The duration of conflict detection is not less than tpersist;
2) The trajectories conflict and TTC is less than tcritical.

Among them, tpersist should be greater than the time of conflict error detection, but less than the time of driver perception. The selection of critical can not interfere with the driver's normal driving, at the same time, it must ensure that the driver has enough time to respond to the alarm to avoid accidents. The maximum value of collision time TTC is usually set as 2S, which is the alarm threshold set by most early warning systems. When the value of TTC is less than 1.5 s, the risk is very high. When TTC is less than 1 s, it is generally considered that the accident can not be avoided. Time gap is a parameter similar to the collision time TTC, which can be used as an early warning threshold. The threshold setting of time gap is usually 2S, that is, when the time gap is greater than 2S, the probability of an accident becomes smaller and smaller; when the time gap is less than 2S, the probability of an accident becomes larger and larger; when the time gap is equal to 0, it means that the accident can not be avoided. The position, speed and acceleration of the vehicle are measured by radar, and the potential collision risk is judged based on the above information.

In reality, because people will refer to other sources of information when making decisions, the problem is more complicated. In addition, people will also consider whether the early warning system alarms when making decisions, which can be seen as a team decision made by people and the early warning system.

3   Experimental Research on Traffic Safety Early Warning Based on Internet of Vehicles

3.1   Data Collection of System Simulation Experiment

The scatter of Newell characteristic reflects the relationship between the speed and the following distance of the vehicle in the real road scene, which also meets the needs of reality analysis, because in the real traffic environment, the vehicle cannot keep the same following distance with the vehicle in front, but there is a safe range of the following distance, and the driver will choose the appropriate following distance within the safe range Distance. Therefore, this study intends to take the straight line calculated by the average value of the parameters of the whole fleet as the standard straight line of vehicle safe driving, which reflects the average level of vehicle following distance changing with speed in the whole road section, and determines the threshold range of safe driving based on the scattered points generated in the real environment, so as to judge whether the vehicle has a dangerous state.

3.2   Generalized Predictive Control Based on Error Correction

The error of dynamic prediction analysis is $y_e(k+j)$ is used as the error compensation to modify the generalized prediction

$$y(k+j) = y_m(k+j) + y_e(k+j) \tag{6}$$

Where, $y_m(k+j)$ is the prediction value of traditional generalized prediction algorithm at k time.

$$y(k+j) = G_j(z^{-1})\Delta\mu(k+j-1) + F_j(z^{-1})y(k) + y_e(k+j) \tag{7}$$

After error compensation, the optimal solution is as follows

$$\Delta U = (G^T G + \lambda I)^{-1} G^T (Y - F - Y_e) \tag{8}$$

The generalized predictive control is used to predict the prediction error, which can be adjusted on-line, changing the situation that the weights of the network are fixed after training in the traditional error correction.

4   Experimental Analysis of Traffic Safety Early Warning Technology Based on Internet of Vehicles

4.1   Simulation Analysis of System Experiment

In this paper, we can see that there is an obvious speed change process in the fleet, which not only conforms to the car following state described by Newell model, but also provides the conditions for calculating the parameters of the model. First, the real-time values of parameters τ and d are calculated by using the wave velocity method, and then the mean values of each parameter are calculated and compared. The experimental results are shown in Table 1.

Table 1. Parameter mean table of Newell car following model

NELL model	Fleet 1	Fleet 2	Fleet 3	Fleet 4
τ/s	1.5157	2.0156	1.4723	1.4566
d/s	7.123	8.6991	6.8068	6.3128

Parameter Mean Value Diagram of Newell Car Following Model, as shown in Fig. 1. Because of the same frequency of each index, the average index mode does not exist. In addition, the discrete coefficient of the arithmetic mean is less than that of the median, so the arithmetic mean is selected as the pair mean to represent the whole statistical array.

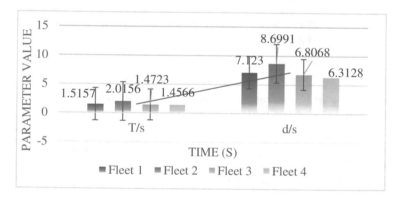

Fig. 1. Parameter mean value diagram of Newell car following model

4.2 Simulation Experiment Analysis

In the simulation system, the second-order object transfer function model is selected, the modified generalized predictive control algorithm is adopted in the controller, the time driven method is adopted in the sensor, and the event driven method is adopted in the controller and the actuator. The experimental results are shown in Table 2.

Table 2. Comparison of average index results of car following distance

Average index	Upper boundary		Lower boundary	
	Mean value	Dispersion coefficient	Mean value	Dispersion coefficient
Arithmetic mean	19.8561	0.3333	1.06512	0.1840
Median	17.1556	0.3801	0	–
Mode	10.5649	0.9120	0	–

On the basis of obtaining the upper and lower boundaries of all vehicle following distance thresholds, the average index and discrete coefficient of the upper and lower boundaries are calculated respectively, as shown in Fig. 2. For the upper boundary, the discrete coefficients of the arithmetic mean and the median are obviously less than the mode, and the discrete coefficients of the arithmetic mean are less than the median, so the upper boundary selects the calculation result of the arithmetic mean as the mean; for the lower boundary, because the mean values of the median and the mode are both 0, it indicates that small data disturbance will also have a huge impact on the discrete coefficients. The discrete coefficient of the arithmetic mean is 0.1840, which is already a small number, indicating that the fluctuation of the whole data is very small. Therefore, for the lower boundary, the calculation result of the arithmetic mean is also selected as the mean.

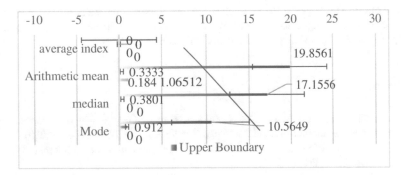

Fig. 2. Comparison chart of average index results of car following distance

5   Conclusions

In this paper, a simulation experiment system is designed, which can carry out repeated experiments on two traffic scenes of road and intersection. The proposed method of dangerous state identification and safety early warning is verified. Firstly, the influence of simulation step size on simulation accuracy is analyzed. The results show that in order to accurately describe the effect of driver's reaction time on vehicle braking distance, the simulation step should be accurate to 0.1 s. Secondly, the impact of early warning system on reducing traffic accidents in two traffic scenarios is compared, which verifies that the real-time early warning algorithm proposed in this study can effectively reduce the occurrence of accidents and reduce the incidence of accidents. In addition, compared with the collision time warning method, the experimental results show that the proposed method has good effect in reducing the false alarm rate and false alarm rate. Finally, the impact of the real-time warning system on reducing the severity of the accident is analyzed, and the average speed of the vehicle in the case of normal driving and the presence of the warning system is compared. The results show that, in the case of failure to avoid the accident, the early warning system can reduce the speed of the collision accident, so as to reduce the severity of the accident and reduce the loss caused by the accident.

Acknowledgement. Fund Project 1: This paper is the mid-stage research result of a new generation of information technology project in the key fields of ordinary colleges and universities of the Guangdong Provincial Department of Education "Research and Application of Traffic Safety Early Warning System Based on 5G Internet of Vehicles" (Project No. 2020ZDZX3096) from Guangzhou Nanyang Polytechnic College.

Fund Project 2: This is the phased research result of the "Research on Security Mechanism and Key Technology Application of Internet of Vehicles" (Project No. NY-2020KYYB-08) from Guangzhou Nanyang Polytechnic College.

Fund Project 3: This paper is the research result of the project of "Big Data and Intelligent Computing Innovation Research Team" (NY-2019CQTD-02) from Guangzhou Nanyang Polytechnic College.

Fund Project 4: This paper is the research result of "Research on Vehicle Collision Warning Method Based on Trajectory Prediction in Internet of Vehicles" (Project No. NY-2020CQ1TSPY-04) from Guangzhou Nanyang Polytechnic College.

References

1. You, X., Chen, Y., Fu, J., et al.: Design and test scheme of expressway bad weather warning system based on Internet of Vehicles. Road Traffic Saf. **019**(003), 28–32 (2019)
2. Lu, J.: Research on vehicle active safety early warning system based on Internet of Vehicles. Technol. Market **7**, 80–81 (2019)
3. Zilong, W., Fen, K.: Research on security architecture of Internet of Vehicles based on blockchain technology. Jiangxi Commun. Technol. **000**(001), 41–44 (2019)
4. Lu, X., Lu, X.: Overview of traffic safety patent hot technologies based on Internet of Vehicles. China New Commun. (17), 17
5. Li, Z., Liu, T., Guo, Z.: Literature review on Internet of Vehicles technology and application. Sci. Technol. Outlook **000**(012), 100–100 (2015)
6. Hong, X.R.: Research on application of Internet of Things on intelligent traffic safety early warning. Appl. Mech. Mater. **263–266**, 2890–2894 (2013)
7. Cheng, J., Cheng, J., Zhou, M., et al.: Routing in Internet of Vehicles: a review. IEEE Trans. Intell. Transp. Syst. **16**(5), 2339–2352 (2015)
8. Juan, B., Wei, C., Zhengtao, X., et al.: Effect analysis of early warning for abandoned object on highway based on Internet-of-Vehicles CA model. Discret. Dyn. Nat. Soc. **2018**, 1–14 (2018)
9. Xie, L.S., Chen, X.Y., Lin, Y.Z.: Bus safety regulatory system design based on Internet of Vehicles technology. Appl. Mech. Mater. **694**, 114–117 (2018)
10. Wu, R., Zheng, X., Xu, Y., et al.: Modified driving safety field based on trajectory prediction model for pedestrian-vehicle collision. Sustainability **11**(22), 6254 (2019)

Application of Virtual Reality Technology in Training of Substation in Coal Mining Area

Yangtao Chuan[✉]

Department of Electrical and Control Engineering, Xi'an University of Science and Technology, Xi'an, Shaanxi, China

Abstract. The substation in the coal mining area is responsible for multiple tasks such as underground substation, power distribution, and protection. How to quickly improve the working ability of electricians in the mining area in terms of system operation and maintenance and troubleshooting during the training process, and regulate the behavior of electricians, so as to ensure the safe and reliable operation of the mining area substation is very important. According to the requirements of the National Mine Safety Administration that coal mine electricians need to strictly abide by the safety and normative requirements of the "Coal Mine Safety Regulations" and "Electrical Safety Operating Regulations", this paper designs and develops a coal mine substation training system based on virtual reality technology. Through 3D modeling, scene building, script logic design, functional module development, etc., training functions such as roaming inspections of mining district substations, equipment learning, simulation exercises, fault simulation, and knowledge assessment have been realized. Tests show that the system provides trainees with an immersive training environment that is easy to operate, vivid, and self-service interaction, so that trainees can master the operation, maintenance and troubleshooting of mining district substations faster and more deeply, greatly improving the efficiency of coal mine electrician training.

Keywords: Virtual reality · Simulation training · Mining area substation · 3D Modeling

1 Introduction

In the coal field, due to its inherent immersion and interactivity, virtual reality technology has also received more and more attention and has carried out experimental applications in some areas, and has achieved more research results. TICHON and others applied virtual reality to the safety training exercises of miners, which effectively improved the level of safety awareness of miners [1]. PEDRAM and others use virtual reality technology to assess the risks of underground work [2]. STOTHARD and others combined virtual reality technology with the concept of sustainable mining, based on real geological data to visualize and simulate a huge mine on a giant screen, and achieved good teaching effects [3]. Based on complete simulation data, AKKOYUN and others have established an interactive visualization environment for teaching and learning related to mining engineering [4]. Mao Shanjun and others initially designed a

© The Author(s), under exclusive license to Springer Nature Switzerland AG 2021
J. Abawajy et al. (Eds.): ATCI 2021, LNDECT 81, pp. 117–123, 2021.
https://doi.org/10.1007/978-3-030-79197-1_15

prototype system of a coal mine virtual environment system [5]. Literatures [6, 7] use Virtual Reality simulation and modeling software such as MultiGen Creator and Vega to show the running process of long wall fully mechanized mining face in a three-dimensional manner, and enables users to complete some simple tasks such as roaming and equipment operation. Interactive function. Literatures [8, 9] research on adding online data to the virtual scene of fully mechanized mining. Meng Long uses high-quality modeling technology to establish rescue models and plans, and studies a multi-person collaborative coal mine rescue exercise system based on the real environment of the mine and the evolution of disaster accidents [10].

Although there is a lot of research on virtual reality in coal mines, the research directions are mainly in coal mining equipment, disaster simulation, safety training, operation procedures of fully mechanized mining faces, etc., and there are few virtual reality simulations on substations in underground coal mines. The mining area substation is the regional power supply center under the coal mine. It is responsible for multiple tasks such as power transformation, power distribution, and protection. It plays an extremely important role in the mine power supply system and safe production [11–15]. This article relies on documents such as "Coal Mine Safety Regulations" and "Electrical Industry Safety Operation Regulations", through 3D modeling, scene construction, script logic design, functional module development, etc., to realize the simulation training of coal mine substations based on virtual reality technology. The system includes training functions such as roaming inspections, equipment learning, simulation exercises, fault simulation, and knowledge assessment.

2   General System Formula

2.1   Introduction of Mining Area Substation

The mining area substation is at the core of the power supply system of the entire mining area. It undertakes the power supply task of one or several mining areas, and changes the high voltage of 10(6) kV from the central substation to 1140(660) V. The low voltage of this mining area is used to supply power to the electrical equipment and other electrical equipment of all mining working faces in the mining area. In addition, part of the high voltage in the substation of the mining area will be directly distributed to the mobile substation in the mining area, and the mobile substation will reduce the voltage and then supply it to the fully mechanized mining face. The main electrical equipment of the mining area substation includes: high-voltage explosion-proof switch, mobile substation, dry-type transformer, low-voltage feeder switch, magnetic starter and comprehensive lighting protection, etc.

2.2   System Demand Analysis

The coal mining area is an important production site underground, and the mining area substation is the center of the mining area's power supply. It is responsible for the power supply, transformation, and distribution tasks of the entire mining area. It is necessary for coal mine electricians to be proficient in the operation and maintenance of related electrical equipment in mining areas. Coal mine electricians need to have the following technical qualities:

(1) Have a certain basic knowledge of electricians, be familiar with relevant regulations and standards of substation operation, and be able to work independently;
(2) It is necessary to understand the power supply system diagram of the substation, be familiar with the power supply of the substation and the load nature, capacity and operation mode of the equipment, be familiar with the characteristics, general structure and working principle of the main electrical equipment of the substation, and master the operation methods;
(3) Master the methods of handling on-site electrical accidents.

In view of the many shortcomings in the traditional coal mine power supply safety training, the training system has developed 5 training modules. Figure 1 shows the functional structure of each module.

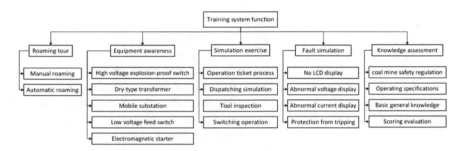

Fig. 1. Functional structure diagram of simulation training system

2.3 General Structure of System

The simulation training system architecture of substation in coal mining area based on virtual reality technology is shown in Fig. 2, which is mainly divided into four parts:

(1) 3dsMax modeling includes the establishment of 3D models of all required scenes and equipment and the production of animations;
(2) Unity development includes four main parts: scene construction, human-computer interaction interface, scene global lighting baking and logic writing. The logic writing part is based on the MVC design pattern. Through layering, the functions of each module are divided, which is easy to maintain and modify the software;
(3) HTC Vive is mainly for interactive development of VR, including character movement and gamepad interaction;
(4) Finally, the system was released on the Window platform. Considering some machines with Low configuration, six modes were set: Very Low, Low, Medium, High, Very High and Ultra. The picture quality was successively improved to adapt to machines with different configurations.

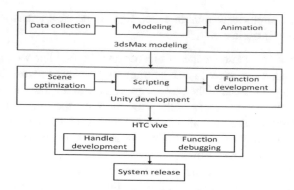

Fig. 2. General structure of the simulation system

3 System Development

3.1 Scene Building

(1) 3D modeling. It mainly includes modeling of electrical equipment in substations of mining area, environment modeling of substations, and environment modeling of coal mine tunnels, etc. The electrical equipment model is an important model of the training system. Because of its complex structure and many parts, it needs refined modeling. Modeling can be simplified for some models that are not demanding in details such as fences, publicity slogans, fire extinguishers, and fire sandboxes.

(2) Scene baking. In order to show more realistic details, use the global illumination technology provided by Unity to set tags for static objects in the scene, create mixed mode lights, and bake the scene. As shown in Fig. 3, it is a comparison diagram of coal mine tunnel scene before and after baking.

(a) Before baking   (b) After baking

Fig. 3. Coal mine tunnel scene before and after baking

3.2   Realization of Key Technologies

3.2.1   Rotation Positioning Function Realization

System interaction involves many rotary operations, such as opening and closing doors, control knobs, isolation switch handles, etc. In order to bring users a more realistic experience, the rotary positioning function is used to track the user's handle position in real time and feedback to the model position data. Change to achieve smooth rotation.

First obtain four position data: the position data of the rotating object axis, marked as A (a1, a2, a3); the handle position data, marked as T (t1, t2, t3); the rotation axis vector of the rotating object, marked as $\vec{r}$, default $\vec{r}$ = (0, −1, 0); the position data of the interaction point, marked as H (h1, h2, h3). Figure 4 shows a three-dimensional vector diagram of the rotation positioning data.

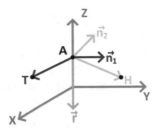

Fig. 4. Three-dimensional vector diagram of rotating positioning data

Calculate the normal vector of the plane formed by the position of the handle, the axis of the rotating object, and the axis of rotation of the rotating object, and unitize it, and marked it as $\vec{n_1}$:

$$\overrightarrow{AT} = (t_1 - a_1, t_2 - a_2, t_3 - a_3) \tag{1}$$

$$\vec{n_1} = \frac{\overrightarrow{AT} \times \vec{r}}{\left|\overrightarrow{AT} \times \vec{r}\right|} \tag{2}$$

In the same way, calculate the direction of the plane formed by the position of the interaction point, the axis of the rotating object, and the axis of rotation of the rotating object, and unitize them, marked as $\vec{n_2}$:

$$\overrightarrow{AH} = (h_1 - a_1, h_2 - a_2, h_3 - a_3) \tag{3}$$

$$\vec{n_2} = \frac{\overrightarrow{AH} \times \vec{r}}{\left|\overrightarrow{AH} \times \vec{r}\right|} \tag{4}$$

After obtaining $\vec{n_1}$ and $\vec{n_2}$, the FromToRotation method under Quaternion provided by Unity can be used to calculate the angle required to rotate the rotating object from the current position to the position of the handle, expressed in quaternion. Finally, follow the rotation of the rotating object through the rotation method under the Transform component.

3.3   Fault Simulation Function Realization

There are two types of components in the main circuit and control circuit of the electromagnetic starter: power connection lines and switches. Establish a digital model of power connection line components, with attributes including: number, live status, fault status, nodes, etc.; establish a digital model of switching components, with attributes including: number, node, on-off status, live status, and fault status, etc.

As shown in Fig. 5, it is the main circuit of the QJZ2-120 electromagnetic starter. By writing and binding the driver between the three-dimensional model and the digital model, the background simulation results are shown in Fig. 6. Among them, green indicates a normal uncharged state, red indicates a charged state in normal operation, and blue indicates a fault state. The color of the components in the figure shows that HK and C-phase disconnection or virtual connection fault occurred between KM.

Fig. 5. Schematic diagram of the main circuit of the QJZ-120 model electromagnetic starter

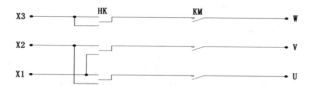

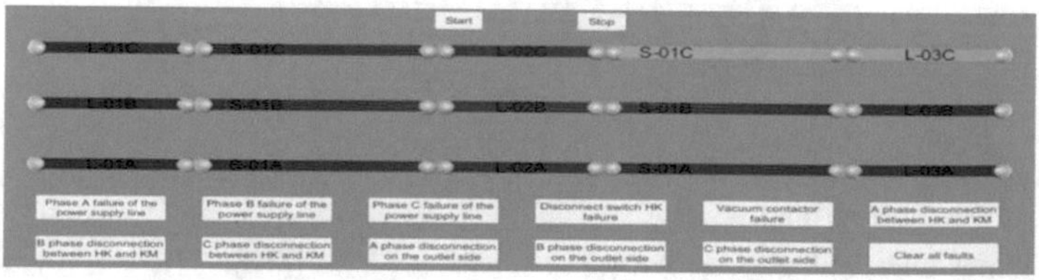

Fig. 6. Background simulation results of the main circuit of the QJZ-120

4   Conclusions

The professional level of electricians in coal mining areas directly affects the safe operation of coal mine power supply systems. This paper designs a virtual simulation training system based on relevant standards and regulations. The system mainly includes five modules: equipment roaming inspection, equipment cognition, simulation exercise, fault simulation, and knowledge assessment. Through virtual reality technology, it brings a more vivid, more realistic and immersive experience to coal mine electricians, increases the enthusiasm of coal mine electricians to learn, and enables coal mine electricians to master the internal structure of the equipment and the entire maintenance process in a short time. The introduction of this system can change the training mode of coal mine electricians, and replace the actual training center with software and hardware systems, provide safety guarantee for coal mine electrician training, save training costs in the coal industry, and have better economic benefits and practicability.

References

1. Tichon, J., Burgesslimerick, R.: A review of virtual reality as a medium for safety related training in mining. J. Health Saf. Res. Pract. **3**(1), 33–40 (2011)
2. Pedram, S., Perez, P., Palmisano, S.: Evaluating the inffiuence of virtual reality-based training on workers' competencies in the mining industry. In: 13th International Conference on Modeling and Applied Simulation, Red Hook, MAS 2014, pp. 60–64. Curran (2014)
3. Stothard, P., Laurence, D.: Application of a large-screen immersive visualization system to demonstrate sustainable mining practices principles. Trans. Inst. Min. Metall. **23**, 199–206 (2014)
4. Akkoyun, O., Careddu, N.: Mine simulation for educational purposes: a case study. Comput. Appl. Eng. Educ. **23**(2), 286–293 (2015)
5. Zhangs, X.: Augmented reality on longwall face for unmanned mining. Appl. Mech. Mater. **40**(6), 388–391 (2010)
6. Zhang, X., An, W., Li, J.: Design and application of virtual reality system in fully mechanized mining face. Procedia Eng. **26**(4), 2165–2172 (2011)
7. Wanl, R., Gao, L., Liu, Z.H., et al.: The application of virtual reality technology in mechanized mining face. Adv. Intell. Syst. Comput. **181**, 1055–1061 (2013)
8. Wan, L., Gao, L., Liu, Z., et al.: The application of virtual reality technology in mechanized mining face. Adv. Intell. Syst. Comput. **181**, 1055–1061 (2013)
9. Tang, S., Wei, C.: Design of monitoring system for hydraulic support based on LabVIEW. Adv. Mater. Res. **989**, 2758–2760 (2014)
10. Meng, L.: Research and development of coal mine disaster prevention and rescue drill platform based on VR virtual reality technology. J. Phys.: Conf. Ser. **1549**, 042060 (2020). https://doi.org/10.1088/1742-6596/1549/4/042060
11. He, Z.: Development and application of virtual reality application platform for common electrical equipment in substation of mine mining area. In: Proceedings of 6th ABB Cup National Papers Competition for Automation System Engineers, pp. 309–313. Chinese Association of Automation (2013)
12. Su, K., Zhang, D.L., Chen, Y.L.: Simulation and implementation of virtual lab circuit based on Unity3d and 3dmax. In: Proceedings of 3rd International Conference on Mechanical and Electronics Engineering. Institute of Management Science and Industrial Engineering: Computer Science and Electronic Technology International Society, 2019, pp. 92–96 (2019)
13. He, Y.: Application analysis of mine QZJ type multifunctional vacuum starter. Mech. Manag. Dev. **34**(06), 75–76 (2019)
14. Guo, X.: Application of fault diagnosis technology for key components of substation equipment in mining area. Hydraul. Coal Min. Pipel. Transp. **4**, 114–115+118 (2019)
15. Zhang, Z.: A mine mining technology on the substation power supply scheme. Value Eng. **31**(30), 103–105 (2012)

Analysis of Emergency Repair Cost of EHV Transmission and Transformation Project Based on Difference Comparison

Kaiwei Zhang[1], Jiangmin Liu[2(✉)], Yi Chen[1], Yan Yan[1], and Wei Liu[1]

[1] Guiyang Bureau, Extra High Voltage Power Transmission Company, China Southern Power Grid, Guiyang, Guizhou, China
[2] China Electricity Council, Beijing, China

Abstract. The emergency repair project of power grid is different from the conventional repair project of power grid and the large-scale repair project of power grid, which determines that it is not suitable for the conventional pricing model and cannot achieve cost control through the standard quota system. Therefore, this paper studies the charging method of emergency repair project of EHV transmission and transformation projects, and further analyzes the pricing process and relevant basis of emergency repair projects from the perspective of difference analysis, which is conducive to solving the problem of unreasonable pricing of emergency repair projects and reducing disputes among all parties, so as to improve investment income and safeguard the legitimate interests of all parties involved in the project.

Keywords: Difference analysis · Ultra high pressure · Emergency repair · Billing method

1 Introduction

With the rapid development of my country's economy, the overall electricity consumption of the society has increased year by year. Driven by the demand for electricity, the power grid has developed rapidly and entered the EHV era, forming a huge grid structure with EHV as the backbone. As an intermediate link between continuous power generation and power consumption, the security and stability of the power grid is of great significance for ensuring normal social power supply, ensuring social and economic development and the normal living order of the people. The power system exists in the natural environment and the social environment, and is extremely vulnerable to challenges from natural conditions (such as earthquakes, ice disasters, typhoons, etc.) and intentional or unintentional damage or destruction from the social environment. In recent years, power grid emergency repair accidents have occurred frequently and are extremely harmful. If they are not handled in time, they will cause significant economic losses to the affected areas.

Whether it is national power grid security or regional power grid security, it is related to the rapid development of the national economy and the safety of people's lives and property. The "National Emergency Plan for Dealing with Large-scale Power

J. Abawajy et al. (Eds.): ATCI 2021, LNDECT 81, pp. 124–131, 2021.
https://doi.org/10.1007/978-3-030-79197-1_16

Outages in the Power Grid" promulgated by the Chinese government in January 2006 listed the emergency plan for large-scale power outages as a national plan. Major power companies have also issued local or regional emergency plans accordingly [1, 2].

The power grid emergency repair project has the characteristics of suddenness, tight construction period, high risk, complicated emergency measures, relatively small amount of main body engineering, and no scale effect. The specific summary is as follows:

(1) Sudden and tight schedule
    Usually emergency repair projects are sudden and the repair time is urgent. In order to ensure the safety of the power grid and the reliability of power supply, the construction period is much shorter than the technical renovation and repair projects carried out according to the normal plan.
(2) High risk and complicated emergency measures
    Accident emergency repairs and fault emergency repairs are affected by the on-site environment and emergency repair guarantee conditions. They are highly dangerous and have a narrow construction area. The measures used to simultaneously ensure continuous power supply and the safety of emergency repair personnel are more complex than conventional projects.
(3) The main body has a small amount of engineering
    Compared with conventional power grid construction projects, power grid technical renovation and maintenance projects, and large-scale natural disaster emergency repair projects, accident and fault emergency repair projects have scattered operating points and small amount of main body engineering, which cannot form a scale effect.

2   Research Status

2.1   EHV Emergency Repair Work Process

The phase division of emergency repair engineering project cycle is the same as that of conventional power grid engineering, which can be divided into: pre-phase, preparation phase, implementation phase and commissioning phase. However, compared with conventional power grid engineering projects, the occurrence of emergency repair projects is uncertain. There is no project proposal, pre-feasibility study and feasibility study, project evaluation and decision-making, and some emergency repair projects do not need to be redesigned. Directly adopt the original design drawings. Therefore, the early stage and preparation stage of the project management of the emergency repair project are mainly reflected in fund planning, framework bidding, emergency start, and standby of related units, which are quite different from conventional power grid projects. The comparison between the emergency repair project cycle and the conventional power grid project is shown in Table 1.

Table 1. Comparison table of emergency repair engineering project cycle and conventional power grid engineering project

Stage name	Emergency repair project stage	Conventional power grid engineering project stage
Early stage	• Fund planning, framework bidding • Establishment of emergency repair team	• Project proposal • Feasibility study report
Preparation Phase	• Emergency startMan-machine dispatch	• Engineering project planning and design
Implementation phase	• Site exploration • Construction of emergency repair works • Completion acceptance	• Construction preparation • Project construction • Trial production and completion acceptance
Operational stage	• Resumption of production	•The project is put into production and operation

2.2   Analysis of the Difference with General Maintenance Engineering

After investigation and sorting, the difference between emergency repair engineering and general maintenance engineering is mainly reflected in the management process, organization method, settlement method, etc. The specific content is shown in Table 2.

Table 2. Comparison table of emergency repair engineering project cycle and general maintenance engineering project

Difference	Emergency repair project	General maintenance engineering project
Management process	• In the event of an emergency fault repair, an emergency report should be made first, and the project implementation should be carried out after obtaining the consent of the relevant department (implementation without the reserve link)	• In accordance with the approval system, the application for initiation, review and approval, material declaration and procurement, the signing of various engineering service contracts, and the acceptance of goods can be completed in sequence in accordance with the approval system. Construction can only be carried out after the application process is completed
Organization	• Need to dispatch personnel and materials urgently, and the coordination work will increase compared with general maintenance projects	• Carry out maintenance tasks in accordance with the established plan and normal organization methods [3, 4]

(*continued*)

Table 2. (*continued*)

Difference	Emergency repair project	General maintenance engineering project
	• The management personnel of the construction unit will organize the construction on site throughout the whole process	• The management personnel of the construction unit concurrently manages multiple maintenance projects, and transfers the risk points and key points of the project implementation to the on-site supervision of the maintenance projects
Project implementation	• Strong timeliness, generally 24 h uninterrupted operation • There are different situations such as holidays, night construction, continuous construction, etc. • There are many uncontrollable factors on site, which are likely to cause insufficient personnel, equipment, and materials, and uncontrollable construction time, etc.	• General maintenance works are carried out according to the power outage plan, or carried out according to the maintenance construction plan. In general, 24 h continuous operation is not arranged • Carry out construction on the scheduled power outage day • The site has been surveyed in advance, the personnel, equipment, and materials are prepared in advance, and the time is controllable
Budget preparation	• Due to the long compilation period, the budget estimate will be supplemented after the completion of the project, or no more compilation	• According to the actual situation of the project, formulate a general routine maintenance plan, and prepare an estimate based on the technical renovation and maintenance quota
Bidding	• The selection of contractors for emergency repair projects usually adopts a framework bidding model. The grassroots units will bid for the emergency repair contractors for the next year based on the repair work that has occurred in the past 2 years • When emergency repairs occur, the grassroots units directly select the emergency repair contractors from the list of emergency repair contractors	• According to the annual regular maintenance work arrangement, in accordance with relevant regulations, bidding and procurement of projects that meet the requirements of bidding and bidding are carried out. After the bidding and procurement, a framework service contract is signed, and the contractor provides engineering services within the time limit stipulated in the contract
Settlement method	• A lump-sum fee (or fixed fee + lump-sum compensation fee) is included in the settlement of some projects, and the fixed amount of maintenance is not applied • The total price contract is adopted for settlement of some projects	• According to the characteristics of the project and the contract, the fixed total price or actual settlement method shall be adopted

2.3   Current Pricing Status and Existing Problems

The 2015 version of the technical renovation and maintenance pre-plans and quotas are compiled under normal geographic and climatic conditions based on the reasonable construction organization design, reasonable construction period and conventional construction machinery equipment required to complete the project content, although it has been adjusted on the basis of the emergency repair project. The number of man-days and shifts is still not fully applicable to the overall cost level of the grid emergency repair project. The actual cost of personnel, materials, and equipment incurred by the construction unit to organize and implement an emergency repair and the actual cost of emergency repair measures is higher than the settlement amount prepared by applying the repair quota to the engineering quantity. Therefore, when companies in various network provinces face different emergency repair projects, the settlement methods will be different. When the cost difference is not large, the maintenance quota is still used for settlement. When the cost difference is large, the cost of the construction unit will be balanced [5]. Actual settlement or a lump-sum fee, so settlement disputes are relatively large.

3   Cost Composition Analysis

In order to determine the cost composition of the emergency repair project [6], this section is mainly based on the analysis of the difference between emergency repair and conventional construction, the analysis of construction efficiency reduction, and the combination of investigation and fund collection and the relevant content of the 2015 version of the maintenance pre-plan. The final cost composition of the emergency repair project includes three parts: construction (installation) project cost, equipment purchase cost and other costs. On the basis of the 2015 version of the pre-regulated maintenance cost structure, there are the following differences.

3.1   Measures Fee

Measure fee refers to the cost of non-engineering entities before and during the construction of the emergency repair project to complete the construction of the project. Combined with the characteristics of the emergency repair project, the measure fee includes: emergency dispatch and coordination fee, construction efficiency reduction fee, and rush work Measures costs, night construction increase costs, winter and rainy season construction costs, temporary facilities costs, safe and civilized construction costs, construction tools and utensils usage fees, construction costs in special areas.

(1)   Emergency dispatch and coordination fee
     Due to the suddenness of the fault, in order to ensure that the fault is eliminated as soon as possible and the safe and stable operation of the power grid is restored, construction personnel and machinery need to be urgently arranged to participate in the emergency repair work. If the emergency repair personnel and machinery in this area are limited, relevant personnel and machinery need to be mobilized from other regions. Participate in the emergency repair work, the 15 version of the

maintenance pre-regulation did not fully reflect the man-machine dispatch work before the emergency repair, so it is recommended to increase the cost of this measure. Survey data shows that the longer the dispatch distance, the shorter the repair time, and the higher the cost of dispatching repair teams and materials.

(2) Construction cost

Due to organizational reasons, construction period reasons, overtime fatigue, etc., the construction efficiency of personnel and machinery is reduced. In the 15th edition of the maintenance pre-regulation, the conventional organization design and construction plan are considered, and the construction efficiency reduction cost is not included. Therefore, it is recommended to increase the cost of this measure.

(3) Rush measures fee

Compared with the normal construction method, the emergency repair project adds rush measures, which causes an increase in construction costs [7]. The 15 version of the maintenance does not include the cost of rush measures, so it is recommended to increase the cost of this measure.

(4) Increased cost of night construction

The purpose of emergency repair works is to ensure power supply, requiring the restoration of power supply in the shortest time, and the workload is relatively concentrated. The work time of emergency repairs in a unit's working day often exceeds 8 h, and there are continuous construction at night and low efficiency at night [8]. According to the needs of the project schedule, the projects that must be continuously constructed at night should be included; this cost does not include the rush cost. In the case of tunnels, tunnels, etc., where lighting is required for construction, additional costs shall be calculated according to the specified plan.

(5) Construction increase in winter and rainy season

Emergency repair works occur all year round, and the rainy season occurs more frequently. In order to ensure the quality of the project, necessary rain and moisture-proof measures are usually taken. There is no difference from the pre-regulated work content of the 15 version of the maintenance, and the workload will increase.

(6) Temporary facility fee

Due to the tight repair period or restrictions on site conditions, the repair personnel can only choose hotels for accommodation, and the cost of accommodation is high.

(7) Safe and civilized construction fee

This fee is a mandatory fee (non-competitive fee) and will not be adjusted [9].

(8) Construction tool usage fee

The emergency repair project is the same as the general repair project, and no adjustment is made.

(9) Construction increase in special areas

The emergency repair project is the same as the general repair project, no adjustment is required.

Through the above analysis, the cost composition of emergency repair engineering measures for the power grid is summarized, which mainly includes six parts: emergency repair assembly fee, post-disaster weather construction increase fee, rush overtime increase fee, construction tool usage fee, temporary facility fee, and safe and civilized construction fee.

3.2   Enterprise Management Fee

Due to the particularity of emergency repair projects, in order to ensure the progress of the emergency repair construction and the construction safety during the emergency repair process, the emergency repair unit will invest more management personnel, thus increasing the investment of enterprise management fees. In addition, the implementation of the relevant national policies and regulations of the "business tax to value-added tax" [10], the "urban maintenance and construction tax", "education surcharge" and "local education surcharge" that were originally included in the tax were added to the enterprise management expenses, which also increased the rate of corporate management fees.

4   Project Case Verification

Three projects including "emergency repair of a 500 kV overhead transmission line" were selected for level testing, and the measure costs, enterprise management fees and project static investment calculated using this result were compared with the original settlement costs of the case project. The results are as follows (Table 3):

Table 3. Level test result

Case study	Total settlement fee (RMB)			Measure fee (RMB)			Enterprise management fee (RMB)		
	Original cost	Estimated cost	Rate of change (%)	Original cost	Estimated cost	Rate of change (%)	Original cost	Estimated cost	Rate of change (%)
Case 1	35538	39381	10.8	2632	5357	103.5	3729	4848	30.0
Case 2	15447	16750	8.4	862	1973	128.9	965	1158	19.9
Case 3	989304	1020833	3.2	105623	116519	10.3	103457	124090	19.9

The deviation rate of the total cost after calculation from the original cost is (3.2%, 10.8%), which can better reflect the actual situation of the ultra-high voltage emergency repair project.

Acknowledgements. This work was supported by the scientific and technological project of China Southern Power Grid Company Limited. The name of the project is the research and application of the pricing model and cost control of the emergency repair project of ultra-high voltage transmission and transformation.

References

1. Toshchakov, P.V., Kotov, O.M., Kostarev, A.F.: Evaluation of versions of electric power grid repair schemes from the results of structural reliability calculations. Power Technol. Eng. **46**(5), 421–427 (2013). https://doi.org/10.1007/s10749-013-0372-y
2. Wang, W., Lou, B., Li, X., Lou, X., Jin, N., Yan, K.: Intelligent maintenance frameworks of large-scale grid using genetic algorithm and K-Mediods clustering methods. World Wide Web **23**(2), 1177–1195 (2019). https://doi.org/10.1007/s11280-019-00705-w
3. Dong, L., Wu, J., Pu, T.J., Zhou, H.M.: Research of the maintenance scheduling optimization considering the grid risk. Adv. Mater. Res. **2116** (2013)
4. Xie, C., Wen, J., Liu, W., Wang, J.: Power grid maintenance scheduling intelligence arrangement supporting system based on power flow forecasting. Phys. Procedia **24**, 832–837 (2012). https://doi.org/10.1016/j.phpro.2012.02.125
5. Anonymous: Households Shouldn't Pay for Power Grid Repairs-Putin. Interfax: Russia & CIS Energy Newswire (2011)
6. Antonyan, O., Maksimchuk, O., Solovyova, A., Chub, A.: The formation of the cost of overhaul of apartment buildings in the budget of the region. In: Proceedings of the Volgograd State University International Scientific Conference on Competitive, Sustainable and Safe Development of the Regional Economy (CSSDRE 2019) (2019)
7. Lee, E.-B., Alleman, D.: Ensuring efficient incentive and disincentive values for highway construction projects: a systematic approach balancing road user, agency and contractor acceleration costs and savings. Sustainability **10**(3), 701 (2018). https://doi.org/10.3390/su10030701
8. Edward Minchin, R., Brent Thurn, S., Ellis, R.D., Lewis, D.W.: Using contractor bid amounts to estimate the impact of night construction on cost for transportation construction. J. Constr. Eng. Manag. **139**(8), 1055–1062 (2013). https://doi.org/10.1061/(ASCE)CO.1943-7862.0000688
9. Ding, L.P., Zhao, T.S., Liu, X.Z.: Study on the system of safe and civilized construction. Appl. Mech. Mater. **1366** (2011)
10. Jin, Z.: Construction of different types of projects after replacing business tax with Vat the impact of VAT tax burden on enterprises and suggestions. Probe Account. Audit. Tax. **1**(1), 1–4 (2019)

Based on Fast-RCNN Multi Target Detection of Crop Diseases and Pests in Natural Light

Mingyuan Xin[1(✉)], Yong Wang[2], and Xiangfeng Suo[3]

[1] School of Computer and Information Engineering, Heihe University,
Heihe, Heilongjiang, China
[2] Institute of International Education, Heihe University,
Heihe, Heilongjiang, China
[3] School of Computer and Information Engineering, Heihe University,
Heihe, Heilongjiang, China

Abstract. Crop diseases and insect pests detection is a necessary means to ensure the healthy growth of crops. With the increase of crop planting area, in order to improve the detection efficiency, the application of deep learning algorithm to crop diseases and insect pests detection has become a research hotspot. However, the accuracy and efficiency of the traditional deep learning model is not high because of the natural concern of crop diseases and pests and the complex background. In this paper, we learn from and improve the fast-RCNN method which performs well in the task of target segmentation. We use cyclegan to supplement illumination and fast-RCNN to extract contour. In order to alleviate the problem of insufficient labeled samples, this paper studies the transfer learning mechanism of fast-RCNN, designs and implements the importance sampling of training data, parameter transfer mapping and other methods. Experiments on real data sets show that the algorithm can better extract the contour of the image and further identify the disease and insect pests in natural light with only a small number of labeled samples.

Keywords: Fast-RCNN · CycleGAN · Object detection

1 Introduction

In the field of computer vision, crop leaf pest image recognition is an important research direction. However, the images of crop leaves obtained in natural environment are usually not ideal, which affects the accuracy of image recognition. First of all, the image has partial overexposure or insufficient light, the position of the light source, the degree of reflection, and the influence on the visibility of leaves and diseases and insect pests may present completely different situations; second, there are land, weeds and other graphics similar to the appearance of the object to be identified in the background; in addition, multiple target leaves exist at the same time and overlap each other, making the leaves present angle Different conditions. These problems greatly increase the difficulty of image recognition of leaf diseases and insect pests.

J. Abawajy et al. (Eds.): ATCI 2021, LNDECT 81, pp. 132–139, 2021.
https://doi.org/10.1007/978-3-030-79197-1_17

2 Illumination Compensation Method for Images Under Complex Illumination Conditions

As one of the most reliable computer vision methods, neural network plays an increasingly important role in this field. The common methods to solve the problem of illumination are histogram equalization, homomorphic filtering, gradient domain image enhancement [1], Retinex algorithm [2], gamma correction [3] and so on. However, most of the papers related to these methods have a long history, and the effect of light treatment is not satisfactory. Combined with the above, various recognition problems under complex lighting conditions are still unsolved topics in the field of computer recognition. Due to the changes of image local pixel values and the distortion of object features and textures caused by lighting, the recognition efficiency and accuracy will be greatly affected, and it is more difficult to further recognize images and videos on this basis. In order to solve this kind of problem, in the fields of face detection [4], skin color detection [5], lane detection [6], some scholars have tried to overcome the influence of illumination and improve the recognition accuracy with various methods.

2.1 CycleGAN

Circularly generated antagonism net is a variant of conventional Gan proposed by Zhu et al. [7]. The discriminator loss and generator loss of one unidirectional Gan are represented by the following Eqs. (1, 2), respectively:

$$L_{CAN}(G_{XY}, D_Y, X, Y) = E_Y\left[\log D_y(y)\right] + E_x\left[\log\left(1 - D_y(G_{XY}(x))\right)\right] \tag{1}$$

$$L(G_{XY}, D_{YX}, X, Y) = E_X\left[\|G_{YX}(G_{XY}(x)) - x\|_1\right] \tag{2}$$

In the above formula, X and Y respectively represent the real input of the two classes to be transformed, gxy and GYX respectively represent the two generators, and Dy represents the discriminator of the unidirectional network. The complete cyclegan network consists of two unidirectional networks, which share two generators and each has one discriminator.

In the whole cyclegan, the loss of the whole network can be simplified as follows:

$$L_{cyc} = E_{x \sim P_{data(x)}}\left[\| F(G(x)) - x \|_1\right] + E_{y \sim P_{data(y)}}\left[\| G(F(y)) - y \|_1\right] \tag{3}$$

Where, G and F represent the two generators, x and y represent the inputs of the two generators respectively, that is, the normal illumination/complex illumination image matrix to be converted, and $Lcyc$ represents the cycle loss of the two generators. The smaller the value is, the more similar the input and output of the generator are.

2.2 Illumination Compensation Steps of Circularly Generated Antagonism Net

By reversing this process, inputting complex illumination images and performing similar operations through another unidirectional network, and making the training cycles of the two unidirectional networks alternate, a complete closed-loop generation network can be formed, and two generators and two discriminators in the network can be trained continuously. In the training, two generator models will be obtained: the model of transforming normal illumination image into complex illumination image and the model of transforming complex illumination image into normal illumination image.

3   Asymmetric Fingerprinting

3.1   Target Detection

R-CNN [7] uses regions generated from selective search [8] or edge box [9], proposes to generate region based features from pre trained convolutional neural network, and uses SVMs for classification. Fast RCNN [10] uses regional proposal network (RPN) instead of selective search. RPN is used to generate candidate bounding boxes (anchor boxes) and filter out the background area, and then another small network is used to classify and regress the bounding box position according to these suggestions. R-fcn [11] uses position sensitive region of interest pool (psroi) to replace the pooling of region of interest in fast RCNN, which can improve the accuracy and speed of the detector. Recently, deformable convolution and variable psroi are proposed in variable convolution network [12], which can further improve the accuracy of rfcn.

3.2   Faster-RCNN

In order to generate the target candidate box on the detection image, the fast RCNN designs the RPN sub network. First, it sets the benchmark anchor on the WXH convolution feature map which is the last output of the backbone network, and then slides the benchmark anchor on the feature map. Then, it maps each sliding window to a lower dimension feature vector, and generally uses a nonlinear activation relu function, Then two full join layers are used to classify (els) and regress (reg) the mapped feature vectors, and the full join layer is shared for all spatial positions on the feature graph. In the training phase, fast RCNN sets a binary class label for each anchor frame generated in the RPN network to indicate whether the anchor frame is a foreground target or a background. The corresponding setting rules are as follows: when an anchor frame has the highest IOU (overlap ratio) with the real annotation box or the IOU of the anchor frame and the real annotation box is greater than 0.7, it is labeled as the foreground target; when the overlap ratio of an anchor frame with all the real annotation boxes is less than 0.3, it is labeled as the background; the anchor frame with the overlap ratio between 0.3 and 0.7 is not very useful for RPN network training, so it is abandoned. According to the above definition, faster RCNN defines the loss function of RPN network as formula (4):

$$L(\{p_i\}, \{t_i\}) = \frac{1}{N_{cis}} \sum_i L_{cis}(p_i, p_i^*) + \lambda \frac{1}{N_{reg}} \sum_i p_i^* L_{reg}(t_i, t_i^*) \tag{4}$$

Where, i is not the index value of a frame in the corresponding small batch, and P_i is the probability that the network predicts whether the i anchor frame is the foreground target or the background. If the anchor box is the foreground target, the corresponding real label box is p_i^* take the value as 1, otherwise it is 0. t_i represents the four parameterized coordinates of the anchor frame, which are the coordinates of the center point of the anchor frame and the width and height of the anchor frame, t_i^* represents the coordinates of the corresponding real dimension box. Lcls is the logarithmic classification loss, Lcls is the border regression loss, $L_{reg}\left(t_i, t_i^*\right) = R\left(t_i - t_i^*\right)$, where R is the smooth L1 loss function. $p_i^* L_{reg}$ is an identifier indicating regression loss only when the corresponding anchor box is positive sample ($p_i^* = 0$) Otherwise, the regression loss term is 0. $\{p_i\}$ and $\{t_i\}$ It represents the output of classification layer (cls) and border regression layer (reg) respectively.

4   Transfer Learning Under the Condition of Lack of Samples

In crop diseases and insect pests images, most of the target annotation data are manually framed, which cannot directly meet the training needs of fast-RCNN. Because the existing target instance segmentation requires that all training instances must be marked with pixel level segmentation mask, the cost of annotating new categories is very expensive. Therefore, in the above scenario, this paper studies the transfer learning under the condition of lack of samples.

4.1   Construction of Migration Sample Dataset Based on Resampling

In order to evaluate the similarity between the training samples from Imagenet source domain and the target domain samples of the power equipment image where the task is located, we need to establish the evaluation function of the target similarity. Considering the accuracy and complexity, the measurement method is as follows:

$$sim(I_1, I_2) = e^{=\|f(I_1)-f(I_2)\|_2^2} \tag{5}$$

Where: I_1 and I_2 represent any two images in the training sample set and the target sample set respectively; $f(I_1)$ represents the feature vector extracted from the image, where the gray level co-occurrence matrix is used to calculate; the similarity of two images is taken as the negative exponential power of the distance between the gray level co-occurrence matrix. The values obtained are between 0 and 1, which can be interpreted as the similarity probability of training samples and target samples. According to this probability distribution, the training data set can be obtained by sampling the training data set. In this data set, the samples with higher similarity are more likely to be selected many times; the frequency of samples with lower similarity is significantly reduced.

4.2   Generating a Few Mask Annotation Images with OpenCV

After generating the mask annotation, the mask information of each pixel in the image is generally represented by 0 or 1, where 0 means that the pixel does not belong to a specific target area, and 1 means that the corresponding pixel belongs to a specific target area. The mask annotation can be generated by opencv API.

5   Experimental Analysis

5.1   Experimental Environment

The processor is 1.6 GHz dual core Intel Core i5, the memory is 8 GB 1600 MHz DDR3, and the graphics card is Intel HD graphics 6 thousand 1536 MB. The development language Python and deep learning framework tensflow are used, and pycharm is the main development tool. The data set uses the open source database as the training set, including 50 thousand pictures, and the disease sample data of Institute of intelligence of Chinese Academy of Sciences as the verification data set.

5.2   Performance Evaluation Index

The common performance indicators of target detection include precision, recall, average precision (AP), mean average precision (map), etc. In order to calculate the values of these indicators, we assume that there are only two types of classification targets after detection: positive and negative. True positives (TP): the number of positive cases determined by the classifier when the label is positive. False positives (FP): the number of negative cases whose labels are judged to be positive by the classifier. False negatives (FN): the number of positive cases whose labels are determined to be negative by the classifier. True negatives (TN): the number of negative cases determined by the classifier when the label is negative. Precision = TP/(TP + sssFP), Recall = TP/(TP + FN) = TP/P. It can be found that the lower the accuracy, the higher the recall rate. If recall value is taken as the horizontal axis and precision value as the vertical axis, the P-R curve can be obtained.

5.3   Result Analysis

In order to improve the final detection accuracy of the model, this paper uses the method of feature extraction and target candidate box generation. Table 1 shows the performance statistics of the two methods on the test data set.

Table 1. The Detection rate and detection time of illumination compensation generation model for circularly generated countermeasure network

Network model	Running time(s)	Simple %	Secondary %	Complex %
YOLOv3	0.18	93.28	91.42	82.78
Faster-Rcnn(VGG)	0.24	89.12	83.61	72.81
Feature exchange-VGG	0.31	94.37	91.61	81.83
Faster-Rcnn (CycleGAN)	0.36	95.22	92.46	83.56

The convolution networks in Table 1 are all based on vgg16 network for feature extraction. It can be seen that the model detection time is increased by 0.07 s and 0.12 s respectively after adopting the cyclic countermeasure network and feature exchange method. For the accuracy of the model, because the feature exchange makes the shallow position information in the final feature image and the multi-scale prediction can also obtain more target features, so the accuracy of the three types of samples with different complexity has been improved. The percentage of improvement according to the simple sample, the medium complex sample and the complex sample is: for the feature exchange is yes 25%, 8.0%, 9.02%; for the cyclic countermeasure network, it is 6.1%, 8.85%, 10.7%. When the cyclic countermeasure network and feature exchange method are used at the same time, the accuracy of the model is further improved. As can be seen from Table 1, compared with yolov3, the accuracy of medium complexity samples and complex samples is improved, especially for complex samples, the accuracy is increased by 0.87%. The results show that Yolo network is more prone to miss detection for occluded or small targets. In order to show the performance of the improved convolution network more clearly, the P-R curves of the original fast RCNN network and the improved convolution network on the test set are compared. Figure 1 and Fig. 2 show the P-R performance curves of the three convolution networks on simple sample and complex sample test sets respectively.

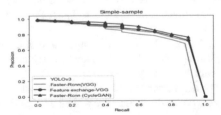

Fig. 1. P-R performance curve of improved convolution network on simple sample test set

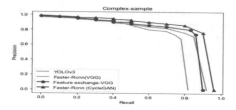

Fig. 2. P-R performance curve of improved convolution network on complex sample test

6   Conclusions

This paper introduces the current situation of crop leaf diseases and insect pests recognition based on fast RCNN algorithm and traditional image recognition methods. In the natural and complex environment, the traditional image recognition accuracy is not high. This paper proposes a target detection method based on light compensation and small sample space. The method is composed of cyclegan and fast-RCNN algorithm. The test method can effectively improve the efficiency of target detection in crop diseases and insect pests images, and get rid of the dependence on manual registration and partition to a certain extent.

Acknowledgements. This work was supported by Heilongjiang Provincial Natural Science Foundation of China: LH2020F039.

References

1. Tivive, F.H.C., Bouzerdoum, A.: A new class of convolutional neural networks (SICoNNets) and their application of face detection. In: International Joint Conference on Neural Networks, svol. 3, pp. 2157–2162. IEEE (2003)
2. Tivive, F.H.C., Bouzerdown, A.: An eye feature detector based on convolutional neural network. In: Eighth International Symposium on Signal Processing and ITS Applications, pp. 90–93. IEEE (2006)
3. Chen, Y.N., Han, C.C., Wang, C.T., et al.: The application of a convolution neural network on face and license plate detection. In: International Conference on Pattern Recognition, pp. 552–555. IEEE Computer Society (2006)
4. Szarvas, M., Yoshizawa, A., Yamamoto, M., et al.: Pedestrian detection with convolutional neural networks. In: IEEE Proceedings. Intelligent Vehicles Symposium, 2005, pp. 224–229. IEEE (2005)
5. Subr, K., Majumder, A., Irani, S.: Greedy algorithm for local contrast enhancement of images. In: Roli, F., Vitulano, S. (eds.) Image Analysis and Processing – ICIAP 2005, pp. 171–179. Springer, Berlin, Heidelberg (2005). https://doi.org/10.1007/11553595_21
6. Land, E.H.: Recent advances in Retinex theory and some implications for cortical computations: color vision and the natural image. Proc. Nat. Acad. Sci. **80**(16), 5163–5169 (1983)
7. Lau, H., Levine, M.: Finding a small number of regions in an image using low-level features. Pattern Recogn. **35**(11), 2323–2339 (2002). https://doi.org/10.1016/S0031-3203(01)00230-8

8. Girshick, R., Donahue, J., Darrell, T., et al.: Rich feature hierarchies for accurate object detection and semantic segmentation. In: Proceedings of the IEEE Conference on Computer Vision and Pattern Recognition, Honolulu, HI, pp. 580–587 (2014)
9. Uijlings, J.R.R., Van De Sande, K.E.A., Gevers, T., et al.: Selective search for object recognition. Int. J. Comput. Vis. **104**(2), 154–171 (2013)
10. Zitnick, C.L., Dollár, P.: Edge boxes: locating object proposals from edges. In: Fleet, David, Pajdla, Tomas, Schiele, Bernt, Tuytelaars, Tinne (eds.) Computer Vision – ECCV 2014: 13th European Conference, Zurich, Switzerland, September 6-12, 2014, Proceedings, Part V, pp. 391–405. Springer, Cham (2014). https://doi.org/10.1007/978-3-319-10602-1_26
11. Girshick R. Fast R-CNN. In: Proceedings of the IEEE International Conference on Computer Vision, Honolulu, HI, pp. 1440–1448 (2015)
12. Ren, S., He, K., Girshick, R., et al.: Faster R-CNN: towards real-time object detection with region proposal networks. In: Advances in Neural Information Processing Systems, SPAIN, Barcelona, pp. 91–99 (2015)

Research and Exploration on the Application of Domestic Password in the Transportation and Water Transportation Industry

Xuekun Sang[1]([⊠]) and Jinying Huang[2]

[1] Research Institute of Water Transport, Ministry of Transport, Beijing, China
[2] China Academy of Space Technology, Beijing, China

Abstract. With the development of domestic cryptography technology, domestic cryptography technology is gradually mature and has the feasibility of application and promotion. Gradually promote the domestic password application system and information infrastructure of transportation and water transport, which will help to resolve the network security risk that the current password system may be cracked, and help the transportation and water transport industry to establish a more secure and reliable support and guarantee system.

Keywords: Domestic password · Transportation and water transportation · Independent and controllable

1 Introduction

With the acceleration of science and information technology, network security has become an important part of national security strategy. In April 19, 2016, general secretary Xi Jinping pointed out at the Symposium on network security and informatization that "to protect Internet Security and national security, we must break through the difficult problem of core technology GB/t22239-2019 basic requirements for classified protection of information security technology network security, issued on May 10, 2019, puts forward specific requirements for the application of domestic commercial password products in e-government and other fields, and password technology has become the core technology and important support of network security [1, 2].

In recent years, with the rapid development of cryptographic technology in China, the State Password Administration has formulated and publicly released commercial cryptographic algorithms and related industry standards with independent intellectual property rights and high security strength, which is of great significance to ensure the autonomous control of cryptographic algorithms and reduce the risk of sensitive information disclosure and information system attack. At the same time, China has basically completed the domestic password application infrastructure and started to provide services. The domestic password industry chain has been mature, creating technical conditions and application environment for each information system, especially for the self-developed information system in the industry to implement password substitution [5, 6].

J. Abawajy et al. (Eds.): ATCI 2021, LNDECT 81, pp. 140–146, 2021.
https://doi.org/10.1007/978-3-030-79197-1_18

The situation of information security problems faced by the transportation industry is relatively severe. The existing international general password algorithm can not meet the current and future application security needs. In the guidance on strengthening password application in important fields issued by the central office of the CPC Central Committee, it is clearly proposed that "transportation is the application scope of key information systems and key industrial control systems, so it is necessary to speed up the application of domestic passwords and protect the network and information security; the existing networks and information systems shall undergo password localization transformation". In 2016, the Ministry of transport issued the master plan for the promotion of password application in important business areas of the transport industry to promote the application of domestic password algorithm in the transport industry [4], strengthen the management of password application, consolidate the cornerstone of the information security of the transport industry, realize the independent and controllable information security of the transport industry, and promote the realization of "big transport" and "four transportations "For important security. On August 22, 2018, the general office of the Ministry of transport issued the notice on the work plan of password application and innovation development (2018–2022) in the field of transport, which clearly states that by 2022, the domestic password will be fully used and fully covered in the field of transport, and the technical standards, application, evaluation, certification and guarantee system of password in the transport industry will be sound and effective, safe, reliable, autonomous and controllable The password security defense line of is basically completed.

2   Current Situation and Problems of Password Application in Transportation and Water Transportation Industry

China's transportation industry is a comprehensive transportation industry, whose fields include: highway, waterway, railway, civil aviation, post, salvage, maritime and supporting management services related to transportation, in which the transportation industry is an important part of the transportation industry. During the "13th five year plan" period, the basic database group of personnel, vehicles, ships, goods, etc. of the transportation industry In this formation, the real-time data acquisition quality of key industries (AIS, VTS, CCTV, etc.) has been steadily improved. Relying on the construction of demonstration pilot projects, new breakthroughs have been made in the intelligent application of smart ports, smart shipping, modern logistics and other fields, and the convenience of information services has been significantly improved. The public information service capacity of the water transportation management departments at all levels has been further improved, the commercialized intelligent transportation information service has been booming, the public information service experience has been improving, and the development environment of the transportation information service industry has been continuously optimized. Therefore, the application of domestic commercial password is very urgent, although domestic password has gradually carried out practical application in information infrastructure, internal administrative office, main business management platform, etc., and achieved some results. However, the overall effect is still relatively backward, and the problems such as weak awareness of network security, weak foundation, insufficient investment, insufficient ability and coverage are prominent, which make it difficult to effectively deal with high-intensity network attacks and adapt to the increasingly severe network security situation.

The main problems are as follows:

(1) Management and service system is not yet sound

The domestic password management organization of the industry is the password management office of the Ministry of transport. Some provincial industry administrative departments have not yet defined the functions and management mechanisms of the domestic password management organization. At present, from the perspective of the industry, the whole transportation and water transportation industry has not yet formed an effective and perfect domestic password management system. In the construction of business system, operation and maintenance, personnel management and other aspects of the industry, there is generally a lack of password related management system, and there is no special password management organization and service system. In the development of domestic password application, there is no reference basis.

(2) The application rate of domestic password is not high and the risk is controllable

At present, most of the established business information systems and infrastructure in the industry adopt international cryptographic algorithms, such as DES, 3DES, RSA, etc. Most of the intellectual property rights of cryptographic algorithms belong to some foreign institutions. It is uncertain whether there are backdoors and loopholes in the content of algorithms. In addition, it is difficult to transform, improve and make deep use of it. Therefore, the security risk of the whole business information system and infrastructure is less controllable.

(3) The standardization of technical specifications is not uniform

The form of security is becoming more and more severe. At present, due to the different construction time, period and manufacturer of different projects in the water transport industry, the standards of cryptographic algorithms used by different cryptographic devices are not uniform. For example, the cryptographic algorithms of CA authentication server, VPN gateway and electronic key depend on software implementation. The security of some cryptographic algorithms is determined by the length of their keys. Facing the increasing computing power of geometric series, the hidden trouble of being cracked is also increased significantly. In addition, individual algorithms are no longer reliable from the point of view of mathematical calculation because of their own design defects, but they are still widely used for historical reasons or because they are temporarily irreplaceable.

3 Research and Exploration on Application of Domestic Password in Transportation and Water Transportation Industry

In order to implement the domestic password application, it is necessary to define the application scope and specific application object. The application scope is mainly the system developed and constructed by the water transport industry itself, which can be gradually expanded to the information system developed by domestic and foreign software manufacturers but with copyright. The main objects should be the current and running public display portal, office system, business information system and infrastructure.

Establish the technical route for the application of domestic passwords in the transportation and water transportation industry, clarify the requirements for the application of domestic passwords in the industry system from the aspects of industry policies, regulations and technical schemes, and gradually promote the application of domestic passwords in the transportation and water transportation industry.

This paper mainly focuses on the following two parts:

3.1 Management and Service System

In promoting the application of domestic passwords, a sound management system is the prerequisite to ensure the standardization and effectiveness of the technical system. The management system is the basis for the systematization, normalization and standardization of the application of domestic passwords. It is also an important restriction and supervision basis to ensure that domestic passwords can be truly implemented in the transportation and water transport industry.

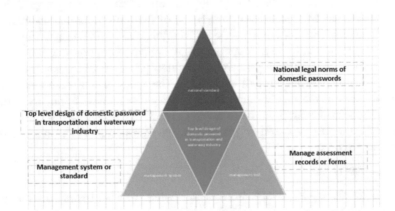

Fig. 1. Domestic password application management system of transportation and water transportation industry [3]

It can be seen from the figure that the domestic password application management system of the transportation and water transport industry includes: (1) relevant laws, regulations, requirements and standards of the national password management organization; (2) top level design of domestic password application of the whole transportation and water transport industry; (3) relevant management system proposed by the water transport industry management organization of the Ministry of transportation and the water transport industry management organization of each province; (4) various internal management systems of the industry Tools for management, implementation, audit and supervision [7, 8].

In addition, in the work of promoting the application of domestic password, the perfect service system of domestic password application consultation, evaluation, operation and maintenance, and support training in the transportation and water transportation industry is the guarantee to ensure the operation of the whole life cycle of domestic password application. It mainly includes: (1) reasonable and scientific scheme design; (2) standardized safety assessment; (3) effective operation and maintenance management; (4) scientific process training [9, 10].

3.2   Applied Research and Exploration

The domestic password shall focus on providing independent and controllable network security guarantee in the aspects of open sharing of key elements such as ships, personnel, goods, vehicles and other information, online coordination of management, convenient interconnection of waterway transportation, etc. The password application of each public service website, administrative office and business application system shall be constructed and implemented in accordance with unified standards, and authentication and authorization shall be carried out in accordance with unified standards.

In view of the fact that the non-autonomous and controllable cryptographic algorithm has been widely used in various systems of the transportation and water transportation industry. Therefore, it is necessary to fully consider the design scheme of the original system resources, the new system or the system under construction to develop the application of the domestic password in the transportation and water transportation industry. According to the current situation of the industry, the domestic password application route of the whole transportation and water transport industry should be carried out synchronously from the following aspects (as shown in Fig. 1: the overall structure of domestic password application in the transportation and water transport industry), as follows:

(1) Public display

The public display is mainly aimed at the portal website, mainly considering the functions of government affairs publicity, information release and service. Domestic password security protection measures can be taken, mainly including trusted site access, identity identification of management personnel, information release responsibility tracing.

(2) Administrative office system

For the office system, mainly considering the work of cooperation, efficiency and information sharing, domestic password security measures can be taken, mainly including user strong identity authentication, electronic seal of official document circulation.

(3) Business management system

For the business management system, mainly considering that the data involves a large number of ships, vehicles, personnel, cargo information, etc., the domestic password security protection measures can mainly include strong identity authentication of system users, mobile device authentication, data transmission security, electronic seal and data security storage.

(4) Information infrastructure

In view of the application security risks of information infrastructure, domestic password security protection measures can be taken, mainly including strong identity authentication of users, safe transmission and safe storage of data.

4 Conclusions

To sum up, there is still a long way to go in the application process of the domestic password in the transportation and water transportation industry, but the domestic password algorithm is safe and reliable, the technology is gradually mature, and has the feasibility of application and promotion. Gradually promote the domestic password application or transformation of public service system, administrative office system, business management system and information infrastructure of transportation and water transport, which will help to resolve the network security risk that the current password system may be cracked, and help the transportation and water transport industry to establish a more secure and reliable support and guarantee system.

References

1. Exploration and practice of the application and transformation of the domestic cryptographic algorithm in Maogaoli information system. Journal of Fujian radio and Television University no. 12016
2. Fu Pengxia Promotes the application of domestic cryptographic algorithm to realize the autonomous control of information system. Scientist (10), 104–105 (2015)
3. Shen, C., Gong, B.: Framework of trusted computing system based on domestic cryptosystem. J. Cryptol. **5**, 381–389 (2015)
4. Circular of the Ministry of transport of the people apos; Republic of China on printing and distributing the general plan for promoting the application of passwords in important business areas of the transport industry (Jiao Ban Fa [2016] No. 53)
5. GM/T 0054-2018: basic requirements for password application of information system

6. Circular of the general office of the CPC Central Committee and the general office of the State Council on printing and distributing the work plan for password application and innovation development in finance and important fields (Hall Zi [2018] No. 36)
7. Shannon, C.E.: A mathematical theory of communication. Bell Syst. Tech. J. (1984)
8. Shannon, C.E.: Prediction and entropy of printed English. Bell Syst. Tech. J. (1951)
9. Bellovin, S.M.: Frank miller: inventor of the one-time pad. Cryptologic (2011)
10. Wang, W.Y.: The development of contemporary cryptography technology and application in data security. Comput. Secur. (2012)

Application of Big Data Technology in the Cultivation Mode of Compound Talents in Civil Engineering

Zhanchun Mu[✉]

School of Yunnan Technology and Business University,
Kunming, Yunnan, China

Abstract. Today is the era of big data. While subverting people's activities, big data technology also promotes the transformation of thinking mode, carrier form, and practice mode in the field of civil engineering. It is an innovative creation of the training model for the application of compound talents in civil engineering. The talent training model is a work that all fields attach great importance to. Only by cultivating professional talents in the field can the development of the field be promoted and the involution of the field can be reduced. It is only in different countries and historical periods that the goal of talent training is, the levels and specifications are different. Big data technology has strong data perception and processing capabilities, which can contribute to the network education work of colleges and universities. But at the same time, the emergence of new things will inevitably bring new thinking. Therefore, it is of great significance to study the application of big data technology to the training mode of compound talents in civil engineering. To this end, this article puts the field of civil engineering under the background of big data, and explores innovative talent training models in the field of civil engineering in the era of big data through research and analysis of its basic connotations and changes in the background of big data. Based on the analysis of the research data, it is concluded that big data technology can bring innovation and transformation to the talent training model of civil engineering, and it can also cultivate high-quality talents in the field of civil engineering.

Keywords: Big data technology · Civil engineering · Talent training model · Compound talents

1 Introduction

With the rapid development of big data technology, it has gradually integrated into various human practical activities. This brand-new change has brought about a double revolution of "technology-thinking" and is reshaping all human life including civil engineering. The training of compound talents in civil engineering application is an important part of the civil engineering profession. The development of big data technology provides a new historical stage for it, which is both an opportunity and a challenge. From the perspective of big data, the research on the compound talent training model for civil engineering applications can provide theoretical support and

J. Abawajy et al. (Eds.): ATCI 2021, LNDECT 81, pp. 147–154, 2021.
https://doi.org/10.1007/978-3-030-79197-1_19

practical enlightenment for the effective development of network education in the field of civil engineering, so that the field of civil engineering can better shoulder the role of talent training.

Big data, as one of the core drivers of current social changes, is an important support for the training of composite talents for civil engineering applications. Under the background of the era of big data, in-depth study of how the training of composite talents in civil engineering applications can be accurately integrated with big data technology, and in accordance with the new characteristics and new laws of the thoughts and behaviors of composite talents in civil engineering applications, promote work innovation in the field of civil engineering, and do a good job civil engineering application compound talents' moral quality and personality cultivation, value concept and life pursuit, professional knowledge and academic ethics education and guidance work, not only the civil engineering application compound talent training work to adapt to the inherent needs of the big data era and advancing with the times, it is the fundamental requirement for the long-term development of the civil engineering profession [1]. Compared with domestic research, foreign research on big data started earlier, and many representative and valuable research results have been formed. International Data Corporation believes that big data still has the characteristic of great data value [2]. The application cases of big data in various fields such as medical care, business, finance, government affairs, network services, family life, etc. profoundly illustrate how big data has changed the way of human production, life, management, and thinking at a speed and way beyond human imagination. Existing state of development [3]. Dr. Eric Siegel combined the application of big data in traffic route planning, supermarket sales, criminal behavior, real estate mortgage, political elections and other aspects in "Big Data Forecasting", highlights how the predictive analysis function of big data affects the development of human behavior [4]. In China, the research on big data and its application in civil engineering has achieved very rich results in theory, and it has also been further deepened in practice [5]. Therefore, in the context of the era of big data, domestic scholars are still in the preliminary stage of exploring the perspective transformation, method reform, logical reconstruction, and path innovation of the collaborative education model of ideological and political work in colleges and universities, and there is a relatively lack of relevant research results.

The rise of a series of intelligent information technologies represented by big data and artificial intelligence has an impact on the training of composite talents in civil engineering applications. It is not limited to teaching, management, decision-making, and cultural construction. It should also integrate these technologies with the field of civil engineering. The research is integrated, and efforts are made to improve the efficiency, level and effect of research work [6]. Based on the current status and limitations of the current big data application in civil engineering application of compound talents, with the big data era as the research background and perspective, this article attempts to deeply analyze the actual situation, existing problems and the causes of the problems in the cultivation of civil engineering application compound talents. Based on this, construct comprehensive, logical, and innovative solutions to form a systematic and complete theoretical system framework and practical plan, which will not only lay a solid theoretical foundation for future related research in this field, but also use scientific practical guidance, prominent problem-oriented and strong support will promote the effective integration of civil engineering application compound talent training and big data technology, and improve the effectiveness of talent training.

2 Method

2.1 Connotation of Big Data

To understand big data, we must first understand what "data" is. Since the time when people are tied and remembered, human beings have forged an indissoluble bond with data. Data are traces of information left by humans in practical activities or that can be captured and recognized by humans. Data in a broad sense refers to all information, and data in a narrow sense is a value, which is obtained by humans through observation, experiment or calculation. As a result, information such as numbers, text, images, etc. [7]. Therefore, large scale is what "data" should mean. Obviously, this is not the core point of distinguishing big data from data. To understand the "big" of big data, we must not only understand it in terms of quantity, but also in terms of function. Before the birth of big data, mankind's ability to process data was limited. After the birth of big data, mankind's ability to collect, store, analyze, and use data has made a qualitative leap. In short, big data is not only a very large-scale data group, but also a new comprehensive ability to use data. Among them, there is a formula named Bayes' theorem formula in big data, and the formula is as follows:

$$P(B_i|A) = \frac{P(B_i)P(A|B_i)}{\sum_{j=1}^{n} P(B_j)P(A|B_j)} \tag{1}$$

Where $P(A|B)$ is the probability that A will occur if B occurs. $A1, \ldots An$ are complete event groups, namely:

$$\bigcup_{i=1}^{n} A_i = \Omega, A_iA_j = \emptyset, P(A_1^2) > 0 \tag{2}$$

When there are more than two variables, Bayes' theorem still holds, for example:

$$P(A|B, C) = \frac{P(A)P(B|A)P(C|A, B)}{P(B)P(C|B)} \tag{3}$$

2.2 Value of Talent Training from the Perspective of Big Data

The value of big data itself is very considerable. With the continuous increase of data dimensions, talent training subjects make it possible to tap the value of the invisible world and solve and avoid the risks of invisible problems through the use of big data. Its value will continue to be enriched and deepened [8]. Therefore, talent training under the perspective of big data has rich value, which can be embodied in the following two aspects: First, it has demonstration value. The application prospect of big data is wide, but there are not many real mature big data application projects. Therefore, talent training under the perspective of big data can accumulate valuable experience for the application of big data in education and other industries, open up a new space, and play a role of demonstration and leadership; on the one hand, talent training under the perspective of big data can be achieved through precise measures. Effectively enhance the individual's political judgment ability, on the other hand, through the data analysis of the education object, timely provide the civil engineering products expected by the education object to meet their political needs; second, it has safety value. The so-called safety value can also be understood as the value of avoiding risks [9]. Big data technology is like the brain and eyes of the education subject. It can help the education subject to better judge the complex behavior and thoughts of the education subject, and improve the ability to take precautions and prevent the slightest failure, so as to effectively avoid risks in a timely and effective manner. To integrate big data technology with education and teaching well, it is necessary to understand the characteristics of big data technology in order to better exert its strengths and advantages. Big data technology can communicate anytime and anywhere between information subjects through various wired and wireless networks. These exchanges inevitably bring about the exchange of information, and the key to reflecting the effectiveness of data and information is to increase speed. The advantage of Big data is that it can have the advantage of speed in processing data.

2.3   Talent Training Theory

The broad sense of talent training refers to the cultivation of talents in society and schools, and the narrow sense of talent training refers to the training of talents in schools in our usual sense. The "talent training" in vocational education refers to the educational activities that make the educated person's body and mind have a purposeful change through the educator's adoption of a certain training plan, so that he can become a person with professional knowledge and skills [10]. In other words, educators need to pre-determine the training goals before launching activities, understand the situation of the training objects, take certain training measures to act on the educated, so that they can get physical and mental development, and have a corresponding guarantee system to ensure the smooth implementation of the entire process.

3   Experiment

3.1   Purpose of the Experiment

This article is based on big data technology, draws on the domestic and foreign theoretical research results, and conducts in-depth research and discussion from the civil engineering application compound talent training model by consulting a large number of books and documents, conducting comparative experiments, data analysis and statistics, and result research analysis. To study the changes and trends of using big data technology in civil engineering application compound talent training model, as well as the advantages brought by big data.

3.2 Experimental Design

This article uses the form of questionnaires. The questionnaire has a total of 20 questions, including 13 objective multiple-choice questions and 7 open-ended questions. The respondents who participated in the questionnaire were mainly students majoring in civil engineering at a certain university, and some teachers were interviewed and asked questions. The content of the questionnaire is based on the composition of the elements of the training model under big data. In this survey, 500 questionnaires were distributed and 350 were collected. After analysis, 300 questionnaires were obtained. The response rate of this questionnaire survey was 60%.

4 Result

4.1 Validity Test of the Questionnaire Design

In this experiment, by interviewing 10 experts' opinions and suggestions, they were asked to evaluate the validity of the questionnaire. The validity is divided into three levels, namely effective, more effective and invalid. The following is their opinion form on the questionnaire. Specific as shown in Table 1.

Table 1. Questionnaire validity survey table

Position	Number	Effective	More effective	Invalid
Professor	3	2	1	0
Associate professor	3	3	1	0
Lecturer	4	3	1	0
Total	10	7	3	0
Percentage	100%	80%	30%	0%

It can be seen from Table 1 and Fig. 1 that among the 10 experts, there are three professors, three associate professors and four lecturers. 80% think the questionnaire is effective, including 2 professors, 3 associate professors, 3 lecturers, and 8 in total; 30% think the questionnaire is more effective, including 1 professor, 1 associate professor, and lecturer 1 person, 3 people in total; no one thinks the questionnaire is invalid. Therefore, the design of the questionnaire is relatively reasonable and can reflect the application of big data technology in the civil engineering method of the compound talent training model, indicating that the questionnaire is effective.

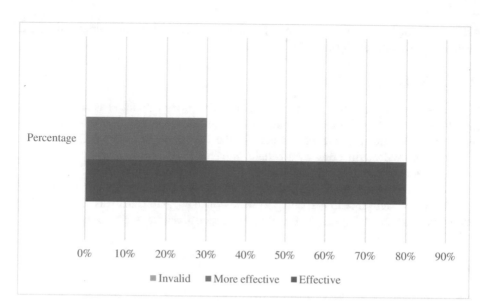

Fig. 1. Questionnaire validity survey chart

4.2   Understanding of Big Data

It can be seen from Table 2 and Fig. 2 that among the 300 questionnaires collected, 261 considered big data technology to be emerging technology, accounting for 87% of the total questionnaires; 191 considered big data technology to be high-tech, accounting for the total number of questionnaires. 63.3% of the questionnaires; 285 respondents believe that the use of big data technology is beneficial in the training of civil engineering application compound talents, accounting for 95% of the total questionnaires; 15 respondents believe that big data technology cannot play a role in the field of civil engineering. Accounted for 15% of the total questionnaires; another 30 questionnaires considered big data technology to be other types of technology, accounting for 10% of the total number of questionnaires. It can be seen from the data that big data is still an emerging technology in the eyes of the majority of civil engineering scholars, and they all believe that it can play an important role in the cultivation of composite talents in civil engineering applications.

Table 2.   Table of understanding of big

Understanding of big data	Percentage	Number of people
Emerging technology	87.00%	261
High-tech	63.30%	191
Good for talent training	95%	285
No benefit to talent training	5%	15
Other	10%	30

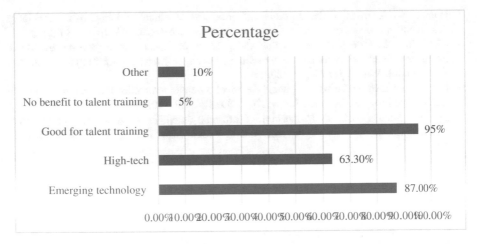

Fig. 2. Figure of understanding of big

5 Conclusion

While the development of big data redefines the human social environment, relational structure, and cultural context, it is creating a new order of collaborative innovation between characters and spaces. Based on the top-level design of the national strategy, docking theoretical guidelines and realistic patterns, carefully considering the fit of big data and civil engineering application compound talent training, and vividly interpreting the value proposition of "Lide Shuren", this is the work of civil engineering talent training in the era of big data.

References

1. Chénruìxìng, yǐnhónglíng, ān dōngshēng, zhāngzhèn léi, suízhìwēi, yáng xuān. Dà shùjù jìshù zài pèi diàn wǎng quán shíxù yùnxíng xiàolǜ fēnxī zhōng de yìngyòng. Gōng yòng diàn **38**(03), 22–30 (2021)
2. Die, F., Yiting, L., Shuang, L.: Analysis of retail industry transformation and upgrade path under category differences. China Collect. Econ. **07**, 81–82 (2021)
3. Chen, L.-K., et al.: Modular composite building in urgent emergency engineering projects: a case study of accelerated design and construction of Wuhan Thunder God Mountain/Leishenshan hospital to COVID-19 pandemic. Autom. Constr. **124**, 103555 (2021)
4. Siegel, E.: RBR enables innovation through responsive OEM partnerships. Ocean News Technol., 1–2 (2019)
5. Gu, X., Angelov, P., Zhao, Z.: Self-organizing fuzzy inference ensemble system for big streaming data classification. Knowl. Based Syst. **218**, 106870 (2021)
6. Valencia-Parra, Á., Varela-Vaca, Á.J., Gómez-López, M.T., Carmona, J., Bergenthum, R.: Empowering conformance checking using Big Data through horizontal decomposition. Inf. Syst. **99**, 101731 (2021)

7. Huang, X.: Analysis of the development prospects of digital and intelligent machinery design and manufacturing. Intern. Combust. Engines Accessories **02**, 167–168 (2021)
8. Li, Z., Li, X., Mou, G.: The research and practice of "specialization and undergraduate integration" application-oriented talent training: taking electrical automation as an example. Sci. Technol. Wind **05**, 175–176 (2021)
9. Liu, J.: Research on the "dual system" electrical automation practice platform based on the real sorting line. Sci. Technol. Innov. Appl. **09**, 128–130 (2021)
10. Lei, H.: Analysis of the application of artificial intelligence technology in electrical automation control. Manage. Technol. Small Medium Sized Enterp. (First Issue) **02**, 173–174 (2021)

Speed Measurement of DC Motor Based on Current Detection

Jian Huang[(⊠)]

XIJING University, Xi'an 710123, Shaanxi, China

Abstract. In order to measure the speed of DC motor effectively, a new method is proposed. By measuring the current magnitude of the driving motor, the corresponding relationship between the current magnitude and the rotational speed is found, so as to realize the accurate measurement of the rotational speed of the DC motor. Compared with traditional Hall method and photoelectric method, this method has the advantages of simple interface circuit, powerful function and high precision. The experimental results show that the method can accurately measure the speed of DC motor, the measurement period is less than 1 s, and the measurement speed range is 100 to 1500 revolutions per minute.

Keywords: DC motor speed measurement · Current detection · Holzer velocity measurement · Photoelectric speed measurement

1 Instruction

In production practice and experiment, in order to effectively control the speed of DC motor, it is necessary to measure it accurately. The commonly used measurement methods are hall velocity measurement and photoelectric velocity measurement. In reference [1], aiming at the disadvantage of poor accuracy of traditional hall velocity measuring device, an improved measure is proposed, which adopts dynamic periodic sampling method to improve the measurement accuracy. In reference [2], hall velocity measurement is improved, and an improved velocity measurement method combining rolling velocity measurement with software frequency doubling is proposed. Reference [3] studies the photoelectric velocity measurement method based on wavelet transform, which effectively removes the interference and improves the measurement accuracy. In reference [4], the angular displacement fitting method is used to obtain the real-time speed of the motor, which effectively improves the performance of the servo system.

The above methods are all based on the traditional speed measurement method, using sensors to directly measure the speed of DC motor. In practical application, if the given voltage is constant, the greater the current, the higher the motor speed. Therefore, this paper proposes to directly measure the current of the motor drive, design the corresponding current amplifier circuit, and convert the current into voltage. After a/D acquisition, it is connected to the microprocessor STM32 for software processing, and the relationship between speed and current is determined through a series of experiments. Compared with the traditional measurement method, this method has the advantages of simple hardware circuit, high measurement accuracy and wide adaptability.

J. Abawajy et al. (Eds.): ATCI 2021, LNDECT 81, pp. 155–160, 2021.
https://doi.org/10.1007/978-3-030-79197-1_20

2 System Design

The system block diagram is shown in Fig. 1. In the figure, stm32f103zet6, a high-performance embedded microprocessor based on Cortex-M3, is used for the main control, and 3.3 V DC power supply is used. There are 112 IO interfaces, of which the SPI interface can be used to drive the 1.44 in. true color display. PWM wave is generated by timer of STM32F103 to drive DC motor. The driving current is connected to ina193 to form a current detection and amplification circuit. The current is converted into voltage value, which is connected to the A/D acquisition input pin of STM32, and the converted voltage value is obtained after software processing [5–8]. In order to get the direct relationship between the voltage value and the speed, the selected motor has its own encoder, which outputs 390 pulses after one revolution. The relationship between the motor speed and the current can be obtained by comparing the two.

3 Hardware Design

3.1 Motor Driver

The supply voltage of DC motor is 6 V, the maximum is 12 V, and the speed is from 100 to 1500 rpm. The motor drive circuit diagram is shown in Fig. 2. In the figure, the DC motor voltage of tb6612 is 12 V, and the logic level is 5 V. The IO output pin PA11 of STM32 is connected to ain1 of tb6612, and PA12 is connected to ain2 of tb6612. These two pins control the motor steering [9, 10]. The pin PA0 of STM32 is connected to the pwma of tb6612, and PA0 outputs PWM wave to control the motor speed. Aout1 and aout2 are connected with DC motor.

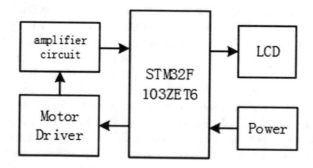

Fig. 1. System block diagram

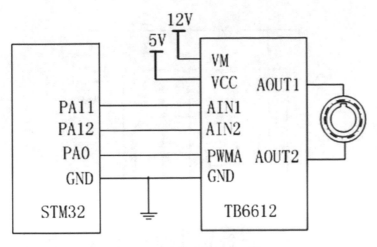

Fig. 2. Motor drive connection diagram

3.2 Hardware Circuit Design of Current Detection

The current detection circuit is shown in Fig. 3. Current detection can select ina193, ina194, ina195, etc. It can accurately detect the current is on the load. The input voltage of ina19x can range from 2.7 V to + 18 V, + 5 V is selected in the test, and the load is DC motor. The size of is will be converted into voltage Vout output, and the calculation method is shown in formula (1).

$$V_{OUT} = \frac{I_S R_S R_L}{5K} \tag{1}$$

If ina193 is selected, RL is 100 kΩ in formula (1), for example, 1 Ω is selected for external resistance rs. Ina193 will amplify the detection current is 20 times and convert it into voltage output. Since the output voltage of Vout is limited to 3.3 V, it should be connected to the A/D acquisition pin of STM32 for processing. Therefore, the current range of is from 0 to 165 ma.

In this design, because the motor speed to be tested is from 100–1500 rpm, the current will certainly exceed 165 ma. Therefore, when connecting ina193, we should first use the parallel resistance R1 and R2 shunt, R1: R2 = 1:20, and take the end of R2 (that is, the current is small) as the load for current detection.

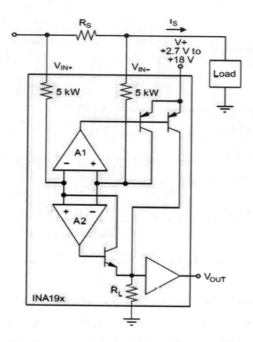

Fig. 3. Current detection amplification circuit diagram

4 Test

The software and hardware are designed according to the above method, and the program is written in C language. After compiling, connecting and debugging in keil mdk5.0, the program is downloaded to stm32f103zet6 development board. Get the following data through the experiment (Fig. 4).

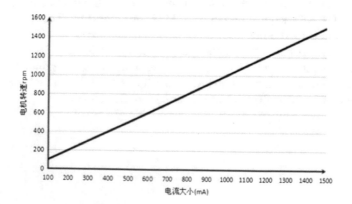

Fig. 4. Corresponding curve between current and speed

Table 1. Relationship between speed and current at 12 V

Speed (rpm)	Current magnitude (mA)
100	101
200	201
300	302
400	403
500	504
600	605
700	706
800	807
900	908
1000	1008
1100	1110
1200	1211
1300	1312
1400	1413
1500	1510

Table 2. Relationship between speed and current at 6 V

Speed (rpm)	Current magnitude (mA)
50	101
100	201
150	302
200	403
250	504
300	606
400	808
500	1010
600	1212
700	1414
800	1615
900	1810
1000	2010
1100	2210
1200	2415
1500	3020

It can be seen from Table 1 that when the DC motor supply voltage is 12 V, the current is almost proportional to the motor speed, and the higher the current is, the faster the speed is. In Table 2, when the DC motor supply voltage drops by half, the proportional relationship between the current and the motor speed remains unchanged, but the motor speed basically drops by half. Using the data in Table 1 to draw the curve in Excel just reflects the proportional relationship between the current and the speed. It is verified that the method of measuring current can be used to measure motor speed.

5   Conclusion

In this paper, a new speed measurement method of DC motor is proposed. Compared with the traditional sensor measurement method, this method is more direct and effective, with simple interface circuit and wider application range. Moreover, the threshold value can be set by software. When the current value is greater than a certain value, the motor can be controlled to stop rotating to prevent burning the motor. This method has a certain practical value.

References

1. Zhang, S.: Research on comprehensive improvement technology of Hall sensor motor speed measurement. Micro Spec. Mot. **46**(5), 31–34 (2018)
2. Han, R., Guo, Y., Zhu, L., He, Q.: Improved speed measurement method of Brushless DC motor based on Hall sensor. Instrum. Technol. Sens. **10**, 115–117 (2017)
3. Du, Y., Song, L., Wan, Q., Yang, S.: Accurate real-time speed measurement of photoelectric encoder based on wavelet transform. Infrared Laser Eng. **46**(5), 1–6 (2017)
4. Wang, H., Hu, J., Wang, S.: Incremental photoelectric encoder angular displacement fitting velocity measurement method. Instrum. Technol. Sens. **10**, 99–101 (2014)
5. Zhao, S., Xiao, J., Guo, Y.: Motor speed data processing method based on Improved Kalman filter. Micro Spec. Mot. **46**(9), 80–82 (2018)
6. Han, T.: Design of DC motor PWM closed loop control system based on MC9S12XS128. Mach. Tools Hydraul. **44**(7), 109–111 (2016)
7. Hua, Q., Yan, G.: Measurement of motor speed based on least square method. Electr. Drive **5**(12), 73–76 (2015)
8. Liu, Q., Zhang, R., Du, Y., Shi, L.: Research on speed measurement of long primary bilateral linear induction motor. Power Electron. Technol. **49**(5), 59–60 (2015)
9. Fu, Y., Sun, D., Liu, Y.: Speed signal estimation of low speed servo system based on Kalman filter. Appl. Mot. Control **42**(5), 17–22 (2015)
10. Lei, W., Huang, C., Li, J.: Indirect speed measurement of DC motor based on morphological filter and center extreme difference. Meas. Control Technol. **34**(3), 17–20 (2015)

Safety Usage of Human Centrifuge

Yifeng Li, Ke Jiang, Minghao Yang, Zhao Jin, Haixia Wang,
Lihui Zhang, and Baohui Li[✉]

Airforce Medical Center, PLA, Beijing, China
liyf8886@sina.com

Abstract. The structure constitute of high performance human centrifuge is made a brief introduction. Then the technology monitoring system is emphasis made stated from basic constitution, fault events source and alarming mechanism and other aspects, explaining that through multilayer protections system can ensure the safety usage of human centrifuge.

Keywords: Human centrifuge · TSMU · Events · Alarming · Safety usage

1 The Brief Introduction of Human Centrifuge

In the field of aviation and aerospace, human centrifuge is the only large device which can imitate lifelike various of flight movement on ground. It can make the subject really realize and feel the influence produced by flight in air for person's physiology and psychology [1, 2]. The research work of human centrifuge in our country begins from the 60's in last century at the earliest stage, while many centrifuges which have been developed successfully do not have the actively dynamic flight simulation function. After many years, through continuous development, a high performance human centrifuge is introduced from Austria's AMST Company in foreign. This centrifuge begins formally working in the year of 2005, and use up to now. This centrifuge has whole balances, three freedom degrees, high +Gz growth rate, and can carry out dynamic flight simulation function etc. [3, 4]. Centrifuge training provides lifelike and safety environment for pilot to complete training with big G load value and big G growth rate, and play a very important role for improving flight person's anti-G load endurance and guaranteeing flight safety. It makes a huge contribution for aviation aerospace business of our country [5–10].

2 The Structure and Constitution of Human Centrifuge

According to each parts of functions of human centrifuge, human centrifuge as a whole is constituted by 8 big systems: machine system, drive system, control system, flight imitation system, medical monitoring system, safety protection system and assistance system, technology monitoring unit. Drive system is constituted by main drive, roll drive, pitch drive, brake unit and servo timing unit. Main drive, roll drive, and pitch drive are used to drive the run of centrifuge cabin, brakes unit realizes machine braking and machine halting braking, and servo timing unit receives order from the control

J. Abawajy et al. (Eds.): ATCI 2021, LNDECT 81, pp. 161–170, 2021.
https://doi.org/10.1007/978-3-030-79197-1_21

system, making sport control of centrifuge and controlling size of acceleration and size of acceleration growth rate. Control system includes data collection, system control, and network communication and other subsystems, which is used for collecting, delivering and recording of various of physiology parameters and engineering parameters, establishing the correspondence connection among each computer, programmable controller, drive and each other performance organizations of human centrifuge, to realize the running control for each subsystem of human centrifuge, such as starting and stopping of subsystem, choice of work mode, choice of control way and ducting of acceleration data etc. Flight imitation system includes manipulation unit and flight imitation unit, to realize truer over load feeling and manipulation experience. Medical monitoring system mainly is to monitor body physiology parameters and each parameter of pilot's anti-load ventilation system, to realize showing, conserving and playback of each signal data. Safety protection system is used for monitoring and judging the status of safety system, adopts to mature and reliable protection measures, and adopts to measures independently making multiple protections for centrifuge device and safety of pilots. Among this the adopted method can be passing video monitoring, as well as passing door- lock, oil pressure and other safety linkage protection measures. Assistance system is composed by power supply subsystem which converts the local supply voltage into work voltage needed by centrifuge running, gas supply subsystem providing compressed air and oxygen, and cooling subsystem which makes cooling for main electrical machine and frequency converter cabin. Assistance system connects with UPS, and under the urgent condition, main arm can be moved to the requested position by operating assistance system, so it has an important function for the normal operation of centrifuge. The safety problem of centrifuge, particularly the personnel's safety problem is heavy medium of heavy, therefore TSMU is specially established in centrifuge system, that is technology system monitoring unit, to real time monitor the whole running of centrifuge and each subsystem. Every station of each station of engineering station, medical station and recording station has TSMU, this system adopts to operation way of touch screen, and by TSMU any subsystem and model needed to be displayed can be selected and made real time monitoring. In this study, safety of device and TSMU system is emphasis made elaboration in detail.

3   The Technology Monitoring of Human Centrifuge

3.1   Total Structure of TSMU

Centrifuge has complicate structure, and system is huge, in order to have an overall monitoring for the status of each system to guarantee safety of equipment and personnel, high performance human centrifuge specially establishes technology monitoring unit to making supervision for each system and part. TSMU can be divided into overview, medical limits, safety matrix, limit switches, driving PLC, safety PLC, bus connection, PC network, alarms, warnings, altogether 10 models. Among them, the overview includes pedestal, gimbal, cabin, low voltage power supply, gas supply system of compressed air etc. altogether 12 parts. The following Fig. 1 explains in detail the structure constitution of TSMU. Centrifuge system is huge, and once appearing fault, it is difficult to seek problems and needs to seek solution by positioning layer by layer. When system is abnormal, will appear a "warning" information and hint an abnormal situation needed to be looked into; While a fault occurs, will appear a corresponding "alarm" information, hint fault occurrence, and induces shutdown of different degree in time, in order to prevent appearing big harm and accident. When appearing "warning" and "alarm", TSMU can give concrete remind at the form of "event", making engineering personnel have a approximate judgment for fault, and engineering personnel can firstly make positioning on big system and subsystem for fault according to the hint, then according to PLC logic status and through looking into all levels electric circuit diagrams and other methods makes sure the position where fault occurrences, and according to fault situation puts forward the corresponding solution. The TSMU unit can greatly raise troubleshooting efficiency in centrifuge using, and the adopted shutting down in time and other measures can greatly prevent and decrease the occurrence of the bad affairs, having closely an important function.

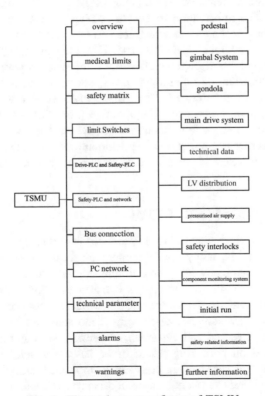

Fig. 1. The total structure figure of TSMU

3.2   The Important Element for Alarming in TSMU-Event

Due to TSMU alarming and the induced different degrees of shutting downs are all expressed out as the form of "event", so its necessary the event-this important element in TSMU is firstly given explanation. TSMU has altogether 103 events, giving explanation and gathering for all kinds of faults situations which may occur in each subsystem. These events may be from peripheral devices, such as event 003 fire alarm building, event 008 INI crane position; one part is coming from drive system, including main drive, roll and pitch drive, such as event 023 CB bearing lubrication OK, this time bear lubricant is needed to add, event 040 CB control voltage main contactors roll/pitch drives, event 045 oil level bearing lubrication; The brake system is very important to the equipment braking and system safety, and events include also the alarming signals coming from three brake systems. Such as event 042 brake main drive warning and fault, event 057 rollbrake not open (left, right), event 067 pitch brake not open, event 068 pressed air tank pitch brake, event_099 not all main brakes are closed etc. Assistance system is also very important for centrifuge, event includes many informations coming from assistance system, such as event 015 Breathing Airsupply ON and Open, event 024 CB auxiliary drive OK, event 026 cooling system fault compressor, event 031 fault compressor 1, event 034 fault compressor 2, event 036 fault air dryer, event 049 UPS system discharged (Warning),event 084 hydraulic power unit oil pressure not OK; The information in events coming from safety protection system, directly gives report for the events which lead to not starting or stopping in running, such as event 048 doors not closed (101, 106, 107, 111), event 075 INI gondola door closed and locked (left and right),event 078 KS Limit reached warning, event 079 KS limit reached STOP, event 083 airtemperature gondola. In addition, there is alarming information coming from safety PLC and drive PLC. Totally to say, in events the events on brake system, assistance system, safety protection system are more. Each event makes a description for fault phenomenon, system fault lies in and the possible reason, and engineering personnel makes positioning of subsystem according to hint, check the provided electric circuit diagram, and makes troubleshooting and maintenance.

3.3   The Safety Matrix Model of TSMU

In TSMU, the events which may induce to produce the same stopping mode is gathered together to form a matrix, many matrix gathers are formed corresponding to normal stop, fast stop, and emergency stop, the different stopping modes. Because it can give important hint and alarming for safety, so it is called as safety matrix. Normal stop is the stopping of equipment in normal operation state, and it does not use brake system in this stopping mode. Fast stop is the stopping mode activated by PLC system when system occurs problems, and roll and pitch movements stop and are fixed at current position by roll and pitch braking. Main brake has a deceleration through control system not through brake system. In emergency stop mode, stoppings of three directions are all by using brakes system, and main drive has a deceleration as the set safety way, roll and pitch movements stop and are fixed at current position, while when emergency stop occurs, it may be PLC system, that is control system gives out order, may be also engineer activates manually emergency stop. The following Fig. 2 is the display interface of safety matrix model. From the figure it can be seen that, with upgrade of stop mode, alarming grade of the corresponding affairs "events" also rises. Such as event048 included in Normal stop, doors not closed (101, 106, 107, 111), hinting one or multi doors centrifuge hall leading to rooms are not closed well; these may be common faults, but these faults can not still induce serious consequence, and leads to ponderance stopping etc. Such as event077 voltage control relay, hinting there is no power supply in cabin; event083 air temperature gondola, hinting the internal temperature of cabin is high. event57 Rollbrake not open (left, right) included in Fast stop, hinting roll brake is not open, and prompting alarming and inducing fast stop; event75 gondola door closed and locked (left and right), hinting cabin doors left and right are not closed and locked, because that cabin doors are not closed and locked in running is very dangerous, the event can prompt alarming and induce fast stop. The faults which are not included in events while are coming from DPLC, give out hint of "Fast stop: from DPLC". Events in emergency stop, such as event4 hardware limit 2 acceleration reached, hinting one of values of hardware limits speed switches on engineering station is over value of level 2, and at this time, the actual speed measured on centrifuge is over the set value, equipment runs as a high speed and faces danger of losing speed, so can induce emergency stop; event42 brake main drive warning and fault, hinting brake pads of main drive occur problem, may have been used up, needing replacing, and at this time, main drive brake faces the risk of being out of order and centrifuge can not brake and stop in time, which is both extremely dangerous no matter for equipment or for the subject, so the event can cause emergency stop. Similarly, the faults which are not included in events while are coming from DPLC, give out hint of "Emergency stop: from DPLC". In the above, each stop mode takes example of two events to make explaining, and in actual application, engineer need make tracking analysis and alarming eliminating.

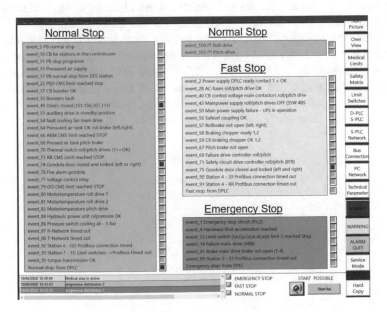

Fig. 2. Safety matrix interface

3.4   WARNINGS Model and Alarm List Model of TSMU

In TSMU, according to the difference of influence affair in events produces for system, the events whose influence is relatively small are united to WARNINGS model, being showed in concentration; the events whose influence is relatively big are united to Alarm list model, being showed in concentration. Influence is relatively small refers to that event in WARNINGS model may be a normal common hint, possibly not inducing system stop; while influence of events in Alarm list model is relatively big, possibly inducing stop of different degrees, also possibly directly inducing system can not start. Events inducing system can not start can be displayed in concentration on interface of "START POSSIBLE", on which the latter in this study has a introduction in detail. Such as in WARNINGS, event048, doors not closed (101, 06, 107, 111), hinting doors are not closed, event17, PB normal stop from DFS station, hinting normal stop produced by operation of pushing button on DFS station, event36, fault air dryer, hinting faults are coming from air dryer; but these can not still induce stops of different degrees and other more severe consequences.

In Alarm list, event022, PED CMS limit reached stop, hinting value measured by strain gauge on pedestal is over limit value of 2 level, event101 roll brake not closed (oil pressure), hinting roll brake is not closed, and the oil pressure switch of roll brake has not response, event102 pitch brake not closed (oil pressure), hinting pitch brake is not closed, and the oil pressure switch of pitch brake has not response; The above events possibly induce stops of different degrees for system, event084 hydraulic power unit oil pressure OK, hinting oil pressure of hydraulic power unit is overtop, and this event may directly lead to system can not start. Engineers should be paid enough attention to the two kinds of alerting information, particularly for the alerting information in "Alarm list" should inspect system in time to carry on fault fixing position and troubleshooting, to ensure the normal running of equipment. The following Fig. 3 and Fig. 4 are separately the interfaces of "WARNINGS" and "Alarm list", and the occurring time of warnings and alarmings, and the detailed content of event and other information are displayed in the figures. In the two models, choosing situations can be set, so as to the needed events can be made fast choosing and inspection in many events.

Fig. 3. WARNINGS interface

Fig. 4. Alarm list interface

3.5   The START POSSIBLE Model of TSMU

It is needed to explain that, there is a special model, "START POSSIBLE" in TSMU, which makes a cover and enumerate for events needed to be examined when centrifuge starts. It includes all affairs "events", and the event appearing problem can be marked out in red. When system can not start, in "START POSSIBLE", the corresponding event which has problem can be marked out in red and displayed, helping engineer fast checking and troubleshooting. The following Fig. 5 and Fig. 6 is the interface of "START POSSIBLE", in the figure, the corresponding events which have problems are marked out in red and the appearing time and the concrete content of event which has problem and other information are displayed. A small window which can be dropped down is fixed at the bottom most of screen, that is no matter which model is selected as current interface, this window is all displayed. In the window, according to time sequence, the each event when occurring warnings and alarmings is listed real time, including occurring time and concrete content and other information, in order to make engineer convenient to in time find fault reason and make a maintenance.

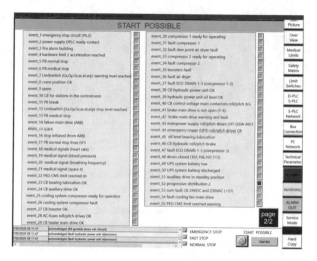

Fig. 5. START POSSIBLE interface 1

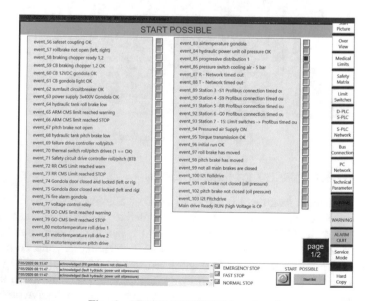

Fig. 6. START POSSIBLE interface 2

3.6   The Other Models and Factors

The above is only to focus on the several models in 10 models which have a close relation with safety. There are other models, such as overview, limits switch, bus connection and other models, which can be displayed as current interface to make checking through touching operation of touch screen. According to the demand engineer checks each model to make system maintain and service. This is guaranteeing safety of equipment running through software and hardware of system. Except for system factor, safety management and control in human factor should also be enhanced, such as before equipment runs, engineers in each working station can ensure to start equipment running only after having carefully checked no personal in centrifuge hall, to avoid inducing accident in human and equipment. TSMU plays a very important role in safety running of centrifuge, but the safety usage of centrifuge needs the joint efforts of system and human factors to get a guarantee.

4   Conclusions

The safety problem of human centrifuge is the most important problem in the usage of centrifuge. Only being acquainted with work principle of centrifuge, structure constitution, system function, improving maintain technical ability, and enhancing safety management and control from maintaining, human factor and other aspects, working efficiency of equipment can be improved, the service life of equipment can be prolonged, occurring of accident can be avoided, in order to guarantee the normal running of equipment and ensure the safety usage of the large-scale equipment human centrifuge.

References

1. Yao, Y., Gaochao, S.: The present situation and development trend of contemporary human centrifuge. Med. J. Air Force **28**(1), 60 (2012). (in Chinese)
2. Lu, H.: Human Centrifuge and Its Application. National Defence Industrial Press, Beijing (2004). (in Chinese)
3. He, Y., Jiang, C., Song, Q.: Survey of the development of centrifuge-based flight simulator. J. Mach. Des. **34**(12) (2017). (in Chinese)
4. Shen, W., Hong, J., Mlin, L., Yang, Y., Wang, X.: Development on large scale centrifuge for aviation equipment. High Dyn. Meas. Control Technol. **34**(6), 85–88, 103 (2015). (in Chinese)
5. Geng, X., Yan, G., Jin, Z.: The study and application of aviation acceleration physiology **32**(5), 189–196 (2004). (in Chinese)
6. Zhao, B., Lv, Y., Sun, X.: The research progress on early warning and protection of cerebral ischemia caused by high performance fighter acceleration. MedJNDFNC **38**(6), 415–417 (2017). (in Chinese)
7. Lin, P.C., Li, S.C., Wang, J., et al.: Measurement of military aircrews' ability to adapt to human-use centrifuge. In: The 40th International Conference on Computers and Industrial Engineering, Awaji City, pp. 1–6 (2010)

8. Jin, Z., et al.: Analysis and discussions on the status of pilots' centrifuge training. Chin. J. Aerospace Med. **28**(4), 249–254 (2017). (in Chinese)
9. Cengiz, Y., Sagiroglu, S.: System safety of human centrifuge and solving angular velocity of main arm with artificial neural network. In: 2017 5th International Symposium on Digital Forensic and Security (ISDFS), pp. 1–6 (2017)
10. Bretl, K., Mitchell, T.R., Sherman, S., McCusker, A., Dixon, J.B., Clark, T.K.: Retention of cross-coupled illusion training to allow for a shorter radius space centrifuge. In: 2018 IEEE Aerospace Conference, pp. 1–7 (2018)

The Application of Computer Simulation in "Yeong Guan Cup" Casting Process Design Competition

Fang Chang[✉], Feng Qiu, Jinguo Wang, and Kai Zhu

College of Materials Science and Engineering, Jilin University,
Changchun, Jilin, China

Abstract. Because of the advancement of science and technology, the application of computer technology has gradually increased. Computers can be used in traditional industrial production. They can increase the rate of industrial production and ensure product quality. They can also be used in teaching, so that students can have a better and systematic understanding of knowledge. "Yeong Guan Cup" Casting Process Design Competition is organized by Foundry Institution of Chinese Mechanical Engineering Society and other institutions. It aims to encourage students to use the casting expertise they have learned to solve casting problems encountered in production, and to improve their practical and independent learning capabilities. The teachers of the colleges participate in the casting competition based on the course design of material forming and control. The casting competition not only requires an excellent casting process design plan, but also needs computer simulation to verify the results. Computer simulation has played an important role in verifying whether the process plan is reasonable. In recent years, computer simulation casting technique has also made some developments.

Keywords: Computer simulation · "Yeong Guan Cup" Casting Process Design Competition · Casting process · Curriculum design

1 Introduction

The casting industry is the basic industry of the manufacturing industry, which promotes the development of the national economy. At the same time, the casting industry is also facing problems such as difficulty in guaranteeing product quality and difficulty in realizing the casting process. Therefore, scientifically controlling the production of castings and designing a reasonable casting process becomes particularly important. Computer simulation of the casting process has only been widely used in industry since the 1980s. Most casting companies have also adopted solidification simulation analysis technology to accurately predict casting defects, adjust process parameters, and improve their yield [1]. With the development of "First-class universities and disciplines of the world" construction, higher requirements are put forward for the quality-oriented education of college students. The purpose of the casting process design competition is to allow the students to apply the casting expertise they have learned to

J. Abawajy et al. (Eds.): ATCI 2021, LNDECT 81, pp. 171–178, 2021.
https://doi.org/10.1007/978-3-030-79197-1_22

solve the production problems of the enterprises. Students complete the topic selection, design and competition under the guidance of the teachers.

2 Application Status of Computer Simulation in the Casting Process

In the casting design process, the factors affecting the quality of the casting should be considered. The changes in the flow field, temperature field and concentration field should be considered during the mold filling and solidification process when the molten metal enters the runner. This is specifically reflected in the selection of casting material, chemical composition, casting temperature and casting time. The temperature field changes during solidification are accompanied by interdendritic convection and shrinkage. Convection and shrinkage will cause microscopic shrinkage, while shrinkage will cause changes in the stress field, which will cause hot and cold cracks. Changes in the temperature field during solidification can also cause shrinkage cavity and shrinkage porosity. When the solidification rate and temperature change are too fast, it will lead to casting defects such as porosity, slag inclusion, insufficient pouring, and cold shut. Therefore, choosing the right casting material, chemical composition, casting temperature and casting time play an important role in controlling the changes of different fields. The main numerical calculation methods used in the casting process are: direct finite difference method (DFDM), finite element method, VEMI, boundary element method, lattice gas method, finite difference method (FDM), of which FDM is the most widely used. No matter which numerical calculation method is adopted, the computer simulation software of the casting process includes three parts: pre-processing, intermediate calculation and post-processing. Among them, the pre-processing part mainly involves model establishment, mesh generation, definition of physical properties, setting of boundary conditions and parameters. The calculation part is mainly about processing settings before numerical computation solution based on fluid mechanics and solidification and heat transfer theory. Post-processing module mainly visualizes and analyzes the results of various numerical calculations. At present, computer simulation mainly focuses on the mold filling and solidification of the casting process, the prediction of shrinkage cavity and porosity, stress analysis and microstructure simulation. Among them, the microstructure simulation is more complicated and is still under research.

At present, there are many kinds of software for computer simulation of the casting process. The foreign software mainly includes Procast, which is mainly suitable for sand mold, die casting, lost foam casting, anti-gravity, centrifuge, continuous, tilt and other casting methods, as well as directional solidification and single crystal investment casting, etc. MAGMA is mainly suitable for sand mold, metal mold, die casting and low-pressure casting. The domestically developed software mainly includes InteCAST, FT-STAR, CASTSOFT, CAST-3D and Z-CAST jointly developed by Shenyang Research Institute of Foundry and South Korea. In recent years, the Unity 3D virtual reality platform has been used for the development of a post-processing display system. Various kinds of software have their own advantages and disadvantages, and Procast is widely used at present. The use of computer simulation can not only optimize the casting process and casting materials, but also design the production process, select the process type, develop the casting system and feeding, design the best riser size and its position, and check the quality of castings and product performance [1].

3   Introduction of "Yeong Guan Cup" Casting Process Design Competition

University discipline competitions are generally a way of group professional competitions for college students in a discipline to examine students' application ability of the basic theoretical knowledge of the discipline, which can fully reflect students' ability to comprehensively use knowledge to solve practical problems. Subject competitions can closely integrate curriculum knowledge with professional practice, discover and solve problems in practice, so as to stimulate the interest and potential of college students. It can not only improve students' learning ability and comprehensive quality, but also help teachers improve their teaching level [2].

"Yeong Guan Cup" Casting Process Design Competition starts in September every year. After the preliminary selection of each school, entries will be submitted in March of the following year. The organizing committee of the competition has a judging committee. The judging committee will mark the content such as modeling methods, understanding and analysis of components, process design, design methods (including three-dimensional modeling, simulation calculation, process drawing, etc.), process documentation, casting cleaning. The judging committee will first judge the third prize and the excellent prize, and recommend the first and second prizes for the final. The final will be conducted in the form of defense. The committee will mark based on the students' defenses, structure, size and material analysis, design level, and production guidance to finally determine the first and second prizes. There are 6 groups of works for selection every year, and each group of works stipulates casting materials. Students can choose the appropriate topic by themselves and complete the work under the guidance of their instructors [3]. *Curriculum Design of Casting Technique* is an important practical teaching link for the major of material forming and control engineering. Its teaching goal is to be able to use the casting theory and process design knowledge to learn and master the casting process and tooling design methods systematically, so that students can formulate reasonable casting technique, and design a reasonable structure of tooling molds. At the same time, through the curriculum design, it also enables students to further improve the design and drawing ability, the basic skills of consulting design materials, and the ability to analyze and solve the actual problems of casting engineering to satisfy the employment demand in the casting industry. However, there are still some shortcomings in the practical teaching. Combining with the casting process design competition for college students, the competition self-learning mode is introduced to cultivate students' self-learning, analysis and solving ability of practical problems. The propositions of the casting competition come from the actual production. The structure of the components is relatively complex, the drawing ability is required to be high, and the casting process design is difficult. All of these require students to comprehensively apply the casting theories, process and tooling design knowledge. The competition focuses on strengthening students' ability to analyze and solve complex engineering problems. The college encourages more professional students to sign up and form teams to participate in the school casting competition and the national competition [3–6].

4   The Design of the Casting Process

After the participating students and teachers have selected the topic, they need to analyze material characteristics of the components, choose suitable modeling and core-making materials. Then they should determine the pouring position and parting surface, select the casting process parameters, design the pouring system, and design the feeding system.

5   Numerical Simulation of the Casting Process and Plan Optimization

A reasonable casting process design plan is a prerequisite for obtaining high-quality qualified castings. Thus, it is necessary to make multiple improvements and optimizations on the basis of the original plan. In the past, the optimization was by trial production. It required repeated operations in multiple steps such as mold manufacturing, modeling, core making, alloy melting, pouring, casting quality analysis, mold modification. The complete casting process consumes a lot of manpower and material resources, and extends the casting process design cycle. Compared with trial production, the numerical simulation technology that has been currently used can simulate the molten metal filling process, analyze the changes in the temperature field and solid phase rate of the metal solidification process, and effectively predict various defects that may occur in the casting process and their causes. It can optimize the casting process plan scientifically, intuitively obtain high-quality castings, and significantly reduce production costs, improve production efficiency, shorten the research and development cycle. It has naturally become the first choice to the market with its high-quality and low-consumption characteristics.

After the casting process is designed in the casting competition, the software CATIA V5 R20 is used for 3D drawing of the casting process. The software Hyper-Mesh is used for mesh generation. During the casting process, the numerical simulation software ProCAST is used for the numerical simulation of the temperature field and solid phase ratio of the filling and solidification process. Based on the data, the casting process is optimized.

5.1   Pre-processing of Numerical Simulation

5.1.1   The Principle of Mesh Generation of the 3D Model

The mesh generation of a three-dimensional model refers to the approximate subdivision of a solid model using several cuboids or cubes. In the computer simulation of squeeze casting, in order to ensure higher calculation accuracy, a smaller step size will be selected to divide the solid model, so that more meshes can be obtained. However, after a large number of meshes are obtained, the judgement of meshes increases the time and space correspondingly, which will inevitably slow down the program, resulting in delayed response or even no response. When performing solid cutting, it should be divided in the order of "surface-line-point". First cutting along the Z axis from bottom to top to obtain several planes, and then cutting the planes along the Y axis to obtain several lines parallel to the longitudinal axis. After that, cutting the obtained lines in the horizontal direction to finally get the required center point [7]. HyperMesh is generally used to divide the mesh. There are two main steps: 1. Using the BatchMesher module to perform mesh generation, and automatically import the software HyperMesh after that to check the quality of the meshes. After confirmation, changing the mesh shape to triangle and check again. 2. Importing the high-quality surface mesh file into the software ProCAST, reading the file into the MeshCAST module in ProCAST. After setting the unit, checking the mesh quality and divide the volume mesh. If there is a mesh quality problem or a cross mesh, repair it in HypeMesh and then divide the volume mesh.

5.1.2 Application of Virtual Mold and Parameter Setting

For sand casting, ProCAST's own virtual mold (Virtual Mold) can be used to save the step of casting mesh. During the simulation parameter setting process, the real flask size cannot be used as the virtual flask size because the outer surface of the virtual flask in this model is defined as adiabatic condition, the flask size must be large enough to avoid large calculation errors.

5.1.3 Setting of Interface Conditions

Due to the relatively complex structure of castings and sand cores, the heat exchange phenomenon at the interface among castings, sand molds and sand cores is obvious. The range of heat transfer coefficient is shown in Table 1.

Table 1. Numerical range of interface heat transfer coefficient

Heat transfer coefficient at different surfaces	h $\{W/(m2 \cdot K)\}$
metal $\sim$ metal	$1000 \sim 5000$
metal $\sim$ molding sand	$300 \sim 1000$
molding sand $\sim$ molding sand	$200 \sim 300$

Set the interface conditions according to the above table:

The heat transfer coefficient of castings and sand mold, casting and sand core: $K = 500W/(m2 \cdot K)$;

Heat exchange coefficient of chill and sand mold: $K = 500W/(m2 \cdot K)$;

Heat exchange coefficient of chill and castings: $K = 2000W/(m2 \cdot K)$.

5.1.4 Initial Condition Setting

ProCAST is used for numerical simulation of the filling and solidification process. Initial Conditions mainly set the pouring temperature, the initial temperature of the sand mold, the initial temperature of the core sand, the molten metal flow, and the filling time/filling speed.

5.2   Three-Dimensional Simulation of the Physical Field of the Filling Process

The physical numerical simulation of the filling process includes the numerical simulation of the flow field, the temperature field and the stress field of the solidification process, the numerical simulation of the structure grain growth and morphology, and the numerical simulation of the structure grain growth and morphology. Intuitive analysis of the single or coupled integrated field quantity in the casting process to find out the inherent causes will help optimize the casting process and obtain high-quality castings [8].

5.2.1   The Flow Field

The flow of molten metal in the mold cavity directly determines the quality of the casting. In addition to the structure of the casting itself, the influence on the flow of the molten metal is also related to the selected pouring temperature. When the pouring temperature is too high, the fluidity is good, but the structure is coarse, and entrainment is easy to occur. When the pouring temperature is too low, it will lead to poor liquidity of the molten metal, and the unfavorable filling situation will cause defects such as cold isolation and insufficient pouring. Therefore, studying the filling process of the molten metal is of great significance for obtaining high-quality casting blanks [8].

5.2.2   The Temperature Field

In the filling process, due to the different structures of castings, the thickness of the castings is different, resulting in the heat dissipation of each part of the castings is different. At the same time, when the contact area between the castings and the sand mold is different, the heat transfer speed is also different. If the temperature field is not well controlled, it will lead to the existence of keyholes, segregation and other defects. Generally, the cooling rate of different parts is controlled by adding chills [8].

5.2.3   The Stress Field

Simulating the distribution law of the stress field in the solidification process of castings can effectively predict the hot crack and deformation of the castings. The current simulation of the stress field in the casting process mainly focuses on cast iron, steel castings and aluminum and magnesium alloy castings. The preheating temperature of the sand mold has a great influence on the residual stress and deformation. The preheating temperature of the sand mold is inversely proportional to the residual stress and deformation of the casting. The pouring temperature has the least effect on the deformation of the castings, and the pouring temperature is directly proportional to the residual stress and deformation. The pouring speed has a least effect on the residual stress, and the pouring speed is inversely proportional to the residual stress and deformation [8].

5.2.4 Multi-field Coupling

At present, the application research of casting simulation has developed from a single field such as stress field, temperature field, and flow field to multi-field coupling integration, focusing on the analysis of the impact of multi-field coupling on the entire casting process, making the entire casting simulation process macrocosmically and microcosmically combined, which can provide active guidance for the development and optimization of casting process plans [8].

5.2.5 Microstructure Simulation of Castings

With the help of three-dimensional numerical simulation technology, it is possible to intuitively analyze the nucleation and evolution mechanism of the solidification structure of castings. The methods mainly include deterministic methods, probability theory and phase field methods. Since they are still far from application, they haven't been applied in the casting competition [8].

5.3 Analysis of Simulation Results Using the Post-processing Module

At present, the post-processing module that comes with the software is mainly used to observe the filling process of the molten metal, the flow state and temperature distribution at different times, and analyze the solid phase ratio and the distribution position of solidification defects. It can provide reference for optimizing and improving the casting process.

6 Conclusion

In "Yeong Guan Cup" Casting Process Design Competition, students can learn the use of software and use their casting knowledge, which will improve their practical ability and reflect the college's quality-oriented education development. Meanwhile, participating in the competition can also win the honor for the college.

Acknowledgement. Thanks to the Chinese Mechanical Engineering Society, Foundry Institution of Chinese Mechanical Engineering Society and major magazines for organizing this competition. Thank you for the support of the college.

References

1. Zhang, J.: Application of computer simulation in casting process and technological optimization. Digital Space **09**, 117 (2019)
2. Liu, H.: Promote the cultivation of practical ability of material control major students under the program of excellence through discipline competitions. J. High. Educ. **21**, 55–56+58 (2015)
3. "Yeong Guan Cup" Casting process design competition announcement of defense and award results. Foundry **66**(06), 645–650 (2017)
4. Cao, Y., Shifang, S.: About the problems in attending casting process design competitions of Chinese college students. Foundry **60**(03), 313–314 (2011)

5. The first "Yeong Guan Cup" casting process design competition award ceremony was held in Huazhong University of Science and Technology. Foundry Technol. **31**(07), 817 (2010)
6. Chen, S.: Teaching reform and practice of curriculum design of foundry technology. Times Fortune, **03**, 67+69 (2020)
7. Zheng, Q., Yang, J.: Application research of computer simulation software pre-processing technology in squeeze casting process// 2017 Metallurgical enterprise management innovation forum collected papers. Hong Kong New Century Cultural Publishing Company **01**, 45–46 (2017)
8. Yang, Z., Qi, H., Guo, H., et al.: Research and application of numerical simulation of casting process in China. Foundry Technol. **38**(09), 2072–2075 (2017)

Application of Information Technology in Building Material Testing

Yufu Li$^{(\boxtimes)}$ and Yinbo Li

Guangdong Construction Polytechnic, Guangzhou, Guangdong, China

Abstract. With the rapid development of the construction industry, the construction industry has brought huge economic benefits and promoted the overall development of related fields. As we all know, the success of construction projects cannot be separated from the improvement of the quality of building materials. After scientific demonstration, the improvement and design response of the material testing information network monitoring platform, more integration of computer network technology, play a positive role in promoting the healthy development of construction projects. This article mainly elaborates the shortcomings in the quality inspection of building materials and the application of information technology in building materials, and puts forward some suggestions for the problems of material inspection.

Keywords: Information technology · Construction · Materials · Detection

1 Introduction

With the rapid development of social architecture, engineering quality has received more and more attention. The data in building material inspection is very complicated. In construction engineering, the role of information technology is very extensive, especially in the quality monitoring of construction engineering. Since the quality of construction projects is closely related to the interests of the people and society, the use of information technology to monitor construction materials can effectively improve its accuracy and ensure the quality of construction projects. Relevant personnel should increase the processing of these inspection data when performing detailed processing [1]. The quality inspection of building materials is very complicated. The quality of building materials is directly related to the quality of the building. In order to ensure the accuracy of the data, we must do a good job of data classification and storage. In the process of data processing and calculation, traditional processing methods are more complicated, mainly based on manual calculations. The use of information technology for data processing is the future development trend [1].

J. Abawajy et al. (Eds.): ATCI 2021, LNDECT 81, pp. 179–185, 2021.
https://doi.org/10.1007/978-3-030-79197-1_23

2 Principles and Characteristics of Information Technology in Building Material Testing

The design framework of computer network application in building material inspection is based on internal management, daily inspection work, establishment of inspection management system, setting of recording inspection status and network monitoring system tracking, and realization of WAN interconnection according to business needs. For the management content described above, if it is to be fully realized, it is necessary to solve the technical problems of the network monitoring system, the detection management system, and the networking between the laboratories. Among them, the testing management system is connected with the testing machine and the terminal computers of the various operating departments in the testing center to form the entire testing operating environment [2]. The testing management system can complete the complete processing of testing and test data, mainly including the collection of original data, the collection and processing of test pieces, and the evaluation and processing of test data in accordance with relevant national standards and regulations [1]. The advantages of the computer network management system mainly include the following aspects: fully implement confidential operation, that is, set permissions for each user, and must have corresponding permissions to enter the operation of each module; realize data collection automation; have laboratory needs the basic management and statistical functions of the system, that is, evaluation results, automatic calculation, printing of paid reports, and powerful statistics and query functions; the operation is streamlined, and the inspection process is completely reproduced [2].

3 The Application of Information Technology in Building Material Testing

3.1 Receive Samples

Generally, it is a commission order with corresponding data issued by the sample receiver, and the inspector conducts the material inspection according to the commission order. In the sampling stage, through the application of computer technology, the entire inspection work is systematized [3]. When the entire work is in an effective working state, each inspection material can be independently numbered when receiving materials, so as to ensure that each The uniqueness of the number avoids the confusion of detection data caused by repeated numbers.

3.2 Use Information Technology for Data Collection

In the actual application of information technology, it is necessary to consider various factors. For the purpose of this article, we need to monitor the testing process. Most informatization testing uses new technologies and adopts new data recording methods [3]. In the process of operating the machine, errors will inevitably occur, so relevant personnel must carry out supervision work and centrally manage the information that cannot be collected. Staff need to increase the management of test data. When inspectors perform inspection work, they need to control the details and increase the analysis and control of the details. When information technology is used for data collection, it is necessary to use related software to check and verify its authenticity [4]. Most traditional data collection work relies on labor, and information technology effectively avoids the problem of manual collection. Information technology can make the data collection process in our building material inspection process more accurate, scientific and reasonable.

3.3   Information Technology for Data Processing

We should use information technology for data processing to make the inspection work more convenient for the relevant staff to perform detailed accounting on the quality of the materials. In addition, when the data is processed manually, the workload is very large, and it is inevitable that people will record, fill in and fill in the data. Mistakes such as processing [4]. Therefore, relevant staff are required to use information technology to help deal with the details of material testing data when processing data. It is necessary to increase the verification of detection accuracy, analysis and verification of the results, the use of information technology can play a great role, and when the detection data is processed, it is helpful to improve its accuracy, as shown in Fig. 1. Data processing is a key part of the entire material testing. The data processing module includes many subsystems, and each subsystem should have the functions of data collection, data calculation, data judgment and generation of test reports. The size and complexity of the data processing module depends entirely on the subsystems. A data processing module should consist of several inspection data sub-items and an overall evaluation module [5]. The building material inspection data processing system consists of several subsystems, each of which interacts with each other. Independence does

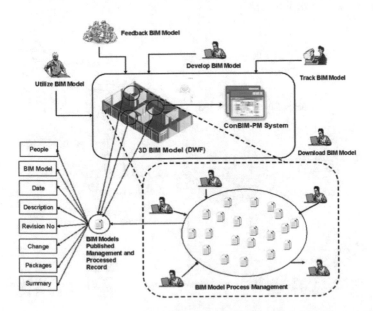

Fig. 1. The concept and framework of the proposed Information technology approach

not affect each other. Each subsystem is composed of a data input module, a data processing module, a test result output module, a database, and a query module.

3.4   Realization of Information Technology in the Supervision Process

In terms of material quality testing, information technology has been very commonly used in building material testing, and advanced scientific and technological information control systems can be introduced based on traditional experiments, which use the real-time recording, shooting and storage functions of the computer itself. It can realize real-time monitoring of all aspects of accurate and fast testing of the entire material during the product quality inspection process, and save it on the inherent disk of the computer. In the process of material testing, material quality data monitoring and review are also carried out by keeping information technology records [6]. Through information technology, it can be directly related to the daily supervision data obtained by the superior. The local testing data will eventually be unified into the previous one. Level database and all upper and lower levels can check the results of related materials in real time, so that problems in material testing and work can be found in time. All aspects of material testing can be monitored in real time, testing problems can be found in time during the testing process or during the testing review process, and the entire testing process can be fully tracked. The network monitoring system mainly takes pictures of the working conditions of the test pieces of some destructive test items such as mortar, concrete, cement, etc., and requires the destruction process of the test pieces to synchronize and correspond to the original data generation curve [6]. Said the advantage is that this will bring a good environment and effect to material quality supervision, and ensure that the higher-level competent department can effectively supervise the accuracy, effectiveness and timeliness of the lower-level department's inspection work in real time (as shown in Fig. 2).

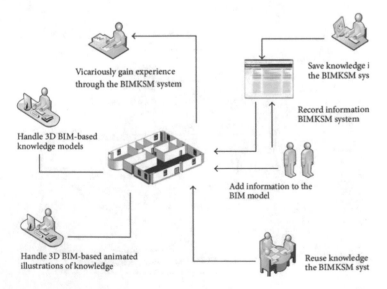

Fig. 2. The application of the Information technology management in construction

3.5   Information Technology for Data Query

In the inspection of building materials, the database system can not only store the basic data generated in the process of sampling materials, but also store important data such as the material quality inspection process and inspection conclusions. Only managers and technicians with access rights can log in to the system and perform operations, which is convenient for querying the results of material testing [7]. The operating system in the database can not only query the classified inspection data of different material quality inspections, but also query the specific inspection date, inspection items, inspection methods, and information of inspectors.

3.6   Information Technology Realizes Data Application

During the quality inspection of materials, the construction material inspection unit should give a detailed introduction and explanation of the details of the work. Inspectors need to conduct data analysis based on past data and experience after the material quality inspection is completed. People build a standard database based on information technology, so that we have directions and standards on the road of future testing [6]. The database will continue to be improved and updated with the development of the times. Building materials are constantly updated every year, and the testing database is constantly expanding [7]. In order to comply with the current society's promotion of energy-saving and environmentally friendly materials, we will increase the quality inspection module for energy-saving and environmentally friendly materials.

4   Measures to Strengthen the Testing of Building Materials

4.1   Establish and Improve the Material Testing Management System

Strict system is an important basis and theoretical reference for standardizing testing behavior, clarifying testing procedures and ensuring material quality. Establish and improve a series of information management systems, such as laboratory job responsibilities, testing instrument use system, equipment maintenance records, testing report issuance system, etc., implement a testing job responsibility system, and achieve a clear division of responsibilities for building materials testing, and testing reports [8]. Once they occur, they should be tracked to the end. In terms of equipment, it is necessary to improve the maintenance and maintenance system of laboratory equipment and the evaluation system of management personnel, and do a good job of regular maintenance, timely calibration and proofreading of testing equipment to ensure the smooth progress of testing.

4.2   Speed up the Development of New Inspection Software

In recent years, information technology has been widely used in various fields. Although the current information technology used in building material testing has brought a lot of convenience to the testing work, when the amount of information is large, information transmission is often unstable, computers and software. For problems such as failures, maintenance and technical processing require waste of time and manpower, and data loss may also occur in the process, which affects work efficiency and data reliability and safety [9]. Therefore, if you want to become a leading company in the industry, it is imperative to strengthen the development of new detection software. The company should increase its efforts based on the actual situation of the unit, and organize front-line employees and scientific and technical personnel to jointly develop new products with high accuracy and strong storage capabilities [8]. Detection software, anticipate possible shortcomings in the work, in the new software, avoid weaknesses in actual operation, increase the transmission speed of relevant experimental data, and ensure information security.

4.3   Strengthen Material Storage Management

During the construction process, the construction parties are complementary to each other, and the purpose is to better complete the project. The construction unit is the main body of the project progress. As the inspection unit, it can kindly remind the construction unit to strengthen the storage management of construction site materials [9]. In terms of testing, some suggestions and precautions are given. According to the characteristics and needs of different materials used in the project, we strictly control the storage conditions of the building materials used on the construction site and store the materials in a suitable environment [10].

4.4   Choose a Suitable Sampling Method

Different materials use different sampling methods. Therefore, the sampling of construction materials should be based on actual conditions. According to the characteristics of the raw materials and the requirements of project acceptance, a scientific and reasonable sampling method that meets the standards and specifications should be selected for the validity of the test results. Provide a reliable guarantee for different materials. The selected detection parameters and detection methods are very different [10]. Therefore, it is necessary to select the best sampling and detection methods according to the characteristics and properties of the actual materials, and let the data speak to reflect the actual situation of the project site.

5   Conclusion

Overall, the application of information technology to the inspection of building materials not only makes the quality inspection of materials more convenient, fast and accurate, but also avoids the mutual interference between people, which can make the results of the quality inspection of building materials better. Reliability, authenticity and effectiveness are guaranteed, which requires us to further study and explore its realization in the future actual work.

Acknowledgments. Project source: Approved Project of Guangdong Provincial Science and Technology Innovation Strategy Special Fund in 2020, Project Number: pdjh2020a0943.

References

1. Zhou, P.: Talking about the testing technology of construction engineering materials. Technol. Market **7**(02), 153–156 (2018). (in Chinese)
2. Zhou, H.Y.: Analysis of the application of information technology in building material testing and laboratory. Urban Construct. Theory Res. **11**(10), 162–165 (2016). (in Chinese)
3. Xue, N., Pan, L., Zhu, G.F.: A trial discussion on the testing and control measures of building materials. Build. Mater. Decorat. **10**, 108–111 (2014). (in Chinese)
4. Liu, P.: Problems and improvements in building materials testing and management work. Jiangxi Build. Mater. **9**(11), 266–270 (2016). (in Chinese)
5. Jiang, X.X., Liu, T., Yang, Z.: Inspection and supervision of construction materials for construction engineering. Sichuan Cement **11**(04), 132–135 (2015). (in Chinese)
6. Xu, X.L.: Analysis of quality control measures for construction engineering materials. Architect. Des. Manage. **34**(08), 50–55 (2017). (in Chinese)
7. Zhou, Y.L.: Explore the key points of building materials testing. China Real Estate Indus. **4**(19), 226–229 (2016). (in Chinese)
8. Yang, L.T.: Talking about the impact of building material testing organization management on testing quality. China Standardization **6**(12), 83–84 (2017). (in Chinese)
9. Li Y.N., Wang Z., Zhao Q.Y.: Application of computer technology in the measurement of dynamic elastic modulus of building materials. Non-destructive Testing **12**(05), 27–32 (2015). (in Chinese)
10. Lin, Y., Yao, Z.: Application of microcomputer system in building material testing institutions. J. Zhejiang Water Conservancy and Hydropower College **12**, 27–32 (2015). (in Chinese)

Intelligent Wind Turbine Mechanical Fault Detection and Diagnosis in Large-Scale Wind Farms

Zhihui Cong[1(✉)], Jinshan Wang[1], Shuo Li[1], Liang Ma[1], and Ping Yu[2]

[1] China Datang (Chifeng) Corporation Renewable Power Co., Limited,
Chifeng, Inner Mongolia, China
[2] Zhongke Nuowei (Beijing) Co., Limited,, Beijing, China

Abstract. In the past few years, with the rapid development of the wind power industry, a large number of wind turbines have been built and deployed. With the development of the wind power industry, more and more scientific researchers pay attention to the stable and safe operation and fault diagnosis of wind turbines. The gearbox is an important component of wind turbine drive chain, which is affected by many factors in operation, such as wind speed fluctuation and loads dynamic change. Because of the bad operating conditions, the gearbox is prone to failure. Once the gearbox fails, it may cause the collapse of the fan drive chain, and then cause huge economic losses. Therefore, the research of gearbox fault diagnosis is of great significance to maintaining the normal operation of the wind turbines.

Keywords: Wind turbine · Fault detection · Fault diagnosis · Algorithmic solution

1 Introduction

With the development of society, people's demand for energy is increasing, and the problem of energy shortage is becoming more and more prominent. Vigorously developing and using new energy is an important way to solve the problem of energy shortage. According to statistics, the renewable energy industry is the fastest-growing energy industry in recent ten years. As a kind of renewable energy, wind energy has a good development prospect and has gradually become an important part of sustainable development in many countries. It can be seen that wind energy is no longer indispensable new energy. The mature wind power generation technology and rapid development momentum make it a new industry.

2 Fault Diagnosis Methods of Wind Turbines

The development of fault diagnosis at home and abroad has experienced the following three stages [1–6] and can be summarized as follows:

J. Abawajy et al. (Eds.): ATCI 2021, LNDECT 81, pp. 186–191, 2021.
https://doi.org/10.1007/978-3-030-79197-1_24

(1) Artificial Diagnosis

The early fault diagnosis technology of mechanical equipment is basically to record the vibration signal of mechanical equipment through special instruments, and then analyze and judge the signal according to the professional knowledge and experience of experts in related fields. At this time, the analysis method is relatively simple, mainly based on Fourier transform, such as spectrum, power spectrum, and so on. The function of diagnosis and analysis in this period is fixed and not easy to expand.

(2) Computer-Aided Diagnosis

With the development of computer technology and signal processing technology, the computer plays an increasingly important role in fault diagnosis. The collection of mechanical equipment status information, signal analysis and database management are all completed by computer. In some cases, the computer can even get the diagnosis conclusion directly and realize the unmanned fault monitoring and diagnosis. At this time, a variety of analysis methods are used, such as wavelet analysis, neural network, and expert system, and so on.

(3) Network Diagnosis

At present, an important development direction of fault diagnosis system is a network. In the composition of the network system, we make full use of the existing network resources of enterprises to realize the intellectualization of the diagnosis system, which not only saves the investment but also facilitates the realization of remote monitoring and diagnosis. At this time, the monitored parameters are more abundant, including not only the traditional monitoring parameters such as displacement, velocity, and acceleration but also the physical quantities such as temperature, pressure and sound, to more comprehensively analyze the running state of mechanical equipment.

3 Fault Frequency Analysis and Fault Detection Mechanism

The rolling bearing is mainly composed of three parts: inner race, outer race, and rolling element. The structure of the three parts is shown in Fig. 1. There are three main rolling bodies: ball, cylinder, and tapered roller, which can be rolled between inner and outer circles. The inner ring of the rolling bearing is assembled with the shaft. In most cases, the outer ring of the rolling bearing does not move, and the inner ring rotates with the shaft. The faults of rolling bearings usually occur on the outer ring, inner ring, and rolling body [7]:

The vibration types of rolling bearings are very complex. According to the different vibration sources, vibration can be divided into three types as follows:

(1) Vibration caused by bearing deformation. When the rolling bearing is running, the bearing will bear a certain load, which will produce elastic deformation and elastic vibration. Such vibration is the inherent vibration of the bearing, and it is not related to whether the bearing has a fault.

(2) Vibration caused by bearing machining error. In the actual processing environment, there will be some machining errors in each part of the bearing, and these errors will also produce certain vibration when the bearing is running.

(3) Vibration caused by bearing failure. The above-mentioned rolling bearing will produce the corresponding vibration in wear, fatigue, fracture, and deformation, which reflects the failure condition of the bearing.

Through the above analysis, it can be seen that the vibration of bearings is random vibration generated by the superposition of vibration sources with multiple frequency components. The vibration frequency of bearing failure is determined by the rotation speed of the shaft, the fault location, and the physical structure of the bearing itself. To get accurate diagnosis results, the interference of the first two vibration sources should be eliminated as much as possible in the fault diagnosis of rolling bearing.

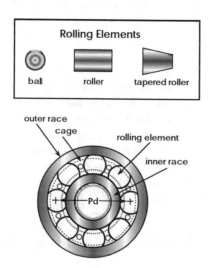

Fig. 1. The rolling bearing of wind turbines

When the rolling bearing has wear and other faults, when it rotates, it will produce a fixed frequency impact, causing periodic vibration. This frequency becomes the fault frequency of the bearing, and the fault type of the bearing can be inferred by using the characteristics of the fault frequency. The fault characteristic frequency of rolling bearing mainly includes inner ring fault frequency (BPFI), outer ring fault frequency (BPFO), ball spin frequency (BSF), and cage fault frequency (FTF). The calculation formula is [8]:

$$BPFO = \frac{nf_r}{2}\left(1 - \frac{d}{D}\cos\phi\right) \tag{1}$$

$$BPFI = \frac{nf_r}{2}\left(1 + \frac{d}{D}\cos\phi\right) \tag{2}$$

$$BSF = \frac{D}{2d}\left[1 - \left(\frac{d}{D}\cos\phi\right)^2\right] \tag{3}$$

$$FTF = \frac{f_r}{2}\left(1 - \frac{d}{D}\cos\phi\right) \tag{4}$$

Where, is the pitch diameter of the rolling element, is the diameter of the rolling element, is the rotation frequency of the bearing, is the contact angle, and is the number of rolling elements. When a certain kind of fault occurs in the bearing, the corresponding spectrum peak of fault frequency will appear in the vibration spectrum. In practice, the actual fault frequency is not exactly equal to the theoretical value calculated by the above formula due to the influence of the objective factors such as the error produced in the bearing processing and the deformation caused by the load during operation. Therefore, when finding the corresponding fault frequency on the frequency diagram, it is generally necessary to find the spectrum peak which is close to the theoretical value.

Similar to rolling bearings, according to the difference of vibration sources, the vibration types of gears can also be divided into three categories as follows [9–11]:

(1) Natural vibration

In the process of meshing, the gears will be subjected to periodic load to produce vibration, and the high-frequency component of vibration is the natural vibration frequency. The natural vibration frequency of gear can be calculated by the following formula

$$f_c = \frac{1}{2\pi} \cdot \sqrt{\frac{K}{M}} \tag{5}$$

Where K is the stiffness of the gear and M is the equivalent mass of the gear. The natural frequency of gear has nothing to do with whether there is a fault or not. Calculating the natural frequency of gear can effectively avoid resonance.

(2) Meshing vibration

In a pair of meshing gear pairs, the big and small gears mesh with each other to produce vibration, and the meshing frequency can be calculated according to the following formula

$$f_m = f_n \cdot z \tag{6}$$

Where f_n is the rotation frequency of the shaft and z is the number of teeth. In addition to the above formula, for planetary gear structure, the calculation formula is relatively complex as follows:

$$f_m = \left(f_s^r - f_p\right) \cdot z_s \tag{7}$$

Where, f_s^r is the absolute rotation frequency of the sun gear, f_p is the rotation frequency of the planet carrier, and z_s is the number of sun gear teeth.

(3) Vibration caused by gear failure

When the gear has wear cracks and other faults, the gear will be affected by the above two kinds of conventional vibration excitation sources and fault excitation sources at the same time, and the amplitude and phase of the gear vibration signal will change, that is, amplitude and phase modulation. In the time domain, the vibration amplitude generally increases and the periodic pulse occurs, while in the frequency domain, the frequency spectrum component increases and the amplitude of specific frequency points increases. In addition to the basic meshing frequency, the fault vibration signal also contains harmonics with two or more times meshing frequency as the main component.

The above analysis shows that the vibration mechanism of gear is consistent with that of rolling bearing. The expressions of the gear meshing vibration model and rolling bearing vibration model can be summarized as follows:

$$s(t) = \sum_{k=1}^{K} S_k(1 + a_k(t))cos(2\pi k f_m t + \phi_k(t)) \tag{8}$$

Where S_k is the k amplitude of the first harmonic, $a_k(t)$ is the k amplitude modulation function of the second harmonic, f_m is the meshing frequency, $\phi_k(t)$ is the k phase modulation function of the second harmonic. Due to the existence of modulation, the gear vibration signal will form a series of modulation frequency bands in the frequency spectrum, which contain rich fault information. The analysis of the form of the sideband is an important basis for judging the type of gear fault. For example, when there is a misalignment fault in the shaft, the sideband of or on both sides of the meshing frequency will be formed in the frequency spectrum of the vibration signal; when the gear has an eccentric fault, the down variable frequency band will appear in the frequency spectrum of the gear vibration signal.

4   Conclusions and Future Work

With the change of the external environment, the operation state of the fan is constantly adjusted to adapt to the environment. The transmission chain of the fan gearbox is the hub to control the operation state of the fan. Once the transmission chain of the gearbox fails, the fan will face the danger of shutdown and collapse. The rolling bearing and gear are the key components of the transmission chain supporting the fan gearbox, which are prone to failure. Rolling bearing fault can be divided into three categories according to fault location, namely inner ring fault, outer ring fault, and rolling element fault. The vibration mechanism of rolling bearing is similar to that of gear, which contains multiple vibration sources. When a rolling bearing or gear fails, its vibration signal will change, and the frequency spectrum of the vibration signal has large amplitude at a specific frequency point, which is called fault frequency.

References

1. Ribrant, J.: Reliability performance and maintenance—a survey of failures in wind power systems. Unpublished doctoral dissertation, XR-EE-EEK (2006)
2. Huang, N.E., Shen, Z., Long, S.R., et al.: The empirical mode decomposition and the Hilbert spectrum for nonlinear and non-stationary time series analysis. In: Proceedings of the Royal Society of London A: Mathematical, Physical and Engineering Sciences. The Royal Society, vol. 454, no. 1971, pp. 903–995 (1998)
3. McFadden, P.D.: Detecting fatigue cracks in gears by amplitude and phase demodulation of the meshing vibration. J. Vib. Acoust. Stress. Reliab. Des. **108**(2), 165–170 (1986)
4. Miller, A.J.: A new wavelet basis for the decomposition of gear motion error signals and its application to gearbox diagnostics. The Pennsylvania State University (1999)
5. Wismer, N.J.: Gearbox analysis using cepstrum analysis and comb liftering. Application Note. Brüel & Kjaer. Denmark (1994)
6. Miao, Q., Makis, V.: Condition monitoring and classification of rotating machinery using wavelets and hidden Markov models. Mech. Syst. Signal Process. **21**(2), 840–855 (2007)
7. Abbasion, S., Rafsanjani, A., Farshidianfar, A., Irani, N.: Rolling element bearings multi-fault classification based on the wavelet denoising and support vector machine. Mech. Syst. Signal Process. **21**(7), 2933–2945 (2007)
8. Felten, D.: Understanding bearing vibration frequencies. Electrical Apparatus Service Association (2003)
9. Yang, Y., Nagarajaiah, S.: Output-only modal identification with limited sensors using sparse component analysis. J. Sound Vib. **332**(19), 4741–4765 (2013)
10. Li, Y., Cichocki, A., Amari, S.I.: Analysis of sparse representation and blind source separation. Neural Comput. **16**(6), 1193–1234 (2004)
11. Gong, M., Liang, Y., Shi, J., Ma, W., Ma, J.: Fuzzy c-means clustering with local information and kernel metric for image segmentation. IEEE Trans. Image Process. **22**(2), 573–584 (2013)

Fault Detection and Diagnosis of Multi-faults in Turbine Gearbox of Wind Farms

Zhihui Cong[1(✉)], Jinshan Wang[1], Shuo Li[1], Liang Ma[1], and Ping Yu[2]

[1] China Datang (Chifeng) Corporation Renewable Power Co., Limited,
Chifeng City, Inner Mongolia, China
[2] Zhongke Nuowei (Beijing) Co., Limited, Beijing, China

Abstract. This paper proposed and developed fault detection and diagnosis solution based on the ensemble empirical mode decomposition and support vector machine (SVM). The collected field signals are used as the input signals that are decomposed into a series of intrinsic mode functions (IMF) by using the empirical mode decomposition. The dominant eigenmode functions are then selected in the proposed solution and their approximate entropy and energy ratio are calculated as signal features. Finally, the feature vector of the signals is used as the input into the support vector machine for model training to classify different mechanical faults of the wind turbine gearbox. The developed algorithmic solution is extensively verified through a set of experiments and the numerical results confirmed the effectiveness of the proposed solution.

Keywords: Fault diagnosis · Wind turbine gearbox · Mechanical faults · EEMD · SVM

1 Introduction

The gearbox of the wind turbine is a key link in the transmission chain of the wind turbine. In the face of the changeable natural environment outside the wind turbine, the gearbox is prone to failure, resulting in a shutdown and huge economic losses. Therefore, it is of great significance to study the fault diagnosis of wind turbine gearbox. Due to the fact that fault detection and system maintenance cannot be carried out in a timely fashion, the fault degree is more and more serious in the operation process, and finally, it is very likely to cause mechanical failure or disastrous results. To avoid this kind of loss, many researchers have put forward a series of advanced technical methods to diagnose the fault type, fault location and fault degree of the gearbox, which greatly improves the utilization rate of wind turbines and reduces the operation and maintenance cost of the wind farm.

To date, there are many methods of mechanical equipment state monitoring and fault diagnosis [1–5]. In practice, the fault detection solutions based on the vibration measurements are the most mature method of fault detection of rotational machinery equipment. The vibration signal processing based solutions are widely used in the practice due to their advantages such as strong applicability, good effect and intuitive signal processing. The existing fault diagnosis methods based on vibration signal

© The Author(s), under exclusive license to Springer Nature Switzerland AG 2021
J. Abawajy et al. (Eds.): ATCI 2021, LNDECT 81, pp. 192–198, 2021.
https://doi.org/10.1007/978-3-030-79197-1_25

analysis can be classified into three categories: the first is time-domain analysis, including various dimensionless parameter indexes and dimensionless parameter indexes; the second is frequency domain analysis methods, such as spectrum analysis, cestrum analysis, envelope spectrum analysis, etc.; the third is frequency domain analysis methods, such as wavelet analysis, short-time Fourier transform and so on. The following three kinds of analysis methods are explained and compared with their similarities and differences, advantages and disadvantages.

(1) Time-domain analysis based method

The fault detection and diagnosis solution based on the time-domain analysis is often calculated from the time-domain signal to obtain a series of parameters. It is the most intuitive and simple analysis method in the vibration signal, which is considered suitable for periodic vibration signal, simple harmonic vibration signal and shock signal. These parameters are divided into dimensionless parameter index and dimensionless parameter index according to whether there is dimension. Dimensionless parameters include kurtosis, margin, deviation, waveform index, peak index and pulse index. Among them, kurtosis, peak, and pulse indexes can effectively reflect the energy of impact components and are sensitive to the strong impact faults, so they can be used for mechanical fault detection and diagnosis. The dimensioned parameter indexes include the maximum value, the minimum value, the peak value, the mean value, the root mean square value, the variance, and so on. Among them, variance directly reflects the vibration intensity, and the root means square value directly reflects the energy of the vibration signal. The dimensioned parameter index is closely related to the working condition of the gearbox, and changes with the difference of the types and sizes of mechanical components. Therefore, the gearboxes are different, and the dimensioned parameters and indexes measured by different gearboxes cannot be directly compared.

In conclusion, the time-domain analysis method is intuitive and concise in principle, but the fault information reflected is limited, and the effect of fault location is poor, and there are many limitations in the field of fault diagnosis.

(2) Frequency domain analysis method

Frequency domain analysis is a series of signal processing methods based on Fourier transform, which plays an important role in the vibration signal processing of rotating machinery, and has been widely used in practice. Through various frequencies, the vibration source can be found, and the causes of the fault can be analyzed. The common methods of spectrum analysis include frequency domain analysis, high-order spectrum analysis, and spectrum adjustment analysis. The following are discussed respectively. For example, the higher-order spectral analysis can be implemented as follows: when mechanical equipment fails, with the increasing of the fault degree, the nonlinear characteristics of the signal are more and more obvious. High order spectrum is a frequency domain analysis tool for nonlinear, non-stationary and non-Gaussian signals. It represents random signals from higher-order probability structure, including phase information, which makes up for the deficiency of second-order statistics (power spectrum). The high-order spectrum has strong noise elimination capability. Theoretically, it can completely suppress Gaussian noise and hence provide efficient performance to exact the fault information through the analysis of vibration signals with a high-order spectrum.

(3) Time-frequency domain analysis method

The spectrum analysis method is mainly suitable for stationary signals and cannot locate the time domain information. The fault signal of the gearbox is mostly non-stationary and nonlinear due to various factors. Due to the fact that the analysis based on the time-domain and frequency-domain cannot extract and characterize the fault features, the combination of time domain and frequency domain analysis was proposed and developed in the literature. The previous studies have confirmed that the time-frequency analysis methods can accurately perform the fault location and characterize the distribution of different frequency components in the time domain to meet the requirements of non-stationary signal analysis.

Linear time-frequency transform mainly includes STFT and wavelet transforms (WT). The selection of short-time Fourier transform window function affects the resolution of time-frequency, but limited by the uncertainty principle, the time resolution and frequency resolution can only be eclectic. Also, the application premise of short-time Fourier transform is that the local signal is stable, and for some signals whose frequency changes rapidly with time, the ideal analysis effect cannot be achieved.

Nonlinear time-frequency transformation mainly includes spectrum, winger Ville distribution, wavelet scale graph, etc. compared with linear time-frequency transformation, nonlinear time-frequency transform is the weighted average of the kernel function of the bilinear product of signal, and the energy of the signal is distributed in time-frequency space, which does not meet linear superposition.

2   Proposed Fault Detection and Diagnosis Solution

2.1   EEMD Method

The energy industrial mode decision (EEMD) is a noise-aided data analysis method for the shortcomings of EMD. EMD algorithm can deal with nonlinear and non-stationary signals and is widely used in fault diagnosis. However, this does not mean that the EMD algorithm is perfect, it still has some shortcomings, which may be magnified in some cases, and the wrong results can be derived. One of the main disadvantages of the EMD algorithm is that the mode aliasing problem is easy to occur. Mode aliasing refers to the existence of different scale signals or the same scale signals in the same Eigenmode function. To solve the problem of mode aliasing, Huang et al. Improved the EMD algorithm and proposed the overall empirical mode decomposition algorithm (EEMD). The main idea of the EEMD algorithm is to use the characteristics of uniform distribution of Gaussian white noise, and artificially add Gaussian white noise to the original signal for multiple decompositions to decompose the original signal into a series of Eigen modal functions of different scales to suppress the phenomenon of mode mixing.

2.2   Algorithmic Solution

The proposed fault diagnosis solution of the wind turbine gearbox is designed based on EEMD and SVM techniques. The collected sensor signals are decomposed into a set of IMF by the use of EEMD and the collection of features are extracted from the IMFs to produce a feature vector. The feature vector is input into the trained support vector machine model to get the final diagnosis result. In this work, the one-to-one and one-to-many training methods are adopted in the proposed algorithmic solution. Firstly, the normal samples are taken as positive classes, and the fault samples are taken as negative classes. One-to-many model training is used to train SVM1, and then the normal samples are removed, and the one-to-one training method is used to train SVM2.

(1) Feature extraction

Feature extraction is an important factor to determine the training effect and diagnosis effect. Selecting appropriate features can greatly improve the accuracy of fault diagnosis. The sensor signal can be decomposed by EEMD to get multiple IMFs. In this paper, the approximate entropy and energy ratio of these IMFs are extracted to form eigenvectors for fault diagnosis. Because the IMF contains more information, the approximate entropy and energy ratio of the first five IMF are extracted to form a 10-dimensional feature vector.

(2) Approximate entropy

Approximate entropy (ApEn) is a statistical measure, which can measure the complexity of a time series. Suppose that a given time series $x(t)$ is artificially divided into N small segments, each segment contains m time points, and the i segment can be recorded as [6]

$$X_i = \{x(i), x(i+1), \ldots, x(i+m-1)\}, \ i = 1, 2, \ldots, N \tag{1}$$

Define the distance between the i segment and j segment as follows:

$$d(X_i, X_j) = \max_{k=1,2,\ldots m} (|x(i+k-1) - x(j+k-1)|) \tag{2}$$

Given any segment X_i, the parameters can be defined as (3):

$$C_i^m(r) = \frac{\sum\limits_{j \neq i} l(r - d(X_i, X_j))}{N - m + 1} \tag{3}$$

$l(x)$ is the indicator function that meets (4) as follows

$$l(x) = \begin{cases} 1, & x \geq 0 \\ 0, & x < 0 \end{cases} \tag{4}$$

Where r represents a predefined constant that is related to the standard deviation of the time series $x(t)$, as given in (5):

$$r = k \times std(x) \tag{5}$$

Where k is a constant and according to the parameter of very segment $C_i^m(r)$, the parameters of the time series of the overall segment can be defined as follows

$$\phi^m(r) = (N - m + 1)^{-1} \sum_{i=1}^{N-m+1} \log C_i^m(r) \tag{6}$$

Therefore, the approximate entropy of ideal state can be defined as [7–9]

$$ApEn(m, r) = \lim_{N \to \infty} [\phi^m(r) - \phi^{m+1}(r)] \tag{7}$$

Generally, the length and the number of segments of the signal are not infinite, so the approximate entropy can also be written as

$$ApEn(m, r, N) = \phi^m(r) - \phi^{m+1}(r) \tag{8}$$

2.3   Experiments and Numerical Results

The experimental data is from the official website of Case Western Reserve University (CWRU) bearing data center [10]. Figure 1 shows the physical diagram of the test-bed built by Case Western Reserve University. The power source of the test-bed is the motor on the left in the figure, which can provide up to 2 horsepower (HP). The middle part of the test bench is a torque transmission device, and the right side is equipped with a dynamometer. The sensor is placed on the magnetic base of the device, and the sampling frequency is 12 kHz or 48 kHz per second. The fault types are divided into inner ring fault, outer ring fault and rolling element fault.

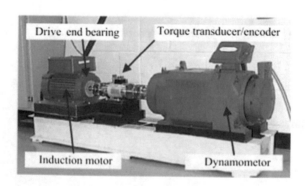

Fig. 1. CWRU bearing testbed

In this experiment, samples are randomly selected from the SKF Bearing Failure test group and normal test group to form a sample set with 150 samples as training data, and each sample contains 4000 sampling points. After EEMD decomposition of each sample point, the first five IMF components are retained and their approximate entropy and energy ratio are calculated. In the normal sample points, three sample points (under different loads) are randomly selected and their eigenvectors are extracted. It is found that approximate entropy is not sensitive to load change, but the energy ratio will change greatly. The advantage of using approximate entropy as a feature vector is that it can balance the impact of load change and prevent some normal samples from being misjudged as fault samples due to load change.

To avoid the selection of training samples, the number of test samples is twice the number of training samples. The test results of sample points under various states are shown in Table 1. It can be seen from the table that all sample points are correctly classified, which verifies the effectiveness of the algorithm.

Table 1. Performance of fault detection accuracy

Bearing condition	Number of test samples	Classification accuracy (%)
Normal	80	98.2
Inner fault	130	100
Outer fault	50	97.1
Ball fault	70	100
Total	140	94.5

3 Conclusions and Discussions

In this work, the multi-fault diagnosis algorithm of wind turbine gearbox based on EEMD and SVM. EEMD is proposed and developed to decompose the sensor signal to mining the deep features of the signal. The energy ratio and approximate entropy of the IMF component are selected. Before the specific diagnosis, the algorithm needs to construct a support vector machine model. The model proposed in this work includes two classifiers. The first classifier is used to judge whether there is a fault in the gearbox, and the classifier is generated according to the one to many methods; the second classifier is used to identify the specific type of fault and is generated according to the one to one method.

It should be highlighted that the establishment of the model takes into account the highest proportion of normal samples in the fault diagnosis sample set. If one to many methods is used to establish the model directly, the training time of the model may be greatly prolonged. Under the condition that the one-to-one method is used to establish the model directly, the number of positive and negative samples will not match, which will affect the classification effect. The proposed method takes into account the imbalance of training time and training samples, and the numerical results obtained from the experiments demonstrate the effectiveness of the proposed fault detection and diagnosis solution.

References

1. Huang, N.E., Shen, Z., Long, S.R., et al.: The empirical mode decomposition and the Hilbert spectrum for nonlinear and non-stationary time series analysis. Proc. Roy. Soc. Lond. A Math. Phys. Eng. Sci. **454**(1971), 903–995 (1998). The Royal Society
2. McFadden, P.D.: Detecting fatigue cracks in gears by amplitude and phase demodulation of the meshing vibration. J. Vib. Acoust. Stress. Reliab. Des. **108**(2), 165–170 (1986)
3. Miller, A.J.: A new wavelet basis for the decomposition of gear motion error signals and its application to gearbox diagnostics. The Pennsylvania State University (1999)
4. Wismer, N.J.: Gearbox analysis using cepstrum analysis and comb liftering. Application Note. Brüel & Kjaer. Denmark (1994)
5. Miao, Q., Makis, V.: Condition monitoring and classification of rotating machinery using wavelets and hidden Markov models. Mech. Syst. Sig. Process. **21**(2), 840–855 (2007)
6. Pincus, S.M.: Approximate entropy as a measure of system complexity. Proc. Natl. Acad. Sci. **88**(6), 2297–2301 (1991)
7. Pincus, S.: Approximate entropy (ApEn) as a complexity measure. Chaos: Interdiscipl. J. Nonlinear Sci. **5**(1), 110–117 (1995)
8. Richman, J.S., Moorman, J.R.: Physiological time-series analysis using approximate entropy and sample entropy. Am. J. Physiol.-Heart Circul. Physiol. **278**(6), 2039–2049 (2000)
9. Jörgen, B., Heiko, R., Andreas, H.: Approximate entropy as an electroencephalographic measure of anesthetic drug effect during desflurane anesthesia. J. Am. Soc. Anesthesiol. **92**(3), 715–726 (2000)
10. Bearing Information. http://csegroups.case.edu/bearingdatacenter. Accessed 2 Mar 2016

Intelligent Condition Monitoring Technology of OPGW Optical Cable Junction Box

Jing Song[1(✉)], Ke Sun[1], Di Wu[1], Fei Cheng[1], and Zan Xu[2]

[1] Zhejiang Huayun Power Engineering Design Consulting Co., Ltd, Hangzhou, Zhejiang, China
[2] Nokia Communication System Technology (Beijing) Co., Ltd, Hangzhou, Zhejiang, China

Abstract. At present, the power communication network has formed a communication network with an optical fiber communication network as the core. In the optical fiber communication network, the structure design and installation position of the junction box has a very important impact on the operation of the communication network. To improve the stability and reliability of the OPGW optical cable junction box, this paper proposes an intelligent monitoring technology, which can comprehensively monitor the environmental temperature, humidity, height, image, internal water immersion and air pressure of the junction box through sensors, and formulate targeted maintenance measures.

Keywords: Intelligent condition monitoring · OPGW · Cable junction box

1 Introduction

Optical fiber composite overhead ground wire is also called OPGW optical cable. It mainly refers to placing the optical cable in the ground wire of the overhead high-voltage transmission line to form the optical cable communication network on the transmission line. The OPGW optical cable has the dual functions of ground wire and communication, especially in the power grid 500 kV, 220 kV, 110 kV voltage level lines is widely used. In the installation process of OPGW cable, many factors need to be considered, including conductor stress, sag and insulation gap, as well as the influence of self-weight. An optical cable junction box is easy to be affected by external forces, weather factors and electromagnetic effects in the process of use, resulting in the loss [1–3]. At the same time, the installation location and structural characteristics will also bring a lot of inconvenience to the later operation and maintenance, seriously affecting the operation quality of the communication network.

The OPGW optical cable is installed on the power tower, and the stainless steel metal shell is used to avoid the influence of external force and weather factors on the cable joint. It is difficult to find the optical cable joint fault from the appearance, and the maintenance is difficult [4]. With the extension of service time, the failure rate of the OPGW optical cable joint box shows an increasing trend, which seriously damages the quality of the communication network. The common faults of the OPGW optical cable joint box are analyzed as follows:

J. Abawajy et al. (Eds.): ATCI 2021, LNDECT 81, pp. 199–205, 2021.
https://doi.org/10.1007/978-3-030-79197-1_26

(1) The structure design of optical cable joint fault box. According to the analysis of the daily operation and maintenance data, the line fault of optical cable communication is more prominent than the equipment fault, and in the optical cable communication line fault, the optical cable junction box fault is an important reason. Because the optical cable junction box is installed on the power tower, it is greatly affected by environmental factors, and there are problems in the design, it is easy to cause the junction box seal is not tight, and the water in the box corrodes the metal At the same time, water ion electrolysis or freezing will also cause damage to the cable ends.

(2) Failure due to the installation process. In the installation process of the cable junction box, if industrial alcohol is used to wipe the optical cable line, the non-volatile alcohol will corrode the coating, causing the coating of the optical cable line to fall off and the fiber core to break. Because the optical cable joint box is installed on the power tower, it is greatly affected by environmental factors, such as long-term strong wind leading to the shaking of the optical cable joint box, resulting in the drop of the optical cable joint box or the failure of the fiber core in the box [5–7].

(3) Lack of optical cable junction box fault sensing means. At present, there is no automatic sensing technology for the optical cable junction box. The inspection of the power optical cable junction box is mainly through the manual regular box opening inspection to repair the faulty junction box and deal with the fault. Due to the particularity of power communication, the power cable is erected with the power line. The conventional joint box inspection is not only time-consuming and labor-consuming but also inefficient. At the same time, in the process of work, there is also the possibility of accidental collision damage caused by the operation of optical cable [8–10].

2 Design of Intelligent Fault Monitoring Technology for OPGW Optical Cable Junction Box

The intelligent fault monitoring technology of the OPGW optical cable junction box is to transmit the operation data of the junction box through the 4G/5G Internet of things chip as the transmission channel. Build a cloud server and view data through mobile applications. The design details are presented and discussed as follows:

Firstly, the sensors are based on the monitoring device. The intelligent fault monitoring technology of the OPGW optical cable junction box is connected to other sensors through serial port 232/485. It includes a water inlet sensor, temperature sensor, wind speed sensor, voltage sensor, and so forth. The monitoring device provides 2–4 232/485 ports and a 12 V power supply. It can connect the alarm signal of the water immersion sensor to the monitoring device, at the same time, it can monitor the outdoor temperature of the junction box in real-time, and collect the monitoring data of the wind speed and wind speed, which is collected every 2 min to form the big data summary of the report. Because the power supply is through solar energy and battery power supply, the junction box monitoring device voltage monitoring sensor is needed to monitor the voltage in a real-time fashion.

Secondly, video monitoring and collection need to be designed and implemented. OPGW optical cable junction box fault intelligent monitoring technology uses a low-power video monitoring chip and video recording algorithm to collect video. When the video is dynamic, the video is stored, which can reduce the power consumption of video and the space of the storage chip.

Finally, the real-time positioning of the connector box is required for the real-time condition monitoring and maintenance system. Through the positioning chip of GPS in the monitoring device, the real-time positioning of the junction box is carried out, and the positioning data is transmitted through the 4G/5G channel so that the real-time position of the junction box can be displayed through the GIS map. The networking scheme of OPGW optical cable joint box fault intelligent monitoring technology is illustrated in Fig. 1.

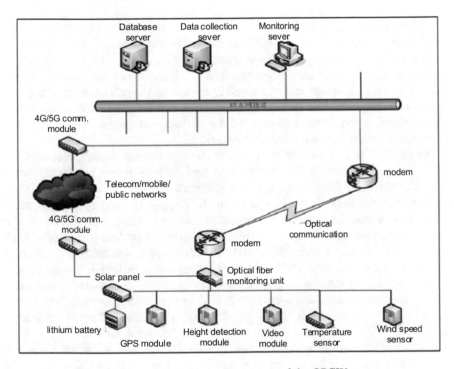

Fig. 1. The condition monitoring system of the OPGW system

OPGW optical cable joint box is an important component of the OPGW line, which is used to connect two sections of the OPGW optical cable to extend the distance of the optical cable and extend the transmission channel. The single-coil length of OPGW cable is generally 3–5 km. In the OPGW optical cable line project, due to the long distance between substations, a line is connected by multiple optical cables, and the connection between two optical cables is completed in the optical cable joint box. To weld two OPGW optical cables, it is necessary to peel the optical cable from the ground wire first, and then weld each fiber core with the optical fiber welding machine. The OPGW optical cable joint box is installed on the power tower, which is inconvenient for daily inspection. To prevent the fiber core from increasing attenuation due to the influence of external environment such as electromagnetic, external force, and weather, the OPGW optical cable joint box uses stainless steel and other metal shells, which is not easy to find problems in appearance. The installation position and structural characteristics have adverse effects on the operation and maintenance of the OPGW optical cable joint box, which makes the OPGW optical cable joint box more stable Optical cable joint box becomes the blind area of operation and maintenance.

3   Fault Detection of OPGW Optical Cable Junction Box

The requirements for OPGW optical cable joint box are good optical cable channel extensibility, that is, the smaller the optical cable connection attenuation introduced by the OPGW optical cable joint box, the better. The connection loss is not only related to the structure of the junction box but also related to the technical level of the construction and installation personnel; the optical cable junction box has a good sealing effect, protecting the fiber core in the box from the influence of the external environment. As the OPGW optical cable joint box is erected on the power tower, it is mostly in the field, river, mountain, and another harsh environment, so it is required to have moisture-proof, water-proof, anti-external force, temperature resistance, and other functions, and has higher requirements for sealing.

The junction box shall consist of the enclosure, internals, sealing elements, and optical fiber joint protection. The shell shall be made of high-strength materials such as stainless steel or aluminum alloy; the sealing elements are used for the sealing of the joint box itself and between the joint box and OPGW, and the sealing methods can be divided into mechanical seal and heat shrinkable seal. The mechanical seal uses adhesive, vulcanized rubber, non-vulcanized self-adhesive rubber, paste rubber packaging mixture, etc., and the heat-shrinkable seal uses the pipe with the inner wall coated with hot melt adhesive heat shrinkable or flaky polyolefin materials are sealed after heating; optical fiber joint protection can be heat shrinkable or non-heat shrinkable.

The installation technology of the OPGW optical cable junction box has an important impact on the quality of the junction box. The sealing effect of the junction box will not be ideal if the construction is not standardized and the technology is not in place, and some will directly damage the optical cable. At present, in most cases, to meet the tightness, once the construction and installation of the OPGW optical cable joint box are completed, it will not be opened, unless the joint box is damaged or the channel is interrupted due to excessive core decay. In case of failure, the maintenance personnel usually replace the junction box directly. For the junction box with serious water immersion, they also cut off the connected optical cable 1–2 m. The length of the remaining cable is shortened. If the length of the remaining cable is not enough, the location of the optical cable junction box should be improved, which is closer to the high-voltage power line. However, the difficulty of maintenance construction is increased, and the safety risk of maintenance personnel is also increased, which increases the electromagnetic field strength of the junction box and worsens the environment. Therefore, it is not an ideal operation and maintenance mode to repair after the failure.

This paper puts forward the joint box technology operation and maintenance detection method of tracking detection auxiliary regular open box inspection. According to the condition-based maintenance requirements of power communication optical cable, routine core detection should be carried out at the initial stage of operation of communication OPGW optical cable, and various state quantities should be collected and evaluated. During the installation and construction of the line project, the desiccant is placed in the junction box. After the construction, the distance information and attenuation data of the junction box on the line are collected as the initial state information of tracking detection. In the first condition-based maintenance evaluation after the OPGW cable is put into operation, OTDR is used to test the attenuation of the OPGW cable joint box, record and compare with the initial state information, select the joint box with a larger attenuation value, open the box to check and replace the desiccant.

Since the fiber core of the optical cable is affected by moisture or water, it cannot be reflected on the optical power loss immediately, and the comparison of the attenuation before and after the test cannot be used as the judgment basis for the open box inspection. Moreover, once the attenuation value increases greatly, it indicates that the optical cable has been irrecoverably damaged, so the damaged optical cable can only be cut off, or even the whole optical cable can be replaced. Therefore, according to the local actual conditions, after the OPGW optical cable line has been put into operation for three years, the joint box should be opened regularly and selectively for inspection every year, and the moisture or water immersion in the box should be observed. At the same time, the desiccant, sealing rubber ring, or silicone rubber ring should be replaced. If it is found that the joint box is seriously affected by dampness or water, the inspection scope should be expanded.

At present, there is no regular open box inspection project for OPGW line operation and maintenance, which leads to the serious water immersion of some joint boxes, but they are not found by operation and maintenance personnel. The sealing ring of the joint box is exposed to the outside for a long time, and it is damaged by the sun and rain, electric corrosion, insect bite and so on. The aging is very fast, and it is easy to have cracks so that the box is affected by moisture, water slowly, and the fiber core of the optical cable is eroded. Therefore, it is necessary to open the joint box regularly and replace the aging sealing ring. The quality of the sealing material is the main reason for the failure of the sealing performance of the optical cable joint box. At high temperatures, the sealing material is easy to soften, and at low temperatures, the sealing material is easy to brittle and loses its elasticity. The airtightness and water tightness of the joint box produced at home and abroad cannot reach 100% reliability.

The statistical data of the bell core show that about 90% of the faults in long-distance optical cable lines will affect all the fiber cores in the optical cable, which provides a basis for using OTDR to test a spare fiber core to determine the fault of optical cable. Every year, the spare fiber shall be used to test the attenuation of the optical cable regularly, and the attenuation curve of the optical cable in the junction box shall be tracked. When the attenuation curve of the junction box is abnormal, the maintenance plan shall be arranged in time, and the box shall be opened for inspection. If possible, an optical fiber on-line monitoring system can be set up to monitor the attenuation of the junction box position on the whole OPGW optical cable line in real-time.

4   Conclusions and Remarks

In recent years, the failure of optical cable junction box presents an increasing trend. The installation position and structural characteristics of the OPGW optical cable junction box have adverse effects on the detection and operation and maintenance of the junction box. This paper focuses on the analysis and comparison of the junction box seals with more problems in practical application and puts forward a technical operation and maintenance detection method of the junction box, which is tracking detection assisted by regular open box inspection. It is suggested that the inspection should be carried out in the bidding and factory. To improve the stable operation level of the OPGW optical cable joint box, the quality control of the OPGW optical cable joint box is carried out.

This paper summarizes and analyzes the problems existing in the current optical cable junction box, especially the problem that the junction box is not well sealed, which causes the box to be immersed in water, and puts forward a technical operation and maintenance detection method of the junction box, which is tracking detection assisted by regular open box inspection.

Acknowledgments. This work was supported by the Science and Technology Development Project of Zhejiang Huayun Power Engineering Design Consulting Co., Ltd.

References

1. Lanier, J.: Design and installation of optical ground wire from a transmission utility perspective. In: 2001 Power Engineering Society Summer Meeting. Conference Proceedings (Cat. No. 01CH37262), Vancouver, BC, Canada, vol. 1, pp. 94–96 (2001). https://doi.org/10.1109/PESS.2001.969991
2. Iwata, M., Ohtaka, T., Kuzuma, Y., Goda, Y.: Analytical investigation on OPGW strands melting due to DC arc discharge simulating lightning strike. In: 2012 International Conference on Lightning Protection (ICLP), Vienna, Austria, pp. 1–5 (2012). https://doi.org/10.1109/ICLP.2012.6344238
3. Serizawa, Y., Myoujin, M., Miyazaki, S., Kitamura, K.: Transmission delay variations in OPGW and overhead fiber-optic cable links. IEEE Trans. Pow. Deliv. **12**(4), 1415–1421 (1997). https://doi.org/10.1109/61.634154
4. Iwata, M., Ohtaka, T., Goda, Y.: Melting and breaking of 80 mm^2 OPGWs by DC arc discharge simulating lightning strike. In: 2016 33rd International Conference on Lightning Protection (ICLP), Estoril, Portugal, pp. 1–4 (2016). https://doi.org/10.1109/ICLP.2016.7791340

5. Günday, A., Karlık, S.E.: Optical fiber distributed sensing of temperature, thermal strain and thermo-mechanical force formations on OPGW cables under wind effects. In: 2013 8th International Conference on Electrical and Electronics Engineering (ELECO), Bursa, Turkey, pp. 462–467 (2013). https://doi.org/10.1109/ELECO.2013.6713885
6. Murata, H.: Application of optical communications for power distribution companies. In: IEEE TENCON 1990: 1990 IEEE Region 10 Conference on Computer and Communication Systems. Conference Proceedings, Hong Kong, vol. 2, pp. 788–792 (1990). https://doi.org/10.1109/TENCON.1990.152720
7. Jie, L., Gang, L., Xi, C.: Study on the thermal stability of OPGW under large current condition. In: 2009 Pacific-Asia Conference on Circuits, Communications and Systems, Chengdu, China, pp. 629–635 (2009). https://doi.org/10.1109/PACCS.2009.103
8. Yuqing, L., Xi, C., Chen, L., Yang, W., Baosu, H.: Study on a new and high efficient OPGW melting ice scheme. In: 2015 2nd International Conference on Information Science and Control Engineering, Shanghai, China, pp. 480–484 (2015). https://doi.org/10.1109/ICISCE.2015.111
9. Ji, J., Gao, Y.: Reason analysis of OPGW breakage caused by ice cover. North-China Electric Pow. 7, 15–17 (2008)
10. Liu, W.-T., He, S., Chen, Y., et al.: Defensive strategy for wide area ice disaster of power grid based on DC deice. Autom. Electric Pow. Syst. 36(11), 102–107 (2012)

Construction Project Management Based on Computer Information Technology

Jingfeng Yue$^{(\boxtimes)}$

Liaoning Jianzhu Vocational College, Liaoyang 111000, Liaoning, China

Abstract. China's traditional pillar industry, especially the construction industry, is faced with great challenges under the promotion of computer information technology innovation. For example, to improve the application of computer information technology in construction project management is a major topic faced by China's construction engineering industry. Based on the development status of computer information technology in Construction Engineering Management in China, according to the characteristics of construction industry, from the necessity of applying computer information technology, this paper puts forward the relevant practices that are helpful to the application of computer information technology in construction engineering management.

Keywords: Construction engineering management · Information technology · Management software · Construction engineering information management

1 Introduction

With the continuous development of modern science and technology, computer, information technology has been widely used in various fields of society. In the modern construction industry, if we continue to use the traditional management mode, it cannot meet the needs of modern social development. Therefore, we need to choose a new and advanced management mode to improve the management level and meet the development needs of modern society. The application of computer information technology is an important means to realize modern construction engineering information management. By applying computer information technology to construction project management, the efficiency and quality of project management can be comprehensively improved.

2 Characteristics of Computer Information Technology

2.1 Resource Sharing

The so-called resource sharing mainly refers to the opening of terminal permissions within the LAN according to the actual needs. This technology should be based on the database technology. For example, in the actual construction project management, the staff should make statistics on the specific material information according to the actual demand. After the information is confirmed to be correct, the actual consumption can be input into the computer database, Enterprise financial management personnel can access the information in the database to understand the specific financial situation of the enterprise. From this work, we can also see that the computer information technology can show the obvious characteristics of resource sharing, strengthen the engineering management effect of construction enterprises to a certain extent, and avoid the excessive consumption of manpower and time cost.

2.2 Precise Management

There are many management contents involved in construction project management, such as technical safety, quality management, cost management, etc. in the development of these management work, the staff should formulate unified management measures according to the actual situation in advance, which virtually increases the working pressure of management personnel, and also prone to management mistakes. Therefore, people can introduce computer information technology In addition, managers can also use the information management platform to maintain the smooth completion of related work [1].

3 The Importance of the Application of Computer Information Technology in China's Construction Project Management

3.1 Application of Computer Information Technology to Realize Real-Time Management

Because of the fixity of engineering construction, it must be used for a long time at the place where it is built, and it is connected with the foundation and connected with the land as a whole. Compared with industrial products, construction products are immovable. In the continuous construction of construction equipment and construction personnel, it is necessary to strengthen the timely management of engineering construction. If the traditional management method is adopted, it is difficult to achieve. The application of information technology can transmit the project management information to the construction manager in real time to realize the real-time management.

3.2 Applying Computer Information Technology to Realize Diversified Management

Compared with other industries, the construction scale of the construction industry is relatively large. In the process of construction, different management modes should be selected according to the area and actual use of the building. Even in the case of the same architectural style, the construction conditions, environment and methods of different construction places will be different, which will lead to different modes of engineering construction management and greatly increase the difficulty of management. With the application of computer information technology, we can have a comprehensive understanding of the construction conditions and environment of various construction areas, so as to adopt corresponding and diversified management.

3.3 Application of Computer Information Technology to Realize the Whole Process Management of Buildings

Due to the large scale and long construction period of the construction project, the project will experience the changes of four seasons, which will not only greatly increase the difficulty of open-air construction, but also greatly increase the difficulty of the whole process management of the project. Therefore, the application of computer information technology for building construction management, through the whole process of scientific management of construction projects, can effectively shorten the construction period.

3.4 Application of Computer Information Technology to Realize All-Round Management of Buildings

Due to the large scale of modern engineering projects, and the construction site conditions and environment are more complex. If we adopt the traditional engineering management mode, it will inevitably consume a lot of manpower, material resources and financial resources, and the management level is relatively low. Therefore, the application of computer information technology, human, material and financial resources can be summed up into a unified whole, through all aspects and links of the project management, in order to coordinate the smooth implementation of various types of work, so as to improve the quality of engineering construction management [2].

4 Application Status of Computer Information Technology in Construction Project Management

4.1 Lack of Coordination Between Software Development and Engineering Management

The traditional construction project management mode is usually based on experience, which cannot achieve the coordination with new technology. Most construction enterprises do not set up special information management personnel. Ordinary administrators are not competent for the corresponding management work without computer professional skills. In addition, in the process of developing management software, software developers blindly emphasize profit maximization and lack of reasonable cognition of construction engineering management. The software products developed are difficult to meet the practical needs of the development of the construction industry, and cannot be effectively used, which affects the full play of its functions.

4.2   Lack of Professionals

The survey shows that most of the construction project managers in our country are lack of in-depth understanding and cognition of computer information technology, and few of them master professional technology. Moreover, many times, managers will be bound by traditional ideas, unable to carry out management work from the perspective of sustainable development of construction industry. In addition, the management of construction enterprises do not realize the role of computer information technology, and think it is unnecessary to employ professional technical personnel with high salary, which will inevitably lead to the deviation of the direction of construction enterprises.

4.3   Frequent Changes in Personnel

In the process of building construction, the environmental conditions will change, which leads to great uncertainty. Moreover, many factors cannot be controlled artificially. The rhythm and frequency of construction are difficult to control, and the demand for management personnel also fluctuates greatly. Due to the frequent mobilization of personnel, many construction enterprises are unwilling to spend time and energy on computer for management personnel Information technology related training.

4.4   Changing Management Environment

Construction production has the characteristics of liquidity, which also makes its management environment changeable. Both visible and invisible factors are in dynamic change. If we continue to use the traditional construction project management mode, it is obviously unable to adapt to the dynamic management environment and affect the level and effect of construction project management [3].

5   Application Strategy of Computer Information Technology in Construction Project Management

5.1   Accelerate Software Development and Utilization

Software developers should base on the practical needs of construction project management, develop management software, and test it in pilot projects, so as to promote the popularization and application of computer information technology in construction project management. Government departments should pay attention to the need, promulgate corresponding laws, regulations and policies, and urge some key projects to speed up the construction of computer information management system and information network through mandatory requirements, so as to provide a good policy and market environment for software development enterprises, and ensure that they can strengthen communication with construction enterprises, Develop management software to achieve mutual benefit and win-win situation [4].

5.2   Attach Importance to the Promotion of Related Technologies

The construction department should pay attention to the coordination work, promote the computer information technology through various ways, so that the construction engineering enterprises can truly realize the important role of computer information technology in the construction project management, and effectively apply the computer information technology to the production and operation management. For large and medium-sized construction enterprises, we can base on their own actual situation, invest a certain amount of money to develop management software, speed up the grasp of modern technology, and popularize computer information technology in enterprises. In addition, from the perspective of ensuring the rational application of computer information technology, enterprises should also pay attention to the training of information management personnel, take the application ability of computer information technology as an important reference for managers to take up their posts, optimize the internal management team of enterprises, and enhance the effectiveness of management [5].

5.3   Building BIM Information System

In the construction project management, the workflow will affect the management effect to a certain extent, and the characteristics of the construction project itself make it have many objective factors in the project approval and bidding link, and the management system is relatively complex. In the data statistics calculation link, it needs to involve a large number of projects. According to this characteristic, BIM can be constructed. The model is used to improve the objectivity and accuracy of data calculation results [6]. Through the reasonable use of computer information technology, it breaks through the limitations of traditional departments and links the relevant contents into a whole. Therefore, it realizes the joint monitoring of various business modules, emphasizes the coordinated operation of all sectors of construction engineering, and ensures the management effect of construction projects.

5.4   Improve the Quality of Management Personnel

In the construction project management, we should emphasize people-oriented and pay attention to the "comprehensive, coordinated and sustainable" development. From the perspective of construction enterprises, we should improve the quality of management personnel and give full play to the role of "people" in construction project management [7]. It should be realized that the application of computer information technology in construction project management largely depends on the cognition of the enterprise management. If the management personnel do not pay attention to it, the application of computer information technology will be out of the question. Therefore, it is necessary to carry out the corresponding training work in the construction enterprise, improve the professional ability and comprehensive quality of enterprise management personnel, combine continuing education with professional qualification certificate system, strengthen the training of architects, engineers, project managers and other personnel, improve their mastery and application ability of computer information technology, and insist on training before taking up posts This is to improve the professional skills and professional quality of management personnel, and realize the reasonable application of computer information technology [8].

5.5   Improve the Adaptability of Project Management

Construction engineering construction is easy to be affected by the external environment, whether geological environment, climate environment or natural environment, will have a significant impact on the resource investment and construction period of construction projects, and may also interfere with the accuracy of various data [9]. In view of this situation, the construction parameter setting function should be introduced into the construction engineering information management system to ensure that the user can customize the relevant parameters according to the environment of the construction project, so that the various mechanisms in the running state can be consistent with the actual situation, do a good job of information feedback, and managers can also optimize the parameters according to the feedback information To adjust [10]. Under the premise of fully ensuring the safety and quality of construction, the management personnel should take the cost and progress of construction engineering as the management core, realize the efficient collection and processing of data information and promote the maximization of management efficiency through the reasonable application of computer information technology [11].

6   Conclusion

Information construction is an in-depth management reform. In the construction project management, we should constantly innovate the way of thinking, fully introduce advanced computer information technology in the strategic planning of information construction, organically combine various information technology and management methods, transform information technology, effectively improve the effective utilization rate of various resources, complete the transformation of intensive management, and finally promote the progress Industry information application process, to maintain the efficient and stable development of construction engineering.

References

1. Yang, J.: Discussion on the application of computer information technology in construction engineering in BIM era. J. Beijing Print. Inst. 28(1), 145–147 (2020)
2. Jin, D.: Analysis of the application effect of computer information technology in construction project management in BIM era. China New Commun. 21(5), 101–103 (2019)
3. Yang, H.: Application of computer information technology in construction project management. Constr. Technol. Dev. 44(11), 66–67 (2017)
4. Chen, Y.: Discussion on project management informatization of contemporary construction enterprises. Sichuan Build. Mater. 46(3), 192–193 (2020)

5. Fang, L.: Research on the application of computer network information technology to improve the management level of construction projects. Constr. Technol. Dev. **44**(23), 87–88 (2017)
6. Liu, H., Liu, L.: Intelligent management of building big data based on digital watermarking fingerprint technology. Sci. Technol. Progr. Countermeas. **35**(24), 98–101 (2018)
7. Waleed, U., Siddiqui, M.K.: Use of ultra wide band real-time location system on construction jobsites: feasibility study and deployment alternatives. Int. J. Environ. Res. Publ. Health **17**(7), 2219 (2020)
8. Jiang, X., Wang, S., Wang, J., et al.: A decision method for construction safety risk management based on ontology and improved CBR: example of a subway project. Int. J. Environ. Res. Publ. Health **17**(11), 3928 (2020)
9. Zhang, M., Cao, T., Zhao, X.: Applying sensor-based technology to improve construction safety management. Sensors (Basel) **17**(8), 1841 (2017)
10. Zhu, Z., Yuan, J., Shao, Q., et al.: Developing key safety management factors for construction projects in China: a resilience perspective. Int. J. Environ. Res. Publ. Health **17**(17), 6167 (2020)
11. Liu, Z., Zhang, A., Wang, W.: A framework for an indoor safety management system based on digital twin. Sensors (Basel) **20**(20), 5771 (2020)

Evaluation Index of Algorithm Performance in Building Energy Saving Optimization Design

Yinghuan Liu[✉] and Chungui Zhou

Software Engineering Institute of Guangzhou, Guangzhou, Guangdong, China

Abstract. With the rapid development of the logistics industry, the data and information of logistics are also growing geometrically. However, the current information system is usually unable to deeply analyze these huge data and information, so as to accurately understand the law of the development of the industry and the direction that should be guided. Therefore, it is necessary to establish a new information system. The traditional database management information system can not make good use of and analyze the large amount of data accumulated in the database. Data mining and data warehouse technology can solve this problem. Data mining is a technology dedicated to data analysis and understanding, revealing the knowledge contained in the data. It is one of the important goals of the application of information technology in the future. It can explore and analyze a large number of data, mining meaningful rules, in order to provide appropriate reference and suggestions for future decision-making. In recent years, data mining technology has been widely studied and applied, and has been successfully applied in the fields of Commerce, finance, medical treatment and so on.

Keywords: Data mining · Logistics transportation · Multimodal transportation · Genetic algorithm · Dynamic programming

1 Introduction

With the rapid development of information technology, enterprises can obtain information more quickly. Customers' requirements for products tend to be complicated, diversified and personalized. The commodity market begins to shift from the seller's market to the buyer's market. In order to obtain and maintain competitiveness, enterprises must constantly shorten the time of product development and research, improve product quality, reduce product cost and shorten delivery cycle. The logistics industry, which originated from the logistics service for the war and is known as the "third profit source", has been paid more and more attention by the public. How to better tap this part of the profits and improve the level of domestic logistics has become the focus of all aspects of research [1].

The professional research on Logistics in China started late, and the development of logistics industry has only been concerned in recent years. China Transport Association pointed out that China's high freight transport costs need to be attached great importance. China's freight transport costs are three times that of western developed

J. Abawajy et al. (Eds.): ATCI 2021, LNDECT 81, pp. 213–217, 2021.
https://doi.org/10.1007/978-3-030-79197-1_28

countries, and logistics costs account for up to 30% of the total cost of goods. In addition, the diversity of customer needs and personalized requirements also provide logistics transportation enterprises with high-level logistics services of multi frequency, small quantity and timely delivery. As well as the fierce competition in the logistics industry, logistics transportation enterprises are required to provide differentiated logistics services at appropriate low cost. In order to meet the needs of customers and gain competitive advantage in the fierce competition, logistics transportation enterprises must constantly introduce various modern information technologies and establish their own logistics transportation system, so as to improve logistics efficiency, reduce logistics cost and improve high service quality.

2 Introduction of Dynamic Programming

Dynamic programming is suitable for solving such a problem: a big problem can be divided into several stages, each stage forms a sub problem, each stage is interrelated, each stage has to make a decision, and the decision of a certain stage often affects the decision of the next stage, thus affecting the overall decision. In the process of solving dynamic programming, the optimal strategy as the whole process has the following properties: no matter what the past state and decision are, for the state formed by the decision, the remaining decisions must constitute the optimal strategy, which is the optimization principle of dynamic programming. By using this principle, the solution process of multi-stage decision-making can be regarded as a continuous recursive process, which can be calculated step by step from the back to the front. In solving, the state and decision in front of each state is only equivalent to its initial condition for the subproblem behind it, and does not affect the optimal strategy of the subsequent process.

3 Application of Data Mining Algorithm in Logistics Transportation System

3.1 The Meaning and Types of Logistics Distribution

Distribution is an important part of modern logistics. It is a comprehensive product of modern market economic system, modern science and technology and modern logistics thought. It is essentially different from the well-known "delivery". Modern enterprises generally recognize that distribution is an important part of business activities, it can create higher benefits for enterprises, and it is an important means for enterprises to enhance their competitiveness [2].

The distribution center is a circulation enterprise specialized in goods distribution, with a large scale of operation. Its facilities and technological structure are specially designed and set up according to the characteristics and requirements of distribution activities. Therefore, it has a high degree of specialization and modernization, complete facilities and equipment, and strong ability of goods distribution. It can not only deliver goods from a long distance, but also deliver goods of various varieties, It can not only distribute raw materials of industrial enterprises, but also undertake supplementary goods distribution to wholesalers. This kind of distribution is the main form of goods distribution in industrial developed countries, and it is the future development of distribution [3].

3.2   Advantages of Genetic Algorithm

Genetic algorithm has obvious advantages for other algorithms, such as saving algorithm:

① The search process is applied to the encoded string, not directly to the specific variables of the optimization problem. The random transformation rules are used in the search, not the certain rules. It uses heuristic search instead of blind exhaustive search, so it has higher search efficiency.

② Genetic algorithm starts to search from a group of initial points, rather than from a single initial point. What's more, it gives a group of optimal solutions instead of an optimal solution, which can give designers more choices. It can search fully in the solution space and has the ability of global optimization.

③ Genetic algorithm has strong parallelism, which can improve the speed of computation through parallel computing, so it is more suitable for the optimization of large-scale complex problems.

3.3   Application of Dynamic Programming in Logistics Transportation System

According to the dynamic programming theory mentioned above, for the multimodal transportation problem with n cities and M transportation modes to choose between each city pair, there are m transportation modes in total, and the dynamic programming method is used to find the optimal solution [4].

Starting from a certain city, the total cost can be expressed as follows:

$$P_{n-1}(k, L) = t_{n-1}(k, L) + qC_{n-1}(L) \tag{1}$$

K is the mode of transportation into the city, L is the mode of transportation out of the city, and l is the unit cost of the mode.

$$P_{n-1}(k, m^*) = \min\{P_{n-1}(k, L)\} \tag{2}$$

From one city to another, choose the best way according to the following equation:

$$P_i(k, L) = t_i(k, L) + qC_{i,i+1}(L) + P_{i+1}(L, m^*) \tag{3}$$

$$P_i(k, r^*) = \min\{P_i(k, L)\} \tag{4}$$

There are n cities on the same route, and each city has m kinds of transportation modes to choose from. The dynamic programming method needs m (n − 1) calculations in total.

4 System Optimization

The optimization of the system mainly includes two goals, one is to reduce costs, the other is to reduce inventory. For finished goods logistics, the general optimization is divided into two kinds: the first kind of logistics network optimization, the second kind of inventory optimization.

The objective of inventory optimization is to determine the reasonable inventory quantity and distribution of each warehouse, and determine the required area of each warehouse. The main objective is to reduce the inventory.

The basic logic of logistics network optimization is integer programming. From n production places to m demand points of products, how many warehouses are needed and the location of warehouses. At this time, the system using linear programming can well meet this point. The daily standard function of linear programming is the lowest logistics cost, and the main constraint is the demand response time of each demand point, that is, the demand response time in the service level, and how long can the order be delivered. Optimization results: determine the number of warehouses, the location of each warehouse, the area covered by each warehouse. Determining the coverage area of each warehouse is the key result. For example, enterprises have warehouses in Shanghai and Guangzhou. For example, in the market of Northern Fujian, should the goods be delivered from Shanghai or Guangzhou? Network planning can solve such problems through detailed calculation.

The cost target of network planning refers to transportation cost (most enterprises now outsource all logistics, only transportation cost and storage cost). Transportation cost accounts for 80–85%, and storage cost accounts for a low proportion. Storage cost is affected by unit storage price and inventory, but not by network quantity and location. So network optimization can take transportation cost as optimization objective function.

For example, according to the optimized parameters, the enterprise only needs 20% of the original inventory to meet the demand. On the one hand, many enterprises are faced with the problem of large inventory, and the inventory is not well controlled, so they can find a lot of optimization space; on the other hand, there is a big gap between ideal and reality, so it is difficult to estimate the optimized parameters in enterprises. For most enterprises, the sales staff are more afraid of out of stock than inventory, The impact of out of stock on performance is greater than that of inventory. Therefore, the huge space found in the optimization results often can not be fully realized, and half of the realization is very good. At this time, the economic significance of data mining technology is obvious.

The logistics transportation system designed in this paper takes modern information technology and logistics technology as the background, takes improving the service quality of distribution and reducing the cost of distribution as the core, focuses on the rationalization of the distribution process, the construction of the business process of distribution enterprises and the system control strategy, combines with the scientific problems in the logistics distribution system planning, and uses the logistics distribution system planning method. Through the external and internal environment of logistics distribution center operation, using data warehouse and data mining technology, through scientific planning, it is an effective way to achieve the best balance between the economy and competitiveness of distribution activities.

5   Conclusion

With the development of information technology and social economy, the world economy is increasingly characterized by globalization, networking, information and knowledge. Enterprises produce a lot of important data and information every day, but only a small part of them will be used in the relevant business analysis. Most enterprises are in the state of "data surplus, information shortage". In order to help logistics distribution enterprises adjust their operation policies, optimize their business processes, and improve their management level and competitiveness with the change of market demand, the author applies data warehouse and data mining technology to logistics transportation system, and combines genetic algorithm with specific application to solve the problem of distribution route optimization, which can help enterprises optimize resources, Make assistant decision for the development of enterprise.

Acknowledgements. Logistics system control, school level research team, number: ST202003.

References

1. Kang, X.: Data Mining Technology Based on Data Warehouse, pp. 13–15. China Machine Press, Beijing (2004)
2. Li, M.: Basic Theory and Application of Genetic Algorithm. Science Press, Beijing (2003)
3. Wang, R.I.S., Xu, B., et al.: Applied Dynamic Programming, 1st edn. National Defense University Press (1987)
4. Jin, S.: Model and algorithm of urban vehicle distribution route design. Comput. Eng. Appl. **22** (2002)

Ant Colony Algorithm in Tourism Route Planning

Haizhi Yu[✉], Jing Wang, and Yuhua Jiang

Hulunbeir College, Hulunbeir, Inner Mongolia, China

Abstract. With the rapid development of China's national economy and the substantial improvement of people's living standards, more and more people will choose tourism, which is beneficial to their physical and mental health, and actively participate in it, enjoying the psychological and physical relaxation brought by a pleasant trip. However, China has a large population, inconvenient transportation, limited travel time for office workers, and the phenomenon that people are tired when they don't visit all the scenic spots according to the plan often occurs. To a certain extent, people's psychology and body can't be relaxed. As a result, people's enthusiasm for outbound tourism is obviously not high, so the development of domestic tourism is relatively backward. At present, in order to achieve rapid development of domestic tourism, it is urgent to study the planning and design of tourism routes.

Keywords: Route planning · VRP · Ant colony algorithm · VRP database

1 Introduction

As one of the important driving forces of global economic development, the tourism industry not only accelerates the international capital flow and the popularization of information technology management, but also further creates an efficient behavior consumption mode and demand value orientation, more and more people will choose tourism, which is beneficial to their physical and mental health, and actively participate in it, enjoying the psychological and physical relaxation brought by a pleasant trip. However, due to China's large population, the transportation is inconvenient, the travel time of office workers is limited, and the phenomenon that people are tired after not visiting the scenic spots according to the plan often occurs. To a certain extent, people's psychology and body are not relaxed, which leads to people's low enthusiasm for traveling. Therefore, the development of domestic tourism is relatively backward. At present, in order to achieve rapid development of domestic tourism, it is urgent to study the planning and design of tourism routes [1].

Tourism route planning is not only a small part of tourism planning and design, but also the core of travel agency management. Whether in China or other countries, there are few achievements on tourism route planning and design, and high-level research is rare. People want to understand some basic characteristics of tourism activities at a higher level, only through the formed and developing tourism system and the standardized and described discipline research results.

© The Author(s), under exclusive license to Springer Nature Switzerland AG 2021
J. Abawajy et al. (Eds.): ATCI 2021, LNDECT 81, pp. 218–222, 2021.
https://doi.org/10.1007/978-3-030-79197-1_29

2   Research on Related Theories of VRP

2.1   General Description of VRP Problem

VRP was first proposed by two experts, G. Dantzig and J. ramser, in 1959. Since then, it has been highly valued by experts of related disciplines. Their research subjects mainly involve combinatorial mathematics, operations research, computer application, graph theory and other disciplines [2]. These experts have done a lot of theoretical research on vehicle routing problem, and made rigorous analysis on relevant experimental data.

This problem is usually described as: in a graph G composed of vertices and edges, namely G = (V, E), where the parameters in the set of vertices are generally expressed as:

$$V = \{V_i | i = 0, 1, 2 \cdots, n - 1\} \tag{1}$$

The set of edges is represented as:

$$E = \{(v_i, v_j) | v_i, v_j \in V, i \neq j\} \tag{2}$$

However, these customers are also a collection.

2.2   Basic Model of VRP Problem

The mathematical model of vehicle routing problem (VRP) is generally expressed as:
Objective function:

$$\min Z = \sum_{i=0}^{n-1} \sum_{j=0}^{n-1} \sum_{k=1}^{m} C_{ij} X_{ijk} \tag{3}$$

For large-scale VRP problems, only intelligent algorithms with strong search ability can be used to solve them. Although genetic algorithm has strong global search ability and fast speed, it is easy to cause early convergence. For the shortcomings of genetic algorithm, ant colony algorithm can make up for them with its flexibility and positive feedback mechanism to solve large-scale RP problems, Therefore, ant colony algorithm is deeply loved by scholars at home and abroad, and has broad application prospects in solving VRP problems. Although ant colony algorithm in solving problems, but as long as the reasonable selection of parameters and the improvement of the algorithm.

3   Research on Ant Colony Algorithm

The inspiration of ant colony algorithm comes from the collective foraging behavior of ant colony in nature. The algorithm is an artificial intelligence algorithm, which is designed by simulating ant foraging behavior. When ants go out for food, they will leave a kind of special substance on the path they pass, and ants communicate with each other through this kind of special substance, and judge their own direction by perceiving its concentration [3]. The schematic diagram of building network weighting is shown in Fig. 1. The block network partition diagram after clustering is shown in Fig. 2.

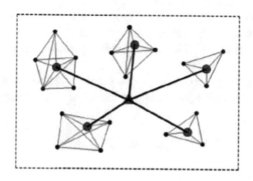

Fig. 1. Diagram of network weighting

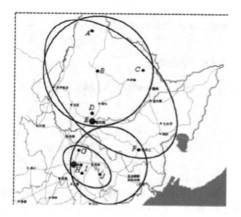

Fig. 2. Block network partition diagram after clustering

3.1   The Principle of Basic Ant Colony Algorithm

which is researched and designed by researchers according to the observation of ant's food seeking behavior. In the process of searching for food source, ant colony determines its crawling direction by perceiving the concentration of a substance called pheromone. When ant passes a certain path, it will leave pheromone on that path. This substance will spread in the air at a certain rate. However, ants communicate with each other through this substance. When ants judge which direction to go, they always try their best to choose the path with more pheromones. From the characteristics of ants, we can see that the collective activity of ants is a process proportional to the amount of information, that is to say, its regulation process is a positive feedback process.

At the beginning, the amount of pheromone is usually a fixed value:

$$\tau_{ij}(0) = C \tag{4}$$

Where C is the constant we set.

3.2   Parameter Setting of Basic Ant Colony Algorithm

Ant colony algorithm is a kind of dynamic algorithm. Different parameter combinations will lead to different results. Therefore, when using, choosing a reasonable combination of accuracy of the solution. However, at present, the ant colony algorithm has not formed a complete theoretical model in parameter setting, that is, there is no unified conclusion in parameter setting. Only through continuous experiments can we finally determine the parameter combination in line with the actual problem. In this way, the ant colony algorithm lacks strict theoretical support in parameter setting.

3.3   Summary of This Chapter

This chapter describes the basics of the initial algorithm, describes the main components in the VP and some of its model model model model patterns in the prosecution, the element and the clement and the behavior of the architecture, It also bears the general meaning of the knowledge of the VRMP, the degrees RMPs have some degrees of ambition at the top of the team's algorithmetic algorithm, parameters of the mathematics of the mathematics, which describes a quantities of Satpossessed with various degrees. Captain, the test results in this chapter are an opportunity to solve the prejudice of ants [4].

4   Conclusion

Nowadays, with the increasing pace of life, tourism has become a good way for most office workers to relax. However, due to the limitation of their time and the number of times they go out, this paper designs a set of personal travel routes for specific personal time arrangement, To make the travel enthusiasts can get the best travel experience in the shortest time and cost. This paper reviews a large number of domestic and foreign information about the design and optimization of tourism routes, combines theory with practice, and summarizes the current situation and future development trend of China's tourism route design. In the planning and design of tourist routes, we should focus on the long-term development of tourism industry and tourism enterprises, and improve the design level of tourist routes to a higher level.

Tourism route planning is a very broad concept, which mainly includes the analysis and prediction of the planned regional resources and market customers, the accurate positioning of its regional planning, the development trend and the necessary development strategy for the planning. At the same time, it is also necessary to know the characteristics and contents of the products and the future development direction of the products, It can accurately put forward the development area of key projects, focus on innovation, characteristics and personality, find the development methods and principles of relevant elements, and outline the ideal blueprint from the macro level of regional tourism development planning.

Acknowledgements. (1) Transverse Project: Study on the Individual Spell Group Model of Travel Agency.

(2) Inner Mongolia Autonomous Region Project of the 13th five year plan of Education Science in 2019: Research on the Construction of Teaching Quality Evaluation System of School-Enterprise Collaborative Education Based on Model SERVQUAL [NGJGH2019003].

(3) Transverse Project: Study on the Optimization of Travel Agency Salary System.

References

1. Jiang, D., Yang, X., Zhou, X.: Genetic algorithm for vehicle routing problem. Syst. Eng. Theory Pract. **1** (2011)
2. Xiao, P · Partheno genetic algorithm for vehicle routing problem. Comput. Technol. Autom. (2010)
3. Cai, T., Qian, J., Sun, Y.: Simulated annealing algorithm for multiple transportation scheduling problem. Syst. Eng. Theory Pract. **18**(10), 11–15 (2011)
4. Wang, X., Li, Y.: Research on multi depot and multi model loading and unloading hybrid vehicle routing problem. Control Decis. **24**(12), 1769–1774 (2009)

Human Error Analysis of Civil Aviation Air Traffic Control Based on Modeling and Prediction Algorithm

Zheng Wei$^{(\boxtimes)}$

Xi'an Air Traffic Control Center, Northwest Area Air Traffic Administration, Xi'an Civil Aviation, Yanliang, Shanxi Province, China
123321981@Sohu.Com

Abstract. As a high-tech industry, civil aviation industry should focus on technology, but accident reports show that at least 3/4 of the accidents are caused by human error. Therefore, in ATM system, human error is a potential weak link. We must take measures to minimize human error and its impact, and optimize the quality of human operation for error detection and recovery. The ultimate goal of the research on human errors in air traffic control is to take protective measures and means against human unsafe behaviors, so the identification and analysis of human errors in air traffic control is very important.

Keywords: Air traffic control · Human error · Knowledge representation framework · Ontology · Rule reasoning

1 Introduction

Human error refers to any human behavior or inaction that has potential or actual negative impact on the system. FAA and Eurocontrol put forward hera-janus human error analysis method after analyzing the theoretical background, concept coverage, analysis method and reliability of HFACS and Hera. Janus starts from the external manifestation of human errors, analyzes the psychological causes of human errors from the outside to the inside, and combines with the environmental factors when human errors occur, forming a set of systematic and scientific human error classification analysis method [1].

The task of air traffic management is to effectively maintain air traffic safety, maintain air traffic order and ensure smooth air traffic. It includes air traffic service, air traffic flow management and airspace management. As the main part of air traffic management, air traffic service includes air traffic control service, flight information service and warning service. As the provider of air traffic control service, the task of controller is to prevent the collision between aircraft and aircraft, prevent the collision between aircraft and obstacles, and maintain and accelerate the orderly flow of air traffic. Based on the particularity of the controller's task, any error may lead to disastrous consequences.

J. Abawajy et al. (Eds.): ATCI 2021, LNDECT 81, pp. 223–227, 2021.
https://doi.org/10.1007/978-3-030-79197-1_30

2 Human Error Model and Knowledge Representation Framework of Air Traffic Control

Air traffic control human error has the characteristics of wide range of knowledge, many factors to be considered and complex analysis process. In view of this situation, this paper explores the effective analysis method of ATC human error, analyzes the existing ATC human error model, and finds that hera-janus and hera-smart model analysis process is more detailed, and the analysis results are more refined, so the above two human error analysis models are deeply studied. On this basis, this paper attempts to introduce the ontology method to solve the problems of unclear concept definition, weak concept logical connection and complicated analysis process in the process of ATC human error analysis by constructing the ontology knowledge base of ATC human error. This paper introduces the domain knowledge representation framework of ATC human error in detail, which provides the basis of knowledge representation for the construction of ATC human error knowledge base.

2.1 Definition of ATC Human Error

In ICAO doc9859AN/474, human error is defined as the negative effect of the system caused by the action or omission of the operator. Air traffic control system is a complex man-machine system, which is used to ensure air traffic safety, maintain air traffic order and promote air traffic fast and smooth. Therefore, human error in air traffic control can be defined as: due to the behavior or inaction of the controller, the negative effect of the air traffic control system occurs, which affects the air traffic safety, is not conducive to maintaining air traffic order and hinders the smooth flow of air traffic [2].

2.2 Human Error Analysis Model of Air Traffic Control

Human error in air traffic control is an interdisciplinary scientific field, involving aviation psychology, aviation ergonomics, aviation medicine, anthropometry and other disciplines. When analyzing the human error of air traffic control, it is necessary to consider not only various factors, but also the influence of the interaction of various factors on the analysis results.

Calculate the angle of satellite horizontal and near point at the observation time:

$$M_k = M_0 + nt_k \tag{1}$$

Generally, the iterative method is used to calculate E:

$$E_{k+1} = M_k + e \sin E_k \tag{2}$$

Substitute the satellite ephemeris into the following formula:

$$\Phi_k = v_k + \omega \tag{3}$$

There are many factors that lead to human error unsafe events in air traffic control, including controller personal factors, organizational factors, training and training factors, airspace structure and flow control factors, team resource management factors, working environment factors, control hardware/software factors, etc. Therefore, it is not feasible to analyze human error by analyzing the role of various factors in unsafe human error events and the interaction between them.

3 Construction of Ontology Knowledge Base of ATC Human Error

Aiming at the problems of imperfect knowledge expression and weak logical connection expression in the process of ATC human error analysis, it is necessary to build a formal conceptual model to organize ATC human error domain knowledge. By referring to hera-janus model, this paper extracts the key concepts in the field of ATC human error and analyzes the logical relationship between them. According to the characteristics of ATC human error analysis, it constructs the ontology of ATC human error. By analyzing the additional relations between concepts, the inference rules of ATC human error are constructed. Combined with human error ontology and human error reasoning rules of air traffic control, the ontology knowledge base of human error of air traffic control is constructed, which provides knowledge base support for semi-automatic analysis of human error of air traffic control.

3.1 Overview of ATC Human Error Ontology Knowledge Base

The knowledge base aims at the needs of analyzing the problems in the field. It uses the way of knowledge representation to store the theoretical knowledge, expert experience, factual basis and other knowledge in the field into the computing platform for management, and uses the knowledge base to perform reasoning, inquiry and other operations [3]. Although the theoretical research on human error in air traffic control is relatively mature, due to the complexity of the research on human error in air traffic control, the research on human error in air traffic control is very complicated, Scholars do not use a unified term in their research, so it is difficult to understand each other, which greatly increases the difficulty of knowledge sharing and reuse. Ontology knowledge base can organize language well, express language clearly and accurately, and be recognized by computer, which is conducive to the storage, sharing and reasoning of domain knowledge. The construction of ATC human error ontology knowledge base includes two parts: the construction of ATC human error ontology and the construction of ATC human error reasoning rules.

3.2 Construction of Human Error Ontology in Air Traffic Control

Ontology structure is defined as o: (C, R, a, t, I), in which class (c) represents the basic concept of domain knowledge; relationship (R) represents the relational attributes between classes; attribute (a) represents the data attributes of class; dataset (T) represents the different data types of attributes; instance (R) represents the individuals contained in class ∞. According to the above definition, the ontology concept will be constructed, the relationship between concepts and the characteristics of the relationship will be determined, the concept data attributes will be determined, and finally the ontology instance will be constructed.

3.3   Ontology Conceptual Modeling of ATC Human Error

Ontology class is an accurate and formal description of domain concept. Based on hera-janus model, this paper uses the top-down construction method to determine four top-level classes: "Janus human error", "work task", "information and equipment" and "risk scenario factors", It is used to describe the types of human errors, the tasks of air traffic controllers when unsafe events occur, the information of air traffic controllers dealing with errors, the equipment involved and the risk scenario factors affecting air traffic controllers. However, the constructed ontology is mainly used for failure protection analysis, and a class needs to be constructed to record the failure protection description, so that the recorded description information can be used for human error type reasoning.

Ontology failure protection description class is used to record the failure of system protection obtained from hera-smart model analysis, so that the information can be used for human error type reasoning. Therefore, the failure protection description class should include all the concepts related to the system failure protection description. Through the analysis of failure protection of multiple unsafe events, 11 related concepts are summarized, including aircraft, control equipment, control information, control instructions, etc. [4].

Ontology task class is used to record the tasks that controllers are performing when human errors occur. Therefore, it must include all kinds of tasks of controllers, Considering that radio communication and command, control room communication, coordination and process list tasks can be divided into more detailed types of tasks, the above four categories are further subdivided.

By analyzing the role of the concepts in the human error ontology of air traffic control in human error analysis, it is found that the object relationship between ontology concepts can be divided into two types. One is the relationship between controllers and human error types, work tasks, information and equipment, and dangerous scene factors: controllers make certain human errors, controllers are engaged in certain work tasks, Controllers deal with some information and some equipment wrongly, and they are affected by some dangerous scene factors. The other is the relationship between the controller and other subclasses in the failure protection description class: the controller fails to make the control plan, the controller encounters dangerous situations, the controller commands the aircraft, the controller fails to communicate with the pilot, and the controller fails to send control instructions. In addition, there is also the relationship between the pilot flying the aircraft and the failure of the pilot to receive/execute the control command.

4   Conclusion

Human error is an important risk source in the operation of air traffic control. It is necessary to study the human error accidents of air traffic control, investigate the causes of the accidents in detail, study the potential factors behind the accidents, and then explore the preventive measures. The existing human error analysis models of air traffic control generally have the problems of imperfect knowledge expression and weak logical connection expression, which need to be supported by semi-automatic analysis tools through corresponding means. To solve this problem, based on hera-janus model, the ontology knowledge base of ATC human error is constructed, which provides a complete domain knowledge expression of ATC human error.

References

1. Lu, L.L., Zhou, Y,, Zhou, M.: Classification and causes of human errors in air traffic control team. Chin. J. saf. Sci. **19**(1), 64–70 (2009)
2. Wang, J., Yang, H.: Ontology modeling analysis for human factors in ATC safety system. J. Civil Aviat. Univ. China **27**(4), 3336 (2009)
3. Yuan, L., Chen, Y.: Several important models of human error in air traffic control. Air Traff. Manag. **1**, 35–37 (2010)
4. Sun, R., Zhao, Q.: Research on ecar model. Chin. J. Saf. Sci. **22**(2), 17–22 (2012)

Testing Based on Edge Agent of Internet of Things

Jianye Huan[1], Yuanliang Fang[1], Aiyi Cao[2], Hengwang Liu[2(✉)],
and Bingqian Liu[1]

[1] Electric Power Research Institute of State Grid Fujian Electric Power Co., Ltd.,
Fuzhou 350007, China
[2] Anhui Jiyuan Inspection And Testing Technology Co., Ltd., Hefei 230088,
Anhui, China

Abstract. With the continuous development of information technology, the equipment on the edge of the power is also constantly updated. As the edge equipment, the power station area fusion terminal has not only improved its intelligence, Applications based on converged terminals can be customized according to different power business needs, Realizing the collection and processing of various information on the edge side not only greatly saves personnel investment, but also improves the processing efficiency of data on the edge side, making business operations more efficient. This paper explores and researches the testing methods of micro-applications of edge IoT agents, aiming to improve the accuracy of data collection and processing of micro-applications at the edge, and to ensure the stability and reliability of daily business operations.

Keywords: Edge-side applications · Intelligent fusion terminals in the platform area · Micro-application testing

1 Introduction

With the rapid development of information technology, traditional cloud computing cannot meet the needs of the rapid development of the Internet of Things, social networks, and intelligent Internet networks. Relying on cloud computing alone is not enough to achieve the goals of high reliability and ultra-low latency in computing and communication in 5G. Most resources have limited computing, communication, and storage resources, and must rely on cloud or edge devices to enhance their functions [1]. Therefore, a new trend is emerging in the computing field, that is, the function of the cloud is more and more toward the edge of the network, the use of mobile edge computing (MEC) to supplement cloud computing, mobile devices migrate tasks to network edge nodes, both It can make up for the shortcomings of long delay in mobile cloud computing, and can meet the needs of the expansion of the computing power of terminal equipment to meet the requirements of mobile applications with high energy efficiency, low delay and high reliability.

While edge-side applications are being implemented and promoted by the State Grid Corporation of China, the reliability of data collected and processed by edge-side applications cannot be verified in time, and the supervision methods are relatively

© The Author(s), under exclusive license to Springer Nature Switzerland AG 2021
J. Abawajy et al. (Eds.): ATCI 2021, LNDECT 81, pp. 228–233, 2021.
https://doi.org/10.1007/978-3-030-79197-1_31

backward. Therefore, research on edge-side micro-application testing technology to improve micro-application business collection The reliability of processing and processing provides support and guarantee for the construction and implementation of the Smart IoT system of the State Grid Corporation of China.

2 The Necessity of Research on Edge-Side Micro-application Testing Technology

2.1 Increasing Demand for Micro-application Testing

Edge-side applications have received extensive attention from all walks of life, and have blossomed in many application scenarios, but there are still many problems that need to be studied in actual applications of edge-side applications. If different levels of edge servers have different computing capabilities, how to solve load distribution problems based on performance issues such as demand, latency, bandwidth, and energy consumption; How to solve the problem of service communication responsibility and deployment dependency under the distributed architecture; how to solve the interoperability problem between different edge-side applications.If edge-side applications are to be successfully implemented and promoted in enterprises, in addition to quoting the framework system of edge-side applications, all supporting technologies for edge-side applications will be improved [2]. Edge-side application detection technology is an effective way to verify the quality of edge-side applications. Real-time application of edge-side application detection technology can detect defects in the research and development process in time, improve product and service quality, optimize product performance, and reduce production failures. Therefore, with the prevalence of edge computing frameworks, enterprises have put on the agenda to improve the edge-side application system and study edge-side application detection technology.

2.2 The Testing of Edge-Side Applications is More Complicated, and the Research of Automated Testing Technology has Become the Focus

Smart converged terminals are emerging terminal devices, and the testing of their micro-applications is quite different from the testing of traditional application software. Whether it is the interface protocol or the internal container technology, the testing technology and method have a greater test. On the other hand, smart terminal equipment, as the Internet of Things proxy equipment on the edge of the power, has a large number. Relying only on manual testing not only reduces efficiency, but also hinders the large-scale implementation of terminals. Therefore, research on the testing technology and methods of edge-side micro-applications greatly guarantees the stability and reliability of the power edge-side equipment.

3   Overall Architecture of Edge-Side Applications

3.1   Overall Architecture of Edge-Side Applications

Edge-side micro-applications are the edge terminals in the three major systems of cloud, edge, and terminal. They are mainly edge IoT agent equipment. Its composition mainly includes: basic platform, container, data management module, data bus module and edge-side micro-application module, as shown in Fig. 1:

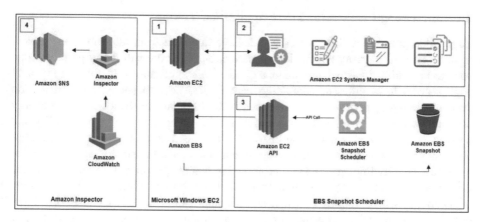

Fig. 1. Overall architecture diagram of edge equipment

3.2   Design Principles and Classification of Edge-Side Applications

3.2.1   Design Principles for Edge Applications

The overall framework design of edge-side micro-applications follows three basic principles:

Micro-applications interact based on the message mechanism to avoid private communication, realize data interaction decoupling, and reduce the complexity of interaction management;

Centralized data management, avoiding the establishment of private databases by micro-applications, ensuring data security performance, and improving data usage efficiency;

The naming, functions, and reserved interfaces of micro-applications are clear to ensure effective management of micro-applications.

3.3   Classification of Edge Applications

From the perspective of micro-application design and management, according to micro-application usage scenarios, it is divided into the following types:

Basic management: used for the management of the terminal itself. For example, data center, UART, LTE, carrier agent, master station agent, such micro-applications are installed by default in smart terminals.

Collection and monitoring: used for equipment information collection and monitoring. For example, communication collection, external remote signal collection, leakage protection collection monitoring, etc.

Application analysis: used to support the company's user services, operation and maintenance management and other businesses [3, 4]. For example, low-voltage topology dynamic recognition, low-voltage fault location, orderly charging of electric vehicles, etc.

4 Test Method and Content of Edge Side Application

4.1 Test Requirements for Edge-Side Applications

Edge-side applications are functional applications for important data collection and management at the edge of power. The accuracy and reliability of its data are the prerequisite and guarantee for the stable development of power business. Therefore, in-depth research and mastery of the testing requirements of edge-side micro-applications are necessary prerequisites for testing. The testing requirements for micro-applications can be summarized as follows:

(1) Verify whether the data collection and management functions of the edge-side micro-applications are correct and reliable.
(2) Verify the correctness and security of micro-application message transmission.
(3) Verification of the interaction capabilities of micro-applications, including: interaction with the cloud and interaction capabilities between micro-applications.
(4) Verify whether the performance of the micro application is stable.
(5) Verify whether the micro-application protocol is reliable.
(6) Verify the compatibility of micro-applications.

4.2 Micro-application Test Module Architecture

(1) The interaction between the edge-side application and the data center and the frame center is to forward information through the MQTT Broker forwarder. The subscription request and sending request of the edge-side application are forwarded to the data center through the MQTT Broker, and then the data center passes through the response information MQTT Broker is sent to the edge side application.

In order to implement performance tests such as concurrency of edge-side applications, the cloud performance platform will mobilize multiple presses to form a cluster system. Each press contains a number of execution containers, which simulate a certain number of virtual users, and jointly generate pressure on the system under test [5, 6]. During the test, each basic test execution container will generate a large amount of performance data, such as the response time of each server request, return code, and so on. Summarize the data of all execution units to get the overall performance measurement of the test.

4.3   Test Process

The tester initiates a test request through the web interface, and is processed by the test management module. The test task is assigned to a suitable set of test machine nodes, and the load test of the system under test is initiated. The metrics are counted in real time from the original test data, together with system monitoring. The data is displayed together in the test dashboard. After the test runs, the resources of the test machine are automatically cleaned up and recycled.

4.4   Testing Method

This article has an in-depth understanding of the edge-side application testing of the State Grid Corporation of China, analyzes the status quo of the edge-side application testing, combines the characteristics of the edge-side structure and the technical requirements and management requirements of the edge-side testing, and proposes a detailed analysis and explanation of the target range of the test. Efficient and orderly program basis [7, 8]. According to the characteristics of the edge-side application architecture, research the edge-side application test technology in the edge-side environment, refine the overall requirements and current situation of the edge-side application test, and form the edge-side application detection method to provide a reference for the edge-side application test [9, 10].

(1) Edge-side application function test: realize the test of the interaction function between edge-side micro-applications and between edge-side micro-applications and the cloud; realize the operation data collection and message verification of edge-side micro-applications; realize the edge-side APP Automated testing of functional interfaces.

(2) Edge-side application performance test: test the response time of micro-applications, concurrent user requests, and the performance test of the MQTT Broker integrated terminal;

(3) Compatibility test of edge-side applications: Test the compatibility between micro-applications, the compatibility of containers, and the compatibility with terminals.

(4) Edge-side application interaction protocol test: Test the agreement consistency and reliability between micro-applications.

5   Summary

At present, the most important thing in the construction of the State Grid Electric Power Internet of Things is the construction of side-end equipment and smart terminals. A large number of electric energy information collection, monitoring, control, metering and other equipment on the power grid side and the power consumption side will usher in comprehensive technological innovation and hardware replacement upgrades. The stability and reliability of the side-end equipment integrating many functions and integration will become power Guarantee of business implementation. This article aims to explore the testing methods and technologies for micro-applications of side-end equipment, improve the reliability of power services, further improve the testing level of side-end equipment, and ensure the quality of online equipment.

Funding. Supported by Science and Technology Project of State Grid Fujian Electric Power Co., Ltd. (Research on Key Technologies and Application Expansion of Smart IoT, 52130419002M).

References

1. Wang, H., Jiao, H., Chen, J., Guo, Y.: Research on the edge division method and application of distribution network based on stable connection. Electric Pow. Inf. Commun. Technol. **2020**(01) (2020)
2. Matt, Y.: Edge computing ushered in the golden 10 years. Comput. Netw. **2020**(04) (2020)
3. Ding, Z.: The application of edge computing in the field of smart manufacturing. Comput. Knowl. Technol. **2020**(15) (2020)
4. Zhong, Y., Wang, H., Lin, Z., Cui, Z., Huang, J.: 5G-based edge computing gateway and its application in the power grid. Autom. Appl. **2020**(06) (2020)
5. Wu, D.: Edge computing empowers smart cities: opportunities and challenges. Internet Econ. **2020**(06) (2020)
6. Shi, F., Liang, S., Jiang, Y., Yan, L., Xu, J., Guo, X.: Research and practice of edge computing in oilfield internet of things system. Comput. Eng. Appl. **2020**(16) (2020)
7. Mu, J., Chai, Y., Song, P., Bi, L., Wu, D.: Development status of edge computing and research on standard system. Inf. Commun. Technol. **2020**(04) (2020)
8. Zhang, H., Fang, B., Guo, J.: Application of edge computing in new smart city. Inf. Commun. Technol. **2020**(04) (2020)
9. Arno, M.F., Robert, B.: From edge computing, fog computing to cloud computing. Mod. Manuf. **2019**(07) (2019)
10. Cao, Z.: Edge computing: framework and security. Confidential Sci. Technol. **2019**(09) (2019)

Data Mining Algorithm in Engineering Quantity Information of Architectural Design Stage

Haiyue Tang[(⊠)]

Beijing City University, Beijing 100083, China

Abstract. This paper analyzes the application advantages of data mining technology in the construction engineering design stage, combined with the key points of engineering quantity audit in the construction engineering design stage, including the parts with relatively large engineering quantity, the parts with high engineering cost, and the construction links that are prone to make mistakes, etc. The purpose is to improve the construction quality of the construction engineering design stage, and then promote the stable development of the industry economy.

Keywords: Data mining technology · Project cost · Construction link · Audit method

1 Introduction

The life cycle of construction engineering is becoming longer at this stage. The main reason for the current situation is that the scale of construction projects is also expanding. The increase of the amount of work in the design stage of construction engineering. It makes people tend to apply new design technology. The traditional audit methods include comprehensive audit method, standard budget audit method, comparative audit method, screening audit method, index audit method, key audit method and so on. Considering the current construction scale, many engineering projects will choose the index audit method as the audit method in the engineering design stage, and θ m (building information model) is a construction method that takes the relevant data of the engineering project as the model parameters and simulates the overall structure of the building with the help of digital technology [1]. It is of great significance to improve the effectiveness of audit results and the utilization rate of construction project resources.

2 Application of Data Mining AdaBoost Algorithm in Architectural Design Quantity Information

In order to enhance the information collection efficiency of the method operation, increase the strength of EB attack information identification, constantly change the data collection frequency, and check the strength of the signal sent out by the element

© The Author(s), under exclusive license to Springer Nature Switzerland AG 2021
J. Abawajy et al. (Eds.): ATCI 2021, LNDECT 81, pp. 234–238, 2021.
https://doi.org/10.1007/978-3-030-79197-1_32

software, find the corresponding signal location according to the signal characteristics, at the same time, assist the attack information query function, accurately find the attack information number required by the system, and store the signal content, according to the collection organization rules, the basic signal collection operation is realized, the data information identification status is adjusted continuously, the ADA boost algorithm is entered into the malware system identification, and the identification formula is set as shown in Eq. (1):

$$K = \int (a+p)dxdy - Q \tag{1}$$

The setting is shown in Eq. (2):

$$P = \frac{S-T}{N} \tag{2}$$

Set the corresponding judgment equation as shown in Eq. (3):

$$J = \sqrt{w+l} \cdot \frac{C-V}{d} \tag{3}$$

3 Application Advantages of Data Mining Technology in Construction Engineering Design Stage

3.1 Information Integrity

The expansion of construction scale increases the total amount of daily data of construction projects. If such data can not be effectively sorted out, it is bound to increase the probability of problems in the construction process of construction projects [2]. The application of daily M technology can effectively avoid the occurrence of such situations, and the technology also has the following application advantages in practical application) information integrity in the process of construction engineering, a large number of data information will be generated every day, including a large number of data information need to be collected in the design stage of construction engineering, such as geological condition information of construction area, hydrogeological condition information, climate data information, etc. The integrity of these data information will directly affect the effectiveness of architectural engineering design. BM technology can collect this kind of information completely, provide accurate information data for square sinus design, and improve the quality of design scheme. Data relevancy in construction engineering, all data information have mutual relevancy. Data mining technology can effectively sort out the relationship and confirm the relationship between all the information, so as to improve the order of the construction process and speed up the subsequent construction efficiency of the construction project.

3.2   Data Consistency

Design error is an inevitable factor. How to reduce design error to the lowest level is what all enterprises need to do. In the process of statistical data, BM technology adopts a unified data measurement unit to process all data parameters, which can effectively reduce the calculation error caused by human factors, and further improve the scientificity of project design scheme. 4) simulation of building related parameters designed in the traditional design stage, it needs to use experience or hand-made model to evaluate the application value of the design scheme, and the time cost is high. BM technology can build equal scale 3D model with the help of corresponding design software. According to the relevant test results of 3D model, we can more intuitively understand the performance indicators of building structure and improve the feasibility of design scheme [3]. After the above recognition operation, classify the identified data, divide the collected signal data according to different classification attributes, standardize the management of data and storage space, select target recognition template, and identify and match with relevant information database. Set the data recognition matching graph, as shown in Fig. 1.

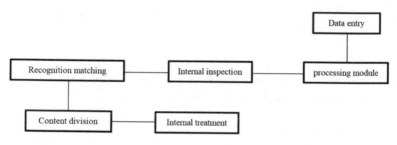

Fig. 1. Data recognition matching diagram

4   Key Points of Quantity Audit in Construction Engineering Design Stage

4.1   The Part with Large Quantities

The large part of the project has always been one of the key points of the project audit, because the large part of the project belongs to the core structure, and its quality will directly affect the overall construction effect of the building. Before the audit of engineering quantity, technical personnel need to understand the construction nature of the construction project. Generally speaking, the construction engineering quantity of the building with the main structure is relatively large. For example, for buildings with actinide reinforced concrete as the main structure, the largest part of the workload is the concrete pouring process. For buildings with brick and stone structure as the main structure, the quantities related to brick and wood are higher than those of other structures. These are the key parts to be audited, including the parameters of materials, material ratio, etc.

4.2   The High Cost Part of the Project

The construction project can be divided into several construction links, each construction link can be called a separate construction project, maintaining the relative independence, but there is a progressive or parallel relationship between each other, and in these decomposed projects, the project with high project cost also belongs to the very important audit content. In construction engineering, the construction cost of foundation engineering, reinforced concrete engineering, metal structure fabrication and installation engineering in steel structure, advanced decoration engineering and other construction engineering links is high. When applying index design method to project audit, it is necessary to conduct a detailed review to find out the problems in project design in time and improve the design value of construction engineering project.

4.3   Error Prone Construction Links

For the parts that are easy to make mistakes, for example, in reinforced concrete engineering, whether the quantities of beam to column and beam to slab joint parts are repeatedly calculated, whether the statistics of the number of components are correct, and whether the concrete foundation beam is applied with belt pile cap; When calculating the work load of wall body in masonry engineering, the quantities of reinforced concrete columns, beams and lintels embedded in the wall body are calculated separately for the inner and outer walls [4]. When calculating the earthwork of foundation trench, according to the regulations, the length of the inner wall is calculated according to the net length of the inner wall. If it is calculated according to the axis, the quantities are more calculated. In earthwork, site leveling and surplus soil transportation are generally not included in the engineering drawings It means that it is easy to miss calculation in the preparation of project budget and settlement.

5   Research on Audit Method of Construction Engineering Quantity in Design Stage Based on Data Mining

Drawing process drawing process is a very basic link in the construction stage of construction engineering, but drawing is also an important guarantee to ensure the orderly development of Construction Engineering in the formal construction process. With the participation of data mining technology, engineering quantity audit needs to focus on the following aspects at this stage.

(1) In the process of drawing, technicians need to review the structural types of construction projects, the total amount of building materials used, the basic size of materials and other contents to determine whether they are in line with the relevant national standards, so as to improve the effectiveness of the application process of construction projects

(2) After the completion of data collection, technicians also need to verify the accuracy of the collected data. They can calculate all the information parameters with the help of expert system, big data technology and other information technology, and at the same time, they need to determine the correlation between the data

(3) In the audit process, once the wrong data is found, it should be modified in time. However, considering the correlation between data information, BM technology needs to calculate the impact of data changes, mark the changed data information, pay attention to the accuracy of the changed data, and reduce the error rate in the drawing process, So as to improve the design quality of the design scheme.

6 Epilogue

To sum up, the increase of the total amount of construction projects increases the work difficulty in the process of daily management, and the application of data mining technology can complete the collection of data information, and can effectively sort out the data information, so as to improve the rationality of the audit results of construction projects in the design stage. In addition, the application of BM technology can provide necessary data support for the optimization of engineering quantity audit method. Ensure the smooth implementation of the construction link, shorten the construction period of the construction project.

References

1. Lin, S., Lin, C.: Several key points to be controlled in the preparation of bidding documents for EPC. Bidding Procurement Manag. **11**, 23–26 (2020)
2. Meng, Y., Wu, N., Zhu, C., Kong, L.: Application priority of BIM technology in landscape architecture design stage. J. Beihua Univ. (Nat. Sci. Ed.) **21**(06), 822–829 (2020)
3. Wang, W.: Exploration and prospect of EPC management mode. China Surv. Des, **10**, 40–49 (2020)
4. Feng, J., Wang, J.: Research on two-stage bidding method for architectural engineering design. Arch. Econ. **41**(S1), 274–277 (2020)

Research and Semiconductor Production Equipment Operation Management System

Hairong Wang[✉]

College of Optoelectronic Engineering, Yunnan Open University,
Kunming 650223, Yunnan Province, China

Abstract. In semiconductor manufacturing enterprises, production equipment is the material basis for semiconductor enterprises to carry out productive business activities. With the rapid development of semiconductor technology in the past 30 years, the industry competition is also increasing rapidly. Therefore, how to manage and use these semiconductor production equipment and improve the management level of production equipment is of great significance to improve the strength of semiconductor production enterprises in the industry competition and promote the progress and development of enterprises. With the continuous development and progress of semiconductor industry, higher requirements are put forward for the research and development of equipment management system. The equipment management system is not only the management of equipment itself, such as equipment asset management, equipment spare parts management, etc., but also needs to be connected with the production and operation of equipment. It is necessary to consider how the management system can improve the utilization rate of equipment in the production and operation process, reduce the occurrence of defective rate, and increase enterprise benefits. And these have gradually become enterprises in the current fierce competition in the semiconductor industry.

Keywords: Semiconductor production · Log file · PCS · HVM · Equipment maintenance

1 Introduction

Equipment management is an important part of enterprise management. Long term stable and trouble free operation of equipment is the key factor of enterprise economic benefit assessment [1]. It can ensure the production plan to be carried out on schedule and respond to the market demand of customers in time. At the same time, it can effectively reduce the production cost of enterprises, such as direct costs (raw materials, water, electricity, etc.), Management costs (capital interest, labor, contract liquidated damages, accounts receivable collection, etc.) can be effectively controlled or reduced. Moreover, it can effectively improve the qualified rate of products and reduce the probability of defective products, and ultimately improve the level of economic benefits of enterprises.

Among the productive enterprises, production equipment is the most important means of labor for enterprises to carry out production and operation activities, and it is

© The Author(s), under exclusive license to Springer Nature Switzerland AG 2021
J. Abawajy et al. (Eds.): ATCI 2021, LNDECT 81, pp. 239–243, 2021.
https://doi.org/10.1007/978-3-030-79197-1_33

also the material basis for enterprises to carry out modern production. Enterprise's production capacity and production scale are closely related to the production equipment owned by enterprises. The advanced nature of production equipment and its stability are reliable guarantees for the market competitiveness of enterprises. The scientific and effective management of production equipment and the full play of the production capacity of the equipment are of great significance to the improvement of the comprehensive competitiveness of enterprises and the stable and sustainable development of enterprises. The structure of modern semiconductor packaging and testing equipment is becoming more and more complex, the precision is constantly improving, and its automation degree is also improving day by day, followed by the rising cost of equipment, the price is high, so how to pursue the most economical cost and the highest comprehensive efficiency of the equipment in the life cycle of semiconductor equipment is particularly important and realistic in the production management of semiconductor enterprises.

2 Overview of Maintenance and Evaluation Methods for Semiconductor Production Equipment

2.1 Characteristics of Semiconductor Production Equipment

With the continuous development of "Moore's law", semiconductor packaging and testing equipment is constantly updated. At the same time, due to the processing accuracy, production capacity, etc., the equipment has high precision, high utilization rate, strict environmental requirements, high equipment precision and complexity, and so on, followed by high equipment prices. Specifically, it is reflected in:

Fast update: the development of semiconductor devices is changing with each passing day, and the semiconductor production equipment is also bringing forth new ones. A new generation of semiconductor production equipment, from R&D to product production, usually takes no more than 10 years [2]. The actual production cycle is about 5 to 7 years.

High price: semiconductor production and inspection industry is a typical capital and technology intensive enterprise. The equipment price is often calculated in tens of millions of dollars. A medium-sized semiconductor packaging and testing factory needs several hundred million dollars of fixed assets investment.

High precision: the design precision of semiconductor equipment is usually in micron level, and some devices even require nanometer level, such as wafer processing equipment.

High load operation: due to the large market demand of semiconductor devices, the production equipment is basically running at high load or full load all day long, and is in continuous operation throughout the year. In addition to the maintenance time, the utilization rate of the equipment can reach more than 80%.

2.2 MA (Machine Availability) Usability Evaluation System

Machine availability, abbreviated as Ma, is the ratio of available time of equipment to total working time. Ma is an important parameter to reflect the working capacity of equipment, and is also a general index 1 for industrial enterprises. The higher the Ma, the higher the utilization rate of the equipment in unit time. Conversely, the lower the Ma, the lower the availability of the equipment in unit time. Ma is also widely used in many fields of production equipment performance indicators. The calculation period of Ma is generally one week (7 days). Equation 1 is the formula definition of Ma.

$$MA = \frac{168 - SDT - USDT}{168} \times 100\% \tag{1}$$

The working time is 168 h, which refers to the whole time of a week (7×24), which is determined by the characteristics of all-weather production of semiconductor enterprises. For semiconductor production equipment, the working time is generally in hours.

This formula reflects the proportion of working time of a single equipment in a week, and also reflects the time of equipment failure and repair. To a certain extent, it is also an index to assess the professional skills of equipment maintenance technicians.

2.3 Equipment a Compliance Performance Index Evaluation System

A confidence is another equipment performance index based on Ma. It is defined as the ratio of the number of shifts (SHT) reaching the Ma target value and the total number of shifts in the unit time (usually one week as a unit time). Equation 2 is the definition of A confidence.

$$A\ Confidence = \frac{Meet\ \ MA\ \ Goal\ \ Shift\ \ Quantity}{Total\ \ Shift\ \ Quantity} \times 100\% \tag{2}$$

A confidence is an indicator used to measure the availability of machinery and equipment. Its purpose is to reduce the impact of regular or irregular downtime of machinery and equipment on production scheduling. It is different from MA in that Ma will be affected by scheduled downtime (usdt) and unscheduled downtime (SDT). If only Ma indicators are taken into account, sometimes Ma fails to achieve the target because of unreasonable regular maintenance arrangement rather than real equipment failure. In order to achieve the goal of a confidence, both usdt and SDT should be small. This requires the equipment maintenance personnel to arrange the periodic maintenance as short as possible, while not affecting the quality of periodic maintenance. Therefore, if we allow the scheduled and unscheduled downtime of the same time and the shortest time every day, it is the most ideal state. It is not difficult to see from the formula that the higher a confidence is, the better it is. The highest is 100%, and the lowest is 0%. The higher the a-confidence is, the more favorable it is for the production planner to arrange production. The higher the a-confidence is, the better the production cycle can be reduced.

3 Requirement Analysis and Design of Semiconductor Production Equipment Operation Management System

In large-scale production enterprises, there are thousands of equipment. Especially for the semiconductor production equipment, because the product needs to go through the test link of the later stage to judge whether it is good or bad. If the equipment fault in the front process can not be detected in time, when the product flows to the later stage test, there are a large number of products or semi-finished products that need to be scrapped in the production line. This will cause serious losses to the economic benefits of enterprises. Even affect the normal delivery, damage the corporate image [3]. It can be seen that timely and effective detection of production equipment problems to prevent product hidden dangers is the focus of the problem to be solved by the system.

For some equipment problems, the same failure may occur repeatedly for a long time due to design defects or other reasons. For equipment technicians, one equipment technician may need to manage several or even dozens of equipment and hundreds of machines. If the maintenance records of each kind of equipment are effectively saved, when the equipment encounters problems, some key words of failure mode can be input into the system to find the historical solutions and methods of such problems. Equipment technicians can take these historical data as reference to get the fastest solution to this problem. This will greatly shorten the maintenance time of technicians, and win more economic benefits for enterprises. At the same time, it can avoid the loss of rabbit technicians and the impact of enterprise production.

Another problem to be considered is how to quickly solve the current equipment failure and resume production if the current equipment failure cannot be solved by technicians. This requires that the maintenance management system be upgraded. The system should encourage the equipment technicians to take the initiative to solve the equipment failure, so as to avoid some cases of not actively repairing the machine due to personal reasons. At the same time, some problems that the technical personnel can not solve can be raised to the equipment experts or engineers timely and effectively, so as to quickly restore the equipment production.

Demand analysis of data volume: the running database of the foreground system can store at least one quarter of data. When the amount of data exceeds one quarter, the system needs to back up the data files in the running database to the server's special backup drive letter. When necessary, these files can be put back into the running database, which not only ensures the traceability of data, but also improves the speed of system operation. This paper only introduces the implementation of data backup in detail. The log file database of the foreground system running database is saved by one folder. The purpose of this is to facilitate the system to analyze the alarm situation every day. As shown in Figure 3-3, the log file in the system running database [4]. For more than a quarter of the log files need to be backed up to the specified server drive letter, and saved to a folder named logarams, and the file name is backed up with the start time of logarams plus log files, so that it is convenient to quickly find the required log file data in the traceability process in the future.

4   Conclusion

In today's fierce market competition, the level of equipment management has become one of the key factors for the success of competition between enterprises. As an important part of the daily management of enterprises, equipment management not only directly affects whether the production and operation of enterprises can be carried out smoothly, but also relates to whether enterprises can obtain economic benefits and long-term development in production and operation activities. In this paper, through the research and collection of domestic and foreign semiconductor industry equipment management theory, combined with the characteristics of semiconductor production equipment, established the system design objectives. According to these requirements, through prototype design, functional module design, program flow design and database design, combined with computer software design related knowledge, the design and implementation of semiconductor production equipment operation management system is finally completed.

References

1. Li, W.: Semiconductor equipment development and application and project management research. Sci. Technol. Commun. 5(1), 25–28 (2012)
2. Li, B.: Review and recommendation of equipment spare parts inventory structure and spare parts management conception. Equip. Manag. Maintenance 06, 6–8 (2004)
3. The role of equipment management in enterprise development. China Text. 09(15), 6–7 (2001)
4. Jing, F.: TPM Management and Practice. Northeast University Press, Heilongjiang (2003)

Genetic Algorithm with Reliability Evaluation Matrix Algorithm for Electrical Main Connection

Shimin Wang[✉]

Shandong Vocational College of Science and Technology,
Weifang 261053, China

Abstract. In order to get a practical reliability evaluation software for main electrical connection, the algorithm must be universal for different main connection types and efficient for calculation and analysis. The minimal path set/cut set algorithm based on adjacency matrix is a new algorithm for the evaluation of electrical main wiring, which is suitable for the reliability evaluation of various main connection types. In this paper, the fast computing technology of adjacency matrix algorithm is deeply studied. Through the sparse processing of minimal path set/cut set matrix, dimension reduction of minimal path set matrix and fast search of minimum cut set, the calculation speed of the algorithm is effectively improved. The calculation efficiency is 6 times higher than that of the conventional matrix algorithm, which can meet the requirement of the rapidity of the evaluation software for the main electrical wiring.

Keywords: Main electrical connection reliability evaluation · Adjacency matrix algorithm · Accelerated calculation technology

1 Introduce

It has been about 30 years of research on the reliability of the main electrical system, and the important research on the reliability of the main power system has been achieved for about 30 years. However, in order to apply the research results to practical projects, the algorithm must meet two requirements: the universality of various main connection lines and the high efficiency of calculation and analysis. The minimum path set/cut set algorithm based on the adjacency terminal matrix uses the adjacency matrix and the terminal matrix to describe the main wiring system, without any special setting for the network structure; at the same time, the matrix represents a multi port network model [1]. It can adapt to the multi input and multi output conditions of the actual system, and solve the problem of universality of the algorithm. Modern hydropower projects are becoming more and more large. Taking the Three Gorges Hydropower Station as an example, it has installed 27 units, 15 AC outgoing lines and 2 DC outgoing lines. In order to develop a practical evaluation software for main electrical wiring, the algorithm must be efficient. On the basis of the literature, this paper studies the accelerated computing technology of adjacency matrix algorithmx.

J. Abawajy et al. (Eds.): ATCI 2021, LNDECT 81, pp. 244–248, 2021.
https://doi.org/10.1007/978-3-030-79197-1_34

2   Basic Algorithm of Main Wiring Evaluation

The minimal path set/cut set algorithm is an algorithm based on network connectivity. Its key is how to find the path from the outgoing line to the source point (generator) and further obtain the corresponding cut set events, so as to enumerate the fault states of the system, analyze the fault consequences and obtain the corresponding reliability index results. The flow chart of the core algorithm minimum cut set method for fault search is shown in Fig. 1.

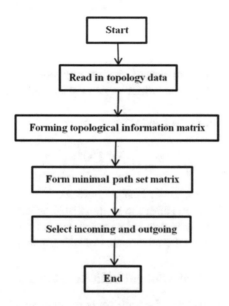

Fig. 1. Algorithm flow of minimum cut set

   After getting the minimum cut set information matrix of the system, the possible fault states and consequences of each cut set event are analyzed one by one, such as permanent fault overlapping outage, permanent fault overlapping outage and planned maintenance overlapping outage, and the corresponding outage time, failure rate and load loss are calculated respectively, and finally the system reliability index is accumulated.

3   Computing Acceleration Technology

It is found that the calculation time is mainly consumed in the matrix multiplication operation of searching the minimal path set and the column vector logical addition operation of finding the minimum cut set from the minimum path set matrix. The amount of calculation is closely related to the dimension of the minimal path set matrix and the number of components in the system. Aiming at these two calculation modules, the accelerated computing technology is studied.

3.1   Sparse Processing of Minimal Path Set Matrix

The application of sparse matrix technology in power system is deeply studied. The difference between sparse matrix and conventional matrix lies in its "zero row storage" [2]. The most commonly used is retrieve cells by row (column) storage format and triangle.

Based on the sparse storage format, we can carry out "zeroing operation" which can effectively reduce the number of cycles of the program and greatly improve the operation efficiency. If the non-zero element in the k-th row of the adjacency matrix is m $m_k (<n)$ and the non-zero element of the j-th column of the end-point matrix is m, $m_j (<n)$, then the number of cycles of the zeroing operation is $m_k \times m_j$ when the elements (k, j) are calculated once, which is obviously less than the cycle times n^2 of the conventional matrix.

If the sparsity is defined as a matrix:

$$p \frac{\tau}{m \times n} \times 100\% \tag{1}$$

The p value of the actual power system network topology matrix is very small, usually less than 50%. Therefore, after the sparse matrix is used, the "zero storage" and "zeroing operation" can greatly improve the operation efficiency. Moreover, the smaller the sparsity, the more obvious the advantages of using sparse matrix.

3.2   Dimension Reduction of Minimal Path Set Matrix

In addition, the number of paths of the path set is not changed. In order to speed up the searching process of minimal cut set, the array dimension of path set moment should be reduced as much as possible. It is found that if the following vectors are the same in the element path set matrix. The effective method of dimension reduction is to merge the same columns in the path set matrix, and finally get a new path set matrix with different column vectors, so as to minimize the column dimension. In this process, two merging principles based on different topological characteristics are included.

3.3   An Accelerated Algorithm for Searching Minimum Cut Sets

The minimum cut set obtained from the minimal path set matrix is realized by logical addition of column vectors. According to the conventional method, for $m \times n$-dimensional path set matrix, searching all the second-order cut sets needs $n(n-1)/2$ cycles, and searching all third-order cut sets needs $n(n-1)(n-2)/3$ cycles. Obviously, there are only a few times in these cycles that can get an effective minimum cut set result. In order to speed up the search process, it is necessary to reduce the number of invalid addition cycles. In addition, the number of rows of the matrix is less than that of the original path set matrix P, and the computation required to determine whether the column vector is a vector is also smaller. Taken together, the efficiency will be greatly improved.

4   Case Analysis

Aiming at the 3/2 connection scheme of a hydropower project, two different methods, the minimal path set/minimum cut set method of conventional matrix and the adjacent terminal point matrix algorithm based on sparse matrix technology, are used for calculation and analysis. The results of the two algorithms are the same and comparable.

There are 152 components in the scheme, including 9 units, 9 step-up transformers, 9 generator circuit breakers, 6 overhead lines, and 23 500 kV intervals, which are divided into 38 sections [3, 4]. Take "HCI" outlet point as an example.

The conventional minimal path set/minimum cut set method always analyzes the network structure of a single output point. When multiple output points need to be analyzed, the similar searching process of minimal path set/minimum cut set must be completely repeated. The algorithm used in this paper is to search the path set matrix of the whole network at one time, and then take the sub path set matrix corresponding to each input and output point to analyze the corresponding minimum cut set, which reduces the repeated route set search process.

The efficiency improvement of this algorithm compared with the traditional algorithm is closely related to the specific electrical wiring mode. Different main wiring schemes have different minimum path length (number of components), different minimum paths associated with each element, sparsity of minimum path matrix and efficiency of corresponding algorithm.

The ratio a reflects the degree of the network structure of the system. If the correlation is close, the dimension of the sub path set matrix of each output point is large, and the efficiency of the algorithm is not improved; if the correlation is loose, the dimension of the sub path set matrix of each output point is small, and the efficiency of the algorithm is greatly improved. But in general, the new algorithm can improve the computational efficiency to a certain extent (the computational efficiency of the example system increases from 6 to 60 times). To sum up, the main factor affecting the efficiency of the algorithm is still the system size (reflected by the total number of components), and the sparsity of the road set matrix is about low, so the proportion of improving the efficiency of the algorithm in this paper is larger.

5   Conclusion

In this paper, the fast computing technology of the adjacency matrix algorithm used in the reliability evaluation of the main electrical connection is deeply studied. Through the sparse processing of the minimal path set/cut set matrix, the dimension reduction of the minimal path set matrix and the fast search of the minimum cut set, the calculation speed of the algorithm is effectively improved. The calculation efficiency is increased by at least 6 times, which meets the requirements of the practical software for the evaluation of main electrical wiring.

References

1. Yongji, P.I.: Principle of Reliability Engineering. Tsinghua University Press, Beijing (2002)
2. Boming, Z., Shousun, C.: Analysis of Higher Power Network. Tsinghua University Press, Beijing (1996)
3. Yongji, G., Xiang, Z.: Reliability Evaluation Software for Main Electrical Connection of Power Plant Substation and Its Application in Longtan Hydropower Station: Project Research Report. Department of electrical machinery, Tsinghua University, Beijing (2002)
4. Zongxiang, L., Yongji, G.: Reliability evaluation of main electrical connection of hydropower station. Power Syst. Autom. **25**(18), 16–19 (2001)

Neural Network Control for Water Injection Circulation

Xin Chen[✉]

Daqing Oilfield Co., Ltd., Second Oil Production Plant, Daqing 163414,
Heilongjiang, China

Abstract. With the continuous progress of oilfield development technology, numerical simulation technology and results have been applied to the development dynamic analysis work, but there is no application precedent in finding inefficient and ineffective water injection circulation field. Therefore, this paper focuses on the application of numerical simulation of movable oil saturation, movable water saturation diagram and single well single layer history fitting results to judge inefficient and ineffective water injection circulation field, which provides the basis for dynamic analysis and development adjustment.

Keyword: Numerical simulation of active water saturation in invalid circulation field

1 Introduction

In the process of water injection development, long-term water injection scour will form a large pore channel for the rapid flow of injected water, and then the injected water will be extracted from the oil production well at a very fast speed, resulting in a large single-layer water injection in the water injection well and a rapid increase in water cut in the oil production well, which will form an inefficient and ineffective water injection circulation field. In the late stage of ultra-high water cut development, reducing production cost and increasing recoverable reserves are important means of reservoir fine management [1]. The traditional method of judging inefficient and ineffective water injection circulation field is to combine all kinds of static and static data to analyze and judge, but the numerical simulation can accurately reflect the mining situation in each period of reservoir. The application of numerical simulation results can guide the development and adjustment measures and the formulation of schemes. Make judgment, prediction and decision more scientific and reasonable.

2 Conventional Method to Identify Inefficient and Ineffective Water Injection Cycle

Through comprehensive utilization of dynamic and static data, combined with dynamic management experience, it is concluded that inefficient and ineffective water injection circulation field has the following characteristics:

J. Abawajy et al. (Eds.): ATCI 2021, LNDECT 81, pp. 249–253, 2021.
https://doi.org/10.1007/978-3-030-79197-1_35

Appear in well-injected sand bodies;

The sand body has good connectivity and high permeability; There are high-yield liquid and high-aquifer in oil production wells, and the water absorption between water injection wells varies greatly. The liquid production and water cut in well formation are increasing rapidly; By analyzing the production condition of oil production well, for the wells with fast rising water content and high liquid production, according to the annulus test data in recent years, if there is a layer with high liquid production ratio and high water content in a single well, the test data such as carbon-oxygen ratio energy logging and wall coring are used to verify the medium and high water-flooding layer.

3   Using Numerical Simulation to Discriminate the Inefficient and Ineffective Water Injection Cycle

The fluid in the pores of the reservoir can be divided into four parts: bound water, movable water, movable oil and residual oil. In the late stage of high water cut development, it is very important to accurately determine the residual oil-active oil and effectively improve the utilization rate of injected water-active water. The numerical simulation results can provide us with the dynamic water saturation map and the dynamic oil saturation map in each stage of the reservoir, which can accurately reflect the washing condition of the reservoir and the distribution of the remaining oil. As an example, the method of judging inefficient and ineffective water injection circulation field by numerical simulation results is analyzed.

The accuracy of the historical fitting results of the west block geological model of Nanliu area after permeability correction is high [2]. Compared with the closed coring well, the coincidence rate of flooding grade is higher, reaching 72.1, in which the coincidence rate of thin layer is 68.8. Therefore, it is feasible to use numerical simulation results for dynamic analysis.

3.1   Range of Movable Oil Saturation and Movable Water Saturation Determined by Phase Permeability Curve

According to the coring data of well 6-4-28 in well Nanliu 6, in high permeability reservoir, the irreducible water saturation is low, generally 15–35%, with an average of 25%, the remaining oil saturation is 24–37%, with an average of 29%, and the oil-water two-phase co permeability area is 35–55%, with an average of 49%, so the maximum saturation range of movable oil and water in reservoir pores is 0–55%; in medium low permeability thin layer, the irreducible water saturation is high, generally 25–42%, with an average of 49% The remaining oil saturation is 21–36%, with an average of 31%. The oil-water two-phase co permeability area is 30–44%, with an average of 39%. The maximum saturation range of movable oil and water in reservoir pores is 0–44. Details are shown in Table 1 below.

Table 1. Average saturations of oil and water in oil and water of all types of oil and water in well No. 4–28

Oil reservoir type	Water saturation (%)	Residual oil saturation (%)	Oil-water two-phase co-permeability range (%)	Water production rate above 90 per cent		
				Water saturation (%)	Dynamic water saturation (%)	Dynamic oil saturation (%)

From the relative permeability curve and coring data, it can be seen that when the water saturation of oil layer reaches more than 45%, the water production rate is above 90%, the reservoir is highly flooded, the movable water saturation is 20–26 and the movable oil saturation is 19–25.

3.2 Application Examples

Through the analysis of the movable water saturation and movable oil saturation diagram of each single sand layer in the west block of Nanliu District, it is preliminarily determined that there may be 23 inefficient and ineffective water injection circulation fields. Through the analysis of production conditions and oil-water well test data,19 inefficient and ineffective water injection circulation fields were determined, and the coincidence rate reached 82.6. In the process of searching for invalid water injection circulation field, for a few circulation fields that have been controlled by water injection, we use numerical simulation to screen the circulating field with high movable water saturation but less than 10% of the whole well water.

In the dynamic oil saturation diagram and the dynamic water saturation diagram, the inefficient and ineffective water injection circulation field has the following characteristics: inefficient and ineffective water injection circulation field is distributed in well area; Dynamic water saturation of main line between 2.2.2 oil wells is generally higher than 20%, and dynamic oil saturation is lower than 25%.

the invalid circulation field is mainly developed in the inner front-edge single sand layer, the branch-tuo transitional single sand layer, the outer front-edge phase I and II single sand layer, the ratio of the inner and outer front-edge single sand layer is similar, and the ratio of the water injection circulation field in the inner front-edge single sand layer is 52.8%, mostly in the connected channel sand,

In the main sand and the non-main sand, in the non-main reservoir, the ratio of channel sand to the water injection circulation field formed by the main sand is the ratio of the main sand to the water injection circulation field formed by the main sand is 42.9. The ratio of non-main sand to non-main sand is 25.0.

3.3 In the Connected Well Group, the Direction of Inefficient and Ineffective Water Injection Circulation Field is Affected by the Type and Thickness of Each Well Point Drilling

Well 6-2-125 in the south, for example, is a primary infill injection well for the extraction of grape flower and Gaotai oil layer, which opens 37 sedimentary units, opens 31.2 sandstone thickness m, and opens 11.4 m. effective thickness 100 m as at

December 2003 368 m per day3. It can be seen from the map of movable water saturation of Group II that the movable water saturation of the four perforating layers in the connected area with the perforating layer of the corresponding oil production well has reached more than 20%, and the corresponding movable oil saturation is less than 20%, so it can be judged that there may be inefficient and ineffective water injection circulation field in the perforation layer of the well.

The a layer of Pu II 8 m, sandstone 0.6 m, of main sand and sandstone with effective 0.5 m Effective 0.2 m of non-main sand, There are three open wells connected around [3]. The south 6-Ding3–122 well m 270 from the injection well is sandstone 1.0 m, Effective 0.5 m main mat sand, The south 6-20-627 well,165 m from the injection well, is sandstone m,1.0 Effective 0.4 m of non-substrate sand, The other well 6-20-626, which is 171 m from the injection well, is a second class off-balance sheet layer of sandstone 0.4 m.

From the dynamic water saturation field, The water saturation of the injection well is 41.8%, Dynamic oil saturation 0.24, The dynamic water saturation of well N6-D3-122 located on main sand is 24.2%, Dynamic oil saturation 19.2%, The dynamic water saturation of oil wells on non-main sand is 28.0%, Dynamic oil saturation 14.0%, The active water saturation of oil wells on the off-balance sheet is 14.4%, The movable oil saturation is 27.6. From the current oil saturation field, The oil saturation of South 6-Ding 3-122 well is 52%, Water saturation 48%, According to the law of phase seepage curve, the water content of this well is over 90%, High aquifer; Besides, According to the results of single layer liquid production fitting, Well S6-D3-122, this layer currently contains 96.5% water, The fluid production accounts for 13.2% of the whole well, Is the main production layer of the well; Well 6-20-627 in South China, 98.5% water, According to the above data, we believe that there are inefficient and ineffective water injection circulation fields between wells 6-2-125,6-Ding 3-122 and 6-20-627.

We verify with dynamic data. On the basis of the water absorption profile data of the well,13 water absorption profiles were tested in the a layer of Portuguese II 8, and 7 water absorption was shown ~ 12 and 26 m, respectively. In 2002, the water absorption ratio was 28.7%, so it is judged that there is an inefficient and ineffective water injection circulation field between the South 6-2-125 well and the surrounding three oil production wells [4]. Water absorption profile, water-finding data, carbon-oxygen ratio energy logging data, well-wall coring data prove that there are 4 perforating layers in this group.

4   Conclusion

It can be seen from the dynamic water saturation diagram of Gaotai sub-layer that the movable water saturation of Gaotai sub-layer section of the well perforation is less than 15% in the area between oil and water wells. According to the water absorption profile of South 6-2-125 well in recent years, it is shown that only the Portuguese II group absorbs water, and the Gaotai sublayer does not absorb water.

For this reason, we have carried out stratified self-regulation of water injection wells, and the distribution of injection volume of Portuguese II1-Portuguese II4 layer is m from 503. Down to 25 m³. From 40 to 30 m, the proportion of injection in the 10th layer of Portuguese II 5-Porto II3 and the daily injection was reduced by 22 m3. The daily fluid output decreased by 28 t, the daily oil output increased by 2 t, and the water content decreased by 1.7 percentage points. It is proved that it is feasible to use numerical simulation technology to analyze inefficient and ineffective water injection circulation field.

References

1. Yuchun, W.: Application of Relative Permeability Curve in the Study of Water Drive Law of Low Permeability Reservoir. Heilongjiang Petroleum Society Reservoir Engineering Association (2001)
2. Lau, E.: Special Topic of Reservoir Numerical Simulation Method. Petroleum University Press, Dongying (2001)
3. Yun, L.: Reservoir Simulation. Petroleum University Press, Dongying (1999)
4. Dakuang, H., Qinlei, C., Cunzhang, Y.: Basis of Reservoir Numerical Simulation. Petroleum Industry Press, Beijing (1993)

Study on the Role of Computer Simulation Software for Polymer Flooding in EOR

Cong Fu[✉]

Daqing Oilfield Limited Company, No.4 Oil Production Company,
DaQing 163511, China
fucong@petrochina.com.cn

Abstract. After polymer flooding, about 50% of the crude oil remains in the ground. The further research on EOR after polymer flooding is highly valued by petroleum scientists and technicians. In this paper, physical simulation method is used to evaluate the oil increasing effect of the technical method of further expanding sweep volume and improving oil washing efficiency after polymer flooding. For heterogeneous reservoirs, the contribution of expanding sweep volume to recovery is greater than that of improving oil washing efficiency. Considering the technical and economic effects, zeolite/polymer solution flooding system has more application value. Computer numerical simulation has been widely used in various sectors of society. To a certain extent, theoretical calculation has replaced the actual experimental process. In the initial simulation analysis process, it can effectively reduce the difficulty of work, shorten the research cycle and reduce the cost of capital. In the field of oilfield development, improving oil recovery has always been a hot topic in research. Great progress has been made in tertiary oil recovery by polymer flooding in recent years.

Keywords: After polymer flooding · High concentration polymer · Polymer flooding · Mathematical model

1 Introduction

Polymer flooding technology has the characteristics of high investment risk and high technology cost, so it means a high success rate in practice. Through the design and optimization of computer simulation software, to achieve the formulation and optimization of polymer flooding technology scheme, predict the implementation effect of oil displacement scheme, and carry out economic evaluation, which can greatly guide the practice and implementation of the technology. Therefore, it is of great significance to develop a set of professional software which can simulate the oil recovery process of polymer flooding in order to meet the actual needs of oil production technology. In view of the existing technology progress, the oil displacement mechanism simulation description and simulation equation algorithm improvement were carried out, which laid a theoretical foundation for the realization of perfect polymer simulation oil displacement software [1].

J. Abawajy et al. (Eds.): ATCI 2021, LNDECT 81, pp. 254–258, 2021.
https://doi.org/10.1007/978-3-030-79197-1_36

2 Numerical Simulation of Polymer Flooding

First of all, the corresponding simulation calculation should be carried out to obtain effective and feasible technical methods for polymer flooding in the same line, and the recovery results should be predicted, and the economic feasibility evaluation of the results of analysis and prediction should be carried out, so as to demonstrate the implementation risk, so as to make the research and analysis process of reservoir development management more scientific, and ensure that the development process of polymer flooding can be carried out under high economic benefits [2].

According to the research area, the following aspects need to be optimized: adding multi molecule separate injection development and single molecule profile control and flooding function. According to the new function, the digital analog algorithm is improved and the software module is designed. Combined with the software architecture, the function of POLYGEL model is selected for the following aspects of numerical simulation.

(1) Single molecular polymer flooding simulation includes core flooding process, polymer flooding process and polymer flow mechanism.
(2) Multi molecular polymer flooding simulation includes separate injection in different layers, separate injection in different wells, and multi molecular flooding process.
(3) Adjust the injection and production profile numerical simulation, including core flooding process, strong gel and weak gel flow process, polymer flooding mechanism.

3 Model Equation and Algorithm Improvement

In order to improve the algorithm, the pressure equation needs to be solved. w represents water phase and o represents oil phase. The pressure calculation equation is as follows:

$$\varphi C_t \frac{\partial P}{\partial t} + diV(\mu_W + \mu_o) = Q_W + Q_o \tag{1}$$

The up-stream sorting algorithm is used to iterate the pressure solving equation to calculate the water saturation Sw. According to So = 1−Sw, the oil phase response saturation is obtained.

$$\varphi \frac{\partial S_W}{\partial t} + diV\mu_W = Q_W \tag{2}$$

According to the calculation of plane layered water flooding, 1000 mg/L low molecular weight polymer is injected into the water flooding to 95% water cut to simulate the oil displacement process. The results are shown in Fig. 1. Compared with the actual situation, the simulation model can basically reflect the oil recovery process of water flooding and polymer flooding.

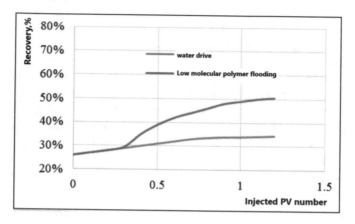

Fig. 1. Simulation results of low molecular polymer flooding

4 Challenges of Polymer Flooding Technology

Polymer flooding has been carried out for nearly 20 years in China's onshore oilfields, and has achieved a lot of theoretical and practical results, which is quite mature technology, which will be of great help to the application of polymer flooding EOR technology in offshore oilfield, and many theoretical achievements and practical experiences and lessons can be used for reference. However, polymer flooding in offshore oilfield has different characteristics from onshore oilfield.

(1) Offshore platforms lack fresh water. The traditional polymer flooding technology uses fresh water to prepare polymer solution, and at the same time produces a lot of sewage. If the waste water is used for injection, the concentration of polymer must be greatly increased. However, it is almost impossible to use fresh water to prepare polymer in offshore polymer flooding, only seawater (salinity 32000 ∼ 35000 mg/L or formation produced sewage) can be considered.

(2) The reservoir conditions are complex. For example, the tertiary oil fields in Bohai Sea have many stratigraphic series, great difference between layers, high permeability, high salinity of formation water and high viscosity of crude oil; the formation environment in Bohai Sea also determines that the salinity of formation water is high, while that of seawater is higher [3].

5 Breakthrough of Polymer Flooding Technology in Offshore Oilfield

The key of polymer flooding EOR technology in offshore oilfield is polymer, which requires polymer to be soluble, highly viscous, salt resistant (calcium and magnesium resistant) and long-term stability; flexible and highly automated injection equipment and process are the guarantee of polymer flooding implementation. Under the condition of meeting the requirements of polymer materials and injection equipment, the field test of polymer flooding was carried out.

5.1 Development and Application of New Salt Resistant Polymer

Through the design of molecular structure, two kinds of new salt resistant polymers, hydrophobically associating polymer and structural composite polymer, were developed for tertiary oil fields in Bohai Sea. Pilot scale and industrial production of these two kinds of products were realized. The basic properties of various products and the properties of their application in Bohai oilfield were systematically studied. Among them, the hydrophobically associating polymer app4 has been applied to polymer flooding test in Suizhong 36-1 oilfield [4].

Hydrophobically associating polymer is the introduction of a small amount of hydrophobic groups into the molecular chain of partially hydrolyzed polyacrylamide. The polymer molecules in solution can form supramolecular structure through the association between hydrophobic groups, which has certain strength but reversible physical action.

5.2 Polymer Injection Equipment Suitable for Offshore Operation

The field test of polymer flooding in domestic offshore oilfields has not been carried out in the past. The successful implementation depends largely on the polymer injection equipment and technology. The polymer injection equipment in onshore oilfield is huge, which is not suitable for offshore platform application, and there are not many reference materials for equipment and process design. For the special requirements of polymer flooding in offshore oilfield, injection equipment is a blank.

According to the conditions of offshore platform, the dispersed dissolving injection equipment for single well and well group polymer injection is designed and manufactured. The equipment has the following features:

(1) Small size, light weight and integration;
(2) High degree of automation;
(3) The anti-corrosion treatment and low shear of valve were taken to keep the viscosity of polymer solution to the maximum extent;
(4) The system is open and interface is reserved. It is convenient to add additives to the system according to the plan.

6   Conclusion

(1) It is a strategic measure to enhance oil recovery by polymer flooding in offshore oilfields in China, which is of great significance to the development of offshore oil.

(2) There are three challenges in the implementation of polymer flooding technology in Offshore Oilfield: lack of fresh water resources, conventional polymer can not meet the requirements; limited offshore platform space, special requirements for polymer injection equipment; difficult adjustment of well pattern, well spacing and formation series.

(3) The field test of polymer flooding in the Eogene oilfield of Bohai Sea has achieved the expected purpose. The solubility and injection capacity of polymer under offshore platform conditions have been verified. The injection equipment and injection process developed by the company have been investigated. The polymer bearing produced liquid has been obtained and analyzed.

References

1. Haifeng, W., et al.: Surfactant research and development direction of ASP flooding in Daqing Oilfield Petroleum. Geol. Recovery **11**(5), 62–64 (2004)
2. Huiyu, Y., et al.: Recognition and training obtained from pilot test of foam compound flooding. Daqing Petroleum Geol. Dev. **20**(2), 108–110 (2001)
3. Yue, Z., et al.: Interfacial properties and oil displacement mechanism of heavy alkylbenzene sulfonates. Acta Physicochemical **21**(2), 161–165 (2005)
4. Jiecheng, C., et al.: Field test of ASP flooding in Daqing oilfield. Daqing Petroleum Geol. Dev. **20**(2), 46–49 (2001)

Data Mining Algorithm for Evaluation Platform of Resistant Polymer Flooding Effect

Xuefeng Li[✉]

Daqing Oilfield Limited Company No. 4 Oil Production Company,
Daqing 163511, China

Abstract. Based on genetic algorithm, on the basis of the performance evaluation of salt resistant polymer such as viscosity increasing, viscosity stability, anti cutting property and viscoelasticity, two salt resistant polymers with the best comprehensive performance were selected. The physical simulation experiment of secondary oil displacement results was carried out with ordinary polymer, and the h2500 salt resistant polymer with the best performance was selected for field test. As of November 2018, the EOR in the test area is 16.7%, and the EOR is 19.6% more than that of conventional polymer flooding under the same injection pore volume multiple. Therefore, the application of new salt resistant polymer is an effective way development effect.

Keywords: Genetic algorithm · Salt resistant polymer · Oil displacement effect · Field test · Development effect

1 Introduction

Nowadays, field exploitation is increasing. With the increasing difficulty of exploitation, many problems are exposed. The increase of water content in oil and gas exploitation addition, the loss of formation energy increases production. In addition, oil and gas wells seriously reduces the development benefit. Therefore, the water plugging effect of temperature and salt resistant field has become increasingly significant [1]. In many cross-linked polymers, polyacrylamide polymer is an effective water plugging polymer, and it is also widely used at present. It is the general name of acrylamide monomer and monomer copolymer containing amide group or other functional groups.

2 Parallel Strategy of Genetic Algorithm

Using the idea of divide and conquer, PGAs (parallel Ge netic algorithms) was first applied to large parallel computers. There are many ways to realize the idea of divide and rule at the same time. At present, there are four kinds of parallel genetic algorithm models.

Implement and is one of the direct parallelization schemes of genetic algorithm. Only one population, population selection, crossover and mutation operations are completed on the master node machine, and the fitness evaluation is completed in the slave node machine. The slave node receives the individual sent by the master node,

J. Abawajy et al. (Eds.): ATCI 2021, LNDECT 81, pp. 259–263, 2021.
https://doi.org/10.1007/978-3-030-79197-1_37

and the master node obtains the individual fitness value calculated by the slave node. This paper designs a reusable master-slave PGAs framework based on mp; applies the master-slave PGAs to the mining algorithm of fuzzy association rules, and the speedup performance ratio is improved by 19.1%. When the model needs a lot of computing fitness work, this parallel scheme can be adopted. However, there are communication delay or bottleneck between master node and slave node, and uneven load, Causes parallelism to fail [2].

The principle of the original speedup is as follows: suppose there are ρ parallel machines, which can form a higher performance parallel running platform. The operation speed (i.e. performance) of a single computing node is 1, and the serial performance of the structure created by P computing nodes is pref (P). The known speedup ratio is the ratio of serial run time to parallel run time:

$$S_p = \frac{T_1}{T_p} \tag{1}$$

This law is mainly applicable to the case of fixed load. For example, in the master-slave model, almost linear acceleration ratio can be obtained; in the island model, when the population size is constant and the number of subgroups is not proportional, the acceleration ratio is the same as that of the master-slave model. At present, this law is seldom used in neighborhood.

3 Oil Effect Evaluation of Salt Polymer

On the basis of previous indoor performance evaluation, two salt resistant polymers lh2500 and GL with relatively best comprehensive performance were selected. Based on the experimental results of common 25 million molecular weight polymer flooding as the basis of evaluation and comparison, the physical simulation experiment of salt resistant polymer flooding effect was carried out.

3.1 Experimental Scheme

The core size is 45 cm × 45 cm × 30 cm. The gas permeability of low and high permeability layers is 300md1200md. The cores were aged in 45 °C incubator for 12 h after oil saturation. First, the water flooding is finished is more than 98% for 2 h continuously; then, 0.5PV and the experiment is finished when the water cut of the produced liquid is more than 98% for 2 h. Chemical flooding technology with partially hydrolyzed polyacrylamide (HPAM) as main agent has become the leading technology of EOR in China. Under the condition of high temperature and high salinity, the viscosity of HPAM solution will decrease greatly due to the loss of hydrolysis, degradation and precipitation. The viscosity of HPAM solution will be improved by increasing the concentration and dosage, which seriously affects the use efficiency of polymer. In recent years, in order to reduce the amount of polymer, domestic and foreign research focuses on the research and development of new temperature and salt resistant polymers. Literature research shows that the main ways to improve the temperature and salt resistance of polymer are to improve the relative molecular weight of polymer and introduce functional monomers with special structure and functional groups on the molecular chain: first, introduce sulfonic acid and other groups with strong λ hydration ability to enhance the solubility of polymer and avoid precipitation and separation caused by high calcium and magnesium ions; The second is to introduce ring structure to increase the steric hindrance, rigidity and hydraulic volume of polymer chain to avoid the entanglement of molecular chain; the third is to introduce long chain alkyl and other hydrophobic groups to increase the steric hindrance of molecular chain and produce intermolecular association effect, which greatly increases the viscosity; the fourth is to introduce λ zwitterion, which shows the salt like characteristics of anti polyelectrolyte in a certain range with the increase of salt concentration.

3.2 Oil Displacement Experiment was Carried Out After Polymer Preparation

From Table 1 that the enhanced recovery rate of lh2500 flooding is 26.7% and 282%, respectively, which is higher than 2.5%–4% of the common polymer with 25 million molecular weight. The experimental results are shown in Table 1.

Table 1. Data table of oil displacement experiment results

Name of polymer	Polymer concentration	Effective permeability	Water drive recovery rate%	Enhanced recovery value of polymer flooding	Final recovery rate (%)
Average 25 million	1300	474	39.7	24.2	63.9
H2500	1200	415	39.5	26.7	66.2
GL	1100	442	39.6	28.2	67.8

3.3 Oil Displacement Experiment After Polymer Solution Aging

The viscosity of different polymer solutions after aging for 60 days is quite different. viscosity stability difference on oil displacement effect, oil displacement experiments of different polymer solutions aged for 60 days were carried out. The experimental results are shown in Table 2.

From Table 2 that under the same other recovery values of the three polymers are in the range of 18.6%–22.2%, which is lower than that of oil displacement experiment after polymer preparation. This is because the viscosity of polymer solution decreases in different degrees after 60 days of aging, and the controlling effect of mobility is weakened, and the oil displacement effect is relatively poor. The EOR of common polymer with 25 million molecular weight, lh2500 salt resistant polymer and GL salt resistant polymer decreased by 5. 8%, 4. 5% and 9. 0% respectively. The enhanced oil recovery value of salt resistant polymer lh2500 is 3.6% higher than that of common polymer, and the oil displacement effect is the best [3].

Table 2. Data table of oil displacement experiment results

Name of polymer	Polymer concentration	viscosity	Water permeability measurement	Water drive recovery (%)	Enhanced recovery rate of polymer flooding
	1300	50	39.3	444	39.3
H2500	1200		41.9	452	39.8
GL	1100		35.7	414	39.5

4 Field Test of Salt Resistant Polymer Flooding

4.1 Overview of Test Area

In order to explore a new way polymer test was at station 3 (formation water salinity 8217.5 mg/L) in the central part of xingliu area with relatively high salinity of sewage, and the target layer is pui2-3 reservoir. In this area, five point pattern is adopted, and the average injection production well spacing is 125 m. There are 71 wells (33 injection 38 production) and 12 central wells. Lh2500 salt resistant polymer with the best comprehensive performance is selected. Combined with the geological characteristics, water flooded characteristics and d data different concentrations of polymer slug step injection mode is adopted. The polymer concentration of slug 1 was 2134 mg/L, and o061 gate V was injected. The slug dipolymer concentration is 1675 MGL, and 0.949 gate V has been injected as of November 2018. All salt resistant polymer systems are prepared in the way of clean and dilute. In order to compare the test results, the No. 1-2 stations in the central part of xingliu block, which are similar to the development status and geological conditions of No.3 station in the central part of xingliu District, adopt the common polymer flooding with the molecular weight of 25 million, and clean and dilute the oil [4].

4.2 Oil Displacement Effect

By November 2018, 101 PV polymers were put into the test plot. All 37 wells are valid in the test strip. The first effective coefficient was 67.6%, and the total substitution polymer of 1-2 class was 43%, the water content of the raw well decreased by 30% or more, and it decreased by 50% or more, and it decreased by 24%. In the same injection volume, the void fraction was 6.6% and the final recovery rate was improved by 19.6% in the same injection volume.

5 Conclusion

(1) Natural core oil displacement experiment shows that lh2500 salt resistant polymer has which can enhance oil recovery by 13.8%, 4.9% more than that of common polymer with 25 million molecular weight.

(2) In consideration of the problems such as high salinity of wastewater, large amount of polymer and rising cost of polymer flooding in Daqing Oilfield, a new type of salt resistant polymer should be selected to minimize polymer consumption and improve oil recovery.

References

1. Lu, J., Tang, J.: Hydrolysis of HPAM polymer in high temperature reservoir. Acta Petrologica Sinica 6(3), 271–274 (2007)
2. Zhu, L., Chang, Z., Li, M., et al.: Preparation of high molecular weight water soluble (acrylamide/acrylic acid/2-acrylamide-2-methylpropanesulfonic acid) terpolymer by partial hydrolysis. Acta Polymer Sinica (3), 315–318 (2000)
3. Wang, H.: Study on Influencing Factors of viscosity of HPAM solution and selection of complexing agent. Northeast Petroleum University, Daqing (2014)
4. Wang, Z., Li, L.: Preparation and performance evaluation of a temperature resistant and salt resistant polyacrylamide. Chem. Eng. Manag. (2), 6–7 (2018)

Prediction and Analysis of Numerical Simulation Algorithm for Return Characteristics of Resistant Polymer

Ling Liu[✉]

Daqing Oilfield Limited Company, No. 4 Oil Production Company, Daqing 163511, China

Abstract. It is of great significance to accurately predict and evaluate the development and management of polymer flooding. Reservoir micro pore structure parameters are important factors affecting the development effect of polymer flooding, and there is a nonlinear and uncertain complex relationship between them. The parameters which have great influence on the EOR of polymer flooding are selected by correlation analysis. The nonlinear and Uncertain Multivariable System is predicted by polynomial regression analysis and BP neural network. The results show that the artificial neural network method has better adaptability, It can better reflect the internal relationship between various micro parameters affecting polymer flooding effect and EOR value, and the prediction accuracy is high. Therefore, it is considered that the application of BP neural network to predict polymer flooding effect is feasible and effective.

Keywords: Salt resistant polymer · Polynomial regression · Artificial neural network · Pore structure · Prediction

1 Introduction

Salt tolerant polymer flooding is mainly the residual oil driven by bottled water, and its micropore structure is an important factor affecting its distribution. Mobile sand storage is the spatial structure of canyon. The pores of rock include the size and distribution of pores, the connectivity of the throat of pores, the geometry of pores and the micro heterogeneity. The composition and occurrence of clay minerals are pore. This is an important parameter to reflect the micro inhomogeneity of the memory. bend. It not only controls the flow and concentration of petroleum gas, but also has an important impact on oil production capacity. It is an important factor affecting the development effect of polymer actuator, including oil consumption, final development of storage oil and Ge factor, development factor and performance of biopolymer itself. The study of macro geological factors is the basic factor of regional development. It also has the characteristics of pore structure as a micro geological factor.

The porosity and permeability of the core of the Portuguese group in Xingnan area of Daqing Oilfield are relatively low, but the polymer recovery value of the polymer in the physical simulation experiment of polymer flooding in the laboratory is not low,

J. Abawajy et al. (Eds.): ATCI 2021, LNDECT 81, pp. 264–268, 2021.
https://doi.org/10.1007/978-3-030-79197-1_38

some of which are even higher than those of Sai group and saIII group with higher porosity and permeability [1]. The difference of experimental results is due to the difference of different reservoir layers in different regions, which indicates that the micro certain relationship with polymer flooding effect.

Through correlation analysis, it is found has a great correlation with porosity, pore throat radius, peak value and other pore structure parameters. In this paper, we try to use pore structure parameters as analysis data, and use artificial neural network and polynomial regression to predict the EOR potential of polymer flooding. It is found that the prediction accuracy of artificial neural network method is high and has certain application value.

2   Synthesis Mechanism of Salt Resistant Polymer

Due to the influence of high temperature degradation and salt sensitive effect, the viscosity of polymer used for oil displacement decreases in the process of use, so it is necessary to introduce functional monomer to improve the moisture resistance and salt resistance of polymer. Block copolymerization method is used to sub structure insensitive sulfonic group of block polyacrylamide, which can improve salt tolerance and increase salt tolerant monomer 2-acrylamido-2-methylpropanesulfonic acid (AMPS group is insensitive to external cations and has high charge density). For example, the introduction of rigid monomer can significantly inhibit the amide group in the polymer, while the introduction of ring structure greatly improves the shear resistance of hydrophobic groups. Hydrophobic association modification is to introduce hydrophobic long chain group hydrophobic monomer into hydrophilic macromolecules to form intramolecular or association. The ratio of physical group to hydrophobic group affects the association structure of hydrophobic long chain alkyl, which can prevent the rapid decline of viscosity and improve its temperature and salt resistance.

3   Evaluation of Temperature and Salt Resistance of Block Polymers

Polymer solutions with different mass concentrations were prepared with different types of polymers and simulated brine (salinity 2410 mgl1) from Daqing Oilfield. The viscosity of solution was observed at 45 °C and shear rate of 734s1. It can be seen from the data in Fig. 1 that the viscosity of polymer solution increases with the increase of mass concentration. When the mass concentration reaches 1000 mgl -, the increasing trend of viscosity becomes more obvious. The viscosity of gl-80 and gl-100 functional polymers is better than that of HPAM. The viscosity of the new block gl-80 polymer solution is higher than that of the new block-100 polymer solution at low mass concentration. When the mass concentration of the solution reaches 1200 mg/L, the new block G100 polymer shows stronger viscosity enhancement, which is better than that of the new block gl-80 polymer. Figure 1 below shows the viscosity increasing diagram of polymer.

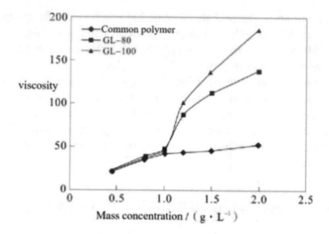

Fig. 1. Viscosity increasing of polymer

4   The Basic Principle of Prediction Method

4.1   Selection of Main Pore Structure Parameters Affecting Polymer Flooding Effect

While the conventional mercury injection method can obtain capillary pressure curve, a number of microscopic parameters can be obtained, which can describe the micro pore structure characteristics of reservoir from different angles. The laboratory physical simulation experiment of polymer flooding can obtain the improved value of polymer flooding recovery after water flooding, and the data obtained by physical model can eliminate the influence of various subjective or macroscopic geological factors in field implementation [2]. Two rock samples are taken from a core in parallel, one is used for mercury injection experiment to obtain the micro pore structure parameters, and the other is used for indoor physical simulation experiment to obtain the enhanced oil recovery value of polymer flooding. The experimental data are used to analyze the influence of micro factors on the development effect of polymer flooding.

The influence degree of formation macro and micro parameters on polymer flooding development effect is analyzed by simple correlation analysis method. The greater the correlation is, the greater the influence degree is. The factors with high correlation are used as the input parameters of ANN and polynomial stepwise regression models.

4.2   Polynomial Stepwise Regression Method

Quadratic polynomial regression analysis method combines orthogonal design and regression analysis. On the basis of orthogonal design and regression analysis, the mathematical model of causal relationship between given variables and dependent variables is established for model prediction or production. Multivariate quadratic polynomial model equation is generally expressed as:

$$y = b_0 + \sum_{i=1}^{m} b_i x_i + \sum_{i=1}^{m} b_{ii} x_i^2 + \sum_{i=1}^{i<j} b_{ij} x_i x_j \tag{1}$$

When processing the data matrix, besides the original data, the quadratic polynomial containing the data is automatically generated (that is, the quadratic polynomial of each independent variable data is also taken as an independent variable factor) [3, 4]. Quadratic polynomial stepwise regression analysis is to retain the items that have significant influence on the dependent variables and delete the insignificant items.

4.3 Artificial Neural Network Method

The known parameters are exactly the same as the polynomial stepwise regression analysis. The formula obtained by artificial neural network method is the following "implicit expression".

$$y = ANN(x_1, x_2, \cdots, x_m) \tag{2}$$

Where ANN is a nonlinear function. This function can not be expressed by the usual mathematical formula, so it is called knowledge base. In this paper, the most widely used error back-propagation neural network (BP network) is used in this paper. This kind of neural network has an input layer composed of many independent neurons, which is used for inputting relevant data and information, and one or more hidden layers for complex intermediate calculation in the process of analysis and simulation, and one output layer for outputting operation and simulation results. Its structure is shown in Fig. 1. The N nodes in the input layer correspond to the input vector composed of N input components, $x = (x_1, x_2, \ldots, x_N)$ The output vector $Y = (y_1, y_2, \ldots, y_L)$ is composed of L output components corresponding to l nodes in the output layer. The hidden node Z_j and output node Y_k are:

$$Z_j = f\left(\sum_{i=0}^{N} w_{ij}^1 x_i\right), j = 1, 2, \ldots, M \tag{3}$$

$$Y_k = f\left(\sum_{i=0}^{N} w_{jk}^2 z_i\right), k = 1, 2, \ldots, L \tag{4}$$

Where w_{ij}^1 and w_{jk}^2 are the weights w_{0k}^2 of input hidden layer and hidden layer output layer respectively, and w_{0j}^1 is the wide value, and $x_0 = 1$, $z_0 = 1$.

5 Conclusion

Through the analysis and comparison of the polynomial regression analysis method and the conventional BP neural network prediction method, it is concluded that the polynomial regression analysis method has the advantages of fast calculation speed and simple model establishment, but its disadvantage is that the prediction equation constructed is more complex and the prediction accuracy is low. The artificial neural network method can be used in the prediction of multivariable system. The neural network has strong information utilization ability and is superior to the conventional multiple linear regression analysis and polynomial regression analysis in the prediction accuracy and trend.

References

1. Wu, M., Deng, S., Wei, F., et al.: Synthesis technology of polyacrylamide and its application in oilfield development. Prog. Fine Petrochem. Ind. **12**(12), 1–4 (2011)
2. Yang, Q., Li, B., Li, Y., et al.: Study on calculation method of polymer sweep efficiency and oil displacement efficiency. Daqing Pet. Geol. Dev. **26**(1), 109–112 (2007)
3. Liu, Y., Chang, Q., Yu, F., et al.: Application of Polyacrylamide in oilfield production. Petrochem. Appl. **33**(4), 9–10 (2014)
4. Li, L., Jiao, L.: Multi factor nonlinear time-varying prediction of oilfield production based on artificial neural network. J. Xi'an Pet. Univ. **17**(4), 42–44 (2002)

Quantitative Analysis of Micro Remaining Oil in Water Flooding Based on CT and Image Analysis Software

Yanping Liu[✉]

Daqing Oilfield Limited Company, No. 4 Oil Production Company,
Daqing 163511, China
liuyanpingdq@petrochina.com.cn

Abstract. There are still a lot of residual oil in the formation after ASP flooding. It is important to clarify the type and quantity of remaining oil for further enhanced oil recovery. Various types of micro residual oil in cores with different permeability after water flooding and weak alkali ASP flooding are quantitatively analyzed by using computer tomography (CT) and image analysis software. The results show that the main types of residual oil after water flooding are cluster, column and blind end; weak alkali ASP flooding has different degree of displacement effect on all kinds of residual oil after water flooding, and the displacement effect of cluster residual oil is the best; the saturation of blind end residual oil increases slightly after ASP flooding. The remaining oil form after weak alkali ASP flooding is consistent with that after water flooding, and the proportion of cluster remaining oil is the highest. It is necessary to further develop this kind of residual oil to enhance oil recovery.

Keywords: Weak alkali · ASP flooding · Micro remaining oil · CT · Image analysis

1 Introduction

By the end of 2017, the average EOR of ASP more than 18%. The research on the basis for subsequent adjustment of development plan and EOR. The research scale of remaining oil can be divided into four volume scales: macro scale, large scale, small scale and micro scale. The research of micro scale is mainly realized by micro visualization model. At present, there are many researches on micro scale residual oil after mechanism of oil displacement, divided into two types: only by adhesion force and by capillary force and viscosity force at the same time [1]. Hu Xuetao et al. Divided the remaining oil after water flooding into three forms: single droplet, double droplet and branched droplet. The core pore structure was generated by random network model to simulate the water drive process, and the distribution proportion of different types was counted.

© The Author(s), under exclusive license to Springer Nature Switzerland AG 2021
J. Abawajy et al. (Eds.): ATCI 2021, LNDECT 81, pp. 269–273, 2021.
https://doi.org/10.1007/978-3-030-79197-1_39

2   Experiment and Analysis

2.1   Experimental Materials

The surfactant was 40% petroleum sulfonate and the weak base was Na_2CO_3 (effective content was 99.99%) [2]. The ternary solution of masonry alkali is prepared from the mother liquor of the above three reagents. The alkali concentration of active agent and polymer is 0.3%, 1900 mg/L and 1.2%, respectively. In order to distinguish the gray value of oil and water in the subsequent image analysis, potassium iodide (effective content is 99.99%) is added into the sewage to prepare 5% potassium iodide solution. The experimental oil is a mixture of degassed dehydrated crude oil and aviation kerosene from a factory in Daqing Oilfield. The working viscosity is 8.5 MPa·s (45 °C), and the experimental cores are artificial sandstone cores with effective permeability of 40 mD and 290 mD respectively.

2.2   Oil Displacement Experiment and Effect Analysis

In the experiment, two artificial cores with effective permeability of 40 md and 290 md were used to carry out water flooding and weak alkali ASP flooding experiments, and the effects of different oil displacement systems under the same permeability were compared. The steps of water flooding experiment are as follows:

Water logging permeability, saturated simulated oil: Inject filtered sewage into the core for water drive at a speed of 0.1 M/min, and stop when the water cut of water drive rises to 98%; calculate the final recovery rate of water drive. Experimental steps of weak alkali flooding: ① inject filtered sewage into rock core at the speed of 0.1 ml/min for water flooding, and stop when the water content of water flooding reaches 92%; ② inject 0.2 pv polymer (protective slug) after injecting 0.3 pv (oore volume) into the core; ③ calculate recovery efficiency and final recovery rate of each stage.

3   CT Scanning Reconstruction and Remaining Oil Type Analysis

The cores of each group after oil displacement experiment were cut and sampled. The real core slices were scanned by skyscan high frequency resolution CT machine skyscan 1172 produced by Belgium skyscan company. The two-dimensional cross-sectional images of core samples were reconstructed by using CT machine with Nepcon software, as shown in Fig. 1. Skeleton of different core samples in the scanning and reconstruction images are processed with appropriate gray level to distinguish the oil-water distribution. The brown, blue and pink areas represent the remaining oil, water and rock skeleton respectively [3]. According to the shape and formation reasons types are mainly divided into three categories: ① column residual oil, remaining oil remaining in narrow and long channels; ② cluster residual oil, whose shape is mostly Y-shaped, H-shaped, or irregular shape formed by several small pieces of residual oil; ③ blind end residual oil, Most of the remaining oil which is not easy to be displaced in the spinel (one end blocked) channel is retained to form blind end remaining oil.

Fig. 1. Scanning reconstruction of core sample

According to Formula 1 described in other parts of the application and U value of water, the correction coefficient of image data can be determined [4]. The U value of water is proportional to the radiation intensity received by the detector. The first radiation intensity can be a constant value. Further description of determining the first radiation intensity is provided in other parts of the present application. The relationship between collimation width and radiation intensity can be described as equal.

$$\frac{1000}{C_w} = c^*(f(w) + P) \tag{1}$$

Where CW is the correction factor corresponding to the collimation width W; P is the first radiation intensity; and C is the coefficient, which can be determined according to Eq. 2.

$$c = \frac{1000}{C_{w1} * (f(w1) + P)} \tag{2}$$

Where w1 is the first collimation width; C_{w1} is the first correction coefficient corresponding to the first collimation width w1; and f (w1) is the scattering radiation intensity corresponding to the first collimation width.

The acquisition unit may acquire image data related to scanning performed at the target collimation width. The image data can be obtained from other components in the CT system, for example, a storage device or a storage module. The image data may be related to radiation scattered through or by a scanned object in a scan performed at a target collimation width.

The comparative analysis of the pictures after water flooding and strong alkali ASP flooding shows that the amount of thin film remaining oil on the pore surface is significantly reduced after strong alkali ASP flooding. After water flooding, there are more thin-film residual oil on the surface of rock particles, while after ASP flooding, the film remaining decreases obviously. The fluorescence images of water drive and ASP drive are shown in Fig. 2 and Fig. 3 below.

Fig. 2. Fluorescence image of water drive

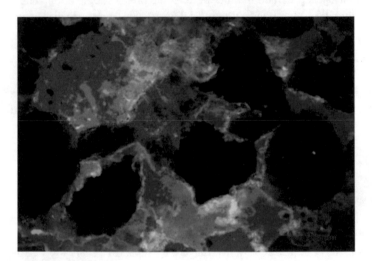

Fig. 3. Fluorescence image of ASP drive

4 Conclusion

CT and computer image analysis technology combined with the advantages of non-destructive and quantitative analysis can accurately calculate the saturation of various types of remaining oil (columnar, cluster, blind end, etc.) Weak alkali ASP flooding can displace all kinds of residual oil to a certain extent after water flooding. The saturation of cluster residual oil decreases by 17.53% and 10.65% respectively in 40 mD and 290 mD cores. The effect of viscosity increasing and oil-water interfacial tension reducing of weak alkali ternary system is the most obvious for cluster oil. However, the proportion of cluster oil is still higher than 50% after displacement, and this kind of remaining oil needs to restart migration to further enhance oil recovery.

References

1. Ma, C., Zheng, H., Song, K., et al.: Numerical simulation of residual oil distribution after polymer flooding in the east-west block of beier. Acta Pet. Nat. Gas **29**(1), 99103 (2007)
2. Guo, W., Li, Z., Jia, G., et al.: Study on quantitative interpretation of micro distribution of residual oil and variation of pore microstructure. Sci. Technol. Eng. **14**(31), 32–36 (2014)
3. Wang, Q., Baofa, Q., Sui, J., et al.: Practice and understanding of tertiary oil recovery technology in Daqing Oilfield. Pet. Geol. Dev. Daqing **20**(2), 1–9 (2001)
4. Guo, W., Cheng, J., Liao, G.: Research status and development direction of tertiary oil recovery technology in Daqing Oilfield. Daqing Pet. Geol. Dev. **21**(3), 1–6 (2006)

The Application of Computer Vision Technology in the Field of Industrial Automation

Qian Cheng[✉]

Hubei University of Technology, Wuhan, China

Abstract. Computer vision technology, as the representative technology of the 21st century industry across the age, has had a significant impact on my country's social and economic development and the quality of life of the people with the rapid development of my country's modernization. The purpose of this article is to analyze the application of computer vision technology in industrial automation from the perspective of computer vision technology under the development of industrial vision. The method is based on the development and changes of the industrial foundation in the field of computing vision technology, and a deeper exploration of its practical application performance in the industrial automation industry.

Keywords: Computer vision technology · Vision technology · Industrial application · Computer · Industry

1 Introduction

Nowadays, as a giant in the computer field, it has branches and leaves. The various computer technologies developed accordingly have better application value in various fields. They have become a necessary part of people's production methods and daily life, Among them, computer vision technology can obtain and transmit a large amount of information. And when this technology is used in the production, processing and design of the industrial automation industry, it can effectively integrate, innovate and control useful information. This technology relies on a variety of high-end technologies to provide support, coupled with the country's overall high-tech industry's continuous development and modernization planning, computer vision technology has shown obvious progress and formal convenience, and is gradually applied to actual production and in life [1].

2 Computer Vision Technology Related Concepts

2.1 Related Concepts

Computer vision technology is connected with computer systems, based on image and signal processing technology, probability analysis and statistics, network neural technology and information processing technology. Used to analyze and process data and

J. Abawajy et al. (Eds.): ATCI 2021, LNDECT 81, pp. 274–279, 2021.
https://doi.org/10.1007/978-3-030-79197-1_40

information. The technical operation is specifically by using a camera to replace the human eye, and the computer body acts as the human brain, enabling it to recognize, judge and remember targets with the support of technology. And replace humans in some production operations [2].

2.2 Working Principle

When the brightness meets the requirements, the camera starts to collect image information of a specific object through the lens, and then carries out internal information transmission. The computer system responds and receives quickly, and uses image processing technology to further process the original image to optimize the image effect. Classify and organize the key information in the image to identify and extract the required accurate and high-level abstract information, and finally store the information in the database, so that the things to be identified can be compared with the stored information in real time, and the technology can be completed efficiently Run the job.

2.3 Theoretical Framework

Since the 1980s, the proportion of computer vision technology in the market has begun to rise linearly. Computer vision technology maintains a "ongoing" tense in terms of its own industrial field technology theory and practical application research strategies. The research level of computer vision technology theory covers many aspects such as computer theory, algorithm and actual realization. If you analyze the visual technology from the perspective of computer theory, you need to use prime maps, dimensional maps, and three-dimensional model representations for further research. In this process, we complete the processing by transmitting effective information and identify other tasks. The multi-layer visual processing is roughly divided into low-level, middle-level, and high-level. It is also based on the complexity of the analysis task, in order to facilitate users to choose by themselves, and achieve quick Complete the function command [3, 4].

3   Industrial Vision

Nowadays, the field of computer industry is highly concentrated and developed, and the technical field has been widely used in the production level of today's society. In the field of industrial automation industry, the image analysis, text recognition and industrial detection functions of computer industry technology can be used to improve industrial production efficiency, and The application of technology in the computer industry in the image processing process can greatly improve production efficiency and accuracy. At the same time, on the creative level of the technical field of the computer industry, the experiment of the non-destructive structure can be realized by both computer vision image processing technology and visual inspection technology and provide quality assurance. Especially in industrial buildings, computer vision technology has changed from the military field to the civilian industrial form. On this basis, computer image processing technology has achieved considerable development and update, and with the continuous update of industrial automation production technology, related technologies in the field of complementary inspection have gradually gained efficiency and accuracy that surpassed the past. As shown in Fig. 1 below:

Fig. 1. Industrial vision

3.1 Promote Industrial Development

In industrial creation, the use of image processing and visual inspection technology can not only achieve non-destructive testing and measurement, but also ensure the quality of products in the production process, thereby reducing the use of manpower and improving the efficiency of industrial automation production. The industrial vision system is mainly embodied in the industrial vision inspection system, which can use vision technology to compare the image and standard of the measured object to ensure the quality of the measured object; the industrial robot system can guide and control in the form of visual measurement and is widely used Industrial production.

3.2 Improve Industrial Production Performance

From the perspective of the technical branch of industrial vision analysis, resolution is a factor that needs long-term attention in the overall relationship of this field. In different industrial operating environments, the resolution requirements are also different under the object size and accuracy. For example, at the spatial change level, only the theorem can be used to ensure that the image resolution meets the specified requirements; at the level of gray-scale changes, the requirements for the light source are particularly high, requiring extremely high sensitivity of the camera system. While ensuring that the measurement form is sufficiently accurate, it is necessary to use strong signals to identify the shape and size of objects, perform related tasks and perform calculations to ensure more accurate image processing. Therefore, the application of computer vision technology can realize the adjustment of various systems and production conditions, so that the entire computer vision technology system maintains good performance.

4   Application of Computer Vision Technology in the Field of Industrial Automation

In the field of industrial automation production, the application prospects of computer vision technology are very wide, and the practical application in the production process is also very prominent, including dimensional measurement, component inspection, guiding robots and other functions. For example, in the processing of parts, the inefficiency and limited accuracy of traditional manual operations may damage the parts during the transportation of the parts. But after applying computer vision technology when performing non-destructive testing on a large number of parts, especially in the repeated inspection process of the same parts and products, the model recognition technology, image processing technology, sensor technology, and computer technology in computer vision technology will play a great role. However, in the actual application process, if the shape of the part changes and the on-site lighting is insufficient, the information collected by the sensor cannot be guaranteed, and the application quality of computer vision technology will be seriously affected. As shown in Fig. 2 below:

Fig. 2. Computer vision technology

4.1   Visual Inspection Technology

Visual inspection technology can usually achieve the following functions. First, collect data in a comprehensive way through various light source changes and sensor devices. The source image input of 256-color grayscale image is a common standard in the practical application of computer vision technology. Then the source image data is fully analyzed and processed to extract relevant data, so that the desired detection result can be output. Under normal circumstances, preprocessing the source image is the prerequisite for the overall improvement of the accuracy of the detection operation, and according to the relevant basic conditions in the preprocessing, the relevant matching technology is used for comprehensive detection. Under the basic conditions of establishing the template, the template is used for comprehensive detection to obtain a better distribution effect. Secondly, the output processing result is the last step of the source image processing. During the entire output process, not only the detection results need to be consistent with the expected results, but also the accuracy of data analysis and processing needs to be improved during the image processing.

4.2   Image Preprocessing

Image preprocessing is the work performed before template matching, mainly to obtain the binary edge map in the image. In the entire image processing process, in the form of image edge detection and data binarization. Among them, the edge detection work is in progress, mainly through the first-order differential operator, the second-order differential operator and the global detection form. Only through these three forms can the preprocessed content meet the actual standards. For example, first-order differential operators are used to implement edge detection behavior. Since the shape of the area near the edge is affected by the wider phenomenon, after the detection result appears, it must be refined. If the second-order differential operator is used, the edge width obtained by using the second-order derivative zero crossing is one pixel. After the detection result appears, there is no need to refine the processing form, thus ensuring the accuracy of the edge effect. Therefore, in the edge detection process, the second-order differential operator is the main application method.

4.3   Template Matching

In computer vision technology, template matching is an important application direction in this technology, mainly for all-round matching and analysis of existing templates and detected objects. Use the similarity value between the two graphics to determine the similarity between the template and the detected object, just like the synthesis of the model and the actual object, so as to ensure the accuracy and validity of the output content. In the process of matching the two, the method used must be able to achieve the invariance of the data model rotation, translation, and scale change, that is, to ensure that the 360° visual perception is the same as the entity. Use two graphics for similarity comparison and matching. This value can not only be greatly convenient for calculation, but also can shape the same shape and size as the detection object. In the calculation of similar values, it is usually necessary to use a distance function for calculation, mainly using the Euclidean distance method. This method can measure the similarity value of two point sets, it does not need to deal with the relationship between different points. Only need to ensure the distance between two finite point sets to achieve this function, especially when processing points with multiple characteristics, the accuracy of the calculation process and the calculation result can be guaranteed at the same time, and the anti-interference ability in the process is also enhanced.

5   Conclusion

As a high-tech specializing in the study of computer recognition functions, Computer vision technology has a variety of technical function branches under the industry field. Although it seems to have a certain degree of complexity, technical functions have strong application value and can bring great convenience to the field of not limited to industrial automation.

"Application" is the biggest driving force in the computer vision system. In the development of all walks of life, computer vision technology has made considerable progress and has gradually formed in market competition. Especially under the continuous improvement of my country's modern economic development model and industry quality awareness, Continuously improving the optimization level of the technology industry can greatly meet the changing needs of the development of the industrial field, and continuously improve the optimization of the technology industry according to the operating standards of computer vision technology, Pay attention to the training of technical personnel. Analyze based on operating technical standards and ideas, and improve the technical integration and fusion application of computer vision technology and industrial automation as a whole with industry development, and regularly evaluate and optimize technical personnel to meet the development needs of computer vision technology in various industries.

In summary, under the modern development of national science and technology, computer vision technology, as a technology derivative of the computer industry, has been widely used in various industries, especially when people's quality requirements for industrial products are increasing. Under the premise of higher, The use of computer vision technology can effectively improve the efficiency and quality of industrial production. Therefore, it is necessary to strengthen the technicality of computer vision technology, determine the operating standards and optimization ideas of computer vision technology, and enhance the development and construction needs of people's production and life. In order to apply this technology scientifically and rationally to improve production efficiency in automated production, it is necessary to further integrate actual production conditions on the basis of a clear understanding of basic theoretical knowledge, explore and persist in practice in the application. Summarize the experiences and breakthrough innovations in the exploration in order to better promote and apply computer vision technology to assist a series of operations such as production and inspection, to better serve the society, serve mankind, and promote social development and technological development.

References

1. Li, J., Dong, S., Gou, L.: Automobile door opening anti collision early warning system based on millimeter wave radar system. Volume 10(6), 339 (2020)
2. Ji, J., Fang, C.: Design and implementation of intelligent obstacle avoidance car based on millimeter wave radar. J. Zhejiang Wanli Univ. 33(1), 85–90 (2020)
3. Jiang, Y.: Research on forward vehicle detection and tracking algorithm based on millimeter wave radar and machine vision fusion. Chongqing University, Chongqing, p. 7 (2019)
4. Wu, R., Jin, Z., Zhong, T., et al.: Vehicle ranging system based on millimeter wave radar. Automot. Pract. Technol. (2), 33–35 (2019)

Risk Identification and Evaluation and Machine Learning in Manufacturing Innovation

Xianke Li[(⊠)]

Guangzhou University Sontan College, Guangzhou 511370, China

Abstract. With the advent of knowledge economy, artificial intelligence (AI) and machine learning play an increasingly significant role in manufacturing innovation. Artificial intelligence (AI) based on computer science makes today's economic society enter into "Digitalization", "Networking" and "Intellectualization". It promotes the industrial upgrading of Intelligent Manufacturing in China's manufacturing industry from three aspects: promoting the production of intelligent manufacturing equipment, promoting the birth of new industrial model, and promoting the application of new business. This is because, on the one hand, (AI) of manufacturing production factors, reduce operating costs, and achieve comparative advantage benefits and economies of scale benefits; on the other hand, it can improve the management efficiency of enterprises, improve their knowledge learning ability, and accelerate the pace of independent innovation.

Keywords: Artificial intelligence · Machine learning · Manufacturing industry · Innovation and upgrading

1 Introduction

Manufacturing power and accelerate the upstream movement of China's manufacturing industry to the global value chain. Intelligent manufacturing is the future transformation trend of China's manufacturing industry. At present, with the maturity of computer technology, AI gradually plays a huge role in many fields [1]. Especially massive data mining in manufacturing industry gives more application scenarios for AI, and machine learning helps. Data is the source of machine learning. In the current Internet economy, massive data is mined and stored, so data mining and learning is the core task. Through the knowledge map analysis, we can clearly see the evolution direction of industrial innovation. Therefore, in the process of manufacturing technology upgrading, it is helpful for researchers to use the results of machine learning for interactive learning.

J. Abawajy et al. (Eds.): ATCI 2021, LNDECT 81, pp. 280–285, 2021.
https://doi.org/10.1007/978-3-030-79197-1_41

2   Research on Artificial Intelligence and Machine Learning

Generally speaking, the application of AI can be roughly divided into three stages: weak AI application stage, strong artificial AI application stage and super AI application stage. The so-called weak AI application stage refers to using the e application stage, the operation and manufacturing of computers break through brain science, and intelligence is very close to human intelligence, which is difficult to achieve with the current technology; In the application stage of super AI, after the great development of brain science and brain like intelligence, AI has become a super intelligent system. Although the current AI is in the weak application stage, its role in the manufacturing industry is constantly being explored.As the core of AI, machine learning is the fundamental way for computer to carry out intelligent analysis [2]. Data acquisition and learning, knowledge and data processing, human-computer interaction.

3   Application of Artificial Intelligence and Machine Learning in Manufacturing Industry

Under the background of the vigorous development of industrial Internet, manufacturing + AI will develop along the three stages of "Digitalization", "Networking" and "Intellectualization". This will provide significant opportunities for the transformation and upgrading of China's manufacturing industry, so as to promote China's manufacturing industry to turn to intelligent manufacturing, promote the emergence of new industries and change the existing business model [2, 3].

(1)   Promoting the development of intelligent manufacturing

With the comprehensive application of AI, the automation of production has reached a new height, machines will replace labor, the proportion of capital production factors in the production investment of enterprises is increasing, and even "unmanned workshop" has appeared. With the improvement of machine learning level, data mining and analysis technology get rid of the limitations of traditional analysis technology and make a new breakthrough. This has greatly improved the level of business intelligence and enabled enterprises to realize real "intelligence" in the fields of risk management, marketing and service.

By cooperating with speech recognition, computer vision and robot technology, machine learning can use a lot of data to improve the level of AI. By cooperating with speech recognition, computer vision and robot technology, machine learning can use a lot of data to improve the level of AI. First, machine learning can predict changes in raw material prices so that finance, operations, and supply chain teams can better manage constraints on plant and demand. Secondly, machine learning is helpful to optimize the production process, realize multi product concurrent production, improve production efficiency and reduce the unqualified rate of products. Thirdly, machine learning can achieve predictive maintenance, which can improve the performance of preventive maintenance, repair and overhaul by providing higher predictive accuracy for components and levels. Finally, machine learning can further optimize the enterprise's supply chain, optimize the input of production factors, make the production always in the optimal stage, and realize the scale economy within the enterprise.

(2) Promote the birth of new industrial model

AI and machine learning have brought new structural opportunities to equipment enterprises, software and service enterprises, communication and solution providers, and factory production processes, etc., and spawned the birth of new industries. At first, the emergence of new hardware equipment enterprises. AI equipment does not depend on human input and output, so a series of new automation hardware equipment enterprises are born. For example, the chip of AI technology, automatic optical detection, automatic logistics vehicle, cooperative robot, etc., bring new product segmentation market for hardware equipment manufacturing enterprises. The second, new opportunities for AI industry software enterprises. Machine learning is widely used in the production and sales of manufacturing industry. The visual analysis of application data needs mature software, algorithm and service enterprises as support to promote the deep integration of application software service companies and manufacturing industry to jointly develop more refined AI. For example, GE and Enel will jointly deploy and optimize GE's asset performance management (APM) software. The third, the upgrading of communication and solution operators. All the applications mentioned above that bring development momentum to equipment enterprises, software and service enterprises all rely on high-speed industrial communication technology and high-level system interconnection. At present, China is vigorously promoting the construction of 5g network and related infrastructure, providing new soil for the upgrading of industrial intelligence.

(3) Promote the application of new business models

Data is the source of machine learning. In the current Internet economy, massive data is mined and stored, so data mining and learning is the core task. Through the knowledge map analysis, we can clearly see the evolution direction of industrial innovation. Therefore, in the process of manufacturing technology upgrading, it is helpful for researchers to use the results of machine learning for interactive learning.For example, the B2C and C2C based e-commerce model based on data analysis and processing has achieved great success compared with the traditional offline retail mode. The development of e-commerce in China has created a lot of data and provided fertile soil for the application of machine learning. On the demand side, traditional industries have gradually realized the power of AI, and began to take AI as the next growth point. The analysis and application based on big data mining has become the key layout direction of China's manufacturing industry; On the supply side, a new industry providing AI technology services and products is gradually formed. There are a large number of computer vision, voice recognition, cloud computing service providers in the market. Industrial Internet has gradually opened up the value chain of b-end manufacturing and consumer services, formed a new business model of C2B and C2M, and promoted the integration of manufacturing and service industries. With the further promotion of industrial intelligence, the future manufacturing industry will realize the transformation and upgrading from the current standardized and large-scale production services to personalized and customized production services.

4   The Impact of Artificial Intelligence and Machine Learning on Chinese Manufacturing Innovation

(1) Control costs and improve production efficiency

With the global industrial transfer, China has a comparative advantage in labor-intensive industries because of its cheap raw materials and cheap labor. Therefore, it undertakes the middle part of the "smile curve" of value-added in the global value chain. Although it will help China to integrate into the international division of labor in the short term, it also falls into the capture type global value chain dominated by developed countries. Artificial intelligence (AI) and machine learning can help Chinese manufacturing enterprises improve the technical content of their products, help them climb to the high end of the industrial chain, enhance the added value of their products, and obtain higher profits. In addition, AI and machine learning can carry out more refined management of production links, reduce management costs, and optimize the input ratio of labor and capital. Especially with the advent of 5G era, communication speed can be further improved, transportation and storage costs can be reduced, repeated links in production can be reduced, and waiting time can be reduced. Through the prediction of future product prices, the strategy of increasing or reducing production in advance can be adopted to optimize the production cost of enterprises.

(2) Intra product division of labor

At present, the production of manufacturing industry has evolved from the global division of labor among industries to the intra product division of labor, and the production processes of products are dispersed to different enterprises, even enterprises in different countries. On the one hand, AI and machine learning help to decompose the complex production processes in the past, and through the standardization of product production processes, the international division of labor within products can be realized. Generally speaking, the factor prices of richer resources in a country are always cheaper, while those of scarce resources are always higher. Therefore, in developing countries, the labor force is relatively rich, so the production cost of labor-intensive products must be low; in developed countries, capital factors are relatively rich, so capital intensive products are relatively cheap. Therefore, in developing countries, the labor force is relatively rich, so the production cost of labor-intensive products must be low; in developed countries, capital factors are relatively rich, so capital intensive products are relatively cheap. Because different countries and regions have different comparative advantages in different resources, AI can help manufacturing enterprises to move their factories to the regions with the most cost advantages, so as to enhance their profit margins and obtain comparative advantages.

On the other hand, AI and machine learning can help enterprises to obtain scale economy benefits. The application of AI algorithm can calculate the level of the production scale of the enterprise to maintain the highest efficiency of resource utilization. At the same time, it helps to change the mode of production, release part of the labor force, make these people free become managers or researchers. With their rich operation in the production field, these workers will significantly optimize the overall production and operation efficiency of the enterprise, and improve the success rate of product research and development. The influence of AI on each link of intelligent production in manufacturing industry is shown in Table 1.

Table 1. The influence of AI on manufacturing industry

Stage	Design	Production	Inspection	Transportation	Storage
Sensor	Digital twins				
Communication industry		Cooperative robot		LOT	
Industrial cloud	Intelligent factory, cross enterprise value chain extension and whole industry ecological construction				
Image recognition	Intellectual property protection			Supply chain management	
Blockchain			Automatic optical inspection		Storage robot
Voice				Voice sorting	
Machine learning	Optimal design	Predictive maintenance		Vehicle cargo matching	

(3) Carry out innovation and enhance product autonomy

Manufacturing industry can continuously produce more abundant data than consumption, and provide abundant "means of production" for the development of AI. Data collection is the basis of data analysis, testing and machine learning. Only when there is enough data base, machine learning can give full play to its effectiveness. Therefore, AI can help manufacturing enterprises upgrade from three aspects of products, services and production. In addition, AI helps to promote equipment innovation in manufacturing industry, thus reducing the dependence of manufacturing automation on technology and equipment. Although China's manufacturing enterprises surpass Germany and Japan in scale, they still rely mainly on Siemens, Ge, Mitsubishi and other German Japanese enterprises in terms of automation equipment, methodology, software and hardware platforms required for product design and production. For example, although Huawei has achieved global competitive advantages in the field of 5G communication, it still relies heavily on foreign enterprises such as the United States in the production of 5G-equipment.

(4) Digital benefit transformation

Digital benefit transformation is the application of digital technology to all aspects of the manufacturing value chain. For example, through the digital twin of R & D, the physical operation is reduced and the iteration is fast in a short time, which can not only save time but also improve the R & D efficiency. Through the R & D data and process management, the application efficiency of existing R & D achievements is improved, the R & D cost is reduced, and the remote collaborative R & D can be realized. Understand user needs through the Internet, and adopt crowdsourcing and other methods in R & D. In the manufacturing process, digital technology is used to optimize the manufacturing process and real-time monitoring, so as to quickly solve the problems in production, adjust the manufacturing progress according to the order, increase the use efficiency of equipment, reduce operating costs, improve product quality and realize the optimization of manufacturing process. In the marketing and service link, we can communicate with customers in new ways, provide new products and services at the same time, establish a community platform, actively understand customers' needs, provide services such as product life cycle maintenance, increase customer stickiness, and tap the "long tail demand" outside the market to form professional collaborative operation, improve efficiency and increase product added value.

5   Conclusions

Will have a great impact on the manufacturing industry, and even promote the comprehensive innovation of manufacturing industry with the application of disruptive technology.Therefore, in the 5G era, AI brings new opportunities to the innovation and upgrading of Chinese manufacturing enterprises. Machine learning will promote the development of knowledge economy, and promote Chinese manufacturing industry to turn to capital intensive and knowledge intensive.

Acknowledgements. "Research on the evolution of enterprise technological innovation network to promote regional collaborative innovation" (**2020GZGJ287**); "Economic opening, innovation vitality of enterprises and coordinated development of regional economy" (**2020WQNCX104**).

References

1. Lv, W., Chen, J., Liu, J.: A quantitative analysis of China's Artificial intelligence industrypolicy from the perspective of policy tools. Stud. Sci. Sci. (10), 1765–1774 (2019)
2. Zhou, H., Li, H., Zhao, L.: Research on the impact of digital transformation of manufacturing industry on green innovation performance – the moderating effect of digital level. Sci. Technol. Manag. **23**(01), 33–43 (2021)
3. Wang, L., Ding, W.: Influencing factors and paths of resource integration in the innovation process of manufacturing enterprises: a multi case study based on enterprise life cycle. China Sci. Technol. Forum (01), 95–105+165 (2021)

Construction of Safety Occupational Adaptability Model of Cloud Based Operation Management and Optimization in Mechanical and Electrical Industry

Yang Liu[1(✉)], Jiaxiang Yi[2], and Lina Mei[1]

[1] Chongqing Vocational College of Light Industry, Chongqing 401320, China
[2] Chongqing University of Science and Technology, Chongqing 401320, China

Abstract. Based on the understanding of the construction of psychological resource model, this paper deeply analyzes and studies the construction content, and based on this, according to the working characteristics of the mechanical and electrical industry, constructs the professional adaptability model of the staff's psychological state.

Keywords: Psychological resources · Electromechanical industry · Occupational adaptability

1 Introduction

With the rapid development of social economy, higher requirements have been put forward for the mechanical and electrical industry. In view of the dangerous characteristics of its work, it is necessary to do a good job of safety management, to prevent and improve the safety problems caused by human factors.

2 Construction of Psychological Resource Model

In the electromechanical industry, it is necessary to construct man-machine system to carry out specific work, but because people have uncertainty and dynamic characteristics, there will be great differences in psychological characteristics and quality. In order to realize the mutual adaptation of man, machine and environment, people need to be improved according to the actual situation. For the safety occupational adaptability, it mainly studies and evaluates the human safety, and constructs the psychological resource model based on it, taking the psychological resource content as the core, which includes the tendency of safety occupational ability, psychological adaptability and the allocation and use of psychological resources to ensure that it has strong safety and adaptability characteristics, and also conforms to the relevant principles [1].

The construction of psychological resource model is mainly based on the theoretical model of psychological load. Its main content covers various ability factors such

J. Abawajy et al. (Eds.): ATCI 2021, LNDECT 81, pp. 286–290, 2021.
https://doi.org/10.1007/978-3-030-79197-1_42

as cognitive ability, motor perception ability, memory and psychological bearing ability of related personnel, and reasonably allocates the psychological resources consumed by their professional ability. Therefore, psychological resources can only be used in work and self-protection.

3   Research on Occupational Adaptability Model of People's Mental State in Mechanical and Electrical Industry

3.1   Psychological Quality Adaptation and State

Distribution is an important part of modern logistics. It is a comprehensive product of modern market economic system, modern science and technology and modern logistics thought. It is essentially different from the well-known "delivery". Modern enterprises generally recognize that distribution is an important part of business activities, it can create higher benefits for enterprises, and it is an important means for enterprises to enhance their competitiveness [2].

The distribution center is a circulation enterprise specialized in goods distribution, with a large scale of operation. Its facilities and technological structure are specially designed and set up according to the characteristics and requirements of distribution activities. Therefore, it has a high degree of specialization and modernization, complete facilities and equipment, and strong ability of goods distribution. It can not only deliver goods from a long distance, but also deliver goods of various varieties, It can not only distribute raw materials of industrial enterprises, but also undertake supplementary goods distribution to wholesalers. This kind of distribution is the main form of goods distribution in industrial developed countries, and it is the future development of distribution [3].

3.2   Advantages of Genetic Algorithm

On the basis of the psychological load model, it is necessary to conduct an in-depth investigation on the psychological load factors of the personnel in the mechanical and electrical industry. In view of the structure, condition and influence of the post psychological load of the electrical operator, the psychological load dimension of the electrical operator is accurately divided, including fatigue degree, task pressure, time pressure, psychological stress degree, etc., and the appropriate data collection method is selected to investigate and collect the psychological state of the personnel, and according to the working content and time of the mechanical and electrical industry, It can effectively get that the degree of psychological load of different post personnel is also quite different. It is in a state of mental sub-health for a long time, and it is easy to reduce the degree of satisfaction and attitude of mechanical and electrical work, and even appear physical and mental diseases.

It is also an important content to adapt to the construction of the model, mainly by comparing the psychological load differences among the different positions in the electromechanical industry. The in-depth analysis of the measurement results of psychological load index can effectively show the relationship between psychological load and psychological resources and provide scientific basis for the construction of psychological state occupational adaptability model. In order to ensure the safety of the mechanical and electrical industry and its overall controllable ability, it is necessary to maintain a balance between the psychological load and the working load of the personnel. On this basis, it is necessary to do a good job of load forecasting, subjective evaluation of the working environment and tasks of mechanical and electrical workers, and testing of their physical, psychological and behavioral tolerance. Therefore, it is necessary to ensure that the evaluation and fatigue analysis accurately judge the degree of psychological load and its relationship with psychological fatigue and resources from a subjective point of view.

3.3   Distribution and Use of Psychological Resources

For the psychological resource allocation and use model, it is an important content of the psychological resource model of safety occupational adaptability, which can be divided into two aspects, namely, the distribution model and the use model. Ability to complete work and self-protection at the same time.

Distribution model. From the point of view of psychological resources input, there is a clear direction, mainly for self-protection resources and work resources supply, but because of its extreme characteristics, in order to avoid the phenomenon of resource depletion, it is necessary to make effective use of the remaining psychological resources, on this basis, it also plays a decisive role in the safety of occupational adaptability. In the process of psychological resource adjustment and recovery, the staff will automatically protect their own psychological function, which leads to a downward trend of work resource motivation, job performance, etc., but if they change their roles and convert them into professional ability and psychological adaptability respectively, It will meet the needs of the actual model construction, improve the quality and level of psychological resource allocation, and meet the actual requirements of safety professional psychological adaptability. In addition, the psychological load of personnel can also clearly reflect the situation of psychological resources, so it is necessary to ensure the accuracy of psychological resource allocation.

Use the model. The construction of psychological resource safety adaptability model is mainly composed of two parts: working resource utilization and safety adaptability. According to the actual situation, it is divided into four parts: low energy high fatigue area, high energy low fatigue area, high energy high fatigue area and low energy low fatigue area. From the point of view of psychological load state, it can also be divided into four forms: first, psychological high load state, mainly due to excessive content and use of mechanical and electrical working resources, which has exceeded the limit value of self-protection resources and ignored the importance of work safety. The main feature is a light workload and short duration.

4   Competency Orientation and Personnel Status

Ability tendency and personnel fatigue state are also important forms of safety professional ability of personnel in mechanical and electrical industry. For ability tendency, it plays a decisive role in psychological load, not surplus resources. In psychological resources, the control of ability is always dominant. Table 1 SWAT Psychological load dimension and division of the assessment method.

Table 1. SWAT Psychological load dimension and division of the assessment method

Level/Dimension	Time	Psychological efforts	Psychological stress
1	More free time, less overlapping activity	The activity automatism is strong, unconscious psychological effort is more	The frequency of panic, danger, anxiety and so on is less. Easy to adapt to environment
2	Occasional space time, more overlapping activities	Have a certain sense of effort, need to give some attention to the activities	Due to the influence of many factors, psychological stress is increased
3	No free time, frequent overlap of activities	Extensive psychological effort is required, activity is complex and requires frequent attention	Higher stress levels require self-psychological control and regulation

First of all, according to the professional characteristics of mechanical and electrical personnel, as well as the characteristics of work between different positions, targeted resource allocation and ability tendency activities, in the process will appear a certain degree of psychological load uneven, warning ability decline, and at the same time, different resource allocation conversion will appear psychological maladjustment, resulting in a certain sense of work fatigue; Secondly, according to the safety characteristics of mechanical and electrical personnel, it is necessary to carry out specific work strictly in accordance with relevant norms and systems, and to strengthen implementation while clarifying their own responsibilities, so as to play an important role in ensuring work safety. Therefore, it is necessary to deepen the cognitive ability of mechanical and electrical staff and to clarify the importance of attention to their specific work; Thirdly, in the process of constructing the safety occupational adaptability model, it is necessary to measure the indexes of the personnel, study the orientation of their ability tendency according to the actual personnel's situation, and make clear their preference for psychological movement ability and operation skill.

Conclusion: to sum up, on the basis of psychological resource model construction, it is an inevitable trend to meet the development of modern society and an important content to improve the quality and level of mechanical and electrical engineering. Therefore, we must do a good job in the psychological construction of mechanical and electrical personnel, strengthen psychological quality and professional ability, conducive to the realization of economic and social harmony.

Acknowledgements. Study on the influencing factors of safety professional ability of mechanical and electrical employees (No: KJ1758497).

References

1. Lu, J., Yang, J.: Study on performance evaluation of subway drivers based on structural equation model. Logistics Technol. **43**(09), 43–47 (2020)
2. Zhao, W.: Research on job burnout of new generation industrial workers based on job requirement-resource model — the role of psychological resilience. Soft Sci. **30**(12), 67–71 (2016)
3. Yang, M., Wen, Z.: Psychosocial security atmosphere in the work requirements-resource model. Chin. Psychol. Soc. 519–521 (2015)

Research on the Application and Practice of Statistical Analysis Methods in Mathematical Modeling

Zhiqin Chen[(✉)]

Jiangxi Teachers College, Yingtan 335000, China

Abstract. Mathematical modeling involves a large amount of real data, and the complex relationship between the data must be processed. Therefore, if the corresponding method is not used, the difficulty of mathematical modeling will increase. The so-called "corresponding method" generally refers to statistical analysis Such methods can provide powerful help to mathematical modeling. Therefore, in order to complete mathematical modeling, modelers should understand the application of various statistical analysis methods. This article takes three commonly used statistical analysis methods in mathematical modeling as examples to discuss the basic principles and application practices of the three methods. The three statistical analysis methods are principal component analysis, factor analysis, and independent component analysis.

Keywords: Mathematical modeling · Statistical analysis method · Method application

1 Introduction

Since the establishment of the concept of mathematical modeling, how to perform statistics and analysis on real data has always been a concern of people. Therefore, a large number of people have conducted research on mathematical modeling statistical analysis methods and proposed a large number of statistical analysis methods. Subsequently, under the competition of various statistical analysis methods, individual methods with excellent performance and significant effects have become common methods in modern mathematical modeling, such as principal component analysis, factor analysis, and independent component analysis. However, there are differences in the application methods and main functions of different statistical analysis methods. Therefore, in order to use statistical analysis methods to complete mathematical modeling, it is necessary to master the correct application methods of each method. It is necessary to carry out relevant research.

© The Author(s), under exclusive license to Springer Nature Switzerland AG 2021
J. Abawajy et al. (Eds.): ATCI 2021, LNDECT 81, pp. 291–297, 2021.
https://doi.org/10.1007/978-3-030-79197-1_43

2   Basic Principles of Mathematical Modeling and Statistical Analysis Methods

2.1   Principal Component Analysis Method

Principal component analysis is a statistical analysis method for extracting transformation variables. The extracted variables are called "principal components" in the theory of this method, that is, principal component analysis mainly relies on orthogonal transformation to transform groups of variables that may have correlations. The variable group is linearly uncorrelated, and the variable group is the principal component. Principal component analysis is mainly suitable for multivariate research. Because there are too many variables in this type of research, it is difficult to conduct research using ordinary methods, but blindly reducing the number of variables will lead to less information. In the early days, people fell into a dilemma in multivariate research. After the advent of principal component analysis, the situation was successfully broken. In principle, people found that there is a correlation between variables in general. In this case It can be determined that there is a certain overlap in information between the two related variables. At this time, the principal component analysis method can be used to extract all the unrelated variables from the related variables. The remaining variables are repeated variables, so you can delete them directly. Variables will continue to decrease, but valuable information will not decrease, so that people can ensure that information will not decrease blindly on the basis of reducing the number of variables, and the difficulty and feasibility of research are guaranteed. In Carl Pearson (the proponent of the principal component analysis method) and subsequent related research, most researchers define the principal component analysis method as a dimensionality reduction statistical method, based on an orthogonal transformation, which promotes the correlation of the variable components the original random vector of is transformed into a new random vector with irrelevant components. Through algebra, this process can be regarded as the transformation of the covariance matrix of the random vector to the diagonal matrix, so that its dimension is continuously reduced. The calculation steps of the principal component analysis method are shown in Fig. 1.

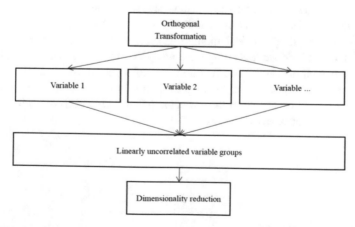

Fig. 1. The calculation steps of the principal component analysis method

2.2   Factor Analysis

The factor analysis method was first proposed by the British psychologist Spearman. In his research, he mentioned that there is a certain correlation between the performance of students in various subjects. Students who have good academic performance in one subject often have scores in other subjects, Very good [1]. On the contrary, the performance of each subject will be poor, indicating that there may be the same common factor among students performance in various subjects. Finding out this common factor can know why the students performance is good and why it is poor, so that teachers can make targeted progress. education. The factor analysis method is gradually formed under Spearmans discovery. This method can find the so-called common factors from many variables and guarantee the representativeness of the common factors. At this time, the essentially identical variables can be used in mathematical modeling. Grouped into the same factor, so that the number of variables is reduced, and the relationship between the variables can be clarified. The scope of application of factor analysis is very wide. Because everything is directly and indirectly related, the common factors of things can be found using factor analysis, but it is worth noting that factor analysis must be based on the source of factors It can only be used on the basis of the same thing, otherwise the factor does not conform to the principle of essentially the same. In addition, the method of finding common factors in factor analysis is generally confirmatory factor analysis. The steps of confirmatory factor analysis are shown in Fig. 2.

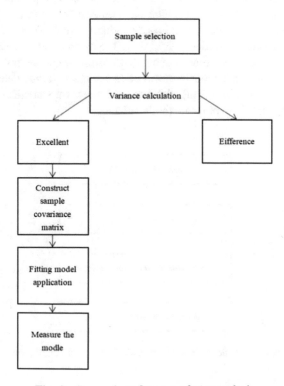

Fig. 2. Steps of confirmatory factor analysis

2.3   Independent Component Analysis

The independent component analysis method is a statistical analysis method based on the above two methods. Its application value is more than that of the first two. It can be used when other methods cannot be used or reliable results cannot be obtained. In the application of independent component analysis, firstly, the mixing matrix can be obtained by calculation, and then the inverse matrix of the mixing matrix is sought. The inverse matrix can help us untie the mixing matrix and understand the actual situation of each independent component in the mixing matrix. In general, the main function of the independent component analysis method is to find the internal factors that support the data object, and separate the factor data information from a large number of data objects to form independent components, and finally the component content can be known according to the data information of each independent component. The calculation steps of independent component analysis are shown in Fig. 3. In the research theory of independent component analysis, the independent component is usually regarded as a function, and the P-dimensional observation data vector is recorded as X. On the basis of obtaining the P-dimensional unknown source signal vector, the matrix can be defined as Mix the matrix and finally find the inverse matrix, and then perform a linear transformation on X to get the output vector. The output vector is the value of the independent component function, and accurate judgments can be made according to the value. In addition, there are general uncertainties in the application of independent component analysis: (1) Uncertainty in the order of the components, that is, in the application of independent component analysis, sometimes it is impossible to know the arrangement of the components, which is not conducive to us. To understand the nature of things, other methods are needed to deal with it to a certain extent; (2) Uncertainty of component volatility; that is, some independent components are not static, they will fluctuate randomly, and independent component analysis cannot determine the magnitude of component volatility. This also has an impact on statistical analysis, which needs to be compensated by other methods [2].

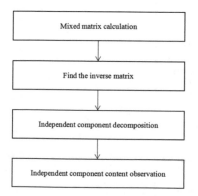

Fig. 3. Independent component analysis method calculation steps

3   Practical Application of Mathematical Modeling and Statistical Analysis Methods

Focusing on the above three statistical analysis methods, the following will analyze the practical application of each method based on actual mathematical modeling problems.

3.1   Application of Principal Component Analysis

In practical applications, firstly, formula (1) is used to obtain the P-dimensional random vector to form the sample matrix, and then formula (2) is used to perform the normalized conversion of the matrix:

$$x = (x_1, x_2, \cdots, x_P)^t \, nx_j = (x_{i1}, x_{i2}, \cdots, x_{iP})^T, i = 1, 2, \cdots, n, n > P \qquad (1)$$

Where x_1, x_2, ..., x_n are the number of variables; T is the variable group; nx_j is the number of samples; x_i is the degree of information overlap between variables; $i = 1, 2,$ ..., n, $n > P$ is the P-dimensional random vector to form the sample matrix.

$$Z_{ij} = \frac{x_{ij} - \bar{x}_j}{s_j}, i = 1, 2, \ldots, n; j = 1, 2, \cdots, P \qquad (2)$$

Where, x_{ij} is the newly generated variable after the transformation of the information overlap degree of each variable; s_j is the coefficient of the matrix.

Secondly, formula (3) is used for conversion to obtain the standardized matrix Z, and then formula (4) is used to obtain the correlation coefficient matrix for the standardized matrix Z. Finally, formula (5) is used to convert the standardized variable index into the main component and evaluate it.

$$R = \left[r_{ij}\right]_P xP = \frac{Z^T Z}{n - 1} \qquad (3)$$

Where r_{ij} is the correlation coefficient.

$$|R - \lambda I_P| = 0 \qquad (4)$$

$$U_{ij} = Z_i^T b_j^o, j = 1, 2, \cdots, m \qquad (5)$$

Finally, the first principal component, the second principal component, and the P-th principal component can be obtained by formula (5), and the weighted summation is performed on this basis, and the variance contribution rate of each principal component can be known, thereby completing the statistical analysis. In mathematical modeling with identification of the main factors of water resources risk, the modeler must not only identify the main factors of risk, but also classify the hazard of the risk. In this case, the principal component analysis method is used and the main risks can be obtained by referring to the above steps. Factor, and then use dimensionality reduction to first standardize the data of each influencing factor, establish a standardized model, and then calculate the principal component to obtain the cumulative contribution rate of each factor. According to the contribution rate of the principal component factor, the water resource risk factor level can be known. According to the result Just make statistics.

3.2   Application of Factor Analysis

In practical applications, the application of factor analysis can be divided into two steps:

(1) Model construction, that is, first assume that the pseudo P-dimensional random vector is consistent with the q-dimensional random variable, namely $q \leq P$. Then there will be a common factor in the X variable, this factor has an impact on all the variable individuals in X, but there is a P-dimensional unobservable random vector in all the variable individuals in X, and the common factor that affects the random vector is defined It is a special factor. For special factors, formula (6) can be used to obtain a factor loading matrix. According to this matrix, all special factors can be listed, and finally calculated to understand the influence of each factor on the individual variable of the X variable;

(2) Validation factor analysis, that is, usually The LISREL software can be used to perform verification factor analysis. The software includes PRELIS, LISREL, and SIMPLIS. The language functions are in order: ① Form a covariance matrix through regression analysis; ② Measure model elements based on the covariance matrix Define the relationship between, and perform model fitting; ③ Continue to fit the model until the model fit is acceptable. Taking the analysis of the main causes of heavy metal pollution as an example, first establish the correlation matrix between heavy metals and the polluted area, use the software according to the matrix to obtain the eigenvalues of the impact factor and the cumulative contribution rate, and then rotate around the cumulative contribution rate of the factor to complete the calculation of the factor loading matrix. Judge the influence of each factor on heavy metal pollution, so that statistical analysis can be completed [3].

$$V_{ar}(X) = \vec{A}\vec{A}^T + \sum \tag{6}$$

In the formula, AA^T is the transformation process when the P-dimensional random vector and the q-dimensional random variable coincide with each other.

3.3   Application of Independent Component Analysis

In practical applications, first, the mixing matrix A must be centered and whitened, and then the inverse matrix B can be obtained after normalization. At this time, the iterative operation can be performed to obtain the characteristic matrix. The characteristic matrix expression is shown in formula (7). According to the characteristic matrix, independent components can be obtained from the matrix, and the influence of each independent component on the target can be understood. Take the project of selecting excellent classification factors for the relationship between genes and tumors as an example.

First, the gene components can be divided into two types: normal group and abnormal tumor group, and independent component analysis method is used for technical analysis. Secondly, the number of gene data matrixes is assumed, and the number of genes in the gene matrix the individual genes are independent of each other. At this time, matrix a is obtained. Finally, the inverse matrix can be obtained under centralization and whitening. After iterative operations, the independent genome can be known. This genome is an excellent factor [4].

$$X = AB \tag{7}$$

In the formula, AB is the factor feature.

4 Conclusion

In summary, this article analyzes the application practice of statistical analysis methods in mathematical modeling. The analysis first introduces three common statistical analysis methods in mathematical modeling, discusses the principles and basic application steps of the methods, and understands the characteristics, advantages, and application conditions of each statistical analysis method. Secondly, the calculation formula of each statistical analysis method is introduced, and the application method of the method is described in combination with actual cases. It can be seen that each method can help people complete mathematical modeling in practical applications and can solve the problem of mathematical modeling.

References

1. Fox, W.P.: Using multi-attribute decision methods in mathematical modeling to produce an order of merit list of high valued terrorists. Am. J. Oper. Res. **4**(6), 365–374 (2014)
2. Waterman, K.C., Swanson, J.T., Lippold, B.L.: A scientific and statistical analysis of accelerated aging for pharmaceuticals. Part 1: accuracy of fitting methods. J. Pharm. Sci. **103** (10), 3000–3006 (2014)
3. Couteron, P.: Quantifying change in patterned semi-arid vegetation by Fourier analysis of digitized aerial photographs. Int. J. Remote Sens. **23**(17), 3407–3425 (2002)
4. Orlov, A.I.: Statistical simulations method in applied statistics. Zavodskaya Laboratroiya. Diagnostika Materialov **85**(5), 67–79 (2019)

Cyber Intelligence for CV Process and Data Mining

Defogging Algorithm Based on the Combination of FCM Clustering Segmentation and Otsu Method for Image Structure Based on Lab Color Space

Changxiu Dai[(✉)]

South China Business College, Guangdong University of Foreign Studies, Guangzhou 510545, Guangdong, China

Abstract. When a single image is de-fogged by dark channel prior, halo is easy to appear at the boundary of the image, and color distortion is easy to be caused in the bright area. To overcome the above defects, this paper proposes an image defogging algorithm based on Lab color space for image structure FCM clustering and segmentation combined with the improved Otsu algorithm. First, convert the foggy image from RGB color space to Lab color space, and decompose the image into structure and texture in Lab color space. Then, FCM clustering is used to calculate the transmittance on the image structure, and the atmospheric light value is optimized in combination with the improved Otsu algorithm. Finally, the structure of the fog-free image is restored through the inverse process of the atmospheric physical model, and the obtained structure is integrated with the texture of the image to obtain the final fog-free image. Experiments show that the algorithm can effectively de-fog, solve the problem of partial distortion in the sky in the dark channel prior de-fog algorithm, can obtain better color vision fidelity, and has stronger applicability.

Keywords: Dark channel prior · Lab · Structure and texture · FCM · Otsu algorithm

1 Introduction

In recent years, with the development of industry, environmental pollution has become more and more serious, thus leading to the generation of haze phenomenon. In the weather view with haze, the image not only attenuates the color seriously, but also reduces the contrast and saturation, which seriously restricts the application of computer vision. Therefore, image defogging technology has important research value.

There are two main types of image defogging algorithms: one is based on image enhancement methods, such as histogram equalization algorithm [1], image enhancement and improvement algorithms based on Retinex theory [2], homomorphic filtering algorithm [3], wavelet Transformation algorithm [4], etc.; the other is based on image restoration methods, such as Tan algorithm [5], Fattal algorithm [6], Tarel algorithm [7, 8], He algorithm [9] and so on.

© The Author(s), under exclusive license to Springer Nature Switzerland AG 2021
J. Abawajy et al. (Eds.): ATCI 2021, LNDECT 81, pp. 301–307, 2021.
https://doi.org/10.1007/978-3-030-79197-1_44

2 Dark Channel Prior

The dark channel prior is a typical algorithm for image restoration and de-fogging. According to the degradation model of fog weather scene imaging, the fog-free image is performed by fog image.

$$I(x) = J(x) t(x) + A(1 - t(x)) \tag{1}$$

I (x): Haze image, J (x): Remove haze image, t (x): transmittance, A: the atmospheric light.

A dark channel priori holds that there are always some pixel points in a fog-free image that are very small and close to 0 in a certain channel. Therefore, the upper formula can be further transformed into the following mathematical model:

$$J(x) = \frac{I(x) - A}{t(x)} + A \tag{2}$$

However, the dark channel prior algorithm has many shortcomings, such as halo and color distortion. Therefore, many improved algorithms have appeared. This paper proposes an image defogging algorithm based on Lab color space for FCM clustering and block calculation of image structure and finding the top 0.1% pixels of the sky area brightness value as the atmospheric light value estimation.

3 Algorithm Improvements

3.1 RGB to Lab

The Lab color space is composed of three components: L (luminance), a and b (color). The addition of fog does not change the color of the image. For L dehazing, a and b will be affected little. Therefore, the first step for image defogging in this article is to convert the image color space, I_{RGB} to I_{Lab}.

3.2 Image Decomposition

Any image is composed of structure and texture. The structure is the larger part of the image, and the texture is the smaller part of the image. The addition of haze noise only affects the structure of the image, and has little effect on the texture of the image. Therefore, the second step of the image defogging algorithm in this paper is to decompose the foggy image, $I._{Lab} = I_{S\_Lab} + I_{T\_Lab}$.

Image decomposition is based on an improved form of total variation, the mathematical model of total variation:

$$\arg \min_S \sum_p \left\{ \frac{1}{2\lambda} (S_p - I_p)^2 + \left| (\nabla S)_p \right| \right\} \tag{3}$$

I: input image I_{Lab}, p: the index of pixels, S: output structure image I_{S_Lab}.

Use formula (3) to obtain the structure I_{S_Lab} of the foggy image I_{Lab}, and then use $I_{T\_Lab} = I_{Lab} - I_{S\_Lab}$ to obtain the texture of the foggy image, and then dehaze the image structure I_{S_Lab} and the image texture I_{T_Lab} Does not go fogging. The image decomposition results are shown in Fig. 1.

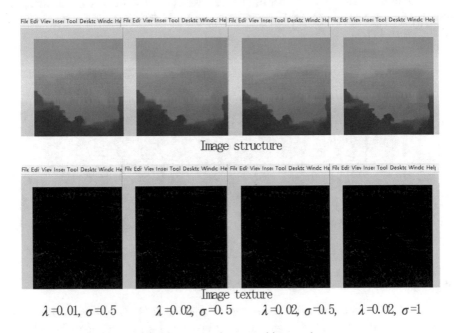

Image structure

Image texture

λ =0. 01, σ =0. 5 λ =0. 02, σ =0. 5 λ =0. 02, σ =0. 5, λ =0. 02, σ =1

Fig. 1. Image decomposition results

3.3 FCM Cluster Image Structure IS Lab Calculates Transmittance

The dark channel prior image de-fogging algorithm induces the larger difference pixels into the same block to calculate the transmittance, which leads to the "halo" phenomenon. Therefore, the image segmentation quality directly affects the image defogging effect. After many years of efforts, researchers have developed many image segmentation methods. There are two-peak method based on similarity detection, region splitting and merging method, edge detection method based on greyness discontinuity detection, edge tracking and so on. These algorithms have achieved good image segmentation results. Since the correlation L,a,b three components in Lab color space is small, the generation of fog will not affect the change of image color, but only affect the brightness of the image L, so it is only necessary to fog up the L component. This paper uses FCM to segment the brightness component $I_{S\_Lab\_L}$ of the foggy image structure and then calculate the transmittance. First take the brightness component $I_{S\_Lab\_L}$, and then FCM divides the n pixel brightness L_i (i = 1, 2, ..., n) into c fuzzy classes. The similarity between pixels classified into the same class is the largest, and the difference between different classes The pixel similarity is the smallest, and the

Euclidean distance is used to measure the similarity between pixels. Its mathematical formula is expressed as follows:

$$J_{FCM}(U, V) = \sum_{i=1}^{c} \sum_{j=1}^{n} u_{ij}^{m} d_{ij}^{2} \tag{4}$$

V: the cluster set, c: the number of cluster centers, d_{ij}: the Euclidean distance, m: a weighted index. Calculated transmittance results FCM the clustering image structure are shown in Fig. 2.

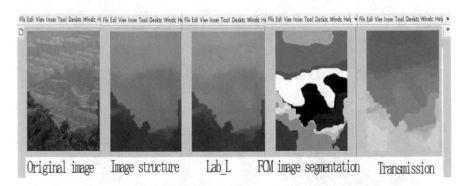

Fig. 2. FCM Calculation of transmittance by clustering image structure

3.4   Optimization of Atmospheric Light Value

Reference [10] selects the pixel corresponding to the 0.1% brightness value of the foggy image as the atmospheric light value. This method is susceptible to the influence of bright light sources in the image, resulting in a large estimate of atmospheric light. Reference [11] uses the quadtree method to estimate the atmospheric light value, which avoids the influence of bright light sources to a certain extent. When calculating the atmospheric light value, the algorithm in this paper first divides the structure map of the foggy image into the sky area and the non-sky area using the improved Otsu method, and then selects the corresponding pixel point of the first 0.1% brightness value of the sky area as the atmospheric light value. The experimental effect is shown in Fig. 3.

Fig. 3. Segmentation results of sky region and non-sky region

4 Experimental Results and Analysis

In view of the de-fogging effect of this algorithm, the following will be evaluated from the subjective qualitative contrast analysis and the fidelity degree of image color respectively.

4.1 Comparative Subjective Analysis

Due to limited space, here is only the comparison of the dehazing effect of Reference [9] and Reference [11] with the algorithm in this paper. The comparison result is shown in Fig. 4.

Original image   reference 9   reference 11   the Algorithm

Fig. 4. Comparison of image defogging effect

From the comparison of experimental results, it can be seen that the method of [9] has obvious effect of de-fogging, which is in the non-sky area with lower depth of field, and has achieved better detail retention effect. However, in the large white or sky area, the method in [9] has the phenomenon of color distortion and edge halo. The method of literature [11] combines color attenuation prior and dark channel prior knowledge. the overall fogging effect of the de-fogging algorithm is better. The image after de-fogging solves the halo phenomenon to a certain extent, but it cannot solve the drawbacks of image color distortion.

The image de-fogging method based on the LAB pattern combined with the sky region and the non-sky region segmentation, in the low depth of field, the details of the dark channel prior de-fogging remain excellent, in the LAB mode, only for the image brightness de-fogging, image color and saturation operation unchanged, effectively solve the color distortion caused by the dark channel prior algorithm. Clustering segmentation image de-fog makes the phenomenon of halo after de-fog significantly reduced. The image de-fogging effect of this algorithm has achieved good results from both detail and color distortion subjectively.

4.2   Fidelity of Post-Fog Images

By observing many images after fog removal using dark channel method, we know that some color distortion in the sky after fog removal is mainly obvious color is not smooth, which violates the prior law of small color change and smooth color in the sky area. After the fog image is removed, the performance of the sky area is still smooth.

5   Conclusion

This paper proposes a defogging algorithm for image structure based on Lab color space. The algorithm decomposes the foggy image into structure and texture in Lab color space, and dehazes the structure. Experiments show that the algorithm in this paper retains the details of the lower depth of field image, solves the problem of partial distortion of the sky color, and improves the dehazing effect and applicability of the algorithm. However, there is still a small amount of color distortion in the sky area. How to completely avoid the distortion phenomenon requires continued research and exploration.

References

1. Reza, A.M.: Realization of the contrast limited adaptive histogram equalization (CLAHE) for real-time image enhancement. J. VLSI Signal Process. Syst. Signal Image Video Technol. **38**(1), 35–44 (2004)
2. Rahman, Z., Jobson, D.J., Woodell, G.A.: Multi-scale retinex for color image enhancement. In: Proceedings of the 3rd IEEE International Conference on Image Processing, pp. 1003–1006. IEEE Computer Society Press, Los Alamitos (1996)
3. Seow, M.J., Asari, V.K.: Ratio rule and homomorphic filter for enhancement of digital colour image. Neurocomputing **69**(7–9), 954–958 (2006)
4. Russo, F.: An image enhancement technique combining sharpening and noise reduction. IEEE Trans. Instrum. Meas. **51**(4), 824–828 (2001)
5. Tan, R.T.: Visibility in bad weather from a single image. In: Proceedings of the IEEE International Conference on Computer Vision and Pattern Recognition, pp. 1956–1963. IEEE Computer Society Press, Los Alamitos (2008)
6. Fattal, R.: Single image dehazing. ACM Trans. Graph. (TOG) **27**(3), 1–9 (2008)
7. Tarel, J.P., Hautiere, N.: Fast visibility restoration from a single color or gray level image. In: 2009 IEEE 12th International Conference on Computer Vision, pp. 2201–2208. IEEE (2009)

8. Guo, F., Tang, J., Cai, Z.: Image Dehazing based on haziness analysis. Int. J. Autom. Comput. **11**(1), 78–86 (2014)
9. He, K., Sun, J., Tang, X.: Single image haze removal using dark channel prior. IEEE Trans. Pattern Anal. Mach. Intell. **33**(12), 2341–2352 (2011)
10. Kim, J.H., Jang, W.D., Sim, J.Y., et al.: Optimized contrast enhancement for real-time image video Dehazing. J. Vis. Commun. Image Represent. **24**(3), 410–425 (2013)
11. Zhuq, Q.S., Mai, J.M., Shao, L.: A fast single image haze removal algorithm using color attenuation prior. IEEE Trans. Image Process. **24**(11), 3522–3533 (2015)

Big Data Technology in Intelligent Translation Algorithm in Computer Science

Hanhui Li$^{(\boxtimes)}$

School of Foreign Languages, Fuzhou University of International Studies
and Trade, Fuzhou 350202, Fujian, China

Abstract. With the rapid development of science and technology, more and more communication and exchanges are transmitting information through the network, which has higher and higher requirements for the difficulty of intelligent translation. On this basis, this article will specifically analyze the main factors that affect computer network information communication, take the computer network information intelligent translation as the starting point, and improve the level of information intelligent translation based on the combination of intelligent translation algorithms. Based on traditional algorithms, this paper can use high-strength artificial fish school algorithms to collect and process the data in the network more efficiently, so as to prevent errors in information translation. The intelligent translation algorithm model model studied in this paper mainly includes information collection and processing and more efficient and comprehensive translation of information. Through the analysis, it can be understood that the intelligent algorithm is to transform the key data and the key string. With the rapid development of big data technology, intelligent translation in computer network information can efficiently and accurately translate confidence. Experimental research results show that in the intelligent translation of computer network information, the use of artificial fish swarm algorithms can effectively improve the efficiency of information translation and the accuracy of translation results, and the speed and accuracy of traditional translation have doubled.

Keywords: Computer science · Big data analysis · Intelligent translation · Algorithm research

1 Introduction

The continuous innovation of intelligent translation technology means that artificial intelligence is being used more and more in translation work. The most notable form of expression is language recognition and the creation of cloud translation platforms. Speech technology has greatly improved the efficiency and quality of translation [1]. In translation work, voice recognition can realize the complete recording and translation of the translation process, and the voice input function can realize the rapid recording of the translated text. After voice recognition, the untranslated text is input into the document, and then translated one by one through some software or program. Due to the development of intelligent translation technology, the error rate of speech recognition and input functions is getting lower and lower. The automatic detection function

J. Abawajy et al. (Eds.): ATCI 2021, LNDECT 81, pp. 308–315, 2021.
https://doi.org/10.1007/978-3-030-79197-1_45

can self-check and adjust the meaning of words according to the context, which greatly improves the efficiency and effect of translation [2]. Secondly, the cloud translation platform provides space for communication and cooperation for translators, and improves the storage space of translation data. Its unique cloud backup function can also save various data generated in the translation process at any time, so as to facilitate reading and inspection personnel at any time, thereby ensuring the accuracy and completeness of the data. The platform will provide space for information sharing, communication and cooperation between employees, and provide an open and safe environment for translators who perform translation projects together but are not in the same space, as well as a platform for people to share experiences and exchanges. And this industry will help to improve personal ability and the prosperity and development of translation, and have a positive impact [3].

Today, driven by information and network technology, translation teachers should get rid of traditional translation concepts. Translation skills are not only translation skills between two languages, but also information skills for processing translated text. The use of information technology to help translation is an irreversible trend [4]. Therefore, when training talented translators, teachers should keep up with the development of the times, timely understand various advanced information translation techniques and translation concepts, and attach importance to training students' ability to use various information technologies to complete translation work in the process of education and teaching. Only in this way can learning be truly realized, because the Internet is the representative of the development of the information age. All types of work in today's society cannot be separated from the Internet, and translation is no exception [5]. Therefore, teachers should be good at guiding students to use network tools to make translation work smarter and more efficient. The Internet plays an important role in preparing, implementing and improving translation. Before starting translation, teachers should encourage students to use the Internet to study the basic knowledge and background knowledge related to the translated text, which is the prerequisite for accurate and fast translation. In the translation process, teachers should guide students to learn how to use the Internet or other software to help translation to solve difficulties and confusion, which is also the key to improving accuracy and efficiency. After the translation is completed, the teacher should guide students to learn how to use the Internet to check and improve the translated text, so that the final product meets the requirements [6].

Since translation work involves multiple fields, and each field studies the different translation capabilities of translators, they can only adapt to the development of this era by continuously improving their work capabilities, including the ability to use language knowledge, translation skills, and accurate selection [7]. This not only means that translators must have the ability to use modern technical information to complete information collection, improve the level of translation and broaden their professional fields, but also must have the ability to use translation technology [8]. Computers and other devices that help translation to improve translation efficiency. The traditional translation method is manual, and individual translation and translation efficiency are relatively low. However, the demand for translation in the era of big data is constantly increasing. Therefore, computer, software and some forms of program-assisted translation work have gradually replaced traditional manual translation, and the translation method is changing from one-to-one to many-to-one, focusing on cooperation and open translation process [9]. Manual translation is gradually transitioning to software translation, thereby improving translation efficiency and quality. Cloud translation platform is a translation platform that combines advanced computer technology, translation technology and cooperation space. The staff involved in translation projects can share information, cooperate and communicate anytime and anywhere, which greatly improves the efficiency and quality of translation work [10].

2   Algorithm

2.1   Artificial Fish School Algorithm

The algorithm is mainly based on the following characteristics: The area with the most fish is usually the place with the most nutrients in the water, so the algorithm can simulate the foraging behavior characteristics of the fish school near the area with the most nutrients to achieve optimization. Collect and process network information through artificial fish school algorithm to obtain the objective function:

$$PER = 1 - (1 - BER) \tag{1}$$

The probability of errors in intelligent translation of computer network information is BER. PER is a measure of the level of intelligent translation of computer network information. It is necessary to use PER as a measure of the translation level of the entire system, which requires the corresponding PER to be obtained from BER.

Np is the number of bits contained in an algorithm, but in fact, the error probability that may be experienced between the bits of the same algorithm is generally not the same. In addition, if the artificial fish school algorithm is used, the error probability between the bits will be correlated, and the hypothesis will be invalid. In the following, for the two cases of artificial fish swarm algorithm and artificial fish swarm algorithm without artificial fish swarm algorithm, we roughly calculate and derive the estimation formula of the intelligent translation probability of computer information. The calculation method of QAM modulation PER under the AWGN channel has been given. Method, first obtain the error probability calculation formula for the transmission mode that does not use the corresponding sequential transmission. Then in the I-order QAM modulation, the error probability of the k-th symbol is:

$$PER_{QAM} = 1 - \prod_j^{\log_2 i}(1 - P(k))^{\frac{Np}{\log_2(i-j)}} - \prod_{x=1}^{\log_2 j}[1 - P(x)]^{\frac{Np}{\log_2(i-j)}} \tag{2}$$

It can be obtained uniformly with the given PER expression. According to the calculation formula of PER, the calculation formula of PER in each mode can be obtained:

$$PER_n(y) \approx \begin{cases} 1, & 0 < y \le y_n \\ a_n exp(-g_n y), & y \ge y_n \end{cases} \tag{3}$$

It is obvious from the intelligent translation algorithm described above that the artificial fish school algorithm greatly accelerates the speed of its translation and the accuracy of the translation results. Through the analysis of the algorithm in the traditional detection system, it can be concluded that during the detection process, the selection of initialization is randomly distributed, so it is easy to trap the algorithm into a local optimal trap, which is difficult to achieve.

3   Intelligent Translation Model Based on Big Data Technology

An information intelligent translation model is established on the basis of big data technology. The establishment and application of this model can promote the communication between different languages in the current information age. This paper proposes a combination of big data technology and artificial fish school algorithm to make fuzzy The idea is to introduce artificial fish school algorithm. The combination of the two forms a better information translation effect. Select the relative threshold number M ($2 \leq M \leq N$) and the exponential weight r ($1 \leq r \leq \infty$). Select the initial segmentation matrix U (0) and the error critical value ε. Initialize cluster centers:

$$di(i = 1, 2, \ldots, M) \tag{4}$$

Based on experience from the sample of input translation information:

$$Xj(j = 1, 2, \ldots, N) \tag{5}$$

Select M samples as cluster centers. Substitute U(t) into formula (5) to calculate fuzzy clustering centers

$$\left\{ \frac{m^{(t)}}{i} = 1, 2. \ldots M \right\} \tag{6}$$

$$U_t \frac{(1/x_j)}{\sum_{t+1}^{m} u} * \sum_i^N (u_t) * U_t * i = 1, 2 \ldots, M \tag{7}$$

The problem of intelligent translation of information is transformed into the class objective function of artificial fish swarm algorithm, and into the least square estimation problem:

$$\{z(t)\} = [U]\{Y(t)\} \tag{8}$$

The response is:

$$\{Y(t)\} = \int_0^t [h(t - \tau)]\{F(\tau)\}d\tau \tag{9}$$

Given the feature vector of intelligent translation, the algorithm formula can be listed as follows:

$$Y_j(t) = -m_j^{i-1} \int_0^t e^{\lambda_j(t-\tau)} \{V_j\}^T \{f(\tau)\} d\tau \tag{10}$$

In order to ensure the reasonableness of the weights, the consistency check of the comparison and judgment algorithm can be expressed as:

$$x_{ob} = \sum_{j=1}^5 u_{bj} Y_j(t) \qquad x_1 = \sum_{j=1}^5 F_{uj} Y_j(t) \tag{11}$$

4 Results

After the big data technology processing of the intelligent translation model, we translate a set of given information, which contains 100 different languages including Chinese, English, French, etc. Through multiple translations of the information, according to the translation result reflects the performance of our algorithm model and what problems need to be improved. The comprehensive performance of the intelligent translation model is obtained by testing the accuracy of the information translation and the coverage of the intelligent translation model of the test information.

Table 1. Performance test

Number of translations	Accuracy (%)	Coverage (%)
1	86.45	78
2	89.23	81
3	91.56	84
4	95.62	91

Table 1 can get the following data: The intelligent translation model of artificial fish school algorithm based on big data technology proposed in this paper is used as follows. The optimization of artificial fish school algorithm has gradually improved the accuracy of information intelligent translation. From 86.45% of the first translation to 95.62% of the fourth translation, the artificial fish swarm algorithm's optimization results for information intelligent translation are constantly improving; the coverage of the algorithm for information intelligent translation is also gradually improving, starting from the first 78% of the first translation increased to 91% of the fourth translation; it can be seen that the performance of the intelligent translation of information based on the artificial fish school algorithm proposed in this article is gradually strengthened, and the performance test based on this algorithm is positive for intelligent translation of information significance.

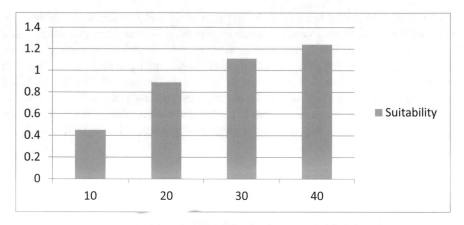

Fig. 1. Algorithm matching degree

From the data analysis in Fig. 1, it can be known that the data during the 40-time algorithm optimization test and utilization process shows that the artificial fish school algorithm has an increasing influence on the intelligent translation of information, and the matching degree of the algorithm in the test process is Keep deepening. The higher the matching degree of the algorithm, the more accurate the intelligent translation of information, and the ability of information translation is also constantly improving. The artificial fish school algorithm is more efficient in the algorithm model. Therefore, the artificial fish school algorithm is used for optimization in traditional information translation, and improving the matching degree of the algorithm and the system can greatly improve the performance of information intelligent translation.

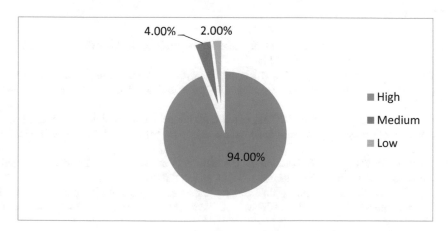

Fig. 2. The use of big data technology in intelligent translation algorithms

From the survey data shown in Fig. 2, we can see the current use of big data technology in intelligent translation. The translation results of this information are processed, and the use of big data technology in intelligent translation is divided into high, medium, three levels lower. According to the data, 94% of the results believe that the current situation of intelligent translation using big data technology is high, 4% of the results believe that the current situation of intelligent translation using big data technology is medium, and 2% of the results believe that current intelligent translation the use of big data technology is low. Through the usage survey, it can be seen that big data technology is indispensable in intelligent translation, and the artificial fish school algorithm combined with big data technology has a unique advantage in intelligent translation.

5 Conclusion

In summary, the computer has become an indispensable tool in people's daily life, and the Internet has become the trend of the times, which has brought a huge impact on our lives. As an important language technology for international communication, translation occupies an irreplaceable place in the era of big data with increasingly close information exchanges. We need to deeply analyze the shortcomings exposed by the current computer information processing technology, and fully integrate the development trend of big data to lay a good foundation for the development of big data technology. Education must be at the forefront of development. The big data environment brings higher requirements for translation work and translation skills, so the training of translation skills must not lag behind. In the era of big data, translation technology tends to be networked and intelligent, so translation teaching should also be adjusted appropriately to prepare for the training of translators needed in the new era.

References

1. Xu, W., Zhou, H., Cheng, N., et al.: Internet of vehicles in big data era. IEEE/CAA J. Autom. Sinica 5(1), 19–35 (2018)
2. Wang, Y., Kung, L.A., Byrd, T.A.: Big data analytics: Understanding its capabilities and potential benefits for healthcare organizations. Technol. Forecast. Soc. Change 126, 3–13 (2018)
3. Wang, X., Zhang, Y., Leung, V.C.M., et al.: D2D big data: content deliveries over wireless device-to-device sharing in large scale mobile networks. IEEE Wirel. Commun. 25(1), 32–38 (2018)
4. Yudong, C., Yuejie, C.: Harnessing structures in big data via guaranteed low-rank matrix estimation. IEEE Signal Process. Mag. 35(4), 14–31 (2018)
5. Gu, K., Tao, D., Qiao, J.F., et al.: Learning a no-reference quality assessment model of enhanced images with big data. IEEE Trans. Neural Netw. Learn. Syst. 29(4), 1301–1313 (2018)
6. Nissim, K., Bembenek, A., Wood, A., et al.: Bridging the gap between computer science and legal approaches to privacy. Harvard J. Law Technol. 31(2), 687–780 (2018)
7. Bravo, G., Farjam, M., Grimaldo Moreno, F., et al.: Hidden connections: network effects on editorial decisions in four computer science journals. J. Inform. 12(1), 101–112 (2018)

8. Ouyang, D., Yuan, L., Zhang, F., Qin, L., Lin, X.: Towards efficient path skyline computation in bicriteria networks. In: Pei, J., Manolopoulos, Y., Sadiq, S., Li, J. (eds.) DASFAA 2018. LNCS, vol. 10827, pp. 239–254. Springer, Cham (2018). https://doi.org/10.1007/978-3-319-91452-7_16

9. Santi, D., Magnani, E., Michelangeli, M., et al.: Seasonal variation of semen parameters correlates with environmental temperature and air pollution: a big data analysis over 6 years. Environ. Pol. **235**, 806–813 (2018)

10. Mariani, D., Martone, J., Santini, T., et al.: SMaRT lncRNA controls translation of a G-quadruplex-containing mRNA antagonizing the DHX36 helicase. EMBO Rep. **21**(6), e49942 (2020)

Establishment and Development of Chinese Language and Literature Database Under the Background of Information

Tongyao Wang[✉]

Shandong Management University, Jinan 250357, Shandong, China

Abstract. Today's world is a world of information globalization. The time and space for people to communicate is getting wider and wider, and the communication between countries is getting closer and closer. Mankind has entered the era of globalization. More and more countries are beginning to attach importance to the important role of language in the game of national culture. As China's comprehensive national strength continues to increase, China's international influence has also increased substantially. Chinese is very popular all over the world. Of course, the unique tool charm and cultural charm of the Chinese make it an important tool for shaping Chinese national culture. The subsystem of Chinese national culture has an irreplaceable function of national identification. An ideal Chinese language and culture will enable China to seize the opportunity to compete with other countries in the world and help China gain an advantage in international competition. In the context of information technology, this article adopts the literature research method and the cross-research method, and draws the following conclusions that the number of people who can use Mandarin accurately and fluently accounts for 40% of the participants; those who can use Mandarin but have some accents account for 35%; Basically, 15% can talk; only 10% can understand but don't speak well, which proves that Mandarin in China is developing towards a good trend.

Keywords: Informatization · Chinese language and literature · Database · Document research method

1 Introduction

National language and culture can be regarded as a field that is rarely contacted. Chinese language and culture are made up of these colorful national languages. But there are not many things that can be directly learned directly from the national language and culture. Therefore, we can only find signs of adaptation in certain related fields, and integrate these parts to construct Chinese language and culture. These fragments are mainly concentrated in the following aspects: national culture is the surface of national language and culture, which defines the concept and constituent elements of national language and culture; the relationship between language and culture is inseparable, and the construction of Chinese language and culture must: Chinese language The relationship with culture The construction of Chinese language and culture is the best international strategy for the promotion of Chinese in the world.

J. Abawajy et al. (Eds.): ATCI 2021, LNDECT 81, pp. 316–324, 2021.
https://doi.org/10.1007/978-3-030-79197-1_46

It is also an issue that we must pay attention to; in addition, some related research fields, such as language design, language resources, language ability, etc. Problems in the scientific field also have enlightening significance for Chinese language and culture.

Tang Xiaoli believes that with economic development, the charm of Chinese language and literature has been recognized by more and more countries. In the context of information technology, how to make a resume about the Chinese language database has become a topic of thinking for many people, and many experts have put forward their own views on this phenomenon [1]. Hirakawa believes that "language promotion is of key significance to the growth of the country's overall strength. "Strong language and culture" can have a greater impact and role in the international community, and it will bring multiple benefits to a country that are difficult to calculate. Not only language and culture, but also the expansion of ideology, and huge economic benefits, the soft power of the country has continued to increase" [2]. Kamath pointed out that "the international promotion of Chinese is a product of the rise of China and the accelerated transformation of the international landscape. It is the result of the dual role of global objective demand and China's independent promotion. It has gone beyond the scope of pure teaching of Chinese as a foreign language, language and culture. Rather, it has risen to a state behavior, which is an important part of China's public diplomacy and even China's overall diplomatic strategy" [3]. Although experts in various fields have made some contributions to the study of Chinese language and literature, they are more concerned about the communication in the context of the information age, and they have not done a good job of Chinese language and literature. In summary, there is no good suggestion for the establishment of the database.

The innovation of this article lies in the context of the information age. Taking modern Chinese ethics design as the main research object, based on the language design information database, starting from the perspective of language design mechanism, through the simultaneous description, analysis and further discussion of the timeliness and trends of its development in different periods of ethnological modern design, Carry out ethical design research and propose specific design research for the development of ethical design [4].

2   Method

2.1   Features of Simplified Chinese Characters

In the process of modern Chinese character simplification, the development and development of Chinese characters are always restricted by Chinese characters. The standard for testing a character system is not whether it is pinyin or something else, but whether it reflects the characteristics of the language it records and whether it can make people feel comfortable in use. A writing system should better serve the corresponding language system. Only when the main purpose of Simplified Chinese is to be accepted by the public, can the purpose of Simplified Chinese be more accurate and accurate in recording the language accepted by the public [5]. The change of Chinese characters is the product of the contradiction between Chinese characters and Chinese. It is not only reflected in the new birth and death of Chinese characters, but also in the changes and development of the structure of Chinese characters. Words are language ontology people created to break the time and space constraints of language communication and serve the exchange of information between languages. Chinese characters can be written and read. People always hope that the simpler the font, the easier it is to write. If it is not simplified, the better, and it is not easy to use. People always want to see the text easier to read, the better, otherwise the text will not achieve the purpose of communication. Simplifying Chinese characters is just for writing and reading. It simplifies the body structure through normal alignment and symbol replacement. By combining certain variants with complex shapes and ambiguous concepts, the requirements for simplifying the number of Chinese characters can be achieved. In short, the simplification of Chinese characters is the product of the changes of Chinese characters, which is the need for the self-development and self-improvement of Chinese characters, and conforms to the laws of Chinese development. The main purpose of simplifying Chinese characters is to better serve the Chinese language. Only by observing this rule can the simplification of Chinese characters be effective. Therefore, when we simplify Chinese characters, we must be fully familiar with and understand Chinese characters. Without divorcing from the reality of the Chinese people, it cannot be simplified at will. Only in this way can the overall simplification and systematic simplification of Chinese characters be achieved. In the work of simplifying Chinese characters, we have always adhered to the policy of "contracting and seeking progress while maintaining stability" and adhere to the mass line. The main feature of this language is convention [6]. From the perspective of language education, students have always been the most important group for learning simplified characters. Whether in class or in life, what students learn, use, write, and see are all simplified characters. If you want simplified characters now, or the structure or outline of some simplified characters recognized by the public is coordinated, it is unrealistic, because students have accepted the current simplified Chinese writing method, and at the same time, learning a set of modified Chinese characters will also Time wasted and the energy of teachers and students. The simplification of Chinese characters is progressing steadily, and optimization and simplification coexist. First of all, in the entire process of simplifying Chinese characters, the simplification of Chinese characters is carried out in batches and batches, and the cycle of drafting, testing, soliciting opinions, reviewing, reviewing, summarizing opinions, and reviewing, until every word list is applied. The regression process applies the "steady progress" policy [7].

2.2 Downplaying the Debate on Grammar Teaching

The introduction of "diluted grammar" has shaken the status of grammar and has continuously simplified the grammar content of the Chinese manual. Affected by this, teachers have gradually neglected the teaching and application of grammatical knowledge in Chinese classes. In the language curriculum model formulated by the Ministry of Education, the requirements for students to learn grammar and rhetoric are also "to learn basic vocabulary and grammatical knowledge with the text, and to understand language difficulties." The result of the test"; and clearly stipulates that grammar knowledge should not be used as the content of the examination. Therefore, in the actual Chinese classroom teaching, grammar teaching is underestimated or even abandoned by many teachers. The direct consequence of this behavior is that students cannot explain some common grammatical phenomena in Chinese learning, and there are more typographical errors and wrong suggestions. In fact, the teaching of grammar and rhetoric was originally a very important content in Chinese teaching, and its role is no less than the teaching of listening, speaking, reading, writing, and memorizing ancient poems [8]. The function of grammar is to allow people to express language correctly and formally, and the function of rhetoric is to allow people to express language appropriately. Learning grammar and rhetoric helps to improve students' language awareness and writing ability, and helps to improve students' English level and reading comprehension. Among the real results of the downgrading of grammar teaching, not only the typographical errors and poor suggestions of school students, but also the publicity of publications, newspapers, television and other media, there have been many violations of grammar rules. In this case, grammar teaching is still very important. Grammar teaching can enable students to understand some basic common logic and the most basic structural laws of their mother tongue, so as to achieve the purpose of gradually improving their language knowledge. Understand the basic logic of Chinese grammar, understand the classification of phrases, understand the rules and proposal writing rules, etc. Although it cannot immediately improve students' language and writing skills, it can cultivate students' understanding of their mother tongue in a subtle way. As a student, using grammar to learn mother tongue should be the most basic quality and knowledge of Chinese [9].

3   Experiment

3.1   Literature Research Method

Collecting a large amount of domestic and foreign documents and making full use of library resources and network resources, we can fully understand the characteristics and influencing factors of language, the relationship between language culture and Chinese international education. Its characteristics and influencing factors are as follows: It provides a reliable theoretical reference for the construction of Chinese language and culture. The relationship between language and culture determines the direction of Chinese language and culture construction. On the basis of a large number of previous research results, the general trend of the development of Chinese language and culture can be predicted [10].

3.2   Cross Research Method

Since Chinese language and culture is a proposition with little contact with countries, there are relatively few documents and experiences that can be directly referred to. Therefore, we should fully learn from the research of social economics, cultural linguistics, communication, semiotics, psychology, cognitive science and other disciplines. Theories and methods, trying to comprehensively analyze the generation mechanism of the Chinese language and culture, especially the questioning points and influencing factors, from a multidisciplinary perspective, in order to construct a Chinese language and culture recognized by the international community.

4   Discussion

4.1   The Impact of Education Level on Language Use

The figure below shows the average frequency of Nanchang and Mandarin spoken by local residents of different "educational levels" for different objects and on different occasions.

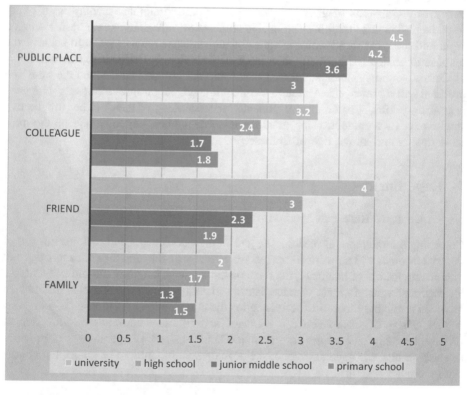

Fig. 1. Differences in the mean of language use by local residents of different education levels

From Fig. 1, we can clearly see such a trend: no matter when, where, or where, local residents with higher educational backgrounds use Mandarin more frequently, and from elementary school to junior high school to undergraduate and above, there is a clear increasing trend. Especially when talking with "friends and colleagues" and in "public places", local residents with college degree or above speak Mandarin in the majority. Only when speaking with family members (elders), local residents with education level above middle school use Nanchang dialect, while those with junior high school education and below use Nanchang dialect. It can be seen that the educational level has a significant impact on the use of Mandarin, and the higher the educational level, the higher the frequency of Mandarin use; the lower the educational level, the lower the frequency of Mandarin use.

4.2   The Language Ability of Residents

Because dialect is their mother tongue, it is a language naturally acquired from birth. If people have normal pronunciation organs and necessary physiological conditions, and they are in a normal language environment after birth, then people's native language ability is not much different. And Mandarin is acquired through various channels the day after tomorrow: some are learned at school; some are learned by watching TV and listening to the radio; some are learned through social interaction, so their Mandarin can be different and different, so We take their Mandarin ability as a language ability to investigate.

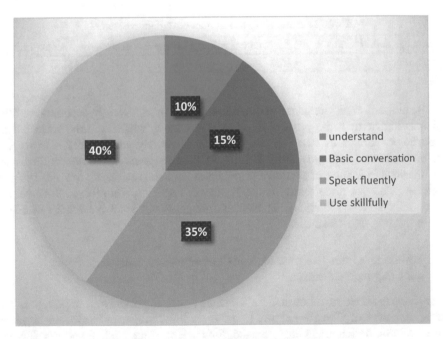

Fig. 2. Residents' mandarin ability map

The pie Fig. 2 above can clearly show the Mandarin ability of local residents. The number of people who can use Mandarin accurately and fluently accounts for 40% of the participants; those who can use Mandarin but have some accents account for 35%; those who can basically talk There are 15%; only 10% can understand but don't know how to say; there are no people who don't understand and cannot speak at all. It can be seen from the statistical data that the comprehension of Mandarin is still very high.

During our investigation, we have not found anyone who cannot understand Mandarin. If we can understand it, it provides favorable conditions for the next step in learning to speak. Therefore, on the whole, the Mandarin ability of local residents is quite satisfactory. Up to 94% of people can use Mandarin. Although they may not be very proficient or have an accent, that is, Nanchang Mandarin, the level of Mandarin is basically the same. Can achieve barrier-free communication with outsiders.

4.3   The Language Ability of Foreigners

Hometown dialect is the mother tongue of the foreigners, and it is the language they have learned from birth. For normal natural persons, people's native language ability is not much different. But Mandarin is acquired through various channels the day after tomorrow, and there is a time difference. Therefore, their Mandarin ability is quite different, so the Mandarin ability of the foreign personnel is also the object of language ability (Table 1).

Table 1. Putonghua ability map of foreign personnel

Language ability	understand	Basic conversation	Can express accurately	Use skillfully
Proportion	10%	16%	19%	55%

The proportion of foreigners who can use Mandarin proficiently accounts for 50%, which is higher than the proportion of local residents, which is only 40%; but the proportion of foreigners who can use Mandarin accurately and fluently is lower than that of local residents, only 19%. It is 42%; the foreigners who can conduct basic conversations in Mandarin are about the same as the local residents, both are 16%; the foreigners who can understand Mandarin but can't speak are similar to the local residents, respectively, 7% and 10%; Neither foreigners nor local residents can neither understand nor speak Mandarin.

4.4   Thinking About the Influence of Contemporary Language Policy on Language Education

The historical development of modern language policy is not only an important part of the development of Chinese culture, at the same time, language is also an important part of cultural heritage. On the one hand, the content of the change and development of language policy is an important part of the content of Chinese education; on the other hand, the change and development of language policy is an important part of the content of Chinese education. Chinese education activities have completed the specific tasks of the language policy. In other words, the language policy needs to realize the practice of language education. Due to the history and specific background of the policy release, combined with many factors, such as specific problems in the implementation process, the impact of language policies on language learning must also be objectively evaluated and investigated. For example, the gap between the purpose of language policy formulation and the actual practice of language education; the particularity of language policy and language curriculum; coordination; the consistency and contradiction between language policy and language itself: the integrity and integrity of language writing and language Educational relationship, etc. Among the many issues, three aspects are worth considering, namely: the constraints of language education and language policy and their value orientation.

5 Conclusion

As a bridge of communication, language plays an extremely important role in the exchange and cooperation of transnational trade and the communication between people. Language is not only a communication tool, but also a carrier of national culture. In the context of market economy and globalization, economic and trade exchanges between different countries and different ethnic groups have become increasingly close. One Belt and One Road strategy and financial cooperation; trade and cultural exchanges between China and countries along the route will become more frequent, and language is a necessary tool for communication. The historical development and development of modern language policy is the inevitable requirement of language development and the inevitable choice of language society. After many years of baptism, these policies have penetrated into the historical process of Chinese education, affecting the development direction of Chinese education, encouraging Chinese education to move towards the standard value orientation, relaxed, beautiful, and accompanied by Chinese education. China's education has also become healthier, richer, more effective, and more dynamic because of these policies.

References

1. Xiaoli, T.: Research on the promotion of English writing ability of junior middle school students in Heilongjiang province by information technology **000**(026), 140–141 (2018)
2. Hirakawa, Y., Chiang, C., Hilawe, E.H., et al.: Formative research for the nationwide promotion of a multidisciplinary community-based educational program on end-of-life care. Nagoya J. Med. Ence **79**(2), 229–239 (2017)
3. Kamath, M.: Prevention of disease and promotion of health: can ayush show the path? Int. J. Adv. Sci. Technol. **29**(4), 1510–1517 (2020)
4. State, V., Tnase, L.C., Petre, R.G.: Study regarding the respect of professional ethics and deontology in the promotion of accounting services in Romania. Valahian J. Econ. Stud. **10**(1), 89–98 (2017)
5. Li, Y.: Reflections on reading promotion service of public libraries in the context of cross-border cooperation **036**(015), 26–28 (2019)
6. Han, C.: Improving practical skills in the teaching reform——taking the major of Chinese language and literature in normal universities for example **034**(003), 89–91,96 (2019)
7. Lihua, Z.: Research on the problems and Countermeasures of Chinese language and Literature Teaching under the new situation%. Manage. Technol. Small Med. Enterp. **000**(001), 83–84 (2019)
8. Luo, Z.: Research on the teaching of Chinese language and literature in modern educational thoughts **002**(002), P.29–33 (2019)

9. Ding, H., et al.: The new biomarker for cervical squamous cell carcinoma and endocervical adenocarcinoma (CESC) based on public database mining. BioMed. Res. Int. **2020**(2020):1–9 (2020)
10. García-Martínez, E., Sanz, J.F., Muoz-Cruzado, J., et al.: A review of PHIL testing for smart grids—selection guide, classification and online database analysis. Electronics **9**(3), 382–382 (2020)

Analysis of Hotel Management Data Based on Fuzzy Logic Mining

Peilin Chen[✉]

China University of Labor Relations, Beijing 100048, China

Abstract. After entering the 21st century, the golden age of hotel industry in China has come, the rapid increase of hotel number and the continuous expansion of hotel scale and business, the competition between hotel industry has become more and more fierce since then. In the increasingly competitive environment, the development of hotels have been threatened and challenged, how to appear in the reverse trend, so that they always maintain the dominant position in the industry, which requires hotels to change the traditional backward business philosophy, marketing thinking and marketing model from top to bottom, and to review the situation. Hotel industry is an important part of tourism in China. According to the survey, the hotel industry creates 16.68 billion yuan of tax revenue for the national finance every year, and provides about 1.51 million employment opportunities for the society every year, which has good social effect.

Keywords: Fuzzy logic · Hotel management · Data analysis · Marketing model

1 Introduction

In recent years, with the rapid development of our national economy and the continuous growth of residents' consumption ability and consumption desire, the service industry has developed rapidly and gradually become the leading industry in modern society. Hotel industry plays an increasingly important role in social and economic life.

The research of hotel management data analysis based on fuzzy logic mining has attracted the interest of many experts and has been studied by many teams. For example, some teams found that people's demand for hotels is increasing, the scale of hotels is increasing year by year, at the same time, the service and grade requirements of hotels are also increasing. According to the China Hotel Market Research and Investment Strategy Research report, China's hotel supply showed an overall growth trend in 2018, and the overall hotel room supply growth rate reached 10.2%. With the increase of consumer demand for accommodation environment, the room supply of middle-end hotels increased rapidly, ranking first among all grade hotels, reaching 15.7% [1]. But from the moment our country joined the WTO, with the implementation of the open and inclusive development policy of the domestic market, some catering tycoons in the international market also entered the Chinese market, to a large extent, it has brought some trouble and crisis to the Chinese hotel industry. In addition, a report released by the National Bureau of Statistics shows that the growth rate of the catering

J. Abawajy et al. (Eds.): ATCI 2021, LNDECT 81, pp. 325–332, 2021.
https://doi.org/10.1007/978-3-030-79197-1_47

industry in China decreasing year by year, and the competition between catering enterprises is becoming more and more white-hot. Therefore, in order to achieve stable growth of operating income and long-term maintenance of hotel brand, hotel enterprises must strengthen hotel marketing. By tapping the advantages of hotel products and services, combining with the internal and external environment, we should fully tap the target customers and potential customers, and constantly optimize hotel marketing strategies to create profits and achieve long-term development in the competition [2]. Some teams found that Qu Gao believes that the development of mobile Internet leads to the change of consumer behavior habits, therefore, the marketing strategy of hotel enterprises should also be changed accordingly. For mobile Internet marketing, Guo Xiaolan believes that in order to make full use of the advantages of mobile Internet in the current hotel marketing activities, hotels should increase capital and technology investment [3]. In particular, it is necessary to ensure that the hotel website can be viewed normally in the mobile terminal equipment, and the loading speed of the website can be guaranteed when needed, so as to facilitate consumers to browse the website better. At the same time, we should increase investment WeChat, the construction of new media marketing platforms such as Weibo in order to make better use of these marketing platforms to carry out related marketing activities. In addition, we should strengthen technical support to ensure that the wireless network of modern hotels in shopping malls can be connected smoothly in order to increase the psychological added value of consumers, better attract consumption and make marketing activities more effective [4]. Other teams found that today, the big data industry is booming, network technology is constantly innovating, the fast-paced information life model has become the mainstream of society, and staying in hotel apartments has become the choice of the public. Because the traditional hotel management mode can not adapt to the new social environment, it is gradually being replaced by intelligent and efficient modern management tools. Although some predictable hotels have purchased hotel management software, most of these software lack big data analysis function, can not provide better service to higher level users through big data analysis. The further development and continuous improvement of hotel management system is of positive significance to the development of domestic hotel industry and positive and important to the development of national economy [5]. Although their research results are very rich, but there are still some shortcomings.

This paper takes Jinhai Tian Hotel as the research object, through fully carrying out the investigation and research on the hotel, starting from the present situation and trend of the whole industry, combining the present situation of Jinhai Tian Hotel marketing strategy and the result of marketing management questionnaire, it provides a strong basis for hotel to formulate the next specific marketing strategy optimization scheme.

2   Method

2.1   Risk Measure

As early as 1693, one of the founders of Halley's actuarial science proposed to use a life table to measure risk, Fisher said "risk refers to the probability that the rate of return is lower than the level of benefit", Fisher's definition is considered the earliest expression of the following risks (downside risk); Thomas and Masgrave believe that investors are most concerned about the probability and extent of loss, so risk should be the possibility of loss and its adverse consequences, so the expected loss rate of investment can be used to measure risk, as shown in formula (1) [6].

$$R = -\sum_{r_i < 0} p_i r_i \tag{1}$$

Pi and ri are the probability and rate of return when the i situation occurs. If the return on investment is lower than the risk-free rate of return, it is considered that the loss has occurred and can be rewritten to formula (2) [7].

$$R = -\sum_{r_i < r_f} p_i (r_i - r_f) \tag{2}$$

From formula (2), it is not difficult to see that risk increases with the increase of expected loss value of investment. The risk measurement index given in formula (2) is very simple and consistent with people's usual thinking habits. But it does not really reflect the impact of volatility on risk measurement.

2.2   Fuzzy Inference Systems

The first fuzzy inference system, the output fuzzy rule is a single point, a value between 0 and 1, the system output is the weighted average of each rule output, where wi is the output of the second half of the i rule, zi is the incentive strength of the second rule, as shown in formula (3) [8].

$$z = \frac{w_1 z_1 + w_2 z_2}{w_1 + w_2} \tag{3}$$

2.3   Bl Neurons

The first class of fuzzy neurons, called fuzzy neurons, is a class of neurons that are energized or standardized to observe or input. it receives discrete values or continuous, determined, or fuzzy unit inputs. The output consists of normalized values determined by fuzzy basic state membership functions, usually in single-input, single-output forms. its input-output relationship is shown in formula (4), where F is a fuzzy function [9].

$$y_i = F(x_i) \tag{4}$$

When the membership function of the basic state of fuzzy variables in the system is complex, the blur of the basic state of fuzzy variables in fuzzy neural networks is often realized by fuzzy network layer. The input-output relationship of fuzzy neurons in network layer corresponds to the membership function of a basic state of fuzzy variables. As shown in (5), the common membership functions are Gao Si, bell, triangle and S [10].

$$y_{ij} = f_j(x_i), j = 1, 2, \cdots n \tag{5}$$

3 Experiment

3.1 Source of Experimental Data

Based on reading a large number of relevant documents, this paper takes Jinhaitian Hotel and its marketing strategy as the research object, analyzes and studies the present situation and existing problems of hotel marketing strategy, combined with the current marketing environment of the hotel. At the same time, questionnaire and statistical analysis.

3.2 Experimental Design

1. Literature research method. 2. Questionnaire method. 3. Comparative research method. 4. Interview method. 5. Statistical analysis.

4 Result

4.1 Analysis of the Hotel Industry

According to the national star hotel statistics report of the Ministry of Culture and Tourism for the first half of 2019, as of the first half of 2019, there are 10284 star hotels, there are 846 five-star hotels, 2542 four-star hotels, 4961 three-star hotels, 1862 two-star hotels hotels, star hotel enterprises in the scale and number of growth. The report also shows that, in the first half of 2019, national star hotel operating income 93.813 billion yuan, among them, catering income 38.215 billion yuan, room income 42.669 billion yuan, other income 12.929 billion yuan, as shown in Table 1 below.

Table 1. Statistical tables of the analysis of the hotel industry

Fivestar Hotel	846 home	Catering income	$38,215 million
Four Star Hotel	2542 home		
Three Star Hotel	4961 home	Miscellaneous receipt	$12,929 million
Two Star Hotel	1862 home		
One Star Hotel	73 home	Room income	$42.669 billion
Amount to	10284 home	Amount to	$93.813 billion

4.2   Analysis on Staff Situation of Jinhai Tian Hotel

The competition of enterprises is the competition of management and talents. There-fore, enterprises should work hard in recruiting and training employees. Jinhai Tian Hotel has 89 employees, including 2 undergraduate students, 12 college students and 45 high school students, as shown in Fig. 1 below. In the local 30 junior high school and below, more than half of the staff in all positions in the hotel, service staff edu-cation is low. From the age structure, the average age of hotel staff is 35 years old, showing a younger trend, the proportion of women in hotel staff is higher, accounting for about 70% of the total number of employees. According to the hotel entry and annual training plan, new entrants must receive 1 month of pre-job training, master hotel knowledge, and pass the training before they can take up their posts. Jinhai Hotel will also hold occasional job skills training and competitions to further ensure that employees are qualified for their positions.

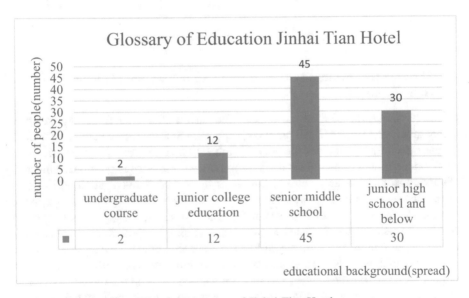

Fig. 1. Education of Jinhai Tian Hotel

4.3   Analysis on the Operation of Jinhai Tian Hotel

Since the opening of the Golden Sea Sky Hotel in late 2015, operating income increased from 4.327 million yuan in 2016 to 11.644 million yuan in 2019, it's growing year by year, food revenue 4.672 million yuan in 2019, an increase of 50.46% over the same period last year; room income $6.972 million, 61.02% year-on-year, as shown in Fig. 2.

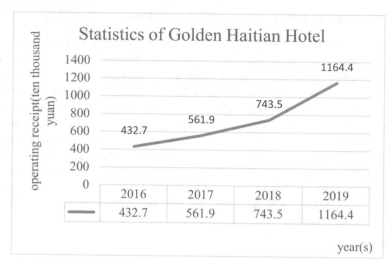

Fig. 2. Statistics on the operation of Jinhai Tian Hotel

4.4   Through the Investigation and Analysis of the Aspects of Jinhai Tian Hotel Products

The results show that 38% of customers think that Jinhai Tian Hotel should appropriately reduce the price of hotel products, especially hotel accommodation, in addition to some agreed unit customers, the price of Jinhai Tian Hotel, especially the price is too high. Compared with other hotels in Honggu District, the price is prohibitive for many ordinary consumers, mainly because the target market of Jinhai Tian Hotel is a high-end commercial market group. Accommodation and other products are priced according to the standard of high star hotel and four star hotel, which is higher than other local hotels, which is related to his simple pricing method and lack of scientific and flexible pricing method. As we all know, reasonable prices have a great impact on the survival of hotels. In a large and competitive environment, they are higher than competitors' price positioning. In special periods, especially in the off-season tourism, Jinhaitian Hotel catering, accommodation prices are not obvious preferential premise, it is difficult to attract consumers to shop consumption, cannot guarantee its competitive advantage in the local hotel market, thus directly affect the hotel operating income, profit space is relatively small. In addition, 20% of customers believe that hotel "green" products are lacking and need to add personalized "green" products.7% of customers report that hotel room space is too small and facilities need to be improved. Hotel products need to be improved valuable advice Jinhai Tian Hotel should actively adopt, improve customer satisfaction, as shown in Fig. 3.

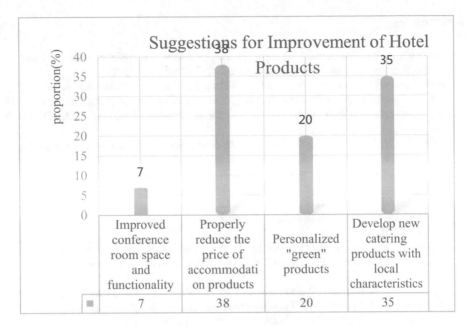

Fig. 3. Recommendations for hotel product improvements

5 Conclusion

The existence of hotel marketing is to meet the reasonable requirements of consumers to the maximum extent, to promote consumption, to meet consumption, so as to achieve the purpose of profit. In this process, various marketing models and marketing strategies promote the consumption of guests with different novelty. In the fierce market competition environment, if hotels, especially star hotels as one of the backbone of China's service industry, want to be evergreen. In a favorable position, we must do marketing work, expand marketing ideas, meet customers' diversified needs and achieve long-term development. Jinhai Tian Hotel, as the only four-star hotel in the local area, has a considerable number of customers in the short years of operation, and has also gained a good reputation and recognition in the industry. However, due to the lack of guidance of modern marketing theory, there is no reasonable marketing strategy.

Acknowledgements. Phased Achievements of the 2021 School-level Scientific Research Project (21XYJS023) of the China University of Labor Relations.

References

1. Grnholdt, L., Martensen, A.: Linking employee, customer, and business results: a study in the hotel industry. Total Qual. Manage. Bus. Excell. **30**(2), 1–9 (2019)
2. Beda-García, M., Claver-Cortés, E., Marco-Lajara, B., et al.: Corporate social responsibility and firm performance in the hotel industry. The mediating role of green human resource management and environmental outcomes - ScienceDirect. J. Bus. Res. **123**(10), 57–69 (2021)
3. Mun, S.G., Woo, L., Paek, S.: How important is F&B operation in the hotel industry? Empirical evidence in the U.S. market. Tour. Manage. **75**, 156–168 (2019)
4. Rockett, G., Yokoyama, M.: A novel approach to inclusion and outreach to be tested at the Hilton Chicago O'Hare Airport hotel. J. Airport Manage. **13**(3), 238–244 (2019)
5. Nurmagambetova, A., Baimukhanova, S., Pukala, R., et al.: Improvement of accounting in the hotel business in the transition to a digital economy. E3S Web Conf. **159**(4), 04019–04020 (2020)
6. Yakushev, O.: Innovative technologies in the management of business processes of enterprises of hotel-restaurant and tourism industries. Restaur. Hotel Consul. Innov. **3**(2), 195–208 (2020)
7. Кузина, О., Kuzina, O., Раденович, М., et al.: Personnel Management in the Hotel Business: the Importance of Training. Management of the personnel and intellectual resources in Russia **8**(1), 56–60 (2019)
8. Sahirova, A.A.: Sahirova, The corporate identity of the hotel business in the tourism marketing system. Reporter Priazovskyi State Tech. Univ. Sec. Econ. Sci. **37**, 182–187 (2019)
9. Blei, I., Tepavevi, J., Stani, M., et al.: Impact of innovations on service orientation of employees in the hotel management. Marketing **50**(1), 13–23 (2019)
10. Povidaychyk, O., Popyk, M.: Methodological approaches to the training of future specialists in the hotel and restaurant business in a high school. Sci. Bull. Uzhhorod Univ. Ser. Pedag. Soc. Work **2020**(1(46)), 106–110 (2020)

Application Value of Big Data Mining Technology in the Field of Financial Venture Capital

Hanfeng Yang(✉)

School of Economics, Shanghai University, Shanghai, China

Abstract. In the context of the rapid development of computer technology, the level of informatization in various industries and fields is also rapidly increasing. In recent years, the scale of big data has continued to expand and has become the backbone of financial venture capital. The market volatility and business complexity brought about by Internet finance have challenged traditional economics and finance research paradigms. This article mainly introduces the application value research of big data mining technology in the field of financial risk investment. This paper uses data mining technology in big data to detect real-time dynamics in the field of financial risk investment and establish an early warning model. The model is solved by the decision tree algorithm, and the data is mined using the typical C4.5 algorithm in the decision tree algorithm to generate a decision tree and transform it into classification rules. Then discover the laws hidden behind financial risk investment to provide a reliable basis for financial investment. The experimental results in this paper show that the decision tree algorithm reduces the occurrence of financial investment risks by 18%, and performs early warning analysis of financial risks. Finally, based on big data, relevant technical analysis is carried out for the financial investment field.

Keywords: Financial risk · Data mining · Early warning model · Decision tree

1 Introduction

At present, the financial market is an important place for financing funds. Both investment and financing needs can be connected in the financial market and can effectively resolve the contradiction between capital supply and demand [1, 2]. The method of analyzing financial risks has always been a hot spot in financial research [3, 4]. With the rapid development of the financial market, investment analysis methods are constantly changing and improving. On the one hand, the application of traditional time series models depends on multiple assumptions, so the application is limited [5, 6].

With the rapid development of big data mining technology, my country's economy and society have made great progress. The application of data mining technology has spread across various fields, changing the work mode and thinking methods in various fields [7, 8]. Salem J's survey report pointed out that users of large-scale systems need to adopt new technologies in order to tap the value of the market and beyond. Use more to create new business growth points for a wide range of parallel processing systems [9]. "Simsek S released a report on the information access tool market, which accounts

© The Author(s), under exclusive license to Springer Nature Switzerland AG 2021
J. Abawajy et al. (Eds.): ATCI 2021, LNDECT 81, pp. 333–340, 2021.
https://doi.org/10.1007/978-3-030-79197-1_48

for 26.6% in the Asia-Pacific region, and predicts that this market will reach 7.2 billion US dollars by 2022 [10, 11]".

The innovation of this article is to propose the application of big data technology and the financial field, and plan and forecast the investment of financial risk management based on big data mining technology. Discussed the division of big data mining technology in the financial field, the role of transaction analysis, researched and analyzed the data mining technology of financial risk investment, and proposed an optimization algorithm for decision tree generation in early warning mode analysis. But because the data scale is not comprehensive enough, the experimental results cannot be very accurate.

2 Financial Risk Investment Under Big Data Mining Technology

2.1 Time Series Model

With the diversified development of Internet finance, various forms of platforms continue to appear. Current mainstream applications include online lending, crowd-funding, third-party payment and digital currency. Since the forms of different applications are very different, it is difficult to summarize all the platforms and mechanisms in this article. Online borrowing and crowdfunding are currently the two most popular forms of Internet finance, and their mechanisms are similar to a certain extent. The same Internet financial platform has different strategic goals and business drivers.

Some platforms are based on helping the poor, while others provide a channel for investment or even shopping. Different strategic goals determine that the platform will adopt different types of returns to maintain the customer base.

The ARMA method is developed on the basis of the ARMA model proposed by Wold combining autoregressive AR and moving average MA. It is the main method of time series analysis, also known as the Box-Jenkins method.

A p-order autoregressive AR(P) model refers to a time series and has the following form

$$X_t = \sum_{i=1}^{p} a_i x_{t-i} + \varepsilon_t \tag{1}$$

Among them is white noise, which can be expressed as ε_T

$$X_t = \sum_{j=0}^{q} \theta_i \varepsilon_{i-j} \tag{2}$$

The autoregressive moving average ARMA (p, q) model is a combination of, and the time series has the following form

$$\alpha(L)X_t = \theta(L)\varepsilon_t \tag{3}$$

2.2 Data Mining Technology

With the rapid development of computer software and hardware technology and its popularization and application, people's ability to collect data and information has been greatly improved. The abundance of information collection methods and the development of data storage technology have accumulated a large amount of data in all walks of life. The collection and processing of information has become an important factor affecting decision-making. Information systems and data mining tools have been widely used to improve the level of scientific decision-making. However, in China's financial market, investors have not yet received the support of effective information acquisition and data mining and other technical methods. This is mainly reflected in the financial data mining technology researchers often have many difficulties in the comprehensive acquisition of financial data. This article focuses on service-oriented the key technology of financial data mining of framework (SOA), researches on several aspects such as the framework and development of financial data mining, data storage and management, and financial data mining technology methods. In the face of rapidly increasing mass data, data analysis and understanding have naturally become people's next goals. However, the original tools for traditional mathematical statistics and data management for data can no longer meet the needs of applications. Therefore, there is an urgent need for powerful and efficient analysis tools to intelligently convert data from massive data into useful knowledge and improve the utilization rate of information.

Designed the architecture of financial data mining, proposed a five-layer architecture model, analyzed the role and function of each layer, studied the distributed deployment and management of financial data mining, gave the formal definition of the main objects of the system and discussed the relationship between them describes the main use cases and their workflow. One of its outstanding advantages is that it can handle linear inseparable situations. At present, this method has become a popular method in the field of data mining, and it is widely used in mining problems such as classification, estimation and prediction.

3 Early Warning Experiment of Data Mining in Financial Risk Investment

3.1 Time Series Data Mining TSDM

A time series database refers to a database composed of a series of sequence values or events that change over time. Time series is the most common and widely used form of time series data representation, and naturally it has become one of the most important research objects of data mining. How to effectively manage massive time series databases, discover the knowledge contained in time series data, and understand more complex time series data is the main task of time series data mining. Time Series Data Mining (TSDM) is time series data mining. The core content of, but divided from the existing research work and the research direction and research objectives in the literature, the concept has a broad and a narrow sense. In a narrow sense, time series data mining TSDM refers to pattern mining from time series. This chapter adopts its narrow definition, and only refers to the pattern discovery of time series. The research methods of pattern discovery of time series are mainly divided into methods based on time series discretization and methods based on phase space reconstruction. The core is to first divide the time series by appropriate methods and map them to points in the high-dimensional feature space, convert the original one-dimensional time series data into a data mining problem for the high-dimensional space, and then perform the high-dimensional space Clustering or classification process to discover temporal patterns.

3.2   Establishment of Financial Security Early Warning Model Structure

This paper chooses a three-layer BP network technology to construct a financial risk dynamic early warning model. The model contains three types of nodes: input nodes, intermediate nodes and output nodes. The Sigmoid function is selected from the input layer to the middle layer, and the Purelin linear function is selected from the middle layer to the output layer.

Modeling process:

(1) Initialization of the model and values; (2) Input the original training samples and expected output results; (3)Output the neurons in the middle layer and the final credit evaluation results: (4) Study the errors of different layers; (5) clarify the correction value of the connection weights of the input layer to the middle layer and the middle layer to the output layer; (6) adjust the connection of different layers Weight; (7) If the established accuracy level or the established upper limit of training times is reached, the training is terminated. At this time, if the error has reached a predetermined level, the training ends, otherwise, return to the second step and start the subsequent training again.

The detailed operation steps are: the first step is to start the MATLAB software at the node, then enter the credit index test set, run the BP neural network program, and use the Map/Reduce module that comes with MATLAB to transfer the operation instructions to the Hadoop set and disassemble it into multiple In the Map job, a Map node obtains a sample of credit index information, completes the corresponding calculation, and then outputs data to the Reduce node, and integrates the data of all Reduce nodes to achieve the final output result. Judge whether the output data is consistent with the expected result, if it does not, the feedback is reversed, and the correction value of the connection weight between different layers is determined again, until the final output is consistent with the expectation, and the BP neural network training ends. Then input the credit information to be evaluated, and perform the evaluation through the trained neural network model to solve the final score. The specific results are shown in Table 1.

Table 1. Based on BP neural network risk prediction model results

Number of hidden layer nodes	2	3	4	5
Relative error percentage	6.34%	4.57%	2.64%	0.24%
Mean square error	0.0095	0.0125	0.0016	0.0007

4 Financial Security Algorithm Analysis Under Big Data Mining

4.1 Financial Security Decision Tree Algorithm Analysis

Decision tree algorithm is applied in the field of financial analysis. It is generally divided into classification tree, which is generally used as decision tree for discrete variables; regression tree, decision tree is divided into the following types: the attribute value of the test attribute may contain a different number, the result each internal node will produce two or more branches. Financial data mining methods based on neural networks have studied the correlation of financial time series, proposed the concept of function correlation, and given the artificial nerve for quickly determining the correlation function. Network model; research on the characteristics of stock price fluctuations, propose the HLP method for stock data preprocessing, and establish an artificial neural network model for stock price fluctuation prediction: research portfolio index tracking stock clustering based on self-organizing neural network model, using this model to analyze the investment portfolio optimization problem shows that a good index tracking effect has been achieved.

The decision tree uses information gain to find the field with the largest amount of information to establish a node, and establish branches with different values of the field, and establish a tree structure to represent the decision set. These decision sets generate rules through the classification of data sets, and can be used for classification and prediction applications. Typical decision tree methods are CART, CHAID D3, C4.5, C5.0 and so on. The results of the financial risk error indicator are shown in Fig. 1.

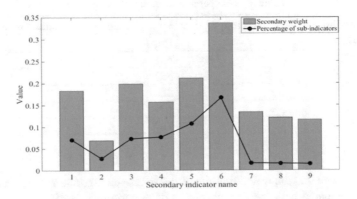

Fig. 1. Error optimization performance of the modified model

When the value reaches 6% points, it reaches a peak value, and the final decision tree generated "leaves thinning". Many branches in the tree reflect abnormal conditions in the training data, which leads to a decrease in the comprehensibility and practicality of the decision tree. "Fully grown" trees require some pruning before they can be used. In other words, this decision tree may be very effective for the original data training set, but the prediction performance is not good when predicting new data, which is the so-called overfitting. Then the decision tree can be pruned to solve the problem.

This article uses the post-pruning method, specifically the cost-complexity pruning algorithm. The process is to first calculate the expected error rate of each non-tree node subtree after being pruned, and then evaluate the branch error rate and the weight of observations, and then calculate the expected error rate without pruning. Compare the two error rates to see if the subtree is retained or subtracted.

4.2   Financial Risk Early Warning Model System Design

The system design of the financial risk early warning agency this result proves the validity and accuracy of the financial risk assessment result. It refers to the process of dividing a collection of data objects into multiple classes or clusters composed of similar objects. It means that the data objects that belong to the same cluster are similar to each other, but the objects of different clusters are different from each other. The difference between clustering and classification is that before the clustering process starts, the category to be divided is unknown, that is, it does not rely on pre-defined classes or training examples with class identification. Clustering itself is an unsupervised learning process that realizes automatic classification according to the similarities and differences between patterns with minimal manual intervention. Cluster analysis is usually based on patterns (patm), and is therefore one of the important methods of pattern recognition. In a narrow sense, a pattern can be understood as a description of an object with a fixed structure that people are interested in, while a pattern class is a collection of similar patterns. Therefore, the data objects and patterns in cluster analysis are logically the same. The specific results are shown in Table 2.

Table 2. Comparison of case matching time

Traditional algorithm matching time (MS)	Improved algorithm matching time (MS)
1025	638
1214	352
1179	386
1233	429
1017	389

It can be seen that the matching time of the improved algorithm is reduced by more than 60% compared with the matching time of the original algorithm.

To test the prediction accuracy of the improved algorithm and the traditional algorithm. 100 cases of the commercial bank are randomly selected as test cases and matched with the case library. The average prediction accuracy of the original algorithm is 84.35%, and the average prediction accuracy of the improved algorithm is 89.47%. The specific results are shown in Fig. 2.

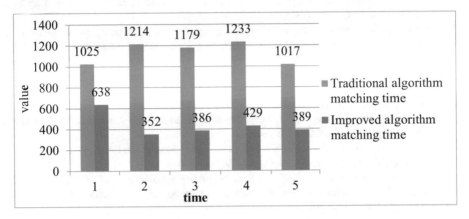

Fig. 2. Comparison of case matching time

5   Conclusions

Although this article has made certain research results on the application value of big data mining technology in the field of financial risk investment, there are still many shortcomings. There are many in-depth researches on the application value of big data mining technology in the field of financial risk investment. There are still many steps in the decision tree algorithm and early warning process that have not been covered due to space and personal ability. In addition, the actual application effect of the improved algorithm can only be compared with the traditional model from the level of theory and simulation.

References

1. Basseri, H.R., Dadi-Khoeni, A., Bakhtiari, R., et al.: Isolation and purification of an antibacterial protein from immune induced haemolymph of american cockroach. Periplaneta americana. J. Arthropod-Borne Diseas. **10**(4), 519–527 (2016)
2. Bruce, T.F., Slonecki, T.J., Wang, L., et al.: Front cover: exosome isolation and purification via hydrophobic interaction chromatography using a polyester, capillary-channeled polymer fiber phase. Electrophoresis **40**(4), NA-NA (2019)
3. Botero, W.G., Pineau, M., Janot, N., et al.: Isolation and purification treatments change the metal-binding properties of humic acids: effect of HF/HCl treatment. Environ. Chem. **14**(7), 417 (2018)

4. Escandón-Rivera, S., González-Andrade, M., Bye, R., et al.: Correction to α-glucosidase inhibitors from Brickellia Cavanillesii. J. Nat. Prod. **80**(1), 233 (2016)
5. Hui, C., Yayue, L., Yang, N., et al.: Polyketides from the mangrove-derived endophytic fungus Nectria sp. HN001 and their α-glucosidase inhibitory activity. Mar Drugs **14**(5), 86 (2016)
6. Chen, S., Liu, Y., Liu, Z., et al.: Isocoumarins and benzofurans from the mangrove endophytic fungus Talaromyces amestolkiae possess α-glucosidase inhibitory and antibacterial activities. RSC Adv. **6**(31), 26412–26420 (2016)
7. Chaichan, M.T., Kazem, H.A.: Experimental analysis of solar intensity on photovoltaic in hot and humid weather conditions. Int. J. Sci. Eng. Res. **7**(3), 91–96 (2016)
8. Franco, A., Fantozzi, F.: Experimental analysis of a self-consumption strategy for residential building: the integration of PV system and geothermal heat pump. Renew. Energy **86**, 343–350 (2016)
9. Salem, J., Champliaud, H., Feng, Z., Dao, T.-M.: Experimental analysis of an asymmetrical three-roll bending process. Int. J. Adv. Manuf. Technol. **83**(9–12), 1823–1833 (2015). https://doi.org/10.1007/s00170-015-7678-x
10. Yan, Z., Wang, H., Wang, C., et al.: Theoretical and experimental analysis of excessively tilted fiber gratings. Opt. Expr. **24**(11), 2107–2115 (2016)
11. Simsek, S., Kursuncu, U., Kibis, E., et al.: A hybrid data mining approach for identifying the temporal effects of variables associated with breast cancer survival. Expert Syst. Appl. **139**, 112863.1-1128631.3 (2020)

Digital Transformation on the Major of Human Resource Management in Big Data Era

Jing Lin[✉]

School of Management, Wuhan Donghu University,
Wuhan 430000, Hubei, China

Abstract. Objective: As the current focus of high attention, the research on the training of new data talents and professional reform under the background of big data era was carried out in this paper. Methods: A questionnaire survey based on 135 respondents was used and the results were analyzed by statistical software. Results: Statistical results showed that the respondents had a high degree of satisfaction with the current talent cultivation model of HRM generally, but they still hoped it can be improved to suit the development of the times and the requirements of enterprises much better. On the other hand, there were differences between students' cognition of the ability requirements and the actual needs of enterprises. Nevertheless, students had high expectation for the digital transformation of HRM, and hoped to join data courses such as data analysis, data-oriented thinking cultivation, and digital platform simulation as soon as possible. Conclusion: The training of complex talents in big data was still in its infancy and exploration stage. It was very necessary for HRM to carry out digital transformation and cultivate big data talents with practical ability and technological innovation ability.

Keywords: Digital transformation · Human resource management · Professional reform · Big data era

1 Introduction

The Outlook on the Development Situation of China's Big Data Industry in 2020 pointed out that the big data industry in China had continued to maintain healthy and rapid development in the past year, at the same time, it had been becoming an important driving force for the innovative development of the digital economy. In 2020, with the rise of developmental upsurge of digital economy, the construction of digital China had been going deeper, and the demand for digital transformation had been unleashed in large numbers. China's big data industry had ushered in a new period of development opportunities and will march towards a higher level [1].

J. Abawajy et al. (Eds.): ATCI 2021, LNDECT 81, pp. 341–347, 2021.
https://doi.org/10.1007/978-3-030-79197-1_49

2 The Significance of Digital Transformation on HRM

2.1 High Demands for Data-Oriented Talents in Enterprises

The report of Digital Transformation of China's Economy issued by the institute of management development and governance joint research center in Tsinghua university united with LinkedIn which is the largest Internet workplace social platform in the world showed that China's talent gap in the field of big data is up to 1.5 million at present and will reach 2 million by 2025, not including the demand for data-oriented talents in non-big data fields, such as human resources, finance, marketing and other fields [2].

The penetration of big data to enterprises had been no longer limited to the professional work of big data or the Internet, more and more related fields had also been affected by the wave of big data. Whoever can attract users by taking advantage of the convenience of Internet and the preference reflected by big data firstly will be able to stand firm in this round of industry reshuffle, which put forward new challenges to human resource management [3]. On the one hand, the human resource management department of the enterprise should select qualified talents to meet the needs of departments in the enterprise more quickly and effectively through rationalized means. On the other hand, the human resource management department also needs data-oriented talents to promote the digitization and informatization of its own work.

2.2 Quick Update of This Major in Universities

The number of universities in China which had been approved to offer majors in big data has increased year by year from 2016 to 2019, which was shown in Fig. 1. According to statistics, the universities approved to offer the major of big data were 3 in 2016, 32 in 2017 and 248 in 2018. In 2019, the number was 196. There had been nearly 500 universities across the country that had been approved to offer big data majors by the end of 2019, accounting for more than one third of the country's undergraduate universities. Big data will become a required major in every undergraduate school in a few years if this trend continues. It can be seen that universities can keep pace with society and update their specialty Settings quickly.

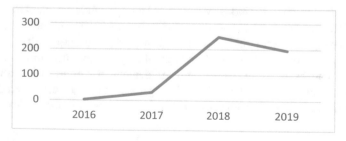

Fig. 1. Number of universities approved to offer major in big data

However, in this process, we should also pay attention to the quality of school and major. On the one hand, With so many universities offering courses in such a short period of time, are the faculty good enough to match the needs of students? Whether the cultivated big data talents can meet the quality requirements of enterprise? On the other hand, the major of big data in practice is mainly about science and engineering, especially computer, Internet and other fields, while the majors of liberal arts and business arc basically not related to it, even the major of new media, marketing and finance which need to be supported by big data.

3   Survey on the Demand of Digital Transformation

135 students in the major of HRM were selected to finish the online questionnaire in this survey aiming on the students' satisfaction with the existing training mode of HRM, their understanding on the talent needs of enterprises and the digital transformation of human resource management. 132 valid questionnaires were collected with effective recovery rate of 97.78%.The results of questionnaire survey were shown as follows:

3.1   Satisfaction on the Existing Training Mode

There were 43 students (32.58%) who were satisfied with the current training mode, 50.76% of them though that the current training mode need to be improved (relatively satisfied or relatively unsatisfied), 2.27% were very dissatisfied. Regarding the aspects need to be improved, students' opinions mainly fallen into the following aspects, as shown in Fig. 2 below. Among them, the high proportion lied in the lack of practical opportunities and the combination between the course and the popular big data technology.

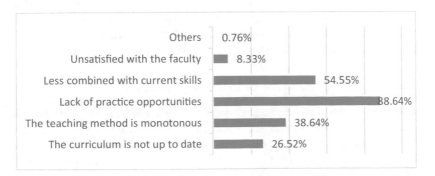

Fig. 2. Dissatisfaction with the existing cultivation model

3.2   Students' Cognition on the Requirements of HRM Practitioners

A clear cognition of the quality and ability requirements of HRM practitioners could help them arrange their time to learn necessary knowledge reasonably and participate in more activities conducive to the improvement of relevant abilities. At the same time, it would also help students to have a profound view on the mode of talent cultivation for their major.

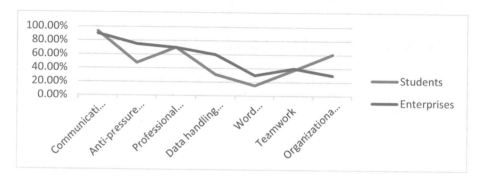

Fig. 3. The cognitive contrast between students and enterprises

As can be seen from Fig. 3, there were great differences between students' cognition and those of enterprises, Among which, the abilities of anti-pressure and data handling were seriously underestimated by students, while the ability of organizational coordinating was overestimated. The reason was the increasing attention on the psychological enduring capacity in recent years with more and more dispute caused by this. With the advent of big data era, data processing ability is favored by enterprises because it can significantly improve work efficiency and change time into money. Therefore, the applicants' data processing ability will also be taken into extra consideration in the recruitment interview. On the contrary, most enterprises do not regard the ability of organizational coordination as one of the essential skills in recruitment, but pay more attention to the candidates' execution.

3.3   Survey on the Demand for Data-Oriented Courses

The majority of students had a strong demand for curriculum reform and hoped to add data courses to enrich the original curriculum system according to the survey. Among all the various courses, the most popular were data-based thinking training course and digital platform simulation course. More than half of the students hoped to offer data analysis course with appropriate difficulty (Fig. 4).

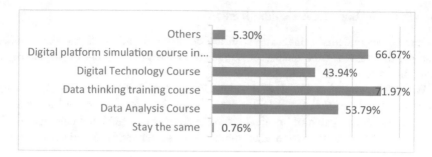

Fig. 4. Survey on the demands of data courses

4   Strategy for the Digital Transformation of HRM

4.1   Changing the Concept and Improving the Quality of Teachers

The shortcomings existing in traditional training mode and the necessity of digital human resource management should have been realized clearly to ensure that universities and teachers would put into more energy in digital transformation to strengthen the investment and research. Meanwhile, universities and teachers should ensure that they have enough resources to meet the needs of students' training. Therefore, they should combine their own advantages to broaden their knowledge. Only on the premise that they have contacted and mastered new knowledge first, can they carried out teaching work in new fields. In this process, universities should play a role in promoting and supporting teachers to go out for study, or inviting enterprises to hold lectures in universities [4].

4.2   Upgrading the Workplace and Building Digital Experimental Platform

The technical support of hardware was also needed in the digital transformation of HRM. At present, part of the enterprises had achieved digital operation, digital or cloud services platform had been used to complete the work of HRM which could greatly shorten the work time, improve efficiency and enhance its automation level.

The attempt of digital operation of HRM in enterprises forced the universities to establish laboratory for students to operate and practice [5]. Therefore, the traditional classroom teaching could not meet the needs of current demands of enterprises in digital transformation any more, instead of that, the digital experiment platform or wisdom classroom would be replaced. Students can exercise and summarize the methods of data application and analysis in the simulation through the actual combat simulation on the platform in order to achieve satisfactory results.

4.3 Building Digital Curriculum System

In view of the students' satisfaction with the current training mode and the change of basic requirements for the quality of human resource management practitioners, it is inevitable to reform the traditional specialty.

With the promotion and development of big data technology, the gap between disciplines has been broken gradually. The universities need to apply big data technology to the construction of practical teaching courses on HRM and cultivate the data thinking for the students so as to adapt the needs of society and enterprises in the future [6]. Although it is difficult for the digital and intelligent reform of HRM, the professionals would not be able to meet the needs of big data era if the traditional training program and major construction had not been reformed. Digital curriculum system need to regard practice ability as the core to realize the combination of theory teaching and practice and the fusion of teaching and research [7]. At the same time, it is necessary to further reconstruct the teaching system, optimize the teaching content, improve the teaching method, standardize the teaching process, perfect the teaching evaluation, and improve the training quality of professional data talents.

5   Conclusions

The training of complex talents in big data is still in its infancy and exploration stage. It is necessary to build and adjust the framework of the practical teaching system by taking the cultivation of digital practice ability and academic innovation ability as the double lead and multi-dimensional integration in the process of digital transformation of human resource management specialty. In addition, it is also necessary to establish two practical teaching chains, which means in-class and after-class, and reconstruct a complete practical teaching system based on the cultivation of innovative ability. The training mode and curriculum system that meet the needs of developing industry in the era of big data will be gradually formed to better serve the construction of regional innovation system and economic.

Acknowledgements. This work was supported by the grants from the 2020 university-level teaching research project of Wuhan Donghu university, a study on the application of big data technology in the reform of management specialty (Wuhan Donghu university teaching [2020] No. 9 Document).

References

1. Research group of big data industry situation: Outlook on the Development Situation of China's Big Data Industry in 2020. China Comput. J. **3**, 012 (2020). (in Chinese)
2. Yong, H.: Big data majors in universities should be prevented from overdoing. China Bus. Newspaper **4**, 2 (2019). (in Chinese)
3. Ran, L.: The digital transformation of human resource management in the information age. Times Fortune **5**, 147 (2020). (in Chinese)

4. Yanli, W.: The digital transformation of human resource management in the information age. Technol. Econ. Guide **28**(30), 20–21 (2020). (in Chinese)
5. Sudi, M.: Digital transformation of human resource management in modern enterprises. Shanghai Inf. **5**, 44–47 (2019). (in Chinese)
6. Fang, W., Fengxiang, J.: The cultivation and reform of human resource management professionals driven by big data. Hum. Resour. Dev. **1**, 48–50 (2020). (in Chinese)
7. Xia, D., Wang, L., Zhang, Q.: On teaching reform and practice exploration of big data application technology course. Big Data Res. **6**(04), 115–124 (2020). (in Chinese)

Application of Data Mining Technology in Construction of Cost Forecast Model in Transmission Line Engineering

Jing Nie$^{(\boxtimes)}$, Nan Xu, Yantao Xie, Bo Zhou, and Yan Song

Economic Technology Research Insitute, State Hebei Electric Power Supply Co., LTD., Shijiazhuang 050021, Hebei, China

Abstract. In view of the large number of problems in technical indicators of transmission line engineering cost, the difficulty of transmission line engineering cost estimation is put forward, and the transmission line engineering cost prediction model based on data mining technology is proposed. Firstly, the technical index of transmission line engineering cost is analyzed through principal component analysis and partial correlation analysis, and the data set is obtained. Later on, the project cost prediction model is created. The method of least squares support vector machine is adopted to make clear of the process of engineering cost. Finally, a simulation experiment is conducted to analyze the proposed method. The simulation results show that this method is characterized with high accuracy in prediction and good effect in application.

Keywords: Data mining technology · Transmission line engineering · Cost prediction

Against the background of continuous development in science and technology and economic globalization, data mining has been widely applied to the whole world and is undergoing rapid development. In this context, due to the development of global scientific information technology, information inflation and data overload were caused correspondingly. Therefore, in this paper, data mining technology is used to process information through new technologies, and valuable and effective parts are screened and extracted [1]. In the utilization of power engineering cost, using neural network technology and data mining technology, principal component analysis and least squares support vector mechanism can be used to effectively realize the construction of scientific cost prediction model so as to improve the accuracy of transmission line engineering prediction.

1 Technical Indicators of Transmission Line Engineering Cost

1.1 Principal Component Analysis

This method can reduce the spatial dimension of high-dimensional variables and data loss, and it is necessary to preprocess data and ensure the precision of data mining. The method of selecting attributes is used to compress the data set and to eliminate the

J. Abawajy et al. (Eds.): ATCI 2021, LNDECT 81, pp. 348–354, 2021.
https://doi.org/10.1007/978-3-030-79197-1_50

irrelevant data. Besides, the data compression can be achieved through quantization processing and transformation. 27 data set attributes were processed and 144 records were obtained. In addition, the mean standard variance is used for data processing, that is, the standard deviation of each indicator is obtained, making the standard deviation to be new sample data [2]:

$$\overline{X}_j = \frac{1}{n} \sum_{i=1}^{n} X_{ij} \tag{1}$$

$$S_j = \sqrt{\frac{1}{n} \sum_{i=1}^{n} (X_{ij} - \overline{X}_j)^2} \tag{2}$$

$$Z_{ij} = \frac{X_{ij} - \overline{X}_j}{S_j} \tag{3}$$

In the equation, X_{ij} refers to the Jth attribute value of the ith project, and Z_{ij} denotes the JTH attribute standard value of the ith project. The data set was analyzed by SPSS software, and the data structure attributes were obtained.

1.2 Partial Correlation Analysis

This method is to analyze the relationship between the two variables after removing the influence of other variables, and SPSS software was used to analyze the above attributes. The limit of partial correlation coefficient was 0.4, and 9 attributes were screened. Table 1 demonstrates the Table for partial correlation analysis. The selected attribute is the power transmission line project cost prediction model, and the obtained data sets are 9 * 144 [3].

Table 1. Table for partial correlation analysis

Attribute no.	Correlation coefficient	The technical attributes
1	1.000	Unit in static/(Ten thousands yuan/km)
2	0.835	Basic rolled steel/(t/km)
3	0.871	Steels for pole shaped tower/(t/km)
4	0.554	The grounding steel/(t/km)
5	0.771	Concrete/(m³/km)
6	0.492	Wire/(Ten thousands yuan/km)
7	0.582	Unit: Tower/(Sill/km)
8	0.565	Unit: Pole and tower/Sill/km)
9	0.621	Earthwork/(m³/km)

2   Construction of Project Cost Prediction Model

2.1   Least Squares Support Vector Machines

Principal component analysis and partial correlation analysis were adopted to obtain the project cost index of transmission line. It was assumed that (x_i, y_i) is the historical samples for transmission line project cost, and the influence factors for project cost in transmission lines is $x_i \in R^n$, that is, the input vector, the cost value of transmission line engineering cost is $y_i \in R$, nonlinear mapping of $\phi(\cdot)$ was used to achieve the mapping of xi so as to create the transmission line project cost performance forecast regression model:

$$y = f(x) = \omega\phi(x) + b \qquad (4)$$

Equation 4 indicates that in the process of establishing the estimation model for transmission line engineering cost performance, the optimal weight vector ω and deviation b should be determined. Based on the principle of minimizing structural risk and slack variable ξ_i, it can be used as an equation constraint optimization problem [4], as follows:

$$\min_{\omega, b, \xi} \frac{1}{2}\omega^T\omega + \frac{\gamma}{2}\sum_{i=1}^{N}\xi_i^2$$
$$s.t. \quad y_i = \omega^T\phi(x_i) + b + \xi_i \qquad (5)$$

γ in the equation refers to the regularization parameter.

Lagrange functions were used for solution of Eq. 5, and the following Lagrange equation was created as:

$$L(\omega, b, \xi, \alpha) = \frac{1}{2}\omega^T\omega + \frac{\gamma}{2}\sum_{i=1}^{N}\xi_i^2 - \sum_{i=1}^{l}\alpha_i[\omega\phi(x_i) + b + \xi_i - y_i] \qquad (6)$$

The $\alpha_i(i = 1, 2, ..., N)$ in the equation signifies the Lagrange multiplier.

For solution of the above mentioned equation, the least square support vector machine (LSVM) transmission line engineering cost estimation model is obtained:

$$f(x) = \sum_{i=1}^{N}\alpha_i K(x, x_i) + b \qquad (7)$$

Figure 1 shows the least squares support vector machine with intermediate nodes corresponding to support vectors.

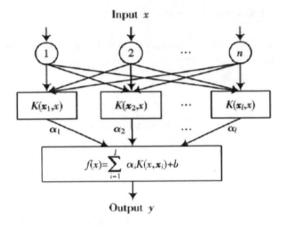

Fig. 1. Least squares support vector machines

By analyzing the modeling process of transmission line engineering cost performance with least square support vector machine, parameter σ^2 and parameter γ will affect the result of project cost estimation. In order to improve the effect of transmission line engineering cost estimation, the particle swarm optimization algorithm is used to determine the optimal value of parameters [5].

2.2   Particle Swarm Optimization Algorithm

It refers to the algorithm to simulate the foraging behavior of birds as they were flying. The particle is used to represent the potential solution of the problem, and the particle determines the fitness function value with the problem solving target, and describes the position of the particle. Supposing that P particles form s = (s1, s2, ..., si, ...sp), the expression for particle in i($1 \leq i \leq$ P) position indicates i($1 \leq i \leq$ P), D refers to the dimension of solution space, and the updated equation for particle state [6] is as follows:

$$\begin{cases} v_{id}^{t+1} = wv_{id}^{t} + c_1 r_1 \left(P_{id}^{t} - x_{id}^{t} \right) + c_2 r_2 \left(P_{gd}^{t} - x_{gd}^{t} \right) \\ S_{id}^{t+1} = S_{id}^{t+1} + v_{id}^{t+1} \end{cases} \tag{8}$$

The inertial weight is W, the particle dimension is D, and r1 and r2 are the interval random numbers at interval of [0, 1].

Inertial weight has an impact on global and local search performance. In order to improve the search ability of the algorithm and ensure the diversity of particles, automatic adjustment of inertia weight is used. The adjustment rule for inertia weight W is:

$$w_i^t = \delta(t)(\varepsilon_i^t + \xi) \tag{9}$$

ε_i^t in the equation refers to the distance between particle i and the optimal particle [7].

2.3 Project Cost Estimation Process

1. For specific transmission line projects, the cost data of transmission line projects are standardized through the collection of historical data of project cost, and the dimensionality of data is unified:

$$x_i' = \frac{x_i - \min(x_i)}{\max(x_i) - \min(x_i)} \tag{10}$$

2. Initializing the particle velocity and position, determining the value range, and mapping the position of least squares support vector machine parameters σ^2 and γ to the particle;
3. Inputting the training sample in transmission line engineering cost estimation into the least squares support vector machine to realize the inverse coding of particle position vector, so as to obtain the parameter value, and obtain the fitness value of each particle by training;
4. Determining the optimal position of each particle and the optimal position of the population;
5. Realizing the updated operation of inertia weight;
6. Updating particle state to generate a new particle swarm;
7. The least-squares support vector machine and training samples were used to calculate the new particle swarm fitness value;
8. Observing whether the end condition is met. If not, go to step 4;
9. Observing the global optimal combination of parameter values based on the optimal population position, and learn the transmission line engineering cost training again to create the transmission line engineering cost estimation model [8].

3 Simulation Experiment

A total of 200 sets of data were collected for the study of transmission line engineering cost data in a city. Figure 2 is the sample in transmission line engineering cost estimation.According to the variation trend of transmission line engineering cost data in Fig. 2, this data are characterized with obvious linear variation and obvious nonlinear variation. Embedding theorem is used to determine the embedding dimension m and delay time τ of transmission line engineering cost data, and the following mathematical model for estimation is created [9]:

$$y(n+1) = [x(n), x(n-\tau), ..., x(n-(m-1)\tau)]^T \tag{11}$$

The embedding dimension and delay time of transmission line project cost data are determined as 7 and 12 by using differential entropy. Before constructing the transmission line engineering cost estimation model by least squares support vector machine, the particle swarm optimization algorithm is used to determine the parameter value, the number of particles is 20, and the maximum iteration times is 500.

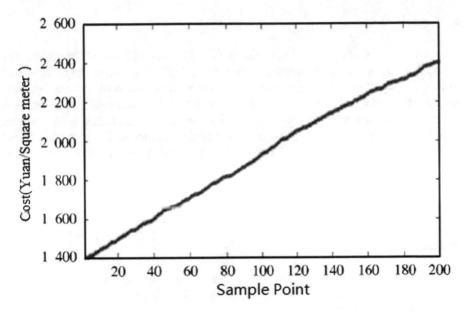

Fig. 2. Sample of construction project cost estimation

Table 2. Comparison results

No.	Static investment (Ten thousand yuan/km)	Regression support vector machines		
		Predicted value (Ten thousand yuan/km)	Absolute error (Ten thousand yuan/km)	Relative error (%)
1	21.32	23.82	2.51	11.5
2	30.52	32.52	1.15	3.82
3	37.52	31.52	−5.4	−15.65
4	26.8	29.45	2.65	10.65
5	22.5	27.11	4.5	18.32

Simulation comparison was carried out through artificial neural network, and the test samples and index parameters were the same as the learning samples. The index parameter is the input neuron, the output index is the unit static investment, the learning sample is trained to design he artificial neural network.

After the training data of neural network model is created, the training output index is close to the actual quality guarantee. Then, the checkout was conducted on the samples. Table 2 shows the comparison results. It can be seen from Table 2 that the artificial neural network can be used to predict the quality of the data, but the stability is not good, and the error of the predicted data is greater than that of the support vector machine. Secondly, it can be seen from the simulation that if the training samples are changed, the test samples will be affected and the output results will be unstable. In this way, the generalization ability and precision of sample prediction of support vector machine theory are better than traditional methods [10] (Advantages of Model).

4 Conclusion

In the cost prediction of transmission line engineering, the difficulty of the project cost is mounted because of the complexity of the influencing factors and the large amount of project. By using partial correlation analysis, principal component analysis and data mining technology, the original index can be merged and dimensionality reduction can be realized, and the comprehensive index obtained is independent. Cluster analysis tools were used to clean the data and create the prediction model. The prediction model designed in this paper can realize training and learning, predict the new project, compare the artificial neural network simulation data, so that the requirements on prediction error in project cost can be satisfied, the feasibility can meet the actual demand.

References

1. Geng, P., An, L., Wang, X.: Construction and implementation of power transmission project cost prediction model based on data mining technology. Modern Electron. Tech. **41**(507 (04)), 165–168 (2018)
2. Zhao, B., Wei, X.: Research on construction of cost estimation model based on data mining. China Real Estate **000**(029), 170–171 (2018)
3. Kong, J., Cao, X., Xiao, F.: Research on transmission line engineering cost prediction model based on support vector machine. Modern Electron. Tech. **41**(04), 135–138 (2018)
4. He, Y., Ju, X., Yong, H., et al.: Construction and analysis of power transmission line engineering cost weight prediction model based on PSO-SSR. Autom. Technol. Appl. **39** (297(03)), 102–106 (2020)
5. Li, L.: Resource sharing of construction cost information under the background of big data. Dev. Guide Build. Mater. (II) **017**(005), 72 (2019)
6. Xin, W.: Construction of big data platform for transmission line engineering cost and application of intelligent analysis control. Guizhou Electr. Power Technol. **021**(011), 8–14 (2018)
7. Yang, J.: Research on the construction of graduates employment direction prediction model based on data mining technology. Comput. Knowl. Technol. Acad. Edn. **15**(04), 16–18 (2019)
8. Hao, H., Zhu, C., Peng, J.: Modeling and simulation of transmission line engineering cost estimation based on small sample data. Autom. Instrum. **241**(11), 163–166 (2019)
9. Sun, G.: Application of BIM technology in power transmission line construction cost management. Intell. City **6**(10), 98–99 (2020)
10. Zhong, G.: Discussion on the countermeasures of construction cost management under the background of big data. Build. Technol. Dev. **46**(05), 89–90 (2019)

Construction of Housing Engel Coefficient Under the Background of Big Data

Lingni Wan[1](✉) and Fang Yang[2]

[1] Department of General Education, Wuhan Business University,
Wuhan, Hubei, China
[2] Department of Business Administration, Wuhan Business University,
Wuhan, Hubei, China

Abstract. Based on the data analysis of per capita consumption expenditure of urban residents in China from 2013 to 2018, this paper puts forward that housing Engel coefficient is used as the measurement basis of house purchase payment ability, and the upper limit of housing payment difficulty is determined by the theoretical upper limit boundary value of housing Engel coefficient, and the degree of housing payment difficulty is divided into five levels.

Keywords: Housing poverty · Housing affordability · Housing Engel coefficient

1 Introduction

The object of housing poverty is the group who cannot afford to buy commercial housing through their own ability, so measuring the family's ability to pay for commercial housing is the key. The housing Engel coefficient can be used as the basis to measure the ability to pay for house purchase. Using this index, the families who can afford commercial housing can be excluded, and the housing difficulties of families can be ranked according to the ability to pay for house purchase.

Engel coefficient was proposed by German statistician and economist Engel Ernst (1821–1896). It can reflect the relationship between basic food demand and household consumption expenditure. Engel coefficient suggests four rules: Firstly, the higher the level of family income, the smaller the proportion of food consumption expenditure in family income; on the contrary, the lower the level of family income, and the larger the proportion of food consumption expenditure in family income. Secondly, no matter whether the family income level is high or low, the proportion of clothing expenses in the family income is small. Thirdly, no matter how much the family income is, the proportion of rent, water and electricity, coal and other consumption expenses in the family income remains unchanged. Finally, the higher the level of family income, the greater the proportion of education, culture and entertainment expenses in family income [1].

J. Abawajy et al. (Eds.): ATCI 2021, LNDECT 81, pp. 355–361, 2021.
https://doi.org/10.1007/978-3-030-79197-1_51

2 Affordability of House Purchase

In 2014, low rent housing and public rental housing were operated simultaneously. In order to truly reflect the degree of housing poverty of families, it is necessary to re determine the purchasing power of commercial housing of families [2]. The purchasing power of commercial housing is shown in Fig. 1.

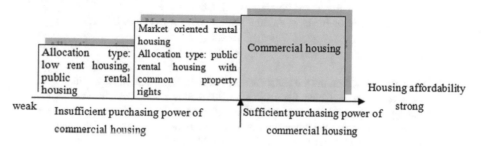

Fig. 1. The ability to pay for the purchase of commercial housing

Figure 1 shows that the ability of families to pay for house purchase is related to the choice of housing product types. When the ability to pay for house purchase is strong, families can solve the housing problem through the commercial housing market. When the ability to pay for house purchase is weak, families need to solve the housing problem through leasing, either through market leasing or applying for indemnificatory housing from the government. Through the interpretation of housing Engel coefficient, this paper aims to explore the trend that the proportion of housing expenditure in total household consumption expenditure changes with the change of income, use the housing expenditure consumption ratio to explain the influence of income level on housing affordability, and find out the feasibility of calculating the upper limit of housing expenditure consumption ratio theory.

3 Calculation of Housing Engel Coefficient

This study assumes that the per capita consumption expenditure of urban residents is all used for housing, food, transportation and communication, education, culture and entertainment [3], that is, $Ct = CHt + CFt + CTt + CEt$, Ctis is the total per capita consumption expenditure of urban residents in t period, CHt represents the per capita consumption expenditure of urban residents in t period, CTt is the per capita food expenditure of urban residents in t period, CTt is the per capita consumption expenditure of urban residents in t period,CEt is the per capita consumption expenditure of education, culture and entertainment of urban residents in t period. Thus, the Engel coefficient of housing can be obtained as follows:

$$\frac{CH_t}{C_t} = 1 - \frac{CF_t}{C_t} - \frac{CT_t}{C_t} - \frac{CE_t}{C_t} \tag{1}$$

It can be seen from Formula 1 that $\frac{CF_t}{C_t}$ is Engel's coefficient, because it is assumed that the per capita consumption expenditure of households is all used for food, housing, transportation and communication, education, culture and entertainment, and the expenditure on clothing, health care, daily necessities, other supplies and services is not considered. This index is actually the theoretical value of the household consumption quota standard for housing.

$$HC_{nt}^{\max} = 1 - e_{nt} - CT_{nt} - CE_{nt} \tag{2}$$

Among them, $HC_{nt}^{\max}$ is the upper limit value of the housing consumption expenditure of the nth family in the whole household consumption expenditure in t period. ent is the Engel coefficient, CTnt is the transportation and communication consumption expenditure of the nth family in t period, CE_{nt} is the education, culture and entertainment consumption expenditure of the nth family in t period.For housing poor families, the ratio of housing Engel's coefficient is:

$$HC_{nt}^{r} = \frac{H_{nt}}{C_{nt}} \tag{3}$$

Hnt is the average housing Engel coefficient of n area in t period. When $HC_{nt}^{r} > HC_{nt}^{\max}$,the burden of housing consumption expenditure is too heavy and housing payment is difficult. Because the Engel coefficient has a certain dynamic, so the housing Engel coefficient will also have a dynamic, so as to solve the bottleneck of the residential poverty index in estimating the type and quantity of consumption expenditure items.

The following is the data of China National Statistical Yearbook 2013–2018 (Table 1) to calculate the housing Engel coefficient.

Table 1. Per capita income and expenditure of urban residents in China, 2013–2018

Project name (unit: yuan)	2018	2017	2016	2015	2014	2013
Per capita disposable income of urban residents	39251	36396	33616	31195	28844	26467
Per capita disposable wage income of urban residents	23792	22201	20665	19337	17937	16617
Per capita disposable net income of urban residents	4443	4065	3770	3476	3279	2975
Project name (unit: yuan)	2018	2017	2016	2015	2014	2013
Per capita disposable property net income of urban residents	4028	3607	3271	3042	2812	2552
Per capita disposable transfer net income of urban residents	6988	6524	5910	5340	4816	4323

(continued)

Table 1. (*continued*)

Project name (unit: yuan)	2018	2017	2016	2015	2014	2013
Per capita consumption expenditure of urban residents	26112	24445	23079	21392	19968	18488
Per capita consumption expenditure of food, tobacco and alcohol of urban residents	7239	7001	6762	6360	6000	5571
Per capita clothing consumption expenditure of urban residents	1808	1758	1739	1701	1627	1554
Per capita housing consumption expenditure of urban residents	6255	5564	5114	4726	4490	4301
Per capita consumption expenditure of daily necessities and services of urban residents	1629	1525	1427	1306	1233	1129
Per capita transportation and communication expenditure of urban residents	3473	3322	3174	2895	2637	2318
Per capita expenditure on education, culture and entertainment of urban residents	2974	2847	2638	2383	2142	1988
Per capita health care expenditure of urban residents	2046	1777	1631	1443	1306	1136
Per capita consumption expenditure of other goods and services of urban resident	687	652	595	578	533	490

Data source: according to the national statistical data of the people's Republic of China.

Table 1 shows that: the per capita consumption expenditure of daily necessities and services, per capita clothing consumption, per capita consumption expenditure of other goods and services, per capita medical and health care consumption account for a low proportion of per capita consumption expenditure of urban residents, and the data ratio in recent six years is relatively stable, without much fluctuation. The per capita food, tobacco and alcohol consumption expenditure and per capita housing consumption expenditure of urban residents ranked first and second respectively, followed by the per capita transportation, communication consumption expenditure and education, culture and entertainment consumption expenditure of urban residents.

According to Table 1, the upper limit value and average housing Engel coefficient of urban residents can be calculated, as shown in Fig. 2.

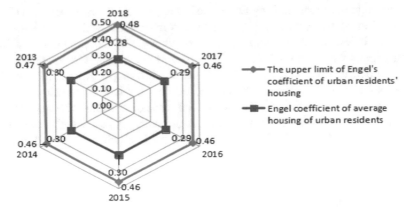

Fig. 2. Housing Engel coefficient of urban residents in China, 2013–2018

As can be seen from Fig. 2 above, the upper limit of Engel's coefficient of urban residents' housing fluctuated little from 2013 to 2018, basically in the range of 0.46–0.48, which decreased from 0.47 in 2013 to 0.46 in 2014–2016, and increased to 0.48 in 2018. The average housing Engel coefficient of urban residents showed a decreasing trend from 2013 to 2018, fluctuating in the range of 0.28–0.30. It was 0.30 from 2013 to 2015, 0.29 from 2016 to 2017 and 0.28 in 2018. This shows that housing Engel's coefficient has a certain dynamic, and will change with the changes of various kinds of expenditure and consumption, which can solve the difference of income quota access of public rental housing caused by regional income difference.

4   The Difficulty of Commercial Housing Payment

Although the upper limit of housing payment difficulty is determined from the upper limit boundary value of housing Engel coefficient theory, the degree of housing payment difficulty should be further divided. For example, The United States considers that the housing expenditure income ratio is greater than 30% as having payment difficulties; It is between 30% and 50% as having medium payment difficulties; It is greater than 50% as having serious payment difficulties. Australia and the United Kingdom consider that the housing rent income ratio is greater than 25% as having payment difficulties hard, Germany believes that the housing expenditure ratio is more than 15% and there is a medium payment difficulty between 15% and 25%. However, most countries have not made a deep division of the degree of housing payment difficulty [4–7].

It can be seen from the policy practice of various countries that the existing research and government intervention on public housing security mainly focus on the housing payment difficulties of low-income groups [8]. Because there is no shortage of purchasing power of basic commercial housing for middle and high income groups. Referring to the Engel coefficient, based on the research hypothesis and the policy practice of various countries in the world, this paper makes a specific division of the difficulty degree of family housing payment (Table 2). Among them, the theoretical upper limit value of urban housing Engel coefficient R1 and the average value of urban housing Engel coefficient R2 in recent six years from 2013 to 2018 are 0.46. If the Engel's coefficient of urban housing is more than 0.46, the proportion of family housing expenditure in the whole family consumption expenditure is too large, the housing burden is too heavy, and there is a serious housing payment difficulty; if the Engel's coefficient of housing is between 0.35 and 0.46, it indicates that the family has moderate payment difficulty; if the Engel's coefficient of housing is between 0.25 and 0.35, it indicates that the family has payment difficulty; if the Engel's coefficient of housing is between 0.25 and 0.35, it indicates that the family has payment difficulty As the lower limit of housing Engel's coefficient, if the housing Engel's coefficient is below 0.25, it is considered to be able to pay for housing. If the housing Engel's coefficient exceeds 0.25, families meet the screening characteristics of family income indicators of public rental housing. See Table 2 for details [9].

Table 2. The division standard of commercial housing payment difficulty

Interval value of housing Engel coefficient	Housing payment difficulty
$R_2 > R_1$	I can't afford it
$R_1 > R_2 \geq 0.46$	Serious payment difficulties
$R_1 > 0.46 > R_2 \geq 0.35$	Moderate difficulty in payment
$R_1 > 0.35 > R_2 \geq 0.25$	Payment difficulties
$R_1 > 0.25 > R_2 \geq 0$	No payment difficulties

Housing Engel coefficient is based on the research method of Kuang Dahei and Ding Yanhao (2018) [10]. Based on the data of urban residents' expenditure and income in China from 2013 to 2018, the theoretical upper limit value of housing Engel coefficient is calculated, with 0.25 as the lower limit value of housing Engel coefficient and 0.46 as the upper limit value. Then according to the policy practice of various countries, the difficulty degree of public rental housing payment can be divided into five grades, namely: no payment difficulty, payment difficulty, moderate payment difficulty, serious payment difficulty, unable to afford, etc.

5   Conclusions

Through the big data observation, it is found that the housing Engel coefficient is more dynamic and regional than other measurement methods in measuring the affordability of family purchase. The value of housing Engel coefficient is a very large value, that is: the larger the value of the family in the index, the lower the income level of the family, the weaker the purchasing power of commercial housing.

The housing Engel coefficient has spatial-temporal and regional differences on the housing payment ability of Chinese urban residents, and low-income groups have different degrees of housing payment difficulties. Therefore, the construction of housing Engel coefficient index can effectively identify the housing payment difficulties of public rental housing families in different regions and different time and space, and scientifically define the family income access standards of public rental housing applicants according to the payment difficulties, so as to ensure that families with different income levels can enjoy different levels of housing security policies according to the degree of housing poverty, and truly protect those in need The housing welfare of the group should not be "one size fits all", but should be managed according to the time and place.

Acknowledgements. This work was supported by Research on the nature and orientation of social insurance agency, a doctoral research fund project of Wuhan Business University. (2017KB010).

References

1. Zhang, Z.: From engel coefficient to tourism engel coefficient: review and application. China Soft Sci. **S2**, 100–114 (2011). (in Chinese)
2. Wu, X., Wang, J., Jiang, Q.: Research on the determination of income line of housing security: a case study of rental housing in Nanjing. Price Theory Pract. **12**, 52–54 (2014). (in Chinese)
3. Li, J., Sun, J., Li, H.: Research on measurement of housing affordability of urban residents from the perspective of surplus income. Res. Technol. Econ. Manag. **03**, 74–77 (2011). (in Chinese)
4. Holmans, A.E.: Housing and Housing Policy in England 1975–2002: Chronology and Commentarya. Cambridge Centre for Housing and Planning Research, London (2005)
5. Ascher, C.S.: The Administration of publicly-aided Housing. New York: International Institute of Administrative Science. Martinus Nijhoff, Leiden (1971)
6. Henderson, J., Race, K.V.: Class and state housing: inequality and the allocation of public housing in Britain. US: Gower Publish (4), 237–278 (1987)
7. King, R.: Dimension of Housing Need in Australia. Australia: In Buchan Allocation, Access and Control, pp. 230–236. Macmillan, London (1973)
8. Liu, G., Chen, L.: Research on access standard of affordable housing from the perspective of housing affordability: ideas methods and cases. China Admin. **04**, 67–72 (2016). (in Chinese)
9. Yi, D., Yin, D.: Statistics of Household Consumption, vol. 71. China Renmin University Press, Beijing (1994). (in Chinese)
10. Kuang, D., Ding, Y.: Spatial and temporal distribution of housing affordability of urban residents in china: an analysis of rent affordability of 35 large and medium-sized cities. Price Theory Pract. **10**, 16–19 (2018). (in Chinese)

Self-recorded Video Segmentation Algorithm Based on Motion Detection

Ling Ma[✉]

Department of Information Engineering, Heilongjiang International University,
Harbin 150025, Heilongjiang, China

Abstract. Content-based video retrieval has been very widely used, which plays an important role in the rapid development of multimedia technology. It also takes effective management and retrieval for large quantities of multimedia data, especially for the increasing number of sports video. This video retrieval has already become hot in the field of multimedia. This paper classifies the state of the self-recorded game video, detects the bottom line by canny operator and separates the zones of service and offensive, detects movements in the corresponding regions. Then, according to the characteristics of difference states of the game, the game station is judged and the video is split. This paper designed the system to achieve the above algorithms. Through the experiments of multiple videos, it comes to an expected result, verifies the algorithm's feasibility and robustness.

Keywords: Video segmentation · Motion detection · Canny operator · Self-recorded video

1 Introduction

In recent years, with the development of electronic information technology, multimedia data storage and transmission technology has made significant progress [1, 2]. Users can access the remote video library anytime and anywhere, such as video on demand, electronic shopping, all kinds of game programs, as well as access to multimedia library. Multimedia information is characterized by large amount of information, which is difficult to describe [3–5]. Therefore, how to organize massive multimedia information to achieve fast and effective retrieval becomes an urgent problem to be solved. In modern life, sports video is a kind of important video. With the continuous improvement of people's quality of life and the rapid progress of science and technology, all aspects of the video game requirements are rising [6, 7]. However, sports games on TV is very long, the corresponding relation of the video data is complex and the huge amount of data, for the audience, really care about and may repeatedly watch just one of a number of exciting part, how to effectively organize the video data, what strategies for sports video data management, and how long the sports video, help to accurately locate the video clips or highlights, provide rich browsing, to give full play to the role of sports video is the key [8, 9]. The structure of sports video content is obvious, such as volleyball technical pause, spike and slow motion etc. Often there are commentators in

J. Abawajy et al. (Eds.): ATCI 2021, LNDECT 81, pp. 362–369, 2021.
https://doi.org/10.1007/978-3-030-79197-1_52

the middle of the game. By analyzing the characteristics of audio, video and subtitles in the sports competition, the automatic recognition of the audience's interesting scenes can save the time for the audience to enjoy the sports games [10].

2   Setting of Service Area

2.1   Common Edge Detection Methods

1) Roberts operator

The principle is: utilize local differential operator to seek edges and get diagonal derivative in 2×2 neighboring area. It is shown in Eq. (1).

$$g(x,y) = \sqrt{[f(x,y) - f(x+1,y+1)]^2 + [f(x,y+1) - f(x+1,y)]^2} \qquad (1)$$

In order to simplify the operation, in practical applications, it is often used the gradient function of the Roberts absolute value or the maximum operator of Roberts. It is shown in Eq. (2) and Eq. (3).

$$g(x,y) = \sqrt{[f(x,y) - f(x+1,y+1)]^2 + [f(x,y+1) - f(x+1,y)]^2} \qquad (2)$$

$$g(x,y) = \max(|f(x,y) - f(x+1,y+1| + |f(x,y+1) - f(x+1,y)|) \qquad (3)$$

Template of Roberts convolution operator is shown in Eq. (4).

$$\begin{bmatrix} -1 & 0 \\ 0 & 1 \end{bmatrix} \quad \begin{bmatrix} 0 & -1 \\ 1 & 0 \end{bmatrix} \qquad (4)$$

2) Laplace operator

Laplace operator has the fundamental idea: it makes second order differential derivative of image function, which is second order differential operator; fetch the form of edges with Laplace operator; calculate the sum of second order partial derivative as Eq. (5):

$$\nabla^2 f(x,y) = \frac{\partial^2 f(x,y)}{\partial x^2} + \frac{\partial^2 f(x,y)}{\partial y^2} \qquad (5)$$

In digital image field, we can use difference method to approximate to differential operation; the discrete calculation like Eq. 6.

$$\nabla^2 f(x,y) = L(x,y) = \{[f(x+1,y)-f(x,y)] - [f(x,y)-f(x-1,y)]\}$$
$$+ \{[f(x,y+1)-f(x,y)] - [f(x,y)-f(x,y-1)]\} \tag{6}$$

3) Canny operator

Canny operator takes this basic thinking for edge extraction [1]: utilize local extreme to detect edges; it has three traits: best detection, best detection precision, one-one correspondence between detection point and edge point; Canny operator theory is carried out in following steps:

Step 1: reduce noise: any edge detection method can't reach good effect in untreated images; so the first is to do smoothing processing of images with Gaussian filter;
Step 2: find gradient of intensity in image: edges in the image can point to different directions; Canny operator detects edges with four filters in horizontal, vertical and two diagonal directions; mark up the maximum of each point and edge direction to generate intensity gradient direction and intensity gradient graph of each point in the original image;
Step 3: track edges in the image: generally, bigger intensity gradient implies more possibly it's edge; Canny operator utilizes hysteresis threshold to limit the value of intensity gradient to determine if it's edge; hysteresis threshold includes high and low threshold; high threshold labels confirmative edges; low threshold tracks fuzzy parts of curves;
Step 4: setting of parameters: on the one hand, during image pre-treatment, setting of smoothing filter's parameters has direct influence on Canny operator; smaller value has less fuzzy effect, and little and obviously changing thin lines are detected; bigger value has more fuzzy effect, good for detecting big and smooth edges; on the other hand, there's a common problem about selection of hysteresis threshold value; lower value helps look more clearly at branch information; higher value leads to lose important information.

2.2   Operator Used in the Paper

Based on introduction and analysis of a few edge detection methods above, we choose Canny operator to detect side lines of volleyball court, because it's a kind of optimizing operator including filtering, intensifying and detection stages, applicable for different places. It can adjust the setting of threshold parameter in accordance to specific requirements and detecting different edge effect. Now that it has good adaptability, the special application in the detection of volleyball court side lines has significance to setting of parameter, better result achieved with suitable threshold. Figure 1 is experimental effect graph. From it, the choice of Canny operator's threshold affects greatly image edge detection. Figure 2 shows too many detected edges; many superfine edges are not necessary. Figure 4 shows very few edges are detected, especially edges of bottom lines are deformed; experimental results are not ideal in the two pictures. Figure 3 gives the best detection effect when operator's threshold is set (25,400) during testing.

Fig. 1. Experimental original   **Fig. 2.** Threshold value (25,200) of the results

2.3   Detection of Serving Area

The detection of serving area is actually setting of serving area, because in self-recording videos there's no particular area used as serving area; we can only observe and analyze to determine one by user's subjective behaviors and then set and demarcate it. The area from the part beyond goal line in shooting direction to the whole image boundary is defined as serving area. Through tests, we get main side lines of the court. Then we need to only find one bottom line from so many edges that we can complete setting and calibration of serving area. From videos we see that in the area from the part underneath court base line to image boundary, base line is the longest curved section. Hence we can get coordinate of base line with this method.

Firstly for edges detected by Canny operator, use RGB (255,0,0) to draw horizontal lines; we draw them because shooting direction is vertical with base line, no value in use of vertical lines. Make settings of Fig. 3 in red lines where Canny operator achieves better detection effect. It is shown in Fig. 5.

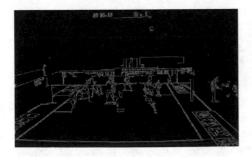

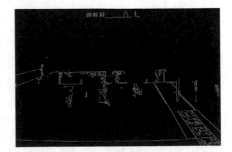

Fig. 3. Threshold value (25,400) of the results   **Fig. 4.** Threshold value (50,550) of the results

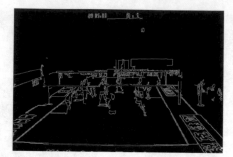

Fig. 5. The results chart on drawing horizontal line

3 Detection of Motion Target

Background subtraction method is a most often used motion detection approach. As its name implies, firstly background needs construct; background frame is unchanging and immobile component in video frame sequence. Since video object used here is about self-recording volleyball match and camera keeps still, so any frame in the background can be background frame. Although the principle of this method background subtraction frame is very simple, the construction and renewal method of background frame is rather critical. Whether background frame is extracted accurately or not affects directly the precision of final detection result. When background image is initialized, it needs to choose a time frame where there are fewer moving objects to get background frame as initialized background frame [3, 4], in the following way: from a longer video, select medium brightness of a single pixel (x, y) of each frame in successive frames as relative brightness of initialized background frame. As we know, background of consecutive frame changes slowly, pixel value in the frame is principal part. Choosing medium value can well reflect background in the time period. In the meantime, to reduce impact of close-up moving target on the background and increase robustness, we can get consecutively a group of medium value $M_i(x, y)$; then take mean value of it by Eq. (7).

$$B_i(x, y) = \frac{1}{T} \sum_{i=0}^{T-1} M_i(x, y) \tag{7}$$

To implement retrieval of content-based self-recording volleyball game videos, the most important is to cut the entire video into different video parts with different contents. Through former analysis, we can realize segmentation by relying on different competition states. Volleyball technician records videos before beginning of the volleyball game. In a few seconds before the match, there is no player at the court. So we can choose video sequence in the first few seconds to calculate initial background frame. Here we select first five frames to get initial background frame with Eq. (7). Next we describe major steps the algorithm includes.

Algorithm Self-recording video segmentation algorithm
Input: self-recording volleyball game video

Output: segmented video clips

    1) Make cycle variable $i = 0$.

    2) Variable $i+1$; by Equation 13, we can get difference image $D(x, y)$ and variable quantity d;

    3) Determine the size of d value; if d<0.4, meaning no big changes between current frame and background frame; current frame replaces background frame; it's believed that it's rest state;

    4) If d>0.8, suggesting lens is covered by object; turn back to (1);

    5) If 0.4<d<0.8, meaning moving target appears; per Equation 14, we can calculate current background frame; it's competition state;

    6) Detect moving target in serving area: get serving area at current frame and background frame; then do background subtraction motion detection; if there's moving target, it is serving state and return to (1); otherwise it's offensive state and go back to (1).

4 Experimental Analysis and Results

Each algorithm needs performance evaluation criterion to validate its usability and robustness. For self-recording videos, different states provide basis for video segmentation. So as long as the algorithm can detect correctly competition state, it can reach ideal effect. This paper gives two definitions, recall and precision, and adaptive double threshold shot boundary detection algorithm evaluation criteria similar. It is shown in Eq. (8) and Eq. (9).

$$B_i(x, y) = \frac{1}{T} \sum_{i=0}^{T-1} M_i(x, y) \tag{8}$$

$$\text{precision} = \frac{\text{Checked the total number of States}}{\text{Traced total number of States}} \tag{9}$$

    To evaluate performance of the algorithm, we need classify different states and analyze recall and precision ratio of each state. That is helpful to make in-depth analysis of the algorithm and know well the detection effect of it about one state. The experimental environment is C++ Visual 6. The experiment object is Tianjin Bundesliga tee across the volleyball team self-recorded video clips. Obviously as seen from Table 1, the recall and precision rate of the proposed segmentation algorithm both reach 90%, on the whole gaining a good effect on the classification of competition video state. The algorithm's analysis requirement is basically reached and its applicability is confirmed.

Table 1. Results of algorithm on self-recording video segmentation

Video clip	Total frames	Serve/find out/check	Attack/find/check	Rest/find out/check	The recall rate (%)	Accuracy (%)
Clip1	24500	28/28/28	25/24/26	1/0/0	95	95
Clip2	35700	27/27/27	28/29/25	2/0/1	90	96
Clip3	25600	24/24/24	29/27/25	1/1/1	93	96
Overall	75600	70/70/70	70/72/74	3/2/2	93	94

But for the detection of rest state, two evaluation indicators are lower, because during rest, there is cleaner or other staff moving back and forth, causing leak detection. We can improve the algorithm from these aspects: reduce as much as possible the irrelevant noise interference; further studies on how to narrow down moving detection area, which would help us focus on the detection of one area as to have more exact judgment of rest state. For self-recorded volleyball video segmentation, the paper designs a regional motion detection algorithm based on the state of the competition, which basically separates the competition state and achieves the desired effect. Using MFC framework for programming implementation in VC++. The overall development interface of this paper is divided into three parts: Control, View and Tree. It is shown in Fig. 6 and Fig. 7.

Fig. 6. The shot browsing on TV

Fig. 7. The shot browsing on self-recording video

Control implements the user's Control function, that is, the bottom row button is the option for video, including video OPEN, PLAY, PAUSE, STOP, EXIT, the two options of RESOLVE1 and RESOLVE2 are used to deal with TV and self-recording video respectively. View implements the user visual function, which is the main area in the middle. When the video lens or fragment on the tree control is not selected, the whole video that is initially opened is played by default. The Tree implements the control function that the user browses to view, namely the right area of the frame. The user can choose the corresponding lens or clip here, and the video lens can be played in the View section. Of course, both options can be used: one is to click video lens or clip to cooperate with the OPEN button, and the other is to double-click the shot or clip.

5 Conclusion

In this paper, in order to achieve the volleyball match state classification based on region segmentation method based on motion detection. All process to remove the rest of the video and other valuable information to facilitate the video operation and storage. Firstly, the video analysis and draw the status category: break, serve, attack three states, secondly, a lot of knowledge and algorithms related to the regional setting and motion detection are introduced, through the Canny operator to detect the bottom line to get the service area, by reduced background algorithm to motion detection, finally, according to the results of the detection of different regions to determine the status of the game on the game, so as to achieve the segmentation of self-recorded video.

References

1. Canny, J.: A computational approach to edge detection. IEEE Trans. Pattern Anal. Mach. Intell. **8**(6), 679–714 (1986)
2. Guan, X.G.: Extraction and update of background in video surveillance algorithm. Microelectron. Comput. **22**(1), 95–97 (2010)
3. Guo, Y.T.: Video traffic monitoring system in the background extraction algorithm. Appl. Video Technol. **5**, 90–93 (2016)
4. Ke, J.: Study on the Analysis Method of Semantic Event Detection based on Video. Jiangsu University (2013)
5. Lipton, A: Local application of optic flow to analyses rigid versus non-rigid motion. In: IEEE international Conference on Computer Vision (1999)
6. Li, H.F.: Multi Level Tennis Video Analysis and Retrieval. Nanjing University of Science and Technology (2019)
7. Meng, G.Z.: Study on the Mechanism of Brain Nerve Movement Decision-making Behavior of High Level Volleyball Players in the Situation, Beijing Sport University (2017)
8. Matthews, R.: On the derivation of a chaotic encryption algorithm. Cryptologia **13**(1), 29–42 (1989)
9. Ma, Y.W., Zhang, W.J.: Based on image segmentation algorithm basketball object extraction method. Microcomput. Inf. **27**(31904), 210–212 (2014)
10. Yu, Y.: The extraction of the audio sensation and video highlights. J. Comput. Aided Des. Comput. Graph. **2710**, 1890–1899 (2015)

Target Detection Algorithm Based on Unconstrained Environment for Video Human Motion

Libo Xing[(✉)]

Department of Information Engineering, Heilongjiang International University,
Harbin 150025, Heilongjiang, China

Abstract. The existing human target detection algorithm can only detect a human body object that is substantially upright in the image. The human body of sport video is in a variety of ways during the movement, and the current detection algorithm is difficult to achieve the desired detection effect. In order to detect various postures from sport video of human targets. The paper proposed a human target detection algorithm without attitude limitation. Firstly, the algorithm finds the human candidate target by shape context matching. Secondly, the hypothetical target is removed by combining the direction gradient histogram features with the support vector machine. The experimental results show that the proposed algorithm not only can detect human targets in various poses, but also has a much higher detection speed than the direction gradient histogram detection algorithm.

Keywords: Sport video · Human action recognition · Target detection algorithm · Video human motion

1 Introduction

Video-based human motion recognition is a very active research field, and the representation of human motion is the difficulty and focus of research. Researchers have proposed a large number of human motion representation methods according to different applications [1, 2]. According to the characteristics of human body tumbling motion in sports video, this paper proposed a method of expressing human motion with constant viewpoint [3]. Firstly, detected the human target from the video sequence. Secondly, applied the target tracking algorithm to obtain the human target sequence in the sequence image. Then, the segmentation algorithm is applied to remove the background and to obtain the silhouette image sequence of the human body, the self-similar matrix diagram of the human body posture is established by the silhouette image sequence to represent the motion of the human body. The detection algorithm of the human target is the basis of all work. Since the human body in the image is very different, it is quite difficult to correctly detect different human targets from the image [4, 5]. Differentiating between different objects in an image can be considered from the perspective of shape, color or texture. The shape of the human body in the image varies widely, mainly due to internal factors and external factors. Internal factors include race, gender and age. External factors mainly include camera shooting angle and distance of focal length [6]. In addition, the human body is a non-rigid body, and the shape of the human body changes with the posture. The color of the human body in the image is different depending on the race of the person and the shooting environment. Texture information cannot be reliably used as a feature of human recognition because of the different dresses of the human body. Most human target detection algorithms are mainly based on shape features, and influence of human shape differences can be eliminated by machine learning. Currently, most human target detection algorithms [7, 8] only can detect a substantially upright posture. However, in the body movements, such as diving, gymnastics and vaulting, the athletes in the video can appear in many inverted postures. Therefore, for a human body in a variety of sports video images, the current detection algorithm is difficult to detect correctly.

2   Human Target Detection Algorithm Without Attitude Limitation

The process of quickly detecting a human target by shape context matching is divided into the following steps: First, the motion Canny boundary operator is used to obtain the boundary in the image, and then the boundary image is sampled to obtain some sampling points, and the shape context descriptors of these points are calculated. Match them with the set of context descriptors in the human body model, and remove the point where the matching distance is greater than the threshold, and a subset of matching points is obtained. Finally, the subset is matched with the shape context in the template to obtain the candidate body in the image [9]. Belongie proposes to obtain a set of points from the internal and external sampling of the contour image of the shape, and the shape of the object in the image can be represented by these discrete points. To describe these points, they propose a shape context descriptor. By comparing the two-point shape context descriptor to compare whether the two points are similar, the similarity between the two shapes is judged by the context descriptor matching of the discrete point set [10].

The human body is hinged and its shape changes as the posture of the human body changes. The posture of the moving human body in the sports video varies greatly. However, the shape of the upper body (including the head, neck and shoulders) of the human body changes relatively little. Therefore, we do not choose the whole body but choose the upper body shape to represent the human body. Seeing from different angles, the shape of the human body is different, for this reason, from the perspective of shooting, the upper body view of the human body can be divided into three categories: left side view, right side view and front (back) side view. We describe the shape of the human body through the shape context. We use the template matching method to detect the human target in the image. First, we explain how to generate three types of shape templates for the upper body of the human body. We collect three images of the human body from the network. Each image set contains 200 human body images with the same angle but different postures, and each image is scaled to 240×160 pixels. Image boundary shape is obtained by applying boundary detection. The Hausdorff distance d between each pair of shapes S is calculated, and then the cumulative distance Di of each shape Si from other shapes is calculated.

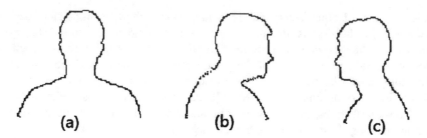

Fig. 1. (a) front (back) view (b) right side view and (c) left side view

We have applied three human body shape templates with different perspectives using the human body image set. For each shape template, we obtain a set of points by sampling, and then calculate the shape context descriptor of each point in the point set, which is taken as the shape of the human body. For each image to be detected, a pyramid image is created according to the image size, and a Canny boundary detection algorithm is applied to each layer of the sub-image to obtain the boundary of the image. Randomly sample from the image boundary to obtain a set S of points of r (about 1/8 of the total boundary pixels). For any point P of the set S, calculate the degree of matching with any point qj in the template shape T.

$$C_{ij}^t = \frac{1}{2} \sum \frac{[h_i(k) - h_j^t(k)]^2}{h_i(k) + h_j^t(k) + \varepsilon} \tag{1}$$

In the background noisy scene, the boundary obtained by applying the edge detection algorithm contains a large number of noise points. In order to reduce the influence of noise, the matching cost function of Eq. (2) is corrected. The distance calculated by Eq. (2) is 85.7, while the distance between the two descriptors is only 10.2, calculated by our modified descriptor 2.4f, since the influence of noise points in the image is removed, which is far less than the former. As a result, through comparative analysis, we can effectively suppress the influence of image noise points on shape context matching by correcting the distance calculation equation.

3   Verify by HOG

Due to the influence of noise, and in context matching, we only select the context descriptor matching of the two points of the target as the judgment basis, in the shape context matching method described above, the candidate body objects detected from the image contain a large number of false positive bodies. Next, the results of the above method detection need to be further verified. We propose a shape feature representation of the rotated direction HOG, and use the support vector machine classifier to verify the candidate human body, delete the false positive body and obtain the final test result.

First, we use a training method similar to Dalal to train a classifier for the upper body. We select 100 images from INRIA static human database. Each picture contains a different body, such as posture, appearance and clothing, and the background of the picture and the light of the shooting are also different. The portion of the images containing the upper body is manually maintained as a positive sample image set. Perform some slight shearing and rotational deformation on each image in the positive sample image set to obtain a new image set containing more than one thousand positive samples with a sample size of 12 times larger. Positive sample deformation can provide generalization performance of the classifier. Like positive samples, negative sample images are also obtained from the 100 images, except that negative sample images do not contain human body. In order to get better classification performance, negative samples must be diversified. Our negative sample images include natural scenes, such as indoors, outdoors, cities, mountains and beaches. Some pictures contain specific objects, such as cars, bicycles, motorcycles, furniture, tableware, and so on. After the positive and negative sample images are collected, the HOG of each sample image is calculated. These the HOGs are input as eigenvectors to a linear support vector machine to obtain a binary classifier (Fig. 2).

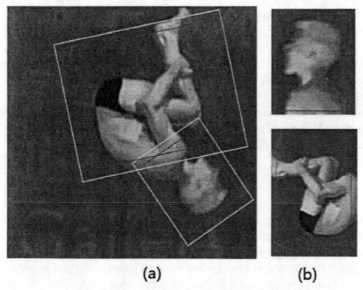

(a)   **(b)**

(a) An image containing two candidate targets
(b) Sub images constructed by candidate targets

Fig. 2. Construction of vertical sub images

Next, using HOG feature, combined with the support vector machine to verify the candidate target process. In the application shape context fast matching method, each candidate body object detected in the image is represented by a rectangular window, and the parameters of the rectangular window are defined by a five-tuple qG. For the i-th candidate body target <Si>, a sub-image containing the candidate body object can be constructed using the rectangular window parameter, and the sub-image is constructed such that the human body object is substantially in an upright posture according to the rotation angle p. Considering the accuracy of the detection, the constructed sub-image size (yellow rectangular area) is slightly larger than the candidate target window, and then HOG feature vector at different positions in the sub-image is calculated, and the calculated feature vector is input into the trained classifier and get the output values of a series of classifiers, the largest value of which is taken as the final value. If the value is larger than one threshold, then accept the candidate human target; otherwise, the candidate target is considered to be a false positive target and remove the candidate target from the result set.

4 Experimental Results and Analysis

First, experimentally observe the influence of different parameters on the performance of our detection algorithm, which is used to determine some parameters of our detection algorithm. Our algorithm mainly has two processes, one is the process of quickly detecting the candidate target, and the other is the process of applying HOG detector to verify. In the first process, there are three parameter values to be determined, and in the second verification process, N. Dalal has discussed the effect of all parameters on the test results, so we will not discuss here.

In the experiment, the detection error trade-off curve is used to evaluate the performance of the detection algorithm. DET curves are widely used to evaluate language recognition methods. DET curve represents the same information as ROC trace does, but DET curve has better discrimination. DET plots the relationship between missed detection rate and false positive rate in double logarithmic coordinates. The value of Y axis represents the missed detection rate MR.

$$MR = \frac{FN}{FN + TP} \tag{2}$$

FN refers to the number of false negative samples, which is the number of the misjudged non-human targets. TP refers to the number of true positive samples, that is, the number of human targets that are correctly detected. The value of X-axis represents the false positive rate, where the number of false positive targets appearing in each detection window is used. In order to facilitate the comparison of the false positive rate of the algorithm, the test image set in the experiment uses the same number of windows, and the total number of detection windows is determined by HOG algorithm. Because the number of detection windows of our algorithm is determined by the fast shape descriptor matching process, the number of windows is related to the image content and cannot be predetermined, and the number of detection windows of HOG algorithm is only related to the size of the image. We draw DET curve under the double logarithmic coordinate system, mainly because the logarithmic coordinate enlarges the small amplitude range and compresses the large value range, which is convenient for comparing the small amplitude change relationship.

In the detection algorithm, three parameters need to be determined for the fast shape context descriptor matching, which are: 1) the sampling rate S of the point. When a shape is represented by a point set, the larger the sampling rate is, the more accurate the shape represents. 2) Threshold parameter TM. Control the proportion of points in the image that match the points on the template. 3) Threshold parameter TD. The relationship between the target size in the image and the target size in the template is controlled. The smaller the TD is, the closer the target size of the target is to the size of the target in the image.

Through the above several experiments, we determine the values of several detection parameters, reflect the relationship between detection time, missed detection rate and parameters. The fast context descriptor matching process only obtains the candidate body, and the result is further verified by HOG feature and the support vector machine classifier. Therefore, the false positive rate is mainly affected by the verification process, but the parameters in the fast matching process are also has an inhibitory effect to the false positive target.

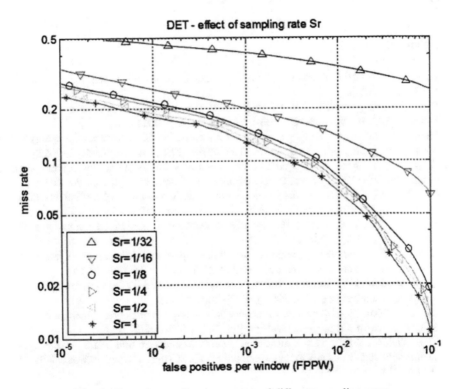

Fig. 3. Detection performance curve of different sampling rates

Figure 3 reflects the influence of the sampling rate S on the detection performance. When the threshold S is less than 1/8 in Fig. 1, the detection performance does not change much. When the sampling rate S increases to 1/4, the detection performance slightly improves by 1/8, but the detection time is doubled. Therefore, considering the detection speed and the detection performance, it is more suitable when the sampling rate S is still 1/8.

5 Conclusion

This paper proposes a human target detection algorithm without attitude limitation. Firstly, the algorithm uses the feature of rotation invariance of shape context descriptors, and combines the template matching algorithm to quickly find out the candidate body targets of various postures. Then, HOG features of the candidate target image regions are established, and the feature vectors are input into the support vector machine for classification and verification. The results show that our algorithm cannot only detect the non-upright posture of the human body target that the original HOG detection algorithm cannot, the detection speed of our algorithm also increases by about 5 times.

Acknowledgements. This work was supported by Heilongjiang Province Educational Science 13th Five-Year Plan (GJB1320332)and (GJB1320333).

References

1. Berl, A., Gelenbe, E., Girolamo, M.: Energy-efficient cloud computing. Comput. J. **53**(7), 1045–1051 (2011)
2. Chou, J.: Energy-Aware Scheduling in Disk Storage Systems, Distributed Computing Systems (ICDCS), pp. 423–433. IEEE, Washington DC (2011)
3. Guenter, B., Jain, N.: Managing cost, performance, and reliability tradeoffs for energy-aware server provisioning, pp.1332–1340. In: INFOCOM, IEEE, Washington D-C (2011)
4. Kliazovich, D.: Green cloud: a packet-level simulator of energy-aware cloud computing data centers. In: 2010 IEEE Global Telecommunications Conference, pp.1–5 (2010)
5. Hao, L., Cui, G., Qu, M.: Cloud computing energy resource scheduling optimization algorithm under cost constraint. High Technol. Commun. **24**(5), 458–464 (2014)
6. Young, C., Albert, Y.: Energy efficient utilization of resources in cloud computing systems. J. Supercomput. **60**(4), 268–280 (2012)
7. Ji, C.C.: Cloud computing resource scheduling optimization based on dynamic trend prediction ant colony algorithm. Sci. Technol. Bull. **32**(1), 187–190 (2016)
8. Lin, W.W., Zhu, Z.Y.: Scalable distributed scheduling method for large-scale cloud resource scheduling. Comput. Eng. Sci. **37**(11), 1997–2005 (2015)
9. Xie, L.X., Yan, Y.X.: Research on service scheduling and resource scheduling in cloud computing environment. Comput. Appl. Res. **32**(2), 528–531 (2015)
10. Yi, P., Wei, C., Yu, P.: Time-shift current balance technique in four-phase voltage regulator module with 90% efficiency for cloud computing. IEEE Trans. Power Electron. **30**(3), 1521–1534 (2015)

The Security Situation Prediction of Network Mathematical Neural Model Based on Neural Network

Ling Sun[✉]

Lanzhou Resources and Environment Voc-Tech College, Lanzhou 730021,
Gansu, China

Abstract. With the gradual development of the modern neural network field, the construction of network mathematical models related to neural networks has become more important, but when using network mathematical models, it is often impossible to achieve strict and accurate theoretical support and data analysis. For the security situation prediction, specific strategies to improve the neural model of network mathematics should be proposed to improve the status quo that network mathematics cannot meet the status quo of scientific research. In the quantitative evaluation of the factors to be considered when selecting the target, the mathematical security situation prediction method will be used efficiently, and the membership of each factor will be determined according to the method and principle of the network mathematical neural model. Through functional analysis and quantitative selection, a mathematical model is established, and the BP algorithm is combined with the neural network algorithm. This paper takes neural network as the theoretical basis of the research, uses BP fuzzy neural algorithm as the main research algorithm, and integrates its important content to analyze and research the optimization of adaptive enhancement algorithm. This paper takes the classical network mathematical neural model as the research object, and optimizes and improves the security situation prediction of the network mathematical neural model. The BP fuzzy neural algorithm in the neural network can be regarded as a kind of node sorting algorithm, so it can be used to construct the network mathematical neural model. The experimental results show that this research uses the network mathematical neural model and the BP fuzzy neural algorithm parameters to estimate the neural network prediction value and calculate the safety situation prediction, and uses the neural network method to test the network mathematical neural model from the safety situation prediction error of the BP fuzzy neural algorithm. The method in the target selection is correct and feasible.

Keywords: Neural network · Network mathematics · Neural model · Security situation

1 Introduction

Network mathematics plays an irreplaceable role in many industries. In the current Internet era, as the top talents in various industries in the future, most network mathematics talents often do not pay attention to mathematics in the research stage, and do

J. Abawajy et al. (Eds.): ATCI 2021, LNDECT 81, pp. 377–384, 2021.
https://doi.org/10.1007/978-3-030-79197-1_54

not understand mathematical knowledge and mathematical calculations. Rarely use some far-reaching mathematical ideas and useful mathematical methods, resulting in insufficient basic mathematical knowledge, unable to analyze and solve problems and unable to gradually study deep technology [1]. With the advent of computerized systems, there are a large number of targets, complex and difficult situations, the use of network mathematics principles, and the rapid requirements of algorithms. Automatic and scientific selection are the main problems in creating network mathematics and neural models, and they are also important for network security. As local colleges and universities pay more and more attention to the promotion of network mathematics thinking and innovation, its effect is not yet clear [2]. The ability to learn network mathematics usually becomes an inevitable requirement for high-tech innovation [3]. Many well-trained scientific and technical personnel are facing huge difficulties, and the scientific and technological innovation in the field of advanced mathematics seems to be insufficient [4]. This article puts forward a lot of researches aimed at improving the security situation prediction of the network mathematical neural model based on the neural network.

Neural network refers to the use of psychological principles and methods to determine how people use digital and information technology. At present, some information collected in neural networks of different scales is relatively vague, and it is impossible to evaluate the state of neural networks [5]. Therefore, it is very possible to introduce network mathematics neural model security situation prediction in the research of neural network, and use its security to predict the situation. The factors affecting network mathematics have been predetermined through data collection and effective verification of information [6]. According to the establishment of neural network algorithm, a specific standard is formulated, and the abstract analysis of the network mathematical neural model can measure the standard formulation [6]. According to this standard, the network mathematical neural model is combined with the correlation analysis between different influencing factors to filter the influencing factors with significant correlation. Finally, these factors are used as clustering variables to conduct the analysis on the topic of interest. Grouping and target positioning, provide suggestions for security situation prediction [7].

Based on this, here, the network mathematics and neural network are organically combined for target selection, that is, when the factors considered in target selection are quantified, and when the target selection decision is made, the network mathematical neural model is used [8]. Multi-dimensional information system analysis is carried out on the abstract data of independent individuals. The basic characteristics are multivariable, multi-level, and strong coupling. Various factors in the system have complex nonlinear effects. The evaluation of the security situation prediction of the network mathematical neural model is more suitable for the use of system analysis to consider the relationship between input and output, combined with the characteristics of the neural network. Due to the generalization function of neural network, the method of network mathematical neural model can be used to transform qualitative evaluation into quantitative evaluation, which will have a better safety situation prediction effect [9]. Compared with other indicators of security state evaluation methods, using network mathematical neural models to describe neural network phenomena has greater deductive and predictive power, and is more convenient for computer simulation. This paper uses neural network and network mathematical neural model to establish a mathematical model to evaluate the security situation prediction, and the verification result is good.

2 BP Neural Network Algorithm

The standard BP algorithm is a gradient descent algorithm using Widrow-Hoff learning rules. The direction of weight and threshold change is the direction in which the editing processing function drops the fastest-the negative direction of the gradient. The calculation formula for each update value is as follows:

$$\Delta w_{jk} = -\eta(\hat{y}_{hk} - y_{hk})\hat{y}_{hk}(1 - \hat{y}_{hk})z_j \tag{1}$$

$$\Delta w_{jk} = \eta(\hat{y}_{hk} - \hat{y}_{hk})\hat{y}_{hk}(1 - \hat{y}_{hk})z_j \tag{2}$$

The number of nodes on the input level corresponds to the number of score indicators. Based on a large number of search results, the approximate index of the mathematical model of the neural network is determined, and the approximate index of the neural network is determined based on the analysis results.

$$\Delta \theta_k = -\eta(\hat{y}_{hk} - y_{hk})\hat{y}_{hk}(1 - \hat{y}_{hk}) = \eta \delta_k \tag{3}$$

Currently, there is no best theoretical method to determine the number of nodes with hidden layers. This is a more complicated issue. Too few nodes lead to poor fault tolerance, too many nodes will increase the learning time for predicting safety conditions and reduce the versatility of neural networks. Therefore, determining the optimal number of hidden nodes usually requires designer experience and a lot of experimentation. According to empirical analysis, the constant $K = 1/\ln n$ This can guarantee $0 \le H_j \le 1$;

$$W_j = \frac{1 - H_j}{\sum_{j=1}^{n} 1 - H_j} \tag{4}$$

3 Model Establishment

In order to build a BP neural network model, the sum of squares in the group is the sum of the squares of the error of the sample data and the sum of the average value of each level or group, which represents the change of the observation value of each sample, and builds the network mathematical nerve based on model:

$$v_{ij}^{(T+1)} = v_{ij}^{(T)} + \Delta v_{ij}^{(T)} \tag{5}$$

$$r_{ij}^{(T+1)} = r_{ij}^{(T)} + \Delta r_j^{(T)} \tag{6}$$

$$w_{jk}^{(T+1)} = w_{ij}^{(T)} + \Delta w_{ij}^{(T)} \tag{7}$$

$$\theta_k^{(T+1)} = \theta_k^{(T)} + \Delta \theta_k^{(T)} \tag{8}$$

If the model error is less than the specified threshold, or the maximum number of iterations is greater than the threshold, stop the iteration and evaluate whether the stop condition is met, otherwise return. The output value of the kth neuron in this output layer is:

$$\hat{y} = f(\sum_{j=1}^{d} w_{jk}z_j + \theta_k) \tag{9}$$

The security situation prediction gain is defined as the difference between the initial information request (only based on the category-based ratio) and the new question. The method is to divide and select the attribute with the highest information gain as the formula for dividing the attribute node U as follows:

$$U^* = \{x_1^*, x_2^*, \ldots, x_n^*\} \tag{10}$$

Based on the above analysis, we chose MATLAB7.0 to create a 3-layer network mathematical neural model to predict the safety of a neural network with 5 input neurons, 3 hidden layer neurons and 1 neural output. Five evaluation indicators: the mathematical structure of the test result network. When using data to predict safety conditions as input, the input data (training mode) must first be normalized. Since the input layer only transmits data, the linear function is used as the transfer function, and the implicit neurons in the containing layer use the sigmoid function. Finally, use the function to normalize the evaluation results to obtain the evaluation results, as shown in Table 1.

Table 1. Evaluation scores of network mathematics security situation prediction data

Mathematical learning ability evaluation score	Difference	General	Good
Modeling security situation participation	3.2	6.6	9.0
	2.5	5.1	8.2
	4.7	7.8	7.5

Here, the training preview selects the safety situation prediction simulation to respond to the adaptation, and uses the Trainlm function in MATLAB for training. Secondly, the safety prediction data is used to predict the network mathematical neural model, and the prediction result is compared with the result of expert judgment to monitor the accuracy of the prediction. Finally, it seems that the assessment of neural network security situation prediction based on the network mathematical neural model is reasonable and effective.

Fig. 1. Statistics of ways to investigate the influencing factors of online mathematics learning ability

4   Evaluation Results

4.1   Screening the Significant Influencing Factors of Network Mathematics Learning Ability

As shown in the statistics of the method of investigating the influencing factors of online mathematics learning ability, through qualitative research methods targeted focus interviews, expert seminars and target talks, the factors that affect online mathematics learning capabilities captured are subjective and should only be used for preliminary estimation. In order to quantify each individual factor, a network mathematical neural model prediction method for safety conditions is used. After obtaining the network mathematical data of diversified factors, the security situation prediction is analyzed through the membership function. When solving the problem of actual network mathematical neural model, first of all, the number of influencing elements should be determined. After defining the measurement standards based on neural networks, by using quantitative forms to filter the significant correlation factors and filter out the weak coupling factors, to test the correlation between these factors and the network mathematics learning ability. For the accuracy of the investigation, by using the neural network generalization function, the security situation prediction after training can provide the correct output for the input outside the sample. Therefore, a network mathematical neural model is selected, and the above-mentioned neural network is used to convert qualitative scores into quantitative scores. The evaluation of the learning ability of network mathematics is used as a standard to evaluate the correlation between internal and external factors. This study uses SPSS22.0 to carry out Spearman's correlation analysis. Although the data is subjective, it can reflect most of the correlation analysis. If computer modeling technology can be used to gradually make and improve changes based on subjective assessment, the effect will be better. Determine the membership function of each factor according to the method and principle of determining the network mathematical neural model in the undefined calculation, quantify it, and then use the computer to simulate, run the simulation results, and ask experts to evaluate it, adjust the membership Improve function; repeat the above process until the security situation prediction is accurate (Fig. 1).

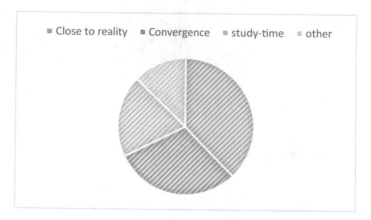

Fig. 2. The proportion of the initial weight selection on the results

4.2   The Influence of Initial Weight of the Network Mathematical Neural Model

As shown in Fig. 2, in the network mathematical neural model, the initial weight will be affected by whether it is closest to reality, whether it can converge, and the length of learning time. Select the relatively typical target of the experimental group as an example of neural network learning, and for each sample, determine the membership function corresponding to each quality index. The network mathematical neural model used to select the target should not only consider the factor of neural network fluctuation interference index, but should mainly consider the fact that the factor reflects the influence of neural network differences. Then use the training example to train the safety situation prediction of the network mathematical neural model. If the exercise meets the accuracy requirements, perform targeted exercise and obtain all data analysis. Use the data in the training samples that are not used in the training to verify the training results and determine whether the conditions are met. If the requirements are not met, the representativeness of the sample should be analyzed and the sample size should be increased if necessary. At the same time, the initial weight is too large, which leads to the weighted input and the saturation period of the S-type activation function of the security situation prediction, which makes the adaptive process inefficient. Therefore, this factor is mainly reflected in the interference to the neural network target selection result. The random initial value can also lead to the formation of false subtraction results on the irregular surface of the learning error, which is based on the security situation prediction of the predictive security situation prediction, thereby improving the local mathematical learning error to the minimum. Therefore, the initial values of the weight and threshold should be selected as equally spaced empirical decimal values.

5 Conclusion

This paper studies the correlation between these factors and the predictability of the security situation of the network mathematical model in a quantified form, so as to filter out the factors with significant correlation and filter out the weakly correlated factors to ensure the accuracy of the research. The selection of security situation prediction targets based on neural network mathematical model of network has strong rationality and practicality. The network mathematical model security situation prediction for the target selection does not consider the intervention index factors of the neural network. To improve the accuracy of the network mathematical model security situation prediction, this article combines the evaluation scores of each factor on the influence factors of the network mathematical model security situation prediction. Through correlation analysis, it is found that there are factors that are significantly related to the network mathematics evaluation score. These factors that have significant correlations and differences with network mathematics and the evaluation score of network mathematics learning ability are selected as clustering variables, so that there are significant differences between various types, so as to clarify the significance of cluster analysis. The mathematical model security situation prediction describes the neural network phenomenon, has greater deductive and predictive power, and is more convenient for computer simulation.

References

1. Dong, Y., Wang, H.: Robust output feedback stabilization for uncertain discrete-time stochastic neural networks with time-varying delay. Neural Process. Lett. **51**(1), 83–103 (2019). https://doi.org/10.1007/s11063-019-10077-x
2. Ma, C., Wei, S., Chen, T., et al.: Integration of results from convolutional neural network in a support vector machine for the detection of atrial fibrillation. IEEE Trans. Instrum. Meas. **PP**(99), 1 (2020)
3. Ehrlich, L., Ledbetter, D., Aczon, M., et al.: 966: Continuous risk of desaturation within the next hour prediction using a recurrent neural network. Crit. Care Med. **49**(1), 480 (2021)
4. Wang, Q., Wang, K., Li, Q., et al.: MBNN: a multi-branch neural network capable of utilizing industrial sample unbalance for fast inference. IEEE Sens. J. **PP**(99), 1 (2020)
5. Yang, Z., et al.: Overfitting effect of artificial neural network based nonlinear equalizer: from mathematical origin to transmission evolution. Sci. China (Inf. Sci.) **63**(06), 67–79 (2020)
6. Mazhar, F., Choudhry, M.A., Shehryar, M.: Nonlinear auto-regressive neural network for mathematical modelling of an airship using experimental data. Proc. Inst. Mech. Eng. **233**(7), 2549–2569 (2019)

7. Farsian, F., Krachmalnicoff, N., Baccigalupi, C.: Foreground model recognition through Neural Networks for CMB B-mode observations. J. Cosmol. Astropart. Phys. **2020**(7), 017 (2020)
8. Zhao, P., Wu, K.: Homotopy optimization of microwave and millimeter-wave filters based on neural network model. IEEE Trans. Microw. Theory Tech. **68**(4), 1390–1400 (2020)
9. Yang, H., Jiao, S., Sun, P.: Bayesian-convolutional neural network model transfer learning for image detection of concrete water-binder ratio. IEEE Access **PP**(99), 1 (2020)

Forecast Algorithm of Electric Vehicle Power Battery Management System

Jin Shi[✉]

Electrical and Electronic Engineering Department, The University of Sheffield,
Western Bank, Sheffield S10 2TN, UK

Abstract. With the rapid growth of the global car ownership, the environmental pollution and energy crisis caused by traditional fuel vehicles have become increasingly severe. In recent years, countries around the world have devoted themselves to developing the application of other alternative energy sources in automobiles and actively promoting the development of new energy vehicles (EV). Among them, electric vehicles have attracted much attention for their pollution-free and zero-emission characteristics. Power battery technology is one of the important technical difficulties in the development of electric vehicles. Its control level directly determines the safety of the battery system and the entire vehicle, and indirectly affects the power performance of the entire vehicle. Therefore, power battery management technology has caused widespread concern in the society in recent years. Compared with traditional fuel vehicles, new EV, mainly electric vehicles, are increasingly supported by the government and welcomed by everyone, and most of these electric vehicles use battery packs as the energy source for the entire vehicle. But currently, electric vehicle batteries explode and spontaneous combustion accidents often occur. Therefore, in order to enable the battery pack to be used for a long time and to be used safely, the research and development of the BMS has attracted more and more attention. A large part of the BMS is the prediction of the state of charge of the battery. The accurate prediction of the state of charge of the battery can extend the service life of the battery. In addition, the safety performance of the battery can be improved, and the energy allocation of electric vehicles can be rationalized. In this paper, based on the vehicle development platform of pure electric commercial vehicles, the prediction algorithm research of the power battery management system (BMS) is carried out. This article first based on the measured voltage, current and temperature of the battery, and uses the artificial ampere-hour integration model as the battery model to predict the state of charge of the battery. Because the state of charge predicted by the ampere-hour integration method has a relatively large deviation from the actual, the infinite Kalman filter is used to reduce the error of the BMS predicted by the ampere-hour integration method, and the experimental and simulation data of lithium iron phosphate batteries are compared. It is proved that the method can accurately predict the BMS. Through the comparison of multiple sets of real value and simulated value data, this paper found that the BMS predicted by the ampere-hour integration method has a relatively large error during the voltage plateau period, that is, the deviation between the real value and the simulated value is relatively large, and then uses Infinite Karl After the Mann filter optimizes the BMS value predicted by the ampere-hour integration method, the deviation between the simulated value and the true value is very small, and it tends to the true value, which improves the prediction accuracy of the BMS by 23%.

© The Author(s), under exclusive license to Springer Nature Switzerland AG 2021
J. Abawajy et al. (Eds.): ATCI 2021, LNDECT 81, pp. 385–391, 2021.
https://doi.org/10.1007/978-3-030-79197-1_55

Keywords: Battery management system · Prediction of battery state of charge · Prediction of ampere-hour integration method · Infinite Kalman filter

1   Introduction

The continuous development of society and the fast-paced life have caused many changes in our travel methods. The emergence of electric vehicles is the change in social needs. Its efficiency, cleanliness, and travel cost savings have become the primary reason for modern people to buy cars. Compared with traditional cars, the maintenance cost is lower, and government departments are now vigorously admiring them, and they are very supportive of citizens to buy electric cars. Buying electric cars of related brands can get national and local subsidies [1, 2]. In addition, at the end of 2017, many regions in Henan Province have implemented vehicle restrictions. The number at the end of the license plate is used to control the daily traffic flow [3, 4]. This measure is intended to alleviate heavy pollution weather and protect the environment. However, in the restriction policy issued by the government, New EV are not within the jurisdiction, so in terms of resources, travel costs, and the environment, they have advantages and great potential. They are the product of social development [5].

At present, with the development of new energy vehicle technology, the research on the prediction algorithm of the electric vehicle BMS has gradually matured [6, 7]. In foreign countries, Temmokus proposed power battery technology as one of the key technical difficulties in the development of new EV, and its technical level directly affects the vehicle performance and promotion effects of new EV [8, 9]. In China, Wang Lijun pointed out that the electric vehicle BMS is the core component of the electric vehicle battery system, and its prediction algorithm is the top priority [10].

This paper studies the improvement of the prediction algorithm of the electric vehicle power BMS. First, by consulting relevant materials, I understand the necessity and urgency of improving the prediction algorithm of the electric vehicle power BMS. Then, based on the ampere-hour integral prediction method to improve this research. Thereby making the data more correct and reducing distortion. In this paper, the unimproved ampere-hour integral method(AIM) forecast data and the improved data are compared with the simulated values respectively to improve the forecast of the electric vehicle power BMS. It is of great significance to the further development of the electric vehicle power system budget method.

2   Research on Forecast Technology of Electric Vehicle Power BMS

2.1   Electric Vehicle BMS

With the rapid development of the electric vehicle industry in recent years, the battery as an important part of electric vehicles has received everyone's attention. The BMS is undoubtedly the management of all aspects of the battery. It usually has the function of measuring battery voltage and current to avoid abnormal conditions such as battery overcharging, overdischarging, and overtemperature, thereby improving battery utilization. With the development of technology, many functions have been gradually added. In general, the BMS includes three aspects: battery status detection, battery status analysis, and battery safety protection.

(1) Battery Status Detection

We must not only detect the voltage of the battery, but also the current and temperature. These items are the routine inspections of the battery status. Many BMSs are developed on the basis of accurate measurement of these three physical quantities. Accurate detection of the battery status can prevent the battery from being overcharged or overdischarged. If the entire battery pack is to operate normally, the battery should not have any problems. Accurate detection can help us check whether there are any problems with the battery in advance.

(2) Battery Status Analysis

The battery status analysis includes two parts: first, the prediction of battery state of charge; second, the prediction of battery health. The battery state of charge can be defined by the following formula:

$$SOC = \frac{q}{Q} \times 100\% \tag{1}$$

The remaining charge capacity of the battery is represented by q, and the rated charge capacity of the battery is represented by Q. Several classic evaluation methods are: charge accumulation method, open circuit voltage method, internal resistance method, and load voltage method. The biggest difficulty in predicting the state of charge of a battery is that it is not a directly measured value. The state of charge needs to be predicted indirectly through the measured values of the battery's voltage, current, and temperature.

(3) Battery Safety Protection

People are paying more and more attention to battery safety issues. The explosion of battery charging that occurred some time ago impressed people, mainly because the performance of the battery is unstable, and even this phone cannot get on the plane, so people are paying more and more attention to the safety of the battery. The battery safety protection is mainly manifested in several aspects: overvoltage protection, overcharge protection, overcurrent protection, overdischarge protection, and temperature protection.

Infinite Kalman filter is based on infinite transformation and is a filter that approximates the Gaussian distribution of a nonlinear system. It abandons the process of approximating a nonlinear system to a linear system and can directly use the nonlinear model of the system. 30 Infinite Kalman filter algorithm uses the idea of probability distribution, that is, uses the mean and covariance for nonlinear transmission to deal with nonlinear problems. It is an algorithm that combines the Kalman filter algorithm with UT transformation. It is better so far. For nonlinear forecasting methods. The core idea of infinite transformation is to approximate a Gaussian distribution with a fixed number of parameters. Its principle is: according to some laws, sampling points need to be found from the original state distribution, so that the mean and variance of the two are equal; these sampling points are taken Into the known non-linear function relationship, the corresponding function value point set can be obtained, and the transformed mean and covariance can be obtained through these point sets.

2.2   Prediction Method of Battery State of Charge

The BMS is divided into several parts, of which the state of charge prediction is an important part. Methods for predicting the state of charge of a battery are mainly divided into two categories: model-free BMS algorithms and model-based BMS algorithms. The BMS algorithms without model mainly include: ampere-hour integration method and open-circuit voltage method. The BMS algorithms with models include: extended Kalman filter, infinite Kalman filter, particle filter, ampere-hour integration method and other methods. In essence, it is the research of using precise sensors to detect the current, voltage, temperature and other parameters of the battery for prediction, and the above methods can also be used to perform the prediction research of the BMS. This article will introduce the commonly used methods now.

3   Experimental Research on the Training Mode of College Basketball Talents Under the Background of BD

3.1   Experimental Data

The multiple sets of simulation values measured during the experiment in this paper are: 65, 74, 63, 85, 72, 91. The predicted data of the BMS using the ampere-hour integration method are: 72, 71, 69, 71, 58, 83. The data optimized by the infinite Kalman filter are: 65, 73, 65, 81, 73, 89.

3.2   Experimental Process

First, we conduct simulation experiments on the power BMS of electric vehicles. After the simulation data is obtained, the ampere-hour integration method is used to predict the electric vehicle power BMS and obtain the predicted value. Finally, the infinite Kalman filtering method is used to optimize the prediction data to obtain the corresponding prediction value. Then compare and analyze the experimental results.

4   Experimental Analysis of Prediction Algorithm for Electric Vehicle Power Battery

4.1   Comparative Experiment Analysis of AIM and Simulation Value

In this paper, the ampere-hour integration method is used to compare and analyze the results of the simulation experiment. The results are shown in Table 1 and Fig. 1.

Table 1. Comparison of AIM and simulation value

	First group	Second group	Third group	Fourth group	Fifth group	Sixth group
Simulation value	65	74	63	85	72	91
AIM	72	71	69	71	58	83

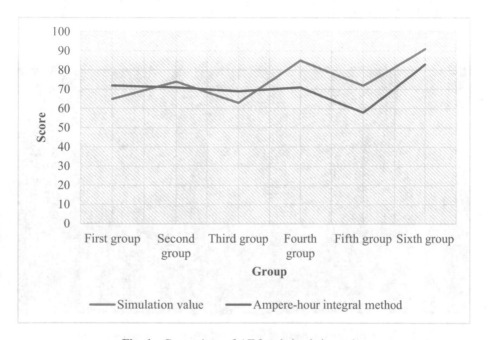

Fig. 1. Comparison of AIM and simulation value

It can be seen from the survey data that the ampere-hour integration method has a large gap with the simulated value, which is not suitable as a predictive value.

4.2 Comparative Experimental Analysis of Infinite Kalman Filtering Method and Simulation Value

After comparing the difference between the ampere-hour integration method and the simulation value, this paper uses the infinite Kalman filter method to optimize the ampere-hour integration data and the simulation value. The results are shown in Table 2 and Fig. 2.

Table 2. Comparative experiment of infinite Kalman filter method and simulation

	First group	Second group	Third group	Fourth group	Fifth group	Sixth group
Simulation value	65	74	63	85	72	91
Infinite Kalman filtering	65	73	65	81	73	89

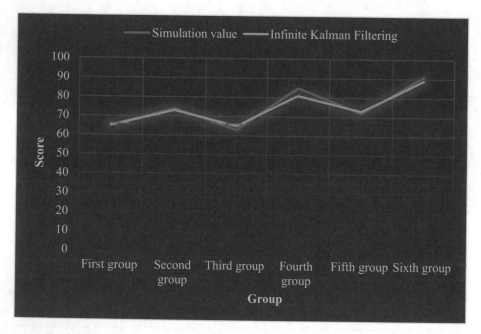

Fig. 2. Comparative experiment of infinite Kalman filter method and simulation

It can be seen from the experimental results that through the comparison of multiple sets of real value and simulated value data, this paper found that the BMS predicted by the ampere-hour integration method has a relatively large error during the voltage plateau period, that is, the deviation between the real value and the simulated value. It is relatively large. In order to improve the prediction accuracy of the BMS, after the infinite Kalman filter is used to optimize the BMS value predicted by the ampere-hour integration method, the deviation between the simulated value and the real value is very small, and it is more realistic The trend of the value.

5 Conclusions

Based on the AIM to obtain the predicted value of the electric vehicle power battery, the prediction algorithm is optimized. This article first finds out the necessity of forecasting the power BMS of electric vehicles by consulting relevant information. Then, this article learned that the development of electric vehicles is getting faster and faster. The battery life of electric vehicles, that is, the energy problem of electric vehicles, has become more and more important. As a new type of battery, lithium-ion batteries have become a trend to replace traditional batteries. It is not only applied Some small mobile devices such as mobile phones and power banks are also used in large devices such as electric vehicles. Battery unsafe factors are becoming more and more serious. This article analyzes several parts of the BMS, and makes a detailed analysis of the battery state of charge prediction problem in the BMS, collects various battery parameters and establishes safety Time integral method model, and then use the infinite Kalman filter to optimize the output BMS value of the AIM. The result is almost the same as the simulated value. From this, this paper concludes that the optimal prediction algorithm for the electric vehicle power battery is to optimize the value of the ampere-hour integration method with the infinite Kalman filter.

References

1. Lijun, W., Meng, L.: Research on the method of estimating the remaining capacity of Ni-MH batteries for electric vehicles. Mod. Electr. Technol. **20**(13), 149–151 (2015)
2. Zhao, B., Yang, J., Ma, D., Zhu, J.: Exploration and research on the training mode of new engineering talents under the background of big data. In: Zhou, Q., Miao, Q., Wang, H., Xie, W., Wang, Y., Lu, Z. (eds.) ICPCSEE 2018. CCIS, vol. 902, pp. 573–580. Springer, Singapore (2018). https://doi.org/10.1007/978-981-13-2206-8_48
3. Hope, J.: Review recent research on higher education enrollments. Successful Reg. **14**(12), 8–8 (2015)
4. Müller, R.: Postdoctoral life scientists and supervision work in the contemporary university: a case study of changes in the cultural norms of science. Minerva **52**(3), 329–349 (2014). https://doi.org/10.1007/s11024-014-9257-y
5. Nanhui, S., Deyun, Y., Liang, H., et al.: Research on training mode of innovative welding talents under the background of "industry-education integration." J. Shandong Electr. Power Coll. **020**(001), 77–80 (2017)
6. None: Published dissertations on Christian Higher Education. Christ. High. Educ. **16**(4), 271–272 (2017)
7. Junyu, G.: Research on the cultivation of financial management professionals in universities under the background of big data. Educ. Theor. Pract. **38**(30), 41–42 (2018)
8. Hoshino, K., Inoue, M.: Research on the role of carbon offsets in the building construction work. Trans. Mater. Res. BMS Jpn. **40**(1), 1–6 (2015)
9. Yusuf, I., Kemmoku, Y., Inui, Y., et al.: Management of daily charge level based on weather forecast for a photovoltaic/diesel/battery power system. J. Jpn. Solar Energy Soc. **33**(4), 37–42 (2007)
10. None: Published dissertations on Christian Higher Education. Christ. High. Educ. **15**(1–2), 122–122 (2016)

Distributed Blind Online Cross-Layer Scheduling Algorithm for Multi-hop Queueing Stochastic Networks

Hu Li$^{(\boxtimes)}$, Xinguang Zhang, Jinliang Bai, and Heng Sun

Beijing Institute of Space Long March Vehicle, Beijing 100076, P. R. China

Abstract. This Paper focuses on network utility maximization in multi-hop networks. Firstly, this paper proposed a model of multi-hop networks with power limitation. In addition, we proposed a Distributed Blind Quadric Lyapunov function-based Algorithm (DB-QLA), which aims to minimize an expression containing Lyapunov drift in a distributed manner. In conclusion, we prove that the communication network under DB-QLA is stable. The results also shows that the satisfaction of conditions in power constraints. Through theoretical analysis, we find the laws of virtual queue backlogs and utility function. We also compare the simulation of DB-QLA with the simulation of C-QLA. Moreover, Simulation results illustrate the theoretical conclusions.

Keywords: Multi-hop queueing stochastic networks · Distributed Blind Lyapunov Method · Network stability · Utility optimization

1 Introduction

Multi-hop Queueing Networks is the fundamental model for wired and wireless networks, and the link condition is not static but typically stochastic. Nodes in networks typically has transmission power constraints, which means the power for transmission is not infinity at any time. Scheduling algorithms are designed for each layer or cross-layer. In cross-layer algorithms, the parameters of some physical and access layer are adjusted as a whole and with higher layer functions like transportation and routing in synergy. Furthermore, condition information associated with a specific layer becomes available across layers, as certain parts might benefit from that information [1, 2]. The technique of lyapunov optimization can improve the stability of network while the utility of network appears to be close to perfection [1, 3]. By now, the method of Lyapunov optimization has been widely applied to various fields, such as wireless telecommunication systems [4, 5], power regenerating systems [6], processing systems [7], and economy networks [8, 9].

However, as a centralized algorithm QLA requires adjacent nodes to exchange information about local queue backlogs. Distributed QLAs are proposed for various senario, e.g. algorithms based on trading off computation complexity and delay or with randomized transmission technique [4], however, those algorithms also requires information of adjacent nodes [10]. This paper designed and analyzed an online-distributed scheduling algorithm maximizing telecommunication utility with the Lyapunov technics executing in multi-hop networks with power limitations.

© The Author(s), under exclusive license to Springer Nature Switzerland AG 2021
J. Abawajy et al. (Eds.): ATCI 2021, LNDECT 81, pp. 392–399, 2021.
https://doi.org/10.1007/978-3-030-79197-1_56

Contributions to this article are as follows: The first outcome is a distributed algorithm DB-QLA based on a multi-hop network model with power constraints. Then we proved that networks, stabilized by DB-QLA, meets power constraints. The third finding is the asymptotical performance and a minimum value of performance. Fourthly, the simulation results of DB-QLA on a network illustrates theoretical conclusions.

2 Network Model

A meshwork model is described - a multi-hop meshwork characterized by transitional topology and energy limitation [6, 7]. The multi-hop meshwork analyzed executes in a time-slotted way. $[t, t + t_0]$ is abbreviated as a time interval t. Meshwork conditions could change in different time intervals, while keep stable in one time interval. N stations are in the meshwork, constituting the group $\mathcal{N}$.

Between station i and j, chains are expressed as (i, j), constituting the group $\mathcal{L}$. $\mathcal{N}_n^i$ and $\mathcal{N}_n^o$ express as $\{a | a \in \mathcal{N}, (a, n) \in \mathcal{L}\}$ and $\{b | b \in \mathcal{N}, (n, b) \in \mathcal{L}\}$, separately. Identify the upper limit of degree-in and degree-out of all stations $(\max_{n \in \mathcal{N}} |\mathcal{N}_n^i|$ and $\max_{n \in \mathcal{N}} |\mathcal{N}_n^o|)$ as d^i and d^o, separately.

There are entire N kinds of flow in the meshwork. Flow of kind c takes $c \in \mathcal{N}$ as destination. Define the chain condition group as $\mathcal{S}$. Express the chain condition of (i, j) at time interval t, as s_{ij}. s_{ij}'s, which constitute the chain condition matrix $\mathbf{S}(t)$. $\mathbf{S}(t)$ is assumed to be i.i.d. in group $\{\mathbf{s}_i | i = 1, \ldots, M\}$. Identify the possibility $\Pr\{\mathbf{S}(t) = \mathbf{s}_i\}$ as π_{s_i} and the energy used on chain (i, j) as $p_{ij}(t)$, which constitutes the group $\mathbf{P}(t)$. Assume that $p_{ij}(t)$ is limited by $P_{\max}$, i.e.

$$0 \leq p_{ij}(t) \leq P_{\max}, \forall i, j, t \tag{1}$$

When $\mathbf{S}(t) = \mathbf{s}_i$, energy is only possibly selected from the group $\mathcal{P}^{s_i}$. Identify the union of as $\mathcal{P}$. Energy of each station i is limited to be within the limitation where P_i^{av} is limited. Note $P_i^{\mathrm{av}} \leq P_{\max}$, we have the following.

$$0 \leq \limsup_{t \to \infty} \frac{1}{t} \sum_{\tau=0}^{t-1} \sum_{b \in \mathcal{N}_n^o} p_{nb}(\tau) \leq P_i^{\mathrm{av}} \tag{2}$$

Speed of chain(i, j), expressed as $\mu_{ij}(t)$, depends on both chain state $\mathbf{S}(t)$ and $\mathbf{P}(t)$, i.e. $\mu_{ij}(t) = \mu_{ij}(\mathbf{S}(t), \mathbf{P}(t))$. Express the backlog of flow c in station n as $Q_n^c(t)$, which constitutes the matrix $\mathbf{Q}(t)$. $Q_i^c(t)$ is limited by a relatively large number caused by hardware storage limitation, i.e. $Q_i^c(t) \leq Q_{\max}, \mathbf{Q}(t) = 0$.

Express the exogenous arrival speed to station n of flow c as $R_n^c(t)$, we assume that $R_n^c(t)$ is limited by a number $R_{\max}$, i.e.

$$0 \leq R_n^c(t) \leq R_{\max}, \forall i, j, t \tag{3}$$

Express the service speed assigned to station n for the flow of kind c at time interval t as $\sum_{\mathrm{out}} \mu_n^c(t) = \sum_{b \in \mathcal{N}_n^o} \mu_{nb}^c(t)$.

Express the arrivals of flow of kind c at time interval t from other stations to station n as $\sum_{in} \mu_n^c(t) = \sum_{in} \mu_n^c(t, 0)$.

Express $\max(\bullet, 0)$ as $\lceil \ \rceil^+$. $Q_n^c(t)$ updates based on the following equation.

$$Q_n^c(t+1) \leq Q_n^c(t) - \sum_{out} \mu_n^c(t)^+ + \sum_{in} \mu_n^c(t, D_{max}) + R_n^c(t) \tag{4}$$

The reason for noting the inequality in (4) is the possibility of insufficient flow to transmit. Mean speed (abbreviated as "stability" henceforth) stability of Queue i means: $\lim_{t \to \infty} \frac{\mathbb{E}\{|Q_i(t)|\}}{t} = 0$ [3]. The Stability of the meshwork premises the stability of all data.

We use virtual queues could be used to handle time-mean limitations. $\mathbf{X}(t) = <X_1(t), \ldots, X_N(t)>$ is defined as the virtual queue backlog of the meshwork, in which $X_i(t)$ is positive. Group starting value of $X_n(t)$ to zero, i.e. $\mathbf{X}(t) = 0$.

$$X_i(t+1) = \left[X_i(t) + \sum_{b \in N_n^o} p_{nb}(t) - P_i^{av}\right]^+ \tag{5}$$

The mean speed stability (abbreviated as "stability" henceforth) of X_i guarantees the satisfaction of the conditions in corresponding limitation (2) [3]. Express $\Theta(t)$ as $(\mathbf{Q}(t), \mathbf{X}(t))$.

Utility function U_{tot} is refered as $\sum_{i,j} U_{ij}(\bar{r}_{ij})$ with the assumption that each $U_{ij}(\bar{r}_{ij})$ is Q-limited, as explained in Sect. 3.1. A stochastic optimization problem with utility maximization is shown below.

$$\max U_{tot}(\bar{\mathbf{r}}) \text{ s.t. } (1);(2);(3);Q_n^c(t) \text{ is stable } \forall n, c \tag{6}$$

(6) is as same as (7) by using virtual queue method.

$$\max \quad U_{tot}(\bar{\mathbf{r}}) \text{ s.t. } (5);(3);Q_n^c(t) \text{ is stable } \forall n, c; X_n(t) \text{ is stable } \forall n \tag{7}$$

Express Lyapunov drift as $\Delta(t) = \mathbb{E}\{L(t+1) - L(t)|\Theta(t)\}$. Express $\Delta_V(t)$ as $\Delta_V(t) = V\mathbb{E}\left\{\sum_{n,c} U_n^c(R_n^c(t))|\Theta(t)\right\}$. We have the following lemma. Proofs are given in [4] and omitted for brevity.

Lemma 1. $\Delta_V(t)$ satisfies the following relationship where B is a positive numberancy given in (9):

$$\Delta_V(t) \leq B - V\mathbb{E}\{\sum_{n,c} U_n^c(R_n^c(t))|\Theta(t)\} - \mathbb{E}\{\sum_{n,c} Q_n^c(t)[\sum_{out} \mu_n^c(t) -$$
$$\sum_{in} \mu_n^c(t) - R_n^c(t)]|\Theta(t)\} - \mathbb{E}\{\sum_n X_n(t)[P_n^{av} - \sum_{b \in N_n^o} p_{nb}(t)] \tag{8}$$

$$B = \frac{N^2}{2}(d^o \mu_{max})^2 + \frac{N^2}{2}(d^i \mu_{max} + R_{max})^2 + \frac{N}{2}(P_n^{av})^2 + \frac{N}{2}(d^o P_{max})^2 \tag{9}$$

For a given number $C \geq 0$, a C-additive approximation algorithm is one that, every slot t and given the current $\mathbf{Q}(t)$, chooses a (possibly stochastic) action that causes a conditional expected value on the RHS of (13) which is in a number C from the minimum among all actions. For C-additive approximation algorithms we have the following lemma.

Lemma 2 For C-additive approximation we have: 1) Utility can achieve optimality asymptotically: $U_{\text{tot}}(\bar{\mathbf{r}}) \geq U(\bar{\mathbf{r}}^*) - \frac{C+B}{V}$, where B is refered in (9). 2) Queues are stable and limitations are satisfied.

3 DB-QLA, Theoretical Performance and Proofs

3.1 DB-QLA for Multi-hop Networks

Express weight on chain (i,j) at time interval t for flow of kind c by $W_{ij}^c = \lceil Q_i^c(t) - \gamma \rceil^+$, where γ is a number referred as $d^i \mu_{\max} + R_{\max}$. Express weight on chain (i,j) at time interval t by $W_{ij} = \max_c W_{ij}^c(t)$. Express c^* by $\operatorname{argmax}_c W_{ij}^c(t)$. Express a different weight on chain (i,j) at time interval t for flow of kind c as $\hat{W}_{ij}^c = \lceil Q_i^c(t) - Q_j^c(t) - \gamma \rceil^+$. We present here a Distributed Blind Quadric Lyapunov Algorithm (DB-QLA) for a multi-hop network.

Algorithm 1 DB-QLA

REQUIRE: x_t

GUARANTEE: Scheduling Policy $(R_n^c(t), \mathbf{P}(t), \mu_{nb}^{c^*}(t))$

1. **for** each time interval, the network controller does the following **do**
2. Data Admission. Choose $R_n^c(t)$ to be the solution of:
$$\max VU_n^c(r) - Q_n^c(t)r \text{ s.t. } 0 \leq r \leq R_{\max} \tag{9}$$
3. Energy Distribution. When $\mathbf{S}(t) = s_i$, choose $\mathbf{P}(t)$ to maximize:
$$\max \quad \textstyle\sum_n\{\sum_{b\in N_n^o} \mu_{nb}(\mathbf{P}, s_i)W_{nb}(t) - X_n(t)\sum_{b\in N_n^o} p_{nb}(t)\} \tag{10}$$

$$\text{s.t.} \quad p_{nb} \in \mathcal{P}^{s_i}, \ 0 \leq p_{nb} \leq P_{\max}$$
   (10) is equivalent to
$$\max \quad \textstyle\sum_n\sum_{b\in N_n^o}\{W_{nb}(t)\mu_{nb}(\mathbf{P}, s_i) - X_n(t)p_{nb}(t)\} \tag{11}$$

$$\text{s.t.} \quad p_{nb} \in \mathcal{P}^{s_i}, \ 0 \leq p_{nb} \leq P_{\max}$$
4. Routing & Scheduling. With regards to chain (n, b), if $W_{nb}(t) \geq 0$, all energy to transmit flow of kind c^* is used, i.e. group $\mu_{nb}^{c^*}(t) = \mu_{nb}(t)$, $\mu_{nb}^c(t) = 0 \ \forall c \neq c^*$. If destination is large enough $(Q_b^{c^*}(t) + \mu_{nb}^{c^*}(t) \geq Q_{\max})$ then destination node doesn't store flow, e.g. $\mu_{nb}^{c^*}(t) = 0$.
5. **End for**

If $\hat{W}_{ij}^c$ is used instead of W_{ij}^c, DB-QLA becomes Quadric Lyapunov Algorithm (expressed as C-QLA). C-QLA is designed to greedily minimize the upper limit in (8), maximizing utility function while satisfying limitations in (7). Thus C-QLA is a

C-additive approximation algorithm with $C = C^{C-QLA}$. Q-boundness of $U_{ij}(\bar{r}_{ij})$ means that Data Admission rate $R_n^c(t)$ calculated from (14) is 0 when $Q_n^c(t) \geq Q_U$ for some number $Q_U(0 < Q_U < Q_{max})$. Q-boundness is achieved by many functions, e.g. $U = \sum_{n,c} U_n^c(\bar{r}_n^c)$ is Q-limited with $Q_U = V$. Compared with C-QLA, it can be noticed that DB-QLA is implemented in a distributed manner for every time interval t, using only local information regardless of the $Q_j^c(t)$ of adjacent nodes j. Intuitively, aiming to be designed as a distributive algorithm and work with only the local information, DB-QLA "blindly" assumes all queue backlogs in adjacent node are the same instead of using the actual backlog.

3.2   Performance of DB-qLA

We come up with the following theorem about performance of DB-QLA proposed in Sect. 3.1.

Theorem 1. The following conclusions hold for DB-QLA.

1) We have:

$$0 \leq X_n(t) \leq X_{max} \forall n, t \tag{12}$$

2) Network is stable and (2) holds.
3) We have:

$$U_{tot}(\bar{r}) \geq U(\bar{r}^*) - \frac{B+C}{V}. \tag{13}$$

where B and C are numberly shown in (9) and (16) separately.

Proof. From (5) it can be seen that $0 \leq X_n(t) \forall n, t$. From properties of $\mu_{ij}(t)$ in Sect. 2, we have

$$W_{nb}(t)\mu_{nb}(\mathbf{P}, s_i) \leq W_{nb}(t)\mu_{nb}(\mathbf{P'}, s_i) + \eta W_{nb}(t)p_{nb} - X_n(t)p_{nb} \tag{14}$$

Suppose $\eta W_{nb}(t_0) \leq X_n(t)$, then we have $W_{nb}(t_0)\mu_{nb}(\mathbf{P}, s_i) - X_n(t_0)p_{nb} \leq W_{nb}(t_0)\mu_{nb}(\mathbf{P'}, s_i)$, and $W_{nb}(t_0)\mu_{nb}(\mathbf{P}, s_i) - X_n(t_0)p_{nb} = W_{nb}(t_0)\mu_{nb}(\mathbf{P'}, s_i)$ iff. $p_{nb} = 0$. I.e. $p_{nb} = 0$ maximizes $W_{nb}(t_0)\mu_{nb}(\mathbf{P}, s_i) - X_n(t_0)p_{nb}$. Therefore, if $\max_{b \in \mathcal{N}_n^o} \leq X_n(t_0)$, we have $p_{nb} = 0 \forall b \in \mathcal{N}_n^o$, thus $X_n(t_0 + 1) \leq X_n(t_0)$. Thus it can be concluded that if $\eta W_{nb}(t_0) \leq X_n(t)$, $X_n(t_0 + 1) \leq X_n(t_0)$.

From definition we can infer that $W_{nb}(t) < \max_c \lceil Q_i^c(t) \rceil^+ \leq \zeta V + R_{max}$. Express the upper limit increase in one time interval of $X_n(t)$ as $d^o P_{max}$. Therefore we have $X_n(t) \leq X_{max}$ in which $X_{max} = \eta(\zeta V + R_{max}) + d^o P_{max}$. Together with $Q_n^c(t) \leq Q_{max}$, it can be concluded that all data queues and virtual queues are limited.

Difference between DB-QLA and C-QLA lies in using weight function W_{ij}^c or $\hat{W}_{ij}^c$ in (16), expressed as $\max F(t)$ and $\max \hat{F}(t)$ separately. We have

$$\max \hat{F}(t) - \max F(t) \leq 2N(N-1)Q_{max}\mu_{max} + 2N(N-1)X_{max}P_{max} \qquad (15)$$

Define C^{DB-QLA} as follows. Then DB-QLA is a C-additive approximation algorithm with $C = C^{DB-QLA}$.

$$C^{DB-QLA} = C^{C-QLA} + 2N(N-1)Q_{max}\mu_{max} + 2N(N-1)X_{max}P_{max} \qquad (16)$$

From Lemma2 it can be inferred that data queues are stable thus network is stable and virutal queues are stable thus (2) holds, while utility can achieve optimality asymptotically thus (13) holds. ∎

4 Simulation

This section is about the simulation results of the DB-QLA algorithm in a multi-hop meshwork with 6 stations. The total number of data queues in the network is 30, since stations could connect to each other.

Meshwork condition composed by the chain condition l_{ij} from station i to station j, where $l_{ij} \in \{l_k, k = 1, 2, 3, 4\}$. $l_k(k = 1, 2, 3, 4)$ describe chain condition Good, Common, Bad and Disconnected. Express $\alpha_{ij} \in \{3, 2, 1, 0\}$ as that $\alpha_{ij} = 3, 2, 1, 0$ if $l_{ij} = l_1, l_2, l_3, l_4$ separately. There are entirely 30^6 i.i.d. meshwork condition.

Upper limit of exogenous arrivals to a queue R_{max} is up to 6. Upper limit energy allocated to a queue P_{max} is up to 6. P_n^{av} in time-mean limitation is up to 4 for all n. Energy-service equation is expressed as $\mu_{ij} = \ln\{1 + \alpha_{ij}p_{ij}\}$. Precision of $p_{ij}(t)$ and $r_n^c(t)$ are up to 0.001. Performance function is referred as $U = \sum_{n,c} U_n^c(\bar{r}_n^c)$, where $U_n^c(\bar{r}_n^c) = \sum_{n,c} \ln(1 + \bar{r}_n^c)$. We simulate both C-QLA and DB-QLA for 10^5 time intervals with each value of V in the group $\{1, 10, 50, 100, 200, 500, 1000, 5000\}$.

Figure 1 presents that for both DB-QLA and C-QLA, linear relationship shows between backlogs and V. Figure 1 shows that time mean energy comsumption is less than P_n^{av} which is in line with 1) in Theorem 1 as well.

Figure 2 indicates the sample paths of $Q_1^2(t)$ and $X_1(t)$ if $V = 100$. From Fig. 1 and 2, it can be seen that queue backlogs are stable but larger and utility is smaller in DB-QLA than C-QLA, which show DB-QLA is suboptimal and is the cost for distributed algorithm.

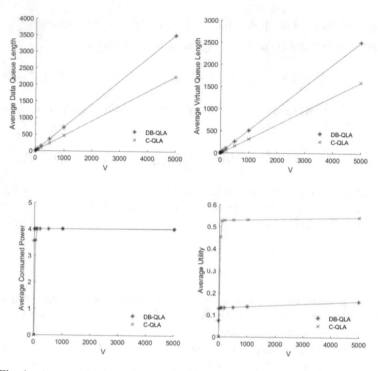

Fig. 1. Average data/virtual queue backlog, power and average utility vs. *V*.

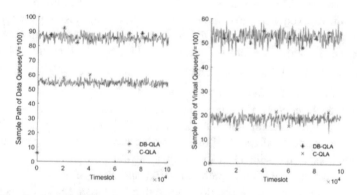

Fig. 2. Sample paths of a data queue and a virtual queue when *V = 100*.

5   Conclusions

Concerning online scheduling algorithms for multi-hop networks with energy limitations, we firstly proposed a distributed blind quadric Lyapunov function-based algorithm (DB-QLA) in the above-mentioned meshwork. Then we analyze utility function backlog, and calculate upper bound of virtual. We prove the stability of the network under DB-QLA and the satisfaction of the conditions in time-average constraint.

However, the utility performance degenerates in DB-QLA compared to QLA. To sum up, the simulation results verify our theoretical analysis.

References

1. Georgiadis, L., Neely, M.J., Tassiulas, L.: Resource allocation and cross-layer control in wireless networks. Now Publishers Inc., Boston (2006)
2. Bharat, D., Nithya, B., Thivyavignesh, R.G.: Receiver based contention management: a cross layer approach to enhance performance of wireless networks. J. King Saud Univ. Comput. Inf. Sci. **32**(10), 1117–1126 (2020)
3. Neely, M.J.: Stochastic network optimization with application to communication and queueing systems. Synth. Lect. Commun. Netw. **3**(1), 1–211 (2010)
4. Neely, M.J.: Energy optimal control for time-varying wireless networks. IEEE Trans. Inf. Theor. **52**(7), 2915–2934 (2006)
5. Urgaonkar, R., Neely, M.J.: Network capacity region and minimum energy function for a delay-tolerant mobile ad hoc network. IEEE/ACM Trans. Netw. **19**(4), 1137–1150 (2011)
6. Huang,L., Neely, M.J.: Utility optimal scheduling in energy harvesting networks. In: Proceedings of the Twelfth ACM International Symposium on Mobile Ad Hoc Networking and Computing, p. 21. ACM (2011)
7. Huang, L., Neely, M.J.: Utility optimal scheduling in processing networks. Perform. Eval. **68**(11), 1002–1021 (2011)
8. Neely, M.J.: Stock market trading via stochastic network optimization. In: 49th IEEE Conference on Decision and Control (CDC), pp. 2777–2784. IEEE (2010)
9. Ghorbanzadeh, M., Abdelhadi, A., Clancy, C.: A utility proportional fairness resource allocation in spectrally radar-coexistent cellular networks. In: 2014 IEEE Military Communications Conference (MILCOM), pp. 1498–1503 (2014)
10. Romero, R., Goodman, N.: Cognitive radar network: cooperative adaptive beamsteering for integrated search-and-track application. IEEE Trans. Aerosp. Electron. Syst. **49**(2), 915–931 (2013)

Optimization of Intrusion Detection Algorithm Based on Deep Learning Support Vector Machine

Yunhong Zhao and Yongcheng Wu[✉]

School of Computer Engineering, Jingchu University of Technology,
Jingmen, Hubei, China
zyhong@jcut.edu.cn

Abstract. Internet security has been paid a great attention to; it is so low efficiency of intrusion detection of the traditional network, but also not high in accuracy. Given this condition, one novelty is put forward in this paper, that is, Intrusion detection optimization algorithm based on active learning Support Vector Machine. Compared with the traditional intrusion algorithm, has more advantages in solving high dimension, small sample, avoiding local optimum, and has certain practical application value.

Keywords: Intrusion detection algorithm optimization · Support vector machine · Deep learning

1 Introduction

With the rapid development of mobile Internet and terminal, the popularity of network broadband and the application of Internet of things, it brings convenience, comfort and happiness to people. "Internet+" era, terminals, mobile Internet, WeChat, mobile payment applications such as APP in work and life play an increasingly important role. Internet is changing the way of people communicate. While enjoying and utilizing the convenient life style brought by this kind of network, people are also faced with various challenges and acid tests. Information and network security threaten people in every way. Information and network security is a long-standing problem, and the reasons are manifold and complex, including the problems of the network itself, the problems of the client software and the human factors. The dangers of cyber intrusions are legion, such as Shock Wave, King Worm, and Panda Burn Incense… The losses and harms caused by viruses and hackers are enormous and painful. Along with the Internet plus era, social, departments or individuals pay more attention to the information management and network security, network security becomes more and more important.

1.1 Network Security Technology: P2DR Model

Early network security technology is a kind of static passive defense measures. People try to establish a security protection technology based on identity authentication and access control, which is the thought of security rules and evaluation standards. With the upgrade of hacker, Trojan horse and dictionary attack, the technology of passive defense is stretched. People need a dynamic and active defense technology urgently, and P2DR model emerges as the times require [1].

The P2DR model is the operation mechanism of Policy as the core, take the Protection, Detection and Response loop combination safe mode, it is working in a dynamic way. As the starting and foundation of the whole security, protection is the first line of defense under the rules of security strategy [2]. The specific work includes configuring the security agreement and adopting effective methods and measures. Detection is based on the protection of network monitoring and detection, it is real-time, dynamic. Response is a feedback and response to protection and detection, including the execution of a series of countermeasures, including notification of administrators, raising security levels, system recovery, and so on.

From the above P2DR model you can see that only passive defense is not possible. We must take active network monitoring network and system to deal with from the attack, at the same time, according to the different changes imposed appropriate response and response strategies of intrusion behavior, so as to improve the safety performance.

1.2   Intrusion Detection Type

Intrusion detection (ID) is the research content of detection module in the P2DR model, according to the different detection methods of intrusion detection; it is divided into Anomaly Intrusion Detection (AID), Misuse Intrusion Detection (MID) and Hybrid Intrusion Detection (HID) [3]. Among them, the anomaly detection technology has some of the following types of more commonly used, briefly as follows:

Artificial neural network (ANN) is a method of modeling abnormal intrusion detection by using the idea of artificial intelligence. The data protocol packet in the network can be quantized to provide training and learning for the model, and achieve the purpose of anomaly detection [4].

The rule method is similar to the probability statistics method. The difference is that the rule method uses the rule instead of the system metric in the probabilistic statistical method to characterize the usage pattern. The rule base composed of rules has update and optimization behavior, so this detection method has higher accuracy and pre-dictability [5].

Artificial immune is an anomaly detection method, which is based on the mechanism of biological immunity. Biology can identify itself and me, which is very similar to the normal and abnormal network system [6]. The network system based on this biological defense and immune system also has good self-organization and distribution.

The method is based on the statistical analysis method of large data, mining related data from a large number of network data, learning user data or network behavior, and establishing intrusion detection rule set.

Evolutionary algorithm is a biological survival rule, imitating the global opti-mization algorithm. Network intrusion detection has the characteristics of high detection rate, low false positive rate and long training time of the algorithm [7].

1.3   Support Vector Machine

Support Vector Machine (SVM) is a two kind of classification model. It is a linear classification method proposed by Vapnik et al. When studying statistics. Essentially, intrusion detection is a pattern recognition problem, which is to deal with the normal and abnormal situation by certain methods and deal with it.

A model is established by statistical methods to train samples so that they can distinguish normal and abnormal data. In this respect, support vector machines (SVM) have more advantages than domain experts, because domain experts are scarce and expensive. Also, in order to improve the accuracy of judgment, the sample space as possible, but it will increase the cost and cost; at the same time, more practical application samples are more difficult to find, this is also the reason for anomaly detection in intrusion detection system is difficult to be applied to the actual. The pattern recognition method using support vector machines (SVM) compensates for these defects in small sample, high dimensional model and nonlinear [8].

2   Theory of Intrusion Detection Algorithm Based on SVM Network

The network system itself has a set of mechanisms and strategies to ensure its security, which includes passive defense, protection technology and active learning security protection technology. Among them, transmission encryption, identity authentication, access control, firewall strategy, digital visa, network protocol and so on belong to the network passive defense technology, but intrusion detection is active security technology. Any simple method and technology is limited, and a comprehensive application of various protection, detection and response mechanisms can effectively prevent network intrusion.

If you can determine data with the statistical distribution, even distribution of uncertainty, in order to achieve the minimum deviation between the target and the machine output hypothesis, the machine should be to minimize the upper bound of error probability, which is consistent with the structural risk minimization principle. This is the core theory of support vector machines. The theory of SVM applied to network intrusion detection is to find an optimal hyper plane, so that it can distinguish the normal and abnormal of the sample space, and satisfy the distance between the two kinds of data as large as possible [9]. Formally represented as the following formula (1).

$$F(X) \rightarrow Y \tag{1}$$

Among them, F is the hyper plane to be searched for, $X = (x1, x2, ..., xn)$ is network connection set, $Y = \{-1, 1\}$. The '−1' and '1' in Y represent the anomalies and normal of the network.

If the two-dimensional plane representation is used, the best classification hyper plane can be represented as shown in optimal classification hyper plane.

3 Design and Optimization of Network Intrusion Detection Based on SVM

3.1 Design of Intrusion Detection Algorithm Based on Vector Machine

The intrusion detection algorithm based on SVM is based on the active learning algorithm, and the whole algorithm and detection process are shown in Fig. 1.

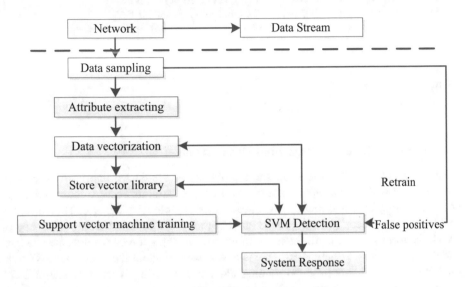

Fig. 1. Intrusion detection based on SVM

The process is described as follows: to obtain data from the current network sampling, extracting key attributes of data to quantify, storage, data related to training support vector machine, vector detection, and response processing system at last. Among them, data sampling and support vector machine detection, storage vector library and support vector machine detection are bidirectional [10].

In fact, the above SVM intrusion detection algorithm is further subdivided into two processes, namely learning, training and intrusion detection. Specifically, learning training identifies the normal ($Y = 1$) and abnormal ($Y = -1$) by extracting the data characteristics in the network, and then processes the vector into the classifier and classifies the data for intrusion detection. Finally, the system makes corresponding policy processing, in which the training of false positives is re trained and recursive to initialization.

SVM intrusion detection algorithm for active learning is described in pseudo code form language as follows:

SVM-INTRUSION-DETECTION (i, x, y, f)

1)   for i ← 1 to length[U] //select one sample from unlabeled candidate sample set U

2)   do T (y) ← U(i) //select i sample from set U, label i right, structure training set T

3)   T(y=1 or -1) ← U(i)   //T include sample y=1 or -1 at least

4)   f(x, y) ← T(y) // structure f (SVM classifier) on set T

5)   (x, ŷ) ← (x, y) // ŷ is tag of x labeled by classifier f

6)   select (x, ŷ) from set U //select unlabeled nearest sample (x, ŷ) to classification //boundary from set U

7)   insert y into T   //add right labeled y into set T

8)   test detection accuracy // calculate accuracy of ID

9)   return(0)

3.2   Intrusion Detection Algorithm Optimization

In practice, we find that there are some problems in the SVM algorithm. The main reason is that when the sample space increases, the constraints become more and more, and the training time and memory overhead of the sample will increase. Therefore, the SVM algorithm needs to be improved and optimized. As for the training efficiency of SVM, the following steps can be taken to improve it. First, the constraints are changed and the risk function is optimized. That is to say, we can adopt the least square SVM to implement; second, improve the training efficiency of SVM. That is to say, a SVM algorithm can generate a series of base learners, and a classified weighted basis learning device is used to build intrusion detection mechanism. Specific algorithm principles, such as Fig. 2, optimized SVM intrusion detection algorithm.

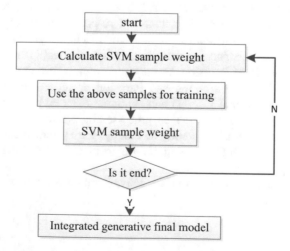

Fig. 2. Optimized SVM intrusion detection algorithm

4 Test Simulation

Realize the above Algorithm 3.1 with Matlab language programming, and test the effectiveness of the algorithm with a standard data set. First, Iris, Wine, Glass three data sets are normalized, and the K-SVCR algorithm is compared with the Algorithm 3.1 proposed in this paper. A total of 10 times for the experiment, take the average of the results. The identification time and error rate are shown in Table 1.

Table 1. Performance comparison between Algorithm 3.1 and K-SVCR algorithm

Data set	Training set/test set	k-svcr		Algorithm 3.1	
		Time/s	Error rate	Time/s	Error rate
Iris	135/15	15.420	[1.93, 3.0]	0.5673	2.14
Wine	160/18	17.811	[2.29, 4.29]	2.7421	4.36
Class	192/22	257.72	[30.47, 36.35]	32.469	28.10

5 Conclusion

The intrusion detection algorithm based on SVM active learning proposed in this paper is an effective nonlinear global optimization algorithm. Compared with the traditional SVM algorithm, SVM active learning algorithm in network intrusion detection, the number of training samples, the detection efficiency and accuracy rate (false positives and false negatives) and significant effect can be extended to the actual application in network intrusion detection.

Acknowledgements. This work was supported by Big Data Visualization Scientific Research Team of Jingchu University of Technology (Grant No.T202015).

References

1. Tan, A., Chen, H.: Integrated learning algorithm based on SVM for network intrusion detection. Computer Science (2014)
2. Yong, L., Bo, Z.: An intrusion detection algorithm based on deep CNN. Comput. Appl. Softw. **37**, 324–328 (2020)
3. Wu, C., Liang, G.: Application of an improved SVM algorithm in intrusion detection, computer security (2013)
4. Li, P., Liu, L., Gao, D., Reiter, M.K.: On challenges in evaluating malware clustering. In: Jha, S., Sommer, R., Kreibich, C. (eds.) RAID 2010. LNCS, vol. 6307, pp. 238–255. Springer, Heidelberg (2010). https://doi.org/10.1007/978-3-642-15512-3_13
5. Khan, L., Awad, M., Thuraisingham, B.: A new intrusion detection system using support vector machines and hierarchical clustering. VLDB J. Int. J. Very Large Data Bases **16**, 507–521 (2007)
6. Li, Z., Gan, Q.: Improved ant colony algorithm to optimize SVM parameters for network intrusion detection model. J. Chongqing Univ. Posts Telecommun. (Nat. Sci. Ed.) (2014)

7. Xiao, G.: Network intrusion detection based on improved ant colony algorithm and support vector machines. Computer Engineering and Applications (2013)
8. Feng, L., Qiang, Z., Zongliang, Q.: A low-complexity detection algorithm for foreign object intrusion in surveillance video. J. Nanjing Univ. Posts Telecommun. (Nat. Sci. Ed.) **1**, 1–8 (2020)
9. Hongbin, L., Sijia, L.: Intrusion detection algorithm in microservice architecture based on feature vectors. Comput. Digital Eng. **47**(12), 3121–3125 (2019)
10. Jian, T., Jifu, G.: Distributed intrusion attack detection system based on artificial bee colony algorithm. Comput. Appl. Softw. **36**(03), 326–333 (2019)

Design Algorithm Based on Parameterization of GC Architectural Design Platform

Hui Li[✉]

Yunnan Technology and Business University, Kunming, Yunnan, China

Abstract. In the information age, computer modeling is used, and then the model is processed to obtain renderings, animation renderings, etc. However, these softwares have so far been a one-way design process rather than an interactive design process. GC adopts a parametric modeling method, which provides a new design method for the construction industry. It is a software mainly used for parametric architectural modeling based on Mscmicrostation v8i, and is a new generation of generative modeling technology. This article mainly introduces the design algorithm research based on the parameterization of the GC architectural design platform. This paper uses the design algorithm research based on the parameterization of the GC architectural design platform, adopts the GC building to design, and reasonably analyzes the feasibility of the platform parameters. The experimental results of this paper show that the design algorithm research based on the parameterization of the GC architectural design platform increases the efficiency of architectural design by 16%. The limitations of the analysis of the parameterized design of the GC architectural design platform provide a good man-machine for the algorithm design application. The methods and approaches of interactive methods are analyzed, discussed and summarized, thereby enriching the academic research results.

Keywords: GC architectural design · Parameterization · Dijkstra algorithm · Algorithm design

1 Introduction

The existing CAD software can no longer meet the requirements of digital architectural design at this stage. The emergence of parametric design software has changed this situation. It can help us explore various new possibilities of architectural forms and make the complexity of the building more complex. It is easy to show [1, 2]. Parametric design quickly obtains different options by adjusting parameters, which improves the efficiency of existing architectural design: by digitizing various constraints [3, 4]. And look for the relationship between the restriction conditions (geometric constraints) and various geometric algorithms and intelligent algorithms. Through program design, complex and efficient architectural forms can be realized. Other information can be encoded and efficiently translated into spatial forms, and can be passed efficient parameter adjustment scheme in response to various changes in the project [5, 6].

With the advancement of science and technology and the rapid development of the Internet, looking at the role of computers in architecture and engineering design, we are

© The Author(s), under exclusive license to Springer Nature Switzerland AG 2021
J. Abawajy et al. (Eds.): ATCI 2021, LNDECT 81, pp. 407–413, 2021.
https://doi.org/10.1007/978-3-030-79197-1_58

in the era of three-dimensional parametric association design in architecture and engineering design. Kirk, D. feel that computers are no longer just tools for architectural expression, but more use of computer-based algorithms to obtain different architectural forms and their deformations [7]. Cairney J. believes that the use of parametric modeling software, combined with various geometric and intelligent algorithms can explore more possibilities of architectural forms under three-dimensional conditions. Parametric design is an inevitable trend in the development of architectural design triggered by digital technology [8]. Its appearance is undoubtedly a revolutionary promotion to the existing architectural design system, and it is more suitable for the complex, changeable and rapid design environment in modern architectural design [9, 10]. It realizes design by establishing a connection between variables and output, forming a derivative relationship. However, there are errors in their experimental process, which leads to inaccurate results.

The innovation of this paper is to propose a design algorithm research based on the parameterization of the GC architectural design platform. Research the parameterized design algorithm of GC architectural design platform, and analyze the effective countermeasures of GC architectural design system. Using GC to parametrically design different roof models. Using GC's parametric design ideas, the real perceptual materials are digitally expressed to generate parametric objects in GC, and the new feature functions of GC are used to realize different structural designs. This part proposes corresponding countermeasures against the problems in GC architectural design, innovating training models, and strengthening the construction of GC architectural design equipment. The aim is to find a new path suitable for the development of the current parametric program system in the architectural design platform through this research.

2   GC Architectural Design Research

2.1   Architectural Design Analysis

During the day, the architects design the concept of the plan, and at night let the computer automatically generate the plan, elevation, and section because the data has been parameterized. It allows architects and engineers to present the design and realization results in a way that was unimaginable before. There have been many design cases generated by GC in countries around the world, such as the London headquarters building of Swiss Re, the British Isle of Wight bus terminal, and the "digital self-regulatory station" designed by the firm. However, there are still relatively few companies using GC in China, and there are not many mature cases. The design method is usually simple modeling of specific shapes, and GC can be used to explore new and creative complex architectural forms. The process of architectural design has changed from searching for results to searching for possibilities. Polygon House I combined with lattice algorithm to carry out parametric design of ancient Chinese buildings. Take GC technical indicators as independent variable X·PAD affective value as dependent variable v, with q dependent variables and p independent variables;

$$E = t_1 p_1 + E_1 \tag{1}$$

$$F = u_1 q_1 + F_1 \tag{2}$$

The unit eigenvector of the largest eigenvalue, the three linear regression equations of E, F, and T, have the following forms.

$$T = t_1 r_1 + E_1 \tag{3}$$

2.2   GC Architectural Design Application Analysis

Analyze the research status of digital architectural design, point out the advantages of parametric design, introduce the parametric associated software GC, and compare it with various computer-aided design software; use GC to realize two different structures of parameter-adjustable roofs Model, combined with lattice algorithm to carry out parametric design of ancient Chinese buildings. It also discusses the parametric modeling method of ancient buildings under geometric constraints. Based on the shortest path algorithm used in architectural design, the shortest path algorithm Dijkstra algorithm is given in the parameterized realization of the GC, and it is applied to the wiring optimization of the building Design, summarize the work of the thesis and give follow-up research suggestions.

In various CAAD software, graphics can be arbitrarily moved, rotated, zoomed, and spliced, and sizes and icons can be automatically marked. In the software, the information of each side and each part of the building can be displayed intuitively and in all directions. Different models are obtained in response to various changes of actual parameters. At the same time, new building function requirements and traditional design methods and tools of building types cannot meet the requirements. The emergence of new intelligent and parametric software provides the possibility for the formation of multiple and hierarchical smart place spaces. Some traditional building types will gradually be replaced by new building types, and new urban forms will appear.

3   Platform Parametric Design of GC Architectural Design

3.1   GC Architectural Design Script Analysis

GC is a parametric CAD software developed by Bently in 2003. It is a software mainly used for parametric architectural modeling based on Ms (MicroStation v8i. At the beginning of 2005, GC applications began to increase, especially in the London construction industry began to be widely used. Commercially released in November 2007. GC is a pioneering building Design tools for the new era. When the design idea changes, you can easily change the entire model only by modifying the logic. Unlike other software, such as CAD, the design concept is completed by drawing lines one by one. Still staying at the stage where the software is only a tool for human use. The principles of GC include the following aspects. GC is a brand-new parameterized and associated design software, which provides users with efficient new ways to consider different designs. GC parameters Way.

3.2   GC Architectural Design in Platform Parameterization

The designer can independently establish the multi-level relationship between the various components, and then can grasp the design intent and design multiple comparison schemes without rebuilding the detailed design model. Through GC, you can use a variety of intelligent and geometric algorithms to explore complex shapes and structures based on intuition.Using the idea of GC parameterized design, various factors in architectural design such as wind, sunshine angle and time have been considered, combined with common polygons and "unfolded book pages" objects in life, and realized polygons and "unfolded book pages" The parametric digital expression of the object, and the use of the new feature function of GC, apply it to the construction of the roof of the building. This type of roof can be used in the design of exhaust walls after expansion and transformation, and it can also be used in the design of building windows after considering the sunlight angle and time factors. The specific results are shown in Table 1.

Table 1. Base information of the item

Data field meaning	Data field name	Type of data	Length
Item number	Item ld	Varchar	20
project name	Item name	Varchar	50
Project scale	Item dimension	Varchar	50

4   Dijkstra Algorithm in GC Architectural Design

4.1   GC Architectural Design Algorithm Analysis

Using the associated parameterized intelligent modeling software GC, after the parameterized relationship between the structure of the ancient building and the various components is clear, the lattice algorithm is combined with the program design, and the appropriate parameters are set. The specific results are shown in Fig. 1. In 2020, my country's total investment in fixed assets will grow at a high growth rate of 9.8%, bringing the total to 56.20 trillion yuan. At the same time, the growth rate of fixed asset investment has almost maintained in the past five years. At this level, the total investment in fixed assets of the whole society has almost doubled.

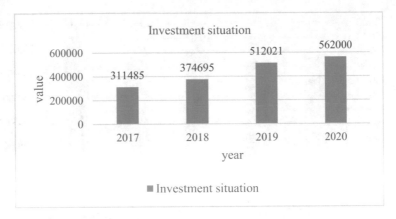

Fig. 1. Social investment in fixed assets

4.2 Dijkstra Algorithm Analysis

The parametric design of ancient buildings is realized by using GC, but the parametric design methods of ancient buildings under various constraints (ie geometric constraints) are not considered. The general trend of my country's urbanization development has led to the development of various construction projects including housing construction, infrastructure construction and commercial construction. As a key link of the construction industry, the architectural design industry is a highly intellectual and technology-intensive industry. This industry is an important way for related scientific and technological achievements to be transformed into actual productivity, and an important driving force to promote technological innovation and the development of related industries. The specific results are shown in Table 2.

Table 2. Contract items of order

Field name	Type of data	Description
Contract no	Varchar (10)	Primary key
Contract name	Varchar (50)	Client's name
Contract sum	Varchar (50)	Amount

The shortest path algorithm application in parametric design is taken as an example, and an application example of the shortest path in architectural design is given. The parameterized design of the building community based on the analysis of building flow lines can be realized with the same algorithm. Dijkstra algorithm is a mature shortest path solution method. Combining relevant constraints such as engineering specifications and practical feasibility, Dijkstra algorithm can be used for wiring. System optimization calculation and programming to explore the best routing method for wiring system design, and use GC's parameterization and graphical design ideas to realize it in GC. The specific results are shown in Fig. 2. It can be seen that GC architectural design companies have a relatively stable and good economic

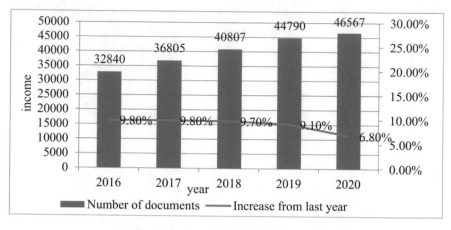

Fig. 2. Data sources of growth rate changes

environment in my country and its provinces, and the development trend of steady growth in investment in park assets has a good macroeconomic environment.

5   Conclusions

Although this paper is studying the design algorithm based on the parameterization of the GC architectural design platform, there are still many shortcomings. The parametric modeling method is based on the shortest path algorithm used in architectural design, and the shortest path algorithm Dijkstra algorithm is realized in the parameterization of GC. The parametric application of the GC architectural design platform not only requires extensive theoretical knowledge, but also a solid theoretical foundation and competence. There are still many in-depth content worthy of research on the design algorithm research based on the parameterization of the GC architectural design platform. There are still many steps in the parameterization analysis of the GC architectural design platform that have not been involved due to reasons such as space and personal ability. The parametric modeling method is based on the shortest path algorithm used in architectural design, and the shortest path algorithm Dijkstra algorithm is realized in the parameterization of GC. The parametric application of the GC architectural design platform not only requires extensive theoretical knowledge, but also a solid theoretical foundation and competence. There are still many in-depth content worthy of research on the design algorithm research based on the parameterization of the GC architectural design platform. There are still many steps in the parameterization analysis of the GC architectural design platform that have not been involved due to reasons such as space and personal ability. In addition, the practical application effects of the shortest path related experiments can only be compared with traditional models from the theoretical and simulation levels.

References

1. Landi, D., Fitzpatrick, K., Mcglashan, H.: Models based practices in physical education: a sociocritical reflection. J. Teach. Phys. Educ. **35**(4), 400–411 (2016)
2. Mckenzie, T.L., Nader, P.R., Strikmiller, P.K., et al.: School physical education: effect of the Child and Adolescent Trial for Cardiovascular Health. Prev. Med. **25**(4), 423 (2016)
3. Wang, J., Shen, B., Luo, X., et al.: Validation of a teachers' achievement goal instrument for teaching physical education. J. Teach. Phys. Educ. **37**(1), 1–27 (2017)
4. Xiang, P., Ağbuğa, B., Liu, J., et al.: Relatedness need satisfaction, intrinsic motivation, and engagement in secondary school physical education. J. Teach. Phys. Educ. **36**(3), 340–352 (2017)
5. Coutinho, D.A.M., Reis, S.G.N., Goncalves, B.S.V., et al.: Manipulating the number of players and targets in team sports Small-Sided Games during Physical Education classes. Revista De Psicologia Del Deporte **25**(1), 169–177 (2016)
6. Ada, E.N., Çetinkalp, Z.K., Altiparmak, M.E., et al.: Flow experiences in physical education classes: the role of perceived motivational climate and situational motivation. Asian J. Educ. Train. **4**(2), 114–120 (2018)
7. Kirk, D.: Physical education, youth sport and lifelong participation: the importance of early learning experiences. Eur. Phys. Educ. Rev. **11**(3), 239–255 (2016)
8. Cairney, J., Hay, J., Mandigo, J., et al.: Developmental coordination disorder and reported enjoyment of physical education in children. Eur. Phys. Educ. Rev. **13**(1), 81–98 (2016)
9. Lodewyk, K.R., Muir, A.: High school females' emotions, self-efficacy, and attributions during soccer and fitness testing in physical education. Phys. Educ. **74**(2), 269–295 (2017)
10. Lander, N.J., Hanna, L., Brown, H., et al.: Physical education teachers' perspectives and experiences when teaching FMS to early adolescent girls. J. Teach. Phys. Educ. **36**(1), 113–118 (2017)

Design of Indoor Path Planning Algorithm for Complex Environment

Hui Li[✉]

Yunnan Technology and Business University, Kunming, Yunnan, China

Abstract. With the rise and rapid development of technologies such as mobile communications, smart terminals and geographic information systems, we are entering the era of mobile Internet. The connection with the environment is gradually strengthening. People's demand for location-based location information services is increasing day by day, and many of these service products have entered people's lives. This article mainly introduces the design research of indoor path planning algorithm for complex environment. This paper uses the design and research of indoor path planning algorithm for complex environment, uses A* algorithm for indoor path planning and design, and reasonably analyzes the feasibility of Dijkstra algorithm and adjacency list model. The adjacency list is a data structure that uses a linked list to store the information of the graph. Each node in the graph has a singly linked list corresponding to it in this data structure. The experimental results of this paper show that the design and research of indoor path planning algorithms for complex environments increases the efficiency of the algorithm by 16%. The limitations of the design and research of indoor path planning algorithms for complex environments provide good applications for ant colony algorithm. The methods and approaches of indoor path planning are analyzed, discussed and summarized, thereby enriching the academic research results.

Keywords: Indoor path planning · A* algorithm · Ant colony algorithm · Dijkstra algorithm

1 Introduction

In the research of internal environmental information service, internal path planning is an important research focus. For different use cases, internal path planning will have different index requirements [1, 2]. For example, in situations such as airport halls or railway stations, users are usually required to plan the shortest path to a specific target [3, 4]. In large shopping malls, users will be required to plan the best route according to their shopping needs. The path map is the data basis and calculation object for path planning. The premise of path planning on the path map is to first express it as data that can be recognized by the computer and organize it in a reasonable way to store it, so that it can be used. For path planning, it is necessary to save storage space [5, 6]. This function of internal space greatly reduces the amount of data that needs to be stored. Defend spatial data: Different from spatial external data, internal spatial data is dependent.

J. Abawajy et al. (Eds.): ATCI 2021, LNDECT 81, pp. 414–420, 2021.
https://doi.org/10.1007/978-3-030-79197-1_59

With the advancement of science and technology and the rapid development of the Internet, the design of indoor path planning algorithms for complex environments has been enhanced [7]. The equipment in a room belongs to the room, and the different rooms in the floor belong to each floor of the building. Bhattacharyya B believes that according to this characteristic, according to the relationship of the internal space, the entity of the internal space can be expressed as a tree structure [8]. Scott JK feels that flexible road features: there is a road concept in the external environment, which restricts the activities of people or vehicles [9]. In most indoor areas, specific moving routes cannot be restricted, and paths can be formed in all areas without obstacles [10, 11]. Therefore, when planning a path, an appropriate path should be formed according to the current environmental conditions and the location of a specific starting point and destination, and the generated path should also be changed when the environment or the location of the change point is changed. However, there are errors in their experimental process, which leads to inaccurate results.

The innovation of this paper is to propose a design study of indoor path planning algorithms for complex environments. Research on the design of indoor path planning algorithms for complex environments, and analyze the effective countermeasures of indoor planning algorithms. This type of algorithm uses the spatial information of the spatial entity in the indoor environment to generate the path. The generated path is closely related to the spatial geometric information of the specific spatial entity in the indoor environment. For example, the size and shape of the room, the width and length of the corridor will all affect the path. The aim is to find a new path suitable for the development of the current indoor path planning system in the complex environment through this research.

2 Indoor Path Planning Algorithm for Complex Environment

2.1 Indoor Path Planning Analysis

From the perspective of the realization method of the storage structure, the adjacency matrix is realized by a two-dimensional array, and the realization methods of other storage structures are all linked lists. Secondly, from the perspective of storage structure, the cross-linked list and adjacent multi-list are more complicated than the other two storage structures. In addition, these storage structures have their own unique advantages and disadvantages. It is easy to judge the relationship between two nodes using the adjacency matrix, and it is easy to obtain the degree of a given node, but the storage space is large. Using the adjacency list can effectively save space, and it is easy to get the out degree of a given node, but it is not easy to get the in degree of a given node, and it is not conducive to judging the relationship between two nodes.

The cross-linked list requires less space and is easy to obtain the out-degree and in-degree of a given node, but its complex structure brings inconvenience to its construction and use. Taking indoor path planning algorithm indicators as independent variables, each child node can only have one parent node, and one parent node can have multiple child nodes. Appropriate application of this function can design an effective algorithm in internal path planning, which has the following form.

$$\begin{cases} I_{A1} = U_A j\varpi C_{01} \\ \quad I_{B1} = 0 \\ I_{C1} = U_C j\varpi C_{01} \end{cases} \tag{1}$$

The calculation value of the indoor route planning algorithm is as follows;

$$3I_{01} = I_{A1} + I_{B1} + I_{C1} = 3U_0 j\varpi C_{01} \tag{2}$$

The adjacency list algorithm test should use the following formula:

$$3I_{02} = I_{A2} + I_{B2} + I_{C2} = 3U_0 j\varpi C_{02} \tag{3}$$

2.2   Indoor Path Planning Algorithm

The Dijkstra algorithm uses the nodes in the graph to represent cities, the edges in the graph represent the roads between cities, and the weight of the edges represents the distance between cities. The Dijkstra algorithm can be used to solve the shortest path problem between given two cities. Dijkstra algorithm the input of is composed of the following content: A weighted graph G, where the two nodes s and e in G represent the starting node and the destination node, respectively. All nodes in graph G form a set V, and each edge in G is an element pair composed of two node elements. The element pair (a, b) indicates that there is an edge connection between node a and node b, and all edges in graph G The composed set is E, and a weight function d is used to define the weight of the edge, that is, the distance attribute of the edge.

In general, the Dijkstra algorithm has a simple structure and is relatively easy to implement. The case suitable for the use of this algorithm is the shortest path calculation in a small range. Its disadvantage is that the calculation amount of the entire search process is relatively large, and the waste of resources is more serious. The following conclusions can be drawn from the path graph and its statistical data: the shortest result path can be obtained using both algorithms, but the number of nodes involved in the search is obviously less than Dijkstra algorithm when using A algorithm.

3   Algorithm Analysis of Indoor Path Planning Algorithm

3.1   A* Algorithm Analysis

Evaluation of the shortest path algorithm First, the Dijkstra algorithm searches for the shortest path in the order of increasing the shortest path distance from the node to the starting node. The entire search process starts from the starting node, and the probability of other nodes in the path graph and the destination node being searched is equal, so all nodes with the shortest path length to the starting node shorter than the shortest path length from the destination node to the starting point will be participate in the search. Secondly, before completing the path planning between the specified nodes, the algorithm will plan the path between it and the starting node for all the nodes participating in the search, and most of the work here is not related to the results of the solution. The occupation grid refers to the adjacency matrix composed of units of the same size according to the law. Under this law, each unit is related to the other eight adjacent units around it.

3.2 Dijkstra Algorithm System

The A algorithm is the same as the Dijkstra algorithm, which also shows that the basis of the A algorithm is the Dijkstra algorithm. Algorithm A can be regarded as a special case of the Dijkstra algorithm. When algorithm A is used, different valuations can be selected according to different usage scenarios.

The algorithm needs to prepare a path map for each layer before use. The path map adopts a knotted model, and uses nodes and edges to represent the entire indoor space. The nodes represent the rooms and the edges represent the corridors. The path diagrams are usually manual. Editing and generating the algorithm is simple in structure and convenient to use, but the generation of the path graph is more troublesome. Secondly, the applicable scene of the algorithm is a simple indoor environment. When the scene becomes more complex, the algorithm will have problems in path representation and path planning. When moving in an outdoor environment, the moving route will be limited to the road. In the indoor environment, there is no spatial entity similar to the road, so it can move freely where there are no obstacles. Using the above two algorithms cannot accurately describe the path in the indoor environment, now another type of algorithm is introduced. The specific results are shown in Table 1.

Table 1. Comparison of several wireless data transmission methods

	Short message	Gsm circuit connection	GPRS
Way of communication	Full duplex	Full duplex	Full duplex
Transmission Rate	26 s/piece	9. 6 kbps	Maximum 171.2 kb
Reliability	Poor	Extremely high	High
Time delay	Large fluctuation range	Small	Medium

4 Indoor Path Planning Algorithm Model

4.1 Internal Path Planning System

The laboratory project in this article proposes a path planning algorithm in space, and briefly describes the path navigation system from the establishment of the map to the optimization and realization of the algorithm. The path planning system and its key technologies have been studied, laying the theoretical foundation for this paper. Secondly, for the indoor map in the indoor path planning system, finite element triangulation is used based on the project requirements, and the indoor path points and indoor information are analyzed. The composition form and the storage form of the indoor map are determined. Then, for the problems in path planning, a single-layer pathfinding algorithm based on the A algorithm and a multi-layer pathfinding algorithm based on the preprocessing road network are proposed, and the data structure of A* proposes an optimization method for the A* algorithm. In the realization of a specific project, the above work results have been verified. Based on the research of maps and algorithms, the indoor path planning module was successfully implemented in the project. The specific results are shown in Fig. 1. The abscissa is the number of synchronizations N, and the ordinate is the absolute value of the error. The test results show that the increase in the number of synchronization makes the error decrease exponentially.

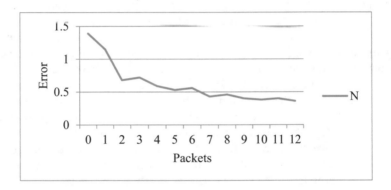

Fig. 1. Mental model work

4.2   Ant Colony Algorithm Model Analysis

Biologists have done a lot of observation and research on organisms at various stages of evolution, and found that whether it is lower organisms such as bacteria, caterpillars, ants, etc., or higher-level species such as fish, birds, and even mammals, their behaviors show this phenomenon, so people began to study whether there is an information sharing mechanism within the biological group that enables individuals to exchange and share information, so as to seek advantages and avoid disadvantages, or cooperate to find food, food transportation, etc. Task, these creatures can therefore rely on the power of the group to survive the cruel competition of nature. One of the classic examples that has been widely studied is the process of ant colonies searching for food sources. After a lot of research and experimentation, people found that they have a very sensitive sense of smell, marking the paths that have been taken, and providing references for other ants. The specific results are shown in Table 2.

Table 2. Statistical table of sample library

	Normal	Ageing	Malfunction
Number of transformers	17	8	8
Total sample	305	160	160
Training samples	204	106	106
Validation sample	101	54	54

The generated path is stored in a graph, and the result path obtained by using the shortest path algorithm in the graph for path planning has the shortest attribute. The indoor path planning algorithm based on the continuous indoor space model board is a representative one of this type of algorithm. It models the spatial plan of a single-layer space to obtain a path diagram. In this diagram, the shortest path algorithm can be used to obtain the shortest path. This method is currently a kind of path planning with better results. Its characteristic result is that the path has the shortest spatial distance. On this basis, the path diagrams between different floors are connected to form a complete diagram. It can also realize the path between different floors. The basic process of planning the algorithm is as follows: Use a continuous indoor space model to generate a path diagram for each floor. The complete path diagram is generated by connecting the corresponding nodes of the elevator and the entry and exit points of the passage in each path diagram. The specific results are shown in Fig. 2. When the number of attributes is the same, as the number of sample data increases, the efficiency of indoor path planning has been significantly improved.

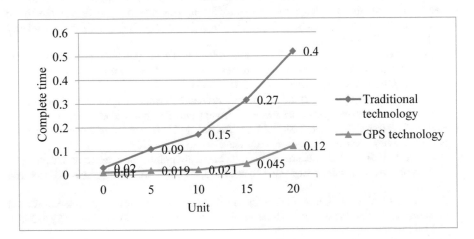

Fig. 2. Questionnaire

5 Conclusions

Although this article is researching the design of indoor path planning algorithms for complex environments, there are still many shortcomings. In the path optimization algorithm, considering the actual user experience, it should be possible to provide multiple paths for selection instead of simply choosing the shortest path. In actual application, the path finding algorithm should be able to combine the merchants and specific users on the path to provide more diversified path search strategies. The design of indoor path planning algorithms for complex environments not only requires extensive theoretical knowledge, but also a solid theoretical foundation and competence. There are still many in-depth contents worthy of research on the design and research of indoor path planning algorithms for complex environments. There are still many steps in the study of indoor path design analysis because of space and personal ability, etc., which are not covered. In addition, the actual application effects of the related experiments of algorithm design can only be compared with traditional models from the level of theory and simulation.

References

1. Li, S., Huang, J.X., Tohti, T.: Fake plate vehicle auditing based on composite constraints in internet of things environment. IOP Conf. Ser. Mater. Sci. Eng. **322**(5), 204–205 (2018)
2. Kumar, N., Rodrigues, J.J.P.C., Chilamkurti, N.: Bayesian coalition game as-a-service for content distribution in internet of vehicles. IEEE Internet Things J. **1**(6), 544–555 (2017)
3. Wang, X., Han, S., Yang, L., et al.: Parallel Internet of vehicles: ACP-based system architecture and behavioral modeling. IEEE Internet Things J. **7**(5), 3735–3746 (2020)
4. Lee, D.G.: A multi-level behavior network-based dangerous situation recognition method in cloud computing environments. J. Supercomput. **73**(7), 3291–3306 (2017). https://doi.org/10.1007/s11227-017-1982-1
5. Jing, M., Jie, Y., Shou-yi, L., Lu, W.: Application of fuzzy analytic hierarchy process in the risk assessment of dangerous small-sized reservoirs. Int. J. Mach. Learn. Cybern. **9**(1), 113–123 (2015). https://doi.org/10.1007/s13042-015-0363-4
6. Chen, H., Feng, S., Pei, X., et al.: Dangerous driving behavior recognition and prevention using an autoregressive time-series model. Tsinghua Sci. Technol. **22**(006), 682–690 (2017)
7. Ruddy, J., Meere, R., O'Donnell, T.: Low frequency AC transmission for offshore wind power: a review. Renew. Sustain. Energy Rev. **56**, 75–86 (2016)
8. Bhattacharyya, B., Raj, S.: Swarm intelligence based algorithms for reactive power planning with flexible AC transmission system devices. Int. J. Electr. Power Energy Syst. **78**, 158–164 (2016)
9. Scott, J.K., Laird, C.D., Liu, J., et al.: Global solution strategies for the network-constrained unit commitment problem with AC transmission constraints. IEEE Trans. Power Syst. **34**(2), 1139–1150 (2018)
10. Liang, H., Liu, Y., Wan, L., et al.: Penetrating power characteristics of half-wavelength AC transmission in point-to-grid system. J. Mod. Power Syst. Clean Energy **7**(1), 123–130 (2019)
11. Liang, H., Liu, Y., Wan, L., Sheng, G., Jiang, X.: Penetrating power characteristics of half-wavelength AC transmission in point-to-grid system. J. Mod. Power Syst. Clean Energy **7**(1), 123–130 (2018). https://doi.org/10.1007/s40565-018-0401-9

Gaussian-Chaos-Enhanced Sub-dimensional Evolutionary Particle Swarm Optimization Algorithms for Global Optimization

Hui Zhi, Lingzhi Hu$^{(\boxtimes)}$, Junhua Hu, and Di Wu

School of Basic Medicine, Shaanxi University of Chinese Medicine,
Xianyang 712046, Shaanxi, China

Abstract. In this paper, a sub-dimensional evolutionary of the particle swarm optimization algorithm based on Gaussian-chaos enhancement strategies (GC-SDPSO) is proposed. The advantage of the sub-dimensional evolutionary particle swarm algorithm is that each dimension of the particle is iterated in turn, rather than a whole. In order to overcome the shortcomings of sub-dimensional evolution algorithm, we introduce the complementary strategy of Gaussian and chaos in the iterative process. Gaussian mutation and chaos perturbation is performed with high aggregation degree to improve the global convergence and to ensure particle diversity. To evaluate the efficiency of the GC-SDPSO algorithm, we also choose nine different types of benchmark functions to test. Compared with SDPSO, GSDPSO, experimental results showed that GC-SDPSO is effective; convergence speed and the quality of the solution are improved.

Keywords: Particle Swarm Optimization (PSO) · Sub-dimensional evolutionary · Gaussian-chaos-enhanced strategy · Gaussian mutation

1 Introduction

Particle Swarm Optimization (PSO) algorithm is a stochastic search algorithm based on swarm intelligence, which has attracted great attention duo to its simple concept, easy implementation, fast convergence speed and less parameter setting. Kennedy [1] constructed and compared several different neighborhood topologies, proposed the use of cluster analysis to improve the performance of the algorithm. Angeline [2] combines the selection operation of evolutionary algorithm to improve the optimization effect in PSO. Natasuki et al. [3] proposed a PSO algorithm based on Gaussian mutation. Liang et al. [4] can keep the diversity of population, and avoid premature convergence of the algorithm in searching process. Lovbjerg et al. [5] introduced the self-organized critical control to the PSO algorithm.

Chaos is a non-linear phenomenon in nature which has characteristics ergodicity, stochastic and sensitive to its initial condition and parameters [6]. Li Bing [7] first used carrier method to map the chaotic variable to the optimization variables to solve the complex nonlinear multi-peak problem. Huang et al. [8] proposed the strategy of chaos-enhanced cuckoo search algorithm (CCS). Jia et al. [9] introduces the PSO algorithm

© The Author(s), under exclusive license to Springer Nature Switzerland AG 2021
J. Abawajy et al. (Eds.): ATCI 2021, LNDECT 81, pp. 421–426, 2021.
https://doi.org/10.1007/978-3-030-79197-1_60

combine chaos and Gauss on local search (CGPSO) for solving high dimensional problems. Turgut et al. [10] introduced the chaotic behavior of quantum PSO algorithm and its application to the design of the heat exchanger. Gandomi et al. [11] applied chaos theory to the PSO algorithm to improve the global search ability and convergence speed of the algorithm. Potter et al. [12] pointed out that the N-dimensional particle can be decomposed into multiple sub-dimensions for spatial search. Bergh and Engelbrecht [13] proposed CPSO-S and further combined it with PSO proposed CPSO-H to achieve the best properties of both algorithms. Xu hui [14] designed the improved PSO algorithm (SDPSO)based on dimension update and applied it to prediction the capacity of coalbed methane production; Ouyang et al. [15] proposed a hybrid harmony search PSO with dimension selection and the experiment proves the high performance of dimension evolution.

2   The basic theory of PSO algorithm

2.1   The standard PSO algorithm

In the standard PSO algorithm, the velocity $v_i = (v_{i1}v_{i2}, \cdots, v_{iD})$ and position $x_i = (x_{i1}x_{i2}, \cdots, x_{iD})$ of the particle is updated according to the following expression:

$$v_{ij}^t = w \cdot v_{ij}^{t-1} + c_1 \cdot r_1 \cdot (p_i^{t-1} - x_{ij}^{t-1}) + c_2 \cdot r_2 \cdot (p_g^{t-1} - x_{ij}^{t-1}), \tag{1}$$

$$x_{ij}^t = x_{ij}^{t-1} + v_{ij}^t, \tag{2}$$

where t is the current iterations number, $i = 1, 2, \cdots, N, j = 1, 2, \cdots, D$; c_1 and c_2 are acceleration factors; r_1 and r_2 are both independent uniformly distributed random variables in the range of [0, 1]; w is the inertia weight. p_i and p_g records the best position it has ever experienced in flight and the best position that all particles have ever experienced. Xu [14] given the mathematical representation of the velocity and position of each sub-dimension is defined as:

$$v_{id}^t = w \cdot v_{id}^{t-1} + c_1 \cdot r_1 \cdot (p_{id}^{t-1} - x_{id}^{t-1}) + c_2 \cdot r_2 \cdot (p_{gd}^{t-1} - x_{id}^{t-1}), \tag{3}$$

$$x_{id}^{t-1} = x_{id}^{t-1} + v_{id}^t, \tag{4}$$

$$v_{id}^t = w \cdot v_{id}^{t-1} + c_1 \cdot r_1 \cdot (p_{gd}^{t-1} - x_{id}^{t-1}) \tag{5}$$

3 Gauss-Chaos-enhanced Sub-dimensional Evolutionary PSO Algorithms

3.1 Gaussian-Chaos Enhanced Strategy

Gaussian distribution is also called normal distribution and often denoted by $N(\mu, \sigma^2)$, where μ is the expectation and σ^2 ($\sigma > 0$) the variance, expressed by.

$$f(x) = \frac{1}{\sqrt{2\pi}\sigma} \exp\left(-\frac{(x - \mu)^2}{2\sigma^2}\right), \tag{6}$$

In this paper, Logistic map is used to generate chaotic variables at current optimal solution, expression by.

$$z_j^{i+1} = 4 \cdot z_j^i \cdot (1 - z_j^i), \tag{7}$$

used following linear mapping to map the chaotic variables to the range of the optimization variables.

$$x_{i,j} = X_{\max,j} - z_j^i \cdot (X_{\max,j} - X_{\min,j}). \tag{8}$$

According to the idea of chaos search, chaotic disturbance is added into the current optimal solution,

$$Z' = (1 - \beta) \cdot \frac{X^* - X_{\min}}{X_{\max} - X_{\min}} + \beta \cdot Z \tag{9}$$

where β is an adjustment parameter and $0 \leq \beta < 1$, Z is the chaotic sequence vector generated by the Logistic map.

3.2 Main Steps of GC-SDPSO Algorithm

The main steps of the proposed algorithm (GC-SDPSO) are presented as follow.

Step 1. Starting: Set the population size, learning factor, the inertia weight, Iterations numbers *iter*;

Step 2. Initialization: Initialize the position and velocity and map it to the range of the variable; Calculating the fitness function value of each particle;

Step 3. Sub-dimensional evolution: Calculate the dimension position value of particle according to the formula (3), (4) and (5), performed the Gaussian mutation operation (11) on half of the particles of high aggregation degree and re-execute. If $iter \geq (2/3)$ Max-FEs, chaotic disturbance is applied to the optimal position of the population according to (9), then continue to step2;

Step 4. Termination condition: If the values of the Max-FEs are meet, terminates the algorithm and outputs the result. Otherwise, go to step 2 for a new iteration.

4 Numerical Experiments and Analysis

In this section, to test the performance of the proposed algorithm and verify the stability of the algorithm. 8 benchmark functions selected from [14–18] according to our purpose. The numerical test functions are listed as follows Table 1. Where $f_1 - f_4$ are unimodal function; $f_5 - f_8$ is a multimodal function, these functions of fitness value are zero. The computational experimental results of PSO, SDPSO [16] and proposed GC-SDPSO are shows in Table 2 and Table 3.

Table 1. Benchmark functions used in experiments

Name	Function	Range
Sphere	$f_1(x) = \sum_{i=1}^{n} x_i^2$	$[-100, 100]^n$
Schwefel	$f_4(x) = \sum_{i=1}^{n} |x_i| + \prod_{i=1}^{n} |x_i|$	$[-10, 10]^n$
Quadric	$f_5(x) = \sum_{i-1}^{n} \left(\sum_{j=1}^{i} x_j \right)^2$	$[-100, 100]^n$
Rosenbrock	$f_6(x) = \sum_{i=1}^{n-1} [100(x_{i+1} - x_i^2)^2 + (x_i - 1)^2]$	$[-5, 10]^n$
Rastrigin	$f_7(x) = \sum_{i=1}^{n} [x_i^2 - 10\cos(2\pi x_i) + 10]$	$[-100, 100]^n$
Griewank	$f_8(x) = \frac{1}{4000} \sum_{i=1}^{n} x_i^2 - \prod_{i=1}^{n} \cos(\frac{x_i}{\sqrt{i}}) + 1$	$[-600, 600]^n$
Zakharov	$f_7(x) = \sum_{i=1}^{n} x_i^2 + (\frac{1}{2}\sum_{i=1}^{n} i x_i)^2 + (\frac{1}{2}\sum_{i=1}^{n} i x_i)^4$	$[-5, 10]^n$
Ackley	$f_8(x) = -20\exp(-0.2\sqrt{\frac{1}{n}\sum_{i=1}^{n} x_i^2})$ $- \exp(\frac{1}{n}\sum_{i=1}^{n} \cos(2\pi x_i)) + 20 + e$	$[-32, 32]^n$

Table 2. Comparisons results of SDPSO, GSDPSO and GC-SDPSO, $D = 10$.

Function		SDPSO	GSDPSO	GC-SDPSO
$f_1(x)$	Mean	2.75E−07	2.15E−07	2.99E−11
	SD	3.05E−02	3.23E−03	3.17E−04
$f_2(x)$	Mean	1.08E−03	9.79E−05	2.37E−05
	SD	3.05E−02	3.89E−04	1.70E−05
$f_3(x)$	Mean	2.54E−05	4.36E−05	2.75E−05
	SD	1.04E−03	2.56E−02	1.15E−04
$f_4(x)$	Mean	4.13E+03	2.35E−02	3.19E−05
	SD	1.63E+03	1.15E−04	1.20E−04
$f_5(x)$	Mean	1.62E+03	3.80E−04	2.69E−05
	SD	1.58E+03	1.35E−01	1.12E−04
$f_6(x)$	Mean	2.18E+03	4.63E−02	2.59E−05
	SD	3.48E+03	2.73E−01	1.22E−04
$f_7(x)$	Mean	1.35E+03	6.21E−01	5.47E−02
	SD	2.04E+03	3.92E−01	2.76E−01
$f_8(x)$	Mean	8.47E+02	5.88E+02	2.93E−02
	SD	1.31E+03	3.58E−01	1.94E−01

Table 3. Comparisons results of SDPSO,GSDPSO and GC-SDPSO, $D = 20$.

Function		SDPSO	GSDPSO	GC-SDPSO
$f_1(x)$	Mean	2.33E+03	2.50E−01	2.61E−04
	SD	3.71E+03	1.43E+00	1.41E−03
$f_2(x)$	Mean	4.52E+02	2.65E−01	1.79E−04
	SD	5.50E+01	1.51E+00	1.20E−03
$f_3(x)$	Mean	1.71E+05	5.06E+01	9.51E−02
	SD	2.57E+03	5.02E+01	7.69E−02
$f_4(x)$	Mean	4.65E+04	4.23E+01	7.08E−02
	SD	2.87E+04	3.61E+01	2.69E−02
$f_5(x)$	Mean	2.69E+03	7.90E−01	2.69E−01
	SD	8.12E+03	2.69E+02	1.84E+03
$f_6(x)$	Mean	8.07E+03	7.75E+01	6.69E+01
	SD	1.53E+04	2.40E+02	1.95E+02
$f_7(x)$	Mean	2.01E+04	7.92E+01	7.92E+01
	SD	2.61E+04	2.42E+02	1.99E+02
$f_8(x)$	Mean	1.67E+03	8.15E+01	6.77E−01
	SD	5.16E+03	1.98E+02	1.97E+00

5 Conclusion

In this paper, we proposed a sub-dimension evolutionary particle swarm optimization algorithm based on the enhanced strategy of Gaussian and Chaos. These algorithms use the idea of "pre division latter merge". Compared with SDPSO, GSDPSO, experimental results showed that GC-SDPSO is more effective and faster to convergence.

Acknowledgements. The authors would like to express their sincere thanks to the responsible editor and the anonymous referees for their valuable comments and suggestions, which have greatly improved the earlier version of our paper.

References

1. Kennedy, J: Small worlds and mega-minds: effects of neighborhood topology on particle swarm performance. In: Proceedings of the Congress on Evolutionary Computation, CEC 1999 (1999)
2. Angeline, P.J: Using selection to improve particle swarm optimization. In: IEEE International Conference on Evolutionary Computation Proceedings, 1998. IEEE World Congress on Computational Intelligence, pp. 84–89 (1998)
3. Higashi, N, Iba, H.: Particle swarm optimization with Gaussian mutation. Proceedings of the IEEE Swarm Intelligence Symposium, Sis, pp. 72–79 (2003)
4. Liang, J.J., Qin, A.K., Suganthan, P.N., et al.: Comprehensive learning particle swarm optimizer for global optimization of multimodal functions. IEEE Trans. Evol. Comput. **10** (3), 281–295 (2006)

5. Løvbjerg, M., Krink, T.: Extending particle swarm optimisers with self-organized criticality (2002)
6. Kiran, M.S., et al: A novel hybrid algorithm based on particle swarm and ant colony optimization for finding the global minimum. Appl. Math. Comput. **219**(4), 1515–1521 (2012)
7. Li, B., Jiang, W.: Optimizing complex functions by chaos search. Cybern. Syst. **29**(4), 409–419 (1998)
8. Huang, L., et al: Chaos-enhanced Cuckoo search optimization algorithms for global optimization. Appl. Math. Model. **40**(5-6), 3860–3875 (2015)
9. Jia, D.L., et al: A hybrid particle swarm optimization algorithm for high-dimensional problems. Comput. Ind. Eng. **61**(4), 1117–1122 (2011)
10. Turgut, O.E.: Hybrid chaotic quantum behaved particle swarm optimization algorithm for thermal design of plate fin heat exchangers. Appl. Math. Model. **40**(1), 50–69 (2015)
11. Gandomi, A.H., et al: Chaos-enhanced accelerated particle swarm optimization. Commun. Nonlinear Sci. Numer. Simul. **18**(2), 327–340 (2013)
12. Potter, M.A., Jong, K.A.D.: A cooperative coevolutionary approach to function optimization. Lecture Notes in Computer Science **866**, 249–257 (1994)
13. Van Den Bergh, F., Engelbrecht, A.P.: A cooperative approach to particle swarm optimization. IEEE Trans. Evol. Comput. **8**(3), 225–239 (2004)
14. Xu, H., et al: Improvement on PSO with dimension update and mutation. J. Softw. **8**(4), 827–833 (2013)
15. Ouyang, H., et al: Hybrid harmony search particle swarm optimization with global dimension selection. Inf. Sci. **10**, 318–337 (2016)
16. Yuan, X., et al: On a novel multi-swarm fruit fly optimization algorithm and its application. Appl. Math. Comput. **233**(3), 260–271 (2014)
17. Ruan, Z.H., et al: A new multi-function global particle swarm optimization. Appl. Soft Comput. **49**, 279–291 (2016)
18. Baykaso, A.L., Ozsoydan, F.B.: Adaptive firefly algorithm with chaos for mechanical design optimization problems. Appl. Soft Comput. **36**, 152–164 (2015)

Thoughts on Criminal Laws and Regulations for Data Security Crimes in the Era of Artificial Intelligence

Sanjun Ma[✉]

School of Business, Hebei Ploytechnic Institute, Shijiazhuang, Hebei, China

Abstract. The accumulation of big data has continuously promoted the development of artificial intelligence. Data has become the core element of the Internet and computer fields. The behavior of using data as the object of crime and the use of data to infringe other legal interests continue to appear, and the importance of data security it is also increasingly recognized by the theoretical and practical circles. "Data is fundamentally different from information and computer information systems, and the main basis for data protection in current Chinese legislation and judicial practice is information security and computer information system security." Therefore, the provisions of China's criminal law have shown limitations and hysteresis. Learn from the provisions of the "European Cybercrime Convention" to discuss how to build China's data security crime sanctions system, and provide a feasible reference for China to effectively combat and control data security crimes.

Keywords: Data security · Artificial intelligence · Criminal law

1 Introduction

In the era of artificial intelligence, the value of data has received unprecedented attention, and data theft and abuse have occurred frequently, which has gradually evolved into types of data security violations, that is, data security crimes. Specifically, data security crime refers to the behavior of "all data processed by big data technology, that is, in digital form, as criminal objects or criminal tools, including accounts and access control data as the core under the technological environment with the rapid development of artificial intelligence, and spread to various unstructured and semi-structured data such as electronic traces, life behaviors, city management, etc., as well as crimes that extend from computer data to multi-terminal data such as the Internet of Things, smart phones, and wearable devices [1, 2]."

J. Abawajy et al. (Eds.): ATCI 2021, LNDECT 81, pp. 427–434, 2021.
https://doi.org/10.1007/978-3-030-79197-1_61

2   Provisions of China's Criminal Law on Data Security Crimes

As the carrier of information, data is essentially not only an information transmission tool, but also the content and necessary part of a computer system. Corresponding to this, the handling rules for data security crimes in China's Criminal Law and judicial interpretations can also be summarized in two ways: one is based on the essential attributes of the data and protecting it as information; the other is based on the technical characteristics of the data, protect it as an integral part of the computer system.

2.1   According to the Essential Characteristics of the Data, Protect the Data As Information

The crime of illegally obtaining computer information system data refers to "violating national regulations, intruding into computer information systems outside the fields of national affairs, national defense construction, and cutting-edge science and technology, or using other technical means to obtain data stored, processed, or transmitted in the computer information system. The circumstances are serious". This crime directly regulates the behavior of using data as the object of crime, and covers the behavior of using data as the object of crime in scope. In the criminal law, the data stored, processed, or transmitted in a computer information system refers to "all meaningful combinations of words, symbols, sounds, images, etc., actually processed in a computer information system [3]." Therefore, it is not so much that this crime protects the data processed, stored, and transmitted on the computer system, as it protects the information carried by the data. It can be seen that the protection of data in China's Criminal Law means protection of information.

The crime of selling or illegally providing citizens' personal information stipulated in the Criminal Law will be included in the crime of suspected selling or illegally providing citizens' personal information when the content contained in the acquired and sold data belongs to citizens' personal information but not operating authority information. When the information obtained involves trade secrets or state secrets, the act will be punished as a crime of infringing on trade secrets or illegally obtaining state secrets. This is also evidence that China's Criminal Law relies on information for protection of "data", and the qualitative nature of the act of acquiring data mainly depends on the nature of the information carried by the acquired data (Fig. 1).

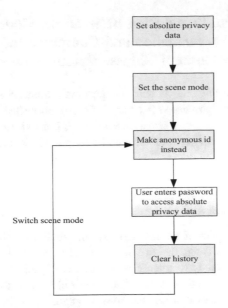

Fig. 1. Data processing

2.2 According to the Technical Characteristics of the Data, Protect it as an Internal Part of the Computer Information System

Data is the processing object of the computer, and its existence and processing must rely on the computer information system. Therefore, China's "Criminal Law" also has another path to deal with data security crimes, which is to treat data as an integral part of a computer system and protect it. The "data" here is limited by its technical characteristics.

At the beginning of adding the crime of illegally obtaining computer information system data, due to the development of technology, most of the truly valuable data can only exist on the computer system, so its scope is basically the same as the boundary of the computer information system [4]. With the development of technologies such as data transmission, storage, mining, and merging, data has gradually become the target of criminal acts. In response to this reality, the judicial organs have expanded the interpretation of the concept of "computer information system". "Computer information system" (and "computer system") is defined as "a system with automatic data processing functions, including computers, network equipment, communication equipment, automated control equipment, etc.". This definition essentially confirms that portable network terminals, including computers, mobile phones, tablets, household smart appliances, and other devices that can automatically process data, have been illegally obtained by equating data with computer (information) systems at the technical level. Computer information systems the protection of data crimes reflects from another aspect that any meaningful content stored, processed and transmitted on these terminals may be identified as "data" and thus protected by the criminal law.

3   The Overlap and Confusion Between the Concept of "Data" and Information and Computer Information Systems in the Current Chinese Criminal Law

In the current criminal law and judicial interpretation, the positioning and meaning of the technical term "data" are not very clear, and the relationship between the two terms "information" and "computer information system" is blurred. In order to cope with the ever-increasing data security crimes and effectively protect data security, legislative and judicial organs have equated "data" with "information" in some laws, and have used "data" with "computer (information) in another legal article" system" is equivalent.

3.1   Data, Information and Computer (Information) System

Although data is the carrier of information, and information is the connotation of data, there are differences between the two in terms of scope and essence. "Data in the meaning of the Criminal Law mainly refers to the object of the crime of illegally obtaining computer information system data, and only refers to the identity authentication information used to confirm the user's operating authority on the computer information system. Other information is protected from illegal acquisition or disclosure. Although it can indirectly protect data carrying relevant information, its scope is still limited [5]." In the era of big data, data is already different in scope from information, especially the information protected by the current Criminal Law. For example, various social platforms, e-commerce platforms, sports apps, etc., collect and accumulate large amounts of data every day. The data stored in portable network terminal devices such as mobile phones and tablet computers can also be used to obtain work and home addresses, communication information, online shopping consumption, travel information, etc. These data can be an important means for the subsequent implementation of criminal acts that violate personal or property safety, but such data are not covered by China's Criminal Law.

Similar to the relationship between data and information, a computer system is the carrier of data, and data is the content of a computer system, but there are essential differences between the two. "Computer (information) system refers to "system with automatic data processing function". The essence of computer (information) system lies in its data processing function, that is, the dynamic processing of data operation, transmission, and storage, while data is static objects processed by computers [6]."

In the era of big data, information largely relies on data to spread, and data is transmitted, calculated, and stored based on computer (information) systems. Therefore, the concept of data cannot be completely separated from "information" or "computer (information) systems." The three are closely related, but independent of each other. Therefore, data security, information security and computer (information) system security are also closely related but independent of each other.

3.2   Data Security, Information Security and Computer (Information) System Security

The "illegal" in the provisions of China's Criminal Law on obtaining data or information actually means that a person who does not know the rights obtains data or information that should be kept confidential, which means that the data or information right holders have broken the confidentiality set by them status [7].

The difference between data security and information security is that in order to protect the "knowable state" of data and information by right holders, information as a completely intangible object only needs to be in a "secret state", that is, not known by others. But for data, in addition to copying, dissemination, and other actions, the perpetrator can destroy the state of the data that can be known by deleting, compressing, and other actions.

Computer (information) system security actually refers to the safe operation of the automatic data processing function of the computer information system. According to the law of the crime of illegally obtaining computer information system data, the function of processing data includes three functions: storage, processing and transmission of data. Therefore, computer information system security means that the computer information system can store, process, and transmit data normally and meet the expectations of the computer information system operator. The computer crime provisions in China's "Criminal Law" protect the ability of computer information systems to store, process, and transmit data, and protect the safety of the operation of computer information systems, which is a dynamic process.

Chin's current regulatory path for data security crimes, whether it is an information-based non-property model or a computer (information) system-based real rights model, is fundamentally an expansion and breakthrough of the existing regulatory path. Compared with the era of artificial intelligence based on the background of big data, this kind of regulation method has obvious lag, and it is necessary to build a more sophisticated and reasonable cybercrime accusation system.

4   The Regulatory Thinking of Data Security Crimes

The shortcomings of China's criminal law on data security crimes are mainly that it only relies on the existing two legal benefits of information security and computer information system security to protect the new legal benefit of data security. Therefore, it is necessary to clarify the relationship between data, information and computer (information) systems and the relationship between the corresponding legal interests, so as to determine the position of data security in the criminal offence system, and establish more accurate data crime charges.

The common manifestation of data security crime is that the perpetrator first invades the target computer system through technical means, and then illegally obtains the target computer data, and after obtaining the data, the obtained data is restored by collecting, cracking, associating, mining, merging, etc. For information and use information to carry out illegal and criminal acts; or through remote control to delete, modify, and compress the data processed on the target computer. According to the nature of the infringement of computer (information) system security, data security, and information security, and the relationship between the three, the information security crime system can be used to regulate subsequent information use behaviors, including computer crimes and data crimes as a dual entry point regulatory system.

4.1 Adhere to the Idea of Protecting Information Security with the Information Security Crime System

In view of the different legal interests of protection, the current criminal law of China's sanctions for information crimes are located in different chapters, mainly including the crimes of stealing, spying, buying, and illegally providing state secrets and intelligence overseas; crimes of infringing on commercial secrets; crimes of infringing on citizens' personal information The crime of illegally obtaining state secrets; the crime of deliberately divulging state secrets; the crime of negligent divulging state secrets; the crime of illegally obtaining military secrets; the crime of deliberately divulging military secrets; the crime of negligent divulging military secrets, etc. The criminal objects involved in these crimes include state secrets, commercial secrets, citizens' personal information, military secrets, etc. These provisions protect the information stipulated by them from being known by people who do not have the right to know. This is the fundamental legal benefit of the information security crime system.

4.2 Establish a Crime System for Computer Crimes and Data Crimes in Parallel

With the increase in the value of data, the dislocation of data security crimes and computer crimes has become more and more obvious, and the disconnect between the regulations of the Criminal Law and data security crimes has become more and more obvious. China's current path of regulating computer crimes and data security crimes should consider separating static data security from dynamic computer (information) system functional security.

Multinational criminal law systems define the meaning of data security as confidentiality, integrity, and availability [8]. Confidentiality corresponds to the information characteristics of the data, that is, the "knowledge state" enjoyed by the owner of the data and the "confidential state" established accordingly to exclude others from knowing; the integrity and the availability of the right owner correspond to the data technical characteristics. Integrity means that the data should remain intact and not be deleted, that is, the data has not been illegally deleted or altered; the accessibility of the data right holder means that the data cannot be illegally compressed or shredded, that is, it cannot violate the data right holder's acquisition or use the right to the data. The dual characteristics of data lead to the two basic crimes of the data security crime system: illegal acquisition, possession, and use of data with data confidentiality as its legal benefit, and illegal destruction with data integrity and the availability of rights holders as its legal benefit, crimes of deleting, modifying, and compressing data.

Crimes of illegally obtaining, holding, and using data. It can be seen from the typical pattern of data security crime that the entire process of data security crime can be divided into three steps: (1) obtaining data; (2) holding data; (3) using data for follow-up activities. Among them, illegal acquisition of data is the so-called act of obtaining data stored, processed, and transmitted on electronic devices through technical means; illegal possession of data means that the perpetrator obtains data through normal business activities, and the law stipulates that it should be deleted without deleting or responding to the data. Encryption to ensure that it is not known to others without encryption; illegal use of data refers to the comparison and mining of unstructured data obtained through multiple channels, or to provide others for comparison and mining, and to perform large amounts of data. The act of integrating and reverting to meaningful information. Obviously, from the perspective of the infringement of the three types of behaviors on citizens and the feasibility of the criminal law system, all three types of behaviors should be included in the scope of regulation of the "Criminal Law".

Crimes of illegally destroying, deleting, or compressing data. China's "Criminal Law" regulates the act of "deleting, modifying, or adding to the data and applications stored, processed or transmitted in the computer information system, with serious consequences" as the crime of illegally destroying the computer information system. In fact, the behavior regulated by this clause is the behavior of illegally destroying, deleting, modifying, and compressing data, which reflects the current situation that China's Criminal Law confuses static data security with dynamic computer system functional security. Regarding this issue, the "Cybercrime Convention" of the Council of Europe (hereinafter referred to as the "Convention") can provide some reference. Article 4 of the Convention makes it a crime to have no right to destroy, delete, modify, or compress data stored on computer information systems, while Article 5 stipulates that contracting states should deliberately affect the normal functions of computer information systems without their rights. The behavior is treated as a crime, and the way to affect the computer information system can be a large amount of input, transmission, modification, deletion or compression of data. Comparing these two articles, it can be seen that Article 4 of the Convention protects static databases [9]. While Article 5 stipulates that crimes can be carried out by destroying data on computers, it protects dynamic databases. Computer information system functions [10]. In other words, based on the provisions of the convention, the perpetrator can violate the security of a dynamic computer information system by destroying the security of a static database, or destroy the security of a static database by destroying the security of a dynamic computer information system. The nature of the behavior is not the same, and it violates different legal provisions. This approach shows the supportive position on the distinction between the dynamic nature of the computer information system's operation and the static nature of data security. Based on the current high volume of data security crimes, this type of legislation in the convention is undoubtedly useful for China and is a feasible path.

5 Conclusion

In response to the judicial dilemma faced by the Chinese Criminal Law in terms of data security protection, the addition of an independent "data security crime should be incorporated into the legislative plan, and relevant provisions in the current criminal law involving data security crimes should be revised in a timely manner. Establish a coordinated regulation model for computer crimes, data crimes, and information crimes. Before the criminal law is amended, the judicial organs can learn from international practices and successful experience, and promptly issue judicial interpretations to explain the penal standards, objective performance, and the definition of this crime and the other crime for data security crimes, so as to make up for this field. The "legal gaps" in the United States have made the prevention and regulation of data security crimes legally based.

References

1. Qianyun, W.: Criminal regulations of data security crimes in the background of artificial intelligence. Legal Forum 2, 10–12 (2019)
2. Zhigang, Y.: Thoughts on sanctions against data crime in the era of big data. China Soc. Sci. 10, 16 19 (2014)
3. Gao, M.: Criminal Law. Peking University Press, p. 536 (2011)
4. Yuanli, L.: Data crime in the era of big data. Chin. Soc. Sci. 6, 22–29 (2015)
5. Xiaobiao, C.: On the criminal law response to web crawlers. J. Henan Procuratorate 10, 55–62 (2020)
6. Xuejun, S.: The origin and logic of the new regulations on Internet credit supervision. Polit. Law Rev. 5, 22–26 (2021)
7. Xiu, Y.: Personal information protection in targeted advertising in the era of big data——? analysis of the industry framework standard for the protection of user information in China's Internet targeted advertising. Int. Press 5, 66–68 (2015)
8. Council of Europe. Cybercrime Convention. European Treaty Series–No. 185
9. Council of Europe. Cybercrime Convention Explanatory Report. Budapest, 23. XI. 2001. Article 60
10. Council of Europe. Cybercrime Convention. European Treaty Series-No. 185

Application of Evaluate Method Based on Universal Design in the Intelligent Computing Product Development of a Medical Lamp

Jianqiu Mao[1]([✉]), Shuguang Sang[1], and Zhiqiang Chen[2]

[1] School of Mechanical Engineering, University of Jinan,
Jinan, Shandong, China
me_maojq@ujn.edu.cn
[2] Shandong Deqin Bidding Assessment Cost Consulting CO., LTD.,
Jinan, Shandong, China

Abstract. Product development is a very complex process, the success of a product largely depends on its early concept development process. In the process of intelligent computing product development, the decision and choice of the product scheme is the key to the success of the product. The general design principle has important guiding significance in the aspect of human-product interaction. This paper focuses on how to help design teams evaluate the impact of different design schemes on human product-interaction during product concept development stage.

Keywords: Universal design · Intelligent computing · Evaluate method · Product development

1 Introduction

The Universal Design were widely used in architects, product designers, engineers and environmental design researchers which was led by the late Ronald Mace in the North Carolina State University [1, 2]. Universal design refers to a product or environment designed for use by all people, regardless of age, disability or other factors. Universal design has its origins in an earlier concept of accessibility. The 7 Principles of Universal Design have a very targeted guiding significance for human-product interaction. Therefore, we designed the evaluation table and the evaluation process to evaluate the medical baking lamp at the different stages through the concept development.

2 Evaluate Process Base on the Universal Design

2.1 Evaluation Indicators and Time Points

This design project is a medical warming lamp used for nursing after hand or foot surgery [3]. It is mainly to keep the temperature for the patient's hand and foot parts. At present, the main for the existing lamp is that it is needs too much space when used, and the storage space also should be too large. Moreover, the human-interaction is not comfortable enough, and the appearance is needing beauty [4] (Fig. 1).

J. Abawajy et al. (Eds.): ATCI 2021, LNDECT 81, pp. 435–440, 2021.
https://doi.org/10.1007/978-3-030-79197-1_62

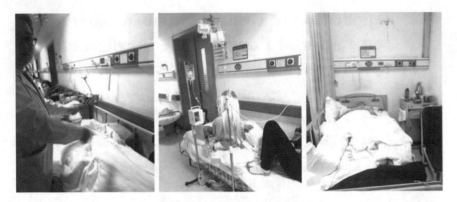

Fig. 1. Usage environments of current medical lamp and using environment

In this project, we focus on the concept development stage (Fig. 2) as a important time point, and the universal design principle is mainly used to help the design team to screen for a more generous scheme.

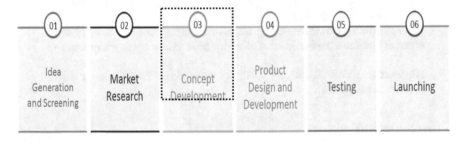

Fig. 2. Product development process

We use the seven principles of universal design as the major indicators to make a seven-star evaluation table, which plays a good role in guiding different design details, especially in measuring human-product interaction (Table 5), (Table 3).

Table 1. The evaluation form for scheme A   **Table 2.** The evaluation form for scheme B

Table 3. The evaluation form for scheme C   **Table 4.** The evaluation form for final

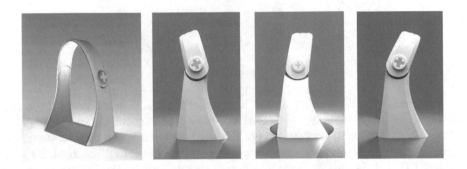

2.2 The Evaluation Analysis of the Schemes

This project evaluates three schemes with universal design principles.

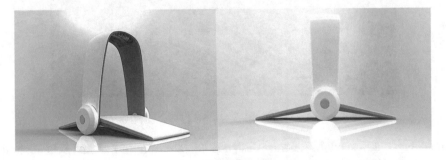

Fig. 3. The design scheme A

Scheme A (Fig. 3) is using on the bed, and part of the upper half is the warming part which can rotate at 270°. The surgical site can be warming from different angles. The overall size of the product is small and easy to store. The forms are simple and beautiful. Evaluate core is 81 that showed in Table 1. We can observe that indicators 2 and 7 are inadequate [5]. The flexibility in Use is restricted because the shape of the product is not conducive to the placement of the patient's injured hand or foot (Fig. 4).

Fig. 4. The design scheme B

The functional characteristics of Scheme B is very similar with Scheme A. They are both placed on the bed and can rotate at 270° [6]. But Scheme B has one more function than A, that the lower part of the lamp structure can support the patient's hands or feet to increasing comfort degree. Evaluate core is 83 that showed in Table 2, it is not difficult to find that the similar disadvantage between B and A is that flexibility in use is defective. The closed circular structure is sometimes insufficient in space for the bandaged hands or feet of the patient. In addition, the height of this lamp is relatively small, and the high temperature may cause burns.

Fig. 5. The design scheme C

Scheme C (Fig. 5) is using on the handrail of the bed. The exposure angle and height of this lamp can easily adjust with one hand. The surgical site can be warming from different angles.The overall size of this scheme is small and easy to store.Aesthetic style comes from a concise form and exquisite structure. The Evaluate core reaches 107 that showed in Table 1. We can observe that all the indicators are not inadequate. The flexibility in Use and the size are suitable [7].

2.3   Final Plan

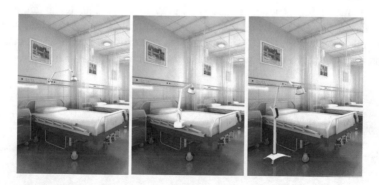

Fig. 6. The final plan

On the basis of Scheme C, modular design is adopted to form the final Scheme in the concept development stage [8]. The Scheme has three ways of use: one is the traditional floor type, but it integrates ECG monitoring instruments [3]. The second is hanging wall type, can be fixed on the wall, does not occupy the ground space; The third is Scheme C. Table 4 shows that the seven indicators have no obvious short-comings, especially in terms of space requirements and convenience of use, which all meet the needs of nurses' operation and the comfort of patients [9]. Most hospitals in China are crowded and the hallways are full of patients, so one of the most important requirements for the product is that it doesn't take up ground space (Fig. 6).

Table 5. The evaluation indicators for final scheme

Major Indicators		Sub-Indicators	Sub-indicator Score
1.Equitable Use	a	Provide the same means of use for all users: identical whenever possible; equivalent when not.	5
	b	Avoid segregating or stigmatizing any users.	5
	c	Provisions for privacy, security, and safety should be equally available to all users.	5
	d	Make the design appealing to all users.	4
2.Flexibility in Use	a	Provide choice in methods of use.	4
	b	Accommodate right- or left-handed access and use.	5
	c	Facilitate the user's accuracy and precision.	4
	d	Provide adaptability to the user's pace.	5
3.Simple and Intuitive Use	a	Eliminate unnecessary complexity.	4
	b	Be consistent with user expectations and intuition.	5
	c	Accommodate a wide range of literacy and language skills.	4
	d	Arrange information consistent with its importance. 3e. Provide effective prompting and feedback during and after task completion.	5
4.Perceptible Information	a	Use different modes (pictorial, verbal, tactile) for redundant presentation of essential information.	5
	b	Provide adequate contrast between essential information and its surroundings.	5
	c	Maximize "legibility" of essential information.Differentiate elements in ways that can be described (i.e., make it easy to give instructions or directions).	4
	d	Provide compatibility with a variety of techniques or devices used by people with	4
5.Tolerance for Error	a	Arrange elements to minimize hazards and errors: most used elements, most accessible; hazardous elements eliminated, isolated, or shielded.	4
	b	Provide warnings of hazards and errors.	3
	c	Provide fail safe features.	3
	d	Discourage unconscious action in tasks that require vigilance.	5
6.Low Physical Effort	a	Allow user to maintain a neutral body position.	5
	b	Use reasonable operating forces.	5
	c	Minimize repetitive actions.	3
	d	Minimize sustained physical effort.	4
7.Size and Space for Approach and Use	a	Provide a clear line of sight to important elements for any seated or standing user.	5
	b	Make reach to all components comfortable for any seated or standing user.	5
	c	Accommodate variations in hand and grip size.	5
	d	Provide adequate space for the use of assistive devices or personal assistance.	4
Suggestion of Judge			

评委签字：

制表人：　　　　　　　制表时间：

3 Conclusions

This paper introduced the evaluation method based on the universal design principles which help to organize the scheme's choice in the concept development stage of the design process. That is, which valuation questions to pose, which methods to choose, and what insights to re-think. Besides, it is also benefit to explore the right design direction, stimulate creative new ideas, and facilitate communication with clients and within design teams. To be relevant to design research and practice, emotion knowledge should be implemented in tools that help design teams to measure, represent, and interpret user emotions.

First of all, the evaluation method of general design principles can effectively evaluate the design scheme rationally. It will help designers to be more rational in the choice of concept. Secondly, it is a validation tool to reflect on each solution and put forward new directions and suggestions for the product. Finally, the evaluation process will be a process of in-depth communication between designers, users, and experts, which is crucial to the success of the product.

Acknowledgment. This work was supported by Science and Technology Project of Higher Education in Shandong Province Foundation. Item No. TJY1803.

References

1. Herbert, L.: Meiselman.: Emotion Measurement, December 2016
2. Kim, J.S., Jeong, B.Y.: Universal safety and design: transition from universal design to a new philosophy, Work **67** (1), 157–164 (2020)
3. Desmet, P.: Faces of product pleasure: 25 positive emotions in human-product. Int. J. Des. **6** (2) (2012)
4. Mao, J., et al.: Application of emotional design to the form redesign of a midwifery training aid, In: International Conference on Design and Technology, Geelong, VIC, Australia, pp. 44–50, 5–8 December 2016
5. http://universaldesign.ie/What-is-Universal-Design/The-7-Principles/ (2012)
6. Norman, D.: Emotional Design -Why We Love (or Hate) Everyday Things. New York: Basic (2004)
7. Grace, R.: Universal design targets products for all ages. Plast. Eng.-Connecticut- **77**(1), 26–31 (2021)
8. Yoon, J.: Escaping the emotional blur: Design tools for facilitating positive emotional granularity, Doctoral thesis, January 2018
9. Yoon, J.: Interactions. Escaping the emotional blur: Design tools for facilitating positive emotional granularity, doctoral thesis, January 2018

Construction of Learner Group Characteristics Model Based on Network Learning Data Perception and Mining

Ping Xia$^{(\boxtimes)}$ and Zhun Wang

Guangzhou College of Technology and Business,
Guangzhou, Guangdong, China

Abstract. With the continuous maturity of online learning methods, a large amount of learning data is stored on various learning platforms. By using big data technology, collecting and analyzing distributed network data, mining learner attribute information, learning behavior and learning results of three types of related data with hidden value. The cluster analysis method is used to extract learner group characteristic parameters, and a six-element learner group characteristic model based on knowledge level, learning motivation, learning attitude, learning style, interest preference and cognitive ability is constructed. The model adopts a multi-level three-dimensional structure, which mainly includes four functional levels: data perception layer, data analysis layer, application service layer, and user visualization layer. Under the effect of the dual mode of learner group characteristic analysis and user group adaptive feedback update, the recommendation function of the hybrid intelligent recommendation engine based on collaborative filtering algorithm and deep learning is used to realize user-visualized personalized recommendation and customized services. Under the function of this model, it meets the actual needs of personalized learning and promotes the development of intelligent learning.

Keywords: Network learning data · Cluster analysis · Learner group characteristic model · Intelligent recommendation engine · Personalized learning

1 Introduction

With the continuous development of Internet technology and the popularization of online learning, massive amounts of learning data are distributed and stored on the Internet. Using big data analysis technology, mining and extracting hid value of network learning data, analyzing the characteristics of learner groups, and realizing personalized recommendation and learning are a potential area worthy of in-depth research. Through the correlation analysis of learning behavior and learning result data, in-depth analysis of the specific performance characteristics of the correlation between the learner's main learning motivation and learning effectiveness. Taking the broad coverage of the group as an entry point, focusing on the universality and diversity of data sources, designing and constructing a model of learner groups characteristics. In the study of this model, the use of data to drive learning kinetic energy is conducive to the analysis of academic conditions and reflects the learner's subjectivity and dominant position. Meet the individualized learning and development needs of learners, promote the scientific transformation of learning concepts and teaching and learning methods, and promote the development of intelligent learning in the information age.

2   Distributed Network Learning Data Collection

As a new learning method that develops rapidly and gradually matures, online learning has significant characteristics such as huge data volume, complex types, wide distribution locations, and different storage methods and storage media. Online learning data is a key factor for analyzing learner characteristics, and it is also the basic data for constructing a group model. Therefore, the comprehensiveness, reliability, accuracy and real-time nature of data sources must be guaranteed. Through the data collection function modules based on synchronous and asynchronous methods, the effective data of various online learning platforms are extracted, classified and stored, providing big data support for the construction of learner characteristic models, as shown in Fig. 1.

Learners as the main objects participating in online learning, after entering the learning system, continue to generate various learning behavior and status data, use real-time synchronous collection to extract this type of dynamic data, and monitor learning process changes and trend development in real time [1]. For the information stored in the learning platform database, asynchronous collection is used to periodically extract such static data, and the static learning data and dynamic learning data are typed and integrated. Use big data technology to achieve preliminary preprocessing of various types of learner data, including data cleaning, screening, classification, aggregation and other operations. Finally, the learner's attribute information, learning behavior data and learning result data are stored in the corresponding data warehouse to realize the collection and preprocessing of distributed network learning data [2].

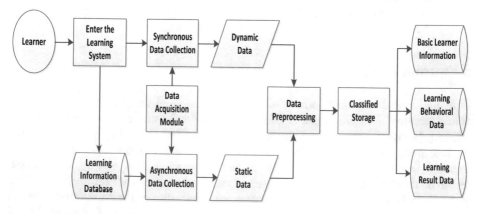

Fig. 1. Flow chart of network learning data collection and preprocessing

3   The Construction of Learner Group Characteristics Model

The learner groups characteristic models to adopt a hierarchical three-dimensional structure, which mainly includes four structural layers: data perception layer, data analysis layer, application service layer and user visualization layer, as shown in Fig. 2. All levels cooperate and assist each other. The lower layer provides services and data interfaces for the upper layer, and the upper layer uses the services provided by the lower layer to achieve higher-level data processing and application service functions. Each level is closely connected, and jointly provides learners with big data-based learning feature analysis and personalized recommendation and customized services.

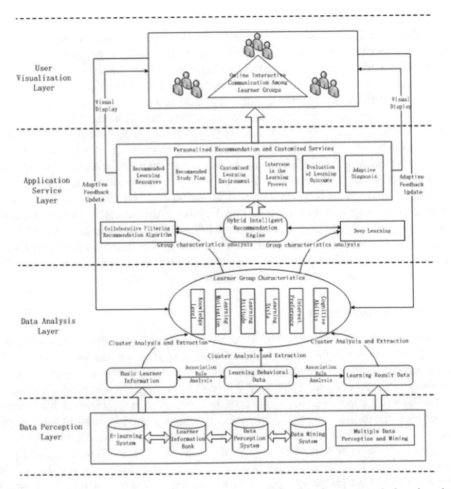

Fig. 2. A model diagram of learner group characteristics based on network learning data perception and mining

3.1 Data Perception and Mining Layer

At the bottom of the model, it mainly realizes the functions of diversified network learning data perception and mining. In the interface interaction with the network learning system, through a series of network learning data collection and preprocessing processes, a learner information database based on individual differentiation is constructed. Based on the basic functions of the information database, it can classify, store and update various learning-related data of each learner in real time, dynamically generate learner electronic files, and construct an information resource network system for learner groups. For diversified massive data, under the function of the data perception and mining system, the data in the learner information database is classified, denoised, sampled and quantified in more detail. By further compressing the volume and structure of the data, eliminating redundant and worthless data, filtering and extracting explicit information with strong relevance, the effectiveness of the original data for feature analysis and the efficiency of data processing is guaranteed.

Each learner's electronic file mainly includes three key modules: learner attribute information, learning behavior data and learning result data. The learner attribute information module mainly stores learner attribute data, including basic static information such as learner gender, age, major, educational background, various skill levels and certificates, awards, hobbies and so on [3]. In the learning behavior data module, the quantitative data of all learning behaviors of learners on the learning system platform is stored, mainly including log-in, browsing, learning dynamics, question and answer, discussion, collection and sharing and other recorded information. In the learning result module, what is collected is various related factors that can reflect learning effectiveness, periodic test scores, learning result levels, ability levels, evaluation and feedback and other representative information. All basic data in the learner information database composed of learner electronic files is the data support for upper-level data analysis and deep-level feature extraction, and is a reliable guarantee for the accuracy and functionality of the learner group feature model.

3.2 Data Analysis Layer

The data analysis layer is the second layer of the model. Under the condition that the lower layer provides explicit data classification and aggregation, the learner's behavior is the central element to analyze the learning effect and the learner's subject. Look for the relationship between the behavioral data collection items that have various effects on the learning effect. For the differences in learning results presented by learners driven by behavior patterns, cluster analysis algorithms are used to collect and analyze data, and learners are classified and hierarchized based on similarity to establish a grouped multi-level similar object set. At the same time, in the scheduling and processing of multi-level data, the implicit factor data that satisfies the analysis of the characteristics of the learner group is extracted from the point of action of the learner's attributes, learning behavior, learning effect, and learner's similar object set. It mainly includes six core characteristic elements of knowledge level, learning motivation, learning attitude, learning style, learning preference and cognitive ability, and traces back the internal factors of learner behavior and group characteristics that have different learning results [4].

In the learner group characteristics model, under the subject concept of scientific accuracy, comprehensive feasibility, and efficient qualitativeness, abstract and quantitative indicators to express characteristic factors are realized, which is convenient for expression, calculation, analysis and storage [5]. Through the presentation of group characteristic data, learners have a deeper understanding of their own learning foundation, learning ability, learning subjective factors, and the positive effect of learning behavior on learning effectiveness, allowing learners to carry out learning activities more efficiently under the driving force of big data.

3.3 Application Service Layer

The more valuable application field of the learner feature model is to design a hybrid intelligent recommendation engine module based on the fusion of collaborative filtering recommendation algorithm and deep learning under the premise of the group feature analysis provided by the data analysis layer to provide users with personalized services. This hybrid recommendation mechanism combines the advantages of the two algorithms, which not only makes up for the shortcomings of a single recommendation algorithm, but also greatly improves the quality of recommended content. At the same time, improve and guarantee the realization of the recommendation and customization functions to better meet the user's personalized learning needs and the development of intelligent learning technology.

Based on the principle of highly highlighting the learner's subjectivity and individuality, the intelligent recommendation engine module mainly implements application functions such as learning resource recommendation, learning plan recommendation, learning environment customization, learning process intervention, learning result evaluation, and adaptive diagnosis. Provide users with a more intelligent, personalized and humanized learning service experience. This module is mainly based on the collaborative filtering recommendation algorithm [6]. According to the calculation amount of the similarity of the user characteristic parameters, the similarity matrix between the group users is established, and the group wisdom effect is exerted to recommend the preferred learning resources and programs for the users in real time [7]. Using deep learning based on artificial intelligence technology as an auxiliary means, according to learner characteristic model data for user portraits, mining hidden features between users and recommended items, and gradually training and feedback to update user portraits. Finally, on the basis of collaborative filtering recommendation, the project resources are quantitatively calculated, resource similarity matching and sorting with user portraits are performed, and the top-ranked data is recommended for users [8].

3.4 User Visualization Layer

The user visualization layer is the direct medium of human-computer interaction between the user and the application system, and is a user-oriented service output interface based on the function of the learner group characteristic model [9]. Through the functions of this level, the personalized services provided by the application service layer are presented to users in a visual and operable mode. Learner groups can use the friendly service interface to realize online interactive communication, discuss and share learning experiences and methods, and promote a popular and personalized learning experience. At this level, users can more intuitively understand the learning progress, learning plan, learning goals, learning process evaluation and learning evaluation results and other information. Users provide adaptive feedback on learning behavior feature parameters and recommendation services, providing a reference data source for the automatic update of learner feature models, and further ensuring the real-time, accuracy and reliability of the model [10].

4 Conclusion

In the context of the rapid development of big data, intelligent learning, and personalized learning, this research is mainly based on the perception and mining of online learning data, and establishes a model of learner group characteristics with six core elements. Through the establishment of a hybrid intelligent recommendation engine, it provides users with personalized recommendation and customized services, helps users' personalized learning and development, and provides theoretical references for the development and research of intelligent learning.

Acknowledgments. This work was supported by Guangdong Province Educational Science Planning Project "Research on the Reform of Computer Network Practice Teaching Based on the Xiyuan Hardware Platform" (2017GXJK206).

References

1. Shi, K.: Research on Related Models in Adaptive Learning System Based on Learner Feature Analysis.: Yunnan Normal University (2020). (in Chinese)
2. Jing, L., Hui, Z.: A flexible network learning behavior data collection and analysis system. Inf. Comput. (Theor. Ed.) **02**, 85–86 (2011). (in Chinese)
3. Nie, L.: Personalize recommendation of learning resources based on behavior analysis. Comput. Technol. Dev. **30**(07), 34–37+41 (2000). (in Chinese)
4. Jiang, Q., Zhao, W., Du, X., Liang, M.: Research on personalized ontology learning resource recommendation based on user model. China Audio-visual Educ. (05), 106–111 (2010). (in Chinese)
5. Wang, M., Xu, J.: Research on the feature mining of learner groups based on online learning behavior data. Softw. Guide, **19**(07), 153–157 (2020). (in Chinese)
6. Yang, W., Han, L.: Research on personalized learning resource recommendation system based on hybrid mechanism——taking the national open university learning network as an example. J. Hubei Open Voc. Coll. **33**(13), 65–66+69 (2020). (in Chinese)
7. Li, Y., Qu, Z.: Analysis of learner characteristics based on analytic hierarchy process. Syst. Simul. Technol. **14**(01), 25–29+48 (2018). (in Chinese)
8. Zhao, J., Sun, S., Guo, J., Zhong, Y., Wang, M., Qin, Y.: Ontology-based network learning resource recommendation algorithm design. Microelectron. Comput. **38**(01), 64–69 (2021). (in Chinese)

 9. Zhang, Q., Meng, Z.: The design and implementation of a distributed student online learning behavior statistics system. Electronic Production, (11), 75 (2015). (in Chinese)
10. Xinjiang Radio and Television University Research Group: Design and implementation of a distributed student online learning behavior statistics system. J. Xinjiang Radio Telev. Univ. **14**(02), 5−9 (2010). (in Chinese)

Scheduling Algorithm and Analysis of Mixed Criticality Real-Time Systems

Juan Xiao, Song Wang, Sheng Duan, and Shanglin Li[✉]

XiangNan University, Chenzhou, Hunan, China

Abstract. An important development trend of modern real-time systems and embedded systems is Mixed criticality real-time Systems, Which is integrating functions with different level of importance into a sharing hardware platform. The biggest research focus of mixed criticality real-time Systems is scheduling problem. At present, the definition and connotation of Mixed criticality real-time system are generally accepted, but there is no unified task model. In this study, according to the research requirements of Mixed criticality real-time system, we first give a task model of mixed criticality real-time system, and define the schedulability of the system. Then it introduces two scheduling strategies based on time and priority, analyzes their advantages and disadvantages, and leads to the classic algorithm EDF in dynamic priority strategy. Finally, the paper focuses on the EDF-VD scheduling algorithm. Based on the task model of Mixed criticality real-time system, an example is given to analyze the EDF-VD algorithm in the Mixed criticality real-time system.

Keywords: Mixed criticality real-time systems · Scheduling algorithm and analysis · EDF-VD (Earliest Deadline First-Virtual Deadlines) algorithm

1 Introduction

Generally, many real-time systems which need to complete work and provide services in time, are embedded systems. In recent years, an important development trend of real-time embedded system design is to integrate different levels of components, which are to specify the level of failure guarantee required by the system components into a common hardware platform [1]. With the development of real-time embedded system, most complex systems are developing to meet the non-functional requirements such as cost, space, weight, heat and power consumption. Non-functional requirements are particularly important for mobile systems. This real-time system with Mixed criticality levels is called Mixed criticality real-time system.

Vestal published the first paper on mixed critical systems in 2007 [2]. In this paper, a fixed priority task scheduling algorithm for mixed criticality system is proposed, and the corresponding response time analysis method is given to analyze the schedulability of the system. At the same time, It is verified that rate monotone (RM) and deadline monotone (DM) optimal fixed priority allocation algorithms are not optimal in the mixed criticality system systems. But, In 2008, Baruah and vestal pointed out that if we do not restrict the system to use fixed priority preemptive algorithm, the algorithm in [2] is not technically optimal, and in fact its performance is incomparable with EDF [3].

© The Author(s), under exclusive license to Springer Nature Switzerland AG 2021
J. Abawajy et al. (Eds.): ATCI 2021, LNDECT 81, pp. 448–454, 2021.
https://doi.org/10.1007/978-3-030-79197-1_64

In 2012, Baruah proposed a series of basic works based on the simplified Mixed Criticality system model. They proved that for such a simplified system and all task instances are released at different times, it is still a strong NP hard problem to verify the schedulability of the system [4].

Although it is only 14 years since vestal first proposed the mixed criticality system real-time scheduling problem in 2007, it has unexpectedly attracted a large number of researchers to pay attention to the research in this field. This paper mainly propose some scheduling algorithms and scheduling analysis of mixed criticality real-time system.

2   Mixed Criticality Real-Time Systems

Generally, mixed critical real-time system is a real-time embedded system that integrates components with different critical levels into a general hardware platform. A mixed criticality real-time system refers to systems with two or more different levels, such as security criticality, mission criticality, low criticality, etc.

2.1   Mixed Criticality Task Model

We found that all previous papers on mixed criticality real-time systems used different task models. Here, we define a simple task model, which has certain universality and can describe the main research results of this paper. As an extension of the traditional periodic task set, the mixed critical real-time task system τ consists of several independent tasks, some other paper call it components.

And each mixed critical task in τ can be represented by a tuple (Pi, Ti, Li, Di):

1) Pi is the small release interval of task τi, also known as period.
2) Ti is the relative deadline of task τi.
3) Li $\in \{1,..., L\}$ is the critical level of task τi, where L is the total number of critical levels of the system. The smaller the number, the lower the critical level.
4) Di = $(D_i^1, ..., D_i^L)$ is the construction execution time vector of task τi, where D_i^s represents the construction execution time of τi when the system is at the critical level S, and we assume that if a < b, then $D_i^a < D_i^b$.

Each task τi will release infinitely many task instances periodically. We use J_i^k to represent the kth instance of task τi. if τi releases an instance at time R, the absolute deadline T of the instance is equal to r + Ti. To simplify the notation, we use Ji to represent the task instance released by τi.

The system starts to run from the lowest critical level. If any task τi fails to complete its execution within the execution time D_i^1, the system will immediately switch to the higher critical level 2. After that, the deadline of all tasks with critical level 1 is no longer required to be met, but for all other tasks with high critical level, they need to be executed according to the more pessimistic execution time D_i^2. Similarly, if any task is not completed in D_i^2 time at this time, the system will further switch to a higher criticality level. In practice, the system may switch from a high-level critical running state to a low-level critical running state due to processor idle. However, from the perspective of modeling, we assume that once the system enters critical level S, all tasks below critical level S will be discarded immediately.

2.2 Definition of Schedulability

We use $\tau(s)$ to represent all task subsets with critical level S, that is, $\tau(s) = \{\tau i| Li = s\}$. Then the schedulability of the system can be defined as follows.

Definition 1. (schedulability of mixed criticality system): a mixed criticality system is schedulable. If and only if the system is running at criticality level S, all tasks $U_{k \geq s}\tau(k)$ can meet their relative deadline.

In order to simplify the representation, when the system only contains two key levels (i.e. $Li \in \{1, 2\}$), we use low key level to represent key level 1 and high key level to represent key level 2.

3 Real-Time Scheduling Strategy

For a real-time system, the main function of the real-time scheduler is to determine the execution sequence of a series of real-time tasks with shared resources (CPU or network bandwidth). The purpose of real-time scheduling is to ensure that the time characteristics of each real-time task in the system can be satisfied. Generally speaking, real time scheduling can be divided into two kinds, one is offline scheduling, the other is online scheduling. Offline scheduling is that all scheduling decisions have been made before the system is executed, and all real-time tasks are executed in the order given by the scheduling table after the system is running. Online scheduling is to make scheduling decisions online according to the time characteristics of the system in the process of system operation.

3.1 Time Based Scheduling Strategy

The scheduler creates a schedule table. Generally speaking, the schedule table is generated before the initial operation of the system (offline scheduling), but some schedulers can also generate a schedule table (online scheduling) during the operation of the system. In the running process of the system, the dispatcher executes each real-time task according to the schedule table, and each task can only be executed within the time point determined in the schedule table.

The biggest advantage of creating scheduling table through offline mode is that it can bear relatively high time complexity (Scheduling table generation and schedulability analysis), which is very difficult for online scheduling mode. Moreover, the real-time system using offline scheduling has very good predictability, because the tasks in the system will be executed strictly according to the pre generated scheduling table. Therefore, offline scheduling is the most commonly used scheduling method in many applications with high security requirements (such as aviation control). But on the other hand, the defect of offline scheduling is also very obvious. Because the scheduling table is generated before the system runs, the flexibility of scheduling is very limited. Once any change occurs in the system running process (such as adding some functionality or changing the hardware environment), offline scheduling can not provide good support, which also promotes the implementation of priority based scheduling strategy Application.

3.2   Priority Based Scheduling Strategy

All scheduling strategies that determine the execution order of tasks in the process of system operation are classified as online scheduling. This kind of scheduling algorithm will schedule the tasks in the system according to the time constraints in the system operation, such as task priority. Scheduling strategies based on task priority are called priority based scheduling strategies.

Compared with time-based scheduling strategy, priority based scheduling can decide which task gets processor resources to execute according to the latest current characteristics of the task, which greatly improves the flexibility of scheduling. Therefore, priority scheduling can cope with the dynamic changes of system workload. For example, as long as the schedulability is guaranteed, the system can add or delete tasks online to achieve flexible functional changes. Priority based scheduling strategy can be further divided into fixed priority schedule (FPS) and dynamic priority schedule (DPS). The difference between the two is whether the priority of the task is fixed in the system operation.

(1) Fixed Priority Schedule, FPS
When the system adopts a fixed priority scheduling, the priority of the task is changed once it is allocated. At any time in the system running, the scheduler selects the highest priority of all ready tasks. According to the different time characteristics of the specific system, there are many different fixed priority allocation strategies. For example, rate monotonic (RM) algorithm has been proved to be an optimal fixed priority allocation algorithm for single processor [5]. RM has more task cycle to assign corresponding priority to it, and the shorter the cycle, the higher the priority.

(2) Dynamic Priority Schedule, DPS
The most famous dynamic priority scheduling strategy is the early deadline first (early deadline first, EDF) algorithm, in this algorithm, the task with the closest deadline at any time is given the highest priority, and is executed by the processor resources. Therefore, the priority of the task is not fixed, but changes dynamically with the operation of the system. EDF algorithm is also proved to be the optimal dynamic priority scheduling algorithm on single processor [6], and compared with the fixed priority scheduling strategy, EDF has higher schedulability. Other dynamic priority scheduling strategies include least laxity first (LLF) [7, 8]. The relaxation time is the time remaining for a task from the deadline to the current time [9]. LLF selects the task with the shortest relaxation time to execute at any time.

4 EDF-VD Scheduling Algorithm and Analysis

4.1 EDF-VD Scheduling Algorithm

The biggest challenge of mixed criticality real-time system is system design. For traditional real-time system scheduling algorithms, such as EDF and RM, will bring huge waste of resources when they are directly applied to mixed criticality real-time system. In recent years, according to the characteristics of different critical levels in mixed criticality real-time systems, researchers have proposed some new scheduling algorithms based on system analysis to improve the utilization of system resources. Among them, EDF-VD (Early Deadline First Virtual Deadlines) has been proved to have better schedulability and runtime efficiency than other algorithms. As the name suggests, EDF-VD is designed on the basis of EDF scheduling strategy. When the system is running at different critical levels, it can balance the schedulability of different critical levels by setting different virtual deadlines for tasks.

4.2 Case Analysis of EDF-VD Scheduling Algorithm

In the traditional periodic real-time system, EDF algorithm is the best. However, for the mixed criticality real-time system described above, EDF may bring lower system performance. For example, we use EDF algorithm to schedule a mixed critical task set with high and low critical levels in Table 1. The task set only contains two tasks with different critical levels, and both tasks release their first task instance at system 0.

Table 1. Mixed-criticality task set

Task	D_i^{L0}	D_i^{H1}	Pi	Li	T_i^{L0}	T_i^{H1}
$\tau1$	8	–	18	L0	18	-
$\tau2$	8	16	20	H1	14	20

As shown in Fig. 1, assuming that the second task instance released by high criticality task $\tau2$ does not complete after the worst execution time $D_2^{L0} = 8$, the system will be triggered immediately at this time (t = 34) to continue to run at high criticality level. After that, although the low critical task $\tau1$ is directly discarded, because the execution time of the high critical task $\tau2$ is $D_2^{H1} = 16$, it still needs 8 time units to complete the execution. Since there are only 6 time units left from its absolute deadline, Obviously, the deadline will be missed. This example shows that in order to prevent the critical level of the system from suddenly switching, when the system is running at a low critical level, even if its deadline is relatively late, it should be given a higher priority, so that it can reserve enough time for the possible critical level transition of the system.

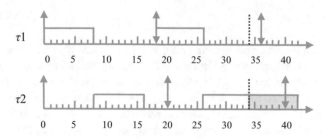

Fig. 1. Mixed-criticality task set for EDF scheduling

On the basis of EDF, EDF-VD algorithm sets different virtual deadlines for each task at different system critical levels, and the scheduler assigns corresponding priority to the released instances according to these virtual deadlines. We use T_i^{H1} and T_i^{L0} to represent the virtual deadline of task $\tau 1$ under high and low critical levels, and T_i^s to represent the virtual deadline of task $\tau 1$ under critical level S. if the critical level of a task itself is less than S, there is no need to set the virtual deadline under critical level S, that is, T_i^s. Now let's look at the example just now. If we set the virtual deadline T_2^{L0} = 14 for task $\tau 2$ (all other parameters remain unchanged), then the updated schedule table is as follows As shown in Fig. 2, the second instance released by $\tau 2$ will have a higher priority than the instance released by $\tau 1$. At this time, if the system switches to the high critical level at the same time, $\tau 2$ still has enough time to complete the worst execution time under the high critical level.

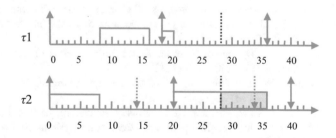

Fig. 2. Mixed-criticality task set for EDF-VD scheduling

By setting such a virtual deadline, we have the opportunity to get extra slack time for those task instances released when the critical level of the system changes, and reduce the workload of high critical tasks brought into the high critical level of the system. On the other hand, we can extend the virtual deadline of some tasks if the system becomes UN schedulable at low criticality level after the virtual deadline is set. In short, to make the system schedulable at all critical levels, it is necessary to set appropriate virtual deadlines for all tasks.

5 Conclusion

For mixed criticality real-time system, what we want to know is whether we can find a way to configure the appropriate virtual deadline for all tasks in the system, so that the system can be scheduled by EDF algorithm at all critical levels without any task missing the deadline. To this end, we need to solve the following two problems, the one is schedulability test, which mainly determines whether EDF can schedule the system at all critical levels under the given virtual deadline configuration. the other is virtual deadline adjustment, which need to give a schedulability determination method, and discuss how to configure the virtual deadline for all tasks to make the system schedulable.

At present, the research of mixed criticality real-time systems focuses on the improvement of scheduling algorithms and related schedulability analysis in various multilevel systems, especially the response time analysis [10, 11].

Acknowledgments. Scientific research project of Hunan Education Department (NO· 17C1487, 17A201).

References

1. Burns, A., Davis, R.: Mixed criticality systems - a review, 4–6 (2019)
2. Vestal, S.: Preemptive scheduling of multi-criticality systems with varying degrees of execution time assurance. In: Proceedings of IEEE International Real-Time Systems Symposium, pp. 239–243 (2007)
3. Baruah, S., Vestal, S.: Schedulability analysis of sporadic tasks with multiple criticality specifications. In: Euromicro Conference on Proceedings of Real-Time Systems, ECRTS 2008, pp. 147–155. IEEE (2008)
4. Baruah, S., Bonifaci, V., D'Angelo, G., et al.: Scheduling real-time mixed-criticality jobs. IEEE Trans. Comput. **61**(8), 1140–1152 (2012)
5. Liu, C.L., Layland, J.W.: Scheduling algorithms for multiprogramming in a hard-real-time environment. J. ACM (JACM) **20**(1), 46–61 (1973)
6. Baruah, S.K., Mok, A.K., Rosier, L.E.: Preemptively scheduling hard-real-time sporadic tasks on one processor. In: Proceedings of 11th Real-Time Systems Symposium, 1990, pp. 182–190. IEEE (1990)
7. Leung, J.Y.T., Whitehead, J.: On the complexity of fifixed-priority scheduling of periodic, real-time tasks. Perform. Eval. **2**(4), 237–250 (1982)
8. Mok, A.K.: Fundamental design problems of distributed systems for the hard-real-time environment, Massachusetts Institute of Technology (1983)
9. Buttazzo, G.C.: Hard real-time computing systems. Real-Time Syst. Ser. **416**, 31–38 (2011)
10. Hussain, I., Awan, M.A., Souto, P.F., et al.: Response time analysis of multiframe mixed-criticality systems with arbitrary deadlines Real-Time Systems (2020) (prepublish)
11. Orr, J., Baruah, S.: Algorithms for implementing elastic tasks on multiprocessor platforms: a comparative evaluation. Real-Time Syst. **57**(1–2), 227–264 (2021). https://doi.org/10.1007/s11241-020-09358-9

Construction of Examination Questions Database of Medical Information Retrieval Course Based on Multivariate Collaborative Filtering Recommendation Algorithm

Lianhuan Li[✉]

Nanyang Medical College, Nanyang 473000, Henan, China

Abstract. With the rapid development of computer networks, the era of big data has come, and fields such as medical retrieval are flooded with massive amounts of data. It is becoming more and more difficult for users to select the knowledge they need from a large amount of knowledge. The personalized recommendation system can quickly recommend the knowledge that people need. Collaborative filtering recommendation algorithm is a hot area of personalized recommendation algorithm, which has been deeply studied and discussed by scholars at home and abroad. The user-based collaborative filtering recommendation algorithm selects neighbors by calculating user similarity, and then recommends neighbors. This is playing an increasingly important role in practical applications. In the current medical information retrieval course, the test question bank for assessment is mainly composed of a large number of test questions, and there is no effective distinction between the test questions under different users, different projects, and different responsibilities. Therefore, the test questions for different personnel are targeted to the medical retrieval course. It is particularly important. In order to solve this problem, this article is based on the existing recommendation system, and through reading a large amount of literature, conducting experimental analysis and other methods to study the role of multiple collaborative filtering recommendation algorithm in the construction of medical retrieval course test question bank. The research results show that the test question system of the medical retrieval course based on the multiple collaborative filtering algorithm is more intelligent, and different test questions can be given to different personnel, which greatly improves the intelligence and efficiency of the medical information retrieval test question database.

Keywords: Multivariate Collaborative Filtering · Recommendation algorithm · Medical information retrieval · Test question database construction

1 Introduction

With the increasing advancement of computer and Internet technology, servers and hosts all over the world have been connected to the Internet. What followed was an exponential growth of web pages, and users had to face a lot of redundant and complex data. The construction of the test question bank in the exam is the sum of all the knowledge points of a course, and it is the complete manifestation of a course. In the

J. Abawajy et al. (Eds.): ATCI 2021, LNDECT 81, pp. 455–462, 2021.
https://doi.org/10.1007/978-3-030-79197-1_65

medical information retrieval course, the test question bank is very large, which covers all the knowledge of the whole medicine. The current test question bank is often completely random when grouping questions, and there is no correlation between different fields. Distinguish, resulting in a complete widening of the course assessment, and the corresponding professionalism cannot be distinguished. Therefore, using the recommendation system of multiple collaborative filtering algorithms to build the test question bank of medical retrieval courses is an urgent problem that needs to be solved in the current medical field.

Applying the personalized recommendation mechanism to the learning field, such as the construction of the test question bank of the medical retrieval course, can improve the quality of service and bring a lot of convenience to users [1]. From the perspective of users, first of all, the personalized mechanism of the test question bank system can reduce the time cost for users to find the knowledge they need, and accelerate the speed at which users find the knowledge they need. Secondly, the personalized recommendation mechanism is just like the teaching teacher can automatically explain to students what they need, and at the same time it can make users "associate" and tap the potential needs of consumers. Users can not always think of everything, especially when new users come into contact with a new field, the user's choice needs to be recommended by the recommendation system [2]. The recommendation system allows users to be inspired to obtain more required knowledge, reduces the user's selection pressure, and improves the user's retrieval comfort. Through related literature research, it can be concluded that the system developed by Xerox in the United States is the first recommendation system and the concept of collaborative filtering is also given for the first time [3]. The recommendation system is mainly used for news recommendation. Users can give feedback on the electronic news they read. If the user likes the current news, they can mark 'like', otherwise, they can mark 'dislike'. When searching for news, users can refer to the label information of users they are familiar with and select news to read. That is to say, the news that users they trust may also like the news themselves. After all, the number of other users that the user knows is limited, and it is impossible to know all users. Therefore, the system does not apply to the scenario of a large user scale in the system. At present, the research of collaborative filtering recommendation algorithm in our country has flourished and rapidly developed in the field of library information and other fields. In recent years, it has gradually increased. After many years of unremitting efforts of scholars, many research results have been obtained in the field of library information [4]. A team proposed an information intelligence service model based on knowledge innovation in its research results, including a professional information intelligence model, a team service model that supports "E-science", and an information intelligence service model that adds value to knowledge. The information intelligence service model has been analyzed in detail [5]. In response to the new requirements of collaborative filtering recommendation algorithms in the online scientific research era, a new knowledge service model for digital libraries that improves knowledge service capabilities from the perspectives of user retrieval and personalized push is proposed; Wang Jixin et al.'s research fully analyzed the semantic grid Applying the advantages of knowledge services, combining ontology and knowledge services related technologies, a knowledge service system model based on semantic grid is proposed, and several important parts of the model are explained in detail, and personalized retrieval is also proposed [6]. It can be seen from the above-mentioned documents that, whether domestic or foreign, a model based on how far collaborative filtering recommendation algorithm is constructed in many fields, it is the future development trend to recommend information in a targeted manner.

This article takes the medical retrieval course as the background and uses medical data to carry out research work on the construction of existing test questions. This article puts forward the concept of preference scoring matrix for a large number of test question data of medical retrieval courses. And on this basis, the How Far Collaborative Filtering Recommendation Algorithm is introduced into the existing medical retrieval course test questions construction, which gives a formal definition of the characteristic attributes of the target object. According to the different behaviors of users, considering the degree of preference of the user for the characteristic attributes of the object under each behavior, the concept of user preference scoring matrix is proposed. After calculating their respective similarities for different behaviors, they are given their respective weights to comprehensively consider the similarities between users, to give different types of test questions, so that the test question assessment of the medical retrieval course is more intuitive and more targeted, thereby improve the overall level of the course.

2 Method

2.1 Multiple Collaborative Filtering Recommendation Algorithms

Since the development of the recommendation system, it has gathered the sweat and energy of countless experts and scholars and achieved great success. New recommendation methods are emerging in endlessly, and the accuracy of recommendation has also been greatly improved. Among these recommendation theories, the most famous and oldest collaborative filtering recommendation algorithm is still attracting attention. Some researchers proceed from the law of people's interest: we may also like items that friends like, and items that we used to be used to, may still be interested in or in demand for similar items. According to this nearest neighbor recommendation strategy, some scholars have proposed a memory-based collaborative filtering recommendation algorithm [7]. The development of big data has highlighted the advantages of data mining technology. Some researchers have introduced data mining related technologies into collaborative filtering, and proposed a variety of model-based collaborative filtering recommendation methods. With the help of machine learning theory, the data sparse problem, expansion problem and cold start problem of the recommendation system can be solved well, and it can also dig out user characteristics and item characteristics, and can establish a recommendation model well. Machine learning methods include: clustering technology, matrix decomposition, dimensionality reduction technology, classification technology, and association rules. Memory-based collaborative filtering has the characteristics of simple principle and clear thinking, and the focus of research is on the quality of neighbors [8]. Whether it is from the concept of similarity or trust, it is to find high-quality neighbors, which can seriously affect the effect of recommendation. Model-based recommendation algorithms are rigorous and efficient. Whether it is a clustering model, a classification model, or a regression model, it can have rigorous theoretical derivation and efficient modeling capabilities. The two methods have their own advantages and disadvantages. The reasonable integration of different types of recommendation algorithms can achieve the complementary advantages of the recommendation mechanism and improve the recommendation effect. On the way to school, people will ask their friends "What are the interesting books?", during vacations, people will ask their friends "what are the fun places?", when chatting with friends, people will also ask their friends: what are they Nice songs, nice movies. There are all kinds of friends with similar interests in life, and they also accept items recommended by friends in many aspects [9]. People will often be recommended by different friends in social interactions. People will prefer items recommended by friends with the same interests. People with the same interests may also like items. Collaborative filtering is derived from the heuristic of similar neighbors. It is proposed to recommend strategies. The following three formulas are the core formulas of the user-based collaborative filtering algorithm.

Formula for calculating similarity:

$$Sim(A, B) = \cos(\vec{A}, \vec{B}) = \frac{\vec{A} \cdot \vec{B}}{|\vec{A}|^2 \times |\vec{B}|^2} \tag{1}$$

Among them, A and B are two different users, and Sim (A, B) represents the similarity between the two users.

Modified cosine similarity formula:

$$Cim(A, B) = \frac{\sum_{i \in I_{AB}} (R_{Ai} - \overline{R_A})(R_{Bi} - \overline{R_B})}{\sqrt{\sum_i \in IAB (R_{Ai} - \overline{R_A})^2}} \tag{2}$$

Among them, Cim (A, B) is the modified cosine similarity, IAB is the set of items jointly selected by two users, A and B represent two users, and RA and RB represent the evaluation values of the two users.

Recommended value formula:

$$P_{Ai} = \frac{R_A + \sum_{j \in N_a} Sim(A,j) \times (R_{ji} - \overline{R_j})}{\sum_{j \in N_a} 1 sim(A,j)} \tag{3}$$

Among them, PAi represents the recommended probability value, Sim(A, j) represents the similarity of two users, and RA, RB represent the average value of the two users.

2.2   Evaluation Index of Recommendation Algorithm

The accuracy of a single recommendation algorithm is not enough to evaluate the quality of a recommendation system. The user's satisfaction, trust, and dependence of the recommendation system should be considered to evaluate the quality of the recommendation system, but these conditions should be obtained through the user's feedback. However, it is difficult to collect reliable samples through the large workload and high cost of surveying user feedback [10]. Therefore, the performance of the recommendation algorithm can also be measured by analyzing the experimental data results of the recommendation system. The main indicators for evaluating the recommendation algorithm are: prediction accuracy and classification accuracy.

3 Experiment

3.1 The Purpose of the Experiment

This article is based on the results of multiple collaborative filtering recommendation algorithms, drawing on the domestic and foreign theoretical research results, using literature, comparative research, mathematical statistics, logical analysis and other methods to analyze the current medical information retrieval course question bank data, and study whether the medical information retrieval course test question database construction under the multiple collaborative filtering recommendation algorithm is more intelligent and more targeted.

3.2 Experimental Design

Most of the data used in this chapter comes from the test question bank of a medical retrieval course in a university, and some of the data comes from the Internet. The main data includes test questions about doctor information, drug information, disease information, treatment methods and other information. In the experiment, different customers are operated with different preferences, and then the test question system recommends relevant test questions according to the user's preference. These experimental data are divided randomly, and the experiment compares and statistics the data by calculating the MAE value.

4 Result

4.1 MAE Mean Value Under Different Similarity Calculation Methods

It can be seen from Table 1 and Fig. 1 that when the number of users is 10, the value of Pearson similarity is 0.804, the value of Cosine similarity is 0.814, and the value of Modified cosine similarity is 0.808; when the number of users is 20, the value of Pearson similarity is 0.81, the value of Cosine similarity is 0.796, and the value of Modified cosine similarity is 0.823. When the number of users is 30, the value of Pearson similarity is 0.806, the value of Cosine similarity is 0.793, and the value of Modified cosine similarity is 0.791. When it is 40, the value of Pearson similarity is 0.789, the value of Cosine similarity is 0.801, and the value of Modified cosine similarity is 0.811; when the number of users is 50, the value of Pearson similarity is 0.814, the value of Cosine similarity is 0.786, and the value of Modified cosine similarity is 0.814, the value of is 0.817; when the number of users is 60, the value of Pearson

Table 1. MAE value under different similarity calculation methods table

	Pearson similarity	Cosine similarity	Modified cosine similarity
10	0.804	0.814	0.808
20	0.81	0.796	0.823
30	0.806	0.793	0.791
40	0.789	0.801	0.811
50	0.814	0.786	0.817
60	0.786	0.799	0.766

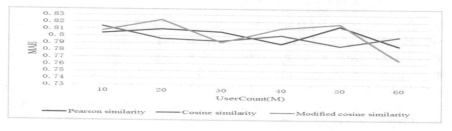

Fig. 1. MAE value under different similarity calculation methods chart

similarity is 0.786, the value of Cosine similarity is 0.799, and the value of Modified cosine similarity is 0.766. The data shows that the MAE values under the three similarity calculation methods all decrease as the number of users increases.

4.2 Comparison of MAE Values in Different Modes

Table 2. Comparison of MAE values in different modes table

	Traditional model	Recommended algorithm mode
10	0.568	0.814
20	0.565	0.796
30	0.543	0.793
40	0.523	0.801
50	0.519	0.786
60	0.505	0.799

It can be seen from Table 2 and Fig. 2 that when the number of users is 10, the traditional mode MAE value is 0.568, and the recommended algorithm mode MAE value is 0.814; when the number of users is 20, the traditional mode MAE value is 0.565, and the recommended algorithm mode MAE value is 0.796; when the number of users is 30, the traditional mode MAE value is 0.543, and the recommended algorithm mode MAE value is 0.793; when the number of users is 40, the traditional mode MAE value is 0.523, and the recommended algorithm mode MAE value is 0.801; when the

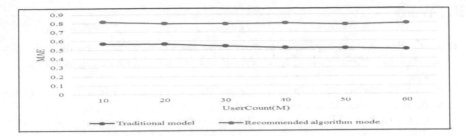

Fig. 2. Comparison of MAE values in different modes chat

number is 50, the traditional mode MAE value is 0.519, and the recommended algorithm mode MAE value is 0.786; when the number of users is 60, the traditional mode MAE value is 0.505, and the recommended algorithm mode MAE value is 0.799. It can be seen from the data that regardless of the number of users, the MAE in the recommended algorithm mode is higher than that in the traditional mode, which can prove that the medical information retrieval course test question database based on the collaborative filtering recommendation algorithm is more intelligent and specific than the traditional mode.

5 Conclusion

With the increasing progress of Internet technology and the advent of the information age, the amount of data has shown an explosive growth trend. Users have to face a lot of redundant and complex data, it is difficult to easily and conveniently obtain the information they want to find, and face many obstacles. However, the traditional medical retrieval course test question bank is randomly organized, and there are many problems. For example, the selected questions have a certain degree of subjectivity and cannot well reflect the actual mastery of all different users. In order to solve these problems, how far the collaborative filtering algorithm is used to realize the management of the test question bank, realize the personalized test paper composition, and analyze different types of users, so that the questions obtained can be more professional and consistent.

Acknowledgements. Project Type: Henan soft Science Research Program Project.

Project Name: Research on the Application of recommendation algorithm based on Multivariate Collaborative filtering in Medical practice qualification examination, Project number: 2124000410192.

References

1. Xi, C.: Research and application of recommendation algorithm based on dimensionality reduction and clustering. Nanjing University of Posts and Telecommunications, pp. 1–5 (2020)
2. Jianmei, C., Yajun, S.: N-dimensional tensor decomposition recommendation algorithm based on user's nearest neighbors. Comput. Eng. **43**(11), 193–197 (2017)
3. Ying, C., Huimin, H.: Collaborative filtering recommendation algorithm based on item attribute preference mining. Comput. Appl. **37**(S1), 262–265 (2017)
4. Ruiqin, W., Yunliang, J., Yixiao, L., Jungang, L.: A collaborative filtering recommendation algorithm based on multiple social trust. Comput. Res. Dev. **53**(06), 1389–1399 (2018)
5. Sun, L., Zhang, Y., Xiao, C.: Collaborative evolution recommendation based on self-attention. Comput. Eng. Des. **42**(02), 382–387 (2021)
6. Jixin, W., Jun, T., Xuan, W., Yitong, W.: Research on "Internet + Localized Classroom" optimization countermeasures based on teaching behavior data analysis. Electron. Educ. Res. **41**(04), 93–101 (2020)
7. Ruixu, Y.: Continuing education personalized intelligent design algorithm based on deep learning. Mod. Electron. Technol. **44**(04), 124–128 (2021)
8. Junjun, L.: The improvement and implementation of user collaborative filtering personalized book recommendation algorithm. Libr. Inf. Guide **6**(01), 38–42 (2021)
9. Manying, N., Guangyi, T.: Research on the thesaurus retrieval of clinical medical nouns based on non-relational databases. Arch. China **12**, 68–70 (2020)
10. Xiang, C., Peibiao, H.: Flipped classroom teaching model of medical literature retrieval based on Chaoxing Fanya platform. Jiangsu Sci. Technol. Inf. **37**(30), 63–66 (2020)

Single Image Region Algorithm Based on Deep Network

Junhua Shao[✉] and Qiang Li

Research Institute, Lanzhou Jiaotong University, Lanzhou 730070, Gansu, China
shaojunhua@mail.lzjtu.cn

Abstract. This paper designed an image region feature extraction and recognition algorithm based on deep neural network model based on deep network as for the problem that it is hard to extract and recognize the eigenvalue in image region, aiming at accurately extracting image eigenvalue while reducing the consumption of time. Through the experimental test of data set, the rationality and feasibility of the algorithm are confirmed, which can effectively reduce the time consumption and improve the accuracy of image processing.

Keywords: Image region · Deep network · Accuracy · Time consumption

1 Introduction

With the rapid development of deep network, especially the great advances in deep neural network in the field of image processing, the efficiency of feature extraction and application of regional images is becoming increasingly higher. In the process of studying and learning the deep network, researchers have found that the deeper the network level is, the richer the image diagnosis that can be extracted, and the more accurate the recognition of regional features in a single image will be, but the network based training will also increase. In the later research, aiming at the difficulty in obtaining the training model in the deep network technology, Kaiming He proposed the residual network [1]. While advancing the network level, the dispersion problem in the training model was constantly solved to improve the feature recognition rate in the single image area. On this basis, VOG network [2] and deep network [3] were proposed to ensure the accuracy and professionalism of image feature recognition.

Based on the deep neural network, this paper completes the extraction and recognition of the regional features of a single image. Through image acquisition, preprocessing, region segmentation and feature recognition, the extraction and recognition of all the regional features of a single image were completed. Finally, the whole processing process of the image was made and the overall flow chart is shown in Fig. 1.

Fig. 1. Processing flow chart

2   Overview of Key Technology

2.1   Deep Network Model

The deep network is equivalent to such a system S: when building a system S with an n-tier structure, the input is I, and the output through system S is O. If the output O is equal to the input I, it can be considered that no information is lost after the input is transformed through the system. This does not exist in practice, because part of the input is always lost if any system is transformed. Therefore, in the setting, the approximate equality of the output O and the input I was taken to make the maximum possible of approximate equality, so that the representation obtained through each level of system S is considered to be a variety of representations of the input I. If a multi-layer deep network is constructed and reached step by step, a multi-level abstract expression of the original input I can be finally obtained, that is, the construction process of the deep learning module is similar to the process of the brain's hierarchical abstract visual acquisition of information.

Deep neural network is, in essence, a deep network model containing more than one hidden layers [4, 5], the calculation of the main unit is neural node that is very similar to the animal's brain neurons. For neurons node in neuroscience, it was found that the input information was processed by mammalian brain cortex by layering process and the function of the visual cortex is not limited to reproduce the image in the retina, and it can extract and calculate perceptual signal. In the same way, the way of the depth of the neural network neurons is very similar to the way of information processing and the organization form of the depth of the neural network is a layer of neurons to process information before transferring the results to the next layer of neurons in the center, the first layer of a layer of neurons mode is similar to mammalian neural network organization in depth superposition manner. Deep neural networks are inspired by the way neurons process information in animals, including humans, by building many neurons into different layers of networks and then adding them on top of each other. The essence of this method is the superposition of multiple hidden layers, and then through large-scale data training, a model that can express super-large scale feature information is obtained, so as to realize the classification and prediction of data samples. In this way, the complex features of data can be learned to a large extent.

2.2   Image Processing Technology

Greenhalgh firstly put forward road signs image recognition method based on the research on image recognition in natural scenes, which can identify the natural scene of road signs characters, the algorithm is mainly made by using the Hough transform and Canny operator to search for images of characters in the candidate area, and then the characters in the image can be recognized. Mammeri A et al. used the HOG algorithm [6] and the SVM classifier to complete the research algorithm of image text recognition under natural scenes, and successfully detected and applied the road sign images in natural scenes. In this literature, A Tesseract method was used to recognize the image OCR.

Literature [7] adopted a "four-point method" to correct the image. Then, a new method of Harris angle points combined with rectangle was proposed to recognize the text in the image. Then, a multi-KNN weak classifier was used to recognize and classify the character data. Literature [8] firstly adopted image morphology method to extract candidate areas of traffic signs, then EOH algorithm was adopted to extract features of candidate areas in images, and then SVM was used to classify characters. In reference [9], Gonzalez A et al. proposed a recognition algorithm to recognize characters in streetview images. In this paper, the text region of the image was firstly obtained through the features of the image color space, and then the same algorithm was used to recognize the char acters as in the literature [10]. The idea of the algorithm is to use MSER algorithm to locate the text, and then use SVM classifier to recognize the characters.

3   Single Image Region Algorithm Based on Deep Network

3.1   Image Acquisition

Aiming at the key eigenvalues in images, this paper takes text data as an example for study. First, a text character module is given. After that, the text module is clipped and the data identification is completed. The specific process is shown in Fig. 2.

Fig. 2. Module identification process

According to the given module recognition process, each character is disassembled, and the character eigenvalues were extracted and recognized based on the deep neural network technology, so that the prediction probability of different eigenvalues can be obtained.

3.2   Image Preprocessing

After the completion of image acquisition, it is necessary to preprocess the acquired image data, aiming at de-noising and extracting key image areas. The process of image preprocessing includes three parts: preprocessing, automatic calculation and interface generation. In the preprocessing part, the principal component analysis method is used to extract the accurate feature value range and the key word information of a region in the image, and then the key information of the image is screened and classified. Moreover, the variance, covariance and mean of regional pixel values are calculated to complete the calculation and matching of correlation, so as to realize the regional feature acquisition, extraction and management of image pixel information. The automatic calculation part is mainly based on the Spark computing framework of open source big data technology, which automatically completes the calculation of the above calculated values instead of the traditional manual calculation method. In order to facilitate the user's understanding and use, the interface generation part mainly displays the feature extraction results to the user in the form of interface.

In the process of image preprocessing, the pixel information is input into the algorithm, and the pixel matrix is processed by the Euclidean distance method. According to these data structures, the variance and mean square error are calculated, and the concentration area of the image eigenvalue is determined by the variance, so as to obtain the key eigenvalue content. According to the covariance, the key eigenvalues are screened out, noise information is removed, and the data information with the maximum positive correlation is retained.

The first is the preprocessing algorithm based on principal component analysis. The specific process is shown in Fig. 3.

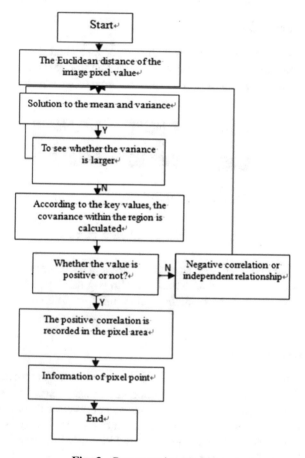

Fig. 3. Preprocessing algorithm

According to the above process, the calculation formula of the mean value is as follows:

$$\bar{x} = \frac{1}{n}\sum_{i=1}^{n} xi \tag{1}$$

$xi(i = 1, 2, ...)$ represents the Euclidean distance of pixels. After that, the variance is calculated, and the formula is as follows:

$$s^2 = \frac{1}{n}\left(\sum_{i=1}^{n} xi - \bar{x}\right)^2 \tag{2}$$

The following is the calculation of covariance, the formula is as follows:

$$\text{cov}(x, \bar{x}) = \frac{1}{n}\sum_{i=1}^{n} (xi - \bar{x}) \tag{3}$$

For the automatic calculation of data, the comparison information of relevant data in the current month was taken as an example to obtain the time sequence diagram, as shown in Fig. 4.

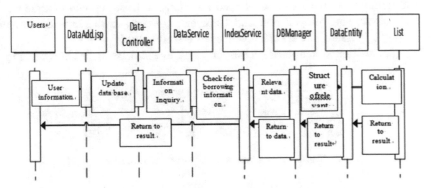

Fig. 4. Timing diagram of data calculation

As can be seen from the figure, firstly, to switch to the interface of Data Add. Jsp, and then the operation was conducted through the input of pixel data. After that, according to the corresponding information in the timing analysis, the display result is saved in the list, and the thread was used to complete the automatic calculation and processing of the data. Finally, the display of the interface is completed.

3.3 Image Area Processing

The eigenvalue range information of a single image region is obtained by image preprocessing, and the deep neural network is used to extract the eigenvalue of the region accurately. The model structure of the deep neural network is shown in Fig. 5.

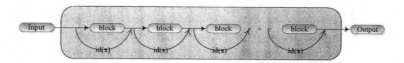

Fig. 5. Structure diagram of deep neural network

As can be seen from the figure, when each layer of the network was chosen to randomly skip a certain layer, and in order to control the current layer ℓ to randomly skip to the next layer, the probability value p_ℓ is set in the residual map, which represents the probability that the current layer ℓ can be directly propagated to the next layer. Then the probability of the next layer being ignored is $1-p_\ell$. Its basic principle is to randomly skip the current layer ℓ through the probability value when training the network, and the control of skipping is realized through a random Bernoulli variable Z, $Pmf(z = 1)$ is the probability mass function of the random Bernoulli distribution, then the probability that the ℓ layer network to be not skipped is $p_\ell = Pmf(z = 1)$. Because a certain number of layers can be skipped randomly, a large amount of time can be saved in the forward propagation and back propagation, so the time consumption caused by the increase of network layers can be well solved.

The composition structure of each layer block in the network is shown in Fig. 6.

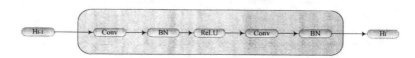

Fig. 6. Block composition structure diagram

In the text eigenvalue of the image, that is, the text model, each block does not consist of a single part but of several basic layers, each of which has the following smaller layers: the convolution layer, the Batch Normalization (BN) layer, and the ReLU activation function. Among them, the main role of BN is to conduct batch normalization of input data, which was mainly to prevent the gradient dispersion problem caused by the uneven release of input data, so that the weight of different levels can be synchronized when updating.

In the process of training, calculation formulas for p_ℓ are as follows:

$$p_\ell = \frac{\ell}{n_{h+v}}(1-c) \tag{4}$$

n_{h+v} represents the total number of network layers, h and v represent the hidden layer and the display layer, and C represents the constant value. The calculation formula is as follows:

$$h_\ell(x) = \begin{cases} \frac{\ell}{n_{h+v}}(1-C)F(x,\{W_\ell\}) + x & ,z=1 \\ [1-\frac{\ell}{n_{h+v}}(1-C)]F(x,\{W_\ell\}) + x & ,z=0 \end{cases} \tag{5}$$

Among which, x represents the input eigenvalue. F demotes the activation function, and the existing Sigmoid activation function is adopted in this paper.

4   Simulation Experiment

In order to verify the rationality and feasibility of the proposed algorithm, a simulation experiment is carried out on the deep neural network based image region eigenvalue processing technology using open data set. Open data sets include Chars 74K, Stanford AI Group and MJSynth data sets, all of which are open image data sets. Open CV is used as the basic tool for image processing, Trensor Flow platform is used as the open source tool for building deep neural network, and all experiments are mainly completed in NVDIA Tesla K80 GPU.

The paper judges whether the algorithm is reasonable and feasible from the accuracy of image processing. The calculation formula of accuracy P is as follows:

$$P = \frac{TP}{TP+FP} \times 100\% \tag{6}$$

TP represents the number of positive samples predicted as positive examples, and FP represents the number of negative samples predicted as positive examples. Assuming that the number of network layers is 50 m, and 2000 images are used in each data set, the calculated results show that the accuracy of images based on deep neural network is 94%, while the result of traditional image processing is 92.9%, and the time is 442 s and 411 s respectively. In other words, compared with the previous algorithm, the algorithm in this paper has improved the recognition accuracy by 1.1%. In terms of the time consumed by character image prediction, according to the above experimental results, the algorithm proposed in this paper consumes more time than the traditional image processing algorithm.

For the 110-layer network model, 2000 and 6000 images from the data set were selected for the experiment, and the experimental results were as follows:

In the image processing technology based on deep neural network, the accuracy of image processing is 96.8%, while the accuracy of traditional image processing technology is 94.9%, and the time is 638 s and 685 s respectively. In other words, the proposed algorithm is 1.9% more accurate than the traditional algorithm, and the time consumption is significantly less.

In conclusion, with the increase of the number of layers and image data, the algorithm has more advantages in processing results and time consuming.

5   Conclusion

In this paper, an image processing algorithm based on deep neural network model is designed for single region eigenvalues in the image. The main steps include image acquisition, image preprocessing and region processing to realize the extraction and recognition of eigenvalues. Through the experiment test, the algorithm in this paper has achieved good application effect.

References

1. He, K., Zhang, X., Ren, S.: Deep residual learning for image recognition. In: Proceedings of the IEEE Conference on Computer Vision and Pattern Recognition, pp. 770–778 (2016)
2. Simonyan, K., Zisserman, A.: Very deep convolutional networks for large-scale image recognition. arXiv preprint arXiv:1409.1556 (2014)
3. Huang, G., Sun, Y., Liu, Z.: Deep Networks with Stochastic Depth (2016)
4. Dongdong, X., Zhixiang, J.: End-to-end speech recognition based on deep optimized residual convolutional neural network. Appl. Res. Comput. 37(S2), 139–141 (2020)
5. Rong, X., Yi, C., Tian, Y.: Recognizing text-based traffic guide panels with cascaded localization network. In: European Conference on Computer Vision. Springer, Cham (2016)
6. Mammeri, A., Khiari, E.H., Boukerche, A.: Road-sign text recognition architecture for intelligent transportation systems. In: 2014 IEEE 80th Vehicular Technology Conference (VTC Fall). IEEE (2014)
7. Baogang, S.: Deep Learning Based Text Detection and Recognition in Natural Scenes (2018)
8. Zhenmao, L.: Research on Rectangular Traffic Sign Detection and Text Extraction Algorithm in Natural Scenes. Beijing Jiaotong University, Beijing (2017)
9. Gonzalez, A., Bergasa, L.M.: Text detection and recognition on traffic panels from street-level imagery using visual appearance. IEEE Trans. Intell. Transp. Syst. (2014)
10. Bergasa, L.M.: A Text Reading Algorithm for Natural Images. Butterworth-Heinemann (2013)

FDTD Method and Data Acquisition for Dispersion of Metal Lens

Min Li[1(✉)], Jingmei Zhao[2], Hairong Wang[2], and Fanglin An[3]

[1] College of Optoelectronic Engineering, Yunnan Open University, Kunming 650223, Yunnan, China
[2] College of Optoelectronic Engineering, Yunnan Open University, Kunming 650223, Yunnan, China
[3] School of Computer Science and Cyberspace Security, Hainan University, Haikou, China

Abstract. In order to overcome the difficulty that the traditional finite difference time domain (FDTD) method cannot calculate the dispersive materials, a dispersion FDTD algorithm based on Drude model is proposed, and the concrete difference formula is derived. The method is applied to simulate the focusing function of metal lens, and the results are in good agreement with the existing theory. The method in this paper is suitable for the analysis of optical waveguides constructed by various dispersive materials whose conductivity is related to frequency.

Keywords: Finite difference time domain (FDTD) method · Dispersion · Metal lens

1 Introduction

Finite difference time domain (FDTD) is a numerical method developed from the finite difference method, which directly uses Maxwell equation to calculate the interaction between electromagnetic field and matter. It was proposed by Ye in 1966. As a full wave analysis method, when trying to simulate the parameters related to frequency, the positive feedback in the time domain will lead to a non physical rapid increase in the computational space and divergence [1]. Therefore, the traditional fdtd can not simulate the material with negative refractive index, such as the metal material in the light field frequency less than the plasma frequency. Many studies have been devoted to this area. In 1990, luebbers et al. proposed to use the recursive convolution (RC) to deal with the frequency-domain dispersive materials based on Debye model, and then used this method to simulate the propagation of electromagnetic waves in plasma. In the same year, Kashiwa et al. Proposed to use ade (auxiliary differential equation) to deal with the dispersive materials based on Debye model and Lorentz model. In 1992, Sullivan proposed an FDTD method for dispersive materials by Z-transform. In recent years, FDTD has been widely used in the field of electromagnetic engineering, and these methods have been well developed. However, these methods are complicated to

© The Author(s), under exclusive license to Springer Nature Switzerland AG 2021
J. Abawajy et al. (Eds.): ATCI 2021, LNDECT 81, pp. 471–475, 2021.
https://doi.org/10.1007/978-3-030-79197-1_67

program. In this paper, we use a simple form to transform the frequency-domain dispersion relation of metal dielectric constant into time-domain by Fourier transform, and then establish the FDTD formula based on this relationship, and finally use it to simulate the focusing function of metal lens.

2 Dispersion FDTD Formula

There is a diffraction limit in the traditional optical theory. With the development of nano manufacturing technology, it becomes the main problem that limits the miniaturization and application of optoelectronic devices. Recently, one of the main methods to overcome the diffraction limit is to use surface plasmon polaritons (SPPs). SPPs are electromagnetic waves confined to the interface of two materials with opposite dielectric constant sign [2]. Usually one of the two materials is metal, the other is dielectric or air. SPPs can only propagate along the interface. In the direction perpendicular to the interface, the field intensity decreases exponentially, so the sub wavelength limitation can be realized.

In nano optoelectronic devices, Au, Ag and other precious metals are used as transport carriers in SPPs based devices. Because the dielectric constant of metal is very different from that of medium, special treatment should be done when FDTD is used for numerical simulation. The dielectric constant and conductivity used in FDTD simulation are both positive real numbers for the medium, while the dielectric constant of metal is a complex number; at the same time, the dispersion of metal is very serious, so the response of metal dielectric constant to frequency must be considered. Since the complex permittivity of metals usually appears in the form of definition in frequency domain, and FDTD is calculated iteratively in time domain, it must be transformed from frequency domain to time domain. The dielectric constant of metal was calculated by Drude model

$$\varepsilon = \varepsilon_0\left(1 - \frac{\omega_p^2}{\omega(\omega + j\gamma)}\right) \tag{1}$$

ω represents the frequency of incident wave, ω_p represents the frequency of plasma, γ represents the collision frequency, and their units are rad/s; ε_0 o is the dielectric constant of vacuum.

TM mode propagates in SPPs based nano photonic devices. The Maxwell equation satisfied by this model is as follows:

$$\varepsilon\frac{\partial E_x}{\partial t} = \frac{\partial H_z}{\partial y} \tag{2}$$

$$\varepsilon\frac{\partial E_y}{\partial t} = -\frac{\partial H_z}{\partial x} \tag{3}$$

$$\mu_0\frac{\partial H_z}{\partial t} = \frac{\partial E_x}{\partial t} - \frac{\partial H_z}{\partial x} \tag{4}$$

3   Numerical Simulation

In order to verify the feasibility of this method, a waveguide array lens with Ag as metal material is used as an example. When the incident wavelength is 632.8 nm, the relationship between the complex refractive index of Ag air Ag waveguide and the air gap is obtained. It can be seen from the figure that the complex equivalent refractive index decreases with the increase of air width, the imaginary part represents the loss, and the real part corresponds to the usual refractive index. According to the optical principle that the electromagnetic wave will deflect towards the direction of high refractive index, a 200 nm symmetrical waveguide array is designed. The width of the Ag strip is 20 nm, and the central air width of the array is 10 nm. The increment of the air width from the center to the two sides is 2 nm. In this way, a region with large central refractive index and gradually decreasing refractive index of both sides is formed. This structure will focus the incident light [3].

Phase distribution of magnetic field. It can be seen that the phase of light wave is modulated by the waveguide array, and the equal phase plane deflects and converges in the waveguide array. After passing through the focal point, the equal phase plane propagates spherical. According to the intensity distribution, the energy flow is also refracted, but the equal phase plane deflection belongs to normal refraction, while the energy flow deflection is a negative refraction phenomenon.

4   Gradient Index Lens Embedded in Metal Nanoslit Array

In this case, the electromagnetic field between adjacent metal slits does not interfere with each other, and each slit can be analyzed and treated as an independent structure. If the thickness of the metal wall is reduced, the electromagnetic fields between the adjacent slits will be coupled with each other. The electromagnetic field transmission mode of the slit array with mutual coupling is different from that of the single slit. Due to the special electromagnetic properties of metals, the coupling coefficient between metal nano slits is negative, which makes the metal nano slit arrays have negative refraction.

The GRIN lens has the property of self focusing. If the GRIN lens is embedded in the metal slit array, its electromagnetic field transmission and focusing properties will be greatly changed. By using this kind of metal slit array structure, the incident light wave can be focused at a very deep sub wavelength scale. In this chapter, we first study the negative refraction properties of metal nanoslit arrays according to the transmission mode equation. Then, we compare the dispersion curves and focusing properties before and after the GRIN lens is embedded in the metal nano slit array. We also study the relationship between the focal length and the structural parameters of the grin metal nanoslit lens by using Hamilton optics principle, And the conditions of light focusing inside and outside the structure.

The skin depth in the metal is about 20 nm. Figure 1 shows the relationship between skin depth and slit width, in which air and dielectric constant of 1.5 are filled in the seam. It can be seen from the figure that the skin depth of spp in the metal decreases after filling with high refractive index medium, which is similar to the result in Fig. 1 (a), but the difference is very small.

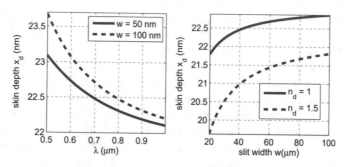

Fig. 1. Skin depth of surface plasmon polaritons in metal/air/metal slits at different wavelengths

Due to the negative refraction property of metal slit array, some traditional phenomena and effects are very different in metal slit array. For example, if the GRIN lens is embedded into a metal nano slit array, its focusing properties will be greatly changed. In this section, we will study the focusing conditions and focusing performance of GRIN lenses embedded in metal slit arrays. This kind of lens structure is called graded index metal nano slit lens [4].

According to the traditional beam propagation theory, the focusing position of GRIN lens is related to the incident wavelength and the gradient coefficient of refractive index change. The steeper the change is, the closer the focusing position is. For the metal slit array lens with gradual refractive index, the situation is much more complicated. The properties of metal materials and the width of dielectric layer and metal wall will affect the focusing position and focusing effect. We use the Hamiltonian method to analyze the focusing position of the GRIN lens.

In addition to the refraction of the light at the exit of the metal lens, the light focusing at the exit of the metal lens is also related to the diffraction effect of the finite aperture. If the aperture of the lens structure is relatively small, the focal position of the lens may shift towards the lens, which is caused by the diffraction of the edge of the finite aperture. Even if the lens does not converge, the diffraction effect of the edge of the restricted aperture may make the diffraction field near the lens have a maximum intensity.

In this chapter, we study the optical transmission characteristics of metal nano slit arrays which can be coupled with each other, especially the cohesive and external focusing properties of metal nanoslit arrays with gradually changing refractive index. According to the wave vector variation law and wave vector matching condition in the dispersion equation, the light refraction direction is analyzed; the light propagation laws of several structural examples are simulated by using the finite element analysis software COMSOL multiphysics; the propagation trajectories of light in the structure and at the exit end are analyzed by using the Hamiltonian method.

5   Conclusion

This paper presents a simple FDTD calculation method based on Drude model and PML absorbing boundary, which can greatly reduce the memory requirement due to less intermediate variables. The numerical results show that the expected function of focusing lens is achieved, and the divergence phenomenon in traditional FDTD is not found. The results are in good agreement with the existing theory. Therefore, this method is suitable for the analysis of optical waveguides constructed by dispersive metal materials in frequency domain.

Acknowledgements. This work is partially supported Hainan Provincial Natural Science Foundation of China (620RC563), the Science Project of Hainan University (KYQD(ZR)20021).

References

1. Inan, U.S., Marshall, R.A.: Numerical Electromagetics: the FDTD Method, pp. 72–112. Cambridge University Press, New York (2011)
2. Yee, K.: Numerical solution of initial boundary value problems involving Maxwell s equationtropic media. IEEE Trans. Ant. Prop. **14**(3), 302 (1966)
3. Jin, L., Baoxue, C., Haima, Y.: Study on symmetric surface plasmon resonance structure excited by planar waveguide. Photoelectron Laser **22**(12), 1821–1825 (2011)
4. Haijun, L.: Near field optical measurement of subwavelength structured nanoholes. Photoelectron Laser **21**(7), 963–965 (2010)

Discussion on Data Mining Behavior of Search Engine Under Robots Protocol

Yong Zhang[1][✉] and Yin Zhang[2]

[1] Liaoning JianZhu Vocational College,
No. 26 Qingnian Street, Liaoyang 111000, China
[2] School of Computer Science and Cyberspace Security,
Hainan University, Haikou, China

Abstract. With the rapid development of cloud computing, the phenomenon of temporary replication is becoming more and more common. The definition of its nature and legal regulation has become the focus of debate at home and abroad. Temporary reproduction belongs to the reproduction in the sense of copyright law, which should be included in the scope of copy right regulation and restricted by joint use system.

Keywords: Cloud computing · Temporary replication · Legal nature · Legislative improvement

1 Temporary Replication in Cloud Computing Environment

Temporary copy refers to the phenomenon that copies appear automatically in the computer memory during the process of reading, browsing, listening and using works by computer. Once the works running or using are shut down or the computer is turned off, such copies will no longer exist. "The global cloud industry has entered a period of rapid development", the network transmission, software operation and online browsing under the cloud computing environment will be temporarily copied. It has the following characteristics.

1.1 Temporary Replication is Technical

Temporary replication is caused by the working principle of computer and network transmission technology. It is the inevitable result of technical operation automatically generated in the process of computer operation, which is the technical collateral of calculation [1]. For example, when browsing a wide web page in the cloud computing environment, the random memory of the computer will automatically generate a copy for the CPU to calculate.

1.2 Temporary Replication is Temporary

In the process of computer program running, once the power supply is turned off or the computer runs new instructions, the copied information will automatically disappear or be replaced. It has certain dynamic replication characteristics, showing randomness,

J. Abawajy et al. (Eds.): ATCI 2021, LNDECT 81, pp. 476–480, 2021.
https://doi.org/10.1007/978-3-030-79197-1_68

temporary and temporary. For example, in the process of temporary replication, data in and out of the transmission device time is as short as microseconds, and once the data is sent out, it will be covered by the new data.

1.3 The Role of LVQ Network Learning Algorithm in Temporary Replication Under Cloud Computing

The steps of LVQ network learning algorithm are as follows: initialize the weight vector of each neuron in the competition layer, assign the small random number randomly, determine the initial learning rate and training times; input samples) find the winning neuron; adjust the weight of the winning neuron according to different rules according to the classification:

$$W_j^i(t+1) = W_j^i(t) + \eta(t)(X - W_j^i(t)) \tag{1}$$

Where t is the number of iterations, is the input sample, and $\eta(t)$ is the learning rate. When the network classification result is consistent with the teacher signal, the weight is adjusted to the input sample direction:

$$W_j^i(t+1) = W_j^i(t) - \eta(t)(X - W_j^i(t)) \tag{2}$$

2 International Definition of Temporary Reproduction

2.1 European Union

Article of the EU Copyright Directive stipulates that "reproduction by any means, directly or indirectly, temporarily or permanently, in whole or in part, shall be prohibited by Member States [2]. Include temporary replication in replication scope." Article 5 provides for exceptions that Member States must implement if temporary reproduction is temporary or incidental and forms an integral and necessary part of the technological process, and its sole purpose is: To make it possible for a work or other object to be transmitted in the network between third parties through an intermediary service provider; and (2) to enable the lawful use of the work or other object, Moreover, if it has no independent economic value, it does not constitute an infringement of the right of reproduction. "It can be seen that the EU has brought temporary reproduction into the scope of the regulation of reproduction rights, and at the same time, it has stipulated exceptional conditions to exclude the temporary replication that meets the conditions from the scope of the regulation of reproduction right.

2.2 U.S.A

The Digital Millennium Copyright Act of the United States does not directly include temporary copying into the scope of the right of reproduction. However, Article 117 of the copyright law is amended in the restricted part of the copyright of computer maintenance or repair, which stipulates that temporary copying of computer programs only for the purpose of maintaining or repairing computers constitutes infringement. In fact, this is to set the exception of replication right for temporary replication under specific circumstances, so it is considered to support the conclusion that temporary replication in RAM belongs to replication. "In the case of cartoon network LP LLP V CSC holdings, LNC, the court of appeal for the second circuit held that a copy constituting a copy within the meaning of the United States copyright law should meet the conditions that the work must be stored in a perceived physical medium and that the copied work must be preserved for a long time rather than for a short time [3]. The data of the plaintiff's program in the buffer is covered by the later data within 1.2 s, which does not meet the second element, does not meet the constitutive requirements of "attachment", and does not constitute illegal reproduction of the plaintiff's TV program. In other words, the United States has included temporary reproduction in the sense of copyright law and excluded some temporary reproduction from the regulation of reproduction right. However, it is difficult to define whether temporary reproduction belongs to reproduction in the sense of copyright law by taking the duration of the copy as the standard, because no definite objective standard can be found for the time limit.

3   Analysis of the Legal Nature of Temporary Replication in Cloud Computing Environment

3.1   Reproduction Right in Copyright Property Right

The right of reproduction is the foundation and core of copyright. Article 10 of China's copyright law interprets the right of reproduction as the right to make one or more copies of a work by means of printing, copying, rubbing, recording, videotaping, remaking, etc. "Computer software protection regulations" defines the right of reproduction as the right to make one or more copies of software. There are two essential elements in the traditional reproduction behavior: (1) the copy must exist in the tangible carrier; (2) it must be stable and lasting.

3.2   Temporary Reproduction Belongs to the Right of Reproduction

With the continuous development of Internet technology and the evolution of reproduction and communication technology, the main use mode of works has also undergone great changes. The new legal phenomenon brought about by this has also made new requirements for the law to keep pace with the times and its application. The provisions and interpretation of the right of reproduction should also keep pace with the times, that is, temporary reproduction should be included in the scope of reproduction right, At the same time, it is supplemented by necessary restrictions. Some scholars hold different views. They think that China's copyright industry at this stage is far less than that of developed countries, and there are a large number of copyright consumer groups. In the issue of temporary replication, the rights of users should be protected as much as possible, otherwise it is not conducive to the spread and development of culture. In the environment of cloud computing, the copyright industry in China is far less than that in developed countries, The cost and difficulty of private reproduction are greatly reduced [4]. The public can easily achieve low-cost and high-quality reproduction, and the interests of copyright owners are vulnerable to infringement.

The author believes that the temporary duplication will damage the protection of the rights and interests of copyright owners, and it is difficult to arouse the enthusiasm of creators. In the context of attaching importance to cultural creativity, if we do not give innovators an orderly communication market, it will hinder the development of cultural undertakings, which is unfavorable in the long run. Therefore, it is necessary to bring temporary reproduction into the scope of traditional reproduction right.

4   Legal Regulation of Temporary Replication in Cloud Computing Environment

As the Internet has developed into the main media for the circulation of works in the information society, and the frequency of temporary replication is very high in the cloud computing environment, therefore, China can not continue to adopt the evasive treatment for the problem of temporary replication, and should actively respond to it.

4.1   It is Clear that Temporary Reproduction Belongs to the Category of Reproduction Right

China's current copyright legal system generally takes a evasive attitude when dealing with the issues related to temporary reproduction, but the opposite positive response is the attitude that should be chosen. Therefore, the author suggests that in the third revision of the copyright law, temporary reproduction should be included in the category of traditional reproduction right, and a clear definition of temporary reproduction should be given, To make up for the loopholes in China's legislation and practice. In spite of the continuous disputes on this issue, however, in the context of the current wave of the international community, China can not be divorced from other countries on the qualitative issue of temporary replication.

4.2   It is Stipulated that Temporary Reproduction that Meets the Restrictions is Reasonable Use

In order to achieve the balance of interests between the copyright owner and the public, while affirming that temporary reproduction belongs to the category of traditional reproduction right, it is necessary to restrict the rights of the obligee, that is, the temporary reproduction in some cases should belong to reasonable use and not constitute infringement. It is suggested that in the third revision of China's copyright law, the provisions on fair use should be added, that is, "temporary reproduction with the following characteristics is reasonable use.". Firstly, it is temporary; secondly, it is an indispensable part of the whole technological process; thirdly, it only aims at the direct processing of software or digital works by computer; fourthly, it has no independent economic significance.

5 Conclusion

With the continuous development of cloud computing technology, the phenomenon of temporary replication is more and more common, which brings great convenience to the public and greatly speeds up the development of science, technology and culture. Although temporary reproduction is classified into the category of traditional reproduction right, which expands the scope of human rights of copyright, the conflict between the interests of copyright owners and public interests can be solved by limiting the principle of reasonable use.

Acknowledgements. This work is partially supported Hainan Provincial Natural Science Foundation of China (620RC563), the Science Project of Hainan University (KYQD(ZR)20021).

References

1. Ying, X.: Legal nature and legislative improvement of temporary replication in cloud computing environment. China Collective Econ. **34**, 91–92 (2015)
2. Haizhong, T.: Detection method of incomplete node data in wireless network under cloud computing environment. Netw. Secur. Technol. Appl. **11**, 96–97 (2020)
3. Xueli, W., Gang, C.: Research on information security risk assessment process under cloud computing environment. Netw. Secur. Technol. Appl. **11**, 93–94 (2020)
4. Weizhong, W., Xin, Z., Dajiang, W., Ke, C.: Network security analysis and solution research in big data and cloud computing environment. Inf. Secur. Commun. Secur. **11**, 102–110 (2020)

Data Collection and Cloud Computing for Local Manager System

Pu Huang[1]([✉]) and Kejie Zhao[2]

[1] Shaoguan University, Shaoguan 512005, China
[2] School of Computer Science and Cyberspace Security, Hainan University, Haikou, China

Abstract. Innovation is an important engine to lead the contemporary economic growth, and human capital, as the main source of innovation ability, plays an indispensable role in the region. In China, there are obvious differences in the distribution and reserve of human capital in different regions, so it has different degrees of impact on the innovation ability of each region. With the development of technology and the arrival of the era of artificial intelligence, more and more close ties have been established between machines and human beings. As intelligent machines with feedback functions such as human-computer dialogue, action and emotion are gradually socialized, the human-computer relationship has also developed and changed accordingly. The social role of the subordinate is both instrumental and developmental. The influence of intelligent machine on interactive objects (people) is not only reflected in the change of human-computer relationship, but also in the impact on human needs. The introduction of artificial intelligence leads to deeper interaction and greater uncertainty in the process of human-computer interaction.

Keywords: Artificial intelligence · Human-computer interaction · Local governance · Regional innovation

1 Introduction

At present, there is no empirical method to explore the relationship among local governance capacity, human capital and regional innovation. In this paper, combined with China's unique national conditions and the actual situation of current development, and absorbing the research experience of scholars at home and abroad, through the description, summary and analysis of the characteristics of local governance capacity, human capital and regional innovation in China, the relationship and mechanism of the three, as well as the endogenous problems among them, are comprehensively and deeply explored and analyzed by empirical methods. This is particularly important for China to continuously realize innovation oriented country and realize economic growth at this stage. It is also of vital significance in optimizing industrial structure, upgrading technology level and personnel training. At the same time, it also makes up for the lack of empirical research on the impact of local governance capacity on Regional Innovation in the existing literature [1]. Therefore, this research not only has academic value, but also has profound practical significance for China's stable and sustainable development.

Based on the era background of artificial intelligence, this research uses interaction design and user experience theory in the field of design in the early stage, with the help of a large number of case studies, and further deepens through quantitative and qualitative user research and research methods. On this basis, the theory and design strategy of human-computer social interaction design are proposed and verified by design practice.

2   Research on the Theory of Human-Computer Social Interaction Design

The biggest collective or shared characteristic of human being which is unique and different from the current computer is self-consciousness. Self consciousness, referred to as "consciousness", refers to the individual's cognition and adjustment of self. Through consciousness, people can reflect and adjust their psychological characteristics, inner activities, behaviors and motives, and the understanding of the relationship between individuals and the outside world. On the level of thinking, people have sociality by means of free consciousness, self-consciousness, self-knowledge and group evolution. Self consciousness is the psychological basis of human social existence. In the past development process, human beings continuously and consciously transform the real world through practice. Human society and human individuals need each other and are closely linked. People and people constitute a team and a collective, and form a collective force, with the advantages of social groups [2].

In order to have feelings and "emotions" in artificial intelligence, it is necessary to study the generation principle of emotions. First of all, it is important to understand the causes of emotion and reasoning. The internal logic program of emotion operation is not equal to the change data of various physiological indexes during emotional reaction. Emotional responses may be external stimuli captured by our senses, internal stimuli that alter homeostasis (body self-regulation) in the body, or our own cognition. Processing stimuli produces changes at the unconscious level of the body state. This is called emotion. If the emotion is strong enough, cognitive, social, contextual, and contextual assessments are made, known as experiencing emotions.

Different from the binary code used in the storage, the real code is used in the path optimization problem, that is, when there are n goods to be picked, the chromosome is composed of 1, 2, 3 … n. It is composed of random disorder sequence of. The fitness of the individual is as follows:

$$fitness = \frac{1}{\sum_{i=1}^{n-1} D_{k_i k_{i+i}} + D_{k_n k_1}} \tag{1}$$

Path optimization requires the shortest path, and according to Formula 1, the path is inversely proportional to the individual's fitness. Therefore, in the calculation process of genetic algorithm, individuals with large fitness value should be selected to retain to the next generation.

The new objective function f (x) is obtained by adding the objective functions of weight coefficients:

$$f(x) = \sum_{i=1}^{n} \omega_i f_i(x) \tag{2}$$

The idea of parallel selection method focuses more on the transformation of genetic algorithm solving process. The idea is to initialize the population according to the standard genetic algorithm, and then divide the whole population line according to the number of objective functions. The division method can be either uniform division or a certain proportion according to the importance of the objective function. After dividing the population, each objective function gets its own initial population, At this time, each objective function independently calculates the fitness of its own population, and then selects the subpopulations. When each branch of the objective function selects the subpopulations, all the subpopulations are mixed to form a combined new species group. The new species groups are fully mixed together for crossover and mutation, and excellent individuals are selected to enter the next generation [3].

Machine social roles need to have independent feedback on information, and their behaviors are not all controlled and predictable. In other words, independent operation and judgment systems are needed. Determine whether the machine may have real emotions. In addition to the fact that people and machines are all made up of matter, machines need to have social attributes similar to human beings when they become social roles. After a lot of research and analysis, the author assumes that the machine becomes a social role, which requires at least four dimensions, namely, "active feedback, certain adaptive ability, active participation ability, and the ability to learn and influence interactive objects as shown in Fig. 1.

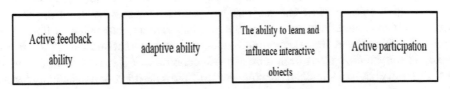

Fig. 1. Four dimensions/conditions of machine becoming social role

The social roles of the four dimensions are explained:

Active feedback: active feedback enables the other party to obtain effective information and achieve a social closed loop. On the contrary, complete passivity is manipulation. Active feedback is the first condition for a machine to become a social role.

Adaptability: the continuous changing interaction process between individuals and various environmental factors. It includes various adaptive changes in behavior, achieving harmony and natural communication.

Active participation: active participation means active social participation. The premise of social interaction is interaction, but artificial intelligence needs to actively participate in social interaction in order to build a certain degree of social relations. "Active participation" is also the biggest difference between active social robots and passive social robots.

3 Description of Variables and Data

Innovation. Because there are no direct indicators to measure technological innovation, the commonly used proxy variables to measure innovation in the existing literature are patent application, patent authorization, new product output value, etc. For example, Zhiyan and Bai Xuejie (2012) measured the innovation performance of enterprises, including the development speed of new products, the number of new products, the number of patent applications per year, and the output value ratio of new products. In accordance with the international standards, the patents in China are divided into invention patents, utility model patents and design patents.

Due to the rise of the theory of human capital, there is an upsurge of research. Becker, minsell and Denison also made outstanding contributions to the theory of human capital [4]. They expounded human capital from different angles. On the basis of deepening the concept of human capital, Becker and minsell studied the economic returns of human capital investment such as education and training by using micro model. On the basis of Schultz's research, Becker made up for the defect that Schultz only analyzed the macro effect of education on economic growth. He made a detailed micro analysis and studied the relationship between human capital and personal income distribution. Becker's research characteristic lies in paying attention to the phenomenon which has nothing to do with economics on the surface, so that it can be connected with economics and analyzed by economic mathematics method. In this paper, a new branch of human capital income distribution is proposed.

At present, the research focus has developed from a single human capital subject to the respective roles of different subjects in specific situations, such as the role of education background, industry experience and innovation qualification in innovation decision-making, behavior selection and enterprise and market success. In short, the theory of human capital has been a branch of economics. If it is applied to the micro enterprise level, it will stimulate the employees of the enterprise, and eventually form a more perfect corporate governance structure, so that enterprises can obtain more benefits and make more contributions to economic growth.

4 Conclusion

As an important power source of economic growth, innovation is the core part of contemporary economy. The relationship between human capital and regional innovation has been widely studied by economists at home and abroad. However, most of the current literature think that human capital can promote regional innovation, and few believe that under specific conditions, the relationship between human capital and regional innovation is uncertain. From the perspective of design, this paper explores the ever-changing and developing interaction between human and machine in the era of artificial intelligence, and studies how to better design human-computer interaction at different levels of cognition, behavior and emotion. Let the machine better serve people, meet the needs of people, and then create a harmonious human-computer environment. Based on the existing technology, this paper studies the design of human-computer social interaction and interaction in the era of artificial intelligence, discusses the deep-seated needs and ultimate goal of human-computer social interaction, and analyzes how the evolving humanoid intelligent robot can meet the dynamic and deep-seated emotional needs of human beings to meet the needs of human beings.

Acknowledgements. Transformation logic of urban grassroots governance in China—An analysis of the relationship between state and society.

This work is partially supported Hainan Provincial Natural Science Foundation of China (620RC563), the Science Project of Hainan University (KYQD(ZR)20021).

References

1. Lin Lin does seven jobs better than robots and artificial intelligence. Comput. Netw. **45**(07), 42–43 (2019)
2. Ying, Z.: Research on the ethical issues of intelligent product design. Design **03**, 49–50 (2018)
3. Kuan, Z., Lingyun, H.: Trade openness, human capital and independent innovation capability. Finance Trade Econ. **40**(12), 112–127 (2019)
4. Dejun, C., Shuming, Z.: High involvement work system and enterprise performance: the impact of human capital specificity and environmental dynamics. Manag. World (monthly) **3**, 86–94 (2006)

Tapping Potential of Water Drive Reservoir in Ultra High Water Cut Stage Based on Computer Image Recognition and Processing Technology

Hongshan Sun[1($\boxtimes$)] and Jun Ye[2]

[1] Daqing Oilfield Limited, Company No. 4 Oil Production Company,
Daqing 163511, China
sunhongshan@petrochina.com.cn
[2] School of Computer Science and Cyberspace Security, Hainan University,
Haikou, China

Abstract. After many years of waterflooding development, the water cut of many old oilfields in China has reached 90% and entered the ultra-high water cut stage. The influence of the vertical and horizontal heterogeneity of the reservoir is intensified. The original stratigraphic division and well pattern deployment are often unable to meet the requirements of EOR. Therefore, it is an important task for decision-makers of Youyang development to carry out the research on formation reorganization and well pattern optimization in ultra-high water cut stage. Computer image processing and recognition play an important role in many fields, so it is necessary to take reasonable methods to improve the image quality. Based on the research and analysis of the current developed technical framework, this paper puts forward the method of improving image processing and recognition, and puts forward the application method of this technology, so as to improve the operation quality of the whole image processing system.

Keywords: Ultra high water cut stage · Heterogeneity · Computer image processing · Identification technology

1 Introduction

In terms of oil and gas development strategy, most of the oil fields in China adopt water injection, and most of them have entered the stage of high or even extra high water cut. Theoretical research shows that for multi-layer sandstone reservoir, when water cut reaches 60% and enters into high water cut stage, 4% of recoverable reserves can be produced, less than half; when 80% water cut enters later stage of high water cut, 627% and less than 23% of recoverable reserves can be produced; therefore, when water cut reaches 90%, only 7% of recoverable reserves can be produced, and more than / 5 of remaining oil can be exploited. It can be seen that even in the ultra-high water cut stage, China's remaining oil reserves are still considerable, and the ultra-high water cut stage is still an important development stage [1]. At the same time, due to the complex geological conditions and frequent adjustment in the long-term development process of each oilfield, there are many problems in the ultra-high water cut stage, such as the change of reservoir properties, the high dispersion of remaining oil on the plane, the low degree of perfection of injection production well pattern, the poor degree of reserve control, the serious disturbance between the vertical upper layers and the large difference between layers, The original formation combination mode and injection production well pattern can not meet the requirements of EOR in ultra-high water cut stage, so it is of great significance to carry out the optimization of formation combination and well pattern adjustment in ultra-high water cut stage.

2 Comprehensive Framework of Computer Image Processing and Recognition Technology

In the computer image processing and recognition technology, the function is to use the database to complete the target recognition based on the high quality image processing. The most basic work is the high-quality image processing. On this basis, the database built in the system can be applied to complete the recognition of the content in the image.

2.1 Computer Image Processing Technology

In computer image processing, camera and other equipment can be used to obtain images, but noise will be generated in the process of signal transmission, so the most basic work in image processing technology is to eliminate such quality factors [2].

In computer operation, there are two working processes in image processing, one is noise removal, the other is super pixel generation. For the denoising work, the idea adopted in this paper is to treat the whole image as a large matrix, and the final image can be obtained by removing the noise in the matrix. The image matrix containing noise is represented as follows:

$$M = C + m \tag{1}$$

Where m is the noise matrix and C is the image matrix. After replacing the image with the image block stack, the equation obtained is as follows:

$$(M_1 M_2 M_3 \cdots M_n) = (N_1 N_2 N_3 \cdots N_n) + (m_1 m_2 m_3 \cdots m_n) \tag{2}$$

This method is to decompose the image into image blocks. Through the time continuity in the video, the existence of noise in the image is analyzed. After finding the noise matrix, the denoised image can be obtained by removing the matrix.

There are two methods for the generation of superpixels, namely QEM algorithm and boundary preserving algorithm. In the boundary preserving algorithm, the equations obtained are as follows:

$$D(x, y) = aB(x, y) + bI(x, y) + chC(x, y) \tag{3}$$

B, I and C functions respectively represent the description functions of image and boundary consistency, intensity uniformity and compactness, a, b and c represent the weight of these three parameters, and h is the compactness parameter. In the algorithm design, this parameter should be set according to the specific image processing requirements.

2.2 Computer Image Recognition Technology

In the application of computer image recognition technology, it is necessary to build corresponding processing module to realize the effective recognition of image content [3]. The sub-systems included in the module are feature point extraction program, database, parameter comparison program, neural network learning program, etc. For rockets, missiles and other spacecraft, the target information and other parameters will be input into the automatic control system. Through the analysis of these information, the system completes the extraction and comparison of various feature points in the image, so as to achieve efficient target recognition.

3 Optimization of Well Pattern Infill Adjustment in Ultra-High Water Cut Stage

3.1 Study on the Policy Boundary of Strata Reorganization in Ultra-High Water Cut Stage

Ideally, the injection rate of each layer should be proportional to the value of the flow coefficient kh/μ. According to the current oil production rate of Shengtuo oil h and Darcy's law, the injection rate of the first layer (permeability $50 \times 10^{-3} \, \mu m^2$) of the model is determined as $Qi = 3 \, m^3/d$, and the injection rate of the second layer under different permeability levels is determined in turn (as shown in Fig. 1).

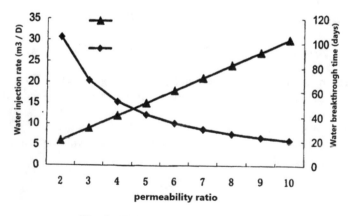

Fig. 1. Effect of permeability gradient

3.2   Pre Water Index

The results show that the interlayer disturbance caused by permeability difference is obvious: according to the flow coefficient kh/μ injection quantity, the water breakthrough time of the second layer becomes shorter with the increase of permeability gradient, and the greater the difference is, the more obvious the difference in water breakthrough time between the two layers is; when the permeability difference is greater than 4, the shortening trend of water breakthrough time is obviously slowed down, and the difference between the two layers is also smaller; the greater the permeability difference is, the shorter the water breakthrough time is, The higher the yield of high permeability layer is. The main reason is that when the oil-water viscosity is fixed, the permeability becomes the main factor affecting the fluid flow. The greater the permeability difference is, the smaller the influence of the high permeability layer is, the faster the leading edge advancing speed is, and the higher the production is, thus the interlayer interference is intensified.

3.3   Index after Water Breakthrough

The theoretical calculation of reservoir engineering shows that: in the case of multi-layer combined production, the thickness of the second layer with high permeability accounts for 50% of the production thickness, but the production accounts for 59% and 82% of the total production; due to the disturbance of the high permeability layer, the production degree of the low permeability layer is small, which restricts the improvement of oil recovery. This effect becomes more and more obvious with the increase of permeability difference, which is manifested in the following aspects: when the permeability difference increases from 2 to 10, the recovery rate decreases from 3396% to 1709%, which indicates that the recovery rate decreases by about 1% only under the interference of permeability.

In the ultra-high water cut stage, interlayer interference is still the main contradiction and factor affecting oil recovery, and it presents a more complex form in the new development stage. Therefore, according to the actual reservoir characteristics of Shengtuo oilfield, a model is established, and the influence of heterogeneous property permeability difference, crude oil viscosity, small layer thickness and flow coefficient on recovery degree is analyzed by using fine numerical simulation method, so as to determine the policy boundary of layer series reorganization, so as to guide the research of strata series reorganization in ultra-high water cut stage [4].

For multi-layer commingled production, waterflooding development, long production interval and different fluid properties, it is not enough to use traditional permeability parameters to characterize interlayer heterogeneity. Thickness and crude oil viscosity will have a significant impact on the effect of combined layer development. Considering the above parameters, the formation flow coefficient, which represents the underground flow of crude oil, is quantitatively analyzed. The former two-layer model is still used in the model, and the distribution of flow coefficient is arranged in different reservoir range, so that the flow coefficient difference can be studied with different values.

4   Conclusion

In this paper, the reservoir engineering method calculation, numerical simulation research and field data analysis are carried out for the main factor affecting the development of multi-layer sandstone reservoir in ultra-high water cut stage, and the field data analysis is carried out. On this basis, the grey correlation analysis and fuzzy cluster analysis are introduced into the study of formation reorganization, In the computer image processing and recognition technology, there are two work contents in the process of image processing, one is to remove the noise matrix in the image, the other is to generate super pixels to improve the accuracy of the image. In image recognition, the application method is to extract the image feature points and feature areas, and then identify the information in the image by parameter comparison. In the application of this technology, we need to decompose the acquired image to generate image blocks, complete the horizontal comparison of each image block on the practice axis, and reduce the workload of the hardware system.

Acknowledgements. This work is partially supported Hainan Provincial Natural Science Foundation of China(620RC563), the Science Project of Hainan University (KYQD(ZR)20021).

References

1. Jishun, Q., Aifen, L.: Reservoir physics, pp. 66–80. China University of Petroleum Press, Shandong (2006)
2. Bo, L., Xianbo, L., Ying, L., et al.: A new method for determining interlayer heterogeneity. Offshore Oil Gas China **19**(2), 93–95 (2007)
3. Yongxia, Z.: Research on Denoising and Super Pixel Generation Algorithm in Image Processing. Shandong University, Jinan (2017)
4. Yanping, F.: Application of computer graphics and image processing technology in copper ore identification. World Nonferrous Met. **23**, 27–28 (2016)

Multimodal Data Fusion Technology in Indoor Environment Art Design

Ran Zhang[1($\boxtimes$)] and Zhen Guo[2]

[1] Shandong Polytechnic College, Jining 272100, Shandong, China
[2] School of Computer Science and Cyberspace Security, Hainan University, Haikou, China

Abstract. The development of interior environment art design has always been the focus of attention. This paper, based on the application of soft decoration materials, expounds the decoration innovation in the interior environment art design, hoping that the research in this paper can provide guidance and reference for the development of interior environmental art design, and also play a role in attracting jade for the research in related fields, Different scores are assigned according to the matching degree, and then it is modeled as an optimization problem with the goal of maximizing the total score of task assignment and constrained by the time and space of workers and tasks. In each time slice, the maximum score problem is transformed into a bipartite graph matching problem.

Keywords: Interior · Environmental art design · Decoration innovation · Task allocation · Prediction

1 Introduction

At present, the popular style of "light decoration, heavy decoration" in domestic interior environment art design makes the advantages of soft decoration materials fully play. There are many kinds of soft decoration materials and flexible forms, which can effectively achieve the designer's purpose and meet the requirements of residents in different styles of indoor environment [1].

Green ecology and sustainable development are the organic whole of mutual cause and effect. Their basic starting point is the "people-oriented" principle. The difference is that green ecological design focuses on contemporary people, while sustainable development focuses on future generations. In 1987, the United Nations put forward the definition of "sustainable development" in the report "our common future", which "satisfies the needs of contemporary people and does not endanger the ability of future generations to meet their needs". The core of this definition is green ecological design. Only in this way can we meet the requirements of the definition, promote each other and form a virtuous circle.

J. Abawajy et al. (Eds.): ATCI 2021, LNDECT 81, pp. 491–495, 2021.
https://doi.org/10.1007/978-3-030-79197-1_71

2 Research on Task Allocation Algorithm

2.1 Algorithm Framework

The difficulty of MSA in spatial crowdsourcing lies in its dynamics. Any arrival order and information about workers and tasks are unknown. In this case, batch processing mode is usually used to solve the problem. The framework to solve this problem is proposed below [2].

In batch mode, the whole time period is divided into several time slices. In each time slice, the task allocation problem is a static task allocation problem. The best way to solve this problem is to transform the maximum score task assignment problem into the maximum weighted bisection matching problem of finding bipartite graph.

Generally, the solution to the problem of minimum cost and maximum weight of bipartite graph is integer programming. The solution process is as follows:

firstly, an indicator variable is defined:

$$x_{j,k} = \begin{cases} 1 & e_{j,k} \in WT_{s_i} \\ 0 & otherwise \end{cases} \tag{1}$$

The sum of the maximum weights obtained by the algorithm is added to the integer programming as a constraint to minimize the cost based on the maximum weight. The objectives and constraints are as follows:

$$\min \sum_{i,k} x_{j,k} c_{j,k} \tag{2}$$

2.2 Shortest Distance First Algorithm

In crowdsourcing, there are differences in the distance between workers and the starting points of different tasks, which will lead to different costs (the cost of workers moving to the starting point of tasks will not be included in the compensation). Although the task workers within the working radius will not be treated differently, it is the workers who prefer to choose jobs closer to each other, which will save unnecessary costs as much as possible; For the task requester, the closer distance means that the task can be executed by the workers as soon as possible; therefore, in the case of ensuring the maximum sum of scores, it is of positive practical significance for workers, tasks and platforms to optimize the sum of distances between different task assignments as far as possible. Therefore, the minimum distance first algorithm (CDP) is proposed.

The location entropy represents the frequency of workers visiting the area. If many workers visit the location in the same proportion, the entropy of the location is higher. On the contrary, if only a few workers visit, the entropy of a location will be lower. Entropy is used to represent the position entropy, and the calculation method is as follows:

$$Entropy = - \sum_{w \in W_t} P_l(w) \times \log P_l(w) \tag{3}$$

Where $P_l(w) = \frac{|O_{w_l}|}{|O_l|}$, Represents the number of times worker w visited location.

3 National Design in Interior Environment Art Design

3.1 The Connotation and Requirements of National Design

The nationalized style in interior design should be based on traditional culture, take its shape, extend its "meaning" and transmit its "spirit", and gradually integrate the formal beauty, implied beauty, wisdom beauty and spiritual beauty of traditional culture and art into modern interior design. This requires designers to have a comprehensive and systematic grasp of the elements of nationalization design. On this basis, the nationalized elements are analyzed and simplified, and they are applied in the interior environment art design, not only copying the original elements, but also copying and imitating the foreign culture.

3.2 National Information Contained in Soft Decoration Materials

The national information contained in soft decoration materials can be reflected from the materials and patterns of soft decoration materials. In terms of materials, the most representative is silk, which has a strong Jiangnan flavor. If a large number of soft decorative materials are used indoors, it will form a strong graceful style.

In terms of soft materials, the decorative patterns can be reflected in soft materials. Taking the dragon and phoenix pattern as an example, it contains clear information of the Chinese nation, has strong Chinese characteristics, and also has the information of good luck. In addition to ethnic information in a broad sense, in China, various ethnic totems with distinct images on soft decoration materials also have strong ethnic flavor, such as tiger totem of Yi nationality, cattle totem of Naxi nationality, horse totem of contentment, etc. these patterns of soft decoration materials can play a role in conveying the identity of the host nation in the room. When guests see these patterns, It can avoid unnecessary troubles [3]. For example, there are large mosque patterns in the room, which indicates that the host is likely to be a Muslim or Muslim. When speaking, it should be forbidden to say words related to ethnic taboos.

4 Humanized Design in Interior Environment Art Design

With the rapid development of economy and the rapid progress of science and technology, people began to pursue leisure, spacious and comfortable indoor space, and began to care about their own physiological, psychological and aesthetic needs. The slogan of "people-oriented" quickly resounded through the whole interior design field. As far as interior environment art design is concerned, only by making full use of soft decoration materials can we truly achieve "people-oriented". Because of its soft texture, good touch, excellent indoor environment and artistic atmosphere shaping ability, soft decoration materials can truly realize "people-oriented".

4.1   Interior Safety Design of Soft Decoration Materials

For soft decoration materials, interior safety design refers to fire protection design. However, for most of the soft decoration materials, fire prevention is impossible, so the focus of the safety design of soft decoration materials falls outside the material itself, specifically to strengthen the fire prevention awareness and fire safety measures of users [4].

Another potential safety hazard that cannot be ignored is static electricity. Most of the soft decoration materials are easy to generate static electricity, which is also the most easily overlooked. Static electricity is basically harmless to adults, but it is a potential hazard for children, which can paralyze the skin in light and damage the nervous system in severe cases [5]. In addition to the direct harm to human body, electrostatic soft decoration materials will absorb a large number of dust and bacteria, forming a potential hazard, and reduce the dirt resistance of materials, increase the difficulty of cleaning. Therefore, designers should pay special attention to the antistatic function of materials when choosing soft decoration materials, especially sofa cushion, cushion, carpet, bedspread, bed sheet, mosquito net and so on.

4.2   Interior Comfort Design of Soft Decoration Materials

In the interior environment art design, we should pay attention to the applicability and comfort. This requires designers to have a good grasp of the soft decoration materials, should choose what material, how much, what pattern, where to use and so on, must be aware of [6, 7]. Good interior space should not only meet the functional requirements, but also be consistent with people's vision and heart ideal. People's perception of indoor space comes from the scale and scale, closed and open, rich and monotonous, cordial and indifferent, artificial and natural, order and chaos, dynamic and static, etc. The coordination of these factors will make people have a good perception of indoor space form, that is, comfort, otherwise it will make people feel uncomfortable or ugly [8].

5   Conclusion

Soft decoration material is not only the "entity" of interior decoration, but also the "soul" of interior decoration. In a narrow sense, the final effect of interior environment art design depends on the selection and application of soft decoration materials to a large extent. In a broad sense, whether the domestic interior environment art design industry can innovate, how much innovation is there, and in which aspects innovation, are closely related to the key factor of soft decoration materials.

Acknowledgements. This work is partially supported Hainan Provincial Natural Science Foundation of China (620RC563), the Science Project of Hainan University (KYQD(ZR)20021).

References

1. Tang, R.: Research on teaching mode of innovative thinking in interior environment art design. Times education (education and teaching edition) (2008)
2. Zhang, Z., Fu, J., Xie, X., Zhou, Y.: Research on crowdsourcing quality control strategy and evaluation algorithm. Acta Sinica Sinica (2013)
3. Wang, L.: Interior environment art design concept analysis. Intelligence (2008)
4. Xin, Y.: Discussion on talent training of architectural interior environment art design. Higher Architecture Education (2008)
5. Wei, J.: Some thoughts on practical teaching of marketing specialty. Rural Econ. Technol. **31** (10), 346–347 (2020)
6. Tan, Y., Han, F., Zhao, Q., Lu, M., Yang, B.: Research Report on practical training of animal husbandry and veterinary specialty in Higher Vocational Colleges of Guizhou Province. China's Livestock Poultry Seed Ind. **16**(05), 3–8 (2020)
7. Wei, Y.: Development and application of practical training package for Construction Engineering Technology Specialty in Higher Vocational Colleges. Vocat. Technol. **19**(05), 60–64 (2020)
8. Xiang, X., Zhou, Y.: Analysis and thinking of practical teaching system of animal husbandry and veterinary specialty in Higher Vocational Colleges. Animal Husbandry Veterinary Today **36**(04), 102 (2020)

Data Mining Based on RNN in Development Technology of Difficult to Recover Reserves

Shilei Zhang[✉]

Daqing Oilfield Limited Company No. 7 Oil Production Company,
Daqing 163000, China

Abstract. With the development trend of large-scale and complex construction projects, safety accidents occur frequently, leading to the original management mode can not solve the more complex and severe problems encountered in the modern construction project management. A new management mode is urgently needed to deal with the contemporary construction projects. In this regard, this paper introduces the concept of construction reliability, which is applied to the construction project management, so as to better achieve the management of construction projects. The construction project optimization model based on four objectives is constructed with the construction system reliability as the constraint condition. The qualitative indexes of quality objective and safety objective are quantified, and the model is solved by multi colony ant colony algorithm.

Keywords: Difficult to recover reserves · Development technology · Countermeasures

1 Introduction

Difficult to recover reserves these resources are low grade and difficult to produce, but they are realistic and reliable replacement resources. Petroleum resource is a kind of pit concept, but it is not static and unchangeable. Our understanding is still very limited, and there is still a lot of room for technological improvement. By emancipating our minds, learning from the experience of low-grade reservoir production at home and abroad, we constantly improve the ability of technology and management innovation, and increase the production of this part of reserves [1]. The three key indicators, namely, reducing block investment, increasing single well production and enhancing the degree of production, are used to liberate these inefficient and difficult to recover reserves by increasing production by technology, improving management efficiency and reducing cost in a comprehensive way. Through further clarifying the ideas and modes of difficult to recover reserves development, strengthening the recognition and re description of sand bodies and faults, and the personalized design of outburst schemes, the development effect is further enhanced, the underground situation is continuously improved, the decreasing amplitude is slowed down, and the development benefit is improved. In the current situation of low oil price, it is necessary to solve the problems of optimization evaluation of remaining undeveloped reserves, technical means of increasing reserves and production, system and mechanism, and rate of return on investment capital.

J. Abawajy et al. (Eds.): ATCI 2021, LNDECT 81, pp. 496–500, 2021.
https://doi.org/10.1007/978-3-030-79197-1_72

2   Connotation of Development of Hard to Recover Reserves

Theoretical innovation may bring broad development space; a technological break-through may bring earth shaking changes. This is "limited resources, unlimited inno-vation". Scientific and technological progress will bring about profound changes and infinite possibilities. We should adhere to the market-oriented and open organizational model. Based on the geological characteristics of the hard to recover reserves, the con-struction of a new management mode for low-grade and difficult to recover reserves is based on the geological characteristics of the hard to recover reserves, through increasing the evaluation and optimization efforts, deepening the geological understanding, increasing the application of new technologies and processes, constantly exploring new mechanisms, new systems and new models for the production of hard to recover reserves, and exploring a new way of economic and effective utilization of hard to recover reserves, so as to realize the steady and steady increase of crude oil production, To select the machine to lay a solid foundation for production [2]. Combined with the geological characteristics of low-grade and difficult to recover reserves, as well as the progress of reserve production technology and the innovation of mechanism and system, the con-struction of effective utilization management mode of low-grade and difficult to recover reserves resources is realized by applying modern management methods and integrating multiple disciplines of reservoir, oil production, surface and management.

3   Calculation Method of Geological Controlled Reserves

The volume method is used to calculate the geological reserves. Calculation of geo-logical reserves of oil reservoirs: when the geological reserves of crude oil are expressed in volume units, formula (1) is used; when expressed in mass units, formula (2) is used. When the geological reserves of dissolved gas are more than $0.1 \times 10^\circ$ m^3 and can be used, formula (3) is used for calculation.

$$N = 100A_0 h \Phi S_{oi}/B_{oi} \tag{1}$$

$$S_{or} = 100 \Phi S_{oi}/B_{oi} \tag{2}$$

$$G_s = 10^{-4} N R_{si} \tag{3}$$

When there is gas cap in the reservoir, the gas cap gas reserves are calculated according to the geological reserves of gas reservoir or condensate gas reservoir. The construction and application of oil and gas control and prediction reserve management system really solves the problem of CNPC reserves management, and information technology plays an important role in the field of reserves management. Through the establishment of oil and gas control and prediction reserve management system, the standardization of reserve data management is realized. According to the characteristics of application software platform and the organization form of multi-disciplinary asset group, reasonable design and implementation are carried out for the installation mode, use mode and related data storage mode of application software.

4   Technical Countermeasures for Development of Difficult to Recover Reserves

4.1   Strengthen Reservoir Evaluation and Research, Plan Production Sequence of Low-Grade and Difficult to Recover Reserves

The difficult to recover reserves have the characteristics of poor grade, poor development effect, high cost and high risk. In order to improve the utilization degree of low-grade hard to recover reserves, the key is to do a good job in the evaluation and research of low-grade and difficult to recover reserves, strengthen the geological research and deepen the geological understanding. Only when the reservoir is more clearly understood can the development and utilization be more targeted. Three dimensional seismic processing and interpretation research was carried out, and 3D seismic and geological data were used for fine structural description, and structural morphology and reservoir characteristics were further confirmed, especially for small amplitude structures and faults with small fault distance and short extension, which laid a foundation for fine reservoir evaluation and research. In the process of 3D seismic acquisition, processing and interpretation, 3D visualization and coherence cube interpretation technology are applied to re implement the structure and oil and gas distribution. In view of the geological characteristics of low-grade and difficult to recover reserves, high-resolution sequence stratigraphy technology and well seismic combined sedimentary microfacies technology are applied to deepen the understanding of reservoir [3]. One is to establish isochronous stratigraphic framework for stratigraphic division and correlation.

4.2   Study on Reservoir Sedimentary Characteristics by Well Seismic Combination

Through core observation and analysis of coring wells, the types and characteristics of sedimentary microfacies association are identified and determined. Logging facies is calibrated by lithofacies, and logging facies model is established. At the same time, the distribution direction, width thickness ratio and length width ratio reservoir heterogeneity of sedimentary sand body are established by fine anatomy of dense well pattern sand body. Plane facies combination is carried out combined with regional sedimentary characteristics to determine the plane distribution shape of sand body. According to the geological conditions of low-grade and difficult to recover reserves, six parameters are selected as comprehensive evaluation indexes, including physical property, number of favorable single-layer areas, effective thickness, oil test production, fluid property and burial depth. Meanwhile, the weight coefficient and evaluation standard are determined by referring to the contribution of each index to the oil-bearing property of the trap. According to different types of difficult to recover reserves, the principle of "overall research, batch production, first fertilizer and then lean, first easy then difficult" is adopted. Reserves to be written off will not be used in a short period of time.

4.3 Using New Technology to Reduce the Production Limit of Low Grade and Difficult to Recover Reserves

The core problem of development and production of low-grade and difficult to recover reserves is to increase the production of single well. Only by increasing the production of single well can we reduce the number of drilling and investment. A small step in engineering technology and a big step in oilfield development. Therefore, in the production process of low-grade refractory reserves, attention should be paid to the application of new technologies at home and abroad, and the effective development of low-grade refractory reserves should be realized by relying on technological progress.

The first is the combination of fracture pattern and optimization of horizontal well pattern. Horizontal well deployment not only considers the matching of well pattern and fracturing fracture, but also considers the injection production mode. The other is horizontal well geosteering technology of real-time curve contrast landing and visual curve continuous comparison to predict well trajectory. Before drilling, the field plan is prepared for the problems that may occur in the drilling process, and the real-time comparison model is established to provide guarantee for smooth geological guidance. During drilling, WD and mud logging data are comprehensively used to judge the reservoir development status, and a visual contrast model is established to formulate countermeasures. The third is the completion technology based on staged perforation and fracturing. When optimizing the perforated well section, the perforating point should be selected according to the physical property and oil bearing property of the reservoir to ensure the conductivity near the wellbore, and the perforation interval and length should be determined according to the direction of in-situ stress to ensure the fracturing effect. In order to solve the problems of short length and small interval of oil layers encountered in the deviation section, the combination of flow limiting and staged fracturing is adopted to improve the degree of reserve production. Fracturing technology can improve the conductivity of special permeable reservoir and increase the productivity of single well [4]. Compared with conventional fracturing technology, large-scale fracturing technology has larger fracturing scale, more proppant types and larger sand addition, so it has better effect on economic and effective production of difficult to recover reserves.

5 Conclusion

At present, the development of hard to recover reserves is facing many major problems that can not be overcome for a long time. There are problems of technology itself, cognition limitation, organization mode and system mechanism. This requires us to break the traditional mindset, innovate scientific research methods, and explore the establishment of a market-oriented and open organizational model.

References

1. Chang, Y.: Analysis of technical countermeasures for development of hard to recover reserves. Chem. Manag. (08), 65 (2018)
2. Xu, J., Liu, Y., Jin, F., Meng, X.: Technical countermeasures and practice of difficult to recover reserves development in Kongnan area of Huanghua sag. Sinopec (07), 19–20 (2017)
3. Shi, H., Fang, K., Xia, B., Qiu, Z., Ding, N.: Technical countermeasures for development of refractory reserves in Block D of Liaohe Oilfield. China Mining Ind. **24**(S1), 352–355 (2015)
4. Wang, L., Ma, L.: Technical countermeasures for development of refractory reserves in Jianghan Oilfield. J. Pet. Nat. Gas (02), 520 (2008)

Legal Issues of SaaS Model in Cloud Computing

Yong Zhang[✉]

Liaoning JianZhu Vocational College, Lioanning China No. 26 Qingnian Street,
Liaoyang 111000, China

Abstract. This paper briefly introduces the changes of software business model
and the general situation of SAS software business model under cloud computing, and discusses the compatibility of the new software use mode with the
current copyright rights in China in terms of authorization mechanism, and the
impact and coordination of the license mechanism on the rights of users of
works, It also analyzes the nature, legal application and format terms of cloud
user agreement involved in online software services, and puts forward suggestions for the above problems. From the perspective of the problems brought by
the business model of SAS software to the current legal system, the author
makes a simple study on the above problems by combining with the introduction
of domestic and foreign academic research status, the evaluation of the current
system, the relevant case study, and the text analysis of agreement terms.

Keyword: Software as a service · Copyright · Software license · User
agreement

1 Classification of Cloud Computing

1.1 Iaas Infrastructure as a Service

Iaas infrastructure as a service is also known as hardware and service, which provides
users with infrastructure such as virtual server, storage space, network equipment,
firewall and so on. Users can choose any operating system and software to deploy. Iaas
enables users to rent virtual computing resources according to their actual needs,
without the cost of purchasing all kinds of computer hardware equipment locally in
advance, as well as the expenses of equipment depreciation, maintenance, updating and
upgrading. At present, Amazon AWS (Amazon Web Service), Google GCE (Google
compute engine), Alibaba cloud and Shengda cloud in China are the main represen-
tatives in this field. For individual users, the most commonly used Iaas service is the so-
called cloud storage or cloud disk.

1.2 SaaS Software as a Service

A popular understanding of SaaS software as a service is online software service,
which transforms tangible sales into intangible services. As a software business supply
and consumption mode, it is far earlier than the concept of cloud computing

J. Abawajy et al. (Eds.): ATCI 2021, LNDECT 81, pp. 501–505, 2021.
https://doi.org/10.1007/978-3-030-79197-1_73

technology. The initial application of SAS can be traced back to the birth of a series of free e-mail services such as Google Gmail, Sina and NetEase. Its business philosophy is to provide users with "software on demand" software services based on the Internet. However, the powerful basic support provided by cloud computing technology makes the service mode of SAS get the promotion in essence [1]. It can run large-scale software, provide online instant service to multi-user and allocate resources on demand. The commonly used SAS services include enterprise oriented online customer relationship management system, enterprise resource planning system, supply chain management system, human resource management system, etc.; the services for individual users include online translation, web game, web map, online personal schedule management, online document editing and other rich application services.

2   Legal Relationship Involved in Business Model of SaaS Software

The business model of SAS software may involve three specific architectures as follows:

(1) SAS software service providers provide users with their own software services directly.
(2) The SAS software service provider is authorized by the copyright owner to provide third-party software services to users.
(3) The SAS software service provider builds the platform, and the third party directly authorizes users to use the software.

The author briefly analyzes the legal relationship and the main issues under the above three frameworks as follows: (1) the same as the network distribution mode, SaaS software business model certainly does not involve the transfer of carrier ownership. (2) However, when a software developer exercises its own copyright, whether it is a software provider or a third-party licensed software, whether it is a third-party software provider or a third-party licensed software, it should have the same relationship with the software provider, After the "copy and distribution" mode of boxed software and the information network transmission mode of network download software, SaaS software business model has further broken through the current copyright bundle. As for this temporary use mode without transferring tangible copies of works, how to find its position in the copyright system needs to be further discussed [2]. (3) The relationship of software license contract has changed. Because the software distribution mode before the cloud era must be accompanied by the user's copy behavior, the license agreement has the function of authorized copy in the sense of copyright law; but in the SaaS software delivery mode, the user's use behavior is not accompanied by relatively stable replication, which leads to the problem of defining the nature of user's use behavior.

In order to meet the needs of users for SaaS application services, SaaS providers need to rent enough virtual machines from SaaS providers to run their own SAS applications According to the change of the mode legal request load and its own judgment, the SaaS provider can adjust the virtual machine leasing policy at any time.

Assuming $T_{i+1} + T_i = T_{period}$. SaaS provider's leasing policy is expressed by formula 1, which determines the pricing method, type and quantity of new leased virtual machines.

$$\pi_{vm} = \left\{ (X_{Ti}^{10}, X_{Ti}^{1R}) \ldots, (X_{Ti}^{j0}, X_{Ti}^{jR}), \ldots (X_{Ti}^{N0} X_{Ti}^{NR}) \right\} \tag{1}$$

Where X_{Ti}^{10} is the number of on-demand virtual machine instances of type $[T_i, T_{i+1}]$ period, and X_{Ti}^{jR} is the number of reserved virtual machine instances with type J to be leased in the $[T_i, T_i + T_{reservation}]$ period. The load handling capacity of SAS application at t $\in$ [T, t + 1] can be obtained by the following formula:

$$totalC_t = \sum_{j=1}^{N} (V_t^{j0} + V_t^{jR}) * C_j \tag{2}$$

3 Research on Legal Issues of User Agreement in SaaS Software Business Model

3.1 Analysis of the Nature of SaaS User Protocol

Different from the traditional software business model, when a software provider provides software as a service to users, its legal relationship also changes from a simple software license to a new type of contract relationship. In addition to the rights and obligations of software licensing, it usually covers multiple rights and obligations such as service, custody, lease and so on. It is very important to analyze the nature of user agreement and clarify the application of law to solve the legal disputes in user agreement.

The agreement between cloud service providers and users belongs to a new type of contract [3]. As far as the contractual rights and obligations involved are concerned, it does not belong to any kind of well-known contract in the current legal system. There is no consensus on the nature of this kind of contract. According to some scholars' investigation on the research status of German academic circles, German scholars generally believe that this kind of agreement is a combination of many different contract types, and has the characteristics of lease, software work license and service contract. The author believes that in addition to the legal relationship of software licensing described in the third chapter, SAS user agreement also has the rights and obligations of software and hardware computing resource lease contract, user information data custody contract and network service contract.

3.2 Analysis on the Legal Application of SaaS User Agreement

Due to the similarity of the service types with the custody contract and consistent with the custody contract, the judgment of service defects not only takes the completeness of service results as the judgment standard of defects, but also takes the appropriateness of service process as the judgment basis of fault. "The author thinks that the relevant provisions of custody contract can be applied to adjust the rights and obligations of both parties in the contract. For example, the user can collect the data information from the cloud service provider at any time; the cloud service provider has the obligation to keep it in person, and shall not transfer the user data information to a third party without agreement; the cloud service provider shall not use or license any third party to use the user data information without agreement; Cloud service providers should be liable for damages if the information and data are lost or leaked due to improper storage of user information and data. In addition, the custody contract stipulates the exemption of non gross negligence for free custody. For the understanding of free, the author believes that even if cloud service providers open free services to users, they still make profits through advertising fees and other means, It is still a commercial operation and should be regarded as a paid custodian.

4   Contract Law Regulation of SaaS User Agreement

In judicial practice, in the case of "Liang Mingyue v. China Eastern Airlines Co., Ltd. over air passenger transport contract dispute", the court held that "in the process of issuing electronic tickets, airlines basically did not list the conditions of the transport contract in the obvious position on the webpage, but expressed in the form of hyperlinks, If the subscriber wants to know its content, it will often be incomplete or unrecognizable due to busy lines and other factors such as transmission, etc., which will deprive passengers of the right to know objectively, and make standard terms become default terms, and passengers have to accept and perform obligations unconditionally, which is not in line with the original legislative intent of the contract law". Therefore, whether the format contract displayed by hyperlink conforms to the express standard or not depends on the specific situation. It can be seen from the above judgment that when judging whether the cloud service provider has fulfilled the obligation of prompt attention under the SAS mode, it should at least ensure the following two points: (1) hyperlinks are displayed in a prominent way on the web page; (2) the access path of hyperlinks is smooth [4]. Moreover, as the arrangement of software services may involve user agreements, service level agreements, privacy policies, statements and other terms and contents of different browsing pages, cloud service providers are obliged to run these browsing contracts through the necessary process of users' ordering software services according to the above standards in the process of users' registration, selection and purchase of services, To ensure that users have full opportunity to review the terms of the agreement.

5   Conclusion

With the promotion of business model of SAS software, both public and private organizations and individuals have begun to migrate to the "cloud". However, this new mode not only improves user experience and reduces user cost, but also brings many legal problems to be solved. Especially with the cloud computing industry becoming more and more important in the strategic development of various countries, compared with the advancement of technology development, the legal system is facing the dilemma of lagging behind in balancing the interests of all parties. The author of this new software business model in copyright protection content licensing mechanism and online software services related to the nature of cloud user agreement, legal application and format terms regulation and other issues, trying to sort out the triple legal relationship involved in this business model, and provide preliminary suggestions for achieving the balance of interests of all parties. As the legal issues related to this business model are still in its infancy, it is hoped that the superficial analysis of this paper will play a role in attracting jade.

References

1. Chen, K., Zheng, W.: Cloud Computing: system examples and research status. Chin. J. Softw. 20(5), 1337–1348 (2009)
2. Alibaba cloud improves the "cloud rendering" operation module and cooperates with cloud shading to draw a blueprint for rendering business
3. Li, J.: research on energy saving task scheduling strategy in heterogeneous cloud computing platform. Nanjing University of Posts and Telecommunications (2014)
4. Huang, L.: Research on cloud computing task scheduling algorithm based on genetic algorithm. Xiamen University (2014)

Improved Camshift Tracking Algorithm Based on Color Recognition

Bo Tang, Zouyu Xie, and Liufen Li[✉]

College of Mathematics and Statistics, Sichuan University of Science
and Engineering, Zigong, China
liliufen@suse.edu.cn

Abstract. For camshift (continuously adaptive mean-shift), the tracking algorithm requires too much target background during target tracking, and it needs to highlight the characteristics of the target. Based on the original algorithm, this paper makes improvements, analyzes the environmental factors in different environments, performs targeted processing on targets with strong light and complex background texture, and performs gamma non-linearization on targets under strong light. Foreground extraction is performed on targets with complex backgrounds. Combine the above algorithms to achieve the purpose of improving the camshift algorithm.

Keywords: Target tracking · Image processing · Camshift algorithm

1 Introduction

Machine vision shines in the science and technology arena of transportation, medicine, criminal investigation, military, etc., and target tracking is one of the frontier directions. In machine learning, there are generative models and discriminative models. Target tracking algorithms can also be divided into these two categories. Among them, the discriminant model is more prominent than the two. The algorithm studied in this paper can be classified as a discriminant model algorithm, with an initial frame, we get the initial frame and compare with each subsequent frame. So as to achieve the purpose of tracking.

In recent years, science and technology have developed rapidly and target tracking technology has become more mature. The emergence of diversity is an inevitable trend. TLD algorithm [3], Kalman filter algorithm [4], particle filter algorithm [5] and other algorithms are all pioneers in this field.

The TLD algorithm solves the problem of target loss during target tracking. When we get the first frame of image, it will automatically perform robust tracking.

Kalman filter algorithm, this algorithm is suitable for linear moving objects, it is very unfriendly to targets such as irregular and nonlinear, and the probability tracking algorithms such as Kalman filter and particle filter have higher requirements for data, so its application Certain limitations.

In 1975, Fukunaga and Hostetler proposed a deterministic tracking algorithm. 20 years later, a team of scientists published a paper to promote the meanshift algorithm. Since then, meanshift has entered the scientific public's field of vision. Then

J. Abawajy et al. (Eds.): ATCI 2021, LNDECT 81, pp. 506–511, 2021.
https://doi.org/10.1007/978-3-030-79197-1_74

based on the meanshift algorithm, Bradski proposed a target motion prediction algorithm based on the color histogram—DP-camshift algorithm.

2   Algorithm Theory

2.1   Meanshift Algorithm

The Meanshift [1] algorithm belongs to the kernel density estimation algorithm. No-parameter estimation is a branch kernel density algorithm in the field of mathematical statistics. This method requires the least prior knowledge and completely relies on data training for estimation. In a large sense, it solves the need for parameter density estimation methods. The problem of the known probability density function.

The Meanshift algorithm does not require any prior knowledge. In the well-known histogram method, after the data is collected, the histogram divides the range of the data into intervals of the same size, fills the data into groups, and calculates the probability through traditional methods. The method calculates the probability value of each group of the histogram. The kernel density estimation algorithm is similar to it, except that a kernel function is added when calculating the data. Kernel function estimation, under certain conditions, more fully converges to any density function, which greatly improves the functionality of the estimation.

For the kernel function, the kernel function used in this algorithm is Epannechnikov, and other kernel functions include Gaussian kernel function and Linear Kernel function.

The basic idea of Meanshift: In a given n-dimensional space Rn, each sample point has a satisfaction

$$F_h(x) \equiv \frac{1}{k} \sum\nolimits_{xi \in S_h} (x_i - x) \tag{1}$$

Where S_h is a set of y points in a high-dimensional sphere with a radius of h that satisfies the following relationship

$$S_h(x) = \left\{ y : (y - x)^T (y - x) \le h^2 \right\} \tag{2}$$

Where $(x_i - x)$ is the offset of each sample to the selected point x, formula 1 is to expect the offset of each sample point to the x point, and the gradient of the non-zero probability density function points to the probability The direction in which the density increases fastest, and the sample points we collected are obtained by using a certain probability density function, then the meanshift vector of each sample point is calculated in this way to point to the gradient direction. When we get enough meanshift vectors, we can get a vector pointing to the most direction, then that direction is the gradient direction of the probability density function. And the final calculation result has nothing to do with the data itself (no need to care about the distance of x), only its position for the x point, but due to environmental factors during tracking, such as illumination, occlusion, object deformation, etc., it is taken away The point closer to the center point x is better.

The Meanshift algorithm obtains the target model and the detection model by calculating the probability of the pixel feature value in the target area and the detection area, and then uses the similarity function to calculate the similarity between the target and the detection model in the initial frame, and selects the one that maximizes the similarity function value. Detect the model and get the meanshift vector about the detection model. The direction that the vector points to is the correct direction of the target from the starting position to the target position. Because meanshift adopts the kernel function estimation method, it has the nature of rapid convergence, so it is finally through continuous iterations. The algorithm will converge to the true position of the target, so as to achieve the purpose of tracking the target.

2.2   Camshift Algorithm

Camshift algorithm is an improvement of the meanshfit algorithm, so it is widely used in the field of target tracking based on continuous target images and color features.

In the color tracking direction, the construction of the back projection map of the color histogram is an important point. The principle is to compare the pixel value of each pixel with the pixel value of the tracking target area during tracking, and finally find similar points, and calculate The probability of finding similar points (only the selected area is compared each time, and the probability of the area outside its range is regarded as 0). Convert the probability to a single-channel gray value (range: 0–255) and finally get the back projection image. The back projection diagram intuitively shows the probability that a certain pixel is the pixel of the tracking target. The greater the probability, the greater the grayscale after conversion, the closer the image is to white, and the brighter the area in the back projection image, the greater the probability that this point is the target tracking pixel.

The Camshift algorithm only improves the target model and detection model of the initial frame in the meanshift algorithm. The core idea has not changed. Camshift will automatically adjust the range of target model selection, and use the detected target model size and position as the next iteration parameter.

The reason why the Camshift algorithm can adaptively select the range is that it compares based on the feature moments before two adjacent frames (usually the zero-order moment and the first-order moment are studied because the centroid needs to be found). Calculate the centroid size according to the characteristic moment:

$$(y_o, x_o) = \left(\frac{M_{01}}{M}, \frac{M_{10}}{M_{00}} \right) \tag{3}$$

The size and position of the window are updated each time according to the zero-order moment M00. The centroid position is the center position of the next frame. At the beginning, we will set a threshold for this tracking. If the threshold is exceeded, then The center of mass and M00 will be recalculated. Similar to the step of finding the next starting point in meanshift. This loop iterates until the distance between the center of mass and the center point is found to be less than the threshold set at the beginning. Finally, the continuous picture sequence is searched and tracked. Because the data of adjacent frames is processed each time, the influence of object deformation on the tracking effect is greatly reduced.

3   Algorithm Implementation

First, the target image is converted from the RGB color space to an image in the HSV space, and the first frame of image is taken out for image preprocessing. The preprocessing is to get the foreground and achieve the purpose of re-tracking after our target moves too fast beyond the originally selected area. Among them, we will analyze the overall target to prevent over-exposure and over-darkness, which will make target tracking fail. The flow chart is as follows (Fig. 1):

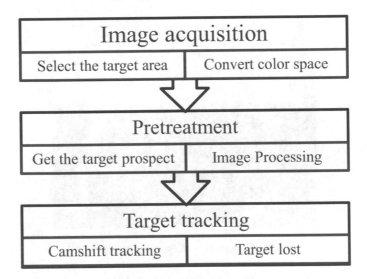

Fig. 1. General flow chart

3.1   Image Preprocessing (Algorithm Improvement)

Select the target, given the selected area, select the first frame of image, extract the foreground of the selected target, and use the grabcat algorithm to separate the foreground and background. Grabcut algorithm [6] uses the texture (color) information and edge (contrast) information in the image to get the foreground and background (Fig. 2).

Fig. 2. Foreground extraction effect comparison

After the foreground is extracted, we will use Gamma correction [2] to avoid target tracking failure due to overexposure and overdarkness of the image. Gamma nonlinearity at this time we use its brightness calculation formula (Fig. 3):

$$L - \sqrt[2.2]{\frac{\left(\frac{R}{255}\right)^{2.2} + \left(1.5\frac{G}{255}\right)^{2.2} + \left(0.6\frac{B}{255}\right)^{2.2}}{1 + 1.5^{2.2} + 0.6^{2.2}}} \tag{4}$$

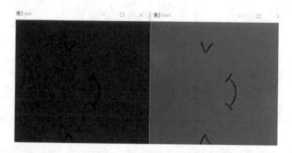

Fig. 3. Gamma correction comparison chart

This is the adaptive Gamma correction used, and the incoming Gamma value is calculated according to the formula and then passed in

$$\text{Gamma} = \frac{\log_{10} 0.5}{\log_{10} \frac{M}{225}} \tag{4}$$

Where M is the global average pixel value of the pixel.

Finally, I will talk about the sum operation between the image of the area and the original image to return the final processed image to achieve the optimized effect. At this point, the pre-processing stage is complete.

3.2 Target Tracking (Result Display)

From the above figure, it can be found that when the target area is lost, the range with the highest similarity probability will be found to be tracked again, and when the target appears again, the target will continue to be tracked (Fig. 4).

Fig. 4. The overall picture sequence of the tracking process

Acknowledgement. This work is supported by the college students' innovation and entrepreneurship training project (S201910622065).

References

1. Aunsri, N., Chamnongthai, K.: Stochastic description and evaluation of ocean acoustics time-series for frequency and dispersion estimation using particle filtering approach. Appl. Acous. **178**(178) (2021). https://doi.org/10.1016/j.apacoust.2021.108010
2. Gharebaghi, N., Nejadrahim, R., Mousavi, S.J., Sadat Ebrahimi, S.R., Hajizadeh, R.: Correction to: the use of intravenous immunoglobulin gamma for the treatment of severe coronavirus disease 2019: a randomized placebo-controlled double-blind clinical trial. BMC Infect. Dis. **20**(1), 1–8 (2020). https://doi.org/10.1186/s12879-020-05507-4.
3. Zhen, X., Fei, S., Wang, Y., Du, W.: A visual object tracking algorithm based on improved TLD. Algorithms **13**(1), 15 (2020). https://doi.org/10.3390/a13010015
4. Jiang, L., Fu, W., Zhang, H., Li, Z., Chi, L.: An Improved robust adaptive Kalman filtering algorithm. Chin. Assoc. Autom. 1109–1113 (2019)
5. Aunsri, N., Chamnongthai, K.: A hybrid algorithm based on particle filter and genetic algorithm for target tracking. Expert Syst. Appl. **147** (2020)
6. He, K., Wang, D., Tong, M., Zhu, Z.: An improved GrabCut on multiscale features. Pattern Recogn. **103**, 107292 (2020)

Construction and Application of 3D Model of Engineering Geology

Liangting Wang, Zhishan Zheng$^{(\boxtimes)}$, and Huojun Zhu

Jiangxi University of Engineering, Xinyu 338000, China

Abstract. with the continuous development of social economy, the rapid development of science and technology, modern society under the background of the rapid development of science and technology in all walks of life are beginning to change, driving force of science and technology in contemporary society is obvious to all, especially the engineering construction field, the rapid development of science and technology had a deep influence for construction engineering industry, construction, exploration technology for improving project quality and construction quality has a direct impact. Especially for engineering geological exploration, with the help of three-dimensional space model construction technology can be more conducive to the construction of labor unions, such as oil exploration, road and bridge construction, water conservancy and hydropower engineering exploration and other aspects play a huge role. Based on the characteristics of 3D space modeling technology, the construction and application of 3D model in engineering geology are deeply analyzed in this paper.

Keywords: Engineering geology · 3D model · Build

1 Introduction

Under normal circumstances, the design of a project to develop before all need to entrust to have the corresponding qualification of surveying engineering project owner enterprises, combines the actual conditions of project engineering construction demand, for the related areas of geology characteristic to conduct a comprehensive investigation, and then based on the survey data to present a construction site and the surrounding area of full and accurate geological evolution process, groundwater activity, rock change, etc., on the basis of these data for the engineering construction scheme design of the original data, and for engineering design unit, can be combined with the data continuously optimizing construction scheme. Under normal circumstances, the construction design units in the two-dimensional space model to reflect the field data of survey, although 2 d spatial data model can relatively comprehensive present a construction site geological conditions, but for two dimensional data processing need to spend a lot of events, some special geological conditions and is not shown in the picture, it is easy to cause geologic conditions to assess risk. Especially as engineering scale increasing, for some complicated geology area, it is particularly important to engineering geological exploration, for these areas must be with the aid of three-dimensional modeling technology, multi-angle comprehensive show the scene actual situation, more comprehensive intuitive visual perception to design unit,

© The Author(s), under exclusive license to Springer Nature Switzerland AG 2021
J. Abawajy et al. (Eds.): ATCI 2021, LNDECT 81, pp. 512–518, 2021.
https://doi.org/10.1007/978-3-030-79197-1_75

comprehensive display of construction site space structure under different geological conditions, provide more perfect geological data for construction design information.

2   General Process of Constructing 3D Model of Engineering Geology

Due to the different emphases of 3D engineering geological modeling, there are many methods, and different design units adopt different methods according to different engineering construction objects. Therefore, in order to avoid generality, the modeling discussed in this paper is a general process, as shown in Fig. 1:

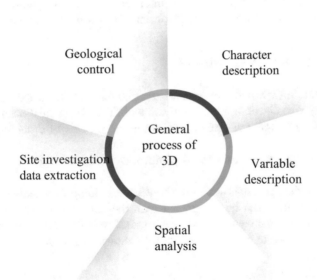

Fig. 1. General process of 3D geological modeling

The general process of 3D geological modeling includes two aspects, the first is the prediction of geological variables, the second is the interpretation of geological features.

The first is the prediction of geological variables. The main function of using three-dimensional model to carry out geological modeling is to predict the spatial changes of one or more geological variables with the help of three-dimensional model. For engineering geological exploration, geological variables are mainly the spatial distribution characteristics of strata, fractures, faults and the corresponding physical and mechanical parameters. For a certain engineering geological variables in the survey, because it could not achieve the continuity of measurement, so often choose some representative observation point measurement, the observation point for the sampling points, and then with the help of different prediction technology, to the whole survey area, the change of the geological variables to speculate, such as by using geological statistics forecast, inverse distance interpolation, direct prediction and condition simulation can achieve this purpose;

Secondly, the interpretation of geological features includes two levels of conditionalization and discretization. Specifically, the engineering geological information is discretized with the control characteristics of geological lithology or rock and soil types as conditions to determine the engineering geological boundary and then describe the relevant features.

Richard is geotechnical investigation CAD software of drilling database stored in various types of geological information, considering the demand of the engineering geological 3 d modeling, mainly using the table drilling, drilling the upper table, using the serial number of the database information to get the drill point, corresponding to the type, drilling point of plane position and thickness of soil layer number, name and information, so that we can provide original data for 3 d model building.

3 Data Structure of 3D Model

The data structure of 3D model of engineering geology directly determines the modeling algorithm, visualization, model analysis algorithm and model accuracy. Therefore, the following factors must be considered in the process of designing the model data structure:

First, design the data structure as simple as possible to facilitate automated modeling;

Secondly, the designed data structure should be as simple as possible in geometry, so that 3D visualization can be carried out with the help of computer software. For example, the surface can be represented by triangulation network, the purpose of which is to clearly reflect the stratigraphic structure, as shown in Fig. 2:

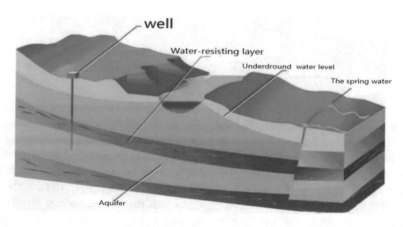

Fig. 2. Geological structure map

The structure of each geological layer is clearly visible in the above figure, which is convenient for construction design.

Furthermore, the designed data structure needs to meet the common operations in the application of 3D geological model such as cutting and subdivision, such as crosscutting or longitudinal cutting of the geological profile.

Finally, the designed data structure must be able to meet the accuracy requirements of the actual construction application.

At present, there are two most common data structures for 3D modeling of engineering geology:

The first is a planar structure, this structure is mainly used to express the level, often use triangulation approach, more clearly reflects the geological structural conditions, but the way the triangulation is not really a 3 d model, but 2.5 d, this kind of structure modeling method is very simple, but is not applicable for the model of the geological condition is complicated.

The second is the volume element structure. This model expression method is an innovation of the opposite structure. It mainly takes the volume element as the basic unit and then gives a detailed description of the internal details of the geology. For the strata in charge, the volume element structure method can minimize the complexity of model operation because it wants to show the distribution of different geological bodies below the surface more clearly.

However, in the process of selecting the data structure of engineering geological 3D model, it is necessary to consider the special phenomena of the special 3D geological model, such as the formation pinching out, convex lens, etc.

After comprehensive consideration of the above factors, combined with the actual situation of the power engineering industry, the posthumous modeling is based on the triprism as the basic unit. Each edge of the triprism represents a borehole, while the upper and lower undersides represent a specific ground plane. There are three specific manifestations as follows, as shown in Fig. 3:

Figure 3 (a) is the most standard triprism volume element, in which the height of the three edges is greater than 0. This volume element structure indicates that the corresponding strata have a certain thickness in the subgrid.

Figure 3 (b) is a variation of the standard tri-prism element, in which an edge has a height of 0, indicating that the corresponding stratum is missing at this point.

Figure 3 (c) shows another special variant, in which it can be seen that the height of two edges is 0, changing from pentahedron to tetrahedron. This situation indicates that the stratum has two points missing in the grid.

And the latter two special volume structures exist in the pinch-out position of the engineering area strata. Different stratigraphic elements have different attributes and characteristics, including the identification number and color of the strata, etc. By recording the structure of these elements, a three-dimensional geological model can be formed.

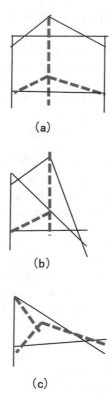

(a)

(b)

(c)

Fig. 3. Geologic 3D modeling of triangular prism geometric types

4 3D Modeling Algorithm of Engineering Geology

The advantages and disadvantages of modeling algorithms directly affect the modeling efficiency and accuracy of 3D model data. This paper focuses on introducing the structural method of triprism model generated by planar structure. The biggest advantage of this modeling method is that the model is built quickly and can meet the accuracy of complex engineering geological application. The specific operation process has the following four steps:

Firstly, according to the point position plane coordinates of the borehole, the Delaunay algorithm is used to generate the corresponding point position plane triangulation network structure, and the connection relationship between the borehole points is determined according to the triangulation network model structure.

Second, all existing soil layer structures are obtained from the data of different borehole locations, and the soil layers are sorted from top to bottom according to the order of different soil layers, and then the list of soil layers of geological structure is generated, and the surface is inserted into the list as the top layer.

Third, attach the depth value of different soil layers to each node of the plane triangulation network structure generated in the first step, and then the planar structure of different soil layers can be obtained. According to the characteristics of the planar structure, if a node does not appear in the corresponding soil layer, the thickness of the coating layer of the soil layer can be taken as the thickness of the soil layer. If there is no coating layer in the previous soil layer, it needs to continue to search upward until the surface.

Fourth, the plane structure generated by the previous link is arranged in pairs in the order from top to bottom, then the triangle network structure of the upper layer and the triangle network structure of the next layer will form a triangular prism structure.

On this basis, the triangular prism structure model can be developed with the help of OpenGL graphics platform, and the three-dimensional geological model can be visualized. Thus, the drilling effect map, the soil layer effect map and the three-dimensional geological effect map can be obtained.

5 Analyze the Body Structure of the Regional Geological Section

Geological profile analytic needs based on the survey work, before the survey, need related working personnel have comprehensive knowledge of engineering geological map specification requirements, in the process of carrying out survey work, must be in strict accordance with the relevant specification requirements needed to collect data, survey process to use the drilling work in the work area on the rock and soil, the surface of drilling process, a detailed record of the relevant data of drilling, timely upload drilling data to computer software, and then generate geological Sect. 3 d figure and columnar hole figure. According to the elements of the three-dimensional model, the corresponding relations among the objects such as points, lines, planes, rings, surfaces and complex bodies are accurately listed to reflect the spatial topological structure relations as accurately as possible. The final presentation mode is usually tree-like structure, which is characterized by strong cohesion and integration.

The second is to generate three-dimensional geological profile. Survey personnel should first process the stratigraphic structure, and then construct the regional geological profile. After all the terrain and geotechnical materials in the survey area are presented, it is indicated that the three-dimensional structural model of engineering geology is formed. In the expression of the volume element of the model, it is necessary to first express the volume element simply, and then form relatively complex volume elements successively. In the process of data processing, logical operation can be carried out on part of the data first, and then the visual expression of the data can be carried out after all the model data are processed.

6 Conclusion

To sum up, the construction of 3D model can provide a more scientific and comprehensive engineering geological situation, especially for construction projects, and can scientifically manage and predict the occurrence of geological disasters. Compared with the traditional geological modeling technology, 3D geological modeling has strong advantages, more in line with the needs of modern engineering construction design and comprehensive management. It can provide engineering geological survey with 3 d dynamic display effect, more clearly reflects stratigraphic structure, with complicated geological conditions, can provide the designer with the clear geological structure and tectonic, and the application of visualization technology for 3 d geological modeling provides a powerful theoretical basis for construction design more convenient design staff.

Acknowledgement. Teaching reform project of Jiangxi Institute "Application and practice of group discussion method in practical training course of engineering survey" of Technology (2020-JGJG-15).

References

1. Yaping, Y., Xiangdong, L.: Construction method and application of 3D model of engineering geology. World Nonferrous Metals **08**, 273–275 (2020)
2. Zhongyang, L., Jin Xianglong, G., Mingguang, H.W., Yinchen, F.: Geological model construction based on MapGIS-TDE three-dimensional platform. Yangtze River **50**(S2), 85–88 (2019)
3. Zhen, M., Yubo, X., Xiaodan, W., Bo, H., Yihang, G.: Geol. China **46**(S2), 123–138 (2019)
4. Sun, B., Wu, X., Lin, G., Zhang, L.: Construction and application of land subsidence model in Dongying City. Shandong Land Resour. **33**(04), 63–66+70 (2017).
5. Zhang, X., Ma, G., Zhu, T., Liu, G.: Research and application of 3D geological application model construction method. Eng. Invest. Surv. **45**(04), 49–54+63 (2017)

Research on the Application of Virtual Reality Technology in College Physical Training

Jian Wang and Dawei Zhang[⊠]

School of Physical Education and Health, Tianjin University
of Chinese Medicine, Tianjin 301617, China

Abstract. Virtual reality technology has the function of simulating human body feeling, and the degree of simulation is very high. Therefore, with the help of technical equipment, people can produce real body feeling in the virtual environment, which makes the virtual reality technology can be applied to many fields related to body feeling, including college sports training. But in reality, the application of virtual reality technology in college sports training is rare, which indicates that colleges and universities have not popularized the technology, and do not know how to play the role of the technology in training. Therefore for the purpose of technology promotion and understanding the role of technology, this paper will analyze the application of the technology in sports training, and mainly discusses the application of virtual reality technology The function, application methods and matters needing attention of physical education training.

Keywords: Virtual reality technology · University · Physical training

1 Introduction

Under the education policy of "moral, intellectual, physical, aesthetic and labor" comprehensive development, colleges and universities pay more attention to the development of students' physical quality, so they begin to vigorously develop physical education. However, in the development process, many problems have been exposed in college physical education, and a considerable part of these problems are reflected in physical training, which hinders the development of college physical education. In the face of these problems, virtual reality technology can help colleges and universities solve these problems, improve the effect of college sports training, and promote the development of students' physical quality. At this time, how to use this technology to solve the problem has become a problem that colleges and universities should think about, so it is necessary to carry out relevant research.

J. Abawajy et al. (Eds.): ATCI 2021, LNDECT 81, pp. 519–525, 2021.
https://doi.org/10.1007/978-3-030-79197-1_76

2 The Role of Virtual Reality Technology in Sports Training

2.1 Help Students Understand Theoretical Knowledge

In essence, physical training is a kind of activity that students carry out training according to theory, and then obtain training results through training. Therefore, the relationship between physical training and sports theoretical knowledge is very close. In this case, if the students do not understand the relevant theoretical knowledge before the sports training, or do not fully grasp the theoretical knowledge, it is easy to make mistakes in the training, resulting in a great discount in the training effect, and may even make themselves injured. It can be seen that this phenomenon not only makes the sports training not play a role in improving the students' physical quality, but also causes the students to bear the physical stress It's a lot of pain. But according to the current situation of sports training, although many students have a certain understanding of sports theoretical knowledge, they do not fully grasp the theoretical knowledge, so the above phenomenon is very common. For example, a student knows that when doing "squat up" training, his feet should be slightly open, but he does not know the specific opening range of his feet. At the same time, considering the different habits of everyone, there is no problem In the end, with his personal habits and some hearsay news, he carried out "squat up" training, but the training did not bring significant effect, and also made the student hurt his knee, which is a typical phenomenon that he did not fully grasp the theoretical knowledge to carry out training. In these phenomena, virtual reality technology can help students safely master knowledge, that is, using virtual reality technology equipment to enlarge the students' body feeling when "squatting up", so that students can feel the body feeling when different "squatting up" posture, and choose the best body feeling from them, so that students can know what the most correct "squatting up" posture is, which will be used in the later training Three positions to do "squat up" training [1]. Figure 1 shows the basic process of theoretical teaching application.

2.2 Cultivate Students' Willpower

Sports training needs to be carried out persistently, and the training intensity needs to be adjusted according to the current physical conditions, at least to ensure that the training intensity is close to the current physical conditions. Therefore, to carry out sports training persistently, people must have strong willpower, and people with weak willpower are easy to give up halfway. In this case, although there is no lack of strong willpower in the student group, most of the students' willpower is still relatively weak, which leads to many students' negative attitude and inner conflict towards sports training, which seriously affects the quality of sports training. For example, a student thinks that every sports training leads to his exhaustion and makes himself very painful, so in the future In physical training, there is a psychological resistance, often do not participate in training, or "cut corners" in training, so physical training can not help the students effectively improve their physical quality. Therefore, how to cultivate students' Willpower has become the teaching difficulty of many physical education teachers, but with the help of virtual reality technology, physical education teachers can

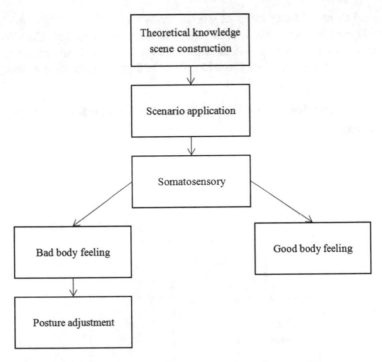

Fig. 1. The basic process of theoretical teaching application

cultivate students' Sports willpower, that is, teachers can use virtual reality technology to build virtual reality scene, organize students to evaluate regularly in the scene, so that students can understand their own physical quality level, and then the teacher group. Organizing students to carry out physical training, and then evaluating after a period of time, can let students know the help that training brings to their physical quality. This feeling will make students gradually like physical training, making them more and more willing to participate in physical training, and in order to improve their physical quality as soon as possible, they will always adhere to training, which is the performance of students' Willpower improvement [2].

2.3   Enhance Students' Interest in Sports

Many students are not willing to participate in sports training, negative response to sports training because they are not interested in sports training, so in order to improve the effect of sports training, teachers should find ways to enhance students' interest in sports, but many sports teachers have nothing to do about this, all kinds of methods to enhance interest have not played a good effect. In this case, teachers can use virtual reality technology to carry out sports training teaching. With the help of technology, teaching can be more in line with students' own interests. Students can also participate in sports training when carrying out activities in line with their own interests, so as to improve their physical quality. That is to say, many modern students like to "play games", and there are many kinds of games, many of which are related to sports for example, There are many "fitness VR Games" in the modern market. The way to play these games is to make players use the equipment to do some sports training, because this teacher can use this kind of game to enhance students' interest in sports [3].

3   Application Method of Virtual Reality Technology Physical Training

3.1   Method Flow

Figure 2 is the basic flow of the virtual reality technology sports training application method.

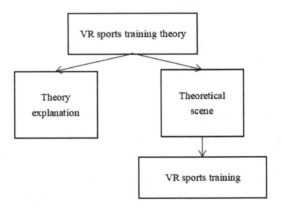

Fig. 2. Basic flow of application method of virtual reality technology physical training

Illustration: (1) VR (Virtual Reality Technology) sports training theory teaching design, that is, teachers should design sports training theory teaching plan before sports training, so as to let students understand the theory with the help of technology; (2) VR sports training theory teaching experience, that is, Teachers should let students enter the virtual reality environment with the help of virtual reality technology equipment to understand the main points of theoretical knowledge and help them understand Students further master theoretical knowledge; (3) VR sports training evaluation, that is, teachers should evaluate students regularly in the virtual reality environment, so as to play the role of virtual reality technology and cultivate students' willpower; (4) VR sports training game teaching, that is, teachers can select some relevant VR games into the classroom, and carry out sports training teaching for students in the process of the game learn to work.

3.2   Application Method

The teaching application method of each link in the process of Fig. 2 is as follows.

VR Sports Training Theory Teaching Design

Take the basic physical training as an example (push up, sprint, etc.), the teacher should design the theoretical teaching plan of each basic training project in advance, such as the correct posture of push up, the help of push up muscles, the help of push up muscles, etc., and then explain these theoretical knowledge in class, so that the students can form a preliminary concept of theoretical knowledge. For example, a teacher designed the theoretical points of long-distance running in teaching, including the physical distribution of long-distance running, the role of long-distance running, and explained them in class, which laid the foundation for the virtual reality technology teaching link [4].

VR Sports Training Theory Teaching Experience

After the students have formed the concept of theoretical preliminary knowledge, the teacher should use the virtual reality technology equipment to let the students experience in the virtual reality environment, in order to let the students master the theoretical knowledge. There are many ways to experience virtual reality technology. This paper mainly lists two ways: ① To demonstrate the correct posture, let students imitate, and feel the body feeling under their own posture. If the body feeling does not match the teacher's description, it means that the posture is wrong and needs to be further adjusted, so that students can gradually master the correct posture; ② To enlarge the students' body feeling, let students adjust the posture and body feeling by themselves experiential sense, which can help students more clearly understand whether they have problems, and then adjust according to the problems. In the two ways, this paper suggests to choose the first one, because it is relatively simple and convenient for teachers to observe the situation of students, so the application effect is relatively higher. Tables 1 and 2 show the efficiency of the two teaching methods.

Table 1. Efficiency of the first teaching experience mode

Project	Result
How long does the student need to master the correct posture	5 min–16 min

Table 2. Efficiency of the second teaching experience mode

Project	Result
How long does the student need to master the correct posture	7 min–22 min

VR Sports Training Evaluation

After a period of training, teachers need to use virtual reality technology to evaluate students' training. They can set up targeted evaluation projects in the virtual reality environment. For example, when teachers want to test the arm strength of students, they can design an evaluation project of "pushing objects" in the environment. With the help of equipment, they can apply the weight of objects. The greater the weight that students can push, the stronger the arm strength of students. According to this, teachers can get the evaluation results. In addition, the evaluation of physical training is not aimed at a certain physical ability of students, but an activity aimed at all physical abilities of students. Therefore, teachers can know students' physical ability strengths and weaknesses through the evaluation, which is convenient for teachers to carry out targeted teaching. At the same time, in order to ensure the reliability of the evaluation, it is suggested that teachers use the system evaluation calculation method to set the standard value first, and then evaluate the calculation The calculation method is shown in formula (1).

$$Bmi = kg/m \tag{1}$$

VR Sports Training Game Teaching

In the past, the teaching content of sports training is relatively single, and the atmosphere is relatively boring, so students can not raise interest. With the help of virtual reality technology, teachers can bring in relevant games to make sports training game, and then start teaching, which can effectively improve students' interest and ensure the quality of teaching. For example, a teacher brought in the "VR long-distance running" game in the long-distance sports training. The students in the game need to keep the long-distance running state all the time, and at the same time, they need to jump with different strength. This process makes the students happy, and at the same time, it also has the effect of sports training. Table 3 shows the students' evaluation of VR sports training game teaching in a university.

Table 3. Students' Evaluation on game teaching of VR sports training in a University (50 persons)

Assessment	Result
Are you satisfied with the game teaching of VR sports training	Yes(48); no(2)
Do you think VR sports training game teaching can raise your interest	Yes(45); no(5)

4 Conclusion

To sum up, this paper studies the application of virtual reality technology in college sports training. Through the research, we know that virtual reality technology has a good role in sports training, can guarantee the effect of sports training, and solve the common problems in the past sports training. This paper puts forward the application method of virtual reality technology in sports training, introduces the implementation mode and effect of each method, and shows that virtual reality technology has good application value in college sports training.

References

1. Jiao, C., Qian, K., Zhu, D.: Application of flipped lassroom teaching method based on VR technology in physical education and health care teaching. IEEE Access **99**, 1 (2020)
2. Li, C., Li, Y.: Feasibility analysis of VR technology in physical education and sports training. IEEE Access (99), 1 (2020)
3. Wang, B., Wu, Z.Z.: Research on application of computer virtual reality technology in sports training. Adv. Mater. Res. **926–930**, 2694–2697 (2014)
4. Li, X.G.: Research on application of computer technology in the virtual reality in sports. Adv. Mater. Res. 1049–1050, 2024–2027 (2014)

Application Research of Graphic Design Based on Machine Learning

Xiangyi Wang[✉] and Hongyang Zhang

School of Art, Xinyu University, Xinyu 338000, China

Abstract. With the continuous development of science and technology, artificial intelligence has gradually become the mainstream direction of today's science and technology development. Machine learning is produced and developed in this background. As a multi-field intersecting learning type, machine learning involves a lot of learning content, such as statistics, algorithm complexity theory. Probability theory and so on, its function is to obtain more new knowledge and new skills through the specialized research of how the computer simulates or realizes human learning behavior. With the continuous development and improvement of machine learning technology, it has been applied to many fields. As an important way in the field of graphic design, image stylization can display the artistic styles contained in different images on the common graph line in some way, and then gradually change the visual characteristics of images. The combination of machine learning and graphic design can, on the one hand, further enrich the image stylization technology in graphic design and give play to the advantages of this technology. On the other hand, it can provide more ordinary users with different image segmentation design convenience. This paper starts with the characteristics and content of machine learning and focuses on the application of graphic design.

Keywords: Machine learning · Graphic design · Stylization of images

With the development of science and technology, all walks of life are beginning to be affected by it. As the mainstream artificial intelligence technology in recent years, machine learning has played a very good role in solving many problems. Of course, as a relatively independent development direction, the development of machine learning in China has been in a period of rapid development [1]. From the beginning of last century, there have already been machine learning algorithm, after one hundred years of development, now the machine learning technology is already relatively mature, especially since the 1980s, the development direction of machine learning has become an independent, in the process related to the machine learning method has a lot of, today's machine learning has gradually got perfect, What we now call machine learning is based on a combination of highly complex algorithms and techniques to study and construct human behavior in an inanimate object and its systems [2]. This technique is now used in many fields, including graphic design. The graphic design described in this paper is increasingly dependent on machine learning. At present, with the image in plane design style of art and design major software technologies such as photoShop, coreldraw, beautiful picture show, these are all relatively professional image processing software, but for those who do not know the art style of ordinary people, with the help of the software to create artistic style you like or need relatively

J. Abawajy et al. (Eds.): ATCI 2021, LNDECT 81, pp. 526–532, 2021.
https://doi.org/10.1007/978-3-030-79197-1_77

complex image, Probably most people will only use Meitu Xiu Xiu to simply process some images, Photoshop, CorelDraw these relatively complex image processing software a lot of people will not use. However, with the development of machine learning technology, especially in the development and application of image stylization technology, the good application of this new technology method in graphic design image stylization can help more ordinary users to create more artistic style images freely [3].

1 Classification and Application Fields of Machine Learning

1.1 The Concept of Machine Learning

First of all, machine learning is not an independent subject content, it is a product of multi-disciplinary work, integrating statistics, complex algorithms, probability theory and other disciplines knowledge content; Secondly, on the basis of these disciplines, with the help of computer integration, we are committed to more realistic simulation of human learning methods, and then divide the existing contents into knowledge structures to maximize learning efficiency [4].

From these two points of view, machine learning is a development stage of artificial intelligence, artificial intelligence is the main research object of machine learning, especially how to gradually improve the performance of corresponding algorithm in experience learning; Simply put, machine learning is the further study of computer algorithms that can be automatically improved by experience, using data and experience to continuously optimize the performance standards of computer programs.

1.2 Classification of Machine Learning Algorithms

Machine learning can be divided into four types according to learning methods, as shown in Table 1:

Table 1. Machine learning is classified according to learning styles

Classification	Describe
Supervised Learning	Training set objectives: marked; Such as regression analysis, statistical classification
Unsupervised Learning	Training set target: no label; Such as clustering, GaN (generating adversarial network)
Semi-supervised Learning	Between supervised and unsupervised
Reinforcement Learning	The agent constantly interacts with the environment to find the best strategy through trial and error

Generally speaking, there are three types of machine learning algorithms, namely supervised learning, unsupervised learning and reinforcement learning. In fact, semi-supervised learning is a combination of the first two and does not have significant characteristics, so I will not make too much introduction.

The first is the supervision algorithm. Since the 1930s, the supervision algorithm has experienced multiple development processes, from the earliest existing discriminant analysis to the perceptron model, and then to the decision tree and distance measurement learning, etc., as shown in Fig. 1:

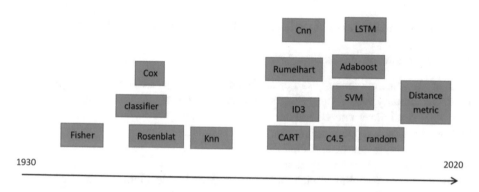

Fig. 1. Development history of the monitoring algorithm

The second is unsupervised algorithm. Compared with the former, this learning algorithm develops relatively slowly, and its content includes two types: clustering algorithm and data dimensionality reduction. Clustering algorithms have emerged since the 1960s, and have undergone the transition development of K-means algorithm, DBSCAN algorithm and OPTICS algorithm.

However, data dimension reduction did not play a significant role in the field of machine learning. It was not until 1998 that the kernel PCA algorithm emerged, which further developed the dimension algorithm as a special nonlinear one.

Secondly, reinforcement learning algorithm. Compared with the previous two algorithms, this algorithm cannot list all the states and actions in a table when solving many practical problems, so it is not widely used. With the emergence of neural network, It brings new development opportunities to reinforcement learning.The most common example of reinforcement learning is international recall, where the agent can decide the next action according to the current state of the board or the change of the environment, as shown in Fig. 2, where the reward is the winning or losing of the game result:

With the emergence and development of neural networks, deep neural networks have been widely used to fit action value functions. This combination enables people to deal with a variety of complex environments with reinforcement learning algorithms, especially in games, robot control and other aspects have been widely used. In addition, the combination of neural network and reinforcement learning algorithm can also directly predict the next operation according to the game screen, the image information of the self-driving car, Go and other image content, accurately.

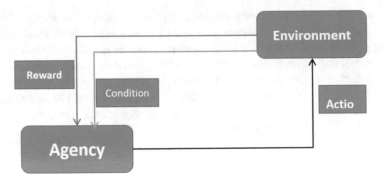

Fig. 2. A special example of reinforcement learning

1.3   Application Fields of Machine Learning

At present, as a technology to help people make decisions, machine learning is widely used in four fields, including data mining, computer vision, natural language processing and robot decision making.

(1) Data mining. Data mining is mainly focused on big data. With the help of machine learning related models, the potential value of information can be mined from big data, and the correlation between different data can be found through information integration mining. For example, the change of housing price can be predicted based on the analysis of big data of the real estate market.

(2) Computer vision. That is, with the help of the computer to see the world, so that the computer can see the world like people, according to the video, images and other information will be identified and classified, which is also the main characteristics of the current graphic design applications;

(3) Natural language processing. Is to let the computer like human ability to understand the language, is no longer a cold machine, can understand the meaning of the text, at the same time according to the connotation of the text to make the corresponding response, for example, the computer can news, books and other contents of classification, or even according to the content of the article and the central idea of the generation of abstract;

(4) Robot decision making. Robots that can think like humans and have the ability to make decisions, such as self-driving car technology, are the result of applying machine learning algorithms.

2   Application of Machine Learning in Graphic Design

2.1   Application of Machine Learning Algorithm in Graphic Design

Supervised Learning. With the help of supervised learning algorithm, a new model is obtained through the training sample learning, and then the algorithm reasoning is carried out with the help of the new model. In graphic design, for example, if you want to identify different images of things, and need to use artificial mark (that is to say to each image is belong to category with beforehand, such as apples, watermelons, grapes, etc.) of the training samples, and then get a model, based on the can use this model to classify things of unknown type judgment. This process is called prediction. If the prediction is only based on the category value of things, it can also be summarized as a classification problem. If you want to predict the real numbers of things, it can be summarized as a regression problem.

Unsupervised Learning. Compared with supervised learning, which needs to build a model for sample training, unsupervised learning has no process of sample training. Unsupervised learning is to realize a given part of sample data, directly classify and conclude these data with the help of machine learning algorithm, and then obtain some knowledge content of these data. The most typical representative of unsupervised algorithm is cluster analysis. For example, 10,000 avatars are taken as samples. With the help of unsupervised learning algorithm, these avatars can be classified according to the criteria of anger, happiness, frustration, joy, etc. In this process, the implementation did not define the category, nor did it implement a well-trained classification model. Instead, it only classified the 10,000 avatars with the help of the clustering algorithm to ensure that each type of avatars was delineated in a theme. When applied to graphic design, it is able to categorize and summarize the same type of images clearly, so as to facilitate better information processing.

Reinforcement Learning. As a special machine learning algorithm, reinforcement learning needs to determine a certain action according to the current environment, and then enter the state of the next stage, repeat the two processes repeatedly, and finally achieve the purpose of maximizing revenue.

2.2   Specific Integration of Machine Learning and Graphic Design

In recent years, machine learning is popular in various fields, especially in graphic design for the realization of image stylization to provide better technology and solutions.

Graphic design image stylization based on machine learning can be divided into four types: image stylization based on brush strokes, image texture synthesis, physical modeling and deep neural network. Among them, brushstroke generation, texture synthesis and physical modeling generation belong to the traditional stylized imaging methods of artistic images. Although these three methods can generate images, they cannot extract the inherent characteristics and higher-level characteristics of artistic images, so the final effect of these three traditional methods is not ideal.

As a branch of deep learning, machine learning is derived from artificial neural network, which simulates the neural senses of human brain to perceive the information of the external world, so as to extract the semantic information and internal features of art image samples. Therefore, in graphic design, the application of machine learning is more to use the method of deep learning convolutional neural network to carry out image stylized processing, that is, to realize the transfer of image style with the help of machine learning convolutional neural network. The specific algorithm is shown in Fig. 3:

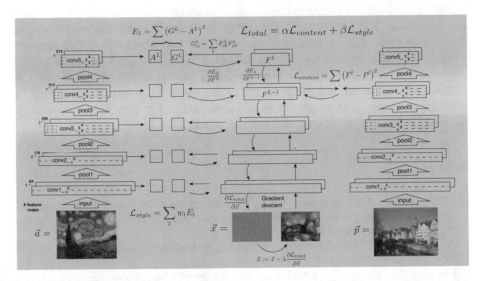

Fig. 3. Flowchart of image style transfer algorithm realized by convolutional neural network

Specifically, the trained convolutional neural network is used to achieve a given style image and an ordinary style image. In the process of the former going through the convolutional neural network algorithm, each convolutional layer will obtain many feature maps, and these feature maps will form a set according to their features. Similarly, ordinary style images will also obtain a large number of feature maps with the help of convolutional neural network. These feature maps will form another set according to their features, and then generate a random noisy image, which will then generate a large number of feature maps with the help of convolutional neural network. Then the final set corresponding to the previous two sets is formed. The final optimization function is that the system adjusts the random noise image to make it look like it not only maintains the content of the ordinary style image, but also becomes a special image with a certain style.

3   Conclusion

In general, the application of machine learning in the field of graphic design can provide better technical support for graphic design. Of course, any algorithm is not perfect. Deep learning convolutional neural network is limited by certain kinds of filters of different artistic styles, so it still has some defects in graphic design. With the help of Deep Dream algorithm, Ostagram image stylized software can ignore the filter template restrictions and can recognize the contents of any two images and transfer the image style of any one image to the other. Of course, the application of this algorithm in graphic design is relatively complex. In particular, it takes a lot of memory resources and events to run. Therefore, in order to give graphic design users a better experience, it is necessary to run processing in the background server at the same time. In this way, on the one hand, it can realize graphic design image stylistic processing faster, and at the same time, it can also avoid occupying large background memory resources. However, in terms of the current application status, the application of mobile terminals is still limited to some extent. In a word, the development and improvement of graphic design image stylization always need the support and technical optimization of machine learning algorithm. Only by constantly optimizing and updating machine learning algorithm can more graphic design users create their own artistic style works and bring better technical experience to more users.

References

1. Clohessy, J.W., et al.: Development of a high-throughput plant disease symptom severity assessment tool using machine learning image analysis And Integrated Geolocation. Comput. Electron. Agricult. **184**, 106089 (2021)
2. Pei, J., et al.: Application of machine learning in apple intelligent production. J. Jilin Agricult. Univ. **12**(04), 1–10 (2021)
3. Kaiyong, L.I.: Image edge detection system for uneven illumination based on machine learning. J. Laser Sci. **42**(03), 130–134 (2021)
4. Feiniu, Y., Zhiqiang, L., Jinting, S., Xue, X., Ya, L.: Image dehazing based on two-stage feature extraction strategy. J. Image Graphics **26**(03), 568–580 (2021)

Cyber Intelligence for Health and Education Informatics

Blended Teaching of Business English Based on Semantic Combination and Visualization Technology

Jinhua Zhang and Fei Li[✉]

Changchun Humanities and Sciences College, Changchun, Jilin, China

Abstract. Because business English plays a great role in international trade, business English teaching methods are getting more and more attention. This article analyzes the problems and their causes of business English teaching in universities in the information age, proposes solutions to these problems, and makes suggestions on the development of business English teaching platforms. This article follows the basic laws of teaching and learning, focusing on how to use semantic combination and visualization technology to build an information-based business English teaching platform. By integrating the process of teaching, learning and practicing, it is hoped that the purpose of constructing a real teaching environment can be achieved through online business English learning.

Keywords: Business English teaching · Semantic combination and visualization · Teaching methods · Reform

1 Introduction

My country's need to integrate into the international market With the continuous deepening of reform and opening up, my country's economy continues to develop and has become an important part of the international market. In this context, the importance of business English has gradually emerged, and my country's demand for business English translation professionals has also increased. Communicating with foreign business partners requires a good command of spoken business English; most business websites use English as the official language, and only those who are proficient in business English can read the content of the website, so that they can keep up with the development trend of world trade. With the development of economic internationalization and global integration, cooperation between countries has become increasingly frequent. As the world's largest developing country and the world's largest trading nation, China's economic advantages continue to increase. The influx of multinational companies has increased the domestic demand for business English professionals [1–3].

With the development of network technology, the network market has also expanded rapidly and has become an important trading place. In the "Internet+" era, Internet trade is an irreversible trend. The famous international e-commerce company Amazon, domestic e-commerce companies such as Taobao and JD.com have had a profound impact. E-commerce not only brings great convenience to people's lives, but

J. Abawajy et al. (Eds.): ATCI 2021, LNDECT 81, pp. 535–540, 2021.
https://doi.org/10.1007/978-3-030-79197-1_78

also plays a major role in promoting the optimization of China's trading environment. In this environment, proficient use of business English is extremely important [4–6].

2   Current Status of Business English Teaching

In the current education system, teachers focus on improving students' reading and writing skills, but ignore the development of listening and speaking skills. The course teaching of college business English still cannot completely overcome this kind of "dumb English", and the teaching situation that emphasizes reading and writing but neglecting listening and speaking still exists. However, the focus of business English teaching should be on the practical application of the language. In business negotiations, both parties must not only understand the terms of the contract and sign the agreement, but they must also be able to communicate in English proficiently. Therefore, it is very important to master the skills of listening, speaking, reading and writing. This requires us to abandon the "dumb English" in traditional teaching and achieve a balanced development of listening, speaking, reading and writing [7, 8].

Business English is a branch of the English subject. As an application of English in the field of trade, business English has its characteristics different from everyday English. First of all, business English not only has its language characteristics as English, but also has characteristics that conform to trade theory and business negotiation. Therefore, the teaching of business English should pay special attention to students' understanding and mastery of business vocabulary. Secondly, business English is not only a language tool for business negotiation, but also a manifestation of business etiquette. If the negotiator does not understand the business culture of the negotiating party during the business negotiation process, the negotiator may offend the other party.

Business English is an applied English, and its basic teaching goal is to cultivate students' ability to use business English in the course of trade. However, most of the current business English teaching lacks a good language environment, students can only learn English in the classroom, they have no opportunity to use business English in social practice, so all the learning is just on paper [9, 10].

3   The Ecological Environment of Classroom Teaching

Classroom is the most important element in the classroom teaching ecological environment. It is an important space that promotes the interaction between the teaching subject and the surrounding environment. With the advent of the information age, smart classrooms have been widely used in colleges and universities. It is a new type of education method and modern teaching method. Based on the Internet of Things technology, it integrates intelligent teaching, intelligent environmental adjustment, video monitoring and remote control, which effectively promotes the further development of information-based classroom teaching and improves the utilization rate of teaching resources. The development of high technology such as the Internet, big data and artificial intelligence provides strong technical support for the construction of an intelligent classroom teaching environment. At the same time, teachers can also popularize the application of various teaching software in the teaching process. For example, business English listening and speaking courses have applied various APPs, superstar learning, fun dubbing, one-click projection and other software are available for teachers and students to use. The teaching process has a dynamic effect, and the teacher-student interaction is diversified.

Classroom layout and seating arrangements have varying degrees of influence on the teaching and learning of teachers and students. The layout and seating of smart classrooms are different from those of traditional classrooms, and most of them are the display mode of round table conferences. It is more convenient to create a group discussion atmosphere and promote various interactive teaching links in such a classroom. The inconsistency of the ecological position of the classroom space will directly affect the construction of an effective informational classroom teaching environment. Therefore, transformation methods promote effective teaching while rationalizing classroom seating arrangements and reducing the distance between teachers and students. Business English courses are a two-way interactive process between teachers and students, rather than a one-way interactive process from teacher to student. Through computer operation, the environment in the classroom, such as posters or other decorations, will also affect students' understanding of the subconscious mind. For example, if business English teaching is conducted in a corporate simulation training room, various entrepreneurial posters on the wall can enhance students' awareness of business English.

4   Business English Teaching Method Based on Semantic Combination and Visualization Technology

To achieve effective listening and speaking of business English, semantic combination and visualization technology should be used, and the interdependent relationship between teaching content and teaching environment should be formed. Only when the teaching process of listening and speaking is balanced in the classroom ecological environment can the teaching goals be achieved and the effective communication between teachers and students, students and teaching materials can be achieved. Constructing an effective listening and speaking classroom teaching environment allows students to boldly express their views and confidently speak their own ideas, so that students can feel the charm of business English language. At the same time, teachers should follow the rules of teaching and pay attention to the individual development of students. In teaching, teachers should guide students to increase their awareness of using business English. The combination of semantics and visualization technology promotes the integration of knowledge and makes classroom teaching more vivid and effective, which not only maximizes information, but also promotes the improvement of students' oral expression ability. The effective communication between teachers and students promoted each other's progress. It not only completed the process of teaching and learning, but also created a win-win situation and created a harmonious atmosphere, from "I want to learn" to "I want to learn". The adaptation of the classroom ecological environment determines the balance of the teaching ecological environment.

The traditional business English courses in Chinese universities are mainly conducted through information teaching methods such as "multimedia network teaching" and "computer-assisted teaching". The traditional teaching classroom model is not suitable for business English classroom. Teachers should use semantic combination and visualization technology to adapt to the information background, integrate teaching models, teaching content, teaching skills, etc., to better guide students to adapt to the classroom teaching ecological environment, and actively integrate into the English listening and speaking teaching process.

In traditional business English listening and speaking teaching, the main teaching tool for teachers is textbooks. Teachers only teach the content of textbooks based on textbooks, and students learn the content of Business English based on the arrangement of textbooks and teachers. In addition, in traditional teaching, the main tool for teachers to cultivate students' English listening and speaking ability is English recording. Students use teachers to play recordings to train their listening and speaking. However, whether it is a textbook or a recording, the learning method is relatively simple, and the amount of knowledge that can be accommodated is relatively small. Therefore, students will inevitably get bored and lose interest in learning. In modern network semantic combination and visualization technology and semantic combination and visualization technology and equipment teaching, business English teaching can make full use of various learning tools, as mentioned above, that is, the smart classroom. Through the use of smart classrooms, business English can use applications, videos, real business meetings, etc., so that students can understand how real business conversations and interactions work.

5   Semantic Combination and Visualization Technology are the Technology Application Trends of Professional Education and Teaching in Various Disciplines Today

Virtual reality can solve many problems such as insufficient teaching resources, limited teaching venues, teaching image, lack of interaction, and dangerous training operations. The animation major of the college cooperates with the business English major to establish scientific research projects to research and develop the application of semantic combination and visualization technology in business English, practical English, and hotel English education and teaching. For example, in the course of professional English course teaching, a realistic simulation of the English scene of the job interview is carried out, and the VR head-mounted display is used to test the English of job candidates in a virtual situation. By using semantic combination and visualization technology, teachers can set a series of concrete English situations and tasks for student applicants, and observe how student applicants deal with tasks and how to solve problems when they encounter problems. During the VR simulation interview process, applicants can move freely in the virtual world with a 720-degree full view, and can use the tracking motion control system to move virtual objects. The teacher observes the students language communication ability through the head-mounted display monitor. Investigation and evaluation.

With the advent of the 5G era, online teaching will be the future education and teaching trend, which can greatly save social and human resources, and truly achieve the ideal of equality of educational resources advocated by Confucius. At present, students lack practical and effective interactive communication, teacher guidance, teaching supervision and evaluation. Virtual reality teaching will combine network technology to solve students' teaching experience, thereby creating a good interactive teaching environment and enabling students to enter the learning situation more vividly. The 5G network solves the technical bottleneck of virtual technology big data transmission, and provides a strong technical guarantee for the network virtual reality teaching of semantic combination and visualization technology such as mobile phones. The virtual screen can be clearer and smoother, and the immersive reality is enhanced. Experience.

Artificial intelligence has played a role in accelerating social development in today's society. Face recognition interactive technology, voice recognition interactive technology, map positioning and navigation technology and other technologies based on big data platforms enable us to work, study, travel, pay, and verify more accurate, faster and safer. The combination of virtual reality teaching and artificial intelligence can optimize the teaching system to recognize the image of students, capture the expressions and movements of the students, analyze and recognize the language and tone of the students, and complete the interaction of virtual reality scenes, images and languages. The most effective evaluation of students cognition and learning level, while analyzing students' personality characteristics and personal interests and expertise, to determine the direction and difficulty of learning for students, so as to truly teach students in accordance with their aptitude, maximize the potential of students, and make the best use of their talents. Make the best use of it.

6 Conclusion

In the context of international Internet commerce, my country's integration into the international market has accelerated, the number of multinational companies has increased, their role has become increasingly prominent, and the scale of cross-border e-commerce has continued to expand. These phenomena have given birth to the business environment in China, and the demand for English-speaking talents is also increasing. However, in most of the current business English classrooms, teachers emphasize reading, writing, listening and speaking, and lack of teaching. Lack of commercial features. In addition, the phenomenon of college English teaching being trapped in the classroom and lack of a social practice platform is also extremely common. The above phenomenon seriously restricts the training of business English talents. Therefore, in the context of "online education", the integration of semantic combination and hybrid business English teaching methods under the background of visualization technology can well solve the above problems and help cultivate high-quality business English talents.

References

1. Lai, Y.: Analysis of the practice of the reform of vocational English blended teaching under the background of the Internet+. Campus Engl. (24) (2018)
2. Lan, G.: Practice and exploration of teaching reform for business English major under the training of applied talents. J. Jilin Radio TV Univ. (6) (2017)
3. Li, D.: Research on SPOC blended teaching model of comprehensive business English course in the Internet age. J. Kaifeng Inst. Educ. (36) (2016)
4. Liu, Z., Liu, S.: Multimedia teaching environment design based on cloud computing. Modern Educ. Technol. (2) (2013)
5. Wei, L.: Business English ABCD integrated practical teaching system based on "three classrooms". J. Hubei Univ. Econ. (6) (2017)
6. Zhang, Y., Liu, Y.: Research and enlightenment of blended learning at home and abroad: from the perspective of language teaching research. J. Lanzhou Inst. Educ. (33) (2017)
7. Zou, L.: Research on the construction of business English online teaching resources. Times Educ. (13) (2018)
8. Xue, Y.: Research on computer aided teaching based on semantic combination and visualization technology. J. Lanzhou Univ. Arts Sci. (Since Nat. Sci. Ed.) 6, 325–329 (2017)
9. Li, X., Xu, J., Pan, C.: Research on the application of semantic combination and visualization technology in college practice teaching. Educ. Teach. Forum (6) (2018)
10. Zhang, J., Li, M., Li, Z., Li, W.: Research on the application of VR technology in actual combat teaching. Comput. Knowl. Technol. (12) (2016)

Detection of Traditional Chinese Medicine Pieces Based on MobilenNet V2-SSD

Yuanyuan Zhu[1], Yuhan Qian[2(✉)], Haitao Zhu[2], and Xuejian Li[2]

[1] Beijing University of Chinese Medicine, Beijing 100029, China
[2] Aerospace Times FeiHong Technology Company Limited,
Beijing 100094, China
yuhan_qian66@sbcmail.com.cn

Abstract. Purpose: The research on automatic detection of image sequence frames of Chinese medicine decoction pieces based on deep learning algorithms has important practical value and could be widely used in medical, production and teaching fields. In the past, the traditional method of extracting the underlying features of the image was used for detection, but this method cannot give a robust detection result under the image condition of a complex background. Therefore, object detection of it requires higher-level image expression methods. Method: After pre-processing the original image of traditional Chinese medicine pieces, this article innovatively proposes a data enhancement method to build a database containing 40 common Chinese medicine pieces (2778 images), as the model training and testing object, using MobileNet v2 replace VGG16 as a feature extraction network to improve the traditional SSD algorithm. Results: Mobilenet v2-SSD could achieve an average recognition accuracy of 78.3% in all the 40 kinds tested. Conclusion: MobileNet v2-SSD is ideal when multiple decoction pieces are shielded from each other and have a complex background, and it has certain application prospects in the future.

Keywords: Traditional Chinese medicine · Object detection · MobilenNetV2-SSD

1 Introduction

Chinese medicine decoction pieces are Chinese medicinal materials that have been processed and processed according to Chinese medicine theories and Chinese medicine processing methods, and can be directly used in Chinese medicine clinics. There are many kinds of decoction pieces of traditional Chinese medicine, and in the process of clinical application and production, it is difficult and workload to quickly identify and classify them one by one. In the actual application of traditional Chinese medicine decoction pieces, its rapid identification mostly relies on the sensory evaluation and empirical judgment of professionals with relevant knowledge of Chinese medicine, though the method is not efficient and still has the risk of misjudgment. With the development of artificial intelligence technology, deep learning technology has also made rapid progress. Due to its strong image expression ability and good generalization, deep learning has attracted wide attention from scholars at home and abroad. At present, the research on the

J. Abawajy et al. (Eds.): ATCI 2021, LNDECT 81, pp. 541–548, 2021.
https://doi.org/10.1007/978-3-030-79197-1_79

picture recognition of Chinese medicine decoction pieces mainly relies on artificially designed underlying image features including shape, color and texture. However, these studies have two main limitations: ① The images of traditional Chinese medicine decoction pieces used are all images with no background and single decoction pieces. This kind of over-idealized experimental scenes and samples is different from the large number of complex backgrounds, multiple decoction pieces, The images of real scenes that are occluded by each other do not match, so previous research is difficult to apply to practical applications; ② Under complex real scenes and backgrounds, low-level features are directly taken from picture pixels without high-level semantic feature information, so these features are easy It changes with the rapid change of the background and cannot be used as a reliable identification feature.

In order to make up for the above shortcomings and improve the accuracy of image recognition and search of traditional Chinese medicine decoction pieces, this article is based on the SSD and uses the advantages of MobileNet v2 to achieve a more complete automatic detection of traditional Chinese medicine pieces. At the same time, establish an image database of common Chinese medicine decoction pieces under complex backgrounds to provide a sample database for future research.

2   SSD Algorithm Network Structure

The convolutional neural network model used for object detection mainly includes: object detection based on Region Proposal [1], including R-CNN [2], Faster R-CNN [3]. Firstly, the method selects the candidate area and then classifies the features, which makes its own network structure complicated and the amount of parameter calculation is large, which affects the detection speed; object detection based on regression, including YOLO [4], SSD [5], etc. By directly performing object classification and bounding box regression on the a priori bounding box generated by the feature map, and outputting the detection result, its detection speed is greatly improved, but its detection accuracy for small object such as Chinese herbal medicine pieces has decreased. SSD absorbs the advantages of Faster R-CNN feature extraction, improves on the YOLO, and draws on the method of reducing the amount of calculation, so that SSD has a certain advantage in detection accuracy and speed. SSD is based on a regression detection model proposed by Liu et al. It uses VGG16 as the backbone network. The SSD model network structure is shown in Fig. 1. The fully connected layers fc6 and fc7 of VGG16 are converted into 3 × 3 convolutions. Conv6 and Conv7 after 1 × 1 convolution, and extract the higher-resolution Conv4_3 and Conv_7 shallow feature maps to detect large object, add four deep feature maps Conv8_2 with rich semantic information behind the backbone network VGG16, Conv9_2, Conv10_2, Conv11_2 to detect large object. Each feature map is independent of each other. The SSD model uses a priori box to extract the predicted values of these six different scale feature maps to classify and regress to achieve the detection of Chinese medicine decoction pieces. However, SSD uses a small number of shallow feature maps to extract the features of small object, resulting in insufficient extraction of detailed feature information for small object such as Chinese medicine decoction pieces, missed detections, false detections, etc., which is resulting in low detection accuracy.

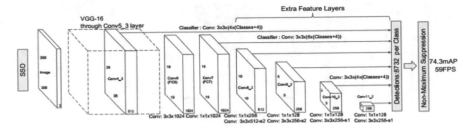

Fig. 1. SSD network structure

3   MobileNet v2

MobileNet v2 is an efficient convolutional neural network, mainly based on the MobileNet network to make two improvements. Firstly, MobileNet v2 draws on the residual structure of ResNet. In ResNet, the residual block first reduces the dimensionality of the input image, convolves, and then increases the dimensionality, while MobileNet v2 first increases the dimensionality of the input image, decomposes the convolution, and then dimensionality reduction, this process is called inverse residual. Secondly, MobileNet v2 improves the deep separable convolution, adding a 1×1 convolution before the deep separable convolution to ensure that the convolution process is performed in high dimensions. In order not to destroy the feature, the activation function after the second 1×1 convolution is removed, which could reduce the damage of the activation function to the feature in the low-dimensional convolution.

3.1   Depthwise Separable Convolution

Deep separable convolution is the main structure of efficient networks. Its main function is to reduce network parameters and speed up network operation. The process of standard convolution [6] is to use multiple convolution kernels with the same depth as the input data, perform convolution operations on them and then sum them to get the result. Depth separable convolution [7] decomposes the standard convolution process into two steps: the first step uses a single-channel convolution kernel to convolve each channel of the input data; the second step uses N kernels with the same depth as the input data. The 1×1 convolution kernel combines the results of the previous step to generate a new result (Fig. 2). The output of the deep separable convolution has the same dimensions as the output of the standard convolution. This decomposed form of convolution greatly reduces the amount of model calculation.

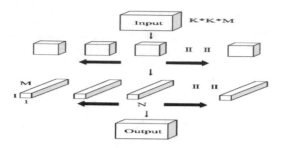

Fig. 2. Depth-wise convolutional filters

3.2   Inverted Residual Block

Random inactivation (Dropout) is a method used by most neural networks to alleviate overfitting, and it can also achieve regularization effects to a certain extent. In a neural network, the interconnection between neurons makes each neuron be affected by surrounding neurons during the parameter update stage, so that the dependence between different trained neurons is enhanced, and there is a complex co-adaptation relationship. Random inactivation can alleviate this effect between neurons to a certain extent, it reduces the dependence between different neurons, and reduces the occurrence of overfitting. The specific operation method is shown in Fig. 3. For each neuron in a certain layer, its weight is randomly reset to zero with probability p during training (the neuron is considered inactive at this time), and this is cancelled during testing. Set it so that all neurons work normally, and multiply the weight by p to ensure that the test phase has the same expectations as the training phase.

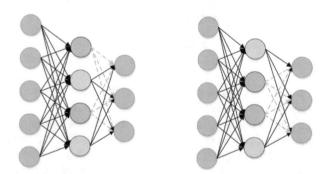

Fig. 3. Schematic diagram of random inactivation

4   MobileNet v2-SSD

In order to improve the resolution of the shallow feature map in the network structure of the improved SSD, the feature extraction of small object detection is strengthened, and the input image size is expanded from the original SSD algorithm 300 × 300 to 512 × 512. Extract the Lay7, Lay14, and Lay19 layers in the MobileNet v2 efficient network, remove the final average pooling layer and 1 × 1 convolutional layer of the network structure, and add the lay20, lay21, and lay22 convolutional layers that have undergone deep separable convolution. Set the corresponding (64 × 64 × 32), (16 × 16 × 96), (8 × 8 × 1280), (4 × 4 × 256), (2 × 2 × 256), (1 × 1 × 256) The output of these feature layers is the bottom-up input structure on the left side of the feature pyramid network in Fig. 4. The improved SSD algorithm network structure is shown in Fig. 4. The extracted feature maps of different resolutions and semantic features are output through feature pyramid network feature fusion, and the predicted values of each fusion feature layer are suppressed by non-maximum values [8, 9] Calculate the test results. In the experiment, in order to compensate for the increase in the amount of calculation caused by expanding the resolution of the input image, the depth_multiplier width factor of the input channel is set to 0.75, and the network model is reduced to improve the detection speed.

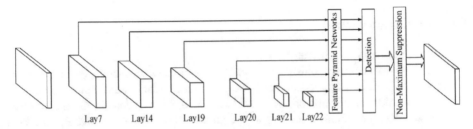

Fig. 4. MobileNet v2-SSD network structure

When using the MobileNet v2-SSD to detect traditional Chinese medicine decoction pieces, the high resolution of shallow feature maps could extract more detailed features to detect small object, and the low resolution of deep feature maps could learn more semantic features. It is used to detect large object. Due to the lack of close connections between the feature layers, the shallow feature layer lacks semantic information to detect the object. Therefore, the feature pyramid network [10] structure is adopted, as shown in Fig. 5, from the top on the right side. Then, the more abstract and semantic deep feature map is up-sampled by 2 times, and then the enlarged area in the Figure is horizontally connected and then upgraded to the previous shallow feature map with the same number of channels through 1 × 1 convolution, and the two features are After image fusion, a 3 × 3 convolution kernel is used for convolution to eliminate the aliasing effect of up-sampling and generate a new feature map. Each layer of feature maps integrates features of different resolutions and semantic strengths, thereby enhancing the semantic information of small object of shallow features to improve the detection performance of small object.

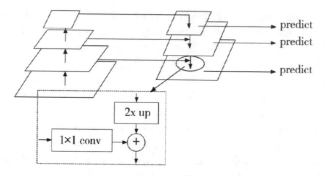

Fig. 5. Feature pyramid network structure

5 Data and Experiment

5.1 Data Augmentation

1) Color jittering (PCA jittering)

The color of an image object may sometimes change with changes in the environment, so in addition to geometric transformation, data could also be enhanced through color transformation, as shown in Fig. 6. First, calculate the mean value and standard deviation according to the three color channels of RGB, and then calculate the covariance matrix for eigen decomposition, and obtain the eigenvalues and eigenvectors for PCA jittering.

Fig. 6. Result of PCA jettering

2) Noise (Gaussian noise)

When a neural network tries to learn high-frequency features that may be useless, overfitting usually occurs, distorting the high-frequency features, which also means that lower-frequency data will gradually become distorted. Gaussian noise with zero mean basically has data points in all frequencies. Adding an appropriate amount of noise can enhance the learning ability. It appears as random black and white pixels spreading in the image. By adding Gaussian noise to the image, a lower level of information distortion is obtained and the robustness of the algorithm is improved. As shown in Fig. 7, the left is the original picture, and the right is the effect picture after introducing noise.

Fig. 7. Result of gaussian noise

5.2   Experimental Results and Analysis

In the experiment, the parameters are batch_size = 24, and the number of training iterations num_steps = 50000. The training set (90%) and the validation set (10%) of the 2778 pictures containing various Chinese medicine decoction pieces after data enhancement are divided into proportions to train the network model, and then the test set is tested on the model obtained through training, and analyze the detection results.

Table 1. Comparison of detection speed and mAP under different algorithm models

Algorithm model	Backbone network	FPS/frame	mAP/%
Faster R-CNN	VGG16	5	70.4
SSD	VGG16	23	72.5
MobileNet v2-SSD	MobileNet v2	28	78.3

Faster R-CNN, SSD, and MobileNet v2-SSD are used on the data set to complete object detection on the traditional Chinese medicine decoction piece image data set, and the comparison and analysis of detection speed and average accuracy rate under different algorithm models are shown in Table 1. The detection speed and detection accuracy of the MobileNet v2-SSD model are better than those of the faster R-CNN and the SSD. The number of frames transmitted per second in the detection speed is increased by 23 frames and 5 frames respectively, and the average accuracy mAP is increased respectively 7.9% and 5.8%.

References

1. Shi, K.H., Chiu, C.T., Lin, J.A., et al.: Real-time object detection with reduced region proposal network via multi-feature concatenation. IEEE Trans. Neural Netw. Learn. Syst. **99** (8), 1–10 (2019)
2. Cai, Z.W., Nuno, V.: Cascade r-cnn: high quality object detection and instance segmentation. IEEE Trans. Pattern Anal. Mach. Intell. **51**(23), 25–33 (2019)

3. Hung, G.L., Sahimi, M.S., Samma, H., et al.: Faster R-CNN deep learning model for pedestrian detection from drone images. SN Comput. **1**(8), 116–120 (2020)
4. Zhang, S., Wu, Y.X., Men, C.G., Li, X.S.: Tiny YOLO optimization oriented bus passenger object detection. Chin. J. Electron. **29**(1), 132–138 (2020)
5. Li, H.T., Lin, K.Z., Bai, J.X., et al.: Small object detection algorithm based on feature pyramid-enhanced fusion SSD. In: IEEE International Conference on Network Infrastructure, vol. 21, no. 73, pp. 336–340 (2019)
6. Lecun, Y., Bottou, L., Bengio, Y., et al.: Gradient-based learning applied to document recognition. Proc. IEEE **86**(11), 2278–2324 (1998)
7. Chollet, F.: Xception: deep learning with depthwise separable convolutions. In: IEEE Conference on Computer Vision and Pattern Recognition, pp. 1800–1807. IEEE Computer Society, Italy (2017)
8. Sandler, M., Howard, A., Zhu, M., et al.: MobileNet v2: inverted residuals and linear bottlenecks. Eprint Arxiv (2018)
9. Song, Y.N., Gao, L., Zhang, B., et al.: Improved non-maximum suppression for object detection using harmony search algorithm. Appl. Soft Comput. J. **81**(34), 54–61 (2019)
10. Chen, J.M., Jin, J., Wang, W.F.: Improved algorithm based on feature pyramid network. Laser Optoelectron. Prog. **56**(21), 165–170 (2019)

Data on Computer-Aided Health in Inner Mongolia Autonomous Region Based on Big Data

Ningping Yuan[✉]

College of Computer and Information, Inner Mongolion Medical University,
Huhehaote, Inner Mongolia 010000, China
20030156@immu.edu.cn

Abstract. It is of great significance to establish a scientific and perfect application model of medical and health big data analysis for promoting the application of medical and health big data. This paper mainly studies the computer-aided medical and health data analysis of Inner Mongolia Autonomous Region based on big data. This paper adopts literature research method, field research method and data science method. Literature research is used to understand big data platforms and their technologies, and to analyze the limitations of data analysis methods in traditional data warehouses. The application process of data analysis based on big data platform is studied by using data science method. The established medical and health data analysis application process based on the big data platform and the described medical and health big data application theme were consulted by using the expert consultation method, and they were modified and improved according to the expert opinions.

Keywords: Big data platform · Computer aided · Healthcare data · Data analysis

1 Introduction

With the continuous advancement of the informatization construction of medical service system, the application of medical and health big data in medical service is becoming more and more extensive. Kankanhalli et al. used support vector machine algorithm to classify and analyze the data sets of five diseases, including obesity, atherosclerosis, hyperlipidemia, hypertension and diabetes mellitus [1, 2]. Tawalbeh etc. Research shows that, at present in the smart devices (wearable sensors, body function monitor) and lifestyle data monitoring (diet, sleep), mobile devices (smartphones, tablets) and clinical data have a lot of work, the emphasis is on using big data technology will all kinds of data integration analysis, contribute to the understanding of a whole way of life not only, also helps to better prevention health care [3].

This paper summarizes and describes the application topics of big data in the field of health care, clarifies the application subjects and scenarios of big data in health care, and provides guidance for the development of big data application in health care. The medical and health data analysis application process based on the big data platform is

© The Author(s), under exclusive license to Springer Nature Switzerland AG 2021
J. Abawajy et al. (Eds.): ATCI 2021, LNDECT 81, pp. 549–556, 2021.
https://doi.org/10.1007/978-3-030-79197-1_80

designed to make the data analysis application process more suitable for the characteristics and application needs of the medical and health big data, so as to provide support for mining the value of the applied medical and health big data and promoting the application development of the medical and health big data.

2 Health Care Big Data Analysis

2.1 Health Big Data Source

2.1.1 Hospital Information System

Hospital information system is an information system that uses modern information technology to assist the medical service and operation management of hospitals. The hospital information system generates the medical database by collecting, storing, processing, extracting, transmitting, summarizing and processing the data generated in each stage of the medical activity [4]. To provide comprehensive, automated management and information services for the overall operation of the hospital.

2.1.2 Regional Universal Health Information Platform

Regional health information platform (hereinafter referred to as the regional platform) data mainly come from medical institutions, and public health institutions (such as the centers for disease control and prevention (hereinafter referred to as the CDC), health supervision institutions, blood center, emergency center, etc.), medical health administrative institutions and grassroots medical institutions, it also with people club department, financial institutions, insurance agencies, public security departments, civil administration institutions, industrial and commercial department, education department, statistics department and so on have further contact. The data of the regional national health information platform is characterized by a wide variety of sources, types, large amount of data, and uncentralized storage [5]. The data of the regional universal health information platform mainly includes three parts: medical service data, public health data and health and family planning management data.

2.1.3 Public Health System

Public health institutions refer to institutions for disease prevention and control, institutions for health supervision and management, institutions for maternal and child health care, institutions for mental health management, 120 emergency treatment centers, blood centers, etc. Public health facility data usually includes disease prevention and control data, health surveillance data, health emergency command data, medical data, women and children's health data, mental health data, blood management data, etc.

2.1.4 Internet Data

With the rapid development of mobile devices and mobile Internet, the data of diseases, health, medical services and consultation generated by major websites are increasing rapidly. Internet data includes data generated by various websites and health testing devices. Health monitoring data include personal health characteristic data generated by various mobile health devices, such as blood pressure, heartbeat, blood glucose, heart rate, weight, ECG respiration, sleep and physical exercise records, etc. [6].

2.2 Health Care Big Data Analysis Application Process

2.2.1 Data Collection

Data collection is realized by the unified data exchange platform in the big data platform, and each collection node uploads the required data by accessing the data exchange platform. Data acquisition and exchange include a variety of access methods, including but not limited to data tables, front-end libraries, real-time communication, files, data distribution, etc. When accessing the new system, the appropriate and best access methods should be selected according to the exchange requirements and char- acteristics of the new system. Data collection is divided into two forms: document - based and library - based. Based on the document collection, the node can upload the document directly, or the original library can be converted and uploaded by the doc- ument conversion tool. Based on the acquisition of the intermediate library, the node writes the data to the front machine and transmits it [7].

2.2.2 Data Integration

Health big data has a large amount of data and heterogeneous, sync (scheduling) and incomplete etc., so the original data of medical health data analysis application has no intention of difficulty, to provide a higher quality for the data analysis phase of the target data set, you usually need to data pre-processing, data preprocessing in general there are four ways: data cleaning, data integration, data code and data transformation [8]. In the original data of the large data for data preprocessing to medical health after the operation to form the target data set, the need to store data set to the big data platform, data size, because of the large medical health obviously does not meet the traditional database storage mode, use of big data distributed storage architecture can be efficient storage of medical health data.

2.2.3 Data Analysis

The core of the application process of medical and health data analysis based on big data platform lies in data analysis. The data analysis module mainly analyzes and calculates the high-quality medical and health data set completed by collection and integration, so as to extract the inherent correlation rules of the data. Data analysis is divided into two parts: data analysis and data calculation. (1) Data analysis: There are many methods to analyze medical big data, which are mainly divided into two types: one is SQL-like queries on medical data sets, such as Hive analysis and Pig analysis. The other is mining and analyzing medical data sets, such as classification mining and text mining. (2) Data calculation: The data computing method of medical and health data is distributed computing. Because of the large amount of medical and health data, the use of distributed computing method can save the overall computing time and greatly improve the computing efficiency. On the basis of designing the data analysis method, we can use the distributed computing framework MapReduce to calculate the data set. The core idea of the architecture is the task decomposition and results of the merge, first by the Map () function to receive the input Key/Value pair (Key, Value), then a break large tasks into smaller shard (Slip), the intermediate results are computed by collection of Key/Value pair (Key, Value), finally by the Reduce () function will all intermediate Value of the same Key Value Value to merge [9, 10].

2.2.4   Data Application Display

The data application display module provides the human-computer interface between the user and the system, through which the user can conduct data analysis and application operations. For example, the hospital management receives the visualization results of the application subject data analysis application, such as the statistical analysis of each data and the sharing and consulting of medical data. The doctor can input the patient data and compare the results of the analysis and calculation of the optimal treatment choice in the system to arrive at the treatment plan recommendation for the patient.

2.2.5   Big Data Basic Algorithm

Mean square error

$$RMSE = \sqrt{\frac{\sum_{(u,i)\in T}\left(r_{ui} - r'_{ui}\right)^2}{|T|}} \tag{1}$$

Mean absolute error

$$MAE = \frac{\sum_{u,i\in T}\left|r_{ui} - r'_{ui}\right|}{T} \tag{2}$$

Accuracy

$$\mathrm{Pr}\,ecision = \frac{\sum_{u\in U}|R(u)\cap T(u)|}{\sum_{u\in U}|R(u)|} \tag{3}$$

3   Data Analysis Practice

3.1   Data Acquisition

Due to the limited ability of the author, the data collection of this application practice is only to collect 1000 data of diabetes patients from the hospital information system of a third-class A hospital in Inner Mongolia through database export.

3.2   Data Preprocessing

The data pretreatment of this application practice is mainly to carry out data cleaning operation. According to factor level statistics, a total of 62 patients with only type 2 diabetes and no complications were obtained in the 1000 pieces of data. A total of 64 hospitalization records; the admission wards are in 10 types of wards; there are 14 attending physicians; there were 236 common treatment schemes; the five categories of treatment plan are drugs, examination, testing, nursing and treatment. Because the other categories are complicated and most of the contents are included in the first six categories, other categories are not considered. The types of examination can be divided into 3 categories, although there are only 3 categories, but the contents of examination are different, which can be withdrawn in the treatment plan; Among the 64 admission records, there were 127 examinations and 474 examinations in total. For missing values in patient data, mean supplement method was used. If the missing value is of distance type, the average value of the existing value of the attribute is used to fill in the missing value, such as hospital stay, etc. If the missing value is non-distance type, according to the mode principle of statistics, the mode of the attribute (that is, the value with the highest frequency of occurrence) is used to complement the missing value, such as the examination number.

4   Data Analysis Results

4.1   K Mean Clustering Method

4.1.1   Clustering is Conducted According to the Four Categories

Table 1. Clustering the number of different samples according to 4 categories

Category	Number of samples
1	4
2	25
3	25
4	10

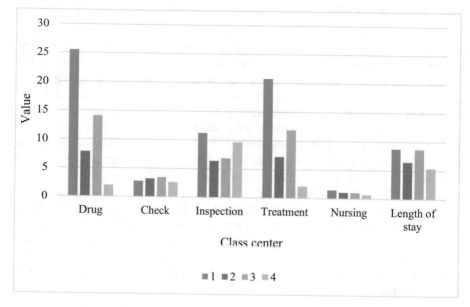

Fig. 1. Cluster the center of each class according to the 4 classes

As shown in Table 1 and Fig. 1, the K-means clustering method was used to cluster according to 4 categories, and the number of samples in 4 categories was obtained. If classified according to four categories, from the point of sample, 2 and 3 category accounted for 78% of the total sample, from the point of such centres, in addition to the examination and hospitalization days in class 1, other projects are significantly higher than the other three kinds of treatment, class 2 and class 3 major difference is that the difference of drugs and treatment project, class 4 check in other projects were significantly lower than other three categories.

4.1.2   Clustering is Conducted According to 5 Categories

Table 2. Cluster the number of samples according to 5 categories

Category	Number of samples
1	4
2	15
3	19
4	17
5	9

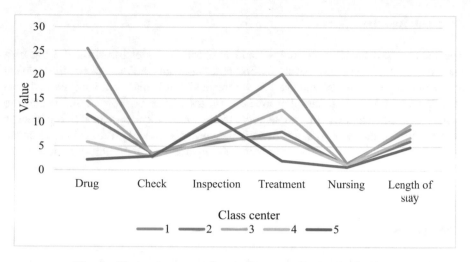

Fig. 2. Cluster the center of each class according to the 5 classes

As shown in Table 2 and Fig. 2, by comparing the sum of deviation squares of clustering variables in each category based on clustering in 4 categories and clustering in 5 categories, it can be found that the sum of deviation squares of clustering in 5 categories is small, so patients in 5 categories are classified in the following part. Patients were divided into 5 categories through descriptive statistics of treatment plans and patient clustering. Although the results of K-means clustering were difficult to explain, the characteristics of patients in the 5 categories were different, and the evaluation of treatment plans could also be obtained by classifying the treatment plans according to the characteristics of patients.

4.2 Analytic Result

The number of frequent item sets with the highest support among the largest number of items under different patient classes was 1, which ensured the uniqueness and consistency of the results. It can be preliminarily regarded as the recommendation of treatment options for different types of patients. For the data analysis and application of 1000 data of diabetes patients in this practice, it can be seen that the data mining method can effectively show the data association rules in medical and health data and assist in diagnosis and treatment. However, for the analysis of medical and health big data with large amount of data, big data analysis technology is also needed to process it, and distributed computing framework and distributed file system are used for medical and health big data analysis and application.

5 Conclusions

The medical and health data analysis and application model based on the big data platform is divided into two parts: the first part is the application topic of the medical and health big data; the second part is the medical and health data analysis and processing based on the big data platform. Firstly, the source of medical and health big data is defined, and then the application theme of medical and health big data is defined. Finally, the medical and health big data is analyzed and processed based on the big data platform to realize the application. Compared with the traditional data warehouse data analysis application model, this paper proposes a medical and health data analysis application model based on the big data platform.

References

1. Zhang, Y., Qiu, M., Tsai, C.W., et al.: Health-CPS: healthcare cyber-physical system assisted by cloud and big data. IEEE Syst. J. **11**(1), 88–95 (2017)
2. Kankanhalli, A., Hahn, J., Tan, S., Gao, G.: Big data and analytics in healthcare: introduction to the special section. Inf. Syst. Front. **18**(2), 233–235 (2016). https://doi.org/10.1007/s10796-016-9641-2
3. Tawalbeh, L.A., Mehmood, R., Benkhelifa, E., et al.: Mobile cloud computing model and big data analysis for healthcare applications. IEEE Access **4**(99), 6171–6180 (2017)
4. Jiang, P., Winkley, J., Zhao, C., et al.: An intelligent information forwarder for healthcare big data systems with distributed wearable sensors. IEEE Syst. J. **10**(3), 1147–1159 (2016)
5. Chen, M., Yang, J., Zhou, J., et al.: 5G-smart diabetes: toward personalized diabetes diagnosis with healthcare big data clouds. IEEE Commun. Mag. **56**(4), 16–23 (2018)
6. Sarkar, B.K.: Big data for secure healthcare system: a conceptual design. Complex Intell. Syst. **3**(2), 133–151 (2017). https://doi.org/10.1007/s40747-017-0040-1
7. Liu, C.Y., Chiou, L.J., Li, C.C., et al.: Analysis of Beijing Tianjin Hebei regional credit system from the perspective of big data credit reporting. J. Vis. Commun. Image Represent. **59**(FEB.), 300–308 (2019)
8. Jun, C., Bo, K., et al.: A big data analysis platform based on the manufacturing specialized library: a case study on implementation of the platform for quality problems. J. Korean Inst. Ind. Eng. **43**(5), 380–387 (2017)
9. Majeed, A., Lv, J., Peng, T.: A framework for big data driven process analysis and optimization for additive manufacturing. Rapid Prototyp. J. **25**(2), 308–321 (2018)
10. Yazdani, M., Kahraman, C., Zarate, P., et al.: A fuzzy multi attribute decision framework with integration of QFD and grey relational analysis. Expert Syst. Appl. **115**(JAN.), 474–485 (2019)

Application of Data Mining Based on Rough Set in the Evaluation of University Teachers' Wisdom Teaching

Dongyan Su[✉]

School of Information Engineering, Jilin Engineering Normal University,
Changchun, Jilin, China

Abstract. In recent years, the country is in an environment of rapid development, vigorously promoting scientific and technological innovation, and continuously increasing investment in university research, and the university itself also attaches great importance to scientific research. In the current background that university teachers pay much attention to teaching evaluation, universities are using Rough set data mining technology to better establish the teaching evaluation system so as to better evaluate their own teaching. In the context of smart education, education big data has been generally valued, and college student classroom behavior data as teaching process data has become a research hotspot in modern education. This paper takes the basic concepts and related technologies of Rough set theory and data mining as the theoretical basis of the research, and integrates its important content to analyze and research the improvement of the evaluation of college teachers' wisdom teaching. This paper takes the data mining of the classic Rough set as the research object, and optimizes and improves the decision number method of teaching evaluation. Data mining through Rough sets can be regarded as an adaptive sorting algorithm. Therefore, the application of data mining technology in smart campuses of colleges and universities can create many opportunities and provide many conveniences for the development of colleges and universities by improving the evaluation of smart teaching of the teacher team. It is also conducive to comprehensively improving the management capabilities of universities. The experimental results show that this research has a better effect on using Rough set and data mining to evaluate the wisdom of teachers in colleges and universities.

Keywords: Rough set · Data mining · Smart education · College teachers

1 Introduction

Data mining, as the name implies, is to extract or "mine" available data from massive data. In many data mining methods, Rough set is a more effective data mining method for using complex systems. Except for the data set, the problem can be solved without providing preliminary information [1]. Rough set theory is a mathematical tool for dealing with fuzzy and uncertain knowledge. By reducing irrelevant data, rules can be derived to make decisions or classify problems. In order to correct the shortcomings of traditional university teacher teaching evaluation, we adopted modern data analysis technology in university teacher teaching evaluation, and used appropriate technology in data mining, and evaluated and summarized university teacher teaching through fair and objective statistical data. Rough set data mining is used to identify potential patterns and help decision makers adjust strategies [2].

University teachers are the main part of academic research and the cornerstone of academic research development. The teaching evaluation of college teachers has become a measure of the development potential of college teachers, and it plays an important role in the evaluation of professional titles, job evaluation, and allowances and benefits [3]. The quality of university teachers directly affects the quality of university education. At present, the research on our country's teacher evaluation system is still in its infancy [4]. From qualitative evaluation to quantitative evaluation, although the scientificity and objectivity of performance evaluation have been improved, there is no algorithmic analysis of the entire university teacher teaching evaluation system. Through statistics of relevant data and information from university teacher teaching evaluation [5], the establishment. The teaching evaluation system of college teachers and the application of Rough set data mining technology to the education management system with complex data is of great significance [6].

Higher education has gone international, and the survival competition between universities is also intensifying. The main competitiveness is the quality of university education and teacher quality education is the cornerstone of higher education institutions [7]. In the information age, it is necessary to analyze and research the key smart campus technologies in universities, and the data mining technology of Rough set is being applied to the smart teaching evaluation system in campus [8]. Thanks to the advantages of Rough set data mining technology, a large amount of complex data can be processed in intelligent colleges and university campuses. According to Rough set data mining technology, management departments can make effective decisions, and the use of this technology is also conducive to promoting colleges and universities to obtain further development. This paper takes the data mining technology analysis of the Rough set as the starting point, and then analyzes the application of the data mining technology of the Rough set to the teacher teaching evaluation in the smart campus of colleges and universities to promote the wide application of the technology [9]. Based on this research idea, this article explores the use of data mining technology to carry out related research around the construction of university teachers' scientific research evaluation indicators, the analysis of factors affecting scientific research performance [10], and the analysis of scientific research input and output efficiency. It not only has a certain theory Research value is also of great significance for further optimizing scientific research incentive policies and stimulating the innovation vitality of college teachers.

2   Rough Set Algorithm

Rough set attribute selection metric is a selection separation criterion, which is a heuristic method to divide the data (marked with class markers) from a specific learning step. The attribute with the best metric score is selected as the global attribute of a specific tuple. The attribute with the highest information gain is selected as the global attribute of node X. This attribute minimizes the amount of information required to classify tuples in the attribute distribution. Recognizable matrix is given a teaching evaluation information system $S = \{U^*, A, V, f\}$, S is a collection of attributes, $C = \{a_1, a_2, a_3, \ldots, a_i\}$. Is a set of conditional attributes and decision-making attributes, the expression formula of its universe is:

$$C_D(i,j) = a_k, a_k \in \{C \wedge a_k(x_i) \neq a_k(x_j)\}, d(x_j) \neq d(x_j) \tag{1}$$

Information gain is defined as the difference between the initial information request (only based on the rate of the category) and the new question (that is, obtained after Part A) by dividing and selecting the attribute with the highest information gain as the formula for dividing attribute node U as follows:

$$U^* = \{x_1^*, x_2^*, \ldots, x_n^*\} \tag{2}$$

Calculate the proportion of the i-th evaluation index of the j-th node in the evaluation factor:

$$p_{ij} = r_{ij} / \sum_{i=1}^{n} r_{ij} \tag{3}$$

$$H_j = -K \sum_{j-1}^{n} P_{ij} \ln P_{ij} \tag{4}$$

Where the constant $K = 1/\ln n$ this can guarantee $0 \leq H_j \leq 1$:

$$W_j = \frac{1 - H_j}{\sum_{j=1}^{n} 1 - H_j} \tag{5}$$

3   Model Establishment

Based on the above formula, establish a database information teaching evaluation system model, $N = \sum_{i=1}^{K} N_k$ is the total number of samples, $m_k = \frac{1}{N_k} \sum_{x=\partial_k} x$. Represents the mean value of the k-th sample, $m = \frac{1}{N} \sum_{K=1}^{K} \sum_{x=\partial_k} x$ is the overall mean. Define the interclass divergence matrix as:

Use the database information teaching evaluation system model to calculate the fuzzy expansion estimation value V conversion formula according to the expansion estimation matrix W and the weight S:

$$V = W \cdot S^T \tag{6}$$

The algorithm is used to analyze the set of all objects whose level value is to effectively reduce the interference between different factors. As a classification loss function, it is used to distinguish each category existing in the corresponding object subset, as shown in the formula:

$$L_{cls} = -\frac{1}{N_{ds}} \sum_i \log[p_i^* p_i + (1 - p_i^*)(1 - p_i)] \tag{7}$$

Among them, p_i is the target in the i node p_i^* is the non-target in the i node. L_{box} it is the regression function of the detection frame, which is mainly used to correct the node coordinates of the foreground to obtain the best detection frame:

$$L_{box} = \lambda \frac{1}{N_{reg}} \sum_i p_i^* L_{reg}(t_i, t_i^*) \tag{8}$$

Among them, p_i^* L_{reg} only if there are foreground nodes $p_i^* = 1$ There is a regression loss when there is no foreground node. $p_i^* = 0$, No return loss, L_{reg} is the regression loss function, and its formula is:

$$L_{reg} = R(t_i - t_i^*) = |t_i - t_i^*| - 0.5 \tag{9}$$

The change of the information source, that is, the statistical measure of the information source $H(n_i)$ decided correct $H(n_i)$. The estimation of the path mainly comes from the product of its distance and the cost δ spent within the unit distance. According to the distance formula $H(n_i)$ estimated method:

$$H(n_i) = \sigma(|x_{Goal} - x_n| + |y_{Goal} - y_n| + |z_{Goal} - z_n|) \tag{10}$$

Put the start node Start into the OPEN list, look for the node i with the smallest F value in the OPEN list as the current node and judge whether it is the end node Goal. If not, move i to the CLOSE list and calculate the neighborhood of node n_i Value of each node in (Table 1):

$$H_c(n_i) = \sigma \max(|x_{Goal} - x_n|, |y_{Goal} - y_n|, |z_{Goal} - z_n|) \tag{11}$$

Estimated method:

$$G_{i,j}(i,j) = 4 \sum_m \sum_{m=-2}^2 G_{i,k-1}\left(\frac{2i+m}{2}, \frac{2j+n}{2}\right) \tag{12}$$

Table 1. The relationship between the EXPAND function and the parameters

Variable	fund	fund-	turn	turn-	pe	pe-	sec	sec-	vol	vol-
Correlation coefficient	0.578	0.345	0578	0.421	0.652	0.425	0.341	0.107	0.248	0.342

4   Evaluation Results

4.1   Application of Rough Set Data Mining Technology in Teachers' Teaching Evaluation

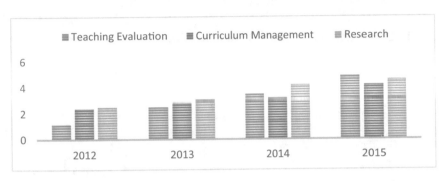

Fig. 1. The application of Rough set data mining technology in teacher teaching evaluation in the past few years

From Fig. 1 from 2012 to 2015, the application of data mining technology in the evaluation of the data mining technology used by college teachers in middle school from 2012 to 2015. In the benchmark database used to evaluate and analyze university education, in the basic information table extracted from teacher teaching evaluation, the teacher may be an academic administrator or administrative staff and does not accept classwork, so there is no teacher in the teacher performance data Achievements, this part of this entry should be deleted. We are studying some students, so some teachers may not have students to be graded, so they can use the data mining technique of the rough set to fill in the homework. When building a smart teaching evaluation system, it is necessary to use data to meet the needs of college teachers, including providing academic research and learning data for teachers and sharing resources between teachers and students. In addition, it can also provide relevant data support for administrators and management data for faculty and staff. The above content needs to be supported by Rough set data mining technology, including system application and perception, network integration and platform management. For the construction of a smart campus based on Rough set data mining technology, it should be carried out from the following three levels. First, smart campus refers to the construction of networking and digitalization. With the support of Rough set data mining technology, college wisdom. The construction of the campus includes teaching, scientific research, management and other content. Teachers, faculty, staff, and students can use information technology to intervene in the campus network to use and share various information resources. This kind of Rough set data mining technology management, which is not restricted by time and space, realizes the informationized campus management mode. Secondly, from the aspect of composition and construction, the main construction of a smart campus includes software and hardware. The hardware facilities include the construction of central processing units, smart sensors, and various networking equipment. The smart sensor refers to the connection of dormitories, libraries, classrooms, etc. on campus in the Internet of Things mode, and then the application software is combined with the Internet through the data mining technology of the Rough set, so as to evaluate the teaching evaluation of teachers in information universities. Implementation of the evaluation.

4.2   The Importance of Teaching Evaluation Grade

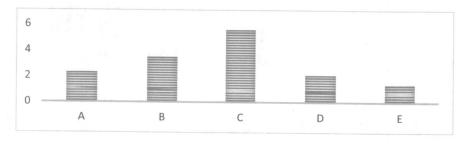

Fig. 2. Level establishment of teaching evaluation levels

From Fig. 2 based on the statistics of college teachers' wisdom teaching evaluation, teaching evaluation grades are established. In order to ensure that the evaluation results of teacher teaching evaluation are accurate and reliable, an indicator system that can reflect the basic quality of teacher teaching is selected. Each rating indicator is divided into five levels: A, B, C, D and E. Use a special method to calculate the corresponding five parts, for the average score, assign different weights to the scores of different components, and then calculate the total score. The data of the evaluation results of university teachers' wisdom teaching is mainly aimed at providing simple statistical reports and information requests to management departments at all levels. In fact, these data also contain more valuable information, such as: the relationship between teacher positions and school occupations and education quality; does the teaching development of teachers contribute to the quality of education? If the intrinsic value of these data can be fully utilized, the wisdom teaching of college teachers will continue to improve. Traditional data analysis technology can not eliminate the association and classification of data. Rough set data mining technology is an important step in the knowledge discovery process, and it is also the focus of modern knowledge discovery field research. In the teacher's wisdom teaching information system based on Rough set data mining technology, the data records in the information represent objects with their attribute values. Use Rough set data mining technology to mine embedded processed data and identify potential data patterns. It can thoroughly analyze the hidden internal relationships between test results and various factors, and extract rules with application value. Make an objective and fair assessment of the evaluation and evaluation of university teachers' wisdom teaching. Moreover, the entire data analysis process in the data mining technology of Rough set is automatic. It has a good open structure and a user-friendly interface, which can provide powerful technical support for the evaluation of college teachers' wisdom teaching evaluation. The application of the data mining technology of Rough set in the evaluation of school teachers' wisdom teaching evaluation has very important theoretical and practical significance for the realization of scientific and effective learning evaluation. It can improve the evaluation of college teachers' wisdom teaching, thereby improving the overall quality of university education.

5   Conclusion

All in all, this paper uses the Rough set theory to reduce the attributes of the data table. After getting multiple possible reductions, use information gain to determine the approximate optimal reduction set. On this basis, the decision tree is further constructed, and used in the comprehensive analysis of teaching evaluation system. This article uses the mining of sample data. Provides a mode for reasonable mining. The data mining technology of Rough set has significant value to the overall construction of college smart campus. Therefore, for colleges and universities, the data mining technology of Rough set should be actively used to build a smart campus, which provides full convenience for the work and learning of teachers and students, and can also assist students in their campus life, using the association rules in data mining. The teaching quality of college teachers is evaluated, and the teaching evaluation system based on Rough set data mining technology is used to mine and evaluate the teaching evaluation data accumulated by colleges and universities for many years. The evaluation results are provided for the decision-making of college teaching management and education administration departments. It is helpful, especially for teachers' teaching work.

Acknowledgments. This work was supported by the 13th Five Year Plan Project of Education Science in Jilin Province in 2019 under Grant ZD19086.

References

1. Begum, S., Sarkar, R., Chakraborty, D., et al.: Identification of biomarker on biological and gene expression data using fuzzy preference based rough set. J. Intell. Syst. **30**(1), 130–141 (2021)
2. Chen, P., Lin, M., Liu, J.: Multi-label attribute reduction based on variable precision fuzzy neighborhood rough set. IEEE Access **8**, 133565–133576 (2020)
3. Sudhakar, T., Hannah, I.H., Senthil, K.S.: Route classification scheme based on covering rough set approach in mobile ad hoc network (CRS-MANET). Int. J. Intell. Unmanned Syst. **8**(2), 85–96 (2019)
4. Wu, X., Kumar, V., Quinlan, J.R., et al.: Top 10 algorithms in data mining. Knowl. Inf. Syst. **14**(1), 1–37 (2008)
5. Witten, I.H., Frank, E.: Data mining: practical machine learning tools and techniques. ACM SIGMOD Rec. **31**(1), 76–77 (2011)
6. Franklin, J.: The elements of statistical learning: data mining, inference and prediction. Math. Intell. **27**(2), 83–85 (2005). https://doi.org/10.1007/BF02985802

7. Yang, C., Li, C., Wang, Y.: Research on the implementing strategies of precise instruction under the environment of smart education. High. Vocat. Educ. J. Tianjin Vocat. **028**(003), 28–32 (2019)

8. Clément, L., Fernet, C., Morin, A.J.S., Austin, S.: In whom college teachers trust? On the role of specific trust referents and basic psychological needs in optimal functioning at work. High. Educ. **80**(3), 511–530 (2019). https://doi.org/10.1007/s10734-019-00496-z

9. Li, X.: Analysis on the responsibility and pressure of college teachers. Int. J. Soc. Sci. Educ. Res. **2**(5), 19–22 (2019)

10. Buttitta, G., William, J.N., et al.: Development of occupancy-integrated archetypes: use of data mining clustering techniques to embed occupant behaviour profiles in archetypes - ScienceDirect. Energy Build. **198**(C), 84–99 (2019)

Surveillance and Disease Control in COVID-19: Big Data Application in Public Health

Muxi Zheng[⊠]

University of California, Irvine, CA, USA
muxi_zheng@yeah.net

Abstract. With the fast development of big data technology, its application in many fields is becoming more common. Big data has played an essential role in public health, and the influence of data usage expands with the rising of technology. This paper would give a brief overview of the current application of big data, specifically in the context of the COVID-19 pandemic and discuss the current issues and problems of data usage in public health related to privacy concerns. While Big Data is looking to be beneficial in many areas, especially in public health, better regulations to ensure data safety and privacy are urgently needed.

Keywords: Big data · Technology · Public health · Collect data

1 Introduction

Big Data is becoming widely applied to many aspects of our lives, including in the arena of public health, which is particularly relevant in the time of the COVID-19 pandemic [1]. The development of Big Data is mainly in three dimensions: volume, velocity, and variety [2]. Such development is beginning to allow better processing of data, including those used in public health. This would benefit not just the individuals but also allow better regulations on the population level, including research and public health surveillance.

1.1 Application of Public Health

On the individual level, Big Data is encouraging the development of personalized medicine. With Big Data's technology, the individual patients' medical examinations and records can be compared to a database with massive data. The data would be integrated to provide personalized medicine and therapy through the processing of the data with artificial intelligence (Fig. 1). This would promote better diagnosis, prognosis, and medical tools and be beneficial in producing solutions to diseases on a broader level, such as developing medicine and general intervention [3]. With such developments, there might be improvements in the general health care level.

© The Author(s), under exclusive license to Springer Nature Switzerland AG 2021
J. Abawajy et al. (Eds.): ATCI 2021, LNDECT 81, pp. 565–570, 2021.
https://doi.org/10.1007/978-3-030-79197-1_82

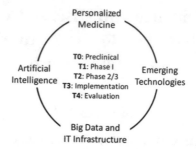

Fig. 1. Big data in the development of personalized medicine (Schork)

On the population level, the access to massive data would be beneficial in public health surveillance, which can include identifying disease trends, so that intervention and regulations can be established to track and control the development of the disease [4]. Research in public health also relies on access to data, and Big Data would promote the research into topics in public health that can be beneficial on the population level [5].

1.2   History of Health Record

The history of keeping data and records in the field of medicine and public health has been very long, with the oldest case report on medical examination report dates back to 1600 BC [6]. As computer systems became widely used, the health records transferred to electronic data from the paper in the 1990s [7]. With the development of technology today, health records are more widely stored on clouds instead of being stored on-site, and this also allows access to the data from more devices [8]. In 2003, the term "electronic health records" was defined, and it refers to any health-related information that is stored in electronic systems "to capture, transmit, receive, store, retrieve, link and manipulate multimedia data for the primary purpose of providing healthcare and health-related services" [2].

1.3   Source of Big Data - Internet of Things

With the development of technology, Big Data sources have expanded, with one of those being the Internet of Things (IoT). The IoT is based on computer chips on intelligent devices, including smartphones, sensors, and other objects like cars and watches, and these together can generate large amounts of health-related information. The healthcare system can then use such information, together with other data from sources, including social media and public data, to monitor and provide real-time public health surveillance [9].

2   The Specific Application of Big Data in Public Health

2.1   Use of Big Data in COVID-19 Pandemic in China

The first outbreak of COVID-19 was in China, and the tracking and controlling of the pandemic in China had employed mainly Big Data and related technologies [10]. In January 2020, the Chinese government and social organizations proposed a prevention and control mechanism, which included wide use of data analysis and mobile personnel monitoring [11]. Later, the use of technology extended to artificial intelligence, cloud computing, and big data, which allowed better tracking, analysis, and control of the pandemic. With 847 million mobile internet users in China, there was a strong foundation for data collection and monitoring, which can provide valuable resources for public health surveillance [12].

One example of data collection is using a QR code to enter public areas, including in public transportation systems. The policy required everyone who enters the area to be registered, which allowed data collection, including contact tracing. The QR code included personal information, including the health status and traveling history, which provided detailed information for public health surveillance.

2.2   Honghu Hybrid System

One of the systems used for public health surveillance in China was the Honghu Hybrid System (HHS) and used in Hubei province, where the pandemic had the first outbreak. The HHS performed Big Data analysis with data collected both from traditional electronic health records, as well as from sources including social media, and performed real-time surveillance on the pandemic (Fig. 2) [13].

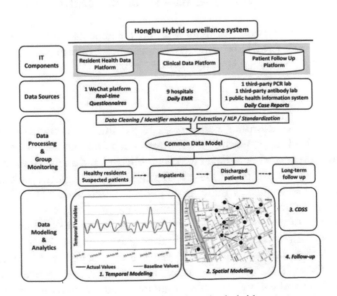

Fig. 2. Use of big data in honghu hybrid system

The Internet of Things has also been used in the process of fighting COVID-19 in China. The data collected from devices had provided massive amounts of information, which contributed to the public health surveillance system in China during the pandemic [14].

As of 3 January 2020, there have been around 100 thousand confirmed cases in China as compared to 29 million cases in the US, which shows that the pandemic has been under relatively better control in China.

2.3   Potential Application of Big Data Related to COVID-19

While Big Data has been chiefly used in the short-term management of COVID-19, there can also be extended applications of the technology that follow-up with the pandemic's impact.

In the medium term, treatments and interventions can be developed by analyzing patient data, while data collection for Big Data extends beyond hospital settings. Drugs and interventions could be assessed using Big Data technology and would promote more effective solutions to the disease [15].

In the long run, the use of such technology could be applied to the development of intelligent cities, promoting the inclusiveness, safety, resilience, and sustainability of cities [15]. This could be achieved through Big Data collection and analysis, which can provide information to policymakers and professionals in public health and thus develop better public regulations. The COVID-19 has shown the importance and effectiveness of Big Data, which calls for extension of the technology application beyond the case of the pandemic [7].

3   Challenges in the Implementation of Big Data

Challenges and criticisms come together with the benefits and convenience brought by the technology of Big Data. One of the most concerning criticisms about the use of Big Data has been the issues related to privacy and ethics.

3.1   Privacy Concerns

With Big Data's characteristics being high in volume, velocity, and variety, data breaches could lead to exposure of a tremendous amount of information. Data breaching has been occurring as data are being collected and analyzed in history, and it could be incredibly concerning in the era of Big Data since every single case of data breaching could lead to data of millions of patients being leaked [16]. While experts have held a positive attitude toward collecting data from sources, including contact tracing apps, most of the population are still reluctant to accept such strategies that can potentially cost their privacy [17]. The IoT is also posing a threat to the public's privacy with devices in lives being monitored and data being used in different applications [18]. These have led to challenges in both legal and ethical challenges in applying Big Data and calling for better regulation and solutions to privacy issues.

3.2   Challenges in Regulation

Solutions to data protection might involve improvements in privacy rules, including the Health Insurance Portability and Accountability Act (HIPAA). There are still gaps in such regulations, including in regulating the controller and processors of data collected. Such gaps have to be addressed to ensure data privacy in third-party entities [19]. One of the important HIPAA regulations includes the deidentification of personal data, but such data could be re-identifiable through processing. On the other side, the "over-protective" nature of HIPAA regulations prevents the use of protected health information for research purposes without the authorization of patients or IRB waiver. This can slow down the data-based innovations as it poses challenges to access data for research purposes [20]. These pose challenges in the development of regulations, which is crucial for data safety.

4   Conclusion

The development of Big Data has shown its benefit in public health through the application of such technology in the time of the COVID-19 pandemic. Big Data has aided health professionals in surveillance, including in contact tracing and health monitoring, which contributed to a wide population's health and safety. The potential use of big data in the medium and long term can also provide a better environment and health. However, privacy and data safety continue to be concerns of the public, and regulations need to develop to promote the safer collection, analysis, and use of Big Data.

References

1. The Best Time to Prevent the next Pandemic Is Now: Countries Join Voices for Better Emergency Preparedness. World Health Organization, World Health Organization. www.who.int/news/item/01-10-2020-the-best-time-to-prevent-the-next-pandemic-is-now-countries-join-voices-for-better-emergency-preparedness
2. Dash, S., Shakyawar, S.K., Sharma, M., Kaushik, S.: Big data in healthcare: management, analysis and future prospects. J. Big Data **6**(1), 1–25 (2019). https://doi.org/10.1186/s40537-019-0217-0
3. Schork, N.J.: Artificial intelligence and personalized medicine. In: Von Hoff, D.D., Han, H. (eds.) Precision Medicine in Cancer Therapy. CTR, vol. 178, pp. 265–283. Springer, Cham (2019). https://doi.org/10.1007/978-3-030-16391-4_11
4. Strachan, M.: Big Data Means Big Benefits for Healthcare Providers and Patients. Trapollo, 25 November 2020. www.trapollo.com/big-data-means-big-benefits-for-healthcare/
5. Hulsen, T., et al.: From big data to precision medicine. Front. Med. **6**, 34 (2019). https://doi.org/10.3389/fmed.2019.00034
6. Gillum, R.F.: From papyrus to the electronic tablet: a brief history of the clinical medical record with lessons for the digital age. Am. J. Med. **126**(10), 853–857 (2013). https://doi.org/10.1016/j.amjmed.2013.03.024
7. What Is the History of Electronic Medical Records? Net Health, 10 March 2021. www.nethealth.com/a-history-of-electronic-medical-records-infographic/

8. Learn The History of EHR Electronic Health Records. ICANotes, 11 December 2020. www.icanotes.com/2019/04/16/a-history-of-ehr-through-the-years/

9. Peeri, N.C., et al.: Defending against the novel coronavirus (COVID-19) outbreak: how can the Internet of Things (IoT) help to save the world? Health Policy Technol. **9**(2), 136–138 (2020). https://doi.org/10.1016/j.hlpt.2020.04.005

10. Coronavirus Disease (COVID-19): How Is It Transmitted? World Health Organization, World Health Organization. www.who.int/emergencies/diseases/novel-coronavirus-2019/question-and-answers-hub/q-a-detail/coronavirus-disease-covid-19-how-is-it-transmitted

11. Coronavirus Disease (COVID-19). World Health Organization, World Health Organization, 12 October 2020. www.who.int/emergencies/diseases/novel-coronavirus-2019/question-and-answers-hub/q-a-detail/coronavirus-disease-covid-19

12. Wu, J., et al.: Application of big data technology for COVID-19 prevention and control in China: lessons and recommendations. J. Med. Internet Res. **22**(10) (2020). https://doi.org/10.2196/21980

13. Gong, M., et al.: Cloud-based system for effective surveillance and control of COVID-19: useful experiences from Hubei, China. J. Med. Internet Res. **22**(4) (2020). https://doi.org/10.2196/18948

14. Rahman, M.S., et al.: Defending against the novel coronavirus (COVID-19) outbreak: how can the Internet of Things (IoT) help to save the world? Health Policy Technol. **9**(2), 136–138 (2020). https://doi.org/10.1016/j.hlpt.2020.04.005

15. Bragazzi, N.L., et al.: How big data and artificial intelligence can help better manage the COVID-19 pandemic. Int. J. Environ. Res. Public Health. **17**(9), 3176 (2020). https://doi.org/10.3390/ijerph17093176.

16. Davis, J.: UPDATE: The 10 Biggest Healthcare Data Breaches of 2020. HealthITSecurity, HealthITSecurity, 16 December 2020. www.healthitsecurity.com/news/the-10-biggest-healthcare-data-breaches-of-2020

17. Perrigo, B.: Will COVID-19 Contact Tracing Apps Protect Privacy?" Time, Time, 9 October 2020. www.time.com/5898559/covid-19-contact-tracing-apps-privacy/

18. Yamin, M., et al.: An innovative method for preserving privacy in Internet of Things. Sensors **19**(15), 3355 (2019). https://doi.org/10.3390/s19153355

19. Bari, L., O'Neill, D.P.: Rethinking Patient Data Privacy In The Era Of Digital Health: Health Affairs Blog. Health Affairs, 12 December 2019. www.healthaffairs.org/do/https://doi.org/10.1377/hblog20191210.216658/full/

20. Price, W.N., Cohen, I.G.: Privacy in the age of medical big data. Nat. Med. **25**(1), 37–43 (2019). https://doi.org/10.1038/s41591-018-0272-7

Innovative University Subject Teaching Mode Under the Information Technology Environment

Wenjing Hu[✉]

Faculty of Management, Hubei Business College, Wuhan, Hubei, China

Abstract. In response to the call put forward by the Ministry of education to "take modern educational technology as a breakthrough to achieve the organic integration of information technology and curriculum", in the long-term teaching practice, guided by constructivist theory, innovative education theory, humanistic learning theory, etc., I analyzes both current situation and misunderstandings during information technology and curriculum teaching integration as now in college teaching practice, then carries out the reconstruction and practical exploration of teaching mode in information technology environment from for dimensions: creating a real teaching situation, building a learning resource platform, providing personalized and differentiated design of independent learning tasks, and evaluating multiple teaching results. It is expected to be able to reform the traditional mode and structure of education and teaching, otherwise to improve students' learning style.

Keywords: Information technology · Teaching mode · Innovate · Research

1 Introduction

Since 1999, the scale of education is expanding rapidly in pace with the continuous expansion of college enrollment. Higher education in China has achieved and advanced into the popularization stage with the speed of running. The continuous expansion of enrollment has also made it difficult for college students to obtain employment. In response to the severe employment situation, there is also a high turnover rate up to 29.4%, some of them have been dismissed because their work skills can't meet the requirements of enterprises, others only due to their own feeling of lack of skills. College education is out of line with social needs, teaching cannot cultivate the ability and knowledge which is required by the society, that's resulting in low monthly income, low professional counterpart rate, high turnover rate, high conversion rate of occupation and industry, and long running in period of workplace. In view of this, "what kind of talents and how to cultivate them" has become the subject of the reform of talent training in universities in the information age.

J. Abawajy et al. (Eds.): ATCI 2021, LNDECT 81, pp. 571–578, 2021.
https://doi.org/10.1007/978-3-030-79197-1_83

2   Objectives

This research is based on the previous research, there are still some left, the characteristics of both higher education and professional courses, select some relevant courses of marketing major in Wuhan ordinary universities, to discuss around the teaching design, learning mode change and information resource platform construction in the discipline teaching of those major under the information technology environment.

3   Methods

This research integrates descriptive, causal and exploratory research to carry out theoretical and practical studying on the innovation of teaching mode, so to create a new learning ecology in the information technology environment. According to the analysis of the requirements of teachers and students, the teaching presentation carrier, the use demand and the expected use effect, It has built a learning resource bank of marketing related courses with promotion value. The specific methods used are: literature research, interview, questionnaire survey, special discussion, experiment, etc.

4   Contents

4.1   The Current Situation of the Integration Between Information Technology and Curriculum Teaching

Through investigation and research, we found that although many colleges and universities are vigorously promoting the above integration, due to the lack of understanding of the essence of the integration, and also the lack of mastering the implementation methods, leading to the following misunderstandings in the specific teaching practice.

4.1.1   Multimedia Courseware and Electronic Screen Replace Blackboard Writing

Through the investigation, it is found that at present, teachers use more in teaching, such as "browsing, searching, collecting materials to prepare lessons", "presenting alternative teaching blackboard with word and PPT", "using video, audio, animation, pictures or demonstration programs to help students understand the course content", which shows that teachers are still in the primary stage of integrating information technology and course teaching. Large screen is now instead of chalk and blackboard, while e-book is instead of teaching materials. On the surface, the whole class has a large amount of information, and the students reflect well. In fact, the original "human irrigation" has been changed to a more efficient "machine irrigation". Compared with the traditional blackboard teaching, the computer screen is more rigid and rigid, especially when teaching some knowledge points that need to be deduced with strong logic, it is not as effective as the teacher's blackboard teaching.

4.1.2   Teaching Mechanization Replaces Teacher-Student Interaction

One class is projected on the screen from introduction to new class teaching, from example to practice. The whole class has become a demonstration class of courseware. What students see is more the continuous switching of images and pictures. What they hear is the teacher's monologue and playful talking, lacking the guidance and communication interaction of students' thinking process. Even if the interactive function of information technology application is more powerful, and the paths and feedback types of courseware preset are more, it may not cover all the problems encountered in students' learning. Reject what is near at hand and seek what is far away, giving up face-to-face communication and pursuing various means of human-computer interaction, that result in one part of students "can't have enough to eat", and the other part of students "can't eat all", the advantages of information technology are not fully reflected.

4.1.3   Insufficient Combination Between Information Network and Curriculum Resources

The combination of information network and high-quality education resources can make students beyond the limits of time and space and realize personalized high-quality learning. However, due to the reasons of funds, technology and ideas, many schools' campus network information resources are poor, and are updated very slow. Some high-quality journals, newspapers, magazines and teaching materials and other resources are not likely to meet teachers and students. On the other hand, curriculum resources of some disciplines are difficult to be developed in the form of Digitalization for teachers and students to share. Even if they are developed, they are only low-level combination and patchwork of network and curriculum resources, rather than integration. Due to the lack of sufficient curriculum learning resources, there is no way for students to explore in groups in class and learn autonomously after class, so it is difficult to change the situation that teachers dominate the classroom and students passively accept knowledge.

4.1.4   Overuse of Information Technology to Replace Students' Actual Group Practice with Simulation Demonstration Practice Teaching

Exploratory and design practice or practical training has been proved to be the most effective way to train students' mind and hand, and cultivate the ability of cooperative exploration and innovation. However, in the current teaching practice, there is excessive dependence on information technology, and to replace students' hands-on practice with simulation demonstration practice teaching, that will make students have dependence and laziness, reduce the opportunities and interests to explore, and miss out the opportunity to cultivate students' ability of scientific inquiry.

4.2   Innovative Research on Teaching Mode Under the Environment of Information Technology

4.2.1   Create Real Learning Situations and Stimulate Students to Be Interested in Learning

Learning environment refers to the social and cultural background directly related to the learning process. Learning is always connected with a certain social and cultural background, that is, "situation". Creating a situation can help students establish the connection between the old and new knowledge, and promote students' thinking development and meaning construction [1]. Taking the marketing course teaching during the research period as an example, when teaching the methods of market investigation, we designed on-site mobile phone micro test, computer-aided personal interview, in-depth interview, observation experiment, lottery investigation and other on-site situations to carry out simulation investigation, so as to increase the interest of learning and stimulate the awareness of participation. In the process of scenario creation and role-playing, it can effectively enhance the memory and application of relevant theoretical knowledge points; stimulate students' interest in studying.

4.2.2   Build Learning Resource Platform and Expand Teaching Capacity

To explore the reconstruction of teaching mode under information technology environment is not same as traditional teaching. in the teaching design, teachers should not only prepare the resources and tools needed in the process of teaching, but also should preset the learning resources and tools needed by students to complete the corresponding learning tasks, as the necessary support for inquiry learning and problem-solving. Learning resources here refer to all tangible and intangible resources used to help students complete their learning tasks, including software resources and hardware systems that support the learning process [2]. These resources and tools effectively help students to break through the difficulties in learning, enhance their studying efficiency, simultaneously, enrich and expand the capacity of teaching, and then realize building and sharing resources together.

4.2.3   Provide Personalized and Differentiated Independent Learning Task Design

The design of information-based teaching emphasizes the process of students' learning and knowledge construction, and highlights autonomy, exploration and reflection in learning. One of the most important tasks for teachers to apply information technology to teaching practice is to design learning tasks that conform to the characteristics of students' knowledge, ability level and psychological age in line with their development goals of knowledge, skill and emotional attitude determined by teaching tasks. These tasks must be able to fully utilize students' initiative and provide them more opportunities to use all knowledge in different situations and form self-reflection based on feedback information.

4.2.4   Multiple Teaching Achievement Evaluation to Guide Students to Love Learning and Enjoy Learning

In pace with the integration of cutting-edge technologies into teaching scene such as cloud computing, mobile internet, data mining and so on, a communication bridge is established between before-class preview, in-class teaching and after-class review to provide information support stand by data and intelligence throughout all teaching evolution. The real realization of teaching evaluation focuses on the assessment of students' knowledge building process rather than results, and focuses on the performance and change in the learning process [3]. In addition to attaching importance to the evaluation of the process, the evaluation of multiple teaching achievements introduces the subject of multiple-evaluation. Students are not only the evaluated, but also the evaluation subject. They should not only participate in the formulation and use of evaluation standards, but also evaluate their own performance and in addition evaluate the performance of their peers.

4.3   Problems and Deficiencies in the Practice of Teaching Model Innovation

4.3.1   Some Teachers' Educational Ideas Lag Behind and Lack of Understanding of Information-Based Teaching

Although education informatization has been included in the important national development strategy, and the education informatization project in Colleges and universities is also in full swing [4], but it is not recognized and accepted by all teachers. Some teachers refuse or resist using advanced information technology in classroom teaching. Some teachers even believe that the applying of information technology to teaching process is a waste of time, distracting students' attention and reducing teaching effect.

4.3.2   Fail to from Two Perspectives as "Teaching" and "Learning" to Realize the Effective Integration of Technology and Curriculum

During the research period, many teachers simply use multimedia or information technology to assist teaching, thinking that this can add color to their own classroom, which represents the advanced teaching concept, without considering whether the subject content and characteristics need to use information technology, how information technology can suit students' psychological characteristics and the curriculum framework.

4.3.3   The School Information Environment Needs to Be Improved

Teaching equipment, multimedia classroom, digital and intelligent multimedia teaching environment which can be controlled and managed remotely are the necessary conditions to advance development of informatization. In the subject research at this stage, there are still many schools' classrooms with outdated equipment so that cannot meet the needs of informatization teaching.

5 Results and Discussion

5.1 Strengthen the Training of Education Concept and Information Technology, and Improve Teachers' Information Literacy

5.1.1 Strengthen Teachers' Learning of Education and Teaching Theories and Teaching Reform Ideas

The educational theory and thoughts are the guiding ideology and important reference basis which are indispensable for the integration of technology to college course. Modern education concepts that should be studied seriously include: quality education, lifelong education, dual subject education, innovation education, EQ based education, four pillar education concepts, etc. Advanced educational theories include: multiple intelligence theory, constructivism learning theory, hybrid learning theory, innovation promotion theory, minimum cost and performance theory [5].

5.1.2 Strengthen Teachers' Learning and Application of Information Technology

The school should first establish a hierarchical and modular information technology training content system, and the training methods can be diversified [6]. For example, the theoretical module mainly uses the reflection mode instead of the inculcation mode which is often used in the previous training. In the reflection mode, after the expert's key explanation and prompt, the teacher will carry out group communication in combination with his own work. The skill module adopts the mode of speaking while practicing. The teaching design module can be combined with some cases for analysis, display and group discussion, so that teachers can participate in the design of teaching and learning training. The training of practice module is based on classroom practice, mainly in the form of observation and discussion, focusing on solving some practical problems in teaching.

5.2 Promote the Integration of Information Technology and Teaching Mode, and Explore the Management of Intelligent Teaching

5.2.1 Achieve Accurate Teaching Design Based on Pre Class Preview Feedback

Teachers should focus on pre class preview, introduce new course content in the form of teaching resource release, pushing micro course video or other forms, and layout preview tasks [7]. Through the assessment and collection of students' Preview trace records and preset problems, to judge the knowledge and technical conditions that students have mastered, then to provide timely, accurate and three-dimensional information basis for preparing lessons. All these help to realize targeted teaching design accordingly, to explain the weak links of knowledge points with supplement, and to analyze the key and difficult points, so students can solve the problems they encounter in new task learning through diversified interactive communication.

5.2.2 Dynamic Evaluation Based on Whole Teaching Process and Learning Quantitative Evaluation

Auxiliary application of information technology makes a lot of process evaluation work can be done by the teacher before the beginning of the course. It can automatically collect the learning behavior of each student in each link, such as class arrival, question answering, discussion, exercise, test, etc., Accompany with the assessment process, it provide the teacher with an index parameter to quantify the learning effect of students by data integration and analysis, so as to provide data support for teachers to accurately evaluate students' knowledge mastery and adjust strategies and progress in time, moreover to promote teachers to free from the heavy work of marking homework statistics, so that teachers can focus more on the improvement of the curriculum, and enhance the flexibility of teachers' mobile teaching.

5.2.3 Create an Intelligent Learning Ecology, Which Based on the Normal Push of Teaching Resources

Every teacher should "be attentive to his/her side and take local materials", should take full advantage of advanced information processing technique, gather, store, process massive information which is collected and realized both in teaching field and real life, then push and release timely by the information technology platform. Thus can make students get the latest learning resources at the first time, deepen theoretical cognition and understanding, learn and realize in life, and finally create a smart learning ecology, so that our classroom is really "fresh" to attract students to fall in love with the classroom.

5.3 Strengthen the Construction of Information Environment to Create Favorable Conditions for Promoting Information Teaching

At present, the construction of information environment still needs to invest a lot of money in the hardware environment and software environment, which is the basic condition to promote the information-based teaching of professional courses in universities. First, improve the construction of hardware environment. Purchase, use, management and maintenance of hardware equipment should be controlled strictly. Second, strengthen the construction of software environment [8]. Using wikis, youtube and other course tools to promote the establishment of high-quality resource base. Subject teachers and information technology personnel should be combined closely to build a cooperative community based on information technology and jointly build teaching resources with high-quality.

6 Conclusions

Through application of technology in innovative practice of teaching mode, we have broken the traditional single teacher-student relationship with teachers as the center, established a teaching mode with schoolteacher as leader and pupil as main body, and realized synchronous or asynchronous teaching communication with the help of technology on the basis of equality and democracy. On the one hand, teachers have changed from traditional "teaching" to "guidance", and their leading role lies in situation creation, information resource provision, cooperative learning organization, research-based learning guidance and independent learning task design; On the other hand, pupils can select appropriate learning contents and forms based on their actual knowledge and ability level, decide their own learning progress, explore and find the required knowledge and information actively, and complete exercises, thoughts and discussions, to ensure that the role of learning subjects can be fully played. Thereby promoting the exertion of their own subjective initiative and realizing the development of individual personality, then to create a good teacher-student relationship, to enables students to return to life, face practice, understand the world, in addition, students' innovative spirit and practical ability are cultivated too.

References

1. Chen, J.: Intelligent learning environment construction. National Defense Industry Press, Beijing (2013)
2. Fang, Y.: Practical exploration of cloud smart classroom teaching mode – taking comprehensive counter business of commercial banks as an example. Sci. Educ. Article Collects **2**, 114–116 (2019)
3. Ge, W., Han, X.: Questionnaire research on teaching ability of university teachers in the digital age. e-Educ. Res. (6), 123−128. (2017)
4. Han, X.: Research on the informatization teaching ability of Chinese university teachers. China High. Educ. Res. **7**, 53–58 (2018)
5. Liu, D.: Online education for twenty years: from education + Internet to Internet + education. Internet Econ. (7), 90−97 (2015)
6. Lu, P.: Research on the construction and teaching practice of hybrid learning resources based on cloud classroom online learning platform. Master's thesis. Central China Normal University (2016)
7. Wu, H.: Design and application of personalized online learning system from the perspective of intelligent learning. China Educ. Technol. **6**, 127–131 (2015)
8. Yuanyuan, Y.: Research on intelligent classroom teaching mode in higher vocational education. China Manage. Informationization **19**, 217–218 (2018)

Design Software Teaching Based on Big Data

Yiru Li[✉]

School of Art and Design, Hubei University of Science and Technology,
Xianning, Hubei, China

Abstract. With the increasingly high requirements of the design industry, Photoshop versions are constantly updated. Although they are powerful, they are also complicated and cumbersome to operate. Without a lot of time of practice, it is difficult to acquire skilled operating skills, and it is impossible to produce wonderful graphic design works. This paper studies the modern network teaching platform based on big data (BD), realizes the discovery of user needs of the teaching platform and the optimization and push of teaching resources, and realizes the full utilization of teaching resources. The teaching effect of "Photoshop Image Processing" for students majoring in art and design can be maximized. New technologies and new means can be effectively used to reform the teaching mode and innovate the teaching methods and means with the help of the power of the BD network teaching platform. The technology of cloud classroom can broadcast the live work of enterprises to the classroom and students' handheld devices to strengthen the connection with the market.

Keywords: Big data · Art design · Professional design software · Network teaching platform

1 Introduction

The emergence of BD technology to solve the problem from the vast amounts of network teaching resources in the discovery and perception of user demand information, provides a method for real-time online teaching to get the data, analyze data, requirements, found and recommended decision provides the foundation base, this topic using BD technology and user portrait method, build the modern network teaching platform based on BD, implementation platform user perception of teaching resource requirements, have the following meaning. Through the analysis of user portraits of users on the platform, we can accurately find teaching resources suitable for users' needs, perceive the needs of users in learning, solve the problem that it is difficult to obtain massive teaching resources and the same cannot be customized, and save users' choice and learning time. To solve the problems that high-quality teaching resources cannot be fully utilized and the management classification is not clear, the diversification of teaching resources mining, persistent data sample collection and multi-dimensional data analysis are realized through BD management, so as to ensure that teaching resources can match users' demands in real time and high-quality teaching resources can be fully utilized [1]. Through the process monitoring, portrait management and other methods, the user's needs and interest points are constantly

J. Abawajy et al. (Eds.): ATCI 2021, LNDECT 81, pp. 579–587, 2021.
https://doi.org/10.1007/978-3-030-79197-1_84

excavated, and the only course that users do not understand or are interested in is found, so as to improve the combination point between the course and the user's interest, so as to realize intelligent online teaching.

With the growing maturity of cloud technology, it provides hardware support for the development of BD. Now, BD has become a hot spot and frontier. From research institutions to government, from enterprises to individuals, BD has been applied in all walks of life. Uskov, points out that can optimize the current teaching mode by using the method of BD, for teachers to provide information on students' learning behavior and learning ways, help teachers understand students area. Areas of interest, and there are problems in some ways, and this kind of analysis can replace the traditional teaching mode of pass an exam to get the analysis results of [2]. Park proposed a clustering algorithm based on Map/Reduce, but the effect was not significantly improved in parallel architectures such as speed increase and scale increase, and it was not widely used [3].

The goal of this study is to take the course "Photoshop Image Processing" as an example to study the application of BD network teaching platform in teaching, which has a theoretical guiding significance for the teaching reform of art design majors.

2   Design Software Teaching Under BD Network Teaching Platform

2.1   BD Network Teaching Platform

(1) Platform Design Standards

When designing the modern network teaching platform based on BD, the following standard principles should be referred to ensure the normal operation of the system.

Principles of system robustness. System development process needs to use the current mainstream technology and stable architecture and related development and auxiliary tools, so as to ensure that the system is efficient, flexible, easy to use, easy to maintain later.

Safety principles. Security principle of the system is an important principle of BD system must be considered strictly follow BD protection in the process of system design specifications, the requirements of user privacy data protection agreement for may threaten point in the system as well as possible to evaluate security vulnerability and risk prevention, to improve the security of data protection and reinforcement measures [4].

The extensibility principle. In the design process, the system must ensure the scalability and portability in the implementation process of the system, the use of standard hardware structure, and the system to provide standard API interface, easy to interact with other business systems.

(2) System Architecture Selection

According to the above principles, combined with the data storage, data cleaning and statistics, data analysis and clustering association of the BD platform may appear performance blocking or reliability problems, the network teaching platform for BD adopts the distributed system architecture based on Hadoop.

Performance Guarantee. The distributed system architecture based on Hadoop is supported by concurrent access and rapid access of massive BD, and real-time database technology and relational database can be integrated [5].

It provides an interface in the form of SQL, supports the dynamic expansion of the cluster to achieve linear performance improvement, and has the capability of parallel complex calculation, statistics and analysis [6].

Support customized development. Customized development can be made according to the business requirements of the system.

(3) Design of Platform Business Logic

User portrait acquisition: the system platform obtains the current learning activities, user attributes, learning tasks, learning objects, learning process and learning roles and other situational elements as well as their attribute values.

User portrait processing: inference and correlation are carried out on the acquired user portrait, and a relatively comprehensive user portrait is obtained and expressed by model.

User profile matching: After completing the modeling representation of the current user profile, the user profile matching requirements are sent to the user demand discovery matching engine.

Resource request for teaching resources: carry out similarity calculation and decision on user portrait model in the historical portrait database, and find the portrait resource record matching with the current user.

Teaching resource return: provide the required teaching resources for the platform users according to the teaching resources matched in the current database.

Teaching resource demand filtering: the teaching resource is screened by integrating the weight threshold of each dimension of the portrait.

Rank teaching resource demand: Rank teaching resource demand according to similarity and demand degree.

Teaching resources push: the final filtered teaching resources are returned to the platform users.

User selects the content to be pushed: User selects the content to be pushed.

User portrait optimization: This is where the current user portrait model is optimized based on user choices and related behaviors.

Users actively search and pay attention to teaching resources: platform users independently search and learn resources.

New demand fusion: Match the current user portrait with the teaching resources independently selected by the user.

(4) Correlation Algorithm

Mean square error

$$RMSE = \sqrt{\frac{\sum_{(u,i)\in T}\left(r_{ui} - r'_{ui}\right)^2}{|T|}} \tag{1}$$

Mean absolute error

$$MAE = \frac{\sum_{(u,i)\in T}\left|r_{ui} - r'_{ui}\right|}{T} \tag{2}$$

Accuracy

$$\Pr ecision = \frac{\sum_{u\in U}|R(u)\cap T(u)|}{\sum_{u\in U}|R(u)|} \tag{3}$$

2.2 Design of Teaching Pattern for Photoshop Image Processing Network Platform

(1) Learning Knowledge

The teaching mode of network teaching platform is to store knowledge materials in the platform and put "learning knowledge link" before classroom teaching. Students study on the network teaching platform in advance. Therefore, knowledge learning is mainly to participate in platform learning activities. On the cloud learning platform, teachers can analyze each student's knowledge mastery through data, which is convenient for teachers to timely set classroom teaching objectives, produce learning materials and upload them to the network teaching platform, so as to facilitate students' online learning.

Courses are offered on the network teaching platform, and learning materials such as micro-lessons, text materials and image materials are provided for students. The teacher is separated from the student on the platform, so the cloud teaching as a virtual classroom is not like the real classroom, the teacher can control the whole. With the network teaching platform supported by cloud technology, its powerful database computing ability can facilitate teachers' teaching management and communication and interaction with students [7, 8]. In addition to the interaction on the cloud platform, teachers and students can also use various technologies to communicate, such as Blackboard, Moodle, etc., or instant communication technologies, such as WeChat, etc., to build a three-dimensional platform. The cloud platform can be set up as several important modules, such as "sharing module", "discussion module", "task module" and so on. In the "sharing module", teachers and students can share graphic materials, micro-lesson materials, etc. In the "discussion module", teachers and students can share their learning experience. In the "task template", the teacher can show the learning objectives and tasks to be completed to the students in the form of a task list.

(2) Analyze the Characteristics of Learners and Determine the Learning Topic

The teaching work of teachers also needs to "know and know". Therefore, teachers must effectively analyze the characteristics of learners [9]. According to the characteristics of secondary vocational students and the teaching needs of "Photoshop image processing", it can be analyzed. The first is the level of knowledge that students have mastered and the cognitive results. Teachers can understand students through analysis and know what they want to achieve.

(3) Design and Make Corresponding Learning Resources According to the Preview Objectives

After designing the teaching objectives, the corresponding preview learning resources can be designed according to the teaching objectives. Teaching video is an important resource for preview. Teaching videos include Flash animation, video lectures, screen videos, etc., among which micro-video teaching is the most widely used [10]. Based on the practical needs of Photoshop image processing, to design and develop the micro course. How students learn. What skills should you have. Taking the micro course of "removing redundant people" as an example, video learning "intelligent filling" is the teaching goal. In the micro course, the ability needs of the job position are firstly analyzed, and then the content is introduced, and the tool explanation and example operation are carried out. Finally, the main steps and tools of "removing redundant people" are summarized, so as to realize the teaching goal in the micro course.

(4) Students' Task List before Class

It consists of two parts: watching micro lessons and practicing independently. In this stage, the coordination between students is very important. Teachers should supervise each study group and grasp their learning status and progress in time.

3 Platform Development

3.1 Platform Development Environment

Through the above analysis, the modern network teaching platform based on BD must give consideration to data processing performance and ontology based BD storage and other capabilities. On the basis of mainstream open-source software as far as possible, the following technologies are selected for systematic implementation.

Table 1. Network teaching platform realizes technology selection

System function module	Development technology and platform
Web access interface	Java + Postgre SQL + Memcached
BD infrastructure module	Java + Python + Shell
Task management	HDFS(Hbase/Hive)
Parallel computing	Apache Hadoop YARN
Parallel ETL	Sqoop
API interface	Thrift
Data storage based on ontology technology	HDFS(Hbase/Hive)

3.2   Infrastructure Module Implementation

Choose distributed architecture based on Hadoop system, according to cost principle, system deployment in a public cloud server, thus ensuring business expansion or inadequate performance can make clastic expansion, from as far as possible to reduce the perspective of system deployment project workload, modern network teaching platform based on the large data using the task scheduling and status monitoring methods such as completed the deployment of the Hadoop framework.

By BD infrastructure module, the system administrator can manage all the nodes in a Hadoop cluster and configuration, at the same time, using the unified control platform, the administrator can be used in the process of BD analysis and management of the open source project to create and configure the environment, such as add or remove the cluster, automated deployment, and each node of the condition monitoring and so on.

(1)   The registry provides the corresponding registration interface for the registration of nodes and scripts. Through the registry, all information can be registered to the full life cycle resource management;

(2)   Manage the workflow through the rule base and provide policy support for the cloud engine;

(3)   The cloud engine will continuously collect information provided by modules, clusters and nodes;

(4)   Scheduling and configuration of resources through unified hardware resource interface;

(5)   Unified management and control of clusters in the system can be carried out through the cluster management interface.

4   Platform Performance Test Results

4.1   Data Loss Rate and Processing Rate

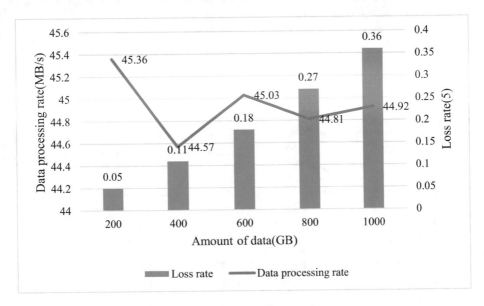

Fig. 1. Data loss rate and processing rate

As shown in Fig. 1, this paper conducted key performance tests on the modern network teaching platform based on BD from two performance indicators: data loss rate and data processing rate. When the data volume is 200 GB, the data loss rate is 0.05%, and the data processing rate is 45.36 MB/s. The data loss rate is 0.11% and the data processing rate is 44.57 MB/s when the data volume is 400 GB. When the data volume is 600GB, the data loss rate is 0.18%, and the data processing rate is 45.03 MB/s. At 800GB data volume, the data loss rate is 0.27%, and the data processing rate is 44.81 MB/s. When the data volume is 200 GB, the data loss rate is 0.36%, and the data processing rate is 44.92 MB/s. It can be seen that the maximum data loss rate is 0.36%, and the data loss rate of the current system is less than 0.5%, which can guarantee the accuracy of data analysis and meet the stability requirements. The data processing rate of the current modern network teaching platform based on BD is within 1000 G, and there is no obvious difference in the processing speed, and the current processing speed can meet the real-time performance requirements of the platform.

4.2   Compatibility Test

According to the design metrics, test the response time of the user login, jump, and load the user's page. The tool simulates a browser POST to submit the user login information, then waits for the 200 status code from the index.aspx page, and gets the

HTML bottom test-specific tag from the returned data stream to determine the load time of the static portion of the page. The tool simulated the request for 5000 times, obtained the minimum response time of 907 ms, the maximum response time of 1637 ms, the average response time of 1274 ms, exceeded the design target (1500 ms) for 3 times. That is, the failure rate is 0.05%, confirmed by the expert group, can be considered to basically achieve the design goal.

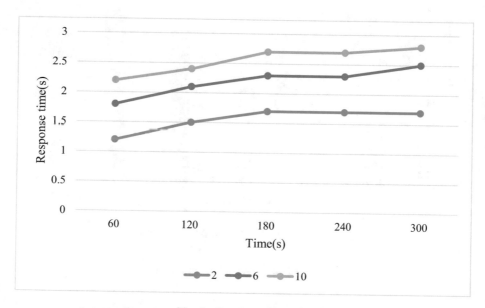

Fig. 2. Pressure test curve

As shown in Fig. 2, according to the system stability requirements, the response time of page login and data query should be controlled within 3 s. The system stability test mainly uses the stress test tool LoadRunner. At first, the test time is set to 5 min, the script is recorded to start the test, and the number of concurrent users is gradually increased from 2 to 10. After 5 min of continuous operation, the number of concurrent users is gradually reduced until all the users exit. As can be seen from the test results, the response time will increase with the increase of test time under different number of users. The maximum response time is 2.8 s when the system runs for 5 min under the number of 10 users, which indicates that the response time of page login and data query in the network teaching platform designed in this paper fully meets the system requirements.

5   Conclusions

With the increasing maturity of Internet technology, online teaching has become the learning method chosen by more and more lifelong learners. However, as there are so many teaching resources, how to accurately find the needs of users has become an urgent problem to be solved. The emergence of BD technology provides a technical basis for personalized customization of network teaching. Combined with BD technology, the construction of a network teaching platform that can accurately discover user needs and make full use of teaching resources has gradually become the core topic of current teaching practitioners and experts. The pre-class knowledge is transferred and practiced and internalized in class by means of graphic courses, micro-classes and flipped classroom. After class, consolidate the core knowledge and take the cloud classroom as the leading. Cooperative and communicative classroom teaching activities foster students' various learning abilities.

References

1. Yu, K., Yue, X.G., Madfa, A.A., et al.: Application of problem-based learning network teaching platform in medical education. J. Comput. Theor. Nanosci. **13**(5), 3414–3417 (2016)
2. Uskov, V.L., Howlett, R.J., Jain, L.C.: Smart innovation, systems and technologies smart education and e-learning. Teach. BD Technol. Pract. Cloud Environ. **59**, 631–639 (2016). https://doi.org/10.1007/978-3-319-39690-3(Chapter 56)
3. Park, Y.E.: Uncovering trend-based research insights on teaching and learning in BD. J. BD **7**(93), 1–17 (2020)
4. Morrison, S.M., Liu, C., Eleish, A., et al.: Network analysis of mineralogical systems. Am. Mineral. **102**(8), 1588–1596 (2017)
5. Maza-Ortega, J.M., et al.: A Multi-Platform Lab for Teaching and Research in Active Distribution Networks. IEEE Trans. Power Syst. **32**(6), 4861–4870 (2017)
6. Marchiori, M.: Learning theway to the cloud: BD Park. Concurrency Comput. Pract. Exper. **31**(2), e4234.1−e4234.17 (2019)
7. Osipov, I.V, Prasikova, A.Y, Volinsky, A.A.: Participant behavior and content of the online foreign languages learning and teaching platform. Comput.Hum. Behav. 50(SEP.), 476–488 (2015)
8. Wang, N., Zhang, Y.: Application Status and Promotion Strategy of Integrated Network Teaching Platform——Taking Northwest A&F University as an Example. Asian Agri. Res. **11**(04), 94–96 (2019)
9. Huang, L.H., Chen, C.Y.: GM(0, N) Model -Based Analysis of the Influence Factors of Network English Learning Platform. J. Grey Syst. **19**(1), 31–40 (2016)
10. García, J., Entrialgo, J.: Using computer virtualization and software tools to implement a low cost laboratory for the teaching of storage area networks. Comput. Appl. Eng. Educ. **23**(5), 715–723 (2015)

Application Research of the Current Situation of College English Online Teaching Model in the Big Data Era

Mei Zhang[(⊠)] and Xiangke Yuan

Zhejiang Tongji Vocational College of Science and Technology, Hangzhou, Zhejiang, China

Abstract. The way of teaching English online in colleges and universities is an important way to promote students' autonomous learning ability. With the continuous innovation of technology, the level of college English online teaching is also constantly improving. In the context of the era of big data, English teaching in colleges and universities has become more intelligent and informatized. This article mainly introduces BP neural network method and hill climbing algorithm. This paper uses big data to analyze the current situation of college English online teaching mode, and establishes a potential BP neural network mathematical model. The model is solved by the BP neural network method, the current situation analysis and application status of the college English online teaching model are analyzed, and the model is revised using historical data to improve the current situation analysis and application status evaluation of the college English online teaching model accuracy. The experimental results of this paper show that the BP neural network method improves the application research effect of college English online teaching mode by 33%. Finally, by comparing the results of online and offline tests, the teaching impact of college English online teaching mode is analyzed.

Keywords: Big data · Online teaching mode · BP neural network method · Hill climbing algorithm

1 Introduction

1.1 Background and Significance

The 21st century is the era of knowledge economy, and the innovation of knowledge acquisition methods will become one of its important contents [1]. With the in-depth study of educational economics, people have a deeper understanding and experience of the inseparable relationship between education and economy [2]. Economics is the foundation of education. Economic development determines the development of education. The development of education also affects the economy and promotes economic development and growth. This is mainly reflected in the increase in the number and quality of educated labor force [3]. As an important carrier of education, teaching plays a key role in promoting the development of education [4]. With the update and iteration of science and technology and the rapid development of the Internet, teaching methods

J. Abawajy et al. (Eds.): ATCI 2021, LNDECT 81, pp. 588–596, 2021.
https://doi.org/10.1007/978-3-030-79197-1_85

are constantly changing. Online teaching began in the 1990s and entered a strong development stage after entering the 21st century [5]. The quality and convenience of online teaching have also had a certain impact on college English teaching.

1.2 Related Work

Zhao J provides a method that can evaluate participatory stakeholder innovation in a complex multi-stakeholder environment to solve essential problems [6, 7]. Based on the principle of common value creation, he proposed an evaluation framework that illustrates the application research process of college English online teaching models in the era of big data. In this process, stakeholders integrate their resources and abilities to assess the university. The current situation analysis and application research of college English online teaching mode in the data age [8, 9]. In order to evaluate this evaluation framework, a number of data have been collected in the study, which represent related issues in the analysis and application research of college English online teaching models in the era of big data [10, 11]. However, because the information collection process is too complicated, the data results are not very accurate [12].

1.3 BP Neural Network Method and Hill Climbing Algorithm

Based on the analysis of the current situation of college English online teaching mode in the era of big data, the application research system of college English online teaching mode is explored and analyzed through big data, and the calculation method of BP neural network method and mountain climbing algorithm is established. The analysis of the current situation of online teaching mode of college English and its application research provide accurate guidance.

2 Status Quo and Application Methods of College English Online Teaching Models in the Era of Big Data

2.1 Bp Neural Network

BP neural network is an advanced multi-layer network with three layers, such as input layer, middle layer, output layer, or unidirectional expansion of three or more layers. Its main feature is that the upper and lower layers are completely connected, and there is no connection between neurons in each layer. Its input-output ratio is a very non-linear mapping relationship. If the number of input nodes is m and the number of output nodes is n, the BP network represents the conversion from m-dimensional Euclidean space to n-dimensional Euclidean space. Among them, n is the connection weight between the input layer and the second layer, the threshold is m, and m is the connection weight between the second layer and the output layer.

The formula is as follows:

$$G = \sqrt{2/O \sum_{O=1}^{Q} (E_Q)^2}, E_Q = 1/3(X - X^{\cdot})^3 \tag{1}$$

Therefore, first standardize the value of each index used to evaluate teaching quality. Then it is used as the input vector of the BP neural network model, the evaluation result is used as the output vector, and when using the BP network, enough samples are used to make it accept the judgment of experts. Understand the index weight, so that the BP neural network model automatically adjusts the value of the weight statement until it reaches the true internal knowledge model, and then adds the specific value of each indicator that has undergone quality learning evaluation to the BP neural network model for evaluation the quality of learning is the result. Finally, the experience and knowledge of experts are used for reference to ensure the accuracy of evaluating teaching quality.

2.2   Hill Climbing Algorithm

The hill climbing algorithm starts from the current node and compares it with the values of surrounding nodes. If the value of the current node is higher, return to that row. As the maximum value, if the surrounding nodes have a larger value, please change the current node to the maximum value node, and then change the value of the surrounding nodes, the hill climbing algorithm will end and the best solution will be obtained.

Since the hill climbing algorithm cannot guarantee the optimal solution, it is not suitable for finding the optimal path in mobile applications. The advantage of the hill climbing algorithm is to remove the minimum value and then loop around the nodes. After many times, the highest point can be obtained. This article uses the maximum-minimum method because the data processing performed by this method is a linear transformation, which can better retain its original meaning without generating information.

The formula is as follows:

$$\varphi = \frac{E - E_{\min}}{E_{\max} - E_{\min}} \tag{2}$$

In the formula, E is the processed neural network input value; E is the unprocessed neural network input value; is the minimum neural network input value.

Because the hill climbing method is based on a unique search method, it can be suitable for solving various optimization problems. The hill climbing method is simple and flexible. Different neighborhood structures and generators are set according to different problems, and the storage complexity is low. When the hill-climbing method is introduced into the association rules, the algorithm takes the classification accuracy as the target, and uses the hill-climbing method to search for appropriate support and confidence thresholds, so as to realize the adjustment of the support and confidence thresholds and ensure the accuracy of the association rules. It realizes the feasible extraction of strong association rules on large data sets, finds out the largest frequent item set in the transaction data set, and uses the obtained maximum frequent items set and the preset minimum confidence threshold to generate strong association rules. In the era of big data, the optimization of college English online teaching mode will have obvious effects.

3   Application Research Experiment of College English Online Teaching Mode in the Era of Big Data

3.1   Experimental Design

Online teaching refers to the form of teaching through the Internet, mobile devices and other communication media. By making full use of various educational resources, online education overcomes the limitations of time and space through network transmission, is no longer limited to traditional offline education methods, and eliminates traditional education barriers. Online teaching allows students to have more choices instead of arranging and setting courses in traditional classroom practice; they can choose more courses that they are interested in or have expertise, and they can choose more from well-known scholars, experts and outstanding Online courses for educators, such as the smart vocational education cloud mobile teaching platform, have a large number of online boutique courses of MOOC colleges. Students can also use the online learning platform for independent and independent learning, and better demonstrate the learners' subjective enthusiasm and motivation. The result is shown in Fig. 1:

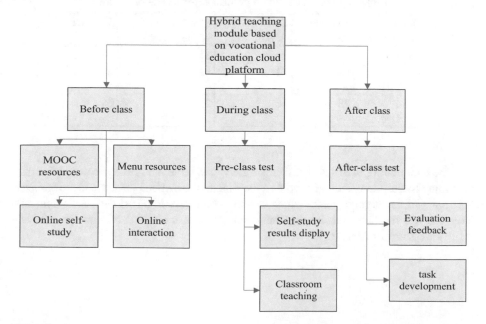

Fig. 1. Online teaching mode of college English based on the vocational education cloud platform

The networked teaching management is operated through the teaching management platform of the computer network, with independent management and interactive processing functions. In the actual network teaching management, such as students' courseware learning, pre-class preview, discussion and answering, online interaction, brain Storm, homework completion, study and examination management, etc., can all be completed through remote interaction through the network, achieving absolute convenience and automation. Therefore, two sets of experiments are designed, one is the control group, the other is the experimental group, the control group is the offline traditional teaching group, and the other is the online platform teaching group. Compare the two groups under the same subject and unit content, under the same conditions Next, what is the teaching effect of the two groups of students in a random situation and the experimental online and offline teaching.

3.2   Collecting Experimental Data in Online and Offline Teaching Mode

In the experiment, the offline and online pre-test was conducted in the first semester. The researchers used a set of final exam assignments from the university to conduct preliminary examinations on two parallel classes in the Department of Water Resources and two parallel classes in the Department of Construction Engineering. This set of test assignments can check students' English skills mastery from different angles. In order to describe the changes before and after the test subjects' English skills, the researchers performed descriptive statistics and independent test T tests on the experimental group and the control group. The test data are shown in Table 1:

Table 1. Online and offline pre-test experimental data table

Online experiment	Average value	Standard deviation%	Standard error of the mean%	Variance
Pre-test1	71	20	31	74
Control pre-test1	69	21	41	89
Mean equation	78	30	23	68

The data mining in this article forms a new data table, and further collects the data according to the name, area, etc., so that the data in the original database is reorganized according to the target demand, which is more conducive to decision-making analysis. You can see the inner relationship between the data more clearly, and the analysis results are shown in Fig. 2:

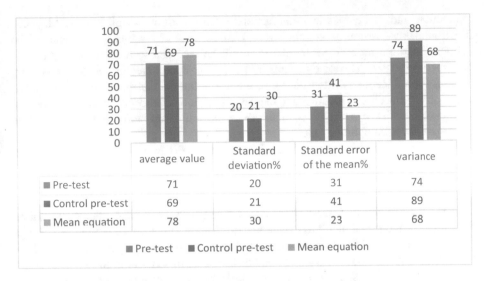

Fig. 2. Data diagram of online and offline experiments

As can be seen from the data in Fig. 2, the control group and the experimental group's other factors are controlled within a certain range, and the average value obtained by using online teaching is 2 higher than the average of offline teaching methods. The difference between the two is not large, indicating that online teaching methods have not played a significant role in improving students' consciousness. However, from the perspective of standard deviation, online teaching methods can give students greater autonomy. Therefore, the use of online teaching also needs to solve the problem of improving students' self-consciousness, and for student autonomy, online teaching has played a greater role.

4   Current Situation of College English Online Teaching Models in the Era of Big Data

4.1   Online Teaching Data of College English

The online teaching model based on the vocational education cloud mobile teaching platform expands the supply of high-quality resources through optimization and integration, and uses the platform to break through the increase in cyberspace resources. The types of resources, the difficulty, and the content of resources are greatly enriched., Different types of resources mobilize the enthusiasm of students and well meet the individualized learning needs of students. With the help of the platform, students can conduct adaptive and personalized learning according to their own learning needs. They can also actively participate in the learning process through teacher-student interaction and student-student interaction, avoiding dumb English. The playback function makes learning more efficient and improves teaching quality. All data of the learner's learning process on the platform will be recorded in the database to form a data set for research. The log data table uses the system log to record various behavioral information of learners on the platform, including operational behavior and behavior trajectory. According to the log data, the basic situation of students' online learning can be counted, which can play a certain role in the follow-up study of students' online learning effects. Supporting role.

The problem with online teaching is that because it is not face-to-face teaching, teachers cannot effectively supervise students, and students are not easy to concentrate. At the same time, teachers cannot teach directly or can only teach face-to-face among a certain number of people. Teachers can neither fully take care of all students, nor can they fully guarantee that every student has listened to or watched the video online during class, and it is difficult to ensure that students will not do other unrelated things during online class. Therefore, it is more to test the consciousness of students. Offline and online testing is conducted at the end of the term. In order to describe the changes before and after the test subjects' English skills, the researchers performed descriptive statistics and independent tests on the experimental group and the control group. The test data are shown in Table 2:

Table 2. Online pre-test experimental data table

Online experiment	Average value	Standard deviation%	Standard error of the mean%	Variance
Pre-test2	88	27	45	89
Control pre-test2	90	22	55	76
Mean equation	79	40	39	87

Through the analysis of Table 2, the data mining in this paper forms a new data table, and further collects the data according to the control group, the experimental group, and the pressure average, so that the data in the original database is reorganized according to the target demand, Which is more conducive to decision-making analysis, and can see the inner relationship between data more clearly. The analysis result is shown in Fig. 3:

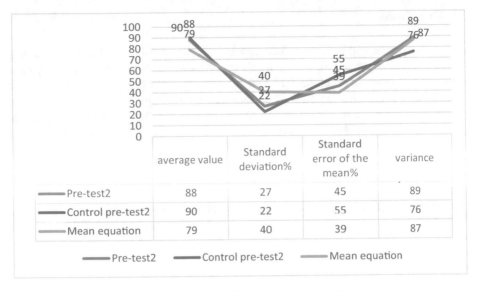

	average value	Standard deviation%	Standard error of the mean%	variance
Pre-test2	88	27	45	89
Control pre-test2	90	22	55	76
Mean equation	79	40	39	87

Pre-test2 ——— Control pre-test2 ——— Mean equation

Fig. 3. Data diagram of online pre-test experiment

The figure shows that the average number of Pre-test2 is lower than that of Control pre-test2. This shows that online teaching can improve teaching quality to a certain extent and increase students' interest in learning, but from a standardized value point of view for students' self-consciousness, offline teaching has advantages over online teaching.

4.2   College English Online Teaching Mode

Online teaching should rely on the Internet, whether it is through a computer or a mobile phone, which places high requirements on the network environment and network speed, especially when many people use it intensively, the network is crowded, the website cannot be accessed, and the teaching cannot be viewed on the platform Video and other issues. In the process of waiting time, time is wasted and efficiency is reduced. Another situation is that the network environment in the area where some students are located is good, but the network environment in the area where some students are located is poor, and network cards may appear during the live broadcast.

Although there are many online course resources, because the teachers are from different schools, the teaching methods used in the works produced by different teachers are also different. The viewing and use of course resources in MOOC will be restricted by the start time. Each course has a start time instead of opening at any time throughout the year.

5   Conclusions

Although this paper has made certain research results on BP neural network method and hill climbing algorithm, there are still many shortcomings. In the era of big data, there are still many in-depth content analysis and applied research methods of college English online teaching models. There are still many steps in the decision-making process that are not covered due to space and personal ability. In addition, the actual application effect of the improved algorithm can only be compared with the traditional model from the level of theory and simulation.

Acknowledgements. The 2020 Higher Education Research Project of Zhejiang Higher Education Association "Research on the Improvement of Higher Vocational English Teaching Quality Based on Dual Platform O2O under the Background of Enrollment Expansion" (No.: KT2020315).

References

1. Xin, W., Huan, J.: Analysis on the current situation of pyramid schemes a-mong college students and the prevention in the new era. Sci. Educ. Literat. Collect. **000**(013), 20–21 (2019)
2. Moldovan, O.T., Constantin, S., Panaiotu, C., et al.: Fossil invertebrates records in cave sediments and paleoenvironmental assessments – a study of four cave sites from Romanian Carpathians. Biogeosciences **13**(2), 483–497 (2016)
3. Xiaowei, Y.: The current situation of college English teaching in the context of the internet age and suggestions for reform. J. Jiamusi Coll. Educ. **000**(007), 408–409 (2016)
4. Mengdie, G.: Research on the current situation and strategy of Counselors' professional development in China in the informatization era. Digit. Educ. **003**(005), 41–45 (2017)
5. Yu, Y., Lu, W.: Research on the current situation of university students' communication in the new media era. Sci. Educ. Wenhui **000**(001), 135–136 (2016)
6. Thyagarajan, C., Suresh, S., Sathish, N., et al.: A typical analysis and survey on healthcare cyber security. Int. J. Sci. Technol. Res. **9**(3), 3267–3270 (2021)
7. Song, Q.: On the current situation and optimization methods of ideological and political education in the micro era%. J. Liming Vocat. Univ. **000**(001): 62–65, 76 (2016)
8. Liu, L., Aremu, E.O., Yoo, D., et al.: Brand marketing strategy of live streaming in mobile era: a case study of Tmall platform. East Asia **1**(1), 65–87 (2020)
9. Yan, H.: Path analysis of college English teaching reform in application-oriented universities in the era of big data. J. Heilongjiang Inst. Educ. **038**(007), 136–139 (2019)
10. Bing, D.: Research on the application of project-based learning in college English teaching in the internet+era. Sci. Educ. Wenhui **000**(006), 173–174 (2016)
11. Xuping, L.: Research on interactive teaching model of college English classroom in the era of mobile internet. J. Jiamusi Educ. Coll. **000**(005), 260 (2017)
12. Zhao, J., Wang, W., Du, S., et al.: A study on the current situation and strategy of image publicity in colleges and universities in the era of self media. Int. Core J. Eng. **6**(5), 116–118 (2020)

Effect Evaluation of PBL Teaching Mode Applied to Information Technology Basic Courses in the Network Environment

Xiaofeng Li[✉]

College of Innovative Management, Valaya Alongkorn Rajabhat University under the Royal Patronage, 1 Klongluang District, Pathum Thani 13180, Thailand

Abstract. The PBL learning content should include the basic knowledge and basic abilities that learners must acquire in the course. This research mainly discusses the effect evaluation of PBL teaching mode applied to basic courses of information technology under the network environment. With the freshman students as the center, the project themes are listed. Each person chooses a project theme. Students who choose the same theme form a group (3–5 people). Using group collaboration methods, through listening to lectures, field observations, literature review, online search, brainstorming, etc., determine the group's research projects and plan the content. Learn to use the multiple functions of the multimedia software PowerPoint, plan creativity, design information presentation methods, and collaborate to complete a presentation. Instructional designers and first-line teachers of the subject participate in the preliminary evaluation of the teaching plan. 88.3% of the students like this kind of group cooperation learning method, and they can cooperate with group members to complete tasks in a tacit way. This research helps to make the implementation of PBL teaching in the teaching of basic information technology courses in colleges more instructive and directional.

Keywords: PBL teaching mode · Network environment · Basic courses of information technology · Multimedia software PowerPoint

1 Introduction

Focusing on learning activities, teachers have carefully organized and designed learning resources and learning tools, but this cannot effectively promote the development of learning activities among students, because learning resources and learning tools only provide external support for the development of students' learning activities, but cannot Provide internal motivation for students to learn.

The development of the entire teaching activities of college information technology courses is based on the three-dimensional teaching objectives in the new curriculum reform. The design and implementation of the PBL model also cannot be separated from the reference of the three-dimensional standards [1, 2]. In the specific application of PBL, teaching activities are generally divided into two modules: First, it is a module to develop students' problem-solving skills. The development of main teaching

© The Author(s), under exclusive license to Springer Nature Switzerland AG 2021
J. Abawajy et al. (Eds.): ATCI 2021, LNDECT 81, pp. 597–603, 2021.
https://doi.org/10.1007/978-3-030-79197-1_86

activities should focus on investigating students to find problems, based on existing information and the collected relevant information conducts a certain analysis of the problem, and proposes the hypothesis of their own problem solution to carry out this link, which enables students to establish the initiative of learning while solving the problem [3, 4]. The second module is to examine and train students' practical operation ability after the problem-solving solution is proposed, cultivate their independent practice, and actively explore the ability to verify the solution [5]. Guide students to have the confidence to learn independently, through continuous exercise, they can improve their practical ability, and can combine knowledge with practical problems for a comprehensive analysis [6, 7]. Throughout the teaching process, the three-dimensional teaching goal is always the positioning criterion for the development of teaching activities, and the final evaluation of teaching activities also refers to the mastery of the three-dimensional goal to complete the evaluation of student learning [8, 9].

The evaluation of the entire learning activity is based on the improvement of students' problem-solving ability, the construction of the knowledge system, the teamwork learning situation, and the student's active learning attitude. This specific teaching process focuses on the information technology mastering allows students not only to collect, process and process information, but also to improve their practical ability to solve problems through the application of information. Questions can be raised by the teacher or raised by the students themselves, but the teacher must control the classroom, and the questions raised by the students cannot be separated from the teaching goals and content.

2   PBL Teaching Mode

2.1   PBL Teaching

With the development of the information society, the capacity of knowledge in various disciplines is rapidly expanding, and the speed of updating is getting faster and faster. Therefore, the learning content of online courses should focus on presenting basic knowledge and cultivating learners' basic abilities. PBL online courses are no exception, and cultivate learners to have stronger adaptability in future study, life and work.

In web-based education and teaching, web courses are ultimately presented to learners in the form of web pages. To a certain extent, the design of courses is to use PowerPoint to design learning content, and using PowerPoint to design learning content is the design process of user interface. Effective use of PowerPoint to design learning content is the key to successfully developing learning content for online courses. In traditional course teaching, the transmission of information is based on the source, and the teacher controls the form and timing of the transmission: in online teaching, the situation is just the opposite. The transmission of information is based on the destination, and learners control the timing of the transmission of information, the length of time that the information is transmitted, the order and form of the information transmission. Therefore, the focus of online courses using PowerPoint to design learning content lies in the integration of learning styles, learning methods, media, etc.

It is necessary for us to find out the elements and methods of learning content design presentation under the network state.

2.2 Network Environment

The teaching design of research learning in the network environment has its own characteristics, which reflect the characteristics of research learning in the network environment, and provide a necessary reference for the design of the teaching process mode. At the same time, these characteristics reveal the difference between research-based teaching design and traditional teaching design under the network environment, and provide a basis for rational analysis of research-based learning teaching design. Because there are many uncertain factors involved in research learning teaching design in the network environment, some unexpected situations may occur at any time. This determines that teachers' teaching design must also be flexible and adapt to changes in environmental conditions at any time. Deal with it accordingly.

The coefficients of the network have been determined and will not change:

$$Q(t+1) = \sum_{j=1}^{n} W_{ij}(t+1)X_j(t) - \theta(t+1) \tag{1}$$

Among them, W_{ij} is the weight coefficient of the connection between the i neuron and the previous j neuron. The convergence speed of the judgment optimization algorithm is faster and the optimization effect is better. It is very suitable for large-scale data or large-scale model problems.

$$m_t = \alpha_1 m_{t-1} + (1 - \alpha_1)g_t \tag{2}$$

$$v_t = \alpha_1 v_{t-1} + (1 - \alpha_1)g_t^2 \tag{3}$$

Among them, m_t and v_t calculate the first and second moments of the gradient.

3 Information Technology Basic Course Experiment

3.1 Teaching Content Design

(1) With the freshman students as the center, the project themes are listed. Each person chooses a project theme. Students who choose the same theme form a group (3–5 people).
(2) Using group collaboration methods, through listening to lectures, field observations, literature review, online searches, brainstorming, etc., determine the group's research projects and plan the content.
(3) Learn to use the multiple functions of the multimedia software PowerPoint, plan creativity, design information presentation methods, and collaborate to complete a presentation.

3.2 Initial Evaluation of Teaching Design

The preliminary evaluation of teaching design is to evaluate the formed initial teaching plan, and make corresponding amendments based on the evaluation to form the final teaching plan to prepare for the implementation of the next teaching plan. This step mainly resolves the practice-oriented part of the plan, that is, the learning activity sequence design part, and makes corresponding modifications according to the actual situation, in an effort to reduce the possibility and intensity of conflicts between the plan and practice. At this stage, it is recommended that instructional designers and first-line teachers of the subject participate in the preliminary evaluation of the teaching plan. Students' understanding of modern education concepts is shown in Table 1.

Table 1. Students' understanding of modern education concepts

Test	Teacher-centered	Student-centered	Pay equal attention to teaching and learning
Pretest data	38%	55%	7%
Post-test data	8%	53%	39%

4 Teaching Effect of Information Technology Basic Courses

4.1 Student Collaboration Ability

In the research, the evaluation of the changes of students' collaboration ability is mainly based on the pre-test and post-test of collaboration ability. At the beginning of the experiment, students conducted a pre-test of their collaboration ability, and at the end a post-test. The measurement of collaboration ability is carried out in the form of a questionnaire, which is designed from two aspects: collaboration attitude and collaboration ability. A total of 54 people in the experimental class completed the pre-test and post-test of collaboration ability.

In the process of research learning in the network environment, the design of learning strategies is generally reflected through the design of learning resources and learning tools. At the same time, for the learning strategies provided by teachers, students may be in a low-level state of unconscious use, and have not risen to the stage of rational use. Therefore, in the actual learning process, teachers should also guide students to rationally use learning strategies through observation or communication with students. It can be seen from the above data that the students' collaboration ability has been greatly improved. In particular, the number of students with "very good" and "excellent" items increased by 26%, and the number of students with "needs to be improved" and "quite limited" decreased by half. The evaluation of student collaboration ability is shown in Table 2.

Table 2. Evaluation of student collaboration ability

Collaboration	Quite limited	Room for improvement	Not bad	Well	Excellent
Pre-test	12%	12%	34%	32%	10%
Post-test	8%	4%	30%	44%	24%

4.2 Teaching Effect

In the classroom, students' interest in learning is relatively high, and their attention is relatively concentrated. This shows that the PBL teaching method used in this lesson can arouse students' interest, improve students' attention and learning initiative, and at the beginning of teaching, The teachers showed that the collection of gourmet websites and blog pages aroused high interest among students. Students used to browse websites made by others. This time they will feel eager to try and eager to make a favorite website by themselves. In the back, the students completed the PowerPoint presentation step by step according to the tasks assigned by the teacher. The whole process was full of enthusiasm. 64% of the students think that they can use the knowledge learned in this lesson to solve the problems encountered in life in the future, but 16% of the students think that they cannot use the knowledge of this lesson flexibly. In addition, this lesson is mainly based on the PowerPoint operation, so the inquisitive thinking has not been greatly improved. However, 81.8% of the students have a good grasp of the content of this lesson, and 79.5% of the students are still willing to continue studying and exploring the content of this lesson. 88.3% of the students like this kind of group cooperation learning method, and they can cooperate with group members in a tacit understanding of the division of labor to complete tasks, and deepen mutual feelings. The learning effect of students is shown in Fig. 1.

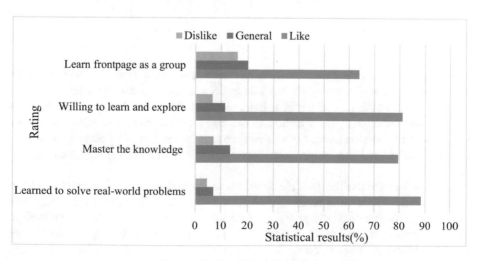

Fig. 1. Students' learning effect

The student's learning attitude is shown in Fig. 2. 78.1% of students like teachers to use this teaching method, and 70.2% of students are able to concentrate and learn enthusiastically in class. There are also a small number of students who jostle in class or whisper not to actively participate in group discussions, and they are not serious in the division of labor to collect information. 82.2% of the students are interested in the problems set by the teacher, and can actively invest in the created problem situation to discover and think about the problem. 88% of the students think that the tasks assigned by the teacher are challenging and can stimulate their desire to explore and seek knowledge, causing cognitive conflict.

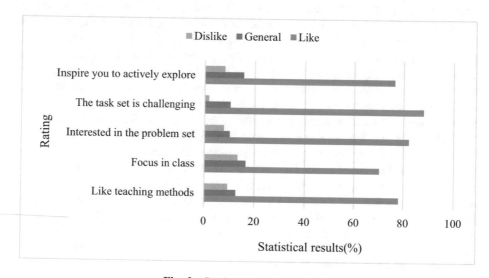

Fig. 2. Students' learning attitude

5   Conclusion

For the problem, creating a situation is its fundamental condition. Teachers combine the actual situation of teaching content and goals to create a situation closely related to their real life, which can make them have a strong interest in learning. For the curriculum, the creation of the situation is the most critical part. The influence of the situation will arouse the interest of students, so that they can actively connect with their thinking and knowledge, which can improve their understanding and handling of problems.

Groups learn to use the multiple functions of the multimedia software PowerPoint, plan creativity, design information presentation methods, and collaborate to complete a presentation. Instructional designers and first-line teachers of the subject participate in the preliminary evaluation of the teaching plan. This research helps to make the implementation of PBL teaching in the teaching of basic information technology courses in colleges more instructive and directional.

References

1. Cho, J., Cho, J.: Development and evaluation of the PBL teaching/learning process plan of 'housing culture and practical space use' for home economics in middle school. Korean Home Econ. Educ. Assc. **32**(2), 59–76 (2020)
2. Tortorella, G., Cauchick-Miguel, P.: Combining traditional teaching methods and PBL for teaching and learning of lean manufacturing. IFAC-PapersOnLine **51**(11), 915–920 (2018)
3. Jiang, Q., Xiao, Z., Ru, C., et al.: Study on the modeling method of knowledge base system in web environment. Int. J. Pattern Recogn. Artif. Intell. 33(9):1959031.1–1959031.15 (2019)
4. Wahyuni, W., Fadryan, E.P., Sitanggang, A.S.: Web-based environmental learning information system in SMA Angkasa Lanud Husein Sastranegara Bandung. J. Comput. Theor. Nanosci. **16**(12), 5360–5363 (2019)
5. Wackett, L.P.: Web Alert: Environmental viruses of prokaryotes: an annotated selection of World Wide Web sites relevant to the topics in environmental microbiology. Environ. Microbiol. **21**(6), 2198–2199 (2019)
6. van der Sloot, K.W.J., Weersma, R.K., Dijkstra, G., Alizadeh, B.Z.: Development and validation of a web-based questionnaire to identify environmental risk factors for inflammatory bowel disease: the Groningen IBD Environmental Questionnaire (GIEQ). J. Gastroenterol. **54**(3), 238–248 (2018). https://doi.org/10.1007/s00535-018-1501-z
7. Childress, V.W.: technology-the extension of human potential. Technol. Eng. Teacher **77**(5), 30–35 (2018)
8. Moon, J., Kim, J., Kim, J., Kim, J., Kim, C., Kim, H.: Roller skiing biomechanical information analysis using GPS, IMU, and atmospheric pressure sensors: a case study. Sports Eng. **21**(4), 341–346 (2018). https://doi.org/10.1007/s12283-018-0278-x
9. Anand, M., Sasikala, T., Anbarasan, M.: Energy efficient channel aware multipath routing protocol for mobile ad-hoc network. Concurr. Comput. Pract. Exper. **31**(4):e4940.1–e4940.13 (2019)

The Construction Mechanism of Lifelong Education Network Gold Course from the Perspective of Artificial Intelligence

Sanjun Ma[✉]

School of Business, Hebei Ploytechnic Institute, Shijiazhuang, Hebei, China

Abstract. Artificial intelligence and informatization have led to accelerated changes in learning methods, and people's learning pathways have gradually expanded. The in-depth integration of lifelong learning concepts with artificial intelligence technology has led to a wider range of learning, and people's demands for lifelong education network quality courses are more urgent. This article uses qualitative research methods to deeply analyze the theoretical source and development context of the network "golden course", explore the application prospects of artificial intelligence technology in the lifelong education network "golden course", and build a lifelong education network "golden course" based on artificial intelligence technology, from the four aspects of building an artificial intelligence lifelong education platform, constructing a precise teaching model, reshaping personalized learning methods, and creating a community-integrated learning environment, it proposes the construction mechanism of a lifelong education network "golden course" from the perspective of artificial intelligence.

Keywords: Artificial intelligence · Lifelong education · Golden course

1 Introduction

In response to the outstanding universal problems of undergraduate education in the new era and new situation, especially the current situation of declining undergraduate teaching quality in Chinese universities, in 2018, the Ministry of Education first put forward the concept of "golden course", emphasizing the reasonable "increasing burden" on college students. Stimulate students' learning motivation and professional interest, and truly turn "water course" into a deep, difficult and challenging "golden course" [1]. The "golden course" design in the information age must first reflect the new knowledge and learning concepts of the information age; second, achieve the deep integration of information technology and education; third, implement a diversified and differentiated evaluation mechanism and cultivate innovative talents. Looking at scholars' understanding of the characteristics of the "golden course", while ensuring the quality, depth and challenge of the course content, the "golden course" also puts forward higher requirements on the teaching process, which are mainly reflected in the improvement of teaching interaction and realization, individuation and precision of teaching, promoting the deep integration of information technology and teaching [2].

© The Author(s), under exclusive license to Springer Nature Switzerland AG 2021
J. Abawajy et al. (Eds.): ATCI 2021, LNDECT 81, pp. 604–611, 2021.
https://doi.org/10.1007/978-3-030-79197-1_87

The above characteristics all reflect the overall goal of "golden course" construction, that is, the overall improvement of curriculum construction and talent training ability, conform to the objective development law of educational resources, and effectively resolve the increasingly prominent contradiction between talent training and curriculum quality. With the rapid development of modern information technology, lifelong education network courses have played an important role in promoting fairness in education and realizing the wide sharing of high-quality educational resources, improving teaching quality, promoting the reform of educational concepts, and cultivating innovative talents with international competitiveness [3]. The construction of a society-oriented society provides indispensable motivation and support.

At present, the new generation of information technology represented by artificial intelligence technology and big data is developing vigorously, constantly changing the style and connotation of traditional education and teaching, promoting the integration of formal learning and informal learning, and building a ubiquitous learning environment for all people [4]. Lifelong learning provides strong support and promotes the deep integration of artificial intelligence technology and teaching to become an important development direction of lifelong education. The application of artificial intelligence education can broaden the horizon and provide ideas for the reform and innovation of various types of education at all levels, and is the most important and effective technical means to achieve educational innovation [5]. Aiming at the lifelong education network "golden course", the integration of artificial intelligence technology can fully meet the lifelong learning requirements of adult learners, while effectively improving the effect of the lifelong education network "golden course" through personalized teaching, precise guidance, human-computer collaborative interaction and other methods provide effective technical support to fundamentally solve the problems existing in the current life-long education network "golden course" construction.

2   The Construction of the Lifelong Education Network Gold Course Under the Artificial Intelligence Perspective

The rapid development of artificial intelligence technology has brought opportunities for learners to use online "golden course" for autonomous learning. Computational intelligence analysis and decision-making, perceptual intelligence human-computer interaction, and cognitive intelligence educational role imitation can create precision for learners sexual, interactive, and personalized lifelong education network "golden course" [6]. Therefore, the research believes that, compared with traditional offline courses and online courses, the support of artificial intelligence technology for the lifelong education network "golden course" is mainly reflected in the three aspects of teaching precision, interactivity and individualization. At the same time, the community should also provide media learning environment guarantee and offline learning support services for the lifelong education network "golden course". On this basis, build a lifelong education network "golden course" model from the perspective of artificial intelligence (Fig. 1).

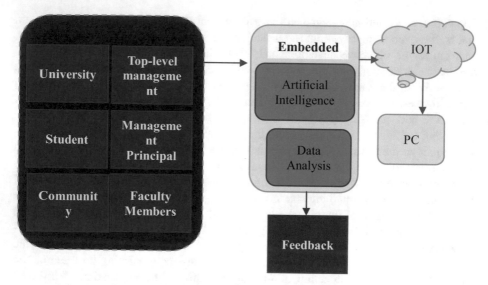

Fig. 1. Lifelong education network golden course model

2.1 Computational Intelligence Realizes Teaching Precision

Although traditional online courses are superior to offline courses in terms of the richness and convenience of teaching resources, the disadvantages of over-focusing on the teaching and instilling of knowledge and skills, ignoring individualization and innovation ability training have not changed, which seriously affects online courses [7]. The online "golden course" supported by artificial intelligence technology can solve this problem through precise teaching methods. Teaching precision is the use of computational intelligence technology to collect micro-data such as individual characteristics, grades, and learning activities of students, analyze different students' learning interests and learning style preferences, predict and effectively warn learning risk points, and use the above analysis results in order to make macro-decisions, formulate personalized teaching plans, adopt different teaching methods and arrange different teaching content for different learners, so as to ensure that students are taught in accordance with their aptitude and provide learners with differentiated and accurate teaching services.

2.2 Perceived Intelligence to Improve Teaching Interaction

Perceptual intelligence refers to the ability of computers to sense vision, hearing, and touch to realize the natural interaction between humans and machines, the core of perception technology is data [8]. The application of perception technology can obtain information from the learner's environment and learning objects, such as voice, emotion, behavior, eye contact, concentration, etc. Instant and real-time data can improve data analysis technology accuracy and truthfulness. The media resources in traditional online courses can only provide learners with visual or auditory stimulation. The perceptual intelligence of artificial intelligence technology can build a virtual environment through language recognition, personalized image recognition, language processing, and virtual reality technology. It can interact through voice input, text input, facial expression changes, etc., which can bring learners "authenticity", "inter-activity" and "immersion" learning experience, eliminate the disadvantages of poor interactivity in the past online courses, and effectively improve the learning effect.

2.3   Personalization of Cognitive Intelligence Service Teaching

The deep integration of cognitive intelligence technology and teaching scenarios is also an important development direction of intelligent education. Cognitive intelligence technology can realize active thinking, learning and decision-making through cognitive reasoning functions, and build a personalized reasoning engine based on subject knowledge graphs and user portraits [9]. Therefore, the realization of personalized teaching can use cognitive intelligence technology to design virtual tutors to imitate educational roles, and quickly and accurately locate learners' knowledge loopholes by analyzing learning behavior data, learning feedback and results, so as to provide adult learners with personalized learning the program, course content and teaching strategies effectively improve the enthusiasm of adult learners in learning and help adult learners to achieve efficient and individualized independent learning.

2.4   Environmental Intelligence Enhances the Efficiency of Community Media Learning Environment and Learning Resources

The construction of the learning environment is the basis for realizing the reform of teaching and learning methods. According to the learning characteristics of adult learners, it provides learners with a more convenient, comfortable and effective community learning environment, explores diversified online "golden course" application models, and builds a network the "golden course" mixed with offline will be an important direction for the development of lifelong education informatization [10]. The media tools in the current community media learning environment can be divided into two parts: explicit hardware resources and implicit cultural organizations. The explicit hardware resources include public media resources and personal media resources. Public media resources mainly include book reading rooms, electronic reading rooms, classrooms, cultural activity rooms, publicity columns, banners, open-air TV, QQ groups and WeChat groups, etc.; personal media resources include televisions, smart phones, computers, tablets, and the Internet. Hidden cultural organizations include community lifelong education centers, community colleges, social celebrity resources, cultural and sports activities, and special training. It is necessary to make full use of various media resources in the community, promote the reasonable allocation of community media resources, and cooperate with artificial intelligence technology to provide effective learning support services for residents' online learning and improve the effectiveness of learning resources.

3   The Construction Mechanism of the Lifelong Education Network Gold Course from the Perspective of Artificial Intelligence

3.1   Make Every Effort to Build a Lifelong Education Platform Based on Artificial Intelligence Technology

In the process of building a lifelong education platform, first consider the construction and update of bandwidth, servers and other hardware resources to ensure that the platform can meet the needs of large-scale online learning, avoid network stalls, server paralysis, slow loading speed, smooth operation of the platform. Second, the platform can integrate new-generation information technologies such as artificial intelligence and big data, provide integration and invocation of "smart technology" for various educational scene applications, and provide centralized output and platform support for functional services [11]. In terms of specific functions, the platform relies on artificial intelligence technology to realize the functions of teaching and learning analysis. As far as teaching functions are concerned, the platform needs to integrate an intelligent question bank and marking system, a personalized teaching resource push system, a school management service system, an operation system, an interactive teaching system, and a teaching evaluation system to establish a unified data standard and break data barriers. Promote the flow of data between various businesses, and ultimately produce well-structured, interconnected, and valuable data. As far as the academic analysis function is concerned, accurate and personalized data analysis is performed on adult learners, a complete credit certification standard and system are established, and big data is used to realize the mutual recognition of credits and learning achievements of adult learners in various teaching platforms. Improve the social recognition of online learning achievements, fundamentally eliminate the chronic problems of online education with high dropout rates, and create a lifelong education network "golden course" platform.

3.2   Build a Precision Teaching Model

With the development trend of diversified, personalized, and high-quality lifelong learning needs, the teaching method of the online "golden course" also urgently needs to complete the transition from extensive to precise, and complete the transformation from purely relying on the analysis of learning results to the online "golden course" learn the transformation of the whole process analysis. Under the big data environment of the lifelong education platform, record and analyze the behavior data of adult learners throughout the learning process, use the analysis results to predict future learning trends, provide reflective clues and teaching warnings, discover tacit knowledge, and realize dynamic monitoring of the learning process [12]. Provide teachers with real-time and dynamic academic analysis services, so as to help teachers adjust teaching strategies and processes, and finally realize a precise network of the whole teaching process including setting teaching goals, classroom exercises, measurement and recording, learning intervention, and teaching evaluation the "golden course" teaching model effectively improves the quality of online teaching.

3.3 Reshape Personalized Learning Methods

With the support of artificial intelligence technology, providing personalized learning methods and educational services for adults, a large-scale and diverse group of needs, is an important goal of the construction of the network "golden course". First, use the adaptive learning system to analyze students in depth and meticulously, accurately draw portraits of adult learners, and accurately push teaching content and learning resources to learners based on the characteristics and learning needs of adult learners, and tailor learning paths to achieve learning visualization of the situation enables learners to dynamically grasp learning progress and differences, and actively adjust learning strategies. Secondly, use eye movement technology to extract and analyze the law of the impact of teaching video display on learners' learning, continuously improve video teaching content and presentation methods, push personalized course content presentation methods according to the characteristics of different learners, and cooperate with virtual reality technology to enhance personalization the interaction between learning experience and curriculum makes the teaching method more in line with the characteristics of adult learners, brings learners a better learning experience, improves the quality of the lifelong education network "golden course" content, and helps adult learners reshape personalized learning the way.

3.4 Create a Community-Integrated Learning Environment

Whether it is media or information technology, it is an extension of man and an intermediary for man to connect to the outside world and obtain knowledge. Converged media refers to the integration of various media forms such as television, newspapers, radio, and mobile phones through the Internet and other information technologies, and the integration of internal and external resources to form a new media form. Compared with traditional media, fusion media has the advantages of integrating resources such as personnel and technology. Integrating regional media resources is conducive to promoting the reform and development of regional media and enabling the effective integration of traditional media and emerging media. The online learning of adult learners, whether it is internal or external influencing factors, or the formation process, depends on the common support and promotion of the community's explicit hardware media resources and invisible cultural organizations. This includes not only traditional media and new media for the residents. At the same time, the learning support services provided by community volunteers and staff to residents, as well as the subtle influence of community training and cultural organizations on residents' learning are also essential. Therefore, the media environment of the community is not limited to the integration of new and old media, but also the orderly and targeted integration of all media resources within the community, which has an impact on the autonomous learning and the whole process of community residents, and strengthens residents' awareness of continuous autonomous learning, to encourage community residents to form an inspiring learning community and establish a vision of sustainable community development. The government and relevant practitioners should fully understand the importance of media integration for the development of community education informatization, strengthen top-level design, coordinate and coordinate the development of community hardware resources and invisible cultural organizations, promote the integrated development of media, and create a media-integrated learning environment for the network "golden course" provides comprehensive offline learning support services.

4 Conclusion

At present, the rapid development of artificial intelligence technology and the concept of online "golden class" provide a good opportunity and technical support for improving the quality of online education in China. We should face the construction of online "golden course" in a rational and objective manner, avoid following the trend and blindly pursue high speed and large numbers. It is necessary to truly meet the learning demands of adult learners, rationally use emerging technologies such as artificial intelligence, establish a network "golden course" that truly meets the needs of all types of people, and avoid duplication of resources. This research hopes that through the above-mentioned strategies and recommendations, it will provide a targeted reference for the construction of the lifelong education network "golden course", and promote the healthy and sound development of the network "golden course" construction and lifelong education informatization.

References

1. Lu, H., Li, Y., Chen, M., et al.: Brain intelligence: go beyond artificial intelligence. Mob. Networks Appl. 23(2), 368–375 (2017)
2. Raza, M.Q., Khosravi, A.: A review on artificial intelligence based load demand forecasting techniques for smart grid and buildings. Renew. Sustain. Energy Rev. 50(10), 1352–1372 (2015)
3. Polikar, R., Shinar, R., Udpa, L., et al.: Artificial intelligence methods for selection of an optimized sensor array for identification of volatile organic compounds. Sens. Actuators B Chem. 80(3), 243–254 (2015)
4. Nandhakumar, N., Aggarwal, J.K.: The artificial intelligence approach to pattern recognition —a perspective and an overview. Pattern Recogn. 18(6), 383–389 (2015)
5. Goyache, F., Del Coz, J.J., Quevedo, J.R., et al.: Using artificial intelligence to design and implement a morphological assessment system in beef cattle. Animal ence 73(01), 49–60 (2016)
6. Hill, J., Ford, W.R., Farreras, I.G.: Real conversations with artificial intelligence: a comparison between human–human online conversations and human–chatbotconversations. Comput. Hum. Behav. 49, 245–250 (2015)
7. Ye, J.J.: Artificial intelligence for pathologists is not near–it is here. Arch. Pathol. Lab. Med. 139(7), 929–935 (2015)
8. Khokhar, S., Zin, A.A.B.M., Mokhtar, A.S.B., et al.: A comprehensive overview on signal processing and artificial intelligence techniques applications in classification of power quality disturbances. Renew. Sustain. Energy Rev. 51, 1650–1663 (2015)
9. Liu, R., Yang, B., Zio, E., et al.: Artificial intelligence for fault diagnosis of rotating machinery: a review. Mech. Syst. Signal Process. 108, 33–47 (2018)

10. Muqiang Z., Chien-Chi, C., Yenchun, W., et al.: The mapping of on-line learning to flipped classroom: small private online course. Sustainability **10**(3), 748 (2018)
11. Gough, E., Dejong, D., Grundmeyer, T., et al.: K-12 Teacher perceptions regarding the flipped classroom model for teaching and learning. J. Educ. Technol. Syst. **45**(3), 390–423 (2017)
12. Gilboy, M.B., Heinerichs, S., Pazzaglia, G.: Enhancing student engagement using the flipped classroom. J. Nut. Educ. Behav. **47**(1), 109–114 (2015)

Wisdom Teaching Method in the Course of Accounting Informationization

Chu Zhang(✉)

Zhanjiang Science and Technology College,
Zhanjiang 524094, Guangdong, China

Abstract. The development of high and new technology makes smart classroom possible. This research mainly explores the application of smart teaching methods in accounting informationization courses. Before the class, teachers carry out teaching design under the concept of smart classroom teaching, and provide supporting teaching resources. Through the rain classroom, the learning content is posted to the rain classroom We Chat platform. Each student completes the learning task and the teacher releases on their own mobile client. After completing the test questions, the teacher needs to analyze the data generated by the rain class to prepare for the students' activities in class. After class, we post homework through Rain Classroom, and rely on Rain Classroom to carry out real-time interaction to solve the difficulties of answering questions in traditional classrooms. Through the pre-class test, in-class test, and after-class test released by Rain Classroom, students can test their mastery of knowledge, skills, and study ability. 30% of students think that through the teaching design oriented to the work process, they can improve their ability to practice on the computer. This research contributes to the intelligent teaching of accounting information courses.

Keywords: Smart teaching method · Accounting informationization course · Smart classroom · We chat platform rain classroom

1 Introduction

Whether it is a national long-term strategic deployment or short-term planning, the in-depth integration of information technology and teaching is the focus and hot issue that needs to be studied. Based on this background, smart classrooms came into being and became a crucial product in the process of educational information reform. In addition, with the rapid development of a variety of new technologies, such as big data and learning analysis technologies provide technical support for smart classrooms, create a smart environment, and make smart classrooms possible.

Wisdom education is the product of the integration of a new generation of information technology and education [1, 2]. Its essence is to comprehensively apply various information technology methods to build a smart teaching environment, introduce smart teaching aids, and build a smart teaching model on this basis, promote learners' smart learning, and cultivate high intelligence and creativity The intelligent talents, improve the intelligence level of the existing education system, and promote the

J. Abawajy et al. (Eds.): ATCI 2021, LNDECT 81, pp. 612–618, 2021.
https://doi.org/10.1007/978-3-030-79197-1_88

innovation and reform of education [3, 4]. In the smart classroom teaching mode, if only a teacher completes these tasks alone, it will greatly increase the workload of the teacher [5, 6]. Therefore, in this teaching mode, teachers need to establish a team, in this team, everyone learns from each other's teaching experience, division of labor to make micro-classes, and encourage each other. Through teamwork, the teaching pressure of teachers is reduced, allowing teachers to spend more time on classroom teaching and caring for students, and improve the teaching efficiency of smart classrooms [7, 8]. In the smart classroom teaching environment, students have richer learning resources and more convenient communication channels. Therefore, in the teaching process, if students are more self-conscious, teachers should arrange class time reasonably and give students more discussion and communication. The opportunity to show learning results, and then cultivate students' innovation ability and independent learning ability [9, 10].

Based on the current actual teaching environment of accounting informatization courses, this article takes teaching practicality as the direction, designs a teaching model based on smart classrooms from the perspective of teachers and students, and obtains the comparison between this teaching model and the traditional classroom teaching model through comparative experiments. Data on the impact of student learning effects and learning experience, and try to explore the impact of the smart classroom teaching model on student learning and the problems of the smart classroom model from multiple perspectives.

2   Accounting Information Course

2.1   Smart Teaching

Smart classrooms need to show the characteristics of personalized learning, and can provide corresponding learning resources suitable for students' academic conditions and learning guidance in line with the current situation of students according to the different learning characteristics of learners. With the development of various technologies, smart classrooms can record and track learners' learning process and learning behaviors related to learners' learning data, so as to analyze learners' learning effects and learning characteristics by exploring data through various technologies. Smart classrooms can provide various forms of learning tools, whether online or offline, to help learners understand and internalize knowledge. These learning tools can promote students to construct meaning of different types of knowledge, and then achieve better learning results. Smart classrooms are equipped with abundant teaching resources and advanced equipment, and under the effective guidance of teaching facilitators, learners actively participate in learning activities and use knowledge to solve problems. In the contextual environment created by the smart classroom, learners actively participate in learning activities and apply knowledge to practical problems efficiently and flexibly.

2.2 Accounting Informationization

The training goal of accounting professionals in colleges and universities has information literacy, can solve the problems in the implementation of informatization for small and medium-sized enterprises, and is a junior accounting talent with accounting qualifications. Accounting talents in higher vocational colleges should have strong practical ability, be able to meet the application-oriented work of the basic accounting of enterprises and institutions, be good at accounting practice, and know basic accounting theories.

The spatial range $V(\varepsilon, T)$ that the perception layer can recognize is as follows:

$$V_B(\varepsilon, T) = \sqrt{\frac{1 - B(\varepsilon, T)}{[1 + B(\varepsilon, T)][1 - 2B(\varepsilon, T)]\rho(\varepsilon, T)}} \tag{1}$$

Among them, T is the space temperature, and ε is the parameter. The energy consumed in each round of monitoring data E_{CH} is as follows:

$$E_{CH}(N/k - 1)lE_{elec}^{Rx} + N/klE_{DA} + \left(lE_{elec}^{Tx} + l\varepsilon_{amp}d_{toBS}^n\right) \tag{2}$$

Among them, l represents the length of the data packet.
Record the extracted time series features as:

$$x = \{x_1, x_2 \cdots x_n\} \tag{3}$$

n is the number of time periods in a sequence. Students trained in colleges and universities should first have the ability of accounting, be able to establish various account books according to the accounting subjects used by the business of the enterprise, and be able to screen various original documents.

3 Wisdom Teaching Application Experiment

3.1 Building a Financial Smart Classroom Model

Before the class, teachers carry out teaching design under the concept of smart classroom teaching, and provide supporting teaching resources. Through the rain classroom, the learning content is posted to the rain classroom We Chat platform. Each student completes the learning task and the teacher releases on their own mobile client. After the completion of the student activity, the teacher needs to analyze the data generated by the rain class to prepare for the student's activity in the class; in the class, the teacher participates in the teaching more as the instructor and the observer, and the entire classroom is based on the students. The main body uses the method of question inquiry to test the students' knowledge application ability. After class, we post homework through Rain Classroom, relying on Rain Classroom to carry out real-time interaction, and solve the difficulty of answering questions in traditional classroom.

3.2   Building a Smart Classroom Learning Performance Model

Through the pre-class test, in-class test, and after-class test issued by Yu Classroom, students can test their mastery of knowledge, skills, and study ability. The class scores measured before the experiment are shown in Table 1.

Table 1. Class results measured before the experiment

Class	Mean	N	Standard deviation
Class 1	72.41	66	13.111
Class 2	79.22	72	6.453
Class 3	78.93	72	6.604
Total	76.98	20	9.595

4   Smart Teaching of Accounting Informationization Course

4.1   Student's Task Completion Status and Other Learning Process Experience

Table 2. Student learning process experience

Survey question	Very much agree	Agree	Don't quite agree	Disagree
Tasks provided can help with pre-class learning	30.56%	51. 39%	9.72%	8.33%
No difficulty in completing before class	43.06%	33.33%	12.50%	11.11%
Actively complete learning tasks	54.17%	27. 78%	15.28%	2.78%
The explanation before class is very detailed	76.39%	13. 89%	9.72%	0.00%
Simple and convenient operation on the Rain Classroom platform	58.33%	41.67%	0.00%	0.00%
Like to adopt the problem-inquiry learning method	44.44%	38. 89%	9.72%	6.94%
Increased interest in learning	47.22%	25. 00%	12.50%	15.28%
Can resolve doubts after class	62.50%	29.17%	8.33%	0.00%
Complete the computer operation task after class	41.67%	36.11%	12.50%	9.72%

The student's learning experience is shown in Table 2. On the whole, the proportion of the choice of consent in the 9 questions is more than 70%, indicating that the students have a better experience of the process of the smart classroom oriented to the financial process. In the process of learning, every time the student's progress is different, the time required to complete the task is also different. 76.39% of the students believe that the amount of pre-class tasks assigned by the teacher is moderate and there is no difficulty in completing it, and 12.50% of the students have a slight completion.

difficult. In terms of learning initiative, 81.95% of students can actively complete the learning task. Of course, some students are forced to complete the task at the request of the teacher. The enthusiasm of this part of the student needs to be cultivated. In terms of platform and resource experience, 90.28% of students believe that the teacher's operation explanation in the micro-class is clear and can be watched repeatedly. All students use the Yu Classroom WeChat public platform to operate and learn easily. In the class, the main way to learn is through question inquiry. 72.22% of students think that the inquiry questions facing the financial process increase their interest in learning, and 83.33% of students like this way of learning. Compared with traditional classrooms, this kind of classroom atmosphere is more relaxed, The students are more active. After class, it is mainly to test the students' learning effect by knowing ability and ability test. 92.21% of students think that the answer to the know ability test part of the objective question test is analyzed in detail, which can solve doubts. 77.78% of students Be able to perform tasks on the computer as required.

4.2   Students' Learning Ability

For the smart classroom teaching model, the solution to operational problems in autonomous learning or after-school exercises is shown in Fig. 1. Through survey and statistical analysis, it is found that the way students solve problems after encountering difficulties is much better than that of front-end academic analysis. From Fig. 1, it can be seen that 12.50% of students choose to directly seek help from teachers through the WeChat platform of Rain Classroom. 8.33% of students are still accustomed to seeking help from teachers in a face-to-face manner, and 20.83% of students will ask for help from classmates. This habit should have a lot to do with the group inquiry in the classroom. 26.39% of students choose to watch again through the platform The operation video pushed by the teacher before class seeks a solution. 15.28% of students choose websites such as Baidu to check the information, reflecting the improvement of students' autonomous learning ability from the side, and 16.67% of students choose to ignore it. This part of the students should be finished There are still difficulties in the after-school tasks, and learning is still relatively passive.

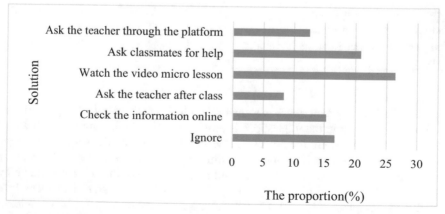

Fig. 1. Solutions to operational problems in self-study or after-school exercises

Figure 2 shows the aspects of improving the ability of the smart classroom teaching model for the financial process. Through investigation and analysis, it can be seen that 10% of students believe that compared with traditional classrooms, work-oriented teaching can increase their interest in learning and are more willing to learn actively; 30% of students believe that work-oriented teaching design can improve their own learning. Computer practical operation ability; the design of pre-class teaching activities is mainly based on self-study. Teachers play the role of instructor in this process. 5% of students believe that their autonomous learning ability has improved. This ratio is higher than expected. Most students are still dependent on learning, and this learning habit still needs to be cultivated.

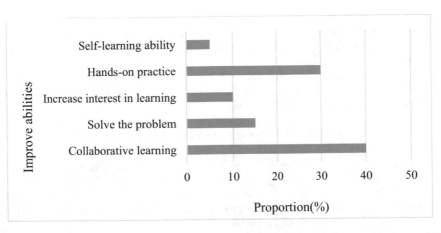

Fig. 2. Aspects of the smart classroom teaching model for financial process to improve ability

5   Conclusion

With the help of smart classroom teaching platform, big data mining and analysis technology is used to effectively solve some problems existing in traditional classroom teaching. For example, on the basis of data analysis, teachers have changed from teaching presuppositions based on personal teaching experience to data analysis of pre-class testing, and changed from monotonous one-quest and one-answer classroom interactions to three-dimensional exchanges and exchanges. The transformation of lagging evaluation feedback into immediate evaluation feedback effectively improves teaching efficiency.

The smart classroom teaching model has changed the teaching concepts and methods of the traditional classroom. This research uses the pre-class test, in-class test, and after-class test issued by Rain Classroom to test students' knowledge, skills, and learning ability. It solves the problems in traditional classroom teaching. This research contributes to the intelligent teaching of accounting information courses.

Acknowledgement. This work was supported by the 2020 Zhanjiang Science and Technology college Research Project on educational and teaching reform: Research on the application of wisdom teaching method in accounting information course (Project No.: jyjx2020013); The 13th five-year Plan of Guangdong Higher Education Institute higher education Research Topics: Research on the teaching reform and application of accounting information course under the background of intelligent finance (Project No.: 19GYB081).

References

1. Asabere, N.Y., Togo, G., Acakpovi, A., et al.: AIDS: an ICT model for integrating teaching, learning and research in Technical University Education in Ghana. Int. J. Educ. Dev. Using Inf. Commun. Technol. **13**(3), 162–183 (2018)
2. Noviana, E., Kurniaman, O., Hermita, N., et al.: Siak culture on local wisdom-based teaching in primary school: a preliminary study. J. Comput. Theor. Nanosci. **24**(11), 8500–8502 (2018)
3. Sunahrowi, S., Anastasia, P., Singgih, K.: Religiosity and local wisdom in teaching cultural science in faculty of languages and arts of semarang state university. IBDA Jurnal Kajian Islam dan Budaya **16**(2), 262–275 (2018)
4. Zhang, S.F., Han, X., Wu, D.H., et al.: Establishment of a new model of regional management of stroke based on the wisdom medical association platform. Fudan Univ. J. Med. Sci. **45**(6), 805–810 (2018)
5. Saputri, S.W., Rohiyatussakinah, I.: The Development of English teaching material base local wisdom at SMKN 1 Cinangka. J. Engl. Educ. Stud. **2**(2), 95–100 (2019)
6. Weisenfeld, L., Mathiyalakan, S., Heilman, G.: Topics for your undergraduate accounting information systems (AIS) course-an exploratory study of information technology (IT) skills and firm size. AIS Educ. J. **15**(1), 58–89 (2020)
7. Garnsey, M., Doganaksoy, N., Phelan, E.: Topics for the accounting information systems course: a dual perspective approach from educators and employers. AIS Educ. J. **14**(1), 36–55 (2019)
8. Moradi, M., Tarighi, H., Hosseinipour, R., et al.: Factors influencing the learning of accounting information systems (AIS): evidence from Iranian students. J. Econ. Adm. Sci. **36**(3), 226–245 (2019)
9. Dierynck, B., Labro, E.: Management accounting information properties and operations management. Found. Trends Technol. Inf. Oper. Manage. **12**(1), 1–114 (2018)
10. Wong, H., Sum, C., Chan, S., et al.: Effect of deliberate practice and previous knowledge on academic performance. Int. J. Bus. Inf. **14**(1), 25–46 (2019)

English Education Model Based on Artificial Intelligence

Mei Yang[✉]

Guangzhou Huashang Vocational College,
Guangzhou 511300, Guangdong, China

Abstract. The development of artificial intelligence will further promote the intelligent revolution of the whole society. This research mainly discusses the design of English education model based on artificial intelligence. Taking constructivism as the basic guiding theory, and at the same time relying on the idea of 4C/ID model. Emphasize the creation of a real learning situation for learning spoken language, use the technological advantages of robots to stimulate students' interest in learning, promote students' learning motivation, and help teachers optimize learning strategies. The educational robot can monitor the student's learning process when the student communicates individually, and establish a personal information database and a feature database for the student to provide personalized learning services. In this learning method, robots help teachers complete repetitive tools and provide more intelligent auxiliary functions, while teachers monitor the teaching process, check for deficiencies, and ensure the orderly progress of teaching activities. In terms of students' language and cognitive ability, the post-test has an average improvement of 1.07 compared with the pre-test. This research contributes to the realization of intelligent classroom teaching methods.

Keywords: Artificial intelligence technology · English education model design · Learning interest · Personalized learning service

1 Introduction

With the advent of the information age and the continuous impact of new technologies, traditional English classroom teaching is facing huge challenges. The disadvantages of traditional English classroom teaching are difficult to solve with traditional methods, and it is urgent to use new perspectives and methods to change. The core of the smart classroom is to use the latest information technology to transform and improve classroom English teaching to create an efficient and intelligent classroom.

In the field of education, language learning is a kind of vague learning, with strong uncertainties. English learning runs through the entire education stage, and the level of English proficiency is directly related to the various assessments of students [1, 2]. In the process of English teaching, how to quantify the learning effect of students has always been a difficult problem [3, 4]. English test questions mostly exist in the form of objective questions, from which it is difficult to draw a definite conclusion on the mastery of a certain knowledge point, which results in the teaching plan cannot be

J. Abawajy et al. (Eds.): ATCI 2021, LNDECT 81, pp. 619–625, 2021.
https://doi.org/10.1007/978-3-030-79197-1_89

adjusted in time due to the mastery of the students [5, 6]. At present, in the process of language knowledge learning, it is a common practice to use exams for practice and exams to promote learning. The answers to a large number of objective questions answered by students after the exam can be regarded as a massive and noisy data collection [7, 8]. Finding key information from massive data is the original intention of data mining technology. Applying it to English teaching can achieve a multiplier effect with half the effort. In today's internationalization, the status of English in international communication is getting higher and higher. Therefore, we must continuously improve English learning ability and reform English teaching to meet the current needs of English learning [9, 10].

The traditional English teaching model has great limitations and can no longer meet our needs. The English teaching reform model mainly uses computer technology and network technology to transform the traditional teacher teaching model into a teaching software, networked, and information-based learning model. The reformed teaching model has largely made up for the limitations of the traditional model and has become more concise and convenient. More and more large and medium-sized enterprises and institutions in our country use multimedia digitization based on networking, intelligence, human-computer interaction, and informatization for English teaching.

2   English Education Mode

2.1   Artificial Intelligence

In the field of education, whether it is the expert system or the intelligent scoring system introduced above, it is a model to imitate teachers or tutors to guide students' learning. There are also neural networks in deep learning, both in terms of pictures and principles, which resemble the way of thinking of the human brain, similar to the reactions of cells and neurons in biology when humans think.

Assume that the standard eigenvalue relative member matrix S and the relative member matrix B_{hj} of all levels.

$$B_{hj} = u_{hj} \left\{ \sum_{i=1}^{m} \left[w_{ij} \left(r_{ij} - s_{ih} \right) \right]^p \right\}^{\frac{1}{p}} \tag{1}$$

By using a neural network to extract valid signals and input them to fuzzy inference, the difficulty of the fuzzy rule process can be greatly reduced. Then after using the neural network to complete the fault diagnosis function, the accuracy of the diagnosis result can be effectively improved.

$$F[M(S, t+1)] = M(S, t) \frac{f(S, t)}{f(t)} \left[1 - P_c \frac{\delta(S)}{L-1} - O(S)P_m \right] \tag{2}$$

Among them, $M(S, t)$ represents a collection of data. If the connection weight between the clarification and the rule layer is δ_i^3, then there are:

$$\delta_i^3 = \sum_{i=1}^{C} \frac{\partial E}{\partial Q_i^4} \times \frac{\partial Q_i^4}{\partial Q_i^3} = \sum_{i=1}^{C} \frac{\partial E}{\partial Q_i^4} \times W_{ij} \frac{\sum_{k=1}^{\delta} Q_k^3 - Q_j^3}{\sum_{k=1}^{\delta} Q_k^3} \tag{3}$$

Artificial intelligence technology mainly studies how to imitate, extend, and expand the human brain to engage in thinking activities such as reasoning, planning, calculation, thinking, and learning, allowing computers or intelligent machines (including hardware and software) to solve complex problems under normal circumstances, and these problems are in In the past, only human experts were able to deal with it.

2.2 English Teaching

Judging from the requirements of the party and the state for the development of education reform in the new era, education should adapt to the requirements of the development of the information society, continue to expand new forms of education, and promote the in-depth integration of information technology with education and teaching. my country needs to continuously explore new educational models that suit its own. So far, there are many research results on smart teaching models in higher education. Compared with the research on higher education, there are relatively few research results on smart teaching models in primary schools, and there are not many researches on smart teaching models in primary school English. Of course, there are relatively few research results and there are certain subjective and objective factors, such as: teachers lack the necessary knowledge and have a weak awareness of the use of this model; students have weak self-control ability and need more teachers to meet the smart teaching model, and some school teachers Insufficient power; some school hardware facilities can not meet the smart teaching model, etc. This makes the theoretical foundation of the primary school English smart teaching model relatively weak, and the amount of research is relatively scarce.

3 English Education Model Design Experiment

3.1 Promote Students' Learning Motivation

Taking constructivism as the basic guiding theory, and at the same time relying on the idea of 4C/ID model. Emphasize that learners are the mainstay, and teachers and robots are the supplementary. Emphasize the creation of a real learning situation for learning spoken language, use the technological advantages of robots to stimulate students' interest in learning, promote students' learning motivation, and help teachers optimize learning strategies.

3.2   Emphasis on Personalized Learning

The educational robot can monitor the student's learning process when the student communicates individually, and establish a personal information database and a feature database for the student to provide personalized learning services. In this learning method, robots help teachers complete repetitive tools and provide more intelligent auxiliary functions, while teachers monitor the teaching process, check for deficiencies, and ensure the orderly progress of teaching activities. The basic situation of the experimental subjects is shown in Table 1.

Table 1. Basic situation of experimental subjects

Class	Nature class	Number of people
Experimental class	1	45
Control class	1	44

4   English Education Model Based on Artificial Intelligence

4.1   Significant Differences

The significant difference analysis is shown in Table 2. Homogeneity analysis is to analyze the difference between the levels of the two classes before the experiment, in order to judge whether the students in the experimental class will adopt the teaching of the English listening and speaking ability training strategy based on the intelligent phonetic system and have significant changes. The English test questions compiled before the experiment are conducted on the listening and speaking tests of the students in the two classes, and the English test results are analyzed to determine whether there are differences in the English listening and speaking abilities of the students in the two classes, which will lay a good foundation for subsequent experiments Foundation. Using SPSS to carry out statistical analysis of data, the test of the experimental class and the control class are carried out with independent sample T test, and according to the results of data analysis, it is judged whether there is a significant difference in the overall mean of the two classes. It can be seen from Table 2 that the sample number of the experimental class is 45, the average English test score is 86.131, the standard deviation is 6.4189, and the standard error average is 0.857; the sample number of the control class is 44, the average English test score is 85.326, and the standard error is 6.3264. The average error is 0.844; it can be seen that the average test scores of the experimental class students are higher than the average of the control class.

Table 2. Significant difference analysis

Class	Number of people	Average (E)	Standard deviation
Experimental class	45	86.131	6.4189
Control class	44	85.326	6.3264

4.2 Comparative Analysis of Students' English Listening and Speaking Ability

During the experiment, SPSS2.0 was used to analyze the distributed student questionnaires and related data, the student questionnaires collected in the experimental class were compared before and after, and the average paired sample T test was performed to observe the English listening and speaking ability of the experimental class students through data analysis Whether it changes before and after the experiment, as shown in Fig. 1. In terms of students' language cognitive ability, the average language cognitive ability of the pre-test is 13.16, the standard deviation is 4.62, the average language cognitive ability of the post-test is 14.23, the standard deviation is 4.14, the post-test is compared with the pre-test average The value increased by 1.07, indicating that through experiments, the language cognitive ability of students in the experimental class has been improved; in terms of students' language understanding ability, the average language understanding ability during the pre-test is 14.55, the standard deviation is 4.26, and the language understanding ability during the post-test The average value is 15.12 and the standard deviation is 4.03. The post-test is 0.57 higher than the pre-test average. It can be seen from the experiment that the language comprehension ability of the students in the experimental class has been improved. In terms of the students' language evaluation ability, the language evaluation of the pre-test. The average ability is 14.11, the standard deviation is 4.51, the average language evaluation ability during the post-test is 14.86, and the standard deviation is 3.77. The post-test is an increase of 0.75 compared with the average of the pre-test, which shows that through the experiment, the language evaluation of the students in the experimental class Ability has been improved.

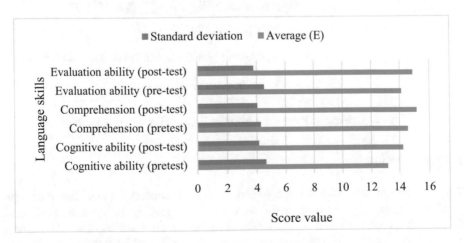

Fig. 1. Changes in students' English listening and speaking ability

The achievements of artificial intelligence have entered various fields of society with a new kind of productivity, allowing human intelligence to be effectively displayed on the auxiliary platform of computers. Computer technology, supplemented, penetrated and promoted by human intelligence, in turn promotes the development of human intelligence. development of. It has expanded a brand-new field for the history of human civilization, brought a large number of industries into a brand-new stage, and enabled the entire scientific community to move forward on a brighter path. It is a remarkable and far-reaching achievement in the history of science. In order to better analyze the changes of students' English listening and speaking interests, the questionnaires related to the students' interest in listening and speaking are analyzed. The changes of students' English listening and speaking interests are shown in Fig. 2. The number of people who chose "very consistent" rose from 19.85% of the total to 24.84%, and the number increased by 4.99%. The number of people who chose "compliant" rose from 30.15% to 38.06% of the total, and the number increased by 7.91%. "The number of people dropped from 5.33% of the total to 3.85%, and the number of people dropped by 1.48%. This shows that in the process of experimental teaching, students' interest in listening and speaking in English is gradually improving, and it also reflects from the side that students' listening and speaking habits are also gradually changing.

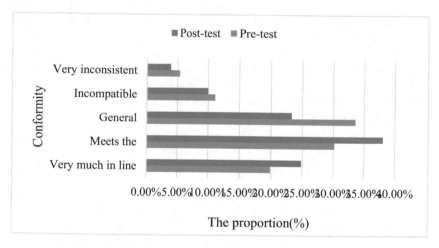

Fig. 2. Changes in students' interest in English listening and speaking

5   Conclusion

This research mainly discusses the design of English education model based on artificial intelligence. Taking constructivism as the basic guiding theory, and at the same time relying on the idea of 4C/ID model. Emphasize the creation of a real learning situation for learning spoken language, use the technological advantages of robots to stimulate students' interest in learning, promote students' learning motivation, and help teachers optimize learning strategies.

The educational robot can monitor the student's learning process when the student communicates individually, and establish a personal information database and a feature database for the student to provide personalized learning services. In this learning method, robots help teachers complete repetitive tools and provide more intelligent auxiliary functions, while teachers monitor the teaching process, check for deficiencies, and ensure the orderly progress of teaching activities. This research contributes to the realization of intelligent classroom teaching methods.

References

1. Uestuenkaya, T.: Artificial intelligence: friend or foe to fashion in consideration of the functionality doctrine? Eur. Intellectual Property Rev. **42**(1), 13–18 (2020)
2. Miao, Z.: Investigation on human rights ethics in artificial intelligence researches with library literature analysis method. Electron. Library **37**(6), 911–926 (2019)
3. Hansen, E.B., Iftikhar, N., Bgh, S.: Concept of easy-to-use versatile artificial intelligence in industrial small & medium-sized enterprises. Procedia Manuf. **51**(7), 1146–1152 (2020)
4. Kim, K.Y., Jung, J.H., Yoon, Y.A., et al.: Designing a performance certification test for automatic detection equipment based on artificial intelligence technology. J. Appl. Reliab. **20**(1), 43–51 (2020)
5. Rong, Z., Gang, Z.: An artificial intelligence data mining technology based evaluation model of education on political and ideological strategy of students. J. Intell. Fuzzy Syst. **40**(5), 1–12 (2020)
6. Adams, R.C., Rashidieh, B.: Can computers conceive the complexity of cancer to cure it? Using artificial intelligence technology in cancer modelling and drug discovery. Math. Biosci. Eng. **17**(6), 6515–6530 (2020)
7. Moon, A.N.: Teaching the history of the English language: a model for graduate students of education. Engl. Lang. Linguist. **24**(3), 47–67 (2018)
8. Gao, B.: Highly efficient English MOOC teaching model based on frontline education analysis. Int. J. Emerg. Technol. Learn. (iJET) **14**(6), 138–146 (2019)
9. Cokun, H.: Description of a model for lesson planning on peace education in teacher training. J. Educ. Cult. Soc. **9**(1), 118–142 (2018)
10. Arsyadana, A., Ahmadi, R.: Learning model-based digital character education in al-hikmah boarding school batu. Didaktika Religia **7**(2), 234–255 (2019)

Analysis of the New Idea of Teaching Management for Universities Under Background of the Big Data

Jingfeng Jiang$^{(\boxtimes)}$ and Ruixue Guo

Department of International Hospitality, Fuzhou Melbourne Polytechnic,
Fuzhou, Fujian, China
jiangjf@fmp.edu.cn

Abstract. In the context of increasing big data, the application of it not only brings great improvement to individuals' work and life, but also has a certain impact on development of education and teaching management in colleges and universities. Combined with the characteristics of the current teaching methodology and management in colleges and universities, integrated into the application of big data, this paper discusses and studies how to reasonably carry out the teaching management in colleges and universities with the development of big data.

Keywords: Big data · Higher education

1 Introduction

At present, with the rapid development of IT, the application of big data is gradually popularized. In people's daily life and production work, big data has strong theoretical value and application value, and by using big data, people's production, life and habits can be used to obtain corresponding accurate analysis.

It is an inevitable trend that big data which is increasingly developed has been incorporated into nowadays colleges and universities teaching management. For the education management of colleges and universities, with the help of massive data, managers are effectively managing related matters in colleges and universities, and can directly analyze and achieve precise management. Furthermore, the application of big data can effectively promote the reform of teaching management, which will lead to the implementation and management of teaching to keep pace with the times, and continue to innovate and develop.

2 Concept and Characteristics of Big Data

The big data refers to the collection of various data and information through the network and other related tools within the set time range, so as to form a collection of user data and information, and conduct data collection of user behavior characteristics based on these data analyses [1].

J. Abawajy et al. (Eds.): ATCI 2021, LNDECT 81, pp. 626–631, 2021.
https://doi.org/10.1007/978-3-030-79197-1_90

Combined with the Big Data application, its features typically include: collecting data rapidly and greatly, rich data categories, quick data dealing speed, and low value density [1, 2]. Throughout them, the so-called data collection refers to the most extensive acquisition of user behavior data information via the using big data processes, which covers all aspects of users' life and work [2]. The categories of rich data, that is to say the acquisition of big data, of information, the information obtained are in fact closely linked to the specific life of the users. In massive user behavior data, information, it contains not only user behavior data, but also user consumption data, and even user learning data [1]. And in presenting data categories, it can be text, words, pictures, graphs, audio, voice and video, etc. Quick receiving and dealing speed means that during analysing and processing of different information, with the corresponding system, it can realize the processing of user information rapidly, and give the corresponding expression for the user's behavior characteristics as soon as possible [2]. Low value density refers to the fact that with big data, the amount of really effective information is very small, which requires the help of certain system applications to extract user behavior characteristics of massive information. Although the amount of value is low, all the extracted information is effective information [1–3].

3   The Source of Big Data in Higher Education Teaching Management

Developing education and training management for colleges and universities, large data can also be used to carry out an in-depth analysis. Combined with the existing large data analysis system, there are three main sources and origins of large data for university educational management.

3.1   Business Management System

Developing education and training in Colleges and Universities cannot be divorced from the appliance and utilization of different business management systems. With the high-speed improvement of information technologies, many universities have also built and utilize information on campuses. Through the application of various business management systems, colleges and universities can fully inform and effectively manage education [3].

For example, in the control of school teachers' attendance, the introduction of an attendance management system; in the management of trainers' teaching behavior, the introduction of performance appraisal system; in addition, for the daily teaching performance management, there are teaching management systems. In conjunction with the use of the different systems above, the acquisition of important data from the management of university education can be achieved through these business management systems [4].

Through the above mentioned comprehensive application of different business systems, such as personnel attendance system, performance appraisal system and teaching management system, the informational of campus teaching management is realized, and at the same time, it also provides an important source channel for teaching management big data [3, 4]. Combined with the above mentioned system, it actually covers related academic information, data, personal data, scientific research data, financial data and so on.The acquisition of these data can promote the school teaching management procedures to have richer and larger data sources, and ensure that after the corresponding data interpretation, we can have a more comprehensive understanding of the evolution for university teaching work.

3.2   Ways of Online Teaching

Equipped with constantly updating information technology, many universities have introduced networked education platforms. Whether from the viewpoint of school education management or from the teachers and students comments, the opening up and use of a networked teaching platform can also make it possible to acquire rich data [5]. Networking is also an important element in university education.The acquisition of teaching data in a network enables a more complete analysis of teaching management work.

In the present implementation of networked teaching in colleges and universities, common means, such as networked classes, networked classes and offline teaching models, form effective supplements. In addition, the network short class mode can be set up, primarily for emphasis on teaching and difficulties in the classroom for a more in-depth interpretation [6].

This type of networked teaching method can not only reduce the distance between faculty and students, but also provide better support for acquiring important data from the management of university education. The summary and analysis of network teaching data can effectively make a clear decision on the development of teaching work and management around colleges and universities [5, 6].

3.3   Teacher-Student Network Behavior

In the construction of the university learning environment, the core goals of teaching management include two groups in trainers and learners. With increasing network technique, the Internet has become a part of people's daily life, which is indispensable. [6]. In the acquisition of large data of university teaching management, the network can be used to obtain the behavior data of university teachers and students [6, 7]. This is likewise another important way to obtain big data for university teaching management. Teachers and students usually use the network or the campus network to browse and collect information during their school days, at the same time, they will also use the network to transmit and interact with teachers in the course of learning [6]. For example, learning on the campus sticker bar or publishing and replying to posts related to campus content; using class WeChat group for class or course content communication, communication, and so on.Through the summary of these network behavior data, it can provide some structured data information for university teaching management.

4 Countermeasure Analysis of Teaching Management of Universities Under the Background of Big Data

In order to better realize the reasonable application of big data, colleges and universities should pay more attention to the introduction of big data information into teaching management in today's rapid development of data. This can provide sufficient support for the development of teaching and the innovation of teaching management [5].

4.1 Realizing the Innovation of Traditional Teaching Management Concepts and the Diversification of Teaching Management Subjects

Enormous amounts of data are made to implement effective teaching management behavior. Based on the analysis of large data, universities are required to base themselves on the development perspective and combine the results of large data analysis to clarify their own problems and deficiencies in the development of teaching management, and actively introduce scientific and rational teaching management concepts to promote and improve the innovation and growing of traditional training or teaching concepts [5].

For school administrators, attention should be given to the innovation of teaching management concepts. Traditional teaching management is under the unified control of the educational administration department. Combined with the results of considerable data analysis, schools have a duty to change their traditional management concepts and gradually turn to the teaching management mode dominated by teachers and students [5, 8, 9]. For example, when a university is carrying out its teaching management work, it will make innovative changes in teaching management work by fully equipped big data, due to the perspectives of teachers and students, and combining the opinions and suggestions of teachers and students on the way and management of teaching management. From the school, to the administrator of the school, to different departments, to teachers and students, all become the main and essential subjects of teaching management, to finally achieve a diversified distribution of teaching management subjects [9].

Fully use of extensive data to strengthen management of specific links in teaching, and at the same time, to innovate teaching management concepts and increase training management effectiveness[5]. In teaching management, a university attaches greatly to the change of educating management concepts, combining with larger data information, so as to realize the ground from the surface to practice of teaching management. In the specific development of teaching management, teaching control should be carried out by different departments and students of different majors based on explicit teaching links.Through this kind of management, not only will be the concept of educating management to be further innovated, but also the effectiveness of teaching management can be fully guaranteed.

4.2 Personalized Teaching Management for Different Requirements of Teachers and Students

Advanced information technology has promoted the reform of teaching management methods and means. In current teaching management developing throughout colleges and universities, besides the customary teaching management system, you can also use the network and mobile devices to carry out the personalized progress of teaching management [5, 9].

In the course of teaching management, through the analysis of the consequences of substantial data, a university emphasizes the realization of personalized management in teaching management, and promotes the practical value of teaching management based on specific practice. In the development of daily teaching, teachers use vast data to personalize the classroom settings with the results of enterprise data for analysis on future talent training [10]. Meanwhile, the huge data association can also promote teachers to have a clear understanding of the acceptability of students' classroom content. Combining the learning ability and learning needs of students at different levels, teachers can teach knowledge in a hierarchical approach when carrying out classroom teaching [9, 10]. In this approach, university teaching management can achieve a sustainable management effect, so that students at different levels can achieve the mastery of theoretical knowledge in classroom learning. In addition, a university also carries out the opening of online classes to personalize and target teaching extension for students with diverse learning needs.In this way, the school has made outstanding achievements in the development of teaching management.

It can be seen that the use of heavy data have a prominent driving significance for the schools teaching management and administration. Utilizing the results of substantial data analysis, universities and teachers will, definitively, have a better target in the organization of training activities, and can effectively integrate learning needs of students to achieve the introduction of personalized teaching methods [5, 6]. With the help of substantial data, we can effectively improve the teaching results in the targeted school, and realize the highest prominence of value of teaching management.

4.3 To Further Improve Teaching Evaluation in Order to Improve Teaching Level

In the process of teachers' teaching effective evaluation, relatively perfect teaching evaluation mechanism can be built with the help of copious data. On the training evaluation, based on the system layout, the teachers, students and teaching offices are set up to evaluate the teaching situation of teachers [11]. By summarizing the data of different evaluations, we can ensure that teachers' teaching evaluation are more accurate and comprehensive.

Currently, many universities begin to build teaching evaluation systems based on large data. The introduction of this system provides a possibility for better teaching management in campus [10, 11]. Last but not the least, with the help of this evaluation system, teachers can better reflect on the teaching work, and make improvements to the shortcomings in order to increase the effectiveness of schoolroom teaching and learning.

5 Conclusion

Under the background of massive data development, university teaching management will inevitably enter a new progress and step. The introduction of data progress in the university teaching management can promote a new change in original management style, promoting fully and further introduction of cutting-edge technology, and implementing with newly teaching management, which will bring about a whole change for the teaching and training management, and promote the innovation. Therefore, universities should pay fully attention to increasing big data technology, effectively using big data to promote the innovation for campus teaching management, and finally achieving the continuous improvement of teaching quality and comprehensive level.

Acknowledgment. This work was supported by 2020 National Vocational Education Research Program (Grant No: 2020QZJ237).

References

1. Dutt, A., Ismail, M.A., Herawan, T.: A systematic review on educational data mining. IEEE Access 15991–16005 (2017)
2. Sagiroglu, S., Sinanc, D.: Big data: a review. In: 2013 International Conference on Collaboration Technologies and Systems (CTS). IEEE (2013)
3. George, G., Haas, M.R., Pentland, A.: Big data and management. Acad. Manage. J. **57**(2), 321–326 (2014)
4. Choi, T.M., Chan, H.K., Yue, X.: Recent development in big data analytics for business operations and risk management. IEEE Trans. Cybernet. 1–12 (2016)
5. Huda, M., Anshari, M., Almunawar, M.N., Shahrill, M., Masri, M.: Innovative teaching in higher education: the big data approach. Turkish Online J. Educ. Technol. **15**(Special issue), 1210–1216 (2016)
6. Chaurasia, S.S., Frieda Rosin, A.: From big data to big impact: analytics for teaching and learning in higher education. Ind. Comm. Train. **49**(4) (2017)
7. Miah, S.J., Vu, H.Q., Gammack, J., Mcgrath, M.: A big data analytics method for tourist behaviour analysis. Inf. Manage. **54**(6), 771–785 (2016)
8. Martin-Rios, C., Pougnet, S., Nogareda, A.M.: Teaching hrm in contemporary hospitality management: a case study drawing on hr analytics and big data analysis. J. Teach. Travel Tour. **17**(1), 1–21 (2017)
9. Grillenberger, A., Romeike, R.: Teaching data management: key competencies and opportunities. In: KEYCIT 2014-Key Competencies in Informatics and ICT (2014)
10. Sigman, B.P., Garr, W., Pongsajapan, R., Selvanadin, M., Bolling, K., Marsh, G.: Teaching big data: Experiences, lessons learned, and future directions. decision line **45**(1), 10–15 (2014)
11. Godwin-Jones, R.: Scaling up and zooming in: big data and personalization in language learning. Lang. Learn. Technol. **21**(1), 4–15 (2017)

An Assessment on the Response of China and America During the Covid-19 Pandemic in Cultural Perspective – Based on Big Data Analysis

Junqiao Liu[✉]

Institute of Psychology, Chinese Academy of Sciences, Beijing, China

Abstract. The level of response to the prevalent covid-19 has varied from place to place. However, culture and big data using rapidly have played a huge role in determining the direction as to which different countries and people respond. A total of 219 people through the internet were sampled randomly for this study during the lockdown time. Questionnaires were electronically distributed through WeChat and other social media platforms through the internet. Descriptive statistics and inferential statistics were using SPSS to analyse the collected data. Through those big data show that Americans from a scale of 1 to 7 had more protection confidence $M = 5$, $SD = 1.22$ than Chinese $M = 3$, $SD = 1.03$. The Chinese had more Pathogenic aversion $M = 5$, $SD = 0.18$ while the Americans scored $M = 4$, $SD = 0.11$ from a scale of 1 to 7. It was noticed that the phenomenon of internal group preference and external group exclusion occurs in outbreaks. The results indicated that the collectivist culture has a higher tendency to think as a group and interdependently corroborate in responding to pandemic outbreaks like the Covid-19 compared to the individualists, which therefore can be attributed to why China a collectivist country has responded better than America an individualist country in dealing with the prevalent Covid-19 pandemic.

Keywords: Collectivism and individualism · Cultural attribute · Big data

1 Introduction

New crown pneumonia (Covid-19) is a novel coronavirus that broke out in China in December 2019. The main symptoms of this new crown pneumonia are fever, dry cough, shortness of breath, etc. This disease suddenly rapidly spread wide, and a large number of people were infected in a short time, followed by a large number of deaths. At the beginning of the epidemic. The epidemic began to spread from China to all parts of the world. According to the World Health Organization's global epidemic data, as of 5:00 pm on November 23rd, the World Health Organization said that there were 58,425,681 confirmed patients in the world, and the total number of deaths due to Covid-19 in the world was 1,385,218. There are 11,972,556 confirmed patients in the United States, and 1,471 Americans died in COVID-19. By the time more data was released, there were 253,931 deaths in the United States, accounting for about one-fifth

J. Abawajy et al. (Eds.): ATCI 2021, LNDECT 81, pp. 632–639, 2021.
https://doi.org/10.1007/978-3-030-79197-1_91

of the global total. In 31 provinces (autonomous regions and municipalities directly under the Central Government) of China and Xinjiang Production and Construction Corps, 92,733 confirmed cases of pneumonia in COVID-19 were found, with a cumulative death of 4,749. The cumulative number of confirmed cases and deaths in the United States ranks first in the world, among which the cumulative number of confirmed cases were about 21% of those in the world, while the population of the United States only accounts for 4% of the world population and the population of China accounts for 17% of the world which is like 4 times more than America. In the face of COVID-19 pneumonia, the gap between China and the United States can be seen to be extremely obvious, but the reason behind such a big difference is worth pondering (See Table 1).

Table 1. The Covid-19 death cases

China	0.03‰
America	7.74‰
Other Countries	2.54‰

2   Methods

2.1   Background Information Comparation Through Big Data Anallisis

The questionnaire was posted in my wechat moment, Facebook and other social medias. Finally a number of 219 ordinary people from the native Chinese and Americans were sampled or the study. The questionnaire test was conducted in stages between Chinese mainland and Westerners from North American countries, and the data were collected at the worst time of the epidemic.

Among the subjects in China, through those data collecting, the age distribution of the selected participants was such that only 1% of the whole sample size among the Chinese was under 18 years of age, 2% were between 18–25 years, 36% were between 26–30 years, another 36% were between 31–40 years, 11% were between 41–50 years, 9% were between 51–60 years, and 4% were above the age of 60. Among the American participants, the age distribution of the selected participants was such that only 2% of the whole sample size of the Americans were under 18 years of age, 17% were between 18–25 years, 29% were between 26–30 years, 33% were between 31–40, 11% were between 41–50 years, 5% were between 51–60 years and 3% were above 60 years of age. From the answering time, we can easily see the younger people will answer faster than older people through the internet questionnaires.

If we define youth as being the age above 30 years, we can notice that the common trend among the two divisions of the Samples analyzed, that is the Americans and the Chinese, is that more than 50% of both the American and Chinese side of the people given questionnaires were above the age of youth meaning both samples had a better understanding of their cultures.

The distribution of education background between the two samples is as shown below in Figs. 1 and 2.

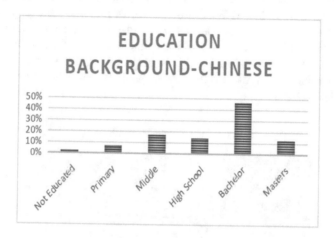

Fig. 1. Education background of the sampled Chinese population

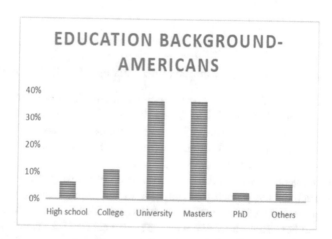

Fig. 2. Education background of the sampled American population

There is a trend observed that most of the Americans sampled for this study were educated at least up to high school. Through those big data collecting from the internet, we see that at least 77% of the Americans had an education level from College, University or bachelor's degree and above while only about 58% of the Chinese had an education level bachelor's degree and above. This paints a picture that since the Americans seem to be more academically educated compared to the Chinese and also taking into account the population of China being bigger than America, the Americans should be able to handle and contain the spread of the Coronavirus better than the Chinese population. The question would be; why has America a typical example of an Individualist culture with more educated people, more medical intellectual power, recorded more cases and death due to Corona Virus as compared to China a typical example of a collectivist culture?

Here, we begin to get the concept that culture and background, sometimes even more than education has a big influence on the behavior of individuals exposed to a similar situations. And can further affect how they respond in a situation of a epidemic.

2.2 Methodology and Analysis of Findings

The measure of what was coined as protection confidence was done on both the sample. The assumption was that the more confident a certain people are in their country being able to handle their problems or rather the covid in this case, the more reluctant they become and hence the more exposed they. As the alternative hypothesis, it is assumed that the difference in protection confidence of the Chinese and the Americans is significant. While the null hypothesis would be that the difference is not significant.

Direct questions were asked to the Americans and the Chinese separately, on a scale of 1 to 7 about how confident they are in their Country being able to fully protect them from the prevalent Covid-19 [1, 2]. On average, the Americans picked a scale of 5 which shows that they were highly confident that their Country would protect them room the virus with $M = 5$, $SD = 1.22$. While the Chinese were less confident in their Country fully protecting them from the virus, on average pick 3 from the scale of 1 to 7, $M = 3$, $SD = 1.03$ from the average value. After preforming a T-test, the P value was $p = 0.0001$ at 95% confidence level. The p value is less that the α value of 0.05, which gives us evidence to support the alternative hypothesis and hence state that the difference in protection confidence of the Chinese and the Americans is significant.

Anxiety was also asset among the two groups. Anxiety is mainly looked at in two directions, there is negative at positive anxiety [3–5]. In this study, negative anxiety was examined using a number of questions. from a scale of 1 to 7, the Chinese and the Americans were asked question that assessed how anxious they were about the Covid, the higher the number from 1 to 7, the less anxious and vice versa. The results were such that the average scale of the Chinese was $M = 3$, $SD = 0.10$ showing a picture that most people among the Chinese picked a level of anxiety close to the average as seen from the standard deviation. While the Americans on average score $M = 4$, $SD = 0.11$. The t test at 95% confidence level gave a p value of $p = 0.0001$ which is less than the alpha value of 0.05 gives evidence that the difference between the anxiety of the Americans and the Chinese is significant. The results were closely spread around the mean. These results communicate that the Chinese showed more negative anxiety compared to the Americans which made them seriously embrace behavioural patterns and practices that helped them keep themselves safe from the pandemic and hence being able to respond better [6, 7].

Pathogenic aversion is tendency of individuals or a certain group to not entertain situation or environments that would potentially cause them be diseased [8]. It is also another culture social behavior that influences how exposed some can be to infectious diseases like the Covid-19. People with high pathogenic aversion will not situation where large clouds come about them, or get so free spirited that they would go about potential disease exposed areas without necessary precaution. Data collected about pathogenic aversion was such that on average, the Americans from a scale of 1 to 7 score M = 4, SD = 0.11 while the Chinese scored M = 5, SD = 0.18. On average, the Chinese had more Pathogenic aversion than the Americans. The difference is proved significant by the t test which produced a p value of p = 0.0001 at 95% significance level which is less than the alpha value of $\alpha = 0.05$.

This suggests that the more people have the negative attitude towards diseased people and disease exposed environments, the less their risk of exposure is to the prevailing disease. The more pathogenic aversion, the less their exposure to communicable diseases like the Covid-19.

Our hypothesis are such that;

Null hypothesis (H0): There is no significant difference between China and America in respect to culture, as to how they have responded to the Covid-19 pandemic.

Alternative hypothesis (Ha): There is a significant difference between China and America in respect to culture, as to how they have responded to the Covid-19 pandemic.

To investigate the hypothesis, an ANOVA test was done using SPSS. Where the variables used met the conditions assumed by ANOVA.

The assumption or the null hypothesis of the ANOVA is that the variance are equal, which when test is done, the p value was found to be less than 0.05 which gives us evidence to reject the null hypothesis and go for the alternative hypothesis which says the variances are not equal.

The results of the ANOVA are in the table below (See Table 2).

Table 2. Analysis of variances.

		ANOVA				
		Sum of squares	df	Mean square	F	Sig.
Protection confidence	Between groups	85.085	1	85.085	80.620	0.0001
	Within groups	226.906	215	1.055		
	Total	311.991	216			
Pathogen aversion	Between groups	29.614	1	29.614	17.443	0.0001
	Within groups	365.013	215	1.698		
	Total	394.627	216			
Covid anxiety	Between groups	21.629	1	21.629	36.932	0.0001
	Within groups	125.911	215	.586		
	Total	147.539	216			

The analysis of variances at a significance level of 0.05 suggests that suggest that there is a statistically significant difference between the group.

2.3   Discussion

2.3.1   Differences in Thinking Mode of Pneumonia in COVID-19

This study set out to investigate the impact of culture on how they respond to the pandemic (Covid-19), that is the collectivism culture, which is found among the Chinese in comparison to the individualism culture which is found among the Americans. Using the variables such as pathogenic aversion, anxiety, protection confidence, education background, we see where each culture stands under the international data comaration. Except that we also know that our governments and research institutions also try their best to get more data to develop insights on these aspects. In order to get rid of covid-19 spread, we started to use location sharing rapidly as well. For Chinese, that is acceptable because of location sharing, it is easy to find the person who had some connection with the potential cases, which is not possible in American. Cause we always consider others more than us. We used to put friends into the first place as well.

2.3.2   Protectionism During the Epidemic

The high confidence that the Americans had in their country being able to protect them from the Covid-19 pandemic, put them in a position to be reluctant and believe that they would somehow not very much be affected as their government machinery would protect them from the virus. But yet this Corona virus has proved to be more powered than expected that it needs more than the efforts of one country, it also needs the efforts of individuals.

The reasoning and perception of the individualism Americans are analytical and pay more attention to the specific characteristics of things; The collectivist Chinese people's reasoning and perception are holistic, and considering things is concerned with the interconnection between things. Right now, the temptation is very strong to do "whatever is necessary" in times of crisis there is an increased need for governments to monitor and control the public, which might make it necessary to limit individual freedom. For Chinese we have confident that our privacy will be hide very well. But American feel uncomfortable with monitor by the social media or government. The data proved this as well. Even for New Year, we have to do covid-19 test to make sure we are healthy enough to travel back and forth. Thanks to the big data and internet convenience, we have the electric paper to prove your health situation as well to save our time, which is hardly to achieve in out counties as well.

2.3.3   Anxiety Index in COVID-19

It is found that after the outbreak of COVID-19 epidemic, individuals with high collectivism have a higher concern for the people in their social group, which in another way can be thought of as anxiety. this put the at a more alert position to want to make sure everyone in their circles are safe and also collaborate to eradicate a problem as seen in the case of the Chinese quickly building the Thunder God Mountain hospital.

Therefore, because of the cultural tendency of collectivism, collectivist countries will have a more obvious sense of protection effectiveness than individualistic countries, so the cultural characteristics of collectivist countries tend to put their people alert and act fast to protect themselves. In the face of the current fight against the epidemic, it is of great significance to build a sense of security and confidence.

2.3.4   China and America Aversion to Pathogens

Like other animals, human beings can use their own sensory organs (such as vision and smell) to detect possible germ threats in the environment and make corresponding behavioral responses. Many researchers believe that aversion plays a very important role in the process of detecting and responding to the threat of germs: as a link of the body's immune system, aversion is used to counter the possible threat of germs and diseases in the external environment. The collectivist Chinese have more pathogen aversion compared to the individualist Americans. This makes the Chinese more inclined and compliant to self quarantine and mask up to protect themselves from the Covid-19, hence this cultural trait can also be attributed to why the China were able to respond better than America in comparison based on cultural inclinations. In order to make sure our society can still go well and our economic can function well, we started to use the internet study to get rid of the bad effect of pathogens. Due to the internet development in China, we could easily restart our life again, within the data, we can easily find students who are willing to started online.

3   Summary

Within the prove of data through the internet, we can find out that: culture has a big influence on how people react to various situations, as noticed from how difference countries powered by difference cultures responded to the Covid-19. The use of location data to control the coronavirus pandemic can be fruitful and might improve the ability of governments and research institutions to combat the threat more quickly. It is important to note that location data is not the only useful data that can be used to curb the current crisis. This study with the data collected is not meant to prove superiority of a specific culture over another, but rather to see how the individualism and the collectivism inclined people responded to the Covid-19 based on their cultural inclinations. Also collecting data can be useful for doctors to searches for vaccines and monitoring online communication on social media which might be good for us to keep our society peace and security. Looking at China which stands at a population of 1.398 billion people as at 2019 and America which stands at a population of 328.2 million people as at 2019 (Worldbank, 2021), we would wonder how China which is 4 times the population of American would manage to contain a pandemic like Covid-19 faster than America. This brings us to the contribution of culture and also the develop of the internet. The traits of interdependent group thinking, and strong inclination to follow orders in the collectivism culture can be copied and embraced and would be useful to help times like these. More studies should be pursued to investigate cultures and see how countries can learn from each other in many ways.

References

1. Kasdan, D.O., Campbell, J.W.: Dataveillant collectivism and the coronavirus in Korea: values, biases, and socio-cultural foundations of containment efforts administrative theory & praxis (2020). https://doi.org/10.1080/10841806.2020.1805272
2. Kim, H.S., Sherman, D.K., Updegraff, J.A.: Fear of Ebola: the influence of collectivism on xenophobic threat responses. Psychol. Sci. 27, 935–944 (2016)
3. Goyal, S.K.: A joint economic-lot-size model for purchaser and vendor: a comment. Decis. Sci. **19**, 236–241 (1988)
4. Hill, R.M.: The single-vendor single-buyer integrated production–inventory model with a generalized policy. Eur. J. Oper. Res. **97**, 493–499 (1997)
5. Kim, S.L., Ha, D.: A JIT lot-splitting model for supply chain management: enhancing buyer–supplier linkage. Int. J. Prod. Econ. **86**, 1–10 (2003)
6. BenDaya, M., Darwis, M., Ertogral, K.: The joint economic lot sizing problem: review and extensions. Eur. J. Oper. Res. **185**, 726–742 (2008)
7. Thornhim, R., Fincher, C.L., Murray, D.R.: Zoonotic and non-zoonotic diseases in relation to human personality and societal values: support for the parasite-stress model original article (2010). https://doi.org/10.1177/147470491000800201
8. Sevastopulo, D., Politi, J., Johnson, M.: G7 countries vow to do 'whatever is necessary' to support global economy. Financ Times (2020). https://www.ft.com/content/571f51e0-67b3-11ea-800d-da70cff6e4d3

Research and Design of Intelligent Detection System in Teaching Manner Training of Normal School Students

Xinyan Zhang[✉]

Jiangxi Teachers College, Yingtan, Jiangxi, China

Abstract. With the progress of society and the increase of computer, network and digital media technology, computer technology has become an indispensable part of modern service industry. The application of the intelligent detection system in the teaching attitude training of normal students has developed in all aspects. As an intelligent detection system, it not only serves teaching, but also serves students. How to use more methods to train students and teaching attitudes is a future job Innovation. This article mainly introduces the research and design of the intelligent detection system in the teaching training of normal students, and uses the intelligent detection system to make a reasonable analysis of the teaching situation in the classroom. This paper uses the research and design of the intelligent detection system in the teaching training of teachers' students, adopts microteaching for classroom analysis, and intelligently and systematically constructs the teaching training of teachers' students to improve the standardization of teaching training in the classroom. The experimental results of this paper show that the application of the intelligent detection system in the teaching training of normal students increases the efficiency of teaching training for normal students by 15% and improves the teaching quality of teaching training for normal students.

Keywords: Intelligent detection system · Normal students · Teaching attitude training · Microteaching

1 Introduction

With the changes of the times, the existing intelligent detection system has been continuously optimized and improved. In the Internet era, streaming media has become a way for the public and students to communicate. The normal teaching department will use the original teaching service system to develop new services [1, 2]. Through the teacher training management network, it can better serve the students' score query, teaching system services and score management, through the network+APP platform [3, 4]. Students and teachers query the results of teaching attitude training online, query information about student educational administration through the platform, and can answer and solve various students' questions online [5, 6].

The current teacher education management only passively collects various data of students. This kind of service to students is one-way teacher performance information and is responsible for relevant teachers and students [7, 8]. Needhi believes that

© The Author(s), under exclusive license to Springer Nature Switzerland AG 2021
J. Abawajy et al. (Eds.): ATCI 2021, LNDECT 81, pp. 640–648, 2021.
https://doi.org/10.1007/978-3-030-79197-1_92

students and teachers can obtain various data and information according to their authorization on the network platform [9]. As an important part of the teaching management system of colleges and universities, relevant information also needs to be connected with the normal educational management system [10]. Arioua L thinks that relevant results should be regarded as an important part of the educational management activities of colleges and universities, and can directly provide feedback to students' teaching effect evaluation [11]. Due to the influence of the long-existing test-oriented education system in history, my country's colleges and universities and the majority of teachers and students do not pay much attention to the teaching of normal courses, and the teaching system aimed at cultivating students' lifelong exercise ability is not complete [12]. However, there are errors in their experimental process, resulting in insufficient accuracy.

The innovation of this article lies in the research and design of the intelligent detection system in the teaching training of normal students. Strengthen the interaction based on the analysis and application of the intelligent detection system and microteaching to teaching training [13]. The impact of constructing a more optimized educational administration management system for teacher training courses in colleges and universities has on students' participation in teacher training, so as to promote the optimization of the management of teacher training courses in colleges and universities, and then provide help for the improvement of the normal education level of students at school. Through the actual participation and Hands-on practice, this article combines practical theory, and uses standard case studies for specific analysis, so as to provide fresh materials and useful attempts for the application of theoretical research on teaching training.

2 Teaching of Normal Students Under the Intelligent Detection System

2.1 Teaching Environment of Normal Students

Our country's current general colleges and universities student curriculum system. A relatively limited number of normal teachers need to undertake the teaching tasks of the whole school, and handle the information about the selection, teaching, examination and other aspects of the whole school. This brings normal teachers far higher than other departments and professional teachers. Realize the optimization of the educational administration system of the normal course, so that the normal teachers can deal with the relevant information of the students more efficiently, and it can also improve the teaching quality of the normal teachers in universities to a large extent. Students in ordinary colleges and universities often need to take different types of courses according to different seasons. The recurring situation of the same student's information in different teaching classes also directly brings greater pressure to the statistics, summary and management of the teaching achievements of normal courses.

The results of normal colleges and universities show diversified evaluations, which will promote the physical health of ordinary colleges and universities. In fact, normal colleges and universities' normal courses involve three subjects with different needs:

teachers, students and school administrative staff. Each link has different authority, teaching management and network platform combine to provide diversified services for college students. Nowadays, there are many blindness in students' self-selection methods, just to get credits for selecting courses, and they don't pay much attention to the improvement and promotion effect that related courses can bring to their own physical fitness. At the same time, they often don't have a deep understanding of the teaching purpose and focus of the curriculum system itself, so they can't understand the real intention of the relevant courses.

At present, the research results of domestic scholars in optimizing the function of the educational administration management system of the designer's model course are mainly concentrated on the framework of related teaching software, the establishment of the process, and the realization of platform security. The realization and construction methods and technologies of related platforms are currently the main research focus of scholars at home and abroad, but the discussion on the construction direction of related frameworks and processes is not rich in the current domestic and foreign research results. Reduce the size of the feature map by down-sampling, use a fixed-size area as a window to slide the value on the input feature map, and select the maximum value in the area as the output result, where h and w represent the height of the down-sampling area Wide, c means the c-th channel.

$$O_{c,m,n} = \max(I_{c,m+1,n+1}) \tag{1}$$

Generally, a nonlinear function is used to perform nonlinear mapping on the input data. The commonly used nonlinear function is shown in the following formula: The formula is as follows.

$$\text{sigmoid(x)} = \max(LdotN, 0.0) \tag{2}$$

Commonly used loss functions such as cross entropy loss for classification and L2 loss for regression, their calculation formulas are as follows:

$$\triangle w_t = w_{t-1} + C_{texture} * L_D \tag{3}$$

2.2   Intelligent Detection System for Teaching Mode

Study the content that the teacher education management system function should have, so as to obtain the innovation of the research perspective. At the practical level, the optimization of the teacher education management system function can better realize the extension of classroom teaching, and provide students with more complete and can reusable teaching resources, and can provide students with much richer teacher knowledge than classroom lectures. At the same time, the optimization of the functions of the educational administration management system of teacher training courses can also directly improve the work efficiency of teachers themselves, and help teachers complete business tasks that are less relevant to the teaching itself. Through richer information transmission, the teaching management system of teacher training courses can also improve the accuracy of teaching, thereby improving the quality of teaching.

When designing a learning project, teachers must first determine the content of learning, set learning goals, and on this basis, further design student learning activities, determine the method of learning, formulate evaluation strategies, and finally generate a learning project day. In order to meet the needs of Japanese-style learning under the information environment, the generated learning projects must be able to be disseminated on the Internet, for example, in the form of web pages or PPT, Word files, etc. Management of student information in project-based learning in an informationized environment, teachers should assume the role of student information management, especially the management of individual learning users and learning groups. 2 Student activity project-based learning emphasizes student-centered and requires students to complete the learning tasks and the final works are formed. These activities can be completed by the students individually or in groups. In the process of project-based learning in an information environment, students' activities include three types: individual student activities, activities within groups and activities between groups.

3   Teacher Training System for Normal Students

3.1   Teaching Training System

The development of teaching attitude training, especially the activities of teachers and students, needs support from all aspects. For this reason, we have designed a learning support module, which is the core of the entire platform. According to their own teaching design, the teachers of the learning project design the educational technology ability training of normal students into many learning projects and publish them in this module. The students learn the project and its basic information through this module, and choose the learning project. The project team selected students of a certain learning project for voluntary combination. Different projects have different requirements for the size of the project team and other requirements. Students form learning groups according to the corresponding requirements and the characteristics of themselves and other students. Teachers have the right to control the learning group. For example, replacing team members and deleting project teams. The group can be discussed in advance and then formed on the platform, or a member can initiate a promotion on the platform and then formed. For ease of management, the creator of the group is the group leader by default.

3.2   Teaching Progress of Normal Students

In order to realize the co-construction and sharing of resources, teachers provide the resources needed for project-based learning and publish them in this module. Students can search, browse, and download resources, and they can also provide excellent resources they think. Although this platform provides teachers with the right to delete resources, teachers are encouraged not to delete the resources provided by students easily, but to give full play to the enthusiasm and initiative of students, and recognize the efforts and dedication of students, except for unhealthy or illegal resources.

Can provide a variety of resources including various examples, software tools and their learning manuals and excellent websites. Project log This module provides students with various forms of logs such as writing project design plans, schedules, progress reports, study summaries and reflections. By writing and managing logs, students are trained to plan and manage the project process. According to the project progress reports and reflection logs written by students, teachers monitor the learning process and progress of students in order to make timely suggestions and guidance. Teachers should guide students to decompose the project work into several tasks, and carry out task division and planning, so as to complete the project according to the design plan and schedule, Download and evaluate the work of other groups. The teacher browses and evaluates the works completed by all students.

The project-based training platform for teaching state training for normal students should have the following functions. Can support the project-based learning process. This is the main difference between the platform and other learning platforms or systems, and it is also the key to effectively improve the educational technology literacy of normal students. Normal students through the platform should be able to easily obtain learning items, develop learning activities, communicate well with other students, and get the help of teachers in time, and finally complete the production of works and evaluate the works. Abundant learning resources can be provided, and the learning resources should be rich and comprehensive, which can meet the needs of completing the training of educational technology ability of normal students, and students can easily obtain learning resources. The final evaluation results of the project works refer to the mutual evaluation among students and the evaluation of teachers. Teachers need to provide work evaluation rubrics when designing projects, and guide students to conduct mutual evaluation. In addition to rating works, students and teachers can add comments and suggestions to works. Project Discussion Students discuss the development of a certain project while participating in project learning. Teachers can also organize students to discuss some key issues and technologies. The specific results are shown in Table 1.

Table 1. Platform development environment

Project	Content
Operating system	Windows Server 2003
Server	IIS
Database	SQL Server 2000
Development language	HML, ASP

4 Microteaching Analysis of Teaching Mode Training

4.1 Microteaching Classroom Analysis

The so-called microteaching is simply: in a controllable teaching system, teachers and students use modern educational methods to train teaching skills and learn teaching skills. During its development, there have been some micro-training modes, such as the University of Sydney mode and the British mode. The practical orientation of micro-teaching will be stronger, otherwise the micro-teaching practice will become a mere formality, just for the sake of micro-teaching, and it is impossible to talk about how to implement the teaching, let alone the teacher students and in-service teachers. The due role of teaching technology training. The specific results are shown in Fig 1:

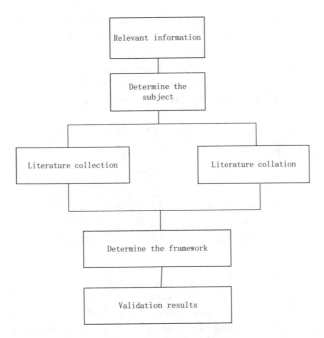

Fig. 1. The optimal capacity allocation

4.2 Artistic Innovation in Visual Media

Among them, the contribution of the University of Sydney model is the development of a complete set of microteaching training materials-"Sydney Microteaching Skills", which explains the six basic skills (imported skills) included in classroom teaching from the perspective of pedagogy and psychology., Presentation skills, etc.); the British model analyzes the relationship between microteaching and teaching theory, and proposes two training models: "cognitive structure model" and "social psychological model". China has 30 Years of development history. At the beginning, there were few words such as "microteaching" or "teaching skills" in the training of teachers in our country. The most basic way for teachers and normal students to improve their teaching skills is to try lectures. Later, the theory and technology of micro-teaching started to be officially introduced to China during a visit abroad at the Beijing Institute of Education. Subsequently, its development experienced three stages: rise, silence and rational knowledge. At present, our country's educators are expanding their research from practical applications to theoretical extensions. For example, designing the guidance training model of microteaching, studying the teaching effectiveness of microteaching, and the evaluation method of microteaching. The specific results are shown in Table 2.

Diversified teaching evaluation research in microteaching Education evaluation research has always been a difficult problem in the education field, and the teaching skill evaluation combined with microteaching is a hot focus area of microteaching research. At present, countries all over the world are actively studying the evaluation of microteaching, but our country has relatively few researches in this area and is not strong enough. This shows that evaluation in microteaching is gradually becoming an important research content of chemistry microteaching. Enrich the theoretical basis of microteaching and diversification teaching evaluation: improve the knowledge and understanding of diversified teaching evaluation of chemistry normal students; improve the development and application of the teaching evaluation system in microteaching; provide research and development opinions for other related research.

Table 2. The performance analysis of 3D model browser

Project	Control group		Test group	
	M	SD	M	SD
Speech flow technology	3.41	842	3.77	3.34
Tone technique	3.28	826	3.65	3.14
Sentence refinement	3.04	838	3.39	2.96

Introduce the definition and significance of microteaching, evaluation and teaching evaluation, focusing on the concept, type, value orientation and development description of diversified teaching evaluation. Theoretical basis combing: combing the theoretical basis of microteaching and diversified teaching evaluation, understanding their ins and outs, and specifying the content and characteristics of these theories. The specific results are shown in Fig. 2. Among the students in our country's colleges and universities, 48% of boys and 31% of girls failed in the test results of teaching attitude training. The number of people who can achieve good and above results only accounts for about 5% of the total number.

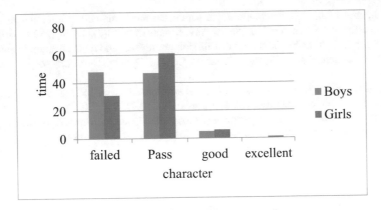

Fig. 2. Teaching evaluation data map

5   Conclusions

Although this article has made certain research results in the research and design of the intelligent detection system in the teaching training of normal students, there are still many shortcomings. Based on the research and design of the intelligent detection system in the teaching training of normal students, there are still a lot of in-depth content worthy of study. There are many steps in the teaching training process that are not covered because of space and personal ability. In addition, the actual application effect of microteaching can only be compared with traditional models from the level of theory and simulation.

Acknowledgements. Project: Science and Technology Research Project of Jiangxi Provincial Department of Education: Research and Design of Intelligent Detection System in Teaching Manner Training of Normal School Students (Project Number: GJJ203205).

References

1. Li, Y., et al.: Driverless artificial intelligence framework for the identification of malignant pleural effusion. Transl. Oncol. **14**(1), 100896 (2021)
2. Hassabis, D., Kumaran, D., Summerfield, C., et al.: Neuroscience-inspired artificial intelligence. Neuron **95**(2), 245–258 (2017)
3. Lu, H., Li, Y., Chen, M., et al.: Brain intelligence: go beyond artificial intelligence. Mob. Netw. Appl. **23**(7553), 368–375 (2017)
4. Ng, T.K.: New interpretation of extracurricular activities via social networking sites: a case study of artificial intelligence learning at a secondary school in Hong Kong. J. Educ. Train. Stud. **9**(1), 49–60 (2021)
5. Camerer, C.F.: Artificial intelligence and behavioral economics. NBER Chapters **24**(18), 867–871 (2018)
6. Hassabis, D.: Artificial Intelligence: Chess match of the century. Nature **544**(7651), 413–414 (2017)

7. Su, J., et al.: Full-scale bending test and parametric study on a 30-m span prestressed ultra-high performance concrete box girder. Adv. Struct. Eng. **23**(7), 1276–1289 (2019)
8. Li, Y., et al.: Estimation of normal distribution parameters and its application to carbonation depth of concrete girder bridges. Discrete Contin. Dynam. Syst. **12**(4 & 5), 1091–1100 (2019)
9. Zhang, G., et al.: Analysis of loss in flexural stiffness of in-service prestressed hollow plate beam. Int. J. Struct. Integrity (ahead-of-print) **10**(4), 534–547 (2019)
10. Needhi, K., et al.: Hybrid testing for evaluation of seismic performance of highway bridge with pier made of HyFRC - ScienceDirect. Structures **20**, 848–865 (2019)
11. Yuanyao, M., Ditao, N.: Effect of vehicle load on the fatigue performance of corroded highway bridge. Sci. Adv. Mater. **10**(6), 845–852 (2018)
12. Arioua, L., Marinescu, B.: Robust grid-oriented control of high voltage DC links embedded in an AC transmission system. Int. J. Robust Nonlinear Control **26**(9), 1944–1961 (2016)
13. Krishnan, B., Ramalingam, M., Vellayutham, D.: Evolutionary programming-based simulation of bilateral real power contracts by optimal placement of flexible AC transmission system devices using contingency analysis. Electric Mach. Power Syst. **44**(7), 806–819 (2016)

The Vocational College English Teaching in the Age of "Internet+"

Shimin Li[✉]

Tianjin Modern Vocational Technology College, Tianjin, China
lishimin1978@sbcmail.com.cn

Abstract. The rapid development of information technology propels the innovative process of vocational college English teaching. It has become increasingly important to explore the transformation and reform of curriculum teaching from traditional to modern teaching by taking advantage of Internet technology in English teaching and by designing an information-based language teaching mode to keep up with the pace of times. This paper expounds the necessity and innovative ideas of vocational college English teaching research in the age of "Internet +", and presents optimization strategies from three aspects: improving teachers' information technology literacy, developing three-dimensional teaching resources and establishing diversified evaluation system.

Keywords: "Internet+" · College English teaching · Optimization strategy

1 Introduction

1.1 The Concept of "Internet+"

What is Internet+ or Internet Plus? Internet Plus is a new model of industries on the basis of the evolution of the Internet. "Internet+" refers to the integration and application process of information technology (including mobile Internet, cloud computing, big data technology, etc.) with Internet technology as the core in various social sectors [1]. It is an incomplete equation where various Internets (mobile Internet, cloud computing, big data or Internet of Things) can be added to other fields, fostering new industries and business development in China [2]. In simple words, Internet is the integration of the Internet and traditional industries through online platforms and IT technology.

One of the main features of "Internet Plus" is: cross-border integration. "+" means crossover, transformation, opening up and reintegration. The Internet is playing an increasingly prominent role in the education industry today. In the age of Internet+, teaching and learning activities are carried out around the Internet. Teachers teach on the Internet and students learn on the Internet. Information flow on the Internet. Offline activities become the supplement and expansion of online activities. However, "Internet Plus" will not replace traditional education, but will give it new vitality.

J. Abawajy et al. (Eds.): ATCI 2021, LNDECT 81, pp. 649–655, 2021.
https://doi.org/10.1007/978-3-030-79197-1_93

1.2   The Benefits that "Internet+" Brings to Teaching

With the arrival of the "Internet+" era, Internet technology has been integrated into all aspects of our life and study, providing us with a lot of benefits in language teaching. The rapid development of information technology propels the innovative process of college English teaching. The intervention of Internet technology provides various forms of interaction for teachers and learners, and the formation of "Internet+" is affecting the trend of language teaching reform and development. It mainly focuses on Internet technologies like mobile Internet, cloud computing, big data technology, etc. Through information technology and network platform, the Internet and the education field is closely integrated so as to provide learners with technical support for the further teaching reform and development, provide learners with a substantial number of three-dimensional, digital, diversified language learning resources, and set up the network platform for the innovative reform and development of English teaching.

"Internet+" breaks the shackles of original knowledge and the boundaries of time and space so that a huge number of superior teaching resources can be shared globally. The globalization of teaching resources promotes the intelligent means, personalized teaching, automatic management, evaluation process, independent learning and diversified development. Learners can conduct independent learning activities such as pre-class preview, question-answering discussion, online testing, and summarizing after class via both online and offline interactive learning platforms to enhance the cultivation of learners' consciousness and independent ability to learn. Informationized teaching methods are used to make up for the deficiency of traditional teaching and more flexible ones are created. The use of digital learning platform is conducive to teachers' strengthening of guidance and supervision on learners' independent learning.

2   The Necessity of Research on Vocational College English Teaching in the Age of "Internet+"

In the age of "Internet+", it has become increasingly important to explore the transformation and reform of curriculum teaching from traditional to modern teaching by taking the advantages of Internet technology in English teaching and by designing the informationized language teaching mode with "interaction", "independence" and "inquiry" as the core. In this sense language teaching can actively adapt to the needs of development of the information age. The integration of modern information technology into classroom teaching and learning is the inevitable trend in the course of college English teaching reform. Many teaching elements, such as teachers' teaching philosophy, course teaching objectives, learners' characteristics, learning strategies and teaching evaluation standards, have begun to change. In this information age, innovating the language teaching mode and learning method is a crucial way and means to promote the reform of college English teaching.

2.1   Change of Teaching Philosophy

The application of Internet technology in the process of language teaching liberates the international top teaching resources, provides a platform to share resources for all learners, thereby giving a new definition for teaching. In the age of "Internet+", language teaching is no longer the "classroom teaching" in the traditional sense. It gives learners more flexibility in learning. Teachers may not necessarily stand on the platform with a book in hand or adopt face-to-face teaching. Students may not necessarily sit in the classroom quietly listening to the teacher, taking some notes once in a while. The traditional classroom teaching model characterized by teachers' giving lectures and students' listening is gradually being replaced by information-based teaching. Through the Internet technology platform, teachers can conduct online teaching guidance, and learners can learn anytime and anywhere. Schools should provide necessary teaching conditions and technical support for learners in the learning process, such as language online courses, online and offline independent learning platforms to cater to the learners with different learning requirements.

2.2   Adjustment of Teaching Objectives

Teaching goal is the concrete manifestation of educational goal and training goal. In the age of "Internet+", the goal of English teaching at college has been changed. The realization of the adjusted teaching objectives should be guided by utilizing the integration of information technology and language teaching to lead learners to positive thinking and action. Cultivate learners' information processing ability to use information means to construct new knowledge. Strengthen their information literacy and cooperation ability. Further improve their language learning ability and communication ability, so as to meet the social demand for innovative technical skills in the information age. Teachers should analyze the subject characteristics of college English course and learners' characteristics. Based on this, formulate specific, clear and executable teaching objectives, determine the teaching objectives of unit learning. Meanwhile, with clear teaching objectives, learners should master the language characteristics and use them correctly through appropriate information teaching methods.

2.3   Learner's New Traits

The children who grow up in the Internet and digital environment are usually called the post-2000 generation. Their way of thinking, learning strategies and learning styles changed obviously. They are quick-minded and resist the traditional way of classroom indoctrination. They pay more attention to visual experience, and are interested in information media. They prefer to use information technology to solve problems; They like challenging tasks, like learning on their own, like to set goals for themselves and use them as motivation to learn. Although the traditional teaching resources with "textbook" as the core can meet the traditional requirements, it is difficult to attract this group of students. It is also difficult to conform to the curriculum requirements of student-centered, interactive and independent learning in the new era.

3 Design Concept of Vocational College English Teaching in the Age of "Internet+"

In the age of "Internet+", teachers and learners must change their ideas and actively adjust to the new mode of digital teaching. When the teaching mode is changed, teachers' teaching mode, authoritative position and teacher-student interaction mode are challenged. Under the new teaching model of "Internet+Teaching", the role of teachers has changed. They are no longer the decision-makers and constructors of teaching activities in the traditional sense, but are becoming the organizers and guides of information learning activities.

Therefore, teachers should take the initiative to adapt to the change of teaching methods, further enhance the literacy of information technology, change the cramming one-way teaching mode into two-way interactive research between the teacher and students, and construct the informationized course teaching with teacher-led and student-centred characteristics. The place of teaching activities should be transformed from a designated classroom to the comprehensive learning environment of "online+classroom+offline" for learners. Learners should improve their learning efficiency, learn to allocate their time reasonably, improve their learning initiative and conduct efficient learning. Learners should strengthen their autonomous learning ability and cultivate their inquiring minds to find the solutions to the problems, which is the new requirement for learners in the network era. The development trend of college English teaching requires integrating traditional language teaching and Internet technology, giving full play to their respective advantages, optimizing the innovative interactive teaching mode focused on learners, and strengthening the application effect of "Internet+" in college English teaching.

4 Optimization Strategies of Vocational College English Teaching in the Age of "Internet+"

As one of the most familiar teaching forms to teachers, traditional classroom teaching has been controversial. In the case of the traditional teaching mode, teachers transfer knowledge and interact emotionally with learners at specific teaching time and place. Its advantage lies in the instant face-to-face communication between teachers and learners, which cannot be replaced by any other teaching mode. Therefore, the necessity of continuation of traditional teaching cannot be denied. However, traditional face-to-face teaching is limited by teaching resources, teaching methods and learning contents, which leads to the emergence of problems such as unsatisfactory teaching effect and low learning efficiency.

In the age of "Internet+", teaching activities are not only a process of teaching and receiving between the teacher and students, but also a process of transferring and circulating information resources. The network information technology promotes the two-way flow of information and realizes the deconstruction and reconstruction of teaching mode and education system. The advantages of Internet teaching and traditional courses cannot be denied. We cannot abandon either of them. Instead, we should combine them organically, complement each other, and promote each other. With the aid of Internet technology, all resources that can be shared and interactive are connected to serve the college English teaching, so as to spread knowledge to the maximum extent. In addition, the advantages of traditional teaching such as face-to-face teaching and instant interaction can be utilized to fully integrate online independent learning and traditional classroom learning, so as to realize the complementary advantages of the two teaching modes and optimize the English course teaching.

4.1 Improve Teachers' Information Technology Literacy

In the age of "Internet+", the design of English course teaching requires teachers to have a vision of the overall pattern of the course, and to have the information technology literacy to have a good idea of classroom teaching design. Information literacy not only includes the basic skills of obtaining, identifying, processing, transmitting and creating information with contemporary information technology, but more importantly, the ability to learn independently, the consciousness of innovation, the spirit of criticism, the sense of social responsibility and participation in the new environment created by contemporary information technology [3]. Teachers' information technology literacy includes information awareness, information acuity, information processing, information sharing spirit and information security awareness [4]. The informationized teaching design of language classroom mainly includes pre-class assignment, collection of learning materials, teaching content, group discussion, language skill training, group interactive display, effect evaluation and summary, etc.

The specific implementation steps are carried out in three parts: online independent learning, classroom focused learning, and offline personal summary. First, the online independent learning can be arranged after class. With the network-assisted autonomous learning platform, students conduct independent online learning and complete interactive and inquiry-based learning tasks such as preview and review. Next, the classroom focused learning should be implemented in the classroom with network environment. Teachers can use cloud class, rain class, super star MOOC platform and other intelligent teaching APPs to organize online interactive activities, such as brainstorming, voting questionnaire, answering questions, group discussion, teaching evaluation, individual tutoring, testing and so on. Lastly, in the offline personal summary process, learners need to sort out knowledge, consolidate key points, complete learning tasks and be able to display homework or personal works on the network learning platform. In class, the network teaching and face-to-face teaching take turns to undertake. There is no fixed proportion on time allocation. Teachers make corresponding adjustment according to the difference of student language level and learning ability. Based on the data analysis of the platform, teachers can timely acquire the learning situation of students on the platform so as to encourage, guide and urge students to study independently.

4.2 Develop Three-Dimensional Teaching Resources

The effect of language learning largely depends on the amount and content of the language the learners are exposed to [5]. In the traditional teaching, textbook is the main way to present knowledge. Textbook has always played an irreplaceable role in classroom teaching. Presenting knowledge through the medium of books is conducive to learners' understanding of the systematization of knowledge, but textbook knowledge often hinders the circulation of knowledge and is usually presented in the form of single static text. With the advantages inherited from the traditional teaching, informationized teaching resources should become the tools to support and expand the thinking process of students. The combination of "traditional textbooks+informationized teaching resources" to develop three-dimensional teaching resources can make up for their respective deficiencies.

The networked teaching environment in colleges is an intelligent teaching environment. It is supported by computer artificial intelligence, virtual reality technology and network technology [6]. By using modern information technology, teachers create language learning teaching scenarios for students in a variety of forms such as text, sound and video, present the extension and analysis of the course content in the form of multimedia teaching resources, and carry out visual information display in class to enrich the input of classroom information. At the same time, information technology resources can be used to create a virtual learning space, build an information-based learning resource sharing platform, and constantly improve and scientifically use important language testing systems such as voice evaluation system and writing evaluation system to meet learners' demands for mobile learning, hierarchical learning and personalized learning.

4.3   Establish a Diversified Evaluation System

Evaluation is one of the indispensable teaching links in the process of teaching, and it is a vital means to test whether students can correctly and flexibly use and process language knowledge information. In the age of "Internet+" age, the organization mode of English course, teaching methods, teaching means, teaching content and so on are quietly changing. Traditional standardized and terminative language skill assessment methods cannot get a foothold in the "Internet+" era, and it needs to adapt to the new diversified, formative and terminative evaluation system. The evaluation should change from the emphasis on the evaluation of teachers' teaching methods, teaching processes and teaching abilities to the evaluation of students' learning strategies, processes and efficiencies.

In terms of evaluation of college English learning, it can be classified into two parts: online learning evaluation and classroom learning evaluation. Learning evaluation should not only focus on students' final examination grades, but also concentrate on students' abilities in information resources collection, processing, independent learning, cooperation and innovation. Teachers evaluate students' independent learning in learning platform and participation in the activity and assess students' performance from the emphasis on "knowledge-based" English course evaluation to the comprehensive consideration of "knowledge ability based". Diversified evaluation forms, including teacher evaluation, student self-evaluation, peer evaluation and online multiparty evaluation, should run through every stage of teaching [7], so as to achieve "everyone makes evaluation at any time and any place". Thus, teachers can understand the latest learning trends of learners in real time, find the problems easily and help their students solve them in time.

5 Conclusion

There is no doubt that the reform of informationized course teaching has improved the quality of college English teaching. In the language classroom teaching with the network environment, the informationized method of language teaching must be aimed at serving the course teaching, and it should be integrated with the teacher's classroom teaching and complement each other. The promotion of informationized teaching must be based on the realization of the teacher-led and student-oriented roles in the course of teaching. The leading role of teachers in education cannot be completely replaced. As far as learners are concerned, they should also get rid of the bondage of textbook knowledge, actively adapt to the new teaching mode, and develop their own innovative and creative ability [8]. The whole teaching process should be focused on learners' needs and organized in the best way. Therefore, in the age of Internet+, vocational colleges should take advantage of the information teaching resources and network technology, innovate the traditional English teaching mode, and promote learners' interactive, inquiry-based and personalized independent learning abilities.

References

1. Qin, N.: Research on the Construction of Blended Teaching Model under the Background of "Internet+", issue: 6, p. 8. Shandong Normal University (2017)
2. Yun, L., Zhao, L., Li, X.: Intelligent Education, Great Transformation of Education in the Internet + Era. Publishing House of Electronics Industry (2016)
3. Zhou, X., Yuan, X.: A survey and analysis of information literacy of college students. Libr. Sci. Res. (2), 84 (2008)
4. Li, Q., Yu, L.: Research on the strategy of improving higher vocational teachers' information technology literacy based on Internet+. Electron. Commer. (11), 87–88 (2020)
5. Spolsky, B.: Conditions for Second Language Learning: Introduction to a General Theory, p. 166. Oxford University Press, Oxford (1989)
6. Sui, X.: Research on the optimization of college english classroom teaching under the network environment. Shanghai Int. Stud. Univ. (12), 59–60 (2013)
7. Xie, Y., Yin, R.: Design and Evaluation of Network Teaching, pp. 341–342. Beijing Normal University Press, Beijing (2010)
8. Zhang, M., Qin, N.: Internet + Education: connotation, problems and mode construction. Contemp. Educ. Cult. (3), 22–28 (2016)

Online and Offline Blended Teaching of College English in the Internet Environment

Xia Wang[✉]

Mianyang Teachers' College, Mianyang, Sichuan Province, China
wangxia1980@sbcmail.com.cn

Abstract. With the continuous deepening of education reform, more and more scientific teaching concepts and novel teaching models have emerged, which have become an effective way to improve the efficiency and quality of modern education and teaching. The effective use of mixed teaching mode in college English teaching can meet the needs of reform and innovation of vocational education teaching, provide students with rich teaching resources, diversified teaching methods, and advanced educational technology services, thereby effectively stimulating college students' interest in learning English. Students learn English knowledge and correctly use English ability to solve more problems to lay a solid foundation and improve the level of talent training in college English education. This article studies the online and offline English teaching in colleges and universities under the Internet environment.

Keywords: Internet · College English · Online and offline · Mixed teaching

1 Introduction

In higher education, English is one of the universal courses, and college students of any major must learn English. College English has the characteristics of grade and comprehensiveness. College students need to take the College English Test Band 4 and 6 during their studies. In addition to learning basic word grammar, they also need to exercise oral English and enhance their reading ability [1]. Therefore, colleges can better help students complete the internalization of English knowledge by using online and offline hybrid teaching models, so that college can use their own fragmented time to complete knowledge consolidation and learning, improve their English application ability, and grow into good high-quality talents with English literacy [1].

2 The Role of Blended Teaching Mode in English Teaching Under the Internet Background

Compared with the traditional teaching model in the past, the "online + offline" teaching model is not only an innovation in teaching methods, but also an innovation in educational concepts, educational resources and even the teaching environment.

J. Abawajy et al. (Eds.): ATCI 2021, LNDECT 81, pp. 656–662, 2021.
https://doi.org/10.1007/978-3-030-79197-1_94

2.1   Combination of Teaching Theories

In the practice of higher vocational English teaching, according to the characteristics of the subject, the English teacher combines online and offline learning theories, and adopts personalized teaching methods in the design of teaching plans and teaching process to adapt to the specific conditions of students and the current teaching environment to meet the needs of students [2].

2.2   Integration of Teaching Methods

Under the requirements of educational informatization reform, higher vocational English teachers should also update their thinking in time, make reasonable use of Internet technology, adopt teaching methods that students love and accept, and closely follow the listening, speaking, reading, writing, and translation links in English teaching [2]. Innovate teaching methods to meet the requirements of informatization development, and further integrate online classroom teaching methods, such as online learning platforms such as MOOC, Rain Class, Flipped Class, etc., to emphasize the dominant position of students and strengthen students' awareness of lifelong learning [3].

2.3   Integration of Teaching Resources

The online "teaching-learning" platform broadens the channels of learning resources and can effectively integrate existing teaching resources. Higher vocational English teachers organize the required English learning resources through the platform and display them in the classroom in the form of micro-classes [3]. Carrying out interesting and contextual English teaching, allowing students to simulate the classroom in terms of pronunciation, semantics, etc., so that the teaching is more professional, the classroom is more efficient, and the students can apply what they have learned.

2.4   The Teaching Environment is Fully Integrated

The integration of online and offline teaching environments is a distinctive feature of the hybrid teaching model. In this way, face-to-face teaching and online learning can give full play to their strengths, complement each other, and complement each other. The fusion of the two makes the teaching benefits maximized [4]. The individual differences of students make it more able to stimulate learning interest, and the learning progress and teaching difficulty are reasonably matched.

3   The Necessity of Online and Offline Mixed Teaching

First, it is proposed in the ten-year development plan of China's educational informatization that the educational development should be guided by through ducational innovation concepts. Innovation in learning methods and education models is the core [4]. Under this plan, vocational colleges are required to use hybrid online and offline teaching to innovate teaching models for English teaching.

Second, with the continuous development of computer applications in the information age, Internet teaching has also been widely used in the education industry, which has also led to the continuous production of advanced teaching technologies. Teachers must change the previous teaching mode of single teaching and use network technology as modern teaching [5]. The main method of English teaching is to use information technology to break the limitation of time and place, formulate a personalized and autonomous teaching mode, promote students' English learning, and provide a technical foundation for online and offline mixed teaching of English [5].

Third, the current situation of college English teaching. English has always been kept as a basic course. Students' English proficiency continues to improve and they have a better foundation in English learning [6]. The phenomenon of college English credits being continuously compressed in vocational schools and English teaching to vocational college students. At this time, it is necessary to train more students with English ability to help improve students' practical application and professionalism. Therefore, it is necessary to continuously change the teaching content in college English courses, from basic English teaching to subject English teaching, and in terms of teaching methods, from traditional teaching methods to a combination of online teaching [6]. Demands also help promote the college students' English self-innovation and development.

4   Problems to Be Paid Attention to in the Online and Offline Mixed Teaching Mode Construction

4.1   Pay Attention to the Improvement of Teaching Evaluation System

The evaluation index of English online and offline blended teaching is diverse. When the evaluation system is established, it adhere to the principle of three-dimensionality and comprehensiveness, and evaluate the comprehensive learning status of students through a quantitative method. The first is process evaluation, which accounts for about 25% of the evaluation module [7]. It mainly provides feedback on students' participation in classroom activities, the frequency of online interaction with teachers, and the timeliness and accuracy of exercise completion. Process evaluation has changed the results-based evaluation method of colleges in the past, focusing on the timeliness of feedback, and can supervise students' daily English learning. The second is performance evaluation, which accounts for about 25%. It mainly records students' classroom presentations and project reports from multiple dimensions, focusing on students' progress in these performance tasks [7]. The main evaluation indicators are language fluency, content richness, and participation enthusiasm. The proportion of this module is about 50%. It is mainly to test the language level of students through a unified examination. In addition to the traditional written test, the evaluation method also includes the oral test.

4.2   Pay Attention to the Optimization of Classroom Teaching

Colleges must first integrate the content of the textbooks according to the requirements of each major and each grade regarding English hours, and formulate their own teaching plans for the characteristics of offline and online classrooms [8]. Specifically, in online classrooms, students need to complete their own previews in a differentiated and hierarchical manner. In this process, students can use preview reports, knowledge notes, etc. to present the results of the preview, and teachers can check the results of the students' preview through project reports and forum questions, as shown in Fig. 1. In offline classrooms, teachers can extract the key points of the course and the difficulties students have in the preview, and then guide students to carry out learning through personal statements, results announcements, course games, content performances, topic debates, etc. Teachers also need to provide targeted tutoring to students online, arrange exercises for students, and help students complete the internalization of knowledge [8].

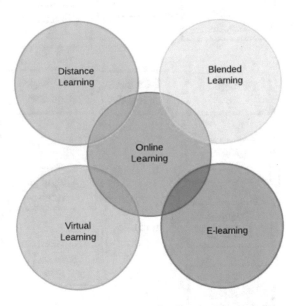

Fig. 1. Features of online learning

5   The Online and Offline Hybrid Teaching Mode Application in College English in the Internet Environment

5.1   New Media + Offline Classroom Mode Application

As a young group, college students pay more attention to new media. Teachers can use new media as the carrier of online mode and the integration point of offline mode to create diversified classroom modes. First, a WeChat public account + offline classroom model can be constructed [9]. In this mode, teachers can publish the key content of the course on the official account, and students can receive this content by opening WeChat, and arrange time for learning according to the amount of content. This method is simple and convenient. Students can learn anytime and anywhere as long as they pick up their mobile phones, as shown in Fig. 2. Second, the MOOC + offline classroom model can be realized. Colleges can comply with the requirements of MOOC and set up "MO Classes" in various colleges to help students learn in different levels and modules through thematic explanations. Finally, a webcast + offline classroom model can be constructed [9]. Teachers can use the live broadcast platform to build an online second classroom, to achieve interaction with students and realize exchanges with each other at any time.

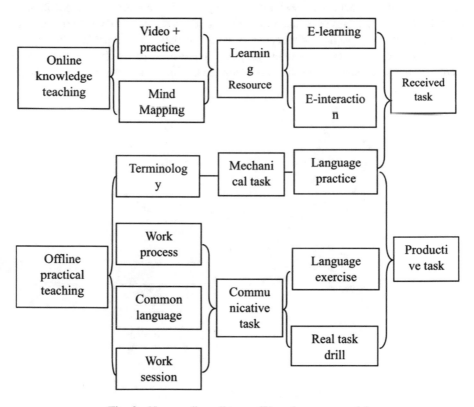

Fig. 2. New media online + offline classroom model

5.2 Application of Learning Software + Offline Classroom Mode

In the rapid Internet technology development, many English learning software have appeared on the Internet. This software are both interesting, informative and convenient, and can be assistants for students to learn online independently. Therefore, teachers can build a comprehensive learning model of English learning software + offline classroom [8]. For example, "Hujiang Dictionary" is an excellent English word memorizing mobile phone software, which can push the content of word learning according to the level of students, and it has strong applicability to students with weak and superior foundations [9].

5.3   Realize the Cooperative Development of Teaching Resources Between Schools and Enterprises

While realizing the online and offline modes combination, English teaching must also realize the integration of English knowledge and professional knowledge. Colleges should strengthen school-enterprise cooperation, pay attention to the provision of online and offline hybrid teaching facilities and the construction of teaching environment, such as the construction of recording and broadcasting training rooms for micro-classes and special classrooms for flipped classrooms. Teachers should use the resources developed by colleges for teaching, and make appropriate adjustments on this basis [10]. It can be said that it is very importance to realize the joint development of resources between schools and enterprises. It can make curriculum resources meet actual teaching and meet the needs of society and positions. In this process, enterprises and universities should understand teaching needs and job requirements through research and teaching research, realize accurate setting, reasonable arrangement, and promote the implementation of cross-over teaching mode [10].

6   Conclusion

In summary, in college English teaching process, cultivating students' autonomous learning ability and subject core literacy is more conducive to promoting students improving their personal development level and comprehensive quality capabilities. In the Internet context, teachers want to further improve the feasibility, comprehensiveness and rationality of the online and offline hybrid-teaching model. Teachers need to be able to analyze and consider relevant data and information, and continuously optimize and teaching content improving, to improve students' comprehensive quality and ability to make full preparations.

Acknowledgments. Project foundation: The First-class Offline Undergraduate Course (English Writing) Project of Mianyang Teachers' College in 2020. Project number: Mnu-JY20228.

References

1. Lin, S.: Exploration of the blended teaching model of college English. Campus Engl. **5**(05), 212–215 (2019). (in Chinese)
2. Cheng, L.Y.: The construction and application of a hybrid teaching model for English majors in colleges and universities. J. Jiamusi Vocat. Coll. **12**, 85–88 (2019). (in Chinese)
3. Hu, C.T.: Research on online and offline mixed teaching mode of college English. Campus Engl. **11**(03), 45–47 (2019). (in Chinese)
4. Liu, F.T.: Research on the construction of online + offline hybrid mode of college English teaching. J. Jilin Radio TV Univ. **9**(09), 57–59 (2019). (in Chinese)
5. Zhang, B.D.: College English blended teaching model and strategy under the background of "Internet+." Seeking Knowl. Guide **10**(07), 115–118 (2018). (in Chinese)
6. Lin, L.T., Zhang, N.: Exploration of college english blended teaching reform from the perspective of "Internet." News Lovers **4**(07), 87–91 (2019). (in Chinese)

7. Zhang, A.M.: The application strategy of the hybrid teaching model in college English teaching under the Internet + background. J. Hubei Open Vocat. Coll. **15**(03), 165–167 (2019). (in Chinese)
8. Zheng, H.R.: Research on college english blended teaching model under the background of "Internet+." J. Hulunbuir Univ. **3**(27), 138–141 (2019). (in Chinese)
9. Chen, Q.T.: Research on college english blended teaching model based on the background of "Internet + Teaching." Sci. Educ. Wenhui **8**(02), 22–26 (2019). (in Chinese)
10. Zheng, Y.T.: An analysis of the college English blended teaching model in the "Internet+" era. Teacher **6**(02), 44–46 (2019). (in Chinese)

Innovative Development Path of Labor Education in the Era of Artificial Intelligence

Xintian Xiu[⊠]

College of Resources and Enviroment, Fujian Agriculture and Forestry University, Fuzhou, Fujian, China
xxt200250@fafu.edu.cn

Abstract. As the main driving force of the new round of scientific and technological revolution, artificial intelligence will inevitably bring a certain degree of threat to the future during the upgrade and transformation, but more importantly, the current new development opportunities. In the era of artificial intelligence, labor education as a socialist qualified builder and reliable successor is facing a profound impact of values and emerging technologies. This article uses qualitative research methods to explore and find a breakthrough in the possible dilemmas of labor education in the era of artificial intelligence, and seek its own development path. This research believes that the development of artificial intelligence cannot change the arduous mission of labor education to train people. Labor education must keep pace with the times and adapt to the rapid development of the technological era. Only when the two resonate at the same frequency can they develop in coordination.

Keywords: Artificial intelligence · Labor education · Students

1 Introduction

Artificial intelligence has undergone three important transformations and has become the main driving force for a new round of technological revolution and an accelerator for the rapid development of the times. However, with the rapid development of artificial intelligence, controversy has gradually increased. Some scholars hold the "machine threat theory", such as the famous scientist Geoffrey Hinton; some experts hold the "development opportunity theory", such as the artificial intelligence scientist Demis Hassabis, who believes that artificial intelligence it can realize the free and comprehensive development of people and even liberate human beings [1]. No matter which viewpoint is a prediction of the future, the final direction still depends on human beings, and it is a manifestation of the joint force in human development. As an important part of the national education system, labor education affects the labor skill level, labor value orientation and labor spirit of socialist builders and successors to a large extent. Emerging technologies and industrial changes have greatly increased social productivity, which not only gave new definitions to labor education, but also triggered great changes in labor forms and methods. Labor education in the era of artificial intelligence needs to be reinterpreted [2]. However, the era changes and connotation changes of labor education cannot be realized immediately. Especially in

J. Abawajy et al. (Eds.): ATCI 2021, LNDECT 81, pp. 663–669, 2021.
https://doi.org/10.1007/978-3-030-79197-1_95

the era of artificial intelligence, it not only puts forward higher requirements on labor education, but also brings many impacts and challenges to labor education. In the era of artificial intelligence, how labor education should go from confusion to transcendence deserves serious consideration.

2   The Dilemma of Labor Education in the Era of Artificial Intelligence

In the era of artificial intelligence, the development of science and technology such as the Internet of Things, artificial intelligence, and big data relies on technological rationality, and the way of thinking is mainly instrumental rationality. The value concept of cutting-edge technology has a profound impact on many aspects of labor education. Labor education is easy to learn. In the face of the inability to implement the technical field, the conflicts between traditional labor value relations and modern technology may exist in a series of difficulties.

2.1   Value Alienation, Technology Dominates

The scientific and technological rationality brought about by artificial intelligence obscures the fundamental value of labor education to cultivate people with its highly materialized tool value, and makes it fall into the value dilemma of marginalization and dislocation. A large amount of labor can be liberated from repeatable and mechanical physical or mental labor, use all kinds of new intelligent tools to complete the labor that should be undertaken, and use the power of technology to meet the needs of material survival and spiritual enjoyment [3]. However, the tool value of technology has liberated the individual's body and brain, but it has also kidnapped the human body and brain. In essence, this kind of liberation is very easy to make students fall into intelligent tool rationality and move towards rigid tool thinking, that is, they believe that tools can help them accomplish the set goals, and they will take care of their time, energy, and personal development plans, and they will gradually be taken care of by them. Technology dominates. Things are not so simple. Faced with the encirclement of artificial intelligence technology, labor education in specific practice often shifts from comprehensive education to one-way cultivation of skills in value dislocation. Artificial intelligence has brought about changes in the content and methods of labor education. In the future, understanding technological products and learning new artificial intelligence technologies to adapt to social needs will become an important part of labor education. These product technologies are the products of labor in a specific social period, and their purpose is to promote students to carry out efficient labor practices and essentially serve people. However, in learning to use digital technology, complex mental work often becomes simple technical work. Teachers and students often focus on smart tools themselves, and develop education around how to use them. Labor education is only skill training for the use of objects for the effective use of new intelligent tools. As a result, the value of labor education will gradually deviate from the original path of cultivating people by virtue, and only reflects the instrumental value of the training of students' skills, and its epochal value for cultivating students' good labor value cognition, labor ability, and labor spirit. The ontological value that promotes the comprehensive and free development of students will gradually disappear (Fig. 1).

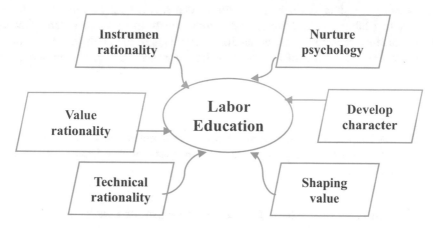

Fig. 1. Labor education under value alienation

2.2   Value is Imaginary, Labor Education is in a Dominant Position

The diversified labor forms in the artificial intelligence era endow labor education with new educational content, but the subsequent simplicity and modelization of labor education content will become an important issue facing labor education in the future [4]. On the one hand, as innovation and entrepreneurship education, leisure education, consumer education, etc. continue to be incorporated into the labor education curriculum system, the content of labor education will be unprecedentedly rich and complex. In view of the fact that labor education occupies very limited class hours in the Chinese education curriculum system, it is even included in the comprehensive practical activity curriculum, and there is no clear class time requirement. The limited teaching time makes the content of labor education can only stay at the level of transmission of shallow knowledge, unable to carry out in-depth education for diverse labor, and fail to subtly infiltrate the soft literacy necessary for students in the future of various labor education content [5]. In particular, it lacks the value shaping of the core labor concept, labor spirit and labor quality and other implicit literacy. To a certain extent, the content of labor education will be summarized in purely knowledge-based teaching, which will reflect the shallowness of labor education thinking, and it will not be able to demonstrate its expected value of nurturing people in the era of artificial intelligence. On the other hand, new labor and traditional labor represented by intelligent labor and virtual labor have their unique value, and both will coexist in labor education curriculum as an important element of labor education [6]. However, the interweaving and collision of the two sides will inevitably lead to new conflicts. Specifically, from the perspective of the content ratio of labor education courses, based on the limited hours of labor education, the update and embedding of new labor education content will squeeze out the proportion of traditional labor education content, and even further compress the traditional labor education content to make it in a dilemma. From the perspective of the implementation of labor education courses, although based on multiple labor values, the main body of education will create unique labor situations and practical methods for students, allowing students to carry out corresponding labor practices freely and autonomously to meet the diverse development of students demand. However, as digital residents in the intelligent age, students are attracted by the instinct of intelligent technology, and may show a preference for new types of labor in their subjective choice of activities. This particular labor practice preference will also aggravate the existing contradiction between new labor and traditional labor, and its labor value will also be in a state of imbalance. The content of the new and old labor education is more important, and how to effectively resolve the conflicts between the two in the arrangement of class hours and the disorder of value will surely be a problem that requires careful consideration of education in the era of artificial intelligence.

3 The Innovative Path of Labor Education in the Era of Artificial Intelligence

3.1 The Value Orientation of Labor Education Must Be Reflected

Artificial intelligence empowers labor education, opening up a new perspective for its development with advanced intelligent technology, but only by demonstrating the purposefulness of artificial intelligence can this empowerment be biased. In fact, technology can serve us well [7]. The problem is that we must understand what we want to achieve with technology and what problems we solve. "Only on the basis of fully clarifying the educational value of labor education to promote the comprehensive and free development of people, can labor education get rid of the technological dependence on artificial intelligence, and gradually realize the value transcendence of scientific and technological rationality [8]. The arrival of the artificial intelligence era means labor, and changes in the form of labor education, but the value orientation of labor education that cultivates students' view of labor, the ability of labor, and the soul of labor will not change, and it will even become more prominent in the complex and changeable social situation. Contemporary students must be in schools and teachers under the guidance of the author, we should correctly understand the educational value of labor education, clarify that labor education in the era of artificial intelligence is cultivating students' key competencies, and develop individual wisdom and exploration spirit that cannot be replaced by artificial intelligence, so as to have both scientific rationality and humanities. Rational labor education enables students to develop into complete human beings and become individualized, innovative, and intelligent new-age laborers who adapt to social needs.

3.2   Labor Education Must Adapt to the Development of Artificial Intelligence

The intelligent technology in the era of artificial intelligence has moved many labor education places to virtual networks, ostensibly narrowing the physical space of labor practice, but labor education can be found in the integration of the online virtual world, the school education world, and the offline life world, and expand the new embodied practice space, and at the same time strengthen the physical input of the body and mind, so that the body can deepen the labor experience in a freer, more open, and more diversified labor space, and obtain a more abundant labor spirit experience [9]. The first is to expand the space for labor education in the online virtual world and encourage students to participate physically and mentally. With the changes in labor patterns, future labor education will inject creative labor content such as smart labor and virtual labor. The main hub of its occurrence is the Internet. Therefore, the virtual labor space based on the Internet is becoming more and more important for the labor education of students. In the Internet field, students mainly perform intelligent virtual labor centered on the application of intelligent technology and the creation of intelligent results. It not only requires the use of hand-brain cooperation to realize the potential and intelligence of the body, but also pays attention to human-computer collaboration and everyone's collaboration, improve the coordination ability of the body and brain and the ability of creative thinking in the practice of intelligent labor. The second is to create a holographic situation in the school education world that relies on VR technology and room networking to satisfy students' physical interaction and emotional communication in labor practice [10]. In the future, smart classrooms that use interactive touch projections as labor teaching tools will use their real-time strong perception and high interaction functions to strengthen the physical interaction and physical participation of students, peers, and teachers in the teaching place, so that the body will return from leaving the

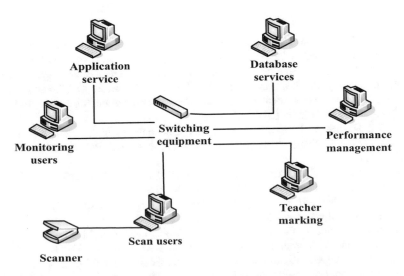

Fig. 2. Labor education adapted to the era of artificial intelligence

field. At the same time, it uses its wisdom to create individualized labor situations for students, such as the productive practice of multi-person collaboration in a distributed virtual reality classroom to improve the educating effect of embodied labor practices (Fig. 2).

4 Conclusion

The tremendous changes in labor forms in the artificial intelligence era mean that labor education is entering a new stage of development. With the emerging industrialization of labor technology, artificial intelligence and labor education will inevitably move toward deep integration. However, the road to integrated innovation is not full of thorns, but full of opportunities and challenges. Faced with the difficulties that labor education may face in the era of artificial intelligence, we should actively respond with an open attitude. Looking forward to the future, with the continuous innovation of productivity and science and technology, artificial intelligence will move from the weak artificial intelligence stage that replaces part of human labor to the strong artificial intelligence stage of omnipotent labor. Labor education will face a more profound revolution of the times. To this end, labor education must focus on artificial intelligence empowerment, grasp the value orientation of labor education to train people, follow the development and education laws of labor in the intelligent era, and innovate the labor education system under the guidance of artificial intelligence technology, fully demonstrate the comprehensive education value of labor education to build morality, increase intelligence, strengthen the body, and educate the beauty, and support the new form of education in the era of artificial intelligence with intelligent and intelligent labor education.

References

1. Villanueva, K.A., Brown, S.A., Pitterson, N.P.: Teaching evaluation practices in engineering programs: current approaches and usefulness. Int. J. Eng. Educ. **33**(4), 1317–1334 (2017)
2. Zheng, H., Perez, Z.: Design of multimedia engineering teaching system based on internet of things technology. Int. J. Contin. Eng. Educ. Life-Long Learn. **29**(4), 293–305 (2019)
3. Ma, L.: Teaching quality monitoring system based on artificial intelligence. Agro Food Ind. Hi Tech **28**(1), 2002–2006 (2017)
4. Liao, H., Hitchcock, J.: Reported credibility techniques in higher education evaluation studies that use qualitative methods: a research synthesis. Eval. Program Plan. **68**(6), 157–165 (2018)
5. Cho, D., Cho, J.: Does more accurate knowledge of course grade impact teaching evaluation? Educ. Finance Policy **12**(2), 1–31 (2017)
6. Spivey, W.A., Caldwell, D.F.: Improving MBA teaching evaluation: insights from critical incident methodology. J. Mark. Educ. **4**(1), 25–30 (2016)
7. Myerholtz, L., Reid, A., Baker, H.M., et al.: Residency faculty teaching evaluation: what do faculty, residents, and program directors Want? Fam. Med. **51**(6), 509–515 (2019)
8. Mohan, L., Kathrotia, R., Mittal, S.: Student feedback in medical teaching evaluation: designing the perfect mechanism. Indian J. Physiol. Pharmacol. **62**(1), 149–155 (2018)

9. Stevenson, J.E., Israelsson, J., Nilsson, G.C., et al.: Recording signs of deterioration in acute patients: the documentation of vital signs within electronic health records in patients who suffered in-hospital cardiac arrest. Health Inf. J. **5**(1), 23 (2016)
10. Ren, P., Xi, L., Jingwei, L.: Research on construction of indicator system for evaluation of the ecological civilization education in Chinese universities. Cogn. Syst. Res. **52**(12), 747–755 (2018)

Teaching Innovation Practice of "Advanced Office Application" Based on Internet Plus

Nan Xie[✉]

College of Information Engineering and Art Design, Zhejiang University
of Water Resources and Electric Power, Hangzhou, Zhejiang, China
xienan@zjweu.edu.cn

Abstract. Advanced office application (AOA) is a general elective course for non-computer majors. It is instrumental and vocational skill operation course. In the face of the "Internet plus" teaching background and the compression of curriculum hours and the undiminished teaching content, it is an important teaching problem to how to solve the phenomenon of students' submission to light theory and copy work and how to use the online learning platform and offline classroom teaching to better cultivate students' innovative ability, independent thinking and problem-solving ability, and improve the efficiency of office work and occupation moral accomplishment of students. The teachers have carried out teaching innovation practice in the following aspects: changing the teaching mode of the course, taking the students' knowledge, skills, operation and quality training as the centre, carrying out the teaching design, integrating into the curriculum, revising the syllabus of the ideological and political elements, and teaching assessment methods, and at the same time, paying attention to the Internet plus students' personalized guidance and assistantship training, forming a good course. The service system mode, with the learning atmosphere of helping more and bringing more, cultivates the students' ability of team cooperation, mutual assistance, innovation and entrepreneurship. Its teaching innovation practice effect is good, and the implementation way has good promotion and application value.

Keywords: Internet plus · AOA online and offline mixed · Curriculum service system · Innovative teaching practice

1 Introduction

At present, general courses such as AOA, C language programming and python programming are generally offered for non-computer major freshmen in undergraduate colleges in China, which include basic computer knowledge, operation and use of office software, C language, python programming and so on. The course teaching basically adopts the independent teaching idea of theoretical knowledge explanation and computer operation and the teaching mode is single, and the teaching methods are mainly teaching method and operation demonstration method. The teaching content is mainly based on operation, supplemented by theoretical knowledge and concepts, especially the use of office software. Sometimes it ignores the universality of students' interest in learning and the cultivation of their sense of inspiration and inquiry [1–3, 8–10].

© The Author(s), under exclusive license to Springer Nature Switzerland AG 2021
J. Abawajy et al. (Eds.): ATCI 2021, LNDECT 81, pp. 670–677, 2021.
https://doi.org/10.1007/978-3-030-79197-1_96

1) Light principle and be careless.
2) Students' cognition of learning courses is insufficient.
3) Pay attention to the imparting of knowledge, ignore the cultivation of ability.

The class explanation and practice cases of this course are based on knowledge. Each course focuses on some knowledge points and ignores the integration of Ideological and political elements. The teaching cases are formulated by teachers according to the teaching content and knowledge points as clues. Students do not have the concept of solving practical engineering problems systematically and their practical ability cannot be improved. In addition, students practice according to the cases designated by teachers, which only reflects their personal ability, lacks the practical training of project team work, and ignores the cultivation of students' innovation ability [4–7].

On May 28, 2020, the Ministry of Education issued the notice of "guiding outline of Ideological and political construction of higher education curriculum" (File No: JG [2020] No. 3), which pointed out that "what kind of people to cultivate, how to cultivate people and for whom to cultivate people are the fundamental issues of education, and the effectiveness of moral education is the fundamental standard to test all the work of colleges and universities." And "the ideological and political construction of curriculum is an important task to improve the quality of personnel training in an all-round way" [1, 2]. Therefore, teachers should combine the teaching mode of "Internet plus" to reform the teaching process, and deeply think about how to improve the quality of teaching and training of the general education. When teachers improve their teaching and teaching abilities, how can they integrate the curriculum elements into the classroom teaching online and offline?

2 Key Problems to Be Solved in AOA Course

Due to the differences in the computer operation level of students in different regions, the course needs continuous teaching innovation and practice to improve the current situation [2–5, 10].

(1) how to change the current single teaching mode, combine the "Internet plus" teaching mode and different teaching methods, so that students can master the advanced OFFICE application skills and skills in a shorter time, so that they can deal with and solve the daily office work more efficiently.
(2) How to integrate ideological and political elements into online platform and offline classroom teaching, so as to better improve students' independent thinking and hands-on ability, team spirit, innovation and entrepreneurship, and create a model of operation general courses that can achieve professional quality and ability training?
(3) How to build a curriculum service system, combining the typical practical application case teaching in the era of "Internet plus", to provide good operational skills and professional ethics for its future occupation, and improve the quality of personnel training.

3 Ways of Innovation and Practice

Combined with several key problems to be solved in the course, the teacher has done teaching innovation practice in the following five aspects [1, 2, 4–6].

3.1 Online and Offline Hybrid Teaching Mode

Combined with the Internet plus and the change of teaching concept in information education, the online and offline teaching mode is adopted. Combined with the syllabus and teaching content arrangement, offline classroom teaching adopts a variety of teaching methods and teaching means, such as task driven, case demonstration operation, heuristic students' computer and work display, student evaluation, etc., and records 26 micro lessons in the way of fragmentation of teaching content knowledge points. Since 2017 grade students began to use online learning platform to assist teaching, the platform construction has been successful The main online open learning platforms are: 1) online development course sharing platform of colleges and universities in Zhejiang Province http://www.zjooc.cn Search the course "advanced application of office software" for self-learning; 2) network teaching platform of Zhejiang University of water resources and hydropower http://i.mooc.chaoxing.com/space/index?t=1610244064916.

3.2 Update of Teaching Design Ideas

In order to better reflect the talent training focus on students' knowledge ability, skill operation and literacy training, the curriculum should be corresponding to "analysis and interpretation of data" in 3.4 of "engineering education certification standard", 3.5 corresponding to the use of modern tools, 3.8 corresponding to professional norms, and 3.9 corresponding to "engineering education certification standard" Individual and team corresponding teaching objectives, teachers in the teaching design has also done ideas update.

The new idea of teaching design mainly adopts six teaching links as shown in Fig. 1. Teachers introduce teaching situation, demonstrate case operation, students imitate, teachers summarize key and difficult points of operation, students' online learning "flipped classroom", students' knowledge summary, students' self-evaluation and teacher-student mutual evaluation.

Fig. 1. Six links of teaching design

The combination of online and offline hybrid teaching mode and six links of teaching design ideas makes teachers' teaching methods and teaching forms more diverse. In the specific offline classroom teaching, the process of "clear teaching object

→ demand analysis → assignment of educational administration → determination of teaching objectives of" knowledge, skills and literacy "to be achieved in each class → teaching design (namely six links) → teaching feedback and evaluation" is used.

3.3 Organization and Implementation of Teaching Content

The selection of teaching content takes technology application as the core, and uses task driven and case-based teaching method to explain knowledge points in class. The course has many basic verification experiments and comprehensive project experiments. The case selection is suitable for students' professional characteristics of post vocational skills teaching content, as far as possible to give every student the opportunity of hands-on operation, work process organization, expression and innovation, so that students can learn to pay equal attention to life and work in the course task learning, making the knowledge system and skill training progress.

The course uses modular case-based knowledge point fragmentation to make micro lessons. Combined with online guidance, interaction and question answering of online development course platform and offline teaching, communication and guidance of traditional classroom teaching, it realizes "flipped" learning of course knowledge. Pay attention to online and offline personalized guidance to ensure students' mastery of key content and advanced learning of knowledge. The specific curriculum implementation strategy is shown in Fig. 2, which consists of four teaching stages.

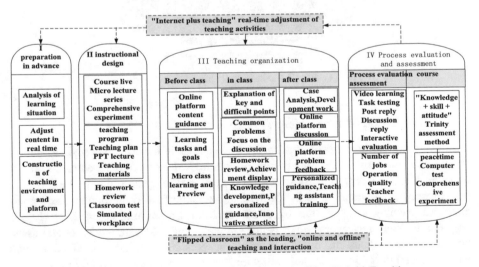

Fig. 2. Implementation strategy of online and offline Hybrid Teaching

Among them, "before class", "after class" in stage III teaching organization and "process evaluation" in stage IV process evaluation and assessment are mainly carried out and evaluated on the online platform. The whole implementation process of the course is dominated by "flipped classroom" and "online and offline" teaching interaction, and the whole teaching content is adjusted in real time.

3.4   Reform of Course Assessment Methods

The course assessment breaks the traditional assessment method of "usual results + final examination", adopts the trinity of "knowledge, skills and attitude" course assessment method, combines the online and offline mixed learning achievements and learning efficiency, and pays attention to the diversity of students' learning process evaluation and assessment indicators, as shown in Fig. 3.

(1) Knowledge is mainly evaluated from the final computer examination, and students' mastery of the course content is evaluated (accounting for 50% of the total evaluation).

(2) Skills are evaluated from the following aspects: the submission of students' project achievements, the quantity and quality of offline practice assignments, online comprehensive exercises and offline classroom tests (accounting for 30% of the total evaluation);

(3) Attitude is mainly from online platform classroom attendance, platform task point, video learning, reply discussion, and offline classroom performance, rush answer, answer questions, peacetime performance and other comprehensive given peacetime performance (accounting for 20% of the total evaluation)

Fig. 3. Trinity assessment chart

3.5   Construction of Curriculum Service System

Following the requirements of post professional ability in the computer age, the course should pay attention to the integration of Ideological and political elements in the revision of the syllabus, and construct the course service system, that is, to condense the collective, pay attention to the individual, pay attention to the personalized guidance and teaching assistant training of students, form a learning atmosphere of "one help many, many lead many", and better cultivate students' teamwork, mutual assistance, diligent research, thinking and independent analysis Innovation and entrepreneurship ability and good computer literacy and professional ethics.

4   Practical Effect

The teaching evaluation and feedback of the course in the past three semesters under different platforms and different teaching methods are shown in Table 1.

Table 1. The next teaching evaluation form in recent three semesters

Implementation semester	Teaching objects	Teaching methods & students	Teaching evaluation and feedback
2019-2020-1	Undergraduate and junior college students; Fraternal institutions	Online and offline hybrid Network platform students: 517	Good effect of flipped classroom and mixed teaching
2019-2020-2	Undergraduate and junior college students; Fraternal institutions; Elective students	Dingding live Offline Teaching Network platform students: 471	Flexible online cross platform learning, willing to recommend to others
2020-2021-1	Undergraduate and junior college students; Elective students	Online and offline hybrid Network platform students: 709	The online platform is free to learn and has good effect in discussion and learning interaction

By comparing the final examination results of grade 14 (pure offline), grade 17 (the first attempt of online and offline mixed teaching) and grade 18 (mixed), and from the Fig. 4, we can see that the excellent rate and average score are greatly improved. The course adopts hybrid teaching and "knowledge + skills + attitude" Trinity course assessment method, which pays more attention to process evaluation, and improves students' learning enthusiasm and hands-on ability. Teachers also need to pay attention to students' learning initiative in real time.

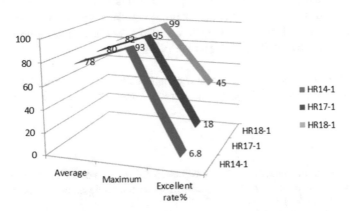

Fig. 4. Comparison chart of final examination results of three students

5 Conclusion

From Table 1 and Fig. 4, it can be seen that the implementation of "online and offline" hybrid teaching effect is obvious, and the average score and excellent rate of final examination of students are greatly improved. Therefore, the next three years of continuous construction and optimization of courses, as well as promotion and application, are mainly taken and planned as follows:

(1) Further enrich the teaching resources of the course, increase the enterprise cases with higher practical and application reference value, and conduct workplace simulation, demonstration operation and video production of teaching cases according to different majors;

(2) The scope and learning objects of shared universities are expanded on Zhejiang Online open platform, super star platform and MOOC platform of Chinese University;

(3) Continue to open up to the society for a long time, ensure good interaction and timely response;

(4) To plan or build a new form and three-dimensional characteristic textbook, which can better meet the needs of online open sharing platform learners;

(5) Strive to regenerate the project results and actively declare the educational reform projects of relevant courses;

(6) To summarize and accumulate teaching experience and write relevant teaching and research papers;

(7) Update the version of the course examination platform, and do the corresponding data update, test and operation and maintenance work.

The office software advanced application has undergone four years of mixed teaching mode reform from 2018 to 2020. Its syllabus, teaching design ideas, teaching content organization and implementation, curriculum assessment methods, etc. are constantly revised and replaced in accordance with the development of the times and the requirements of the state for talent training. The course is online and offline mixed teaching mode and personalized guidance for students It has carried out bold teaching innovation practice. From the students' computer practice and peer teachers' feedback information, it is found that the online and offline hybrid teaching mode is suitable for the operation courses, which has a good application value.

Acknowledgments. This work was financially supported by the 2021 First-class undergraduate course construction project of Zhejiang University of Water Resources and Electric Power.

References

1. Sun, L., Xiong, S., Yang, X., et al.: Construction and practice of online teaching mode based on intelligent vocational education platform under the background of epidemic situation. J. Wuhan Polytech. **19**, 9–12 (2020)
2. Zhang, Y.: Research on the application of "cloud classroom + informatization" in Higher Vocational Computer Teaching. Ind. Sci. Tribune **21**(19), 175–176 (2020)
3. Xiaojian, W.: Hybrid teaching practice of Higher Vocational English based on "intelligent vocational education cloud" platform. J. Wuhan Metall. Manage. Coll. **9**, 69–71 (2019)
4. Weitang, S.: Construction of innovation and entrepreneurship teaching resources based on Intelligent vocational education cloud platform. Sci. Technol. Innov. Daily **35**, 238–241 (2019)
5. Wang, Y., Pan, Y.: Flipped classroom teaching mode and practice based on micro class. Knowl. Econ. 144–146, 148(9) (2018)

6. Hongshen, Yu.: Exploration of teaching methods with the help of micro class flipped classroom – review of flipped classroom and micro course teaching method. Sci. Technol. Chin. Univ. **11**, 106 (2019)
7. Jing, L.: Design and effect research of learning mode based on smart classroom. Comput. Prod. Circ. **4**, 183 (2019)
8. Yang, X.: Teaching plan and implementation of "Internet plus" intelligent classroom. Teach. Manage. **10**, 34–36 (2019)
9. Lihai, L.: The process of feeling, the law of acquisition, the wisdom of development – the wisdom classroom of Chinese in senior high school and its construction strategy. China's Campus Educ. **6**, 142 (2019)
10. Fan, Y., Xu, X.: Research on the "Internet Plus" teaching model for Advanced Application of Office Software. Comput. Era (8), 58–60, 63 (2017)

Employment Risks of Artificial Intelligence Tape and Education Response

Xingzhao Yuan[✉]

Department of Physical Education, Shandong Management University, Jinan, Shandong, China

Abstract. Artificial intelligence technology continues to make major breakthroughs, which is profoundly changing people's production and lifestyles, and it also has an impact on current and future employment. At present, people seem to pay more attention to the application of artificial intelligence technology and the changes it brings to social development, and often ignore the risks that artificial intelligence technology brings to employment. In order to cope with the risks brought by artificial intelligence technology to employment, education in the artificial intelligence era shoulders an important mission. This question has found through research that it is necessary to strengthen the basic scientific research of artificial intelligence, optimize the talent training model of universities, and transform the knowledge production model of universities. Establish the concept of lifelong learning and update the concept of employment education to actively respond to the impact of artificial intelligence on employment.

Keywords: Artificial intelligence · Employment · Education response

1 Introduction

Artificial intelligence technology is the focus of competition among countries in the world today. It is not only an extension of the industrial revolution, but also a starting point for a new revolution. The development of artificial intelligence technology and robots has promoted the continuous and accelerated deepening of the information technology revolution, and has also brought a huge impact on the labor market. The impact of artificial intelligence technology on employment is a hot and focused issue of social concern. Especially in the face of the sudden the "COVID-19", the employment situation has encountered huge challenges. To further promote the digital transformation and development of employment forms, it is urgent to integrate emerging industrial technologies to enhance its ability to cope with the challenges of great changes in the external environment. Exploring the impact and risks of artificial intelligence technology on employment is not only conducive to the promotion of employment for all people, but also an exploration of the major issue of national economic and social development, and is one of the current and future important directions in the field of people's livelihood. To this end, this article analyzes the risks

J. Abawajy et al. (Eds.): ATCI 2021, LNDECT 81, pp. 678–684, 2021.
https://doi.org/10.1007/978-3-030-79197-1_97

of artificial intelligence technology to employment from a global perspective and Chinese trends, and tries to propose relevant strategies from the perspective of education response.

2   Employment Risks Brought by Artificial Intelligence Technology

It is an emerging driving force leading the development of the world, and will make important contributions in promoting economic growth, accelerating technological progress, promoting social harmony, and ensuring national security.

2.1   Artificial Intelligence Technology Reshapes Job Roles

The emergence of artificial intelligence technology will change the traditional manufacturing and production methods. Artificial intelligence technology promotes the intelligentization of production, and intelligent production based on big data improves the feasibility of mass customization in response to the individual needs of the market, and production and business models will also change. The application of artificial intelligence technology in the medical field mainly focuses on medical robots, smart drug development, smart diagnosis and treatment, smart image recognition, and smart health management [1]. Therefore, artificial intelligence technology poses a challenge to medical staff. Medical staff need to rethink their roles and use artificial intelligence technology to formulate a personalized medical plan for each person. The application of artificial intelligence technology in the financial field mainly uses machine learning, voice recognition, visual recognition and other methods to analyze, predict, and identify transaction data, price trends and other information, so as to provide customers with services such as investment and financial management, equity investment, and avoid financial risks provide financial supervision and control. The financial investment of artificial intelligence and big data is the direction of the future development of financial services, which can provide users with data-driven personalized, precise and intelligent integrated financial services [2]. The application of artificial intelligence technology in the transportation field mainly focuses on unmanned driving, traffic behavior monitoring, and traffic risk warning. As far as the current common smart car-hailing software is concerned, artificial intelligence technology can dynamically dispatch and route planning, and can also conduct research and judgment on traffic safety, etc., reshaping the transportation system. Artificial intelligence may affect employment, reduce the proportion of the population directly engaged in productive labor, and eliminate the dependence of related industries on labor [3].

2.2   Artificial Intelligence Technology Reshapes a New Form of Employment

The new form of employment in the future will rely on major technologies such as artificial intelligence technology to promote profound changes in the job market and resource allocation methods. In addition to the deepening of the original employment pattern and the promotion of market mechanisms, artificial intelligence technology will also affect the allocation of resources and the strategic layout of corporate forms, production, and operations in the reshaping of the human resources market, thereby reshaping new forms of employment. Artificial intelligence technology will replace some occupations, such as telemarketers, typists, bank clerks, etc., and will also give birth to some new positions, such as data scientists, machine learning engineers, etc. [4]. At the same time, some new industries will become the main forms of employment for future development. The service industry will become the main driving force of my country's economic development. The development of artificial intelligence technology will promote the intelligent development of the manufacturing industry. Manufacturing companies have widely introduced artificial intelligence to make the production process more "smarter" and more efficient. From January to June 2017, industrial robot manufacturing output increased by 52.8% year-on-year, and the growth rate was 24.6 percentage points higher than the same period last year. Some new industries have grown substantially. The output of new energy vehicles increased by 51.2% over the previous year, smart TVs increased by 3.8%, industrial robots increased by 81.0%, and civilian drones increased by 67.0% [5]. The in-depth application of artificial intelligence will further transform the existing logistics system, inject strong vitality into the adjustment of the employment structure, and reshape the employment pattern.

2.3 Human-Machine Cooperation Has Become a New Development Trend

In the future, to learn to coexist and cooperate with intelligent robots, the future humans must be a new type of human-machine combination, and humans will become more powerful with the help of artificial intelligence technology. It will not only replace some of the existing jobs, it will also change the work paradigm and may create new jobs. Artificial intelligence will reshape the employment structure, and "human-machine collaboration" will become the mainstream work model. The risk of employment in the era of artificial intelligence is contemporary. As enters the era of artificial intelligence in the future, society will also become an intelligent society. The formation of an intelligent society is a long-term process. Therefore, the employment risks of artificial intelligence technology will continue throughout the era of intelligent revolution. It is a real risk and also the potential risks in the future.

3   Educational Response to the Risks that Artificial Intelligence Technology Brings to Employment

The impact of artificial intelligence on education has gradually emerged and will continue to increase. In order to adapt to the impact of the new round of technology on education and employment, China should actively adopt educational response strategies to adapt to the changes brought about by artificial intelligence technology.

3.1   Strengthen Basic Scientific Research on Artificial Intelligence

It will become the endogenous driving force of education reform. The future direction of education development must adapt to the changes brought about by new technologics, especially in the era of artificial intelligence. A fair and quality education system should be constructed to help foster comprehensive and free development, and personal development. Strengthen the basic scientific research, combine artificial intelligence, psychology, brain science, cognitive science, computer science, pedagogy and other disciplines, deeply study the scientific laws of education, and reconstruct the education system to adapt to the artificial intelligence technology. The educational changes, such as changes in teaching methods, and the roles of teachers and students. Through the basic scientific research of artificial intelligence, strengthen the possible influence of artificial intelligence on education, medical treatment, society and other fields to strengthen the avoidance of possible risks and ensure the positive effects of artificial intelligence [6].

3.2   Optimize the Talent Training Model of Colleges and Universities

The key to the growing of artificial intelligence technology lies in talents. The core of talent training lies in universities. Artificial intelligence is a highly intersecting subject, and China is still in its infancy in the cultivation of artificial intelligence talents. Therefore, Chinese universities should actively connect with artificial intelligence, vigorously cultivate artificial intelligence-related talents, and introduce high-level artificial intelligence innovating teams and optimizing the talent training model of colleges and universities, especially in the setting of various professional courses, should be actively matched with the needs of applied talents to achieve the precise training of artificial intelligence talents in China to cope with the changes in technology and the times. In the era of artificial intelligence technology, higher education must accelerate its own reforms to adapt to thc changing needs of skills. The new teaching content needs to focus on the cultivation of students' comprehensive ability, cultivate creative students, stimulate students' cross-border thinking, and promote students' all-round development. New teaching methods need to build an open and shared education platform, and at the same time provide students with personalized and accurate education services with the help of big data and learning analysis technology.

3.3   Transform the Mode of Knowledge Production in Colleges and Universities

With the growing of technologies such as artificial intelligence and cloud platform, universities have faced great challenges and impacts in the knowledge production model. The characteristics of knowledge in this period increasingly tend to be uncertain, globalized, and competitive. On the one hand, artificial intelligence technology has led to profound changes in the knowledge production model. The knowledge production model in universities is being reconstructed [7]. Universities are no longer the only producers of knowledge, and knowledge will become more open and shared. On the other hand, in the context of the globalization and popularization of higher education, and the impact of artificial intelligence technology on higher education, the teaching methods, curriculum content, education goals, education management, etc. of colleges and universities will all change. The traditional university knowledge production model cannot adapt to the goals to be achieved by the integration of artificial intelligence technology and universities, and a new knowledge production model must be produced, while at the same time redefining the educational content, learning methods, learning resources, and education evaluation of universities. Colleges and universities are academic organizations with knowledge as the core. Knowledge production is one of the important functions of colleges and universities, and scientific research in colleges and universities is also a knowledge production activity. It has a profound impact on the knowledge production mode of universities, showing the characteristics of the knowledge production mode 2.0. In the era of artificial intelligence, university teaching should have corresponding characteristics: it emphasizes the integration of multiple disciplines in higher education, pays more attention to the cultivation of students' "computational thinking" ability, and uses interdisciplinary knowledge to cultivate students' communication skills, creativity, critical thinking, and complex problem solving. Ability, writing and learning ability, etc.; pay more attention to the use of artificial intelligence and other technical means to provide students with personalized education services and realize learners' smart learning. The universities should actively carry out the reform of the "artificial intelligence + education" talent training model in order to prepare for the cultivation of talents suitable for the future era (Fig. 1).

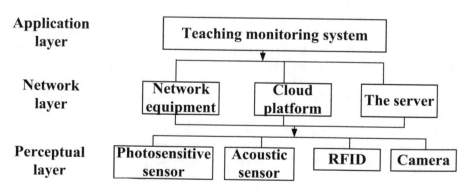

Fig. 1. The mode of knowledge production in universities under artificial intelligence technology

3.4   Establish the Concept of Lifelong Learning

Educational informatization has entered the 2.0 era. In order to adapt to the growing of the artificial intelligence era, it is necessary to establish the concept of lifelong learning. Lifelong learning will become an important way for people to acquire professional skills and adapt to the growing of the times in the future. Lifelong learning puts forward new requirements and great changes to the current education system. For example, it is necessary to cultivate students' interest-oriented and innovative-based education and teaching methods to provide continuous motivation for lifelong learning. Through educational informatization methods and new teaching methods, such as the MOOC learning platform, train students to adapt to future education and cultivate qualified "digital citizens". When artificial intelligence brings more opportunities and challenges to people, in order to better understand and develop themselves, the concept of lifelong learning may be an important choice [8].

3.5 Update the Concept of Employment Education

In recent years, China has made significant progress with regard to employment issues, especially employment education for college students, but it cannot be ignored that the core issue of employment is that the concept of education has not been resolved [9, 10]. In the era of artificial intelligence technology, updating the concept of employment education and preventing various risks in employment are key issues in the current training of higher education talents. On the one hand, in view of the unemployment problem caused by artificial intelligence, big data analysis technology can be used to continuously track and analyze the job market. Colleges and universities should establish a professional early warning mechanism as soon as possible to continuously enhance their own awareness of employment risk prevention. On the other hand, the rapid development of artificial intelligence in education has led to reforms and innovations in the education field. The possible development of artificial intelligence should be taken into consideration when training talents, and the enrollment plans of various majors should be dynamically adjusted in time to avoid excessive positive competition with artificial intelligence.

4 Conclusion

The rapid growing of artificial intelligence technology has changed the original form and mode of employment, which puts forward higher requirements for education to adapt to the development of the times. The education field shoulders the important task of cultivating talents. Only by continuous self-innovation and advancing with the times can we better adapt to artificial intelligence technology and seek new breakthroughs in the talent training model. Employment is the foundation of people's livelihood, and the in-depth integration of artificial intelligence technology and employment is still the focus of future research. This research only puts forward some simple insights, and in-depth and detailed research will continue in the future.

References

1. Hussain, M., BaniMelhem, H.: Impact of human resource management practices on employees' turnover intention in United Arab Emirates (UAE) health care services. Int. J. Inf. Syst. Serv. Sect. **10**(4), 21–41 (2018)

2. Ai, H., An, S., Zhou, J.: A new perspective of enterprise human resource management. Hum. Resour. Manag. Serv. **001**(001), 1–7 (2019)
3. Seo, J.: Board effectiveness and CEO pay: board information processing capacity, monitoring complexity, and CEO pay-for-performance sensitivity. Hum. Resour. Manage. **000**(3), 373–388 (2017)
4. Belizon, M.J., Morley, M.J., Gunnigle, P.: Modes of integration of human resource management practices in multinationals. Pers. Rev. **45**(3), 539–556 (2017)
5. Liu, X., Yue, X.: Design and implementation of hospital human resource information management system. Autom. Instrum. (1), 117–118 (2016)
6. Hao, L.: Talking about hospital human resource information management. Hum. Resour. Manage. **116**(05), 224–225 (2016)
7. Lu, T., Wang, J., Cheng, L.: Thoughts on the basic work of enterprise human resource information management. Sci. Inf. Technol. **001**(014), 135–136 (2017)
8. Qin, N.: Design and analysis of BS institution personnel file information management platform (human resource information management platform). Hum. Resour. Manage. **07**(129), 361–362 (2017)
9. Yun, Z.: Research on human resource information management system based on collaborative filtering algorithm. Electron. Des. Eng. **025**(003), 23–27 (2017)
10. Guo, H.: Analysis and design of human resources information management system for natural gas enterprises. China New Telecommun. **021**(004), 104–105 (2019)

The Application of Virtual Reality Technology in College English ESP Oral Teaching

Xuehong Yuan[✉] and Dongxu Li

Air Force Aviation University, Changchun, Jilin, China

Abstract. The effect of traditional Chinese oral English teaching is not ideal, and the students' oral communication ability cannot meet the actual needs. Language teaching and learning must be carried out in a certain environment. The language environment based on virtual reality (VR) technology plays a pivotal role in modern language teaching. This article first proposes the theoretical basis of VR technology applied to college oral English teaching, and then analyzes The characteristics, advantages and disadvantages of VR technology applied to college oral English teaching are discussed, and finally, the current use of VR technology for classroom oral teaching is proposed.

Keywords: Virtual Reality (VR) · Virtual technology · College English · Oral ESP · English teaching

1 Introduction

For a long time, traditional Chinese teaching concepts believe that the learning process of spoken English is the spoken English knowledge process, that is, the abstract and de-contextualized knowledge in spoken English books is taught by teachers [1]. However, that spoken English is a very social subject. It is produced in actual communication and will also be used in actual communication. Therefore, learning spoken English knowledge in abstraction from specific is bound to enter the passive learning mode of rote memorization, and because the knowledge formed by short-term memory cannot be communicated in real time and effectively in real situations, the learning effect is always not ideal [1].

2 VR Technology is Applied to the Theoretical Basis of College Oral English Teaching

Teaching with VR technology as a medium is mainly based on constructivist theory. Constructivist learning emphasizes in real situations, that is, in the real social. In traditional oral English teaching, it is often impossible to create a natural, meaningful, and authentically English learning environment. VR technology has been completely improved [2]. To overcome this defect, as shown in Fig. 1.

J. Abawajy et al. (Eds.): ATCI 2021, LNDECT 81, pp. 685–691, 2021.
https://doi.org/10.1007/978-3-030-79197-1_98

Fig. 1. VR oral English teaching class

Behaviorist theory is also one of the theoretical foundations of VR. Behaviorism believes that learning is a process of trial and error. With the gradual increase in learners' correct responses, a solid stimulus and response connection will eventually be formed [2]. In a more real and safe virtual environment, learners can continue to be stimulated by various situations, without having to estimate the results of mistakes made in the real situation, and bold trial and error. This is also one of the advantages of VR in oral English teaching.

In addition, the experiential learning theory emphasizes the individual needs of learners. In the network-based VR database, the huge number of tasks and rich task types can meet the self-needs of learners; it creates an environment for oral English learning close to real and natural. Providing learners with opportunities for communication, interaction and meaning negotiation has undoubtedly become the theoretical basis of VR [3].

3   The Application of VR Technology in Oral English Teaching

Immersiveness means that in a virtual environment, participants get a variety of experiences consistent with the real environment, such as sight, hearing, touch, and even smell, so that they can immerse themselves in the virtual world [4]. Interactivity means that through input and output devices, participants can interact with various objects in the virtual world in real time, thereby influencing each other. Conceiving refers to the fact that participants can reproduce the real environment in VR, or they can conceive an objectively non-existent environment at will. Participants do not passively receive information, but generate new ideas and ideas, and actively explore information.

Based on the three characteristics of VR, VR technology has three main positive effects on oral English teaching, namely, improving the interest of learners in oral English learning, enhancing learners' language learning motivation, and enhancing learning autonomy [4].

4   The VR Series of Courses are Applied to Oral English Teaching

From the perspective of the application of the series of VR courses to oral English teaching, the VR course + oral English teaching presents six technical characteristics and advantages, namely, multi-language, multi-scene, multi-function, multi-course, and multi-mode And multi-dimensional evaluation.

4.1   Multilingual

The VR series of courses support multi-language learning and can provide iconic scenes in different countries according to different languages. For example, different national scenes or iconic scenes of the country can be made according to different languages. Students can interact in these scenes and use different languages to explain. In the course of teaching, the system provides panoramic videos to help learners understand the current country's general situation, cultural values, social customs and etiquette, communication styles, business culture and conventions and other knowledge content [5]. Teachers can use tokens to control the perspective in the panoramic video, presenting the full picture of the current country or landmark building in 360 degrees, so as to increase students' cross-cultural knowledge reserve and enable students to more intuitively understand the cultural differences of different countries.

4.2   Multiple Scenes

VR series courses can provide a variety of business scenarios, allowing teachers to customize courseware based on the teaching content to generate their own VR curriculum products that is, using VR technology to develop the system, a large number of business scenarios can be set in the system, including offices and companies [6]. The front desk, the company's centralized office area, offices, meeting rooms, hotels, coffee shops, airports, etc., have a large amount of business communication corpus built in. Students can learn and practice business English in different scenarios, as shown in Fig. 2.

Fig. 2. VR English course method

4.3 Multi-function

VR series courses provide single-person training and multi-person training products to meet different teaching and training requirements, such as: VR English oral speech contest situational teaching training courses, learners can choose single-person training mode to develop topics Speech or impromptu expansion speech, you can also enter into multi-person training to conduct comprehensive knowledge response [7]. In the training scenario, learners can experience immersive information such as the crowded hall, the audience's response, the judge's questions, etc., to enhance the learner's cognitive load on the interference credibility, in speech recognition, emotional control, psychological Photo albums, perception and non-verbal representations create highly simulated experiences for learners, thereby effectively improving practical language knowledge and skills [7].

4.4 Multiple Courses

The VR series courses cover multiple language learning courses, which can be oral English in the workplace, oral business English, oral English, oral English interpretation, oral English speech, transnational cultural experience, etc.; it can also be workplace situations and business in different languages Situations, oral practice situations, interpreting, speeches, etc. Teachers can develop and set up more VR language learning courses according to professional needs, and create personalized school-based language courses [8].

4.5 Multi-mode

The VR series courses provide a variety of learning modes, including self-study mode, follow-up mode, challenge mode, etc. Students can perform sentence learning and training in return visits, hard-listening, and barrier-breaking modes; the system has a large amount of oral English learning content built in. Carry out learning and practical training [8]. According to different training methods, it can be divided into two modes, one is learning mode and the other is practical training mode. The learning mode includes a variety of training modes, such as: follow-up mode, role-playing and other modes, as shown in Fig. 3. The practical training mode can be divided into different levels of difficulty [9]. The system automatically records the latest scores and the best historical scores. Students can choose to replay, re-listen, and learn and train sentences in the past mode.

1) Audiovisual Resolution & Fidelity   2) Audiovisual Immersion & Spatiality

3) Non-Audiovisual Senses   4) Input & Effectors   5) Live & Social

Fig. 3. VR oral English learning multi-mode

4.6 Multi-dimensional Evaluation

The VR series of courses provide intelligent voice evaluation technology to score learners' pronunciation in real voice, intonation, fluency, speaking speed and accuracy [9]. You can also add peer ratings, teacher ratings, learner self-evaluation and other functions to improve the objectivity and effectiveness of the evaluation. In addition, learners can also adjust the personalized learning curve based on personal score feedback, and carry out targeted training and improvement of weaknesses.

5   The Disadvantages and Prospects of VR Technology Applied to College Oral English Teaching

VR technology has been applied to the field of education for more than ten years, but the application of oral English teaching in colleges and universities is still in its infancy [10]. Although a group of pioneers, such as Hangzhou Normal University, Dalian University, Dalian Foreign Languages University, Tianjin Foreign Studies University, Chongqing Technology and Business University, Southwestern University of Finance and Economics, Shenzhen Vocational and Technical College of Information Technology, and China Agricultural University, have actively responded and carried out practice, no matter how From the perspective of breadth and depth, they have not yet reached the stage of popularization. The analysis of factors affecting the VR technology in oral English teaching mainly includes the following three levels:

5.1   School Level

The price of VR technology equipment is high, and the maintenance cost is not low. To realize the manpower deployment, there are greater financial difficulties for ordinary colleges and universities. In addition, oral English courses are generally not regarded as important subjects in colleges and universities [10]. With limited funds, many colleges and universities will inevitably choose to invest in what they think is more important.

5.2   Technical Level

The current VR technology is not yet fully mature, the relatively complete VR oral English learning task library is less, and there is a certain gap between the publicity and actual use of many equipment. Judging from the application situation of existing colleges, the application of VR mostly stays in the "experience" stage [11]. There is still a long way to go to achieve seamless connection with teaching.

5.3   Teacher-Student Level

The majority of teachers and students are not capable of applying modern information technology, and their knowledge and application of VR technology are even rarer. Many teachers think that VR technology is just a gimmick and has no real effect in actual operation. Instead, it wastes a lot of time on equipment debugging. Some students believe that VR will bring a series of physical discomforts [11]. After wearing VR equipment, they feel dizzy and unbearable, and they cannot adapt to a series of health problems.

6   Conclusion

In short, the application of VR technology to college oral English teaching is an innovation and reform of traditional college English teaching, with unprecedented advantages. However, VR technology is still in its infancy, and its hardware and software need to be improved and upgraded to meet the actual needs of teachers and students in oral English teaching.

References

1. Li, L.: Research on the application status in the field of oral English education. Chongqing Norm. Univ. **12**, 76–78 (2017). (in Chinese)
2. Ma, Ch.Y., Chen, J.L.: Computer-assisted language teaching based on virtual reality. Audio Vis. Foreign Lang. Teach. **9**(02), 28–31 (2011). (in Chinese)
3. Ma, H.: The application of virtual reality technology in oral English teaching. Intell. Inf. Technol. Appl. Soc. **7**, 138–141 (2014). (in Chinese)

4. Ji, H.Q.: Talking about the application of virtual reality technology in subject teaching. China Mod. Educ. Equip. **12**, 47–49 (2016). (in Chinese)
5. Cui, L.: Overview of virtual reality technology and its applications. Fujian Comput. **10**(09), 27–35 (2018). (in Chinese)
6. Zheng, Y.Q.: Virtual reality technology and language teaching environment. Teach. Chin. World **12**(04), 43–45 (2019). (in Chinese)
7. Zhao, R.Q., Dong, G.D.: The application of virtual reality technology in college oral English teaching. North. Lit. **11**, 95–96 (2019). (in Chinese)
8. Liu, W.Q.: The application of virtual reality technology in higher vocational English teaching under the background of information technology. Campus Engl. **10**(03), 194–197 (2019). (in Chinese)
9. Fang, Y.N.: Using virtual reality technology in college oral English teaching. Shan Hai Jing **8**(21), 20–23 (2016). (in Chinese)
10. Lan, Y.K.: Research on the application of 3D virtual reality technology in higher vocational business English teaching. Campus Engl. **12**, 55–58 (2018). (in Chinese)
11. Wang, X.H.: Exploration of elementary english teaching supported by virtual reality technology. Educ. Theory Pract. **11**(06), 86–88 (2018). (in Chinese)

Application of Front-Projected Holographic Display + Artificial Intelligence in Theory Teaching

Yan Zhang[✉]

School of Management, Guangzhou College of Technology and Business, Guangzhou, Guangdong, China

Abstract. With the advancement of science, teaching methods and teaching tools have also experienced tremendous development. However, compared with practical teaching, the innovation and development of theoretical teaching methods and teaching tools have been slower. The development of modern science and technology provides technology and possibilities for innovative research and exploration of theoretical teaching methods and teaching tools. Based on the development results of front-projected holographic display and artificial intelligence technology, this paper proposes a new theoretical teaching application model that combines the two technologies, and discusses the future challenges that this model may face and the technical problems that need to be solved.

Keywords: Front-projected holographic display · Artificial intelligence · Theory teaching

1 Introduction

With the continuous advancement and development of science and technology, teaching content, teaching methods and teaching tools have been greatly improved and improved. However, compared with practical teaching, theoretical teaching tends to be slower in the innovation and development of teaching methods and teaching tools. Theoretical knowledge is generally the beginning and foundation of professional learning. However, theoretical knowledge is often abstract and boring, and there is a large gap between it and actual life, which makes it difficult to be understood. Simple rote memorization and cramming-style theoretical teaching are proved easily make student have learning burnout. At the same time, if the teaching methods and teaching tools are backward, and the teaching process is boring, it is difficult for the teachers to teach, and the teachers are also prone to burnout [1–3]. Therefore, for relatively abstract theoretical knowledge content, its teaching methods and teaching tools still need to be continuously researched, explored and improved.

The development of modern science and technology provides technology and possibilities for innovative research and exploration of theoretical teaching methods and teaching tools. In the research of exploring theoretical teaching methods and

© The Author(s), under exclusive license to Springer Nature Switzerland AG 2021
J. Abawajy et al. (Eds.): ATCI 2021, LNDECT 81, pp. 692–697, 2021.
https://doi.org/10.1007/978-3-030-79197-1_99

teaching tools through modern science and technology, Yang Juanjuan studied the application of the "Internet + Chinese Classics" model, and proposed that relevant materials can be searched through the Internet, and multimedia videos can be used to broadcast to students in the class, in order to enhance the attractiveness of the class from the audio-visual angle [4]. Wang Lixue and Wang Xianghong explored the path of using online short videos to carry out ideological and political education of college students. They believe that college students have a strong demand for visual culture as a new way of cultural communication, so it is better to use online short videos for college students' ideological and political education [7]. Based on the "Internet +", Liu Xue explored the "micro-ideological and political" mode of colleges and universities, and proposed that the Internet can be used to establish an ideological and political education platform. In this way, ideological and political theory education and teaching can be carried out through media such as pictures and videos, thereby promoting ideological and political education and teaching quality in colleges and universities [8]. Zhu Huaxi, Tian Min, and Zhao Fang discussed how to use head-mounted 3D glasses to break through the limitations of space and time based on VR technology to increase students' interest in learning [5]. Hong Jiang and Li Zuokun believe that the "5G + front-projected holographic" model can further promote the development of education informatization in primary and secondary schools. They believe that the development of 5G technology can greatly increase the speed of the Internet and bring hope to the progress of front-projected holographic display, which is the key to realize the reform of the future theoretical teaching [6]. In 2018, Imperial College London promoted their front-projected hologram teaching method of their business school to the world. Students and teachers can interact with each other and have a good sense of presence. In recent years, many teachers have explored the application of artificial intelligence in teaching [10–12].

It can be seen that from the initial use of the Internet to collect teaching materials, to the use of online short videos to enhance the teaching effect. From the application of Internet platforms, websites and other complexes, to the application of VR technology, front-projected holographic display and artificial intelligence in teaching. The application of modern science and technology in teaching is gradually developing, and the technology is becoming more and more advanced and mature [9]. These studies and explorations have important reference value for improving the teaching methods, teaching tools and teaching effects of theoretical teaching. However, the specific application model of front-projected holographic display and artificial intelligence in theoretical teaching, as well as the challenges and problems to be solved in the future, have not been fully explored. No scholars have combined front-projected holographic display and artificial intelligence to design and research the application model of theory teaching. Therefore, this paper aims to design and study the application model of theoretical teaching that combines front-projected holographic display and artificial intelligence, which can give play to the advantages of the two technologies, and discuss the technical challenges and problems it may encounter, as well as the technical problems that need to be solved in the future.

2 Analysis of Application Mode of Front-Projected Holographic Display + Artificial Intelligence in Theory Teaching

Front-projected holographic display, also known as virtual imaging technology, is a type of 3D technology. It refers to a technology that uses the principle of interference to record and reproduce the true three-dimensional image of an object. It can break through the limitations of time and space, and achieve face-to-face communication and immersive effect. Compared with the VR technology with glasses, front-projected holographic display is a naked eye 3D technology, which does not require wearing glasses which can make the teaching arrangement easier. At the same time, it has a stronger sense of three-dimensionality that makes the situation more realistic, and the color and clarity almost indistinguishable. Artificial Intelligence (hereinafter referred to as AI), as a branch of computer science, can, to a certain extent, replace human intelligence to complete some complex tasks. Combining front-projected holographic display and artificial intelligence, and applying it in theoretical teaching, can realize immersive, more realistic, and more experiential human-computer interaction. Among them, front-projected holographic display is responsible for restoring and presenting real people and scenes, and AI is responsible for real-time interaction and communication on site. This interaction and communication mainly refers to human-computer interaction and communication. If the characters interacting and communicating are virtual, unconnected or historically restored, then, AI can be used to achieve interaction and communication that cannot be achieved in reality. Whether you are an ancient or modern person, a pioneer in theory or a master of scientific research, whether you are in the same city or far away in a foreign country, you can achieve zero-distance interactive communication. For characters, face-to-face interactive communication can be realized. For the environment, interactive communication between people and the environment can also be realized. Through experiential education and immersive education, teaching interest and teaching effect can be improved and deepened. the sense of experience when interacting can be deepened.

Based on front-projected holographic display and AI, this paper attempts to innovate conventional teaching methods in theoretical teaching, and a new teaching mode realized by new technologies. If thinkers such as Marx and Engels can personally describe their thought development process, explain each theoretical knowledge point and its thought dynamics, and can give explanations and answers to the audience's questions at any time, and realize on-site communication and interaction, this will greatly subvert the original teaching model, which will have an important impact on the improvement of students' interest in learning and the effect of teaching. The technical path to realize this theoretical teaching model is shown in Fig. 1.

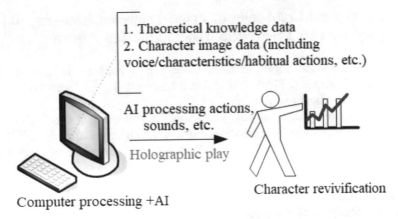

1. Theoretical knowledge data
2. Character image data (including voice/characteristics/habitual actions, etc.)

AI processing actions, sounds, etc.

Holographic play

Computer processing +AI

Character revivification

Fig. 1. The path of front-projected holographic display + AI to achieve interaction with historical figures

In this model, the first step is to input and store all useful information of theoretical knowledge founders, pioneers or other historical figures (including theoretical knowledge data, character image data, voice, character characteristics, habit actions, etc.) in the computer, and model by computer and AI technology to truly show the appearance and behavior characteristics of historical figures, the language and actions of theoretical explanations and the realization of human-computer dialogue. After that, using the front-projected holographic display to broadcast the historical figures through the medium, which can realize the teaching mode of the historical figures personally explain the theoretical knowledge. As shown in Fig. 1, through the combination of the two technologies, the founders, pioneers and important developers of theoretical knowledge are truly showed. They can narrate face-to-face of theoretical knowledge and exchange and interact, which perfectly realized the virtual dialogue with students.

The model of the show-up and interaction with historical figures realized by front-projected holographicdiplay + AI is mainly to improve the methods and tools of theoretical teaching, and improve students' enthusiasm and teaching effect. Of course, this model can also be applied to other areas of education and teaching, such as distance education. At present, the realization of distance education mainly relies on the form of remote meeting or flat video. Based on front-projected display and AI technology, it can realize remote teaching models such as remote situational teaching, remote answering questions and face-to-face communication. With the help of man-machine communication of AI, remote teaching and Q&A communication with no limit on the number of people and classes can be achieved. It is the teaching models that are very worth promoting and exploring. They will be much better than remote meetings or flat video teaching.

3   The Technical Challenges of Front-Projected Holographic Display + AI in Theoretical Teaching in the Future

Although the above-mentioned teaching mode brings hope to the innovation and development of boring theoretical teaching, at present, it still faces problems and challenges in theoretical teaching.

First of all, front-projected holographic media still needs technological breakthroughs. According to the ideal situation, projection with air as the medium is the most simple and true. Unfortunately the technology can't use air as the medium for projection. It needs media such as water mist, holographic film, most of which cannot be used or maintained in teaching situations. At the same time, front-projected hologram with media often have requirements for the viewing angles. Many viewing angles cannot realistically display the projected image, and the projection effect is greatly affected.

Second, The restoration and interaction of theoretical knowledge founders, pioneers, or other virtual characters requires simulation or reshaping of appearances, sounds, etc. The degree of fidelity will affect the experience of theoretical knowledge teaching. The behavioral and vocal characteristics of real figures' speech are often fresh and unforgettable for a long time. Therefore, the appearance, sound, etc. that need to be recorded and restored must be authentic, which poses new challenges to AI technology, and has higher requirements for the collection and storage of character information.

Third, Human-computer interaction issues. Although the current artificial intelligence technology can realize man-machine dialogue, it is still relatively blunt. For the set dialogue content, the communication is relatively smooth, but it is impossible to adapt to the characteristics of the figures.

The last but not the least, storage and transmission. Holographic images need to store and transmit much more content and information than flat pictures and videos. Similarly, AI also needs a lot of information to realize character restoration and human-computer communication. The storage and transmission of the information has higher requirements on hardware devices. If the speed of the corresponding computer cannot be timely or stable, then smooth "face-to-face" communication cannot be achieved.

4   Conclusion

Human beings are progressing, and education is also progressing. How to make boring theoretical teaching more interesting and how to make education and teaching work more efficient is a problem that the world has been pursuing and exploring. With the help of science and technology, to make the original abstract theoretical teaching more vivid and more attractive, it requires continuous experimentation and innovation. I hope that the theoretical teaching model of front-projected holographic display + artificial intelligence proposed in this paper can bring more new inspiration and discussion.

Acknowledgement. Project Foundation: 2020 school-designed ideological and political curriculum "Four One" construction pilot project of Guangzhou College of Technology and Business (project number: KCSZ202001).

References

1. Liu, X.: The analysis of the problems and countermeasures of cultivating interest in ideological and political theory courses in colleges and universities. Think Tank Era **02**, 34–36 (2019)
2. Zhu, C.: Analysis of the causes of college students' learning burnout and countermeasures. Young Soc. **30**, 108–109 (2020)
3. Jiang, J.: Academic debate and enlightenment of "indoctrination theory" – reflections on the educational method of marxist ideology and theory. Party Build. Ideol. Educ. Sch. (High. Educ. Vers.) **09**, 44–46 (2019)
4. Yang, J.: The ideological and political education model of college students from the perspective of "Internet + Chinese classics". Yangtze River Ser. **36**, 24–25 (2020)
5. Zhu, H., Tian, M., Zhao, F.: Integrating red culture into ideological and political education study in colleges and universities with VR technology. J. Jiujiang Vocat. Tech. Coll. **04**, 46–48 (2020)
6. Wang, H., Li, Z.: "5G+ front-projected Holographic display" promotes the informatization of primary and secondary education. Educ. Inf. Technol. **10**, 03–08 (2020)
7. Wang, L., Wang, X.: Exploration of the path of using short online videos to carry out ideological and political education of college students. J. Hubei Inst. Soc. **04**, 71–74 (2020)
8. Liu, X.: Exploration of the "micro ideological and political" mode in universities against the background of "Internet +". Mod. Vocat. Educ. **01**, 06–07 (2021)
9. Wang, X.: Analysis on the reform and application of online teaching in colleges and universities based on curriculum ideology and literacy education. Mod. Vocat. Educ. **01**, 130–131 (2021)
10. Fan, Z., Liu, Y.: The application of artificial intelligence in the teaching of ideological and political theory. J. Guizhou Educ. Inst. **36**(01), 60–64 (2020)
11. Liao, Z., He, M.: Research on improving the teaching ability of ideological and political theory teachers in the era of artificial intelligence. J. Heilongjiang Vocat. Inst. Ecol. Eng. **33**(04), 154–156 (2020)
12. Han, C.: Research on statistics teaching mode under the dual influence of big data era and artificial intelligence. Course Educ. Res. **40**, 31–32 (2019)

Research and Development of Network Teaching Platform for Rock Climbing Course in Colleges and Universities

Jing Huang$^{(\boxtimes)}$, Yuan Ping, and Mingquan Zhu

School of Physical Education, China University of Geosciences,
Wuhan, Hubei, China

Abstract. In view of the development of information analysis technology and a number of advantages of network teaching, this article studies the main draw lessons from modern education information technology principle and multi-disciplinary theory, according to the actual need and development trends of university teaching climbing, using the research methods of literature study, logical analysis methods and software compiling method, prepared to develop a set of suitable for rock climbing teaching network teaching platform. It is convenient for teachers to teach and students to study independently, and further promotes the promotion and popularization of rock climbing in colleges and universities in China.

Keywords: Information technology · Network teaching · Rock climbing course

1 Introduction

Human society has entered the information age, computer technology has been widely promoted and used, computer and other multimedia technology at an amazing speed into thousands of families and affect people's life and learning style, it has become an essential tool in people's life communication and learning. As an important part of educational reform, educational informationization has attracted extensive attention from all over the world, and also attracted the attention of colleges and universities. In the traditional physical education teaching, similar to other traditional teaching, the teaching mode of teacher as the center is adopted, and the leading role of teacher is overemphasized. The teaching mode of teacher's explanation, demonstration and students' practice is still the traditional model of physical education [1]. Although this teaching mode can well reflect the leading role of teachers, it deviates from the leading role of students' active learning. Students passively accept the teacher's "command" in class, which seriously affects the play of students' subjective initiative, restrains the development of students' innovation ability and practical ability, and limits the improvement of students' independent learning ability [2]. The reform of teaching information has broken through the traditional form of teaching and therapy to some extent, and multimedia network teaching has been launched as a new form of teaching.

J. Abawajy et al. (Eds.): ATCI 2021, LNDECT 81, pp. 698–704, 2021.
https://doi.org/10.1007/978-3-030-79197-1_100

As a highly efficient and interactive educational information media, network courseware has been widely used in the teaching of various subjects in colleges and universities, and has gradually developed into one of the most important modern educational technologies. Network courseware with its digitalization, diversification, visualization and many other advantages, has greatly promoted the reform of college physical education. Obviously, the network courseware makes up the deficiency in the traditional physical education teaching mode, the course relies on the network as the platform, flexibly uses the network teaching technology, thus updates the teaching section, changes the teaching method, improves the students' independent learning ability and cultivates the students' interest in learning.

Rock climbing is one of the expanded sports in colleges and universities [3]. It is a new sport emerging in China in the 1980s. It can broaden students' horizons, promote students' physical and mental health and social adaptability, and has been favored by the majority of college students in recent years. Rock climbing has always been in a marginal state in the development of colleges and universities because of the limitation of its site equipment and teachers. At present, only a few schools in more developed areas set rock climbing as a sports teaching project. Although rock climbing has been widely carried out in colleges and universities in recent years, there is still no unified standard rock climbing course materials in colleges and universities in China, let alone vivid and vivid network teaching courseware for students to study by themselves after class. This situation, to some extent, hinders the popularization and development of rock climbing in colleges and universities [4]. In view of the development of information technology and many advantages of network teaching, according to the actual needs and development trend of university rock climbing teaching, this research is to develop a set of network courseware suitable for the use of Chinese university rock climbing course teaching, so as to facilitate teachers' teaching and students' independent learning and use.

2 Guiding Ideology and Basic Principles of the Development of Network Teaching Platform

2.1 Guiding Ideology for the Development of Network Teaching Platform

This project network teaching courseware in the design, strive to make the teaching content is rich, make a teacher classroom teaching auxiliary courseware, can provide students and sports lovers independent study research type of multi-functional network teaching courseware [5]. Due to the late start of rock climbing in China, there is no unified teaching material for reference in colleges and universities. There are very few college teachers with relevant level coaches. Many school teachers of rock climbing are self-taught in the late stage, and they teach only after a short training. They are not familiar with the theory and practical operation of rock climbing. Therefore, when collecting the content of the teaching platform, we should try to refer to a variety of climbing textbooks and materials to ensure the accuracy and richness of the content of the platform.

Considering that this network platform is mainly used by the academic institutions of higher learning in our country, so the platform is designed to be colorful, easy to use, beautiful interface, and practical. Cultivate students' interest in learning through beautiful pictures and wonderful climbing competition videos. Rock climbing course is a course that combines theory and practice closely. When designing and dealing with the layer structure of web pages, special attention should be paid to the design of learning environment and atmosphere [6]. A good learning atmosphere can help learners to improve their interest and enthusiasm in learning. Rock climbing is different from other general sports. It belongs to the high-altitude project in the extension project. Many operations need to be completed on the high rock wall, which has certain risks. Climbing requires high accuracy and safety in the operation, so the design of platform content should be as detailed and meticulous as possible.

2.2   Basic Principles for the Development of Network Teaching Platform

The first is the scientific principle. The network teaching platform of rock climbing course in China's ordinary colleges and universities should accurately reflect the characteristics of rock climbing [7]. The knowledge system presented by the platform must be scientific and cannot violate the basic principles of rock climbing teaching, and the content reflected by the courseware is vivid and intuitive. The selected material is in line with the objective reality, and the video or picture of the technical action in the platform must be standardized and accurate. Text, pictures, video, audio and other media use reasonable, appropriate collocation, clear hierarchy. The second is the principle of effectiveness. The principle of effectiveness is one of the important principles of designing and making courseware. It is the starting point and destination of designing and making courseware. It requires full consideration of teaching content and learners' actual situation when designing and making courseware. To this end, first of all, it is necessary to make clear the teaching objective of rock climbing course, prepare to grasp the teaching content, especially the key points and difficulties in teaching [8]. The courseware produced can meet the needs of students and has definite effectiveness. The third is the principle of operability, the purpose of making network teaching platform is to help teachers to assist teaching or students to conduct independent learning, so the design of the platform must be operational. So that teachers in the classroom teaching process, will not occupy too much time to carry out complex operations, improve the efficiency of teaching. Students can also easily complete self-learning through clear navigation and eye-catching buttons during class time. The fourth is the principle of openness. After the design of the network teaching platform is completed, teachers can change, replace and update the courseware content in time according to the needs of teaching in the process of operation. Network teaching is a dynamic process of communication, teachers can change the teaching content and methods at any time according to students' understanding of the classroom content and students' feedback. Make the courseware can meet the student's request better, improve the teaching quality. Five is the principle of inspiration, the design and production of rock climbing network teaching platform should closely around the teaching objectives, give full play to the leading role of teachers and students. In the platform, vivid and intuitive pictures and video materials are used to stimulate students' association and imagination, so that students can be inspired while learning. Activate the student's brain, train the student's innovative thinking.

3 Research and Development of Network Teaching Platform for Rock Climbing Courses

3.1 Basic Conception of the Development of a Network Teaching Platform for Rock Climbing Courses

The online platform divides the climbing course into four parts. The first two parts include the theoretical part and the practical part [9]. The content of these two parts is planned and designed respectively in the platform, and the overall layout of the platform is fully considered. In addition, the part of picture appreciation and video appreciation is designed for the teaching needs, so that learners can have a deeper understanding and understanding of rock climbing. According to the content of the page to present, to determine the types of subcontracted materials, specific content and layout of the page scheme, and finally write the design of the script. Therefore, the network teaching platform mainly considers the overall design of courseware content and page design.

3.2 Demand and User Population Analysis

Due to the late start of Chinese rock climbing, there is still a big gap between China and developed countries. At present, there is no unified teaching material for reference in the teaching of rock climbing in colleges and universities, and there is a lack of rock climbing teachers in colleges and universities. Up to now, the Chinese Mountaineering Association has begun to accelerate the training of climbing teachers in colleges and universities in order to make up for the shortage of climbing teachers. Teachers came from all over the country, most of whom had no knowledge of rock climbing and were from other disciplines such as track and field, gymnastics and so on. Through a brief training, the teacher can only obtain relatively shallow weak theoretical knowledge and climbing skills, through the survey some of the university teachers eager to have a set of network teaching platform about climbing reference for them in the teaching, ordinary university students also want to through the network platform in the form of learning and understanding of the knowledge and skills about the rock climbing. In view of the many advantages of network teaching and the needs of teachers and students in China"s colleges and universities, in this form, in order to meet the needs of teachers and students in China's ordinary colleges and universities, we hereby develop a rock climbing network teaching platform. The use of the crowd is the first task to determine the production of the platform, in view of the current China's rock climbing movement is mainly carried out in colleges and universities, network teaching platform as an auxiliary teaching materials, the main use of the object is China's ordinary college students, but also can be college teachers and rock climbing enthusiasts.

4   Selection of Content, Main Content and Framework Design of Rock Climbing Network Teaching Platform

4.1   Selection of Platform Content

Rock climbing is one of the emerging projects in China. Since it was introduced into China in 1987, with the rapid development of the times, it has developed rapidly in China in recent years. Because of the restriction of the site, rock climbing has not been popularized and carried out in China's mass sports for the time being. At present, there are very few papers and teaching materials about rock climbing in China. There is no unified teaching materials for rock climbing in Chinese colleges and universities, and many teaching materials are created and compiled by themselves [10]. Since rock climbing is derived from mountain climbing, most of the content about rock climbing is contained in books about mountain climbing. It was not until 2006 that China's Higher Education Press published a rock climbing teaching book compiled by the New Century Physical Education Textbook Compiling Committee. This textbook is divided into four chapters, mainly including the overview of rock climbing, rock climbing techniques, rock climbing training methods, and prevention and treatment of common rock climbing injuries. In 2009, South China University of Technology Press published a book of climbing skills and training compiled by Professor Han Chunyuan. The book mainly introduces the main equipment of rock climbing, rock climbing rope knot technology, basic rock climbing technology and so on, the introduction of rock climbing is relatively comprehensive. Generally speaking, the teaching materials of rock climbing in China are not comprehensive and unified enough. In 2010, China Mountaineering Association compiled and compiled a rock climbing professional appraisal textbook, which is mainly used for the training of social instructors and coaches. This textbook introduces the content about rock climbing systematically. The content of this network platform mainly refers to the above three textbooks and other related books and materials related to rock climbing, the content of the textbook is screened, summed up the essence of each textbook and materials as the main content of the teaching platform, so that the platform is more systematic and comprehensive.

4.2   Design of Main Content of the Platform

This network teaching courseware mainly includes four parts: theory part, practice part, rock climbing picture appreciation, rock climbing video appreciation. The theory part also includes the overview of rock climbing, the features and functions of rock climbing, the classification and difficulty level of rock climbing, the basic equipment and use methods of rock climbing, and the common knot technology of rock climbing. The practice part also includes the basic climbing technology of rock climbing, the setting of the protection station and the descending technology, the top rope climbing and the upper protection technology, the pioneer climbing and the lower protection technology, rock climbing and the protection technology of rock climbing. The picture appreciation part collects pictures about rock climbing from the Internet to strengthen learners' understanding of rock climbing, and the video appreciation part intercepts wonderful competition videos to cultivate learners' interest in rock climbing.

The content of theory and practice covers the related knowledge of rock climbing, which is also the basic content generally involved in the teaching of rock climbing courses in colleges and universities in China.

4.3   Overall Framework of the Platform Content

The structure of the network platform, in fact, is the organization and expression of the network teaching information, it defines the network platform in each part of the teaching content of the relationship and the way of connection, reflects the whole network courseware framework structure and the basic style. Under the network environment, the structure of the network platform design make full use of various information resources to help learners learn, so that the learners in the network environment according to their own needs freedom of choice of learning content, the network teaching platform will be set up navigation on every page, learners can by clicking on the corresponding navigation related knowledge (Fig. 1).

Fig. 1. Network courseware development flow chart

5   Conclusion

According to the teaching content of rock climbing course in colleges and universities and the basic requirements of making network courseware, this paper has preliminarily completed the development of network teaching platform of rock climbing course in colleges and universities. Rock climbing network teaching platform can provide a reference learning platform for college teachers and students as well as rock climbing enthusiasts, and can also provide reference basis for future rock climbing network teaching platform makers. This paper is only the preliminary idea of platform research, in the future research should be added to the platform after the use of the evaluation index system, the rock climbing network teaching platform should be applied to the specific teaching practice, and the platform should be further modified and improved according to the user's feedback and evaluation of the platform.

References

1. Yuexing, L.: Research on the factors and countermeasures affecting the development of rock climbing in colleges. Contemp. Sports Technol. **30**, 129–130 (2019)
2. Baoshan, Q.: SWOT analysis of micro-course teaching applied to rock climbing course in colleges. J. Harbin Inst. Phys. Educ. **5**, 81–84 (2019)
3. Bingwen, H.: Analysis of the development status of outdoor recreational sports in my country. Sports World **3**, 49–51 (2020)

4. Qi, W.: Opportunities and challenges faced by my country's sports development in the era of big data. Sports Sci. **1**, 75–80 (2016)
5. Jiujian, C.: Analysis of the development strategy of the sports industry under the background of big data. Electron. Test. **24**, 47–48 (2015)
6. Qi, L.: Development strategy of competitive sports system in the era of big data. J. Capital Inst. Phys. Educ. **3**, 156–159 (2015)
7. Qiaoxing, L.: Analysis and optimization of the system structure of the sports big data industry. Sports Sci. Technol. **1**, 100–105 (2020)
8. Lifeng, Z.: Application management of outdoor fitness equipment for all people based on big data of Internet of Things. J. Shazhou Vocat. Inst. Technol. **1**, 17–20 (2019)
9. Urakov, A.L., Ammer, K., Dementiev, V.B., et al.: The contribution of infrared thermal imaging to designing a "Winter Rifle"-an observational study. Thermol. Int. **29**(1), 40–46 (2019)
10. Kobelkova, I.V., Martinchik, A.N., Keshabyants, E.E., et al.: An analysis of the diet of members of the Russian national men's water polo team during the sports training camps. Voprosypitaniia **88**(2), 50–57 (2019)

Decision Tree Algorithm in College Students' Health Evaluation System

Maoning Li$^{(\boxtimes)}$ and Hongxia Yang

City College, Xian Jiaotong University, Xian 710018, Shaanxi,
People's Republic of China

Abstract. Decision method. This paper introduces the theory of decision tree, analyzes the structure of decision tree, and discusses the idea of C5.0 algorithm and its advantages and disadvantages. At the same time, symptoms and factors affecting college students' mental health, C5.0 algorithm is applied to college students' mental health evaluation data. According to the mining results, students' mental health problems can be more deeply understood, It is of practical significance to carry out mental health education for college students.

Keywords: Data mining · Decision tree · C50 algorithm · Mental health

1 Introduction

With the accelerating pace of life and the increasingly fierce social competition, college students are facing more and more pressure of study, life, emotion and employment. The resulting psychological problems are becoming increasingly prominent, which directly affect the healthy growth of students and the stability of the campus, The purpose of this study is to make up for the lack of traditional analysis and statistics of College Students' psychological problems, using data mining technology to study college students' psychological problems, so as to find the main psychological symptoms and factors that affect college students' psychological health, To provide a more scientific decision-making basis for school psychological counseling, to provide treatment programs for students with mental health problems, and to provide new methods of early prevention and intervention for college students' mental health, so as to make the school's mental health education more reasonable [1].

2 The Concept and Construction of Decision Tree

2.1 The Concept of Decision Tree

Decision flow chart, which is used to represent a series of judgment process for people to make a certain decision. This method is used to express rules such as "under what conditions, what results will be obtained".

J. Abawajy et al. (Eds.): ATCI 2021, LNDECT 81, pp. 705–710, 2021.
https://doi.org/10.1007/978-3-030-79197-1_101

2.2   Construction of Decision Tree

The main purpose is to prune those branches that affect the accuracy of pre balance [2]. The data in the test data set divided by data preprocessing will play its existing value at this time, and verify the preliminary rules generated in the process of building decision tree from training set data, The smaller the size of the decision tree, the better, because it has predictive ability, the smaller the tree, the stronger, so we should try to build a small tree as much as possible.

3   Decision Tree Algorithm C5.0

The earliest decision tree algorithm was put forward by hunt CLS (concept learning system). Later, Jr Quinlan put forward the famous ID3 algorithm in 1979. It mainly aims at discrete attribute data.

3.1   Advantages and Disadvantages of the Algorithm

C5.0 algorithm can deal with a variety of data types, such as date, times, timestamps and so on. The speed of data processing is faster and the performance of memory occupation is greatly improved. Due to the use of boosting method, the decision tree generated is smaller and has higher classification accuracy; Generally, it does not need a long training time to estimate; cs.0 model is easier to understand than some other types of models, and the rules derived from the model have a very intuitive explanation; powerful technology is provided to improve the accuracy of classification. However, C50 algorithm is more difficult to predict the continuous fields.

3.2   C5.0 Algorithm Selection

The standard C50 decision tree algorithm of decision tree branches uses the information gain of attributes to determine the standard of decision tree branches and find the best grouping variables and segmentation points. Let s be a sample set, the target variable C has k classifications, freq (CI, s) denotes the number of samples belonging to CI class, s denotes the number of samples of sample geometry s, then the information entropy of set s is defined as:

$$Info(S) = -\sum_{i=1}^{k} \left((fred(C_i, S)/|S|) \right) \tag{1}$$

If an attribute variable t has n classifications, the conditional entropy of the attribute variable t is defined as:

$$Info(T) = -\sum_{i=1}^{k} \left((|T_i|/|T|) \right) \times Info(T_i) \tag{2}$$

The information gain brought by attribute variable t is as follows:

$$Gain(T) = Info(S) - Info(T) \tag{3}$$

4   Application of C5.0 in College Students' Mental Health Evaluation System

4.1   Data Preprocessing

The data of this paper comes from the mental health scale of 5065 students of grade 2012 in a university in Fujian Province, including 2263 boys and 2802 girls. The purpose of this paper is to judge the psychological status of college students by answering 104 preset questions of psychological symptoms, Data preprocessing is a very important step in the process of data mining, The data set processed by data mining usually contains not only massive data, but also a large number of noise data, redundant data or incomplete data, For example, students' missing or irregular filling in the test will lead to a large number of noise data in the database, so it is necessary to preprocess the data, which generally requires 70% of the workload in the mining process.

4.2   Establishment of Decision Tree Mining Model

The most famous commercial data mining softwares include spsscementine, etc. SPSS Clementine 20 is selected as the platform for the model establishment and analysis. There are four decision tree algorithms in Clementine: C50, cart, quest and CHAID algorithm. Cart and quest are two branch decision trees, C5.0 and CHAID are multi branch decision trees. The target variables of cart and CHAID algorithms can be continuous or discrete. The target variables of C50 and quest algorithms can only be discrete. Select C50 to build the decision tree model, as shown in Fig. 1.

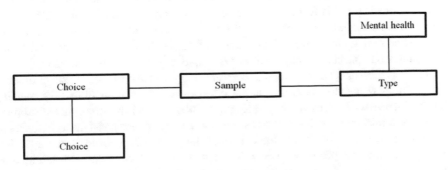

Fig 1. Decision tree C5.0 mining data stream.

5 Problems in the Survey and Evaluation of Students' Mental Health

5.1 Related Problems Due to the Establishment of Institutions

For many years, our psychological counseling work has been attached to the College of humanities. Since 2007, the student Department has set up a counseling center for college students' development, which undertakes part of the psychological counseling and psychological census work. However, due to the scattered personnel and untimely information communication, it is difficult to give full play to the team strength of psychological counseling. Therefore, the integration of resources related to psychological counseling has become an urgent task.

The course of psychology science popularization should be set up. At present, our school offers six courses for students to choose: healthy personality psychology, interpersonal communication analysis, social psychology, psychological hygiene, film text growth psychology and college students' mental health [3]. Through the study of psychology course, college students can strengthen their self-awareness and grasp, and understand the most basic knowledge of mental health. They can not only realize the importance of mental health education, but also understand how to improve their mental health level in college life.

5.2 Due to the Institutional Setup

In terms of mobilizing students, understanding of census data by students and consultants, interviews, and understanding of census work by colleges and students themselves, there are some problems, such as insufficient understanding of census work, and unable to work in a timely and in-depth manner for students. The specific performance is as follows: (1) it is still difficult to form a complete system for the survey and evaluation of students' mental health, which includes computer-based survey, data statistics, result feedback, appointment interview, tracking guidance and so on, Data statistics, result feedback, interviews with key personnel and tracking after interviews are all important contents of the survey and evaluation of students' mental health. At present, there are imperfections in testing, feedback, interview and follow-up counseling, which need to be adjusted. For example, it is difficult to do without missing inspection at present. The missing students are likely to have mental health problems. In addition, there are also loopholes in the interview of key personnel and the tracking after the interview. At present, it is difficult to interview all the important personnel. The author thinks that the first reason is that these students are resistant to defense, so they avoid contacting with psychological counseling institutions. Second, the interview time is not arranged properly. There is no system for follow-up counseling after interview, and there is no corresponding appointment, record and referral method. (2) the screening results need to be further implemented [4]. At present, our school provides a 15 min interview service for the first class of students as soon as the statistical results come out. There are still problems in this process. First of all, there is no guarantee that all kinds of students can be interviewed. Secondly, the screened students are lack of further psychological counseling.

6   On the Perfection of the General Survey and Evaluation System of College Students' Mental Health

The improvement of the survey and evaluation system of students' health status and the effective development of mental health education not only need the change and renewal of educational methods and concepts, but also need the management departments at all levels of the school to look at the work from the perspective of development, overall situation and science. Combined with the actual situation, the author thinks that we should start from the following aspects to improve the survey and evaluation of students' mental health.

6.1   Strengthen the Cooperation Between Departments

The general survey and evaluation of Freshmen's mental health is a systematic project, which requires the close cooperation of the psychological counseling center with all departments, departments and even teachers. In the stage of testing, interview and psychological counseling, all departments need to cooperate with each other. The mental health problems of former college students have aroused widespread concern of college teachers. Most teachers can pay attention to the freshmen's psychological survey and interview, and give support to the related work. However, some teachers do not understand psychological survey and interview, or do not understand psychological counseling and counseling at all, and think that professional knowledge and skills are more important than everything else. In this case, it is difficult to attach importance to and support psychological survey and interview. 3. Strengthen the propaganda of the significance of the survey and evaluation of Freshmen's mental health. The survey of Freshmen's mental health aims to understand the overall mental health of freshmen, provide corresponding mental health education on this basis, and pay attention to students with potential problems, It is a good opportunity for mental health education to provide corresponding services for freshmen, so we should strengthen publicity.

6.2   Institutionalization of Screening Implementation

Freshmen's psychological survey and screening interview is an important routine work of the psychological counseling center. This work must be institutionalized. The actual work must have a process, from the beginning of publicity to the computer testing, data statistics, result feedback, appointment interview and then to the second inspection, follow-up interview, case supervision and other links to form a system. Each task needs to be taken care of by a special person, especially the weak links such as appointment interview, secondary inspection and case supervision of follow-up interview.The interview must be implemented in a systematic way. The implementation of institutionalization should include the following contents: determining how long for the students screened out from the initial interview to have a second interview, appointment by a special person, interview records and referral system, 6. Build a green channel for medical security of College Students' psychological problems. For students with psychological diseases, they should go to the hospital in time for medical treatment, so as to avoid aggravation of the disease. This requires the establishment of a smooth medical mechanism for students' psychological problems in the school, and the referral system should be established between the psychological counseling center and the medical institutions of the school, so that students can be smoothly referred for medical treatment.

In recent years, the psychological counseling center has invested a lot of energy in the "525° mental health publicity month" every year, hoping to make it a window for college students' mental health education through this platform. In a word, psychological census, evaluation and health education are very important basic work in the whole process of the transformation of higher education thinking and mode, And can really implement into the specific work, will make students and their families, schools, society become the ultimate beneficiaries of higher education.

7 Conclusion

This paper analyzes the structure of decision tree, discusses the idea of C50 algorithm and its advantages and disadvantages, and uses Clementine to construct C50 mining model of decision tree, and carries out data mining on College Students' mental health data. According to the mining results, this paper analyzes the main psychological symptoms and factors affecting college students' mental health, and gives some suggestions, It is very helpful to guide the relevant departments and personnel of mental health to make correct counseling plan, assist decision-making, and pave the way for the development of students' physical and mental health.

References

1. Xiaogu, S., Jiacan, W.: Discussion on the early warning and intervention system of College Students' mental health crisis. Xiangchao (second half) **03**, 32–33 (2010)
2. Chunyan, Z.: Research and design of College Students' psychological evaluation system. University of Electronic Science and Technology (2010)
3. Wang, Y.: Some thoughts on the improvement of college students' mental health survey and evaluation system . Mall Modernization **20**, 142–143 (2009)
4. Liang, Y.: Research on college students' psychological archives and evaluation system. Southwest University (2008)

Decision Tree and Data Classification Evaluation for Private Vocational College

YuFang Chen[✉]

Shandong Xiehe University, Jinan 250109, China

Abstract. As an advanced teaching management system, credit system fully respects students' individual differences, teaches students in accordance with their aptitude, emphasizes students' autonomous learning and pays attention to students' personality development, which embodies the educational concept of people-oriented, In addition, the flexibility of course selection, the guidance of learning process, the flexibility of flexible educational system, the broadness of learning content and the adaptability of talent cultivation, etc., have been rapidly promoted and developed, and have been adopted by colleges and universities all over the world and welcomed by master students. The credit system in private colleges and universities is faced with many difficulties, such as lack of teachers and teaching resources. With the help of information technology, making full use of MOOC, a modern teaching and learning mode, is an effective way to solve the problems faced by private colleges and universities in implementing the credit system.

Keyword: MOOC credit system in private colleges and universities

1 Characteristics of MOOC

The International Olympic Committee is a combination of information technology, network technology and high-quality educational resources. The learning platform of the Ministry of education has been the most serious technological change in the field of higher education in the past 500 years. It has chosen learning places, distance learning courses, learning time, learning routes and learning contents all over the world. There is no time and place limit. These characteristics greatly meet the requirements of diversified development of higher education. Compared with the traditional teaching methods of higher education in China, MOOC has the following characteristics: large scale, openness, personalization, etc.

The primary characterization of MOOC is its "massive". There are four connotations of large-scale, one of which is the large scale of students. MOOC platform has broken the barriers of traditional teaching in time and space. MOOC course learning has gone beyond the limit of venue and number of learners. This kind of online classroom without number limit is unprecedented. Second, the scale of the course is large. MOOC platform has a large number of high-quality curriculum resources, providing numerous high-quality courses worldwide. Third, there are more participating institutions of higher learning. Since 2012, more and more colleges and universities around the world have joined the MOOC alliance and participated in the use

© The Author(s), under exclusive license to Springer Nature Switzerland AG 2021
J. Abawajy et al. (Eds.): ATCI 2021, LNDECT 81, pp. 711–715, 2021.
https://doi.org/10.1007/978-3-030-79197-1_102

and construction of MOOC. Most of the elite universities in the United States have joined in the construction of MOOC. European universities have also responded positively [1]. At present, the number of MOOC courses in European universities has accounted for one third of the total number of MOOC courses in the world. In China, the number of colleges and Universities Participating in the construction of MOOC continues to grow. After Peking University, Tsinghua University and Fudan University, colleges and universities have opened characteristic courses on MOOC platform. Fourthly, the number of teachers participating in MOOC is large. Most teachers participate in MOOC construction or course teaching in a team way, and there are many participants. Item bias, which represents the general situation of credit system in private colleges and universities. The formula of prediction scoring model based on formula (1) is as follows:

$$P_{ui} = \mu + b_i + b_u + p^T{}_u q_i \tag{1}$$

Therefore, the new loss function formula to be optimized is as follows:

$$\min \sum_{ui} \left(r_{ui} - \mu + b_i + b_u + p^T{}_u q_i \right)^2 \tag{2}$$

2   The Characteristics of MOOC and Its Influence on the Credit System Reform of Private Colleges and Universities

One is equality. First of all, learning resources are treated equally to all learners, and there is no difference due to age, race, etc. The learning resources provided by MOOC are open, and everyone has the right to choose and learn from these resources. The equality of MOOC is also reflected in the two-way choice of MOOC learning, taking into account the individual needs of students. Education in the traditional sense is "Teacher centered" to a large extent, lacking the freedom of students to choose teachers and courses independently. Under the condition of MOOC, students can choose "MOOC" according to their own hobbies, interests and plans. Teachers should strive for more students. Only by improving their teaching ability and level can they have students to choose from, which fully reflects the equality between teachers and students.

The second is interactivity. MOOC learning attaches great importance to network interaction, emphasizes the communication between learners and teachers, between learners, and learners and knowledge system. The interactivity reflected in MOOC meets the personalized needs in varying degrees. At the same time, the interactivity of MOOC also covers its real-time, timely realizes the learning feedback, enhances the enthusiasm of learners to participate, so as to effectively improve the learning effect of students.

3 The Influence of MOOC on the Credit System Reform of Private Colleges and Universities

The trend of MOOC sweeping the world has been changing the traditional higher education. The influence of MOOC on the development of each university and educational informatization, and the extent to which MOOC can trigger the reform of educational system and mechanism in Colleges and universities depend on whether the educational administrative departments and schools deny MOOC teaching and how to deal with these MOOCS. It is gratifying to note that both the Ministry of education and the provincial education authorities attach great importance to the butterfly effect of MOOC [2]. The opinions especially emphasize promoting the credit recognition and credit management system innovation of online open courses, and encourage colleges and universities to carry out credit recognition, credit conversion and learning process identification in various ways, such as online learning, combination of online learning and classroom teaching. In this context, it is necessary for private colleges and universities to re-examine their own values and missions, face up to the trend of credit system reform.

4 Design of Credit System Based on MOOC

4.1 Guiding Ideology

The guiding ideology of the credit system design of private colleges and Universities Based on MOOC is to conscientiously policy, adhere to the education service for the respect students' right of "learning freedom", and improve students' ability of academic choice. Specifically speaking, in the process of designing and implementing the credit system, private colleges and universities should put morality and talent cultivation in the first place, implement education integration, innovate teaching management mode and talent training mode, optimize the allocation of teaching resources, promote the cultivation of students' personalized development and innovation and creativity, guide students to build their own knowledge system, optimize their knowledge, ability and character structure, and improve the quality of talent training.

4.2 Design Scheme

The credit system scheme of private colleges and Universities Based on MOOC is based on the establishment and improvement of the credit system management system such as course selection system, tutor system, make-up examination and re examination system based on MOOC; it takes the improvement of teaching management, student management and logistics management as the guarantee; takes the construction of modern credit system teaching management information system as the platform to form a dynamic teaching operation mechanism. Make full use of modern information resources, some courses with insufficient teaching resources, make use of the advantages of online education, improve the curriculum system, use MOOC courses, expand the proportion of elective courses, and provide more opportunities for students to choose their own. The credit system is used to measure the students' learning status [3, 4]. Considering the four-year system as a reference, under the premise of graduation with full credits, each student's length of schooling is allowed to change between 3–6 years. Full credits can be advanced to 3 years of graduation, and those without full credits can be postponed to 6 years. The core of the credit system is the course selection system, and the focus of course selection is mainly on general education elective courses. The elective space of public compulsory courses and professional courses is small, and colleges and universities have basically matched these courses and teachers, so the resources and courses in this respect are not very lack. The credit system scheme of private colleges and Universities Based on MOOC is based on the establishment and improvement of the credit system management system such as course selection system, tutor system, make-up examination and re examination system based on MOOC; it takes the improvement of teaching management, student management and logistics management as the guarantee; takes the construction of modern credit system teaching management information system as the platform to form a dynamic teaching operation mechanism. Make full use of modern information resources, some courses with insufficient teaching resources, make use of the advantages of online education, improve the curriculum system, use MOOC courses, expand the proportion of elective courses, and provide more opportunities for students to choose their own. The credit system is used to measure the students' learning status.

5 Conclusion

The rapid development of MOOC also forces Chinese colleges and universities to take the connotative development path with quality improvement as the core. It requires us to emancipate our minds, update our concepts, adhere to the concepts of "openness" and "sharing", focus on students' learning and development, pay attention to the improvement of new information concepts, concepts and capabilities in the Internet era, and dare to break traditional practices, We should reshape the curriculum system, teaching content, teaching methods and methods, reform and even subvert the traditional teaching mode according to the curriculum conditions, constantly build consensus, and comprehensively improve the teaching level and talent training quality.

Acknowledgements. 2020 Vocational Education Innovation and Development Highland Theory Practice Research Special Education Scientific Research School-level Project: Under the background of MOOC, the design of credit scheme of private vocational colleges and universities is studied, Item number: XHXY202008.

References

1. Newman, J.H., Shining, G., et al.: Translated the concept of University. Peking University Press, Beijing (2016). (British)
2. Harvard Committee: Harvard general education Redbook. Li, trans. Peking University Press, Beijing (2010)

3. Feng Huimin general education in modern Chinese universities. Wuhan University Press, Wuhan (2004)
4. Report on the development of online open courses in China, 2013–2016 m. Higher Education Press, Beijing (2017)

Multi-objective Evaluation Software Algorithm for Timbre

Wenji Li[✉]

Baoshan College, Baoshan City 678000, Yunnan, China

Abstract. This paper first studies the universal standards of subjective timbre evaluation in vocal music performance, and then discusses how to quantify and code these standards to input into intelligent objective evaluation software. This paper briefly describes the working principle and implementation method of the core algorithm in the software, as well as the final test results. The core of this paper is to establish the relationship between subjective evaluation and machine objective evaluation, so that the intelligent evaluation results of vocal music can meet the subjective evaluation standards, so as to further promote the accuracy and practicability of the intelligent evaluation system of vocal music.

Keywords: Vocal performance · Timbre evaluation · Subjective evaluation · Objective evaluation

1 On Multiculturalism

Multiculturalism was first proposed in the 1980s and originated in the United States. At that time, there were more immigrants in the United States, which greatly enriched the native culture of the United States and made it present more diversified characteristics. With the continuous development and integration of the cultural field, the economic and cultural environment all over the world have achieved a close interactive state, and the rise and continuous development and application of the network also provide convenient conditions for the cross regional communication of culture. Therefore, under the background of multiculturalism, more harmonious cultural exchanges have been realized among all ethnic groups and regions. Music is an important part of culture and a special way to express culture. People can experience the cultural and emotional content conveyed in music through the creation or appreciation of music, so as to understand the cultural connotation of a country or nation. Pluralistic culture is bound to be more open and inclusive, and the forms of musical works produced in this environment are also more diverse [1]. Music works can be enriched in cultural connotation and become more cultural connotation. With the continuous integration and development of culture, people pay more and more attention to the value of music. Music is gradually changing to modernization and diversification, which correspondingly changes the way of music appreciation and evaluation in the past, and makes the appreciation target change.

2 Subjective Evaluation Criteria of Timbre in Vocal Music Performance and Its Application

At present, the evaluation process of vocal music is mostly artificial, and the evaluation results may have their own preferences and personality characteristics. However, while art is accepted by the public, it also has common aesthetic standards to a great extent. With the development of vocal music singing specialty, Aesthetic identity is obvious. Therefore, in fact, a relatively unified standard has been formed to a considerable extent. Not only the subjective aesthetic of most evaluators tends to such a unified standard, but also the public orientation also influences and recognizes such a standard. Such identity is the inherent characteristic of art, and it is also the feasibility premise of carrying out the unified subjective standard in objective evaluation. Therefore, the purpose of this part of the research is very clear, that is, before establishing the objective evaluation system, we should make clear that the standards of subjective and objective evaluation methods are unified, and they are all based on the basic theory of vocal music and in line with the common aesthetic standards of traditional vocal music singing. In the evaluation system of vocal music, tone, timbre and rhythm are the main reference indicators. If we regard sound as a time series, such a reference index is the most easily materialized characteristic parameter of time series [2]. In the current objective evaluation methods, intonation and rhythm have been studied more, and software has been developed to apply to some entertainment activities or competitions. However, due to the difficulty of modeling the timbre index, there is not a more accurate evaluation software.

3 Phonation Practice

Usually, through the correct voice training, vocal learners with normal voice conditions can complete the corresponding difficult works. However, everyone has his own unique voice. The difference between the congenital voice conditions and training process of each person will cause more or less differences in timbre. For example, the voice with wide and powerful vocal cords and the voice with narrow vocal cord and soft strength, even after the same training process, the final timbre may be greatly different. In the general aesthetic, powerful and brilliant timbre can attract more favor, while the fine and dark timbre is not so popular. This is one of the rules considered in the construction of subjective evaluation criteria model. Second, the same person singing notes in different regions also have different sound quality. The sound quality of the same sound area is more similar. In addition to the congenital conditions, different larynx positions and different degrees of breathing will produce different timbre effects. Generally, the timbre with low and stable larynx position is wider and thicker and rich in metal sense, while the voice with high and unstable laryngeal position is thin and dim. The sound with good respiratory support is stable and firm, while the voice with shallow respiratory support is weak, shaking and unstable. The above two points are the two main aspects of the evaluation of sound quality in this paper.

4 Principle of Objective Evaluation Algorithm

Preprocessing prepares the data for us. Then it enters the signal processing process, which uses a very specialized algorithm. It consists of three parts: model construction, core algorithm and post-processing algorithm. In order to extract timbre features more accurately, we need to conceive the running model first. The main purpose of building the model is to classify and extract different timbre features, so as to give the evaluation value of timbre. This part is also called embedding layer operation. The representation of each timbre is a vector, which may be a combination of single description or multi-dimensional description, corresponding to single feature and composite feature respectively. The function of this part is to map the timbre index to a low dimensional voice vector representation. It is essentially a voice vector table that we learn from the data, and its expression is an embedded matrix, which is obtained from the data training process.

The main purpose of post-processing is to smooth the results so as to avoid the rabbit paying too much attention to a certain feature. This step is the most commonly used method to regularize convolutional neural networks. The principle is to disable the firing of some neurons according to a certain probability. This method can prevent neurons from adapting to a certain timbre feature together and force them to learn more useful features alone. In order to control the error in the whole process, we can define a loss function to measure the error. If we regard the whole iterative process as an optimization process, this function is the objective function of optimization minimization. In this paper, we use cross entropy loss function for the standard loss function of classification problem.

5 Experimental Effect

5.1 Evaluation Software

In the process of our experiment, we have basically realized the preliminary timbre evaluation in vocal music singing, especially the overtone vibration effect with obvious characteristics, and the evaluation results are more consistent with the subjective evaluation. Samples with severe sound jitter and large amplitude generally scored lower, while those with high sound position, small overtone amplitude and strong sound penetration generally scored higher. In the sample, there are singers' voice samples and students' samples. They sing the same pitch and the same language. The voice with high score in the objective evaluation score is not necessarily a famous singer, but may be an ordinary vocal music learner. If the sample used by the famous singer is not in good condition, or the singer's singing level is relatively ordinary, Only because of other reasons and fame, the score will be very low. These are the advantages of objective evaluation. The key problem is that the results of software evaluation are consistent with the subjective evaluation results of vocal music major.

5.2   Image Analysis Algorithm

Before the training of input samples, the precision evaluation value should be normalized. Normalization processing is to convert the data into the number in [0, 1], so as to eliminate the error caused by large difference of magnitude and accelerate the convergence of network training. The maximum minimum method is used to normalize the data, and the function form is:

$$x_m = (x_m - x_{min})/(x_{max} - x_{min}) \qquad (1)$$

$$PID = r_{pup}/R_{iris} \qquad (2)$$

Where: x_m is the data to be normalized; x_{min} is the smallest number in the data series; x_{max} is the maximum number in the sequence 12 samples are selected from data, and sorted and numbered according to their quality, and the precise evaluation indexes of these samples are calculated. Finally, the index values obtained are compared with the sample quality. The results of the two models are compared with those of the GA-1 model [3, 4]. The prediction value of GA-BP neural network model fits well with the expected value. It shows that GA algorithm has good global search ability in optimizing neural network, and provides more accurate weights and thresholds for BP neural network. The fine index analysis results of different quality samples are shown in Fig. 1.

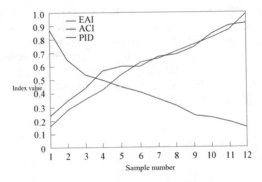

Fig. 1. Analysis results of fine index of different quality samples

5.3   Software Evaluation Effect

In the experiment, the evaluation of the singer's own noise condition is not stable, and the input standard is consistent with the public aesthetic standard. In the experiment, the singing state and vocal singing technology will cover up some of the original timbre characteristics, that is to say, targeted training can effectively make up for the deficiency of natural voice conditions, such as the natural fine voice, The sound penetration and fullness can be enhanced through the adjustment of resonance cavity and strong breathing support, and the score of objective evaluation will be very high. In addition to the systematic evaluation of emotion and style, there will be no systematic evaluation of emotional features.

6 Conclusion

This paper discusses the subjective evaluation standard of vocal music performance and how to use it in the objective evaluation software algorithm in real time. This paper briefly describes the construction process of the simulation neural network in the software algorithm. The software successfully completes the extraction of basic timbre features of sound samples. Through sample training and actual measurement, the software can classify and quantitatively evaluate the main characteristics of different timbres, such as overtone quality, sound saturation, brightness, penetrability, etc. the actual test proves that it can make a more consistent evaluation of the timbre of vocal music singing with the subjective evaluation standard. This software is expected to give birth to the practical products of objective evaluation of vocal music, which will be applied to the daily vocal music teaching and become an important supplement to the vocal music performance evaluation system.

References

1. Aro, H., Mang, L.: Application of timbre evaluation criteria in objective evaluation software algorithm in vocal music singing. Voice Yellow River (11), 61–63 (2017)
2. Yueqin, C.: National vocal music teaching strategies in Colleges and Universities under the multicultural environment. Grand View (Forum) 10, 111–112 (2020)
3. Zhaozhe, H.: Research on Chinese traditional music teaching in multicultural environment. Comedy World (Second Half Month) 09, 14–15 (2020)
4. Jie, G.: Research on the development of Shaanxi art songs in the multicultural environment. Tomorrow Fashion 17, 95–96 (2020)

Risk Assessment of Campus Network Loan Platform Based on BP Neural Network

Lifen Xu[✉]

Yunnan Vocational and Technical College of Industry and Trade,
Kunming 650300, China

Abstract. After the model training and simulation, the data accuracy reaches the expected requirements, which proves that the BP neural network is effective and convenient for the risk evaluation of online loan platform, and proves the effectiveness of the modern evaluation methods such as BP neural network applied to the evaluation of online loan platform, which provides a broader research prospect for the future research. In addition, from the perspective of national regulators, platform itself and investors, this paper describes the use scenarios of the risk assessment method in this paper, and puts forward corresponding suggestions for the control and risk identification of online lending platform.

Keywords: Network loan platform · BP neural network · Risk evaluation

1 Risk Concept

As for the definition of risk, scholars have different definitions of risk due to different starting angles and research fields. However, generally speaking, risk has the following characteristics:

(1) uncertainty. If the results of events can be accurately predicted, then the corresponding countermeasures based on the predicted results will no longer exist, so there is uncertainty about the risk [1]. According to H. Mowbray (1995), risk is the uncertainty of the future, and March & Shapira (1992) thinks that the uncertainty of risk is shown in the non uniqueness of the results of things, and the possible situation of risk can be determined by variance.

(2) Loss, risk has large and small, its result is uncertain, but the final foothold of the size of risk is loss.

(3) objectivity, risk is objective, no matter whether people can feel the risk, risk is not transferred by people's will. Risk can not be eliminated, we can only take various measures to reduce the risk, so that the corresponding loss of risk is minimum.

J. Abawajy et al. (Eds.): ATCI 2021, LNDECT 81, pp. 721–725, 2021.
https://doi.org/10.1007/978-3-030-79197-1_104

2   Characteristics and Functions of Risk Evaluation of Online Lending Platform

To evaluate the risk of online lending platform, we should sum up the characteristics and nature of risk evaluation of online lending platform on the basis of risk assessment combined with its own characteristics. Online lending platform is based on the Internet, which belongs to the scope of website risk assessment. Through the second chapter of the overview of online lending platform risk, this paper summarizes the characteristics of online lending platform risk evaluation as follows: objectivity, reliability, comprehensiveness, timeliness and systematicness. The larger the area occupied by the feature, the more important the feature is in the risk evaluation of online loan platform.

2.1   BP Neural Network

Neural network, as the name implies, is a kind of biological mechanism function of feedback through the form of simulating neurons, imitating the way of thinking of human beings, processing information in the nervous system, simulating classification, and realizing feedback. BP neural network is composed of input layer, some hidden layer and output layer. There is connection relationship between layers, but there is no connection relationship between neurons in the same layer. BP neural network does not need to determine the mapping and weight equation between layers in advance, but constantly adjusts the threshold and weight between layers through the independent learning of BP neural network, until the best effect is achieved [2]. BP neural network belongs to feed-forward neural network. Its learning rule is gradient descent method. The data first propagates forward. The data from the input layer, through the hidden layer, reaches the output layer. After calculating the error, it carries on the back propagation, from the output layer to the input layer, and constantly adjusts until the output of the optimal network model.

2.2   Risk Assessment of Platform Based on BP Neural Network

The input data in the input layer is based on linear combination. The values obtained from the combination are used as the output values of the next layer, which are progressive in turn. The whole process is linear transformation. However, the prediction of many things is not only a linear function. At this time, it is necessary to introduce an excitation function, which can make up for the defects of the linear function and add nonlinear factors to the model, That is, before the linear combination of input data in the input layer, the desired effect can be obtained more accurately through the excitation function. There are two kinds of excitation functions commonly used at present.

Sigmoid function:

$$y = 1/(1 + e^{-x}) \tag{1}$$

Tanh function:

$$y = \tanh x = 2/(1 + e^{-x}) = 2sigmoid(2x) - 1 \tag{2}$$

2.3   Calculation Steps of BP Neural Network

Before BP network training, it is necessary to initialize the BP neural network. It includes initialization input layer, hidden layer, output layer node number, initialization weight vii, initialization input layer and output layer threshold. Set the maximum training times, learning rate and minimum accuracy. BP network calculation process is shown in Fig. 1.

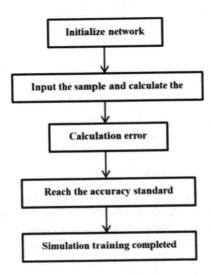

Fig. 1. BP network training process

3   Data Preprocessing

3.1   Positive Treatment of Indicators

Before the model test, the data should be preprocessed to meet the requirements of model evaluation. If the data preprocessing is not carried out, the dimensions of different indicators are different, which will cause the evaluation results to be affected by the indicators with great changes and can not reflect the actual situation. In order to eliminate the adverse effects and make the indicators in the same dimension, it is necessary to normalize the data before normalizing the data. The first step is to normalize the indicators.

In the comprehensive evaluation of things by using multiple indicators, the selected indicators have different characteristics [3]. Some indicators are called positive indicators, and some indicators are the smaller the value, the better the evaluation effect.

This kind of index is called reverse index. There is another kind of index, which is not the bigger the better, not the smaller the better, but the closer to a certain value, the better the evaluation effect, This kind of index is called moderate index. In the process of practical application, we often convert the reverse index and moderate index into positive index, and the process of conversion is also called index.

3.2   Non Dimensional Treatment of Indicators

There is still one step operation after the normalization of indicators, that is, dimensionless processing. Dimensionless data processing is a very important step in data processing. Different evaluation indicators and different indicators have different data. Therefore, if the indicators are not dimensionless, the conclusions drawn from the analysis are meaningless. The dimensionless index can make different indexes change in the same interval, which is convenient for the comparability of different indexes and is conducive to comprehensive evaluation. Compared with the positive indicators, there are more methods to deal with non dimensional indicators, such as comprehensive index method, range change method, high school range change method and so on. In this paper, the most commonly used range change method is selected to eliminate the dimensionless treatment of indicators.

4   Risk Assessment and Grading

4.1   Index Entropy Weight Method

Before BP neural network training, we need to assign the explained variables to the existing data, that is, the expected output results. The 16 indicators selected in this paper are all input variables. The selection of output variables is no longer the method that scholars used the ratings given by online lending platform as output variables, but based on the risk index evaluation system of online lending platform established in this paper [4]. The entropy weight method is used to give weight to each index objectively. Based on the weight of the index, the comprehensive score of 240 online loan platforms is calculated, and the risk level is divided according to the comprehensive score.

4.2   Grading

According to the scores of each online lending platform obtained by entropy weight method, this paper classifies the risk level of 240 online loan platforms. The higher the score is, the better the risk control of online lending platform is, and the lower the risk is; the lower the score is, the greater the possibility of potential risks exists in online lending platforms. In this paper, according to the rating of the corresponding online lending platform Tianyan and the online lending home, and consulting experts in the corresponding field, the risk level of the online loan platform is divided, and the score >0.55 is defined as platform a. the platform with excellent risk control and small risk can be assured of investment; the platform with score >04 is A-level platform, which has good risk control ability and is the best choice for investment; The platform with a score of 0.3–04 is rated as a B-level platform, which is considered as a good risk management and control platform; a platform with a score of 0.15–0.3 is rated as a level C platform with poor risk control; and a platform with a score less than 0.15 is rated as D, which means that it has a large potential risk, and investors are not recommended to invest.

5   Conclusion

In this paper, based on the current risk situation of P2P online lending platform, in-depth study of the causes of the formation of online lending platform risk, build an evaluation index system, and use matlab to simulate BP neural network based on crawling data, and the experimental accuracy reaches the expected effect.By defining the risk of online lending platform, referring to the Roca rating method commonly used in traditional banking industry, and combining with the risk composition of online lending platform, this paper constructs the risk evaluation index system of online lending platform, which includes 5 first-class indicators and 16 s-class indicators, The index evaluation system includes risk control index, product and operation index, compliance index, asset quality index and user evaluation index. Combining with the characteristics of online loan, the index system covers all aspects of risk formation of online loan. The data in this paper comes from the data of online loan home and online loan Tianyan, which is different from the situation of dozens of samples selected in previous studies. In this paper, we increase the sample size and capture 240 online loan platform data for research, so as to increase the sample size and increase the credibility of the model results.

References

1. White paper on China's P2P lending service industry of the first finance and economics new financial research center. China Economic Press (2013)
2. Jie, L., Lu, L.: Research on the influencing factors of default risk of borrowers in BP network lending business research (9), 4554 (2018)
3. Li, C., Tong, Z.: Cao xiaojue credit risk assessment of P2P online loan market based on BP neural network. Manag. Modernization **35**(4) (2015)
4. Li, X., Pu, Z.: Competitiveness evaluation of PP online lending platform, regional finance research (114246) (2017)

Research on the Blended Teaching of College Business English Based on Unipus Platform

Yanyan Xin[✉]

Maanshan Teachers College, Maanshan 243000, China
yanyanxin81@sina.com

Abstract. Unipus platform is a commonly used online teaching platform in english teaching. It has rich english teaching functions and can be applied to college business english teaching. However, it is not advisable to rely solely on online teaching platform for teaching, because there are some defects in online teaching platform itself, so it will cause some negative effects on teaching when it is applied alone. At this time, classroom teaching should be used to make up for the defects of online teaching platform, the combination of the two can form a complementary relationship and effectively guarantee the quality of teaching. In this case, the hybrid teaching mode of college business english should be developed. Based on the unipus platform, this paper discusses the basic concept of the teaching mode, the advantages and disadvantages of online/offline teaching, and the application methods of hybrid teaching.

Keywords: Unipus platform · College business english · Blended teaching

1 Introduction

The hybrid teaching mode is composed of "offline classroom teaching" and "online teaching". The former is the common teaching mode in the past, while the latter refers to a large number of online teaching platforms in the modern network environment. The unipus platform is one of them, and the unipus platform is designed for english teaching. Therefore, the unipus platform can be used as the hybrid teaching mode of college business english part of the teaching model. However, in order to give full play to the effectiveness of the hybrid teaching mode of the unipus platform, teachers must have a full understanding of the unipus platform, and at the same time, they should connect the unipus platform with classroom teaching well to form a complementary relationship between advantages and disadvantages. Therefore, it is necessary to carry out relevant research.

2 Basic Concepts of Mixed Teaching Mode on Unipus Platform

In the mixed teaching mode, unipus platform is mainly responsible for online teaching, and it is connected with the classroom teaching which teachers are responsible for, and carries out business english teaching for students with the cooperation of each other.

J. Abawajy et al. (Eds.): ATCI 2021, LNDECT 81, pp. 726–732, 2021.
https://doi.org/10.1007/978-3-030-79197-1_105

With good connection between online and offline teaching, students can obtain learning materials more easily through unipus platform, and then carry out self-study. Meanwhile, teachers can also give targeted guidance in the process of students' self-study, help students to digest knowledge faster and better, and train students' business english thinking and cultural concepts, so as to make students have both internal and external english literacy and excellent english literacy In the online teaching, teachers can organize students' practical training, train students' knowledge application ability, and also supervise the students' learning situation, guide them to correct the deviation of thinking and bottleneck of students' thinking in time, and finally make them become an excellent business english talent.

3 The Advantages and Disadvantages of Online/Offline Teaching

3.1 Advantages and Disadvantages of Online Teaching

First, the advantages of online teaching are as follows: (1) the basis of online teaching is the Internet, and the level of modern internet technology in China is different from the past. The information transmission rate of the network is very fast. Therefore, online teaching in the Internet environment can guarantee the teaching efficiency. If students want to study themselves, they can download the learning materials directly from the online teaching platform, and then according to the Learning materials can carry out self-study activities. If problems are encountered in self-study, they can communicate with teachers for the first time, and there will be no too much obstacles during the period, which indicates that students can complete self-study in a shorter time; (2) online teaching is not limited by physical space and space, that is, online teaching relies on the internet, and the internet does not occupy real space, so the real space will not create the operation of the internet The restriction influence, at the same time, the internet information transmission has two modes, namely online transmission and offline transmission. These two modes enable teachers and students to send information to each other at any time, and the other party can reply at any time, indicating that time will not limit the self-activity of teachers and students; (3) online teaching resources are more abundant and access is more convenient, that is, online teaching learning resources not only come from textbooks or other textbooks, but also videos recorded by teachers themselves, or some cases that happen in reality. Therefore, online teaching resources are more abundant. At the same time, the acquisition of these resources is convenient, and it is possible to search directly in the network or shoot videos; (4) online teaching has the function of learning analysis, that is, online teaching itself is a platform software at present, such platform software basically has the function of information recording, which can record the behaviors of students on the platform online. The information teachers generated by behaviors can know what students like, what their abilities are, or what are their weak points. Then they can provide targeted teaching services for students, which is conducive to the teaching quality [1].

Secondly, the disadvantages of online teaching are as follows: (1) online teaching is easy to lead to the hard relationship between teachers and students, that is, teachers and students are not actually meeting in online teaching, so teachers can not show their affinity to students, and students are also difficult to grasp their attitudes towards themselves, which will lead to the difficult development of the relationship between teachers and students, even backward, so the relationship between teachers and students becomes rigid; (2) online teaching It will weaken the teacher management function, that is, the main learning activities of online teaching students are self-study. Teachers can only passively provide help to students and can not actively understand the students' learning situation. Therefore, they can not carry out management targeted. In this case, students may indulge themselves; (3) online teaching is not conducive to teaching practice. Taking business english as an example, the student has It is very practical and needs students to have good knowledge application ability. However, the knowledge learned by online teaching students is basically theoretical knowledge, and there is little practical opportunity. Therefore, online teaching is not conducive to teaching practice and will affect the quality of business english [2].

3.2 Advantages and Disadvantages of Offline Teaching

In fact, the advantages and disadvantages of offline teaching are completely opposite to online teaching, that is, the actual meeting between teachers and students in the offline teaching in the aspect of advantages, so teachers can show affinity, build a good relationship with students, give full play to the management ability, or organize students to carry out practical activities, which indicates that the disadvantages of online teaching are exactly corresponding to the advantages of offline teaching. In terms of the disadvantages of online teaching, there is obvious inefficiency in the communication between teachers and students, and physical space and space will restrict both teachers and students. Teaching resources are also mainly from textbooks, which is not abundant. Meanwhile, in the analysis of learning situation, teachers need to collect a large number of data to analyze, but teachers as artificial ability is limited, which can not guarantee the integrity of data The analysis of the quality and efficiency of the online teaching will be the main problem. The advantages of online teaching correspond to the disadvantages of offline teaching. In this case, the hybrid teaching mode has grasped the advantages and disadvantages of the two, and the mixture of the two forms a complementary relationship between the two. Reasonable use of the teaching model can bring many benefits to the business english teaching [3].

4   Hybrid Teaching Application Method

The application method of hybrid teaching under unipus platform can be divided into three steps, which are: online/offline teaching process planning, online/offline teaching and online/offline teaching management. The specific contents of each step are as follows.

4.1   Make Online/Offline Teaching Process Planning

The first step of mixed teaching of unipus platform is to combine online teaching with offline business english class of University. The way to realize the combination is to plan the overall teaching process. According to the advantages and disadvantages of online/offline teaching, the corresponding links in the process are allocated to unipus platform or offline classroom, aiming to give full play to the advantages of both and improve the effect of each teaching link and the integration of online/offline teaching is realized by the integration of teaching process. According to the advantages and disadvantages of online/offline teaching, teachers are advised to assign relevant links of theoretical teaching to unipus platform, for example, to convert the knowledge theory of university business english into electronic version, and then upload it to unipus platform. Students can download and learn at any time, or teachers can also use the homework layout function of unipus platform to arrange homework online, distribute it to students and let students let them do the problem, and send it back to the teacher for review after completion. In the field of online teaching, teachers can design practical activities around theoretical teaching content, and organize students to carry out the practice activities in the offline classroom. Table 1 and 2 are the teaching effects of the process planning before and after the mixed teaching application of unipus platform in a university [4].

Table 1. The teaching effect of the process planning before the application of the hybrid teaching of unipus platform in Business English in a university

Index item	Result
Students' learning efficiency	Chapter 11d/1
Students' learning state	Assessment result: average

Table 2. The teaching effect of process planning after the application of Hybrid Teaching of unipus platform in Business English of a university

Index item	Result
Students' learning efficiency	Chapter 7d/1
Students' learning state	Evaluation result: excellent

4.2   Online/Offline Teaching

For the online teaching part of the unipus platform, teachers should use online and offline teaching methods. Online teaching generally refers to live teaching, that is, teachers can enter the platform at a specific time through the live function of the unipus platform and use the network to teach students. This kind of teaching is real-time, students can ask any questions directly, and teachers can pass the online teaching Live video first time answer, help students understand knowledge. The offline teaching is realized with the help of the upload/download function of the unipus platform, which allows teachers to upload teaching videos or other teaching materials to the platform, and students can download them at any time and place. If they encounter problems, they can leave messages for teachers, and teachers will recover at the first time after seeing them. In addition, it is suggested that teachers should introduce the comprehensive evaluation function into online/offline teaching, so that students can self check and find out their own problems. The logic of the comprehensive evaluation function is shown in formula (1).

$$A_1 \times w\% + A_2 \times w\% \cdots + A_n \times w\% = C \tag{1}$$

Where A is the test factor (1, 2... N is the number of test factors) and $w\%$ is the weight of A.

4.3   Online/Offline Teaching Management

In the unipus platform of college business english blended teaching, teachers must do a good job in teaching management, and the work is carried out in different ways, so how to do a good job in online/offline teaching management is a problem for teachers to think about. First of all, in online teaching, it is suggested that teachers should use the classroom management functions of unipus platform, such as online roll call function in online teaching and information recording function in offline teaching. These functions can know whether students are "hanging up". If they find the "hanging up" behavior, teachers should get in touch with students and make clear the original situation before dealing with it. Secondly, in offline teaching, teachers can observe and understand students' learning situation. If they find that students' learning situation is poor, it means that there are problems in students' learning activities. At this time, teachers should start to manage. In addition, in the aspect of learning situation management, teachers can use the online classroom management function to understand what types of materials students often see, the time period or duration of viewing materials, etc., so that teachers can provide students with targeted teaching, which is also a major point in teaching management. Figure 1 and Fig. 2 show the process of online and offline teaching management respectively.

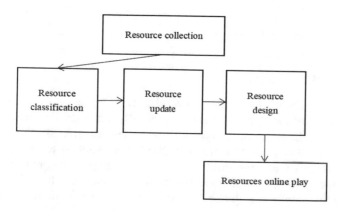

Fig. 1. The process of online teaching management

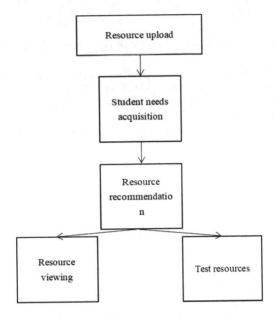

Fig. 2. Process of offline teaching management

5   Conclusion

To sum up, this paper studies the blended teaching of college business english based on the unipus platform. Through the research, the basic concept of the hybrid teaching mode of unipus platform is understood. This mode is born from the complementary relationship between online teaching and offline teaching. With the help of unipus platform, the pertinence of the teaching mode is improved, which can meet the needs of college business english teaching. This paper puts forward the application method of unipus platform for college business english blended teaching, and discusses each step of the method. It can be seen that under the effect of the three steps, the unipus platform and offline classroom play a full role, which is conducive to the improvement of college business english teaching efficiency and quality.

Acknowledgement. The 2020 Humanities and Social Science Key Project of the Department of Education of Anhui Province (SK2020A0677). In 2019, Anhui Province School-Enterprise Cooperation Demonstration Training Center "Ma'anshan Teachers College Hefei Shamanla E-Commerce Co., Ltd. School-Enterprise Cooperation Demonstration Training Center" (2019xqsxzx04). The 2019 Anhui Province boutique offline open course "English Interpretation and Translation" (2019kfkc184). The 2020 school-level teaching and research key project (2020xjzdjy03). Research project of excellent young backbone talents in Colleges and universities of Anhui Province at home and abroad in 2021.

References

1. Smit, R., Engeli, E.: An empirical model of mixed-age teaching. Int. J. Educ. Res. **74**, 136–145 (2015)
2. Chen, H., Ge, M., Xue, Y.: Clustering algorithm of density difference optimized by mixed teaching and learning. SN Comput. Sci. **1**(3), 1–18 (2020)
3. He, F.: Application research of mixed teaching mode based on Wechat applet. J. Phys. Conf. Ser. **1486**, 032006 (2020)
4. Kang, Z.: Exploration and practice of hybrid teaching mode integrating online and offline teaching resources-taking advanced algebra as an example. Creative Educ. Stud. **09**(1), 49–52 (2021)

Small and Medium-Sized Enterprise Management Strategy Research Based on Big Data Analysis

Zhishan Zheng, Huojun Zhu[✉], and Liangting Wang

Jiangxi University of Engineering, Xinyu 338000, China
huojunzhu@aliyun.com

Abstract. In recent years, the Internet has been widely used in all walks of life, made the global economy gradually buy the big data era. Facing the era of big data, the influence of modern enterprise facing the increasingly fierce market competition, and in this case, the enterprise wants to keep good competitive situation will have to pay attention to the application of large data and analysis, especially for small and medium enterprises, the advent of the era of big data for these enterprises produced great influence, traditional decision management mode is not suitable for the external environment change caused by market competition. With this as the background, this paper first makes a brief analysis of the characteristics of the era of big data and the problems existing in the management of small and medium-sized enterprises. On this basis, it focuses on the improvement of the management strategies of small and medium-sized enterprises.

Keywords: Big data · Small and medium-sized enterprises · Management strategy

In recent years, with the rapid development of artificial intelligence, such as cloud computing technology, the influence of these technologies, the era of big data, global big data in recent years the size of the market is becoming more and more big, global big data size of the market for $5.49 billion in 2018, this figure rose to $6.63 billion in 2020, up 11.24% from a year earlier, Fig. 1 shows the world from 2012 to 2020, large scale of data change:

As can be seen from the chart above, with the emergence and development of technologies such as artificial intelligence and cloud computing, the global big data market is getting bigger and bigger [1]. Facing more and more huge data information, began to be multiple influences on the development of the modern enterprise, especially for small and medium enterprises, the advent of the era of big data and global big data size increases gradually, small and medium-sized enterprise traditional management decision model has been unable to external market competition of the market demand, so perfect enterprise management must be combined with characteristics of big data, to better adapt to market competition needs [2].

J. Abawajy et al. (Eds.): ATCI 2021, LNDECT 81, pp. 733–739, 2021.
https://doi.org/10.1007/978-3-030-79197-1_106

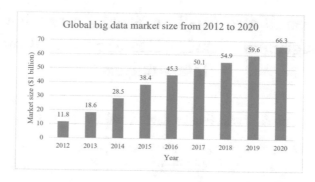

Fig. 1. Change of global big data scale from 2012 to 2020

1 Concept and Characteristics of Big Data

1.1 Large Data

The so-called big data refers to a kind of data whose scale extends to various fields such as information acquisition, storage, management and analysis, and even exceeds the diversity of data types and lower value density of traditional data transmission. From its characteristics, the arrival of the era of big data is not only a challenge, but also an opportunity for small and medium-sized enterprises. The characteristics of the era of big data urgently require small and medium-sized enterprises to make adjustments in management decisions [3].

1.2 Features of the Big Data Era

On the whole, the characteristics of the big data era can be summarized as 5V (Volume, Variety, Value, Velocity and Veracity), as shown in Fig. 2:

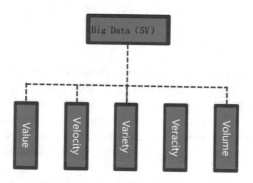

Fig. 2. Features of the big data era

Volume. The primary feature of big data is that the data group is very large. From MB in the earliest MAP3 era to TB, or even PB and EB, the level of big data is getting higher and higher as time goes by. And with the continuous development of network information technology, network information data to show explosive growth, intelligent mobile terminal network, communication network tools, such as social media platform used by billions of user login every day, every day have reached more than 300 terabytes of data information, the vast data cluster if just rely on the traditional method is not possible, must depend on large data processing platform technology, statistics and analysis of huge data cluster and real-time processing.

Variety. In the era of big data, data types are characterized by diversity. Both huge data clusters and diverse data types play their respective roles in network platforms. At this stage with the aid of jingdong, taobao, today's headlines APP can search related data information platform, the platform will also each time the search profile of users collection and analysis, thus to provide more users might like and interested in content, such as after we open the jingdong, taobao will appear "might like" plate, this is the role of big data, has made our daily access records a structured data information.

Velocity. Big data relies on the Internet platform to disseminate, and makes use of the characteristics of the Internet to realize rapid data dissemination. Contemporary society, people have more and more inseparable from the Internet, the Internet is more and more strong, the dependence of the us in the use of the Internet query information, the process of online shopping, virtually will produce large amounts of data information, and timely and processing these data with the help of the Internet, but for a search engine, the analysis of the data processing and storage needs is not only a simple storage, must use big data information processing speed, only in this way can the real-time analysis of huge data clusters. In this case, which platform data information processing speed is faster, who will occupy the absolute advantage in this environment.

Value. Value is the core feature of big data. Compared with the traditional sense of the data information, big data information generated by the greatest value is to be able to deal with the functions of all kinds of data, through its classification, so as to dig up the data for some industries use information, then with the analysis of artificial intelligence, prevent to useful content in mining data information, and the content used in medical, financial, industrial and other fields, for the social governance, improve the efficiency of enterprise production will produce larger role.

Veracity. In the era of big data, data value density is relatively low, but with the continuous increase of data clusters, data value density will also increase. Compared with big data, traditional data information is mostly structured data. Every word and paragraph in paper version of information is valuable, so the value density of traditional data information is relatively high. A book, for example, in which everything is "intentional" and every word has value. However, in the era of big data, all data information is basically unstructured or semi-structured. For example, when we use search engines to search, most of the access logs generated are of no value, and the information with real usable value is less. Therefore, in the era of big data, although the data cluster has increased a lot compared with traditional information, its value has not increased too much. In short, in the era of big data, the corresponding value generated by unit data information is relatively low.

2   Problems of Big Data Analysis in the Management of Small and Medium-Sized Enterprises

2.1   Decision-Making and Management Still Follow the Traditional Mode

For an enterprise, the way of decision-making management is directly related to the future development direction of the enterprise. Especially in the era of big data, the distribution of industrial chain clusters is more obvious, as shown in Fig. 3:

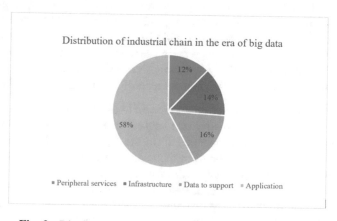

Fig. 3. Distribution of industrial chain in the era of big data

Even in this environment, small and medium-sized enterprises still do not pay enough attention to big data, and still use the traditional decision management mode. Small and medium-sized enterprises still adopt the method of management, policy makers meeting resolution, this way of decision most depend on small and medium-sized enterprise leadership experience to do, not according to the data, so this needs leadership have strong forward-looking, so the management decisions of fault-tolerant rate is relatively low, so the enterprise's operation and development are easily affected by the decision.

2.2   It is Difficult to Identify Data with Economic Value

Due to the relatively low application of big data technology in small and medium-sized enterprises at the present stage, the overall quality of cloud data acquisition is relatively low, coupled with the large amount of information, so small and medium-sized enterprises are easy to make wrong conclusions when processing relevant data and information, which seriously affects their decisions. On the other hand, due to the low value density of big data, it is difficult for enterprises to identify valuable economic data information in the face of huge data clusters. The era of big data in the data explosion growth has far beyond the small and medium-sized enterprises for information processing power, directly increase the difficulty of the enterprise data analysis, combined with small and medium enterprises in the aspect of data information filtering ability is limited, in the increasingly competitive market competition, small and medium-sized enterprise is difficult to grasp the market opportunities [4].

2.3 Important Internal Information of the Enterprise Is Easy to Leak

If a modern enterprise can correctly use big data, it can bring more business profits for the enterprise. However, due to the fast data and information circulation speed in the era of big data, the difficulty of enterprise information management is bound to increase. As small and medium-sized enterprises in the management of confidential information is not perfect, it is easy to appear the problem of information leakage, which directly causes irreparable economic losses to enterprises, affecting the promotion of market competitiveness of enterprises [5].

3 Strategies for Applying Big Data Analysis in the Management of Small and Medium-Sized Enterprises

3.1 Increase the Use of Big Data Analysis to Serve Enterprise Decision-Making and Management

For small and medium-sized enterprises, they must give full play to the characteristics of big data, gradually transform the previous experience decision-making into data-based decision-making, give full play to the characteristics of the era of big data, and find out the core competitive points suitable for their own development. To have an in-depth understanding of big data, learn unique customer models, purchase better products or services, and better manage customer services, as shown in Fig. 4:

Fig. 4. How to leverage the advantages of big data

First of all, small and medium enterprises to use cloud computing platform, collect user information related to enterprise management, the user information to conduct a comprehensive analysis, a comprehensive understanding of enterprise customer preferences, buying the age distribution of products and services, customer base for precise positioning, and then combined with customer demand to develop products, clear enterprise sales direction and goals. On the one hand, it can avoid vicious competition and blind competition in the market, and on the other hand, it can help enterprises to make clear the development direction.

Secondly, the small and medium-sized enterprises can use big data platform for comprehensive understanding of its own operating status, income, expenditure, personnel distribution, etc., help enterprises to a more comprehensive understanding of all kinds of enterprise employees, to help enterprises improve their management level on the basis of further improve staff work efficiency, and gradually improve the market competitiveness of enterprises.

3.2   Improve Enterprises' Ability to Manage and Screen Big Data Information

In today's society, no matter an industry, or an enterprise, or every individual is in an era of information explosion. Any individual or organization can obtain a large amount of data and information with the help of the Internet and various types of mobile terminal devices. Such as jingdong, taobao shopping APP can be excavated in the huge user base to the enterprise's most valuable data and information, and then using the powerful data processing ability of data information integration, such as taobao, jingdong and ability of information processing is a new challenge for enterprises, in the face of the challenge, if the company can reasonable use of the database can be masses of huge data integration out valuable information data.

In the era of big data, enterprises are faced with both structured data and unstructured data. According to different data, different analysis tools are needed to conduct in-depth analysis on the data before it can be used by enterprises. For example, enterprises can according to the characteristics of the structural data can be compiled, must use structured data analysis tools can make the data report can be used for the enterprise specially, this data can be said to be specific to sales management as the core to develop the strategic development goal of small and medium enterprises, these enterprises is market-oriented enterprises.

For unstructured data, and for those who cannot be quantified data information can edit with the help of multimedia, the information have visual characteristics, especially suitable for technology-oriented enterprises, give full play to the technological advantage of data information, a more comprehensive understanding of the latest technical information, help the enterprise to develop new products.

3.3   Permission Control Shall Be Carried Out for Important Data Information Within the Enterprise

No matter in any period, enterprises play two decisions in the process of information dissemination. On the one hand, enterprises are the receivers and users of various data and information; On the other hand is for the benefit of all kinds of information, especially in the era of big data information technology, enterprises on the one hand, with the aid of various network platform, offline traditional platform to launch a product or service widely, promote enterprise's products or services, create the whole image of the enterprise and brand, from the level of enterprise is information disseminator; On the other hand, enterprises can use various information channels to collect data to help them understand the market demand. From this level, enterprises are the recipients of data. Therefore, for modern enterprises, information has become a necessary tool for survival and development. Especially in the face of the impact of the era of large data, small and medium-sized enterprise must strengthen internal management, according to the hierarchical classification to the importance of the internal data information, only people who have a certain management authority conditional access this information, at the same time to strengthen the internal information exchange platform construction, strict limits for company confidential information.

4 Conclusion

In general, the era of big data for any enterprise is more or less influence, particularly for the small and medium-sized enterprises, more should seize the opportunity of the era of large data, with the aid of a wave of big data era, change enterprise's development from the aspects of internal management philosophy, continuous innovation enterprise management model, to help enterprises in the fierce market competition ahead.

References

1. Guo, J., Ning, N.: Discussion on the promotion strategy of enterprise management and human resource management in the era of big data. Enterp. Reform Manag. **28**(04), 72–73 (2021)
2. Cao, M.: Analysis on the problem of enterprise management in the context of big data. Technol. Market **28**(02), 169–170 (2021)
3. Cui, S.: Challenges and countermeasures of enterprise management accounting in the era of big data. Fortune Today (China Intellect. Property) **20**(02), 164–165 (2021)
4. Zhou, J.: Discussion on the necessity of establishing management accounting for small and medium enterprises in the era of big data. Market Wkly. **41**(10), 118–119 (2019)
5. Xiao, M.: On the financial management innovation strategy of small and medium enterprises in the era of big data. Shangxun **26**(6), 84–85 (2019)

Research on Security Strategy Based on Public Security Big Data Security Service System

Zhi-ning Fan[1,2,3(✉)], Xiong-jie Qin[1,2,3], and Min Zhou[1,2,3]

[1] Public Security Department of Jiangxi Province, Nanchang, China
[2] Jiujiang Public Security Bureau, Jiujiang, Jiangxi, China
[3] The Third Research Institute of the Ministry of Public Security, Shanghai, China

Abstract. The security strategy of public security big data security service system is analyzed and studied from three aspects: security policy service configuration, business security policy control and security protection policy control. Policy control is the core research content of big data security strategy, which mainly realizes the aggregation of security alarm, risk, security situation and other information, and carries out association analysis, intelligent reasoning, analysis, judgment and decision-making, forming security protection strategy and business security strategy. Based on the decision-making results, services are arranged, scheduled and configured, including but not limited to linkage, blocking, isolation and access At the same time, it supports flexible expansion mechanism to achieve collaborative protection and linkage.

Keywords: Big data security · Security services · Security policy · Policy control · Trustworthy

1 Introduction

Big data is now the most concerned about the domestic and international information hot spots. McKinsey points out that big data will be the next cutting edge of innovation, competition, and productivity [1]. The US government launched the "Big Data Research and Development Program" in 2012, proposing "through the collection, processing of large and complex data information, from which access to knowledge and insight, to enhance the ability to speed up the scientific and engineering areas of innovation, strengthen the US Homeland Security, Changing education and learning patterns" [2]. The Chinese government 2015 will formally big data into the national security strategy, put forward the "implementation of national big data strategy to promote data sharing and sharing" initiative [3]. In 2017, the Ministry of public security of China issued the "13th five year plan" overall technical framework of public security informatization, which gives the top-level technical design, constructs the "three horizontal and three vertical" technical framework, guides the construction of public security informatization, and focuses on the construction of data policing and intelligent public security. Due to the value characteristics, high confidentiality and confidentiality of public security business, it is necessary to build a security service guarantee system based on public security big data under the premise of ensuring the

J. Abawajy et al. (Eds.): ATCI 2021, LNDECT 81, pp. 740–748, 2021.
https://doi.org/10.1007/978-3-030-79197-1_107

convenient use of business applications, so as to realize the construction of defense in depth system and trusted access service system, and form a scientific and practical systematic security protection capability. The establishment of practical, efficient and economic security access control model service is an important and necessary condition for the construction of public security big data security service guarantee system [4].

2 Public Security Big Data Security Service System Issues

The overall logical framework of public security big data security service system takes data security as the center [5], takes security infrastructure as the support, takes security big data intelligent analysis as the starting point, constructs in depth from six dimensions of "cloud, data, application, network, boundary and end", analyzes and studies unified security management, and constructs a "safe, reliable and compliant" big data intelligent security three-dimensional defense in depth System and trusted access authentication system. Among them, data security is the center. In order to deal with all kinds of high threat security risks that data may face, we can comprehensively improve the data security defense capability level. Due to the lack of overall security planning and management system, in the process of data integration and sharing application, due to the concern of data security problems, it can not gather, dare not serve, and do not allow sharing, which is difficult to meet the urgent demand of public security big data service for information sharing.

2.1 Trusted Access Service System

The main content of the public security big data service system is the ability to build the access system based on zero trust service. It meets the security access service, realizes authentication service, environment awareness service, business security policy control service and audit service, and then completes the authority management and approval service based on the progress of application transformation, so as to establish the public security big data trusted access service system.

2.2 Defense in Depth System

The basic goal of the public security big data service system is to build a secure, credible and compliant defense in depth system based on cloud computing, big data and public security network. The big data security strategy should serve the whole network security situation awareness, the whole process of rapid detection and disposal of security threats, so as to ensure that the whole process of big data is known, controllable, manageable and searchable, and change static into dynamic In order to provide strict security guarantee for the construction of public security big data. Around the network, infrastructure, platform, data, services, applications, etc., to construct an in-depth and efficient big data security policy service mechanism.

3 Public Security Big Data Security Strategy Analysis

Aiming at the problem of public security big data security strategy mentioned in Sect. 2 of the article [6], the article mainly puts forward the following three aspects to study big data security strategy from the perspective of system technology: security policy service configuration, business security policy control and security protection policy control. Policy control is the core research content of big data security strategy [7], which mainly realizes the aggregation of security alarm, risk, security situation and other information, and carries out association analysis, intelligent reasoning, analysis, judgment and decision-making to form security protection strategy and business security strategy, and arranges, schedules and configures services based on decision results, including but not limited to linkage, blocking and isolation At the same time, flexible extension mechanism is supported to realize collaborative protection and linkage. This paper analyzes and studies the security strategy construction of public security big data security system from the above three aspects, and puts forward specific and feasible design ideas and solutions.

3.1 Security Policy Service Configuration

Public security big data security policy configuration generally includes security service policy configuration such as creation, optimization, distribution and approval, including but not limited to network device security configuration, security device security configuration, operating system security configuration, database system security configuration, middleware security configuration, cloud platform security configuration, etc.

1) Policy creation

According to the protected object and security protection ability, the group definition is carried out, the relevant security policy is established, the corresponding security services are selected through the security policy, and then the corresponding security resources are selected to protect the protected object, and the unified life cycle management of the security policy is carried out.

(1) Protected object management

Protected objects include host, application, terminal, server, network equipment, security equipment, application system, data, etc. Register all kinds of assets to the security management center through the asset discovery tool. Group management according to the protected object properties.

(2) Policy configuration

The security policies suitable for the protected objects are displayed in the form of a list, and the corresponding security protection policies are selected according to the security requirements of the protected objects.

(3) Rule configuration

According to the security policy, the corresponding security rules are configured for different security resources.

2) Strategy optimization
Based on security events, security situation and security big data, the security strategy is analyzed and evaluated for different scenarios. According to the results of security policy evaluation, security policy optimization suggestions are put forward.

3) Strategy distribution
Through the way of task, the security policy is distributed to the security identification service, security protection service, security detection service and security response service, and then the security rules are distributed to the security resources through these security services.

4) Policy approval
The distribution of some important strategic mechanisms that seriously affect the security assurance system of big data needs hierarchical approval to achieve the purpose of deterministic security. The approval task generally includes at least the ID of the applicant, the task name, the task type, the applicant unit, the approval content and other information.

3.2 Business Security Policy Control

As a collection of business security policy control capabilities in public security big data security system, business security policy mechanism is responsible for risk aggregation, trust assessment, linkage notification, security instruction distribution and external service system linkage in business security. Business risk sources can be environment aware services, rights management services, audit services and authentication services, as well as security protection policy control; in business security policy control, comprehensive trust assessment is carried out, and security instructions are generated and issued. The execution point of receiving and executing security instructions can be authentication service, authority service or specific execution point in other external system platforms.

Business security policy control provides the whole process of policy control before, during and after the event, and provides functions or capabilities such as risk aggregation, trust assessment, instruction issuance, linkage notification and external linkage.

The business security policy control capability is shown in Fig. 1:

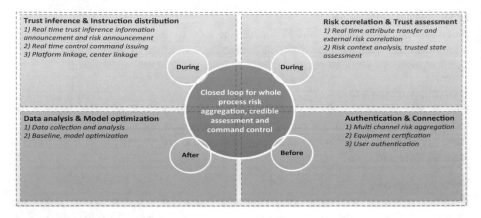

Fig. 1. The Security policy control capability

1) Risk convergence

Business security policy control should have the ability of risk aggregation, which should support the linkage with environment aware service to receive environment aware service information; support the linkage with authentication service to receive token and risk information; support the linkage with authority management service to receive risk information; support the linkage with audit service to receive risk information; support the interaction with external security access and security protection policy Control service linkage and receive risk information. According to the generation time, it is divided into access risk and protection risk. The access risk is mainly generated in the process of users accessing big data, and the source is the terminal through which the business access data flows, the security access platform, and the zero trust centers. The latter is generated by the security protection system, and comes from the threats and events found by the security components in the protection platform.

2) Trust assessment

Business security policy control should have the ability of trust evaluation, support association analysis of multi-dimensional risk information; support comprehensive evaluation based on trust evaluation model; support generation of control instructions based on security policy and evaluation results. Trust evaluation needs to infer and judge based on users and context sensitive data. The evaluation process needs to form historical trust data according to historical data of users, terminal devices, business applications and interfaces. According to historical evaluation, comprehensive inference can be made based on historical data and real-time data.

The business security policy controls the trust inference process, as shown in Fig. 2.

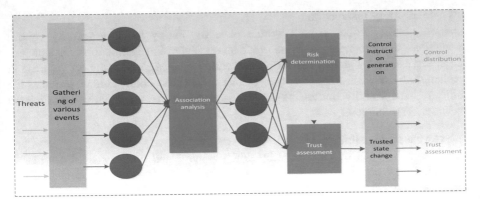

Fig. 2. The trust inference process

3) Linkage notification

The business security policy control should have the ability of risk linkage notification, and support the direct query of risk information through the query interface; the risk notification should support the notification status verification mechanism to avoid information forgery; support the failure retransmission of risk information notification to avoid information loss. Business security policy controls the ability to provide linkage with external services, which is mainly divided into input and output linkage. The input linkage mainly comes from the status and notification of environmental information of environmental perception service, and it comes from the input of various risk information of authentication service, authority management service, business audit service and external system; the output linkage mainly refers to the output of policy control instructions of authentication service and external system.

Business security policy control linkage distribution, as shown in Fig. 3.

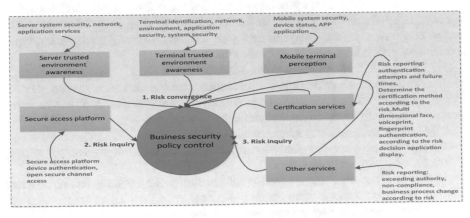

Fig. 3. The Policy control linkage distribution

4) Issue of safety instructions

Business security policy control should have the ability to issue security control instructions, and the ability to issue order board revocation instructions for authentication services and authority services; instruction issuance should support verification mechanism to avoid instruction forgery. Business security policy control provides the ability to issue security control instructions, and policy execution points issue related execution instructions.

5) External service linkage

The linkage between business security policy control and other external services is shown in Fig. 4 below:

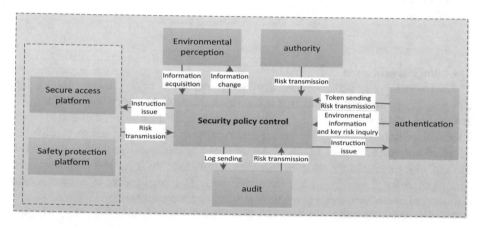

Fig. 4. The Linkage with other external services

(1) Linkage with certification services.

It supports receiving user token and application token from authentication service, receiving risk information from authentication service, sending control instructions to authentication service, and querying hardware features, hardware environment types and terminal risk information of authentication service.

(2) Linkage with environment aware services.

It can synchronize environment information from environment aware service, receive terminal risk information from environment aware service, and receive information change notification from environment aware service.

(3) Linkage with authority management service.

Receive and deliver risk information from rights management service.

(4) Linkage with audit services.

Upload or synchronize the system log information to the audit service; receive and transfer the risk information from the audit service.

(5) Linkage with security policy control service.

Receive the risk information from the security protection policy control service, and transfer the risk information to the security protection policy control service.

3.3   Security Protection Policy Control

Security protection services provide security threat defense capabilities, and provide data content protection, access control, security access, intrusion prevention, security reinforcement and other security capabilities through security resource services, so as to narrow the network attack area, enhance the attack threshold, reduce the impact of security events, and realize all-round three-dimensional protection of cloud platform, data, application, network, boundary and terminal.

1) Alarm and risk information aggregation

Alarm and risk information aggregation supports receiving alarm and risk information generated by security identification service, security protection service, security detection service and security response service, receiving risk and threat situation information of security situation awareness, and receiving risk information of business security policy control. Alarm and risk information aggregation provides standardized information collection interface, and supports unified access of multi-source and heterogeneous data.

2) Analysis and decision making

Analysis and decision-making refers to the process of obtaining the correct scheduling by studying and reasoning the security big data based on the security configuration baseline. In this process, we can add and combine the data from different sources, specify the corresponding security configuration baseline as the judgment condition, and automatically get the desired decision results. Analysis and decision-making include the following main functions: Transform, associate and analyze the original data, automatically synchronize the changes of baseline data of security configuration, update and associate the existing scheduling scheme as the decision result.

3) Scheduling

Scheduling integrates security identification service, security protection service, security detection service and security response service, and decomposes the security capability into reusable steps, such as linkage, blocking, isolation, forensics, blocking, etc. these steps are added and concatenated to form threat protection measures for specific analysis and decision results. When a network event occurs, after analysis and decision-making, the corresponding scheduling scheme is started, and the threat protection measures will be converted into security policy control tasks, which will be distributed to the specific resources carrying the basic security capability for linkage disposal. The whole process can be based on the existing automatic script or user-defined new ones. The whole disposal process is a process Closed loop, we can get a final disposal report, including the collection, enrichment, analysis, decision-making and disposal of the network events.

4) Linkage disposal

Linkage disposal is a security policy control task executed by security identification service, security protection service, security detection service and security response service. By defining a standardized interface, the control task can be executed in all security resources associated with the service without paying attention to the differences between these security resources. Linkage disposal focuses on the inter-action with security resources. After issuing tasks to security identification service, security protection service, security detection service and security response service, it is necessary to monitor and track the security resources of the actual tasks. In case of failure, the task can be re issued to try again. When the retrial reaches the threshold, the task can also be automatically terminated. And feedback the results of task imple-mentation to the disposal report. In addition to distributing tasks to security resources, linkage disposal can also synchronously collect new risk information from security resources and send it to the security management center to drive new security pro-tection strategies. In addition to issuing tasks through security identification service, security protection service, security detection service and security response service, linkage disposal also supports disposal actions through cloud management platform or network management system, and obtains unified disposal report.

4   Conclusion

Starting from the environmental characteristics of the public security big data platform and the requirements of the public security big data service system, this paper analyzes and studies the security policy under the public security big data security service system from three aspects: security policy service configuration, business security policy control and security protection policy control, and carries out the prototype analysis and design through technical means. To do a good job in the control research and technical design of security protection strategy is the basic guarantee ability and necessary condition for the realization of defense security and the safe operation of big data security protection system. It is also necessary to establish a multi pronged mechanism and system to maintain and guarantee the security strategy of public security big data service system.

References

1. Manyika, J., et al.: Big data: the next frontier for innovation competition, and productivity (2011)
2. Whitehouse: Big Data is a Big Deal[EB/OL]
3. http://www.whitehouse.gov/blog/2012/03/29/big-data-big-real. Accessed 29 Mar 2012
4. Xinhua News Agency. Learning China "national big data strategy - Xi Jinping and" thirteen five "fourteen strategy"
5. http://news.xinhuanet.com/politics/2015-11/12/c_128422782.htm
6. Hu, J.: Application of network security technology, November 2017
7. Police technology by Lu Hongbo and Yan Jinduan, issue 5 (2019)
8. Hu, Y.: Wireless Internet technology, issue 6 (2020)
9. Deng, W.: China management informatization, issue 22 (2017)

Research and Design Based on Public Safety Big Data Evaluation Platform

Shaozhi Wang$^{(\boxtimes)}$, Jie Dai, Ming Yang, and Bo Zhao

IoT Tech R&D Center, The Third Research Institute of MPS, Shanghai, China

Abstract. With the continuous advancement of the technological wave, the informatization construction of the public safety industry is also undergoing continuous iterative upgrades in technology and business. All provinces, cities and localities are carrying out informatization reforms and building big data platforms. The platform has accumulated a large amount of police data, such as bayonet, video, criminals, social collection, basic data, case incidents and other data are growing rapidly, and the public security industry has new requirements for big data platforms. How to build a stable, reliable, and safe system has become a core requirement. With the introduction of the concept of new requirements for the construction of big data systems, the previous computing operating platforms and computing architectures are no longer competent in the face of such a huge data scale. A new big data evaluation system is needed to evaluate the big data system from all aspects. The construction situation. At present, some manufacturers have proposed industry benchmark tests and provide corresponding test software packages. However, starting from the versatility of big data and the true flexibility of testing, most of the benchmark test suites can only meet part of the requirements of big data evaluation. Some cannot cover the typical areas involved in big data. The big data evaluation software system takes these requirements as the basis for development, and has developed representative big data benchmark evaluation software that meets the characteristics of big data.

Keywords: Big data · Evaluation · Police data · Public security

1 Introduction

With the exponential increase in the number of big data applications or application scenarios in various countries in the world, the knowledge required to test big data applications and the needs of big data test engineers are also increasing simultaneously. According to data from some IDC institutions, as of 2020, the size of the big data market has reached US$50 billion.

Nowadays, big data not only serves enterprises, but has also become a part of social infrastructure. Like water, electricity, and highways, it is indispensable in people's lives. But the role of big data is not limited to these two, it has begun to play an important role in the field of social public security [9]. Government departments at all levels and various types have accumulated hundreds of millions of big data on public services in social governance [5]. These data are effectively processed and analyzed,

J. Abawajy et al. (Eds.): ATCI 2021, LNDECT 81, pp. 749–755, 2021.
https://doi.org/10.1007/978-3-030-79197-1_108

and responsibility deviations are calibrated step by step, so as to continuously improve real-time dynamic monitoring and real-time early warning capabilities, and promote risk prevention. Scientific and refined control work, knowing all possible risks and their causes, prescribing the right remedy, and responding to the problem, can improve the accuracy and targeting of social governance, and ensure that social risks are resolved at the source. So as to continuously promote the realization of refined social governance [7]. How to achieve refined governance has become a core issue. In response to this problem, it is necessary to evaluate the characteristics of the big data system from all aspects of the big data system to discover problems and iterate the big data system continuously to meet the governance of complex and diverse public safety data.

2 System Architecture Design

2.1 System Hardware Environment Structure

The project uses multiple computing servers to network to build a big data evaluation environment. Including: simulation big data application cluster, evaluation application server, evaluation data server, tester computer terminal and automatic report output equipment, etc.

2.1.1 Simulate Big Data Application Cluster

The test object simulation system is built on a distributed virtual environment composed of 5 computing servers, and an application cluster composed of multiple virtual machines [8]. The cluster is deployed on the hadoop system, with HDFS distributed data management, and data processing algorithms such as Bayes, Kmeans, Pagerank, etc. are configured.

2.1.2 Evaluation Application Server

The hardware part of the evaluation application server is composed of an ordinary computing server, which is connected to the evaluation data server, the tester's computer terminal, and the simulation big data application cluster to be deployed in the same network.

The evaluation application server is installed with a public safety big data system evaluation platform, a public safety view data preprocessing algorithm evaluation tool, and structured, semi-structured, unstructured and other evaluation data generation plugins [14]. The network provides web-based client evaluation application services.

2.1.3 Evaluation Data Server

The evaluation data server is mainly installed with distributed database software to store unstructured data such as videos, images, texts based on mainstream public safety applications and their labeled samples, semi-structured data such as log, xml, and json, and based on the text characteristics of video images Describe the structured data generated.

2.1.4   Tester's Computer Terminal

The tester's computer terminal uses a portable workstation or a laptop to access the web [12] services provided by the evaluation application server by connecting to the evaluation system network and assigning IP addresses, and then to operate the public safety big data evaluation platform, view preprocessing evaluation tools, and self-generation of test data Software, etc., connect the big data system of the test object, establish test tasks, and carry out specific evaluations.

2.2   System Software Structure

2.2.1   Big Data Information Service Standard Evaluation Platform Architecture

The big data information service standard evaluation platform is divided into data generation and import module, big data system performance evaluation module, big data system basic characteristic evaluation module [11], big data system data preprocessing algorithm evaluation module, big data system data management evaluation module, and big data The system security evaluation module, the big data system acquisition evaluation module, the big data system data processing evaluation module, and the front-end interaction module consist of 9 functional modules.

2.2.2   The Logic Flow of the Big Data Information Service Standard Evaluation Platform

The following figure shows the entire process of the evaluation, from the creation of the evaluation task to the evaluation of all aspects, to the output of the evaluation results, and the export of the test report.

2.3   System Network Structure

Introduction to Architecture.

Externally send commands to the Manager Service server through the host (the upper left part of the figure) to control the installation and deployment of the entire cluster (the installation package is obtained from the installation package warehouse).

3   System Function Realization

3.1   Working Principle of Evaluation

The overall Shuffle process [3] includes the following parts: Map side Shuffle, Sort stage, Reduce side Shuffle. That is to say: The shuffle process spans both ends of map and reduce, and includes the sort phase in the middle, which is the process of data output from map task to reduce task input [1].

Note: sort and combine are on the map side, combine is reduced in advance, and you need to set it yourself.

The picture in the official document is: It is divided into map side shuffle and reduce side shuffle [6]. This is the overall process diagram [13]. The specific details of the legend need to be subdivided in the map and reduce stages. See the overall process diagram for details (Fig. 1):

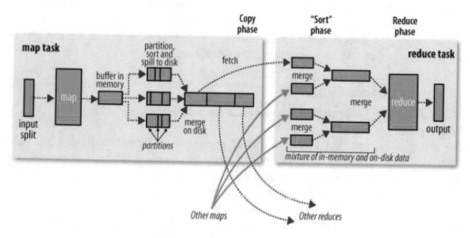

Fig. 1. Working principle diagram

In a Hadoop cluster, most map tasks and reduce tasks [4] are executed on different nodes. Of course, in many cases, Reduce needs to pull the results of map tasks on other nodes across nodes. If the cluster is running a lot of jobs, the normal execution of the task will consume a lot of network resources inside the cluster. As for the consumption of necessary network resources [2], the ultimate goal is to minimize unnecessary consumption. Also in the node, compared to the memory, the impact of disk IO on job completion time is also considerable. From the most basic requirements, for the Shuffle process of MapReduce job performance tuning, the target expectations can be:

1. Partition process.

The output of the mapper is a key/value pair, but the map side only does the +1 operation, and the result set is merged in the reduce task. The method that determines which reducer of the current mapper's part is given is: The Partitioner interface provided by mapreduce hashes the key, then takes the modulo of the number of reducetasks [10], and then sends it to the specified job.

Reduce side Shuffle

Before the reduce task, it continuously pulls the final result of each maptask in the current job, and then continuously merges the data pulled from different places, and finally forms a file as the input file of the reduce task (Fig. 2).

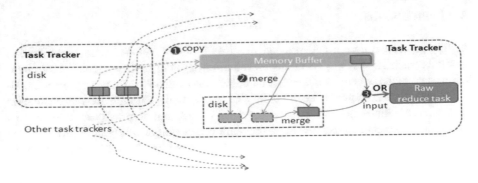

Fig. 2. Reduce side shuffle

3.2 Big Data Evaluation Platform Evaluation Function

A. Data import function: structured data import, semi-structured data import, unstructured data import;
B. Data quality evaluation function: import the structured data, semi-structured data, and unstructured data of the target system to verify the integrity, consistency, and validity [15];
C. Big data system performance evaluation function:
   - In Sort, Wordcount, PageRank, Kmeans, Bayes mode, the system time-consuming feedback under different data processing scales (Fig. 3);

System performance evaluation report

value / data / type / norm	sort			wordcount		
	1024	2048	3072	10	20	30
filenumberofbytesread	436963	343213		2874	2874	2874
filenumberofbyteswritten	14687670	11744000		588944	588944	585880
filenumberofreadoperations	0	0		0	0	0
filenumberoflargereadoperations	0	0		0	0	0
filenumberofwriteoperations	0	0		0	0	0
hdfsnumberofbytesread	587342483	434101946		1700	1700	1700
hdfsnumberofbyteswritten	587342483	434101946		813	813	813
hdfsnumberofreadoperations	3665	2825		13	13	13
hdfsnumberoflargereadoperations	0	0		0	0	0
hdfsnumberofwriteoperations	1490	1150		6	6	6
mapinputrecords	974	2029		15	15	15
mapoutputrecords	974	2029		152	152	152
mapoutputbytes				1431	1431	1431
mapoutputmaterializedbytes				1230	1230	1230
inputsplitbytes	1180	1180		117	117	117
combineinputrecords				152	152	152
combineoutputrecords				103	103	103
reduceinputgroups				103	103	103
reduceshufflebytes				1230	1230	1230
reduceinputrecords				103	103	103
reduceoutputrecords				103	103	103

Fig. 3. Performance evaluation comparison

4 Conclusions

In this paper, we design and provide multiple types of multi-directional evaluation methods, such as system performance evaluation, system basic characteristics evaluation, data management evaluation, data collection evaluation, system security evaluation, etc. It can improve the reliability of evaluation. Through us the evaluation method can reduce the professional level of testers and improve work efficiency. Strengthen the security risk level and prevention capabilities of public safety big data platforms, assist in the construction of high-quality public safety big data platforms, and solve the massive information generated in criminal activities. The data is large and scattered, with complex composition and other public safety big data issues. The use of big data technology to promote public safety governance is of great significance to maintaining the public safety environment in China.

Acknowledgements. This work is sponsored by the National Key Research and Development Program of China (Project No. 2018YFB1004605).

References

1. Zhang, J., Fan, Z., Zhao, Y., et al.: Hadoop Practice of Big Data Analysis and Mining. China Machine Press (2015)
2. Yu, Y.: Software Performance Testing and LoadRunner in Practice. POSTS & TELECOM PRESS (2014)
3. China Electronics Information Industry Development Research Institute, Testing and Evaluation Technology for Industrial Big Data. POSTS & TELECOM PRESS (2017)
4. Cai, L., Wu, X., Liu, Z.: Big Data Technology and Application Series. Shanghai Science Press (2015)
5. Yildiz, O., Ibrahim, S., Antoniu, G.: Enabling fast failure recovery in shared Hadoop clusters: towards failure-aware scheduling. Future Gener. Comput. Syst. **74**, 208–219 (2017)
6. Glushkova, D., Jovanovic, P., Abelló, A.: Mapreduce performance model for Hadoop 2.x. Inf. Syst. **79**, 32–43 (2019)
7. Zhong, P.: Government supervision research in the era of big data. In: First Symposium on International Information Construction (2). Xuri Huaxia (Beijing) International Academy of Science and Technology, January 2016
8. Liu, Z.: Research on knowledge discovery model based on big data. In: 16th Annual Conference of China Association of science and technology: Papers of International Symposium on Technology Information Dissemination and Standardization. China Association of Science and Technology, People's Government of Yunnan Province: Academic Department of China Association of Science and Technology, May 2014
9. Li, Y., Liu, X., Sun, T.: Research on public security intelligence analysis system based on big data. In: Proceedings of the Second China Command and Control Conference 2014 (Part 2). China command and Control Society, May 2014
10. Wu, Y.: MapReduce performance evaluation model for Hadoop 2.x. Comput. Syst. Appl. **30**(02), 219–225 (2021)
11. Zhang, G., Ye, M., Wang, Z., Zhou, T.: Comparison and implementation of core technologies of big data Hadoop framework. Lab. Res. Explor. **40**(02), 145–148+176 (2021)

12. Kovalchuk, S.V., Zakharchuk, A.V., Liao, J., Ivanov, S.V., Boukhanovsky, A.V.: A technology for BigData analysis task description using domain-specific languages. Procedia Comput. Sci. **29**, 488–498 (2014)
13. WenTai, W., Lin, W., Hsu, C.-H., He, L.: Energy-efficient Hadoop for big data analytics and computing: a systematic review and research insights. Future Gener. Comput. Syst. **86**, 1351–1367 (2018)
14. Ishwarappa, Anuradha, J.: A brief introduction on big data 5Vs characteristics and Hadoop technology. Procedia Comput. Sci. **48**, 319–324 (2015)
15. Yeh, T., Huang, H.: Realizing integrated prioritized service in the Hadoop cloud system. Future Gener. Comput. Syst. **100**, 176–185 (2019)

Short Paper Session

Management Optimization of Information Push of Online Shopping Platform from the Perspective of Marketing

Runfa Li[1], Huan Ye[2], Ou Wang[2], and Xianghua Yue[3(✉)]

[1] Guangzhou College of Technology and Business, No. 5 Guangming Road,
Haibu Village, Shiling Town, Huadu District, Guangzhou 510850,
Guangdong, China
[2] Guangzhou International Economics College, No. 28, Dayuan North,
Shatai Middle Road, Guangzhou 510540, Guangdong, China
[3] Xiangnan College, Xiangnan University, 213A, Teaching Building 2,
No.889 Chenzhou Avenue, Suxian District, Chenzhou 423043,
Hunan, China
hntyue@xnu.edu.cn

Abstract. With the rise of online shopping, more and more online shopping platforms will push information with the help of intelligent technology, hoping to bring users a better shopping experience. However, in the actual situation, some online shopping platforms still have some prominent problems in information push, which not only does not improve the user experience, but also leads to the loss of some users. This paper focuses on "the characteristics of online shopping platform information push under the marketing perspective", "the problems existing in the online shopping platform information push", "the management optimization of online shopping platform information push from the marketing perspective", hoping to further optimize the information push management mechanism of online shopping platform based on the marketing perspective.

Keywords: Marketing perspective · Online shopping platform · Information push · Management optimization

1 Introduction

With the increasing demand for online shopping, there are more and more emerging online shopping platforms. In order to stand out from many online shopping platforms and attract more users, some online shopping platforms begin to push information with the help of intelligent technology. In this process, some online shopping platforms have achieved certain success, attracting a large number of loyal users, while some online shopping platforms have encountered various difficulties. Based on the marketing perspective, to explore the information push management mechanism of online shopping platform is to use information push technology more scientifically and reasonably, so as to better serve online shopping platform and online shopping users.

J. Abawajy et al. (Eds.): ATCI 2021, LNDECT 81, pp. 759–764, 2021.
https://doi.org/10.1007/978-3-030-79197-1_109

2 Characteristics of Information Push of Online Shopping Platform from the Perspective of Marketing

2.1 More Information Push Forms

In the marketing perspective, at present, the form of online shopping platform information push is becoming more and more "rich". Specifically, on the one hand, on the online shopping platform, information push forms include text, pictures, and various combinations of words and pictures; on the other hand, on the online shopping platform, information push forms include moving pictures, live broadcast, small video, etc. [1]. These different information push forms have different impact, which can bring users a novel, interesting, aesthetic experience, to meet the diverse shopping needs of different users.

2.2 More Personalized Information Push

With "user demand" as the core, the information push mechanism of online shopping platform is more "personalized". Based on the marketing perspective, each user's aesthetic, hobbies and requirements are different. When pushing information on the online shopping platform, it is necessary to combine the personalized needs of different users to push accurately. Compared with the traditional information push mechanism, personalized information push mechanism has many advantages, specifically: (1) saving user search time; (2) strengthening user purchase intention; (3) upgrading user shopping experience; (4) improving information push conversion rate; (5) reducing overall information push cost of online shopping platform [2]. At present, the "personalized" trend of information push mechanism of online shopping platform is more and more obvious.

2.3 More Diversified Information Push Channels

Now, online shopping platform information push channel is more "diversified". In the traditional online shopping platform information push channels, there are mainly official websites or official apps, such as Jingdong, Taobao, Tmall, Weipinhui, Pinduoduo, etc., as well as some outdoor advertisements and TV advertisements. Nowadays, Tiktok, Jitter, WeChat official account, Micro-blog, and Little Red Book have begun to have powerful information push function as the new media is becoming more and more popular and in-depth [3]. Due to the large number of active users of these platforms, the exposure and effect of information push are also relatively good. Generally speaking, the diversification of online shopping platform information push channel reflects the marketing pattern of online shopping platform in the new era.

3 Problems in Information Push of Online Shopping Platform

3.1 It Has Caused Some Troubles to Some Users

Although the multi-channel information push expands the marketing scope and enhances the marketing effect, at the same time, it is easy to bring trouble to some online shopping users. Specifically, on the one hand, in the process of information push, the multi-channel information push mode increases the operation cost of online shopping platform. In order to make the operation cost have better feedback, the online shopping platform will carefully plan various information push forms to attract users' attention with new, strange and special features. Some users may be interested and some may reject it. Moreover, due to the technical control, some information push without legal person is closed. As long as users browse information, they will be "viewed" and "marketed" [4], which causes a lot of inconvenience to users. On the other hand, in the process of information push, multi-channel information push mode is easy to cause "multiple interference" to users. For example, a message push is read by a user, which will be showed on the top of the Tiktok and followed by Little Red Book too. This kind of high-density marketing form makes users easy to have a sense of resistance.

3.2 Users Are Tagged by the Information Push Mechanism

At present, on the online shopping platform, it is not uncommon for users to be labeled by the information push mechanism. The main performance is as follows: (1) the online shopping platform classifies users according to their browsing traces, search demand, consumption ability and other factors [5]; (2) the online shopping platform pushes different levels of products and services for different categories of users. The core purpose of information push mechanism to label users is to accurately obtain customers and accurate marketing. In this process, it is easy to have these problems: First, the analysis of users' information by online shopping platform is only based on some information materials of users, which is not comprehensive and easy to lead to deviation; second, the information analysis of online shopping platform for users is relatively fixed, and it is unable to analyze and update users in real time. In the case of constantly change of user's information, the online shopping platform may not judge the user accurately; third, the online shopping platform analyzes the information of users, classifies and labels the users. Once the scale and mode are not well grasped, it is easy to give users a bad feeling. Users will feel that they are treated differently and unreasonably, which is easy to stimulate some social contradictions and affect the harmonious development of online shopping platform.

3.3 Serious Homogenization of Information Push Content

How to understand the homogenization of information push content? For example, after a user browses a product or searches for a product on the online shopping platform, it will automatically push similar products and services [6], and repeat marketing to users. Moreover, once these products and services promote activities, price reduction and other situations, the platform will automatically push to users. Such information push mode can meet the needs of users in a short time, but after a long time, it is easy to cause trouble to users, because the purchase demand and purchasing power of users are not invariable. With the change of users' purchase demand and purchasing power, this "homogenization" information push will be out of date, affecting the user's shopping experience. How to avoid the homogenization of information push while accurately pushing is a problem that needs to be considered deeply by online shopping platforms.

4   Management Optimization of Information Push of Online Shopping Platform from the Perspective of Marketing

4.1   Avoiding Causing Trouble to Users

From the perspective of marketing, in order to avoid causing trouble to users, we can improve from these aspects: on the one hand, in the process of information push, the online shopping platform cannot blindly hold the mentality of "spreading the net" blindly [7]. We should carefully conduct channel research, analyze the user types of different channels, and carry out accurate delivery. In this way, not only can the operation cost be reasonably controlled, but also the interference to the inaccurate users can be reduced, so as to achieve better marketing effect; on the other hand, in the information push process, in order to avoid causing "multiple interference" to users, the online shopping platform should optimize and upgrade the information push form. For example, the information push mechanism needs to further open the "independent closing right". When users see the information they have browsed, they can easily close it manually. Many information push fees are related to user click and user view. When the user has the right to shut down independently, it can not only save part of the information push cost, but also improve the user experience, so that the user will not be troubled by the "forced push" information.

4.2   Improving User Label Management

How to improve user label management? Specifically, first, the online shopping platform should make full use of big data technology to systematically analyze the comprehensive information of users, forming a more comprehensive analysis result, so that users feel that they are not being "labeled" [8], but enjoying "intelligent" shopping services; second, online shopping platforms should pay attention to update the information analysis of users, and according to the purchase demand and purchasing ability of users in different periods, the online shopping platform should classify users again, adjust the content and mode of information push to meet the shopping needs of users at different stages [9]; third, online shopping platform should grasp the scale and mode of information analysis for users. Online shopping platform should consider not only the needs of "marketing level", but also "user psychology". For example, the online shopping platform can carry out "personification thinking" on label management, to carry users' ideas and service users' shopping as the starting point, so that users can see not only the information content they need, but also some information content they want and pursue [10]. For example, online shopping platforms can also hold some "user research activities" to understand the shopping aspirations of different users, and constantly improve the information push mechanism of online shopping platforms.

4.3   Solving the Problem of Information Push Content Homogenization

In view of the homogenization of information push content, online shopping platform should be attached great importance. Specifically speaking, first, in the process of information push, online shopping platform should pay attention to the real-time update and analysis of user information, so as to avoid the lag and lag of information push mechanism. In fact, the content of information push is homogenized, which is often caused by the untimely update of information. Second, in the process of information push, online shopping platforms should pay attention to the flexibility and humanity of the push mechanism. For example, users search for a 100 yuan kettle, and the push mechanism can not only push the kettle with the same price, but also from the diversified push such as style, color and performance to meet the needs of users at different levels and help users make comprehensive judgments; third, in the process of information push, online shopping platforms should pay attention to the psychological needs of users. For example, users search for clothes with a price of 80 yuan, but this does not mean that users do not consider clothes of 50 yuan or 200 yuan. The push mechanism can start from the perspective of cost performance and popular styles, so that users can see not only 80 yuan clothing, but also cheaper and more beautiful clothes with different levels. This is not only from the perspective of marketing, but also from the perspective of shopping psychology, every user hopes to buy products with good quality and affordable price.

5   Conclusion

From the perspective of marketing, online shopping platform information push mechanism is indeed a kind of progress and innovation, so that more and more users have a better shopping experience. But the management of information push mechanism should not only consider the marketing level, but also pay attention to user privacy. Some online shopping platforms take "service users" as the slogan and "information push" as the path, which damages the privacy of some users. This not only can't shorten the distance between online shopping platforms and users, but also cause many unnecessary contradictions, leading to the loss of users. In the future, the management of information push mechanism of online shopping platform should comprehensively consider various factors, always take "users" as the core, and protect the privacy of each user while trying to improve the shopping experience of users.

References

1. Li, C.: Analysis of E-commerce personalized recommendation system based on big data. Bus. Econ. Res. **2**, 69–72 (2019)
2. Zeng, F., Zou, Z., Tao, R.: Is personalized marketing bound to cause privacy concerns: based on the perspective of personified communication. Nankai Manage. Rev. **21**(5), 83–92 (2018)
3. Wei, L.: Establishment time of online shopping contract: empirical study, current legislation and due position. Soc. Sci. **12**, 91–93 (2018)
4. Zhang, H., Gu, R.: research on the influencing factors of college students' choice of online shopping platform after "95." J. Changchun Univ. Technol. (Social Science Edition) **3102**, 106–112 (2018)
5. Shi, F., Yang, L.: Lanzhou Academic Journal **10**, 139–147 (2018)
6. Zhu, Y.: Research on consumer conformity behavior in online shopping platform. Times Financ. **26**, 286–287 (2018)
7. Liang, Q.: An analysis of influencing factors of Chinese consumers' willingness to choose online shopping platforms. Huazhong Univ. Sci. Technol. **16**, 148–149 (2017)
8. Li, H., Shi, X., Ma, Y., Hang, Y.: Research on the influencing factors of college students online shopping consumer behavior. Mark. Weekly **03**, 72–73 (2019)
9. Wu, M.: Management Optimization of Online Shopping Platform Information Push from the Perspective of Marketing. Chinese Market **8**, 135–136 (2020)
10. Cui, W.: Design and development of app based on price comparison system and preferential information collection. Knowl. Econ. (5), 53–54,57 (2019)

Construction of Aesthetic Theory System of Multimedia Picture Language

Tingwei Pan[✉]

Primary Education College, Zaozhuang University, Zaozhuang,
Shandong, China

Abstract. The application range of multimedia technology is more and more extensive, and many people pay attention to the aesthetic problems in the design process of multimedia works. How to combine color matching, picture layout, graphic matching and audio video, to present different media around the theme of works naturally, harmoniously and effectively, and bring viewers beautiful audio-visual enjoyment has become an important content of multimedia aesthetics research. This paper starts with multimedia picture language. This paper analyzes and ponders the shortcomings in the aesthetic design of multimedia picture language, puts forward the concept of multimedia picture language aesthetics, and preliminarily puts forward the idea of constructing the theoretical system of multimedia picture language aesthetics. Firstly, by analyzing the characteristics of multimedia picture language, the grammatical structure of multimedia picture language is analyzed. Secondly, through the combination of qualitative research and quantitative research, traditional cognitive behavior experiment and modern eye tracking technology, the research content, design strategy, target meaning and other issues of multimedia picture language aesthetics are summarized. The research hopes to help people find the key to open the aesthetic design of multimedia pictures, improve people's aesthetic ability in experience, speculation and dialogue, and always look at the world with a beautiful eye.

Keywords: Multimedia picture · Multimedia picture language · Aesthetics of multimedia pictures · System info

1 Introduction

The emergence of information technology broadens the channels for people to acquire knowledge. Both electronic information and network information are presented in the form of multimedia. They are all "written" by using multimedia picture language.

With the continuous development of network technology and computer technology, the application of multimedia technology is more and more extensive, mainly because multimedia is intuitive, dynamic, interactive, repeatable and large-capacity, etc. Through multimedia, the original abstract and boring learning content can become intuitive through graphics, animation and other forms of expression, which improves the interest of viewers and improves work efficiency [1]. However, at present, most multimedia works, due to lack of knowledge of multimedia picture aesthetics and lack

J. Abawajy et al. (Eds.): ATCI 2021, LNDECT 81, pp. 765–770, 2021.
https://doi.org/10.1007/978-3-030-79197-1_110

of awareness of beauty design, only talk about the application of functions in multimedia production or teaching, and seldom design from an aesthetic point of view, or turn a blind eye to aesthetic content; Moreover, all kinds of textbooks introducing the use of multimedia are everywhere, all of which focus on the introduction of multimedia functions and operations, and rarely involve the application and strategy of picture aesthetics. This situation leads to the embarrassing situation of multimedia application. Even if people seriously study the use of multimedia software such as Flash and PowerPoint, they simply stack various materials and pay attention to the realization of functions without aesthetic feeling. The main manifestations are as follows: (1) The art design is understated, the design is rough and single, the surface form is pursued, and the actual artistic effect is ignored; (2) The designer's aesthetic ability is not strong, and the artistic factors in multimedia picture language are not fully displayed; (3) Multimedia picture aesthetics lacks systematicness, effectiveness and artistry.

2 Multimedia Picture and Multimedia Picture Language

2.1 Multimedia Picture

Multimedia picture is the basic unit of multimedia technology, a new information picture type after the advent of multimedia, and a comprehensive expression form of visual and auditory media based on digital screen. Multimedia picture is a window for communication between people and information. Multimedia picture is a picture based on screen display, which is different from paper page presentation. Because TV pictures and computer pictures are also displayed on the screen, this is the common attribute of the three. Compared with paper pictures, multimedia pictures are dynamic and have the characteristics of diversity, interactivity and integration. It can stimulate people's senses from multiple angles and achieve the purpose of transmitting information [2].

2.2 Multimedia Picture Language

The concept of "Language with Multimedia" was put forward in 2002 by Youze Qing, a famous educational technology expert in China. This theory makes multimedia design, development and application have rules to follow, and further promotes the improvement of multimedia application effect under the information situation. Multimedia picture language is a new language type, which is different from written language [3]. It mainly relies on "form" to express "meaning", that is, to express knowledge and thoughts or convey artistic aesthetic feeling of audio-visual perception through pictures, texts, sounds, images and other media and their combinations. It can also optimize the communication process and promote people's cognition and development of thinking through interactive functions. Multimedia picture linguistics is an innovative academic born and grown in China, and it is a new research field formed in the information age [4].

3 The Theoretical System of Multimedia Picture Language Aesthetics

3.1 Research Content

Through the characteristics of multimedia picture language, this paper studies the design of multimedia picture aesthetics. It mainly includes: First, starting with Mayer's cognitive theory, the aesthetic characteristics of multimedia picture language are regulated theoretically; Secondly, analyze the psychological process of aesthetics and cognition, and find out the factors that promote the aesthetic improvement of learners by beautiful multimedia pictures; Thirdly, the theoretical category, method and significance of implementing aesthetic design in the process of multimedia screen design. Fourthly, perfect the aesthetic idea, expand the aesthetic theory and improve the evaluation mechanism in multimedia screen design, and form the support and guarantee system of multimedia picture language aesthetics.

3.2 Research Methods

3.2.1 Combination of Qualitative Research and Quantitative Research

At the initial stage of research, qualitative research methods such as philosophical speculation, experience summarization and concept deduction are mainly adopted. For example, the aesthetic design principles of multimedia pictures are summarized and refined by appreciating a large number of multimedia courseware, and the method of summing up experience is adopted. The core concepts of the theory, such as "multimedia picture", "multimedia picture language" and "multimedia picture language aesthetics", adopt the method of conceptual deduction.

On the basis of positivism, we make quantitative research, analyze and measure or obtain data and information by quantitative means, and draw conclusions from it, including experimental research, investigation and content analysis, etc. Among them, the experimental research method is especially suitable for the research of multimedia picture language aesthetics [9].

3.2.2 Combination of Traditional Cognitive Behavior Experiment and Modern Eye Tracking Technology

Traditional cognitive behavior experiments include reaction time measurement, cognitive load test and accuracy test. In this kind of experiment, the subjects are usually randomly assigned to the experimental group and the control group, and the two groups of students use different learning materials to study. After completing the prescribed learning tasks, the learning effect is tested through retention test and migration test, so as to know the quantity, quality and psychological reaction of the test subjects [5].

Eye tracking technology, also known as line-of-sight tracking technology, records people's eye movements when they complete different cognitive tasks, and analyzes eye movement indicators such as gaze time, gaze point, gaze times, saccade times and pupil size, so as to study the individual's internal cognitive process [6]. Eye tracking has an important influence on learning effect. Eye tracking determines whether learners can further remember and think about the content and whether real learning can take place. Eye tracking technology makes an investigation on the aesthetics of multimedia picture language from the visual and perceptual factors. In addition, the research also uses a variety of research methods, such as questionnaire survey and interview, to ensure the rigor and generalization of the research conclusions.

3.3   The Construction of Aesthetic Theory System of Multimedia Picture Language

3.3.1   Grammatical Composition of Multimedia Picture Language Aesthetics

The grammar of multimedia picture language aesthetics can be designed from four functions: picture, text, sound and image. Drawing lessons from the research results of aesthetics, semiotics, psychology, multimedia technology and other related disciplines, this paper makes an in-depth exploration and demonstration of the construction of multimedia aesthetic characteristics from the theoretical framework, research contents, research methods and Mayer's cognitive theory of multimedia learning, hoping to provide a reference framework, ideas and methods for multimedia aesthetic research and aesthetic application in multimedia design. Make multimedia picture art an aesthetic tool to follow in the design and development of multimedia learning materials, thus effectively improving the quality and aesthetic feeling of multimedia pictures in the new media era [7].

3.3.2   Design Strategy of Multimedia Picture Language Aesthetics

First of all, adhere to the design principles of composition beauty, color beauty, animation beauty and sound beauty of multimedia pictures; Secondly, pay attention to the construction rules, semantic rules, pragmatic rules and their matching rules of multimedia picture aesthetics, so as to apply them to the design of multimedia learning resources and the analysis of learning process, and achieve the ultimate goal of optimizing multimedia screen design and improving the application of new media [8]. Finally, starting from the viewer's psychology and serving the new media technology, this paper discusses the realization of the aesthetic application field of multimedia pictures, raises the attention to the beauty of pictures in thought, and looks for strategies to enhance the beauty of pictures and serve the beauty of new media pictures. Establish aesthetic concepts, expand aesthetic theories and improve aesthetic techniques in multimedia applications, and form a support and guarantee system for aesthetic design of multimedia pictures.

3.3.3   The Significance of Constructing the Aesthetic Theory System of Multimedia Picture Language

Aesthetic research with multimedia pictures as the main carrier not only gives people subject knowledge, but also cultivates people's ability to know, experience, feel, appreciate and create beauty through multimedia art design, thus improving people's artistic accomplishment. Multimedia technology shoulders the historical mission, and the development of the times proves the importance of aesthetics. Aesthetics of multimedia picture language is a science, and aesthetics is the need of people's own development. The research will help people find an aesthetic palace to open multimedia screen design, improve people's aesthetic ability in experience, speculation and dialogue, and train people to always look at the world with a beautiful eye [10]. Through the aesthetic education contained in multimedia, the new generation of practical talents with applied beauty, practical beauty and creative beauty can be trained to build a harmonious and stable socialist situation.

In a word, the essence of multimedia aesthetics is restricted by the characteristics of beauty, which points to people themselves and reflects people's hope for a better future. The aesthetic elements in the multimedia screen design are the application of multimedia technology to understand multimedia from an aesthetic perspective, giving a visual angle in the sense of multimedia aesthetics, giving people the motivation to surpass themselves and pursue beauty, giving people direction guidance, and finally achieving moral perfection and aesthetic richness. The aesthetic design of multimedia pictures is carried out by examining the subject content and mining the subject implication, so as to improve people's aesthetic ability.

4   Conclusion

Facing the opportunities and challenges of new media, the use of multimedia technology should take the initiative to innovate the path, integrate aesthetics and art design, and strive to create a new situation of multimedia picture art creation, providing technical support for the deep integration of multimedia picture art design and aesthetics. On the basis of predecessors, the research hopes to systematically explore the aesthetic characteristics of multimedia application, the dilemma and causes of artistic design, and at the same time, study the way out to solve the dilemma. Starting from the current situation of ignoring artistry in multimedia design, this paper studies the aesthetic characteristics of multimedia pictures under the new situation, and explores the theoretical system of multimedia picture language aesthetics in combination with the characteristics of people's psychology and multimedia communication.

References

1. Teaching "Elements" Research Status and Reflection. Li Zelin, Shi Xiaoyu, Wu Yongli. Contemporary Education and Culture 2012(03)
2. Scientific System of Multimedia Learning and its Historical Position. Zheng Xudong, Wu Bojing. Research on the modern distance education 2013(01)
3. Zeqing, Y.: Grammar of multimedia picture language. Inf. Technol. Educ. **12**, 79–80 (2002)
4. Wencheng, G.: On the basic way of aesthetic teaching reform from the visual culture perspective. J. Hunan Inst. Humanities Sci. **02**, 95–97 (2013)
5. Cao, X..: Study on the Influence of Learning Resource Picture Color Representation on Learning Attention. Tianjin Normal University (2020)
6. Zhijun, W., Xue, W.: Research on the construction of multimedia picture linguistics theory system. Audio-visual Educ. China **07**, 42–48 (2015)
7. Qilan, X.: Thinking about the advantages of multimedia teaching in kindergarten. Sel. Essays Next **009**, 1 (2019)

8. Qingzhang, C.: Discussion on media aesthetics in multimedia application system. Electron. Publ. **09**, 9–15 (1997)
9. Chen, L.: Contemporary Construction of Digital Technology Film Aesthetics. Nanjing Art Institute (2020
10. Xinhua, Z.: Some aesthetic thoughts on multimedia works design. J. Mudanjiang Teach. Coll. (Natural Science Edition) **02**, 7–8 (2013)

The Development of Innovating Enterprise Management Mode by Using Computer Information Technology Under the Background of Internet

Ying Zhang and Shasha Tian[✉]

Business School, Central South University, Management Building,
Lu Shan Nan Lu 932, Yue Lu District, Changsha 410083, Hunan, China

Abstract. With the development of the times, computer technology has also undergone rapid changes, which gave birth to many management methods based on the development of science and technology, the most prominent of which is the innovation of enterprise management mode. The previous management model has been unable to keep up with the pace and development speed of the times. If the company wants to develop better, the company management model must be innovated, transformed and upgraded.

Keywords: Internet · Computer information technology · Innovation · Enterprise management mode

1 Introduction

The innovation of enterprise management mode is one of the most important ways and methods to realize the development of the company. If the enterprise cannot actively carry out innovation and transformation and upgrading, it will not only lose its social status in the fierce economic market competition, but also can not realize the long-term development of the enterprise. With the continuous development of the times, the arrival of the Internet provides opportunities for the management innovation of the company Based on the characteristics of enterprise development under the background of Internet, this paper studies its internal management mode, and puts forward the innovation mode of enterprise management.

2 Overview of Enterprise Management Innovation

2.1 Business Management Innovation Direction

Economists have carried out multifaceted and multilevel research on the development, transformation and upgrading of companies, and have made a new definition of enterprise management innovation. In his point of view, the essence of entrepreneurs in the company is to innovate the management mode. Innovation is actually a relational production function of production, in which the combination of production factors and

J. Abawajy et al. (Eds.): ATCI 2021, LNDECT 81, pp. 771–775, 2021.
https://doi.org/10.1007/978-3-030-79197-1_111

conditions of the company is realized, and then the new company management combination is put into the production system. In the process of enterprise management innovation, managers should first investigate the market, use new management thinking mode and Internet technology to establish a management system in line with their own development [1]. This form of innovation is more in line with the development of the times, and can better realize the long-term development and long-term interests of the company.

2.2   Important Significance of Enterprise Management Innovation

Under the dual situation of the growth of the Internet and the impact of the epidemic situation, the development of e-commerce is undoubtedly a major problem in the form of corporate management in China. The close combination of e-commerce and network development has further promoted the transformation of China's companies, among which Tencent, Alibaba and other large enterprises have done better in e-commerce. These companies have well developed and realized the application of computer technology in different kinds of Companies in the society. The innovation of these companies in the company management has opened the door to the scientific and technological development of the company and started the company management innovation in the Internet era [2]. In the era of fierce market competition, the previous management mode cannot achieve the rapid exchange of scientific and technological information, while the modern company management innovation has achieved the courage to try, with distinct development ideas.

3   Enterprise Management Mode Under the Background of Internet

With its excellent characteristics of rapid information dissemination, continuous technology update and wide application in society in the Internet era, the previous company management mode has been constantly subverted. In recent years, with the positive development, the Internet plus mode has emerged. This strategy has brought forth new ideas in the process of company management, and is a new economic form in the Internet era [3].

Virtual Economy: Virtual economy is not an invisible economic situation, its essence is an online economy. For example, news, games and other products established by the network in the market, their cost is very low, but the income is relatively high, which can realize the replacement of the real industry.

Experience Economy: The web has great advantages in terms of customer experience with products. In this way, customers' feedback information is collected, and customers are actively encouraged to participate in the design of the product to achieve the actual experience of customers on the product.

Platform Economy: there are many platform economy models in the network. For example, app stores, games and so on, these platforms are easy to establish, and their development model is not constrained by the environment and economic development system. On the one hand, it can reduce the cost of customers, on the other hand, it can meet the actual needs of customers [4].

4 Current Situation of Traditional Enterprise Management

4.1 Enterprise Management Mode Is Backward

One of the important problems to be solved is that the enterprise management mode is not advanced enough. Although the Internet era develops rapidly and is widely used in the economic market, the traditional ideas are not easy to change when our country's company management. Many use the network management mode, although in the form of modern management, but its internal did not make the corresponding changes, that is, the company management surface, did not realize the essence of Internet enterprise management, just simply imitate the development strategy of some foreign advanced countries, can't start from their own company development status, so, in our country's scientific and technological development There will be some cases of learning to walk in Handan in the management of technology and chemical companies [5].

4.2 Digital Marketing of Enterprise Management Lags Behind

With the rapid development of modern science and technology, Internet technology is widely used in modern company management. The operation of the network makes the company's commodity information spread rapidly, and users can know the company's product information when they read the information. In today's information age, users are disgusted with spam on the Internet. Many companies often use this way in brand promotion, but in fact, this way cannot improve the satisfaction of users, so how to carry out the company's digital marketing has become the key direction of company management innovation under the network background. Internet is a kind of Internet, which can play the role of propaganda and development in the rational use of economy and society. But if it cannot be used reasonably, it will limit the management innovation of enterprises. In the new era, company management innovation must rely on the development of the Internet to achieve sincere communication with users, and establish brand loyalty among users [6]. Network brings a lot of business opportunities to the development of the company, but in the network promotion activities must be accurate positioning, otherwise it cannot achieve the innovation of company management.

5 Innovate the Strategy of Enterprise Management in the Internet Era

5.1 C2B Business Model Application

If the company wants to achieve management innovation in the market competition, it must first realize the communication with users. In the context of the Internet, C2B mode is derived, which truly realizes the development of the company with users as the core and consumers driving the development of the company. Customer first is the essence of the company's development. This mode has been extended to various industries. For example, the Internet mode has been used in catering, and online booking and group buying have been realized. Meituan.com and Baidu group buying have done better in this aspect. This way has changed the previous way of catering telephone reservation, and realized the convenience of consumers' dining. The Internet age has brought great changes to people. It has not only changed the organizational form of the company, but also innovated the company team, making the distance between the company and users closer, and promoting the development of the company. In the previous company management thought, managers took products as the core, did not pay attention to the flow of products in the market, and did not pay attention to the word-of-mouth management of products in the society [7].

5.2   Enterprise Brand Management Innovation

If we want to realize the management innovation of the company, we must pay attention to the management of the company brand. In the process of rapid development in the Internet era, many companies have changed their internal management mode. In today's society with increasingly fierce market competition, it is a wise move for a company to develop a business management model that carries out marketing and brand building at the same time. First of all, the management of the company should be based on brand building and based on marketing strategy [8]. The specific manifestation of this approach in the Internet era is: on the one hand, we pay attention to the building of the company's brand, improve the publicity of the company's brand in the market, and specifically protect the position of the brand in the hearts of customers. The company's brand building is divided into two aspects, one is the construction of corporate culture, the other is product satisfaction.

5.3   Training of Scientific and Technological Talents

In the process of enterprise management, scientific and technological talents are the key elements of the company's development. Talent is not only the bridge and link between network technology and traditional management, but also plays a decisive role in the dissemination of advanced ideas. At present, the rapid development of social platforms such as microblog and wechat fully shows the essence of the Internet, and also realizes the communication and information sharing between company users and company culture [9]. From small aspects to individuals and the whole society, information sharing based on the Internet is affecting our way of life. And the transmission of these information in the company must be realized by talents with scientific and technological ideas. Therefore, in order to better carry out scientific and technological management, we must pay attention to the cultivation of talents in the company. Social management is the way to achieve timely communication. In this link, it realizes the two-way management of talents and timely communication in the company management. At present, many companies have used talent management software, such as Jingdong, Suning, etc. have begun the social era of talent management. This way not only solves the uncontrollable problems in talent management, but also realizes the communication and cooperation between companies and talents. In the company's talent management, the company needs to pay attention to the transfer of positive energy of enterprise development between employees, and promote talents to make better use of the convenience of network communication in the new era, so as to realize the interaction with the market economy [10].

6 Conclusion

With the rapid development of the Internet, information technology is also developing, and the network plays an indispensable role in company management. In recent years, network has been applied at many levels and in many aspects in the innovation of corporate management. It has made great contributions to the promotion of product information, brand building and talent training of the company. In the information publicity, it plays a timely communication with consumers and gets consumer feedback, which is beneficial to the company to improve its shortcomings. Therefore, this paper studies the management innovation of the company, proposes the management mode of the enterprise under the background of the Internet, and puts forward specific innovation strategies according to the current situation of the company management.

References

1. Editorial Department of Enterprise Management. Management change in the internet era. Bus. Manage. (5) (2012)
2. Xuegong, Z.: Research on Management Innovation of Smartphone Industry in Mobile Internet Era. Wuhan University, Wuhan (2013)
3. Tao, D.: On the construction of enterprise group financial management system in the internet era. Theoretical Edition, New West (22) (2013)
4. Chuan, J.: Innovation analysis of enterprise management mode under the environment of big data. Bus. News. **16**, 74–76 (2019)
5. Lei, Z.: The application and innovation of computer information technology in modern enterprise management. Econ. Res. Guide **25**, 15–16 (2017)
6. Xiang, Y.Y., Xin, W.J., Qin, Y.H., Ying, Z.Q.: Applying computer information technology to innovate nursing management mode. Contemp. Med. **5**, 105–106 (2007). (Academic Edition)
7. Liang, W.H.: Computer electronic information technology and project management model thinking research. Comput. Knowl. Technol. **17**(4), 249–250 (2021)
8. Ping, T.Y.: The significance and application of computer information technology in archives management of public institutions. Farm Staff **18**, 189 (2019)
9. Kun, Y.: Application of computer information technology in modern enterprise management. Comput. Prod. Circ. **9**, 4 (2018)
10. Qiang, W.: Application analysis of computer information technology in modern enterprise management. Sci. Technol. TV Univ. **4**, 7–8 (2015)

Analysis on the Sustainable Development of Internet Economy Under Big Data

Lingyan Meng[✉]

Faculty of Management and Economics, Universiti Pendidikan Sultan Idris (UPSI), Darul Ridzuan, 35900 Tanjong Malim, Perak, Malaysia

Abstract. Against the background of big data, comprehensive conditions with low transaction cost, multiple data types and high transaction efficiency were presented in construction of Internet economy. It serves as significant transaction type in social and economic construction at present. However, there are still some problems in the development process of economy model of the network data. It is necessary to conduct research from various aspects to avoid the disadvantages of the data economy to create a healthy and green environment for progress of the Internet economy, and finally realize sustainable development of the economy in Internet under background of big data.

Keywords: Big data · Internet economy · Sustainable development

1 Introduction

As the advancement of IT and development of algorithm model, Internet economy has become one of the main paths of economic development. Big data algorithm provides technical support for progress of the internet economy. The convenience of network makes the digital economy model present beneficial characteristics such as low trade cost, high transaction efficiency and multiple data types, which greatly promotes the efficiency of economic development. But there are still some problems to be solved in the process of the development of network economy. To promote sustainable development of the internet economy, platform based trading can be made available with strict access mechanism, accurate screening of effective information, innovation of a variety of network economy development models. Besides, the economic development should be trapped in high efficiency and low capacity model by adopting the development requirements of the green economy system in the economic development.

2 Characteristics of Internet Economy Based on Big Data Era

2.1 Low Transaction Costs

Digital economy is the new driver to promote the high-quality development of economy in China [1]. A good use of the digital economy model can contribute to effective development of economy. Under model of economy in internet, big data provides

J. Abawajy et al. (Eds.): ATCI 2021, LNDECT 81, pp. 776–781, 2021.
https://doi.org/10.1007/978-3-030-79197-1_112

effective trading reference for construction of digital economy. Flexible use of big data technology could effectively reduce operating costs and transaction costs in the operation process [2]. The sellers could select big data to predict buyer's interest in the transaction, accurately control the inventory, and effectively control the amount of capital input.

2.2   Variety of Data Types

Under the internet economy model, trade information is numerous and complex, and transaction information sources continue to emerge in big data era, which offers new opportunities and challenges to enterprises [3]. While the network platform provides the communication opportunities for the various transaction subjects, the complex data model has also become an important feature of economy in internet. As far as buyers are concerned, network economy provides more choices for them to purchase goods, and there may be dozens of merchants for the same kind of goods.

2.3   High Operating Efficiency

Platform economy is conducive to the creation of diversified business models, and the application and development of big data also promote the improvement of operation efficiency of internet economy [4]. Taking the online shopping model for example, big data can provide consumers with the most efficient reference for purchasing in the shortest time with the Internet economy model. With the help of objective data such as sales volume and evaluation, consumers can determine their purchasing goals more quickly and complete a consumption activity in a short time.

3   Problems in Construction of Internet Economy

3.1   Difficulty in Guaranteeing Information Quality

Although big data provides more trade reference data for both parties in internet transaction, the information source of these data is still characterized with subjectivity to a certain extent. It is difficult to guarantee the quality of the data and information that both parties can refer to. Some incidents in information and data fraud, such as transaction evaluation, have also impacted the sustainable development of the internet economy.

3.2   Low Access Standards

Due to advancement in the internet economy, diversified value creation subjects were presented [5]. Against the background of big data, value subjects of network economy are numerous and complex, and the access threshold of trade is low. Most legal persons, unincorporated organizations and natural persons can become the subject of platform trade through simple registration, which leads to the mixed development of the appropriate transaction subjects and the improper subjects in economic development, increasing uncertainty for trade security.

3.3 Big Data and Protection in Privacy

Optimizing big data processing is an inevitable trend of the development of internet economy [6]. In the process of internet economic transactions under the background of big data, users' transaction information is collected and sorted by the platform into information materials that help facilitate transaction efficiency. The platform will recommend corresponding commodities to users according to their consumption habits to promote user consumption. However, in the process of big data collection, the protection of this part of privacy information data still needs to be further promoted. In the process of data collection, the privacy of users has also encountered challenges.

4 Paths to Promote the Sustainable Development of Internet Economy

4.1 Strict Access Mechanism

The sustainable progress of economy in internet can be made to ensure eligibility of transaction subjects, and credit investigation of subjects is the first step for risk identification in risk control and management [7]. Transaction on the internet platform can effectively reduce trade costs. Most small and medium-sized enterprises tend to create "online stores" for transaction on the network platform. Although the main body of these enterprises is qualified at the time of registration, they may lose the ability to prevent and control accidents in the later trade process due to business risks. For these small and micro enterprises, real-time data detection in the process of trade is also very necessary except for strict market access mechanism. For these small enterprises with insufficient capital reserves and incapability to deal with business risks, they can make use of big data to monitor their credit information in real time during the transaction process. In context of big data, expanding the online platform can also be achieved based on the data source to train and optimize the credit model [8]. Through the data model, real-time credit investigation monitoring is carried out on the subjects of economic transactions, and the change of corporate credit information is paid close attention to while the platform trading access mechanism is strictly performed, so as to ensure the security of transactions to the maximum extent and reduce the risks caused by improper subjects in the process of transactions.

From the perspective of practical operation, transaction subjects of internet economy usually use multiple platforms to conduct trade activities, and strict control of user credit investigation can also be realized through data interconnection of various platforms. Big data provides convenient conditions for the integration of user information, and its connotation lies in the availability of data. Data in credit and transaction data is in a relatively open state under the background of big data. Besides, acquisition, review of user data by platform is very convenient under the mode of internet information exchange. In order to facilitate the access process while strictly enforcing the access qualification, cooperation should be further strengthened in internet economy platforms and "green channel" for the transaction subjects with qualified credit data should be made available to speed up the audit, and avoid the repeated review of qualified subjects.

4.2   Improving Internet Economy System

Applying big data to the internet economy can further optimize and improve the internet economy system [9]. The real economy serves as basis of network economy, and the construction of internet economy is based on development of the real economy. To guarantee the vitality in development of internet economy, perfect network economy system is the key point that needs to be adjusted and grasped. To establish good internet economy ecosystem, big data model can be utilized to effectively predict the trend of the real economy, which adjusts structure of internet economy through data analysis and reasoning, and standardizes and optimizes the system of internet economy by means of adapting to the structure of real economy to realize sustained development of Internet economy.

In addition to adapting the internet economic system to the structure of the real economy, it is also an important part to achieve sustainable development by using big data to adjust the direction of economic progress. In light of the function of big data, data model can provide more targeted suggestions for progress in internet economy and enable logical structure of economic development to adapt to the demands of our times in progress. Analysis and detection function of big data can also be used to monitor the monopolistic behavior in market transactions. Big data is used to conduct scientific analysis on business behaviors of enterprises, and timely and effective supervision is carried out in case of monopolistic behaviors or unfair competition behaviors of enterprises, so as to maintain the fairness and security in internet based market transactions. In addition, economic consumption, under the big data logic, is proved to be characterized with predictability. For some small-micro enterprises that keep abreast of the development of the times, the governments can take advantage of the the results calculated by big data to provide funds and policy support to increase the share of "sunrise industries" that are beneficial to the social development in boosting internet economy, providing green business environment for development of emerging Internet industry on equal footing to stimulate the vitality in the development of the internet economy.

4.3   Progress in Green Economy

In addition to commercial attributes, internet economy is also endowed with social attributes. To maintain sustained development, the use of big data could effectively develop social attributes of the internet economy, promote the development of the overall social economy based on development of the internet economy, and people's living standards can be improved. From the actual situation, with its advantages in low cost, high efficiency and the reserve of large number of customers, internet trade has become an important means to implement targeted poverty alleviation and accelerate poverty alleviation under the vigorous promotion and advocacy of the state [10]. Some poverty-stricken counties have achieved "poverty alleviation" by means of "live streaming commerce", and the internet based economic and trade model has provided new economic development opportunities for these areas.

From the perspective of the social function of economic development, the fundamental aim in enhancing progress of internet economy is to improve living standards of people, and the starting point and foothold of development should be based on improving people's living standards. Therefore, to achieve the sustained development, the focus of the development of the internet economy should also include use of big data to improve people's lives, taking the full advantage of social function of the internet economy, and helping the poor counties and cities to develop the economy with data-oriented and low-cost online transaction mode. In development of the internet economy, the operation difficulty in network platform should also be correspondingly adjusted.In order to attract more people to take their part in boosting the sustainable development of the Internet economy, operation difficulty can be adjusted, the transaction process can be simplified, and the qualification examination of the trading subjects can be realized in a more clear and convenient way when designing the software of the trading platform.

Generally speaking, the sustainable development of internet economy can be made with the optimization of data model, the strict market access mechanism in economic transactions, and the qualification examination of transaction subjects based on credit investigation of big data. Due to complexity of information sources in big data era, control of data information quality also needs to be strictly standardized. In addition, in the process of constructing the network transaction structure. Integrating the internet economy with the development of the real economy based on the development needs of the real economy, and building a green and healthy development model of the Internet economy based on the real economy are also necessary.

References

1. Zhenzhen, S.: Research on the path of high-quality development to empower digital economy in the new infrastructure perspective. Manage. Adm. **12**, 145–149 (2020)
2. Yuxiang, W.: Thinking on the application of big data in internet economy. PR Mag. **7**, 281–282 (2020)
3. Song, X., Lv, G., Du, N., Liu, H.: Research on the use behavior of information sources in competitive intelligence network of small and medium-sized enterprises under big data. J. Mod. Inf. **41**(1), 88-93+110 (2010)
4. Jia, L., Lei, L.: The platform economy model and development strategy in the Internet 3.0 era. Enterp. Econ. **1**, 64–70 (2021)
5. Jingkun, B., Ya, Z., Sihan, L.: Research on the relationship between knowledge governance and value co-creation in platform-based enterprises. Stud. Sci. Sci. **38**(12), 2193–2201 (2020)
6. Meng, Z.: The essential analysis of high-quality development of internet economy. Econ. Res. Guide. **29**, 30–32 (2020)
7. Jie, L.: Research on risk control system of big data of P2P lending in China. Manage. Adm. **1**, 161–165 (2021)
8. Xuejun, C.: Internet consumer finance: technology application, problems and regulatory countermeasures. Contemp. Econ. Manage. **42**(7), 83–91 (2020)

9. Qiong, W.: Sustainable development of internet economy under the background of big data. Trade Fair Econ. **14**, 24–26 (2020)

10. Xiufeng, L.G., Yating, W., Xianming, K.: Internet trade development and income distribution gap. J. Ind. Technol. Econ. **39**(10), 107–115

On the Application of Video Monitoring System in Public Security Work in the Era of Big Data

Xuming Zhang[✉]

Department of Public Security Management, Liaoning Police University,
Dalian 116036, Liaoning, China

Abstract. With the rapid development of Internet information technology, the society has entered the era of big data, video surveillance system has been widely used in public security work, and has realized the leap forward development of public security work. However, there are still some problems in its operation, such as the immature application of intelligent video analysis security technology, the difficulty in the construction of big data public security video monitoring system, the selection of some cameras, and the unreasonable structure. Therefore, it is necessary to discuss the application countermeasures, such as the selection of safe standard video monitoring equipment, and the construction of large database by public security and major enterprises.

Keywords: Big data · Video surveillance system · Application · Public security work

1 Introduction

With the development of network technology, now has entered the era of big data, today's City video monitoring system is more and more mature, through the use of computer technology in the public security work has played a key role, but the public security organ's video monitoring system still has some defects. In order to increase the efficiency of public security work and crack down on all kinds of criminals, the public security department should pay attention to the problems existing in the current video monitoring system, deeply optimize the video monitoring system, and provide strong technical help for all kinds of work.

2 Background of Video Surveillance System Based on Big Data

The wide application of information technology in public security work in the era of big data has added fresh blood to public security work. The application of big data in public security work has initially achieved obvious results. For example, Hubei Anyun, established by Hubei Provincial Public Security Department, has made use of the information convenience of big data to achieve the integration of public security and

© The Author(s), under exclusive license to Springer Nature Switzerland AG 2021
J. Abawajy et al. (Eds.): ATCI 2021, LNDECT 81, pp. 782–787, 2021.
https://doi.org/10.1007/978-3-030-79197-1_113

social control, criminal investigation, technical investigation, network investigation and visual investigation, and cracked up to 150000 criminal cases. From April to June 2015, Shanghai public security organs used video network to supervise and control Control technology has cracked more than 14400 theft cases, with a year-on-year growth of 22.5%. The key reason is that the application of big data practice system has improved the comprehensive research and reconnaissance ability of some public security information to a certain extent, thus enhancing the accuracy and initiative of the public security department in attacking criminals. According to the current data application platform, monitoring data storage, and the face of "hacker' attack and a series of problems [1], how to apply the technology to the actual combat of public security organs is the urgent problem to be solved.

3   Application Status of Video Surveillance System in China's Public Security Work

The establishment and application of video monitoring system in public security work is playing a very important role in arresting crimes and protecting social harmony and stability. The video monitoring system in the public security work often needs to deal with a large amount of data and storage, needs a large number of personnel to deal with the technical maintenance of the monitoring system, and more monitoring equipment is installed, and a lot of manpower and material resources are invested in the maintenance equipment installation. Nowadays, the video monitoring system has become a key tool in the public security work. It often needs to retain a large number of video image content. However, it is difficult to process the information and requires high installation of the monitoring system. There are still some problems in video monitoring system, such as information resource sharing, video monitoring system building, video management system, video monitoring equipment coverage, video monitoring management team building and so on [2]. In any case, the video monitoring system is constantly upgrading, cooperating with major video monitoring enterprises, constantly improving the video monitoring system, so as to make it play the largest role in the public security work.

4   Main Problems in the Application of Public Security Video Monitoring System

(1) The application of intelligent video analysis security technology is not mature
   With the establishment and popularization of ultra clear monitoring system, the video image becomes clearer and the video resources that can be called become richer. In particular, the continuous enhancement of intelligent video analysis technology has given fundamental help to the application of video big data. However, all big data applications must aim at data security. In the era of large-scale networking, a large amount of information and artificial intelligence, the security hazards and methods of video surveillance system have changed greatly [3]. However, most of the video network systems always lag behind in the field of

security. In recent years, the promulgation and implementation of national security standards, the cooperation between the company and public security organs, and the replacement of technology and products will well promote the implementation of public security video monitoring system in the era of big data.

(2) The construction of big data public security video monitoring system is difficult

The work of public security organs is to maintain social stability. Most of what they learn are the knowledge and skills of how to catch criminals and how to crack cases. In the era of big data, they can learn how to apply the video monitoring system. However, for the production and maintenance of video equipment, especially the storage of big data, the public security organs can't do everything. The establishment of this department wastes a lot of police resources. With regard to the creation and development of security standards for video surveillance system, the public security organs naturally can not do it by themselves. They can only cooperate with relevant monitoring data science and technology research institutions and monitoring product development companies in the market [4], and devote themselves to the research in this aspect, so as to effectively implement the norms of big data use and application into the product development and processing scheme.

(3) The selection and structure of some cameras are unreasonable

The traditional video monitoring equipment of public security department includes general gun camera, ball machine, face camera, micro bayonet camera, snap camera, some monitoring equipment that can structure the front-end video stream, and some simple and intelligent monitoring equipment. With the development of the times and the continuous replacement of video monitoring technology, the styles of monitoring equipment are increasing. Because there are many choices, the personal skill standards of the created staff and users are higher. Usually in this important step, due to the urgent start of the project and other reasons, the front-end selection and processing are hasty and random. In addition, the allocation of video monitoring equipment is unreasonable. In some densely populated areas, the placement of video monitoring equipment is unscientific, and many cameras are set around the streets in many urban areas. However, there are few cameras in places with dense traffic and high crime incidence, such as underground parking lots and public squares in downtown areas. The location is simple, and only one-way monitoring can be implemented [5], which will lead to criminals taking advantage of the opportunity to work here What kind of video surveillance dead corner crime.

5   The Assumption of Perfecting the Application of Video Monitoring System in Public Security Work

(1) Standard equipment must be used in public security work to ensure the security of video surveillance.

For the first time in China, based on the technical requirements of video monitoring network information security, the technical standards for the security maintenance of video monitoring network video information and its control signaling data in public security are formulated, which are in line with the information security scheme design, system detection and related equipment development and detection of video monitoring system in public security. In the video surveillance network work is very hot, the network security situation is also slowly showing. Under the influence of the construction idea of ``focusing on construction, focusing on application, and neglecting security', when establishing and using the network video monitoring system under big data, the assets are not clear, the management is not coordinated, the front-line equipment is easy to replace under abnormal conditions, the instructions are simple, and the system security problems appear, which are easy to be cracked and used [6]. As long as the video monitoring system is invaded by others, it is very likely to cause the leakage of video information. Gb35114–2017 is the first technical requirement based on the information security level of video surveillance network in China. It is the first time to propose a detailed standard for the information security of public security video surveillance. It is a technical reference for improving the information security of public security video surveillance as a whole.

(2) Public security and major companies jointly build big data and cloud services.

Dahua Co., Ltd. and Jiaozhou public security bureau established the police big data application laboratory. In mid June 2018, Jiaozhou Public Security Bureau of Shandong Province and Dahua Co., Ltd. formulated a strategic cooperation agreement. Both parties are willing, fair, win-win, and create a strategic partnership based on serving the establishment of Ping'an. Therefore, they created a "police big data application laboratory" to jointly develop the development path of police enterprise cooperation.

Ruian technology and qiniuyun join hands to create big data products. On October 12, 2018, qiniuyun and Beijing Ruian Technology Co., Ltd. signed a strategic cooperation treaty. Comprehensive and long-term strategic cooperation has been achieved in security, network information security, big data services and other important fields. Through in-depth integration, Ruian technology will work with qiniuyun to create the top network big data products in China [7].

The three research institutes of the Ministry of public security and Wangsu science and technology work together in strategic cooperation. At the 2018cdn technology integration and Application Forum in Shanghai, the third research center of the Ministry of Public Security (national network and data system security product quality supervision and inspection center) and Wangsu technology company formally signed an integrated strategic cooperation treaty [8]. According to the treaty, the two sides will implement overall and in-depth cooperation in network security, explore network security together, especially the formulation and implementation of cloud security industry requirements, and support the creation of network security.

(3) New generation video monitoring system video structure development and Application.

The addition of artificial intelligence technology (hereinafter referred to as AI) makes the whole security industry step into a top-down innovation. In the past, the public security industry has been slowly stepping into intelligence. The security industry pays more attention to the high connection of technology and application. AI is not only the addition of camera technology. New security system, the industry will all apply AI technology, so that security products to achieve the ability of automatic identification of criminals, security process from passive to active implementation change. In the process of public security monitoring, video monitoring has already realized the intelligent monitoring, dynamic face control, face recognition and capture of the whole city, but at this time, how to quickly find important information in a large number of video information is obviously an important topic [9, 10]. But there is only one way to deal with it, that is, video structure.

(4) Improve the video monitoring management system of public security work.

Although the domestic public security departments have perfect management system, the management system of video monitoring system is still not mature enough, so we need to improve the management system of video monitoring. Public security departments should work out a set of real and effective video monitoring management system suitable for their own development in accordance with the scientific outlook on development and the relevant national laws. And provide video monitoring management workers with the opportunity to learn and train professional knowledge.

6 Conclusion

Today's Internet era is a revolution, which is "a war that has not started but has started". Public security organs should make full use of cloud platform resources, integrate various fields with public security organs' video monitoring system, establish a new policing mode of cross-border resource integration management and attack, and establish data monitoring centers in public security organs all over the country, Integrating information resources, from data management to cracking down on new crimes, realizing the big data and integration mode of public security organs at each level, strengthening the police with science and technology, cracking down on three-dimensional integration, and integrating the coordination ability among various kinds of police, the strategic cooperation between public security organs and science and technology research and development companies will be further deeply integrated and implemented in an all-round way.

References

1. Cheng, J., Pan, P.: Application of cloud computing technology in public security video surveillance system. Comput. Knowl. Technol. **05**, 244–246 (2019)
2. Zhang, J.: Application of smart security in public security video surveillance. Smart City **06**, 4–7 (2019)
3. Zhao, W., Zhao, Y., Cheng, S.: Overall architecture analysis of public security video image information application system. China Secur. Certification **02**, 15 (2018)
4. Ma, Y.: Interpretation of 2018 global terrorism index report. Int. Res. Ref. **2019**(2), 37–43, 55 (2019)
5. Zhang, X.: Research and application of face recognition technology in public security video surveillance. China Secur. Technol. Appl. **01**, 31–36 (2019)
6. Hu, J.: New direction and aspiration of UAV technology application. China Equipment Eng. **02**, 137–140 (2017)
7. Zhen, W.: Big data era and intelligent video analysis technology. Commun. World **03**, 49–53 (2019)

8. Zhang, J., Zhang, K., Wang, J.: Current situation and development trend of low altitude anti UAV technology. Progress Aviation Eng. **9**(01), 1–7, 34 (2018)
9. Zhang, C., Hu, C.: Research on the construction of public security technology specialty under the background of intelligent policing. Public Secur. Educ. 2019, (12) (2019)
10. Wang, T., Qi, X.: Detailed design of intelligent police system architecture. Comput. Network **42**(11), 3–4 (2016)

Fungal Industry Supervision System Based on Zigbee Technology

Lifen Wang, Hongyu Zhu, Lingyun Meng, and Mingtao Ma(✉)

Jilin Agricultural Science and Technology College, Jilin 132000, Jilin, China

Abstract. In order to improve the production efficiency of the fungus industry, save the planting cost of the fungus industry and build an efficient production mode of the fungus industry. We are based on Internet plus's "smart bacteria industry" and supported by Internet of Things technology. The fungus intelligent greenhouse is constructed, and the ZigBee protocol stack and networking are used to realize the crop environmental monitoring system. The accurate acquisition and stable transmission of environmental temperature, Humidity, gas concentration, light intensity and other data are formed, and the remote monitoring and management system for agricultural production is constructed.

Keywords: Zigbee · Sensor · Fungal industry

1 Introduction

ZigBee is a two-way wireless communication protocol. Compared with other communication technologies, ZigBee has many advantages such as stable communication, low power consumption, low price, low delay, etc. [1]. Users can remotely monitor the data in the intelligent greenhouse in real time through the mobile terminal and the PC terminal. ZigBee technology has many advantages in the detection of intelligent bacteria industry, which can more effectively meet the requirements of accurate acquisition and stable transmission of environmental temperature, Humidity, gas concentration, light intensity and other data in the modern bacteria industry production process, construct a remote monitoring and management system of bacteria industry production, create a good environment for the growth of fungi crops, further enhance agricultural production efficiency, reduce labor force, save agricultural costs, realize the overall upgrade of agricultural production mode, and add a new impetus to the development of modern agriculture in our country [2, 3].

2 Overall System Design

This system is a smart bacteria industry detection platform based on ZigBee protocol stack and networking. The system is mainly composed of a coordinator and a terminal node monitoring center. We connect each sensor through ZigBee networking between the coordinator and the terminal node. The sensor nodes are composed of temperature and humidity sensors, light sensors, gas sensors, etc. and CC2530 modules [4, 5]. The coordinator and the terminal node are networked in a star topology. The terminal node

J. Abawajy et al. (Eds.): ATCI 2021, LNDECT 81, pp. 788–792, 2021.
https://doi.org/10.1007/978-3-030-79197-1_114

is responsible for reading and uploading the data of each sensor to the network. The coordinator is responsible for collecting the data collected by each terminal node and sending the data to the monitoring center through WiFi. The monitoring center computer uses the visual graphical interface developed by VS tool, which can receive, process and display all the data sent from each monitoring area. In addition, the fungus crop data of the monitoring center can be viewed in real time through the mobile phone APP, so that users can view the crop information anytime and anywhere, and the number of terminal nodes can be appropriately increased according to the crop growth environment to increase the reliability of the data (Fig. 1).

Fig. 1. Overall system design

3 System Design

3.1 System Hardware Design

ZigBee networking system is composed of a plurality of terminal nodes and a coordinator. The terminal nodes are composed of a CC2530 module and a variety of sensors. The terminal nodes can realize the data acquisition of crop environment. The MCU in our terminal node uses the CC2530 chip of TI Company [6–8]. As a chip for ZigBee networking, it has lower power consumption and higher performance than the previous generation chip, and is suitable for the overall adaptation of IEEE 802.15.4 protocol and ZigBee software IAR. CC2530 has an intelligent sleep mode compared with ESP8266 using WiFi of the same grade, in terms of battery usage, CC2530 is about 100–200 times longer than ESP8266. The most classic application of such timers as CC2530 is to act as a real-time counter, or change it to a wakeable timer to jump out of power supply mode 1 or power supply mode 2. CC2530 is also much larger than other chips in terms of low power consumption, so we use it as a chip for bacteria detection (Fig. 2).

Fig. 2. CC2530

Table 1. ZigBee compared with other technologies

	WLAN	Bluetooth	ZigeBee
Applied range	Video	Short distance instead of Wirelin	Monitoring and control
Battery use	1–5	1–7	100–1000
Number of Network nod	30	7	255
Broadband	11000	1000	20–250
Distance	1–100	43475	1–75
Merit	Strong adaptability	Easy to operate	Low reliability cost

3.2 System Structure Design

In order to ensure the stability of ZigBee network, we optimized the ZigBee network structure on the premise of ensuring the system performance. In the various networks of ZigBee system network, we finally chose the centralized network topology structure. Star topology network has the advantages of relatively easy maintenance and management, flexible reconfiguration, easy fault isolation and convenient detection, ensuring the stable operation of ZigBee network (Table 1).

3.3 Sensor Module Design

For the environmental data needed to be collected in the current bacterial production, it is necessary to update and optimize the sensor module in real time, so as to ensure the low error detection of sensors such as temperature sensor, light sensor, Humidity sensor and so on, and to ensure the real-time acquisition of crop growth environmental data.

3.3.1 Setting of Temperature and Humidity Module

In the process of setting up the temperature and humidity module, we need to optimize each node that collects data. In the optimization process, the first thing to ensure is that the working voltage of the module should be within the range of 5V ± 0.3 to ensure the service life and accuracy of the sensor. The ZigBee wireless data channel is used to upload the data collected by the sensor in real time to realize the real-time monitoring of temperature and humidity information. Therefore, we use DHT11 temperature and humidity sensor to collect temperature and humidity.

3.3.2 Setting of Light Sensor

The light sensor is mainly a photosensitive module, which is mainly used to collect the light intensity. When the light intensity changes, the photosensitive module will transmit the data to ZigBee network through the network at the simulation port and carry out real-time monitoring.

3.3.3   Gas Sensor

Gas sensor (MQ-2) has good sensitivity to various harmful gases and combustible gases in a wide concentration range. Due to the different internal structures of various substances, it determines their selective absorption to light of different wavelengths, i.e. substances can only absorb light of a certain wavelength. The absorption relation of matter to light of a certain wavelength obeys Lambert2Beer's absorption law. Taking CO_2 analysis as an example, the infrared light source emits infrared light of 1–20 μm. After absorbing through a certain length of gas chamber, passing through a narrow band filter of 4.26 μm wavelength, the infrared sensor monitors the intensity of infrared light of 4.26 μm wavelength to express the concentration of CO_2 gas [9, 10].

4   Conclusion

Internet of Things (IOT) is a new type of industry in China's future technology. Agriculture is the most important thing in China's development. ZigBee protocol stack and networking are new breakthroughs in the development of bacteria industry. In this scheme, ZigBee protocol stack is combined with networking and Internet of Things, and ZigBee is used to collect data on the growth environment of crops. Compared with other WiFi methods, this scheme has more convenient networking, faster transmission rate and more accurate data transmission. Compared with the traditional agriculture, the intelligent bacteria industry of this plan has realized precision and intelligence, fundamentally saved financial, human and material resources, has broad application prospects, and is of great significance for the realization of our intelligent bacteria industry.

Acknowledgements. This work was supported by. Jilin provincial department of education project support project (JJKH20190975KJ).

References

1. Zhu, J.: Research on intelligent fungal industry monitoring system based on wireless sensor network. Hunan Agric. Mach. **3**, 77–79 (2014)
2. Xi, Z.: Applied research of internet of things in smart agriculture. Baosteel Technol. **2**, 46–48 (2018)
3. Zhao, X., Liu, K.: Smart agriculture real-time acquisition and remote control system based on zigbee technology. Agric. Eng. Technol. **6**, 35–36 (2017)
4. Chen, B., Jie, J., Zhang, W., et al.: Design and implementation of intelligent agricultural system 7 based on Lora. Comput. Measur. Control **26**(10), 128–131+136 (2018)
5. Li, X., Li, S.: Research on intelligent monitoring system model of agricultural greenhouse based on Internet of Things technology. Inf. Syst. Eng. **2018**(7), 121 (2018)
6. Liu, S., Zhang, W., Wei, C.: Design and research of intelligent agricultural monitoring system based on android. Internet Things Technol. **7**(04), 79–80 (2017)
7. Chen, X., Liu, S.: Design of agricultural environmental data collection system based on Internet of Things. Lab. Res. Explor. **37**(7):66–68+105 (2018)
8. Fu, C., Lei, W., Zhu, W., et al.: Intelligent integrated table design based on Internet of Things. Lab. Res. Expl. **36**(6), 133–137 (2017)

9. Ge, Q.: Greenhouse environmental monitoring and data separation detection system under the environment of Internet of Things. Inf. Commun. **1**, 158–160 (2019)
10. Fu, C., Tian, A.: Design of an intelligent agricultural Internet of Things system. Lab. Res. Explor. **36**(12), 129–132 (2017)

Reform of Mathematics Teaching Mode from the Perspective of Artificial Intelligence and Big Data

Shanshan Gao[1(✉)] and Rui Jia[2]

[1] Department of Information Engineering, Liaoning Institute of Science and Engineering, Jinzhou, Liaoning, China
[2] Department of Marxism, Bohai University, Jinzhou, Liaoning, China

Abstract. With the rapid development of artificial intelligence and big data, the wide application of artificial intelligence and big data has promoted the reform process of traditional teaching mode. Traditional mathematics teaching makes students lack interest in mathematics, and the ability of applied mathematics in practice, what is more, the assessment of students is too single, so it is imperative to carry out mathematics teaching reform. In the teaching reform, artificial intelligence can stimulate students' study interest, exercise students' application ability, and enrich examination methods.

Keywords: Artificial intelligence · Big data · Mathematics teaching model

1 Introduction

Science and technology is one of the constituent elements of the productive forces and plays a vital role in the development of society. Science and technology are one of the constituent elements of productivity and play a vital role in social development. The development of artificial intelligence promotes the continuous progress of science and technology. It not only improves productivity, but also makes great changes in the way of life of human society, and promote the all-round development of people [1]. With the rapid development of science and technology, artificial intelligence and big data have gradually been applied to human production and life, which has promoted the process of social development, and has gradually been applied to the field of education and teaching with its innovative and contemporary characteristics, and has promoted the traditional teaching model Reform process. Now many schools have introduced artificial intelligence and big data technology as auxiliary teaching methods in their teaching. Artificial intelligence provides a new impetus for teaching reform [2]. It can break the traditional teaching mode, stimulate students' interest in mathematics through various means, and realize the research and innovation of university mathematics curriculum theory and practice through computer.

J. Abawajy et al. (Eds.): ATCI 2021, LNDECT 81, pp. 793–797, 2021.
https://doi.org/10.1007/978-3-030-79197-1_115

2   The Disadvantages of the Traditional Mathematics Teaching Model

College mathematics is an important public basic course of higher education and an important part of general education. Learning mathematics well plays a very important role in cultivating and improving students' thinking quality, innovation ability, and comprehensive use of mathematics to solve practical problems. However, the traditional mathematics teaching model has certain drawbacks.

2.1   The Rigid Teaching Makes the Students Lack Interest in Mathematics

Traditional college mathematics classroom teaching is based on theoretical explanations. Teachers blindly teach the knowledge content specified in the syllabus and mechanically introduce mathematical theoretical knowledge. Classroom teaching is immersed in a large number of mathematical proofs, mathematical derivations, and mathematical calculations, which brings students a sense of boring and lack of interest in mathematics. Many professional theoretical knowledge of college mathematics is very abstract, and it is a discipline with strong logical thinking. If relying solely on the explanation of mathematical principles and mathematical theories, this single teaching method cannot stimulate students' interest in mathematics. Students lack the interest in active research, they only memorize the principles and formulas mechanically in the process of learning, and cannot form a systematic grasp of the knowledge they have learned.

2.2   Students Lack Practical Application Ability

Students passively accept what the teacher says; they have developed a learning habit of accepting existing knowledge and experience and lack the spirit of questioning. At this stage, most college mathematics teaching in colleges and universities is mainly based on theoretical explanations. Teachers are only teachers and students without connecting with reality and paying no attention to application. Most students only know how to calculate math problems, but they don't know the practical application of such mathematical calculations, especially for their own majors. Therefore, many students don't like to learn mathematics, and they can't learn mathematics well. They feel that learning mathematics is useless.

2.3   The Assessment Method is Too Single

The mathematics assessment method in most colleges is the final written examination, which does not conducive to assessing students' practical application ability. To learn mathematics courses well, students need to have practical application ability, so in the teaching process, students need to cultivate various abilities such as inductive analysis, logical reasoning, and innovative practice. The current assessment method is likely to cause students' `assault' learning situation, making students feel that the learning process is loose and tight, the knowledge is superficial, and it is difficult to integrate theory with practice. From the perspective of the long-term development of students, because students' theoretical knowledge is not strong, they will not turn knowledge into abilities, which will affect their career development and further study in the future.

3 Mathematics Teaching Mode Reform

Artificial intelligence and big data technology have become a new driving force for the development of various industries, and the use of artificial intelligence and big data to promote teaching reform and innovation has become a new topic for many schools. At present, there are two main forms of applying artificial intelligence in education and teaching: one is to use network media for online learning that transcends time and space, and its temporal and spatial compression properties change the field where education occurs; the other is to use artificial intelligence equipment to assist teaching in the school education field. Its non-fatigue attribute and information wide-ranging characteristics promote the transformation of teaching methods [3]. Artificial intelligence in teaching reform can stimulate students' interest in learning, exercise students' application ability, and enrich assessment methods.

3.1 Use Artificial Intelligence and Big Data to Stimulate Students' Interest in Learning

Artificial intelligence is an open system that can not only perform autonomous operations, build models, but also interact with the external information environment. Artificial intelligence can simulate, assist, replace, extend, expand and enhance the initiative of students [4]. Artificial intelligence equipment can provide students with vivid virtual practice materials, realize specific functions, and provide applied stimulation to stimulate students' initiative in learning. Professor Hu Xiangen, an expert on intelligent guidance systems, once said: "The learner can optimize the knowledge structure in the best educational environment" [5]. For example, in a mathematics classroom, teachers can use the intelligent learning system to call up information resources such as images, sounds, application examples, and video materials that support classroom teaching at any time, so that mathematics classroom teaching has both pictures and texts, complete audio and video, and can save writing time on blackboard and improve the efficiency of classroom teaching allows teachers to teach more teaching content within the limited classroom teaching time, and teach students richer mathematical knowledge. Artificial intelligence equipment can mobilize students' interest in learning and keep students enthusiastic about learning in math classes.

3.2 Use Artificial Intelligence and Big Data Technology to Develop Students' Application Ability

Students' application ability needs strong theoretical knowledge as a foundation, so students need teachers to answer questions for them anytime and anywhere during the learning process. Facing the massive amount of information, human teachers have limited ability to obtain and process information, and it is also difficult to control the continuous change of information. To adapt to this change, it is necessary to use external smart devices as a cognitive outsourcing of the human brain for information processing and thinking, and AI teachers become important and excellent assistants [6]. Artificial intelligence and big data technology can answer questions for students, and intelligent robots can be a good helper for students' daily learning. When students encounter difficulties in learning, they can use the intelligent learning system to find answers, and the intelligent system supports multiple expressions such as text, voice, and pictures, which improve the efficiency of students' learning. Artificial intelligence and big data technology, can not only enrich students' theoretical knowledge, but also exercise students' practical application ability. For example, artificial intelligence can be used to establish a simulation laboratory, and students can rely on the simulation laboratory to verify the correctness of mathematical formulas. The simulation laboratory has the characteristics of safety and lifelike, which is conducive to students' understanding and application of mathematical abstract formulas.

3.3   Use Artificial Intelligence and Big Data Technology to Enrich Assessment Methods

Modern teaching methods driven by artificial intelligence and big data can enrich the ways to assess students' learning outcomes. In the daily assessment of students, the artificial intelligence face recognition system can achieve accurate attendance; the intelligent system can also display the number of students, class time, course name and other content. For students' online learning situation, artificial intelligence can count the number of student logins, the number and time of text reading, homework completion and other information, and make accurate analysis of student learning based on this information, and can realize automatic correction of homework and score statistics And other functions. From the big data of smart devices, teachers can find learners' interest preferences, attitude habits, emotional personality, and learning styles, so as to develop personalized learning plans for students. At the same time, learners can know themselves more accurately and choose targeted learning content, so as to improve learning efficiency [7]. Therefore, the application of artificial intelligence and big data enables teachers to diversify the assessment methods of students. Artificial intelligence can track students' learning for a long time, and can provide accurate and objective factual basis for teachers' assessment of students.

The wide application of big data and artificial intelligence has a positive impact on the reform of university mathematics teaching mode. We believe that artificial intelligence and big data technology can play an important role in teaching reform.

References

1. Jia, R.: A Marxist perspective on the impact of artificial intelligence on social development. Think Tank of Sci. Technol. **7**, 47–49 (2020). (in Chinese)
2. Sun, Z., Tian, H.: The "change" and "guard" of teaching reform in the new era from the perspective of artificial intelligence. Comput. Era **9**, 80–83, 87 (2019). (in Chinese)

3. Feng, C., Chen, X.: The role of artificial intelligence in teaching reform and its rational examination. J. Soochow Univ. (Educ. Sci. Edition) **1**, 25–32 (2020). (in Chinese)
4. Cai, B.: Artificial intelligence empowers classroom revolution: essence and philosophy. Res. Educ. Dev. **39**(2), 8−14 (2019). (in Chinese)
5. Liu, K., Hu, J.: The theoretical framework of artificial intelligence education application. Open Educ. Res. **24**(6), 4–11 (2018). (in Chinese)
6. Yu, S., Wang, Q.: Analysis on the development of "AI+teacher" cooperation path. E-Education Res. **40**(4), 14–22 (2019). (in Chinese)
7. Jia, J.: Artificial intelligence empowers education and learning. J. Dist. Educ. **1**, 39–47 (2018). (in Chinese)

Factor Analysis of Professional Ability of Fitness Clubs Membership Consultants in the Era of Big Data

Kun Tang[✉]

Department of Physical Education, Liaoning Institute of Science and
Engineering, Jinzhou, Liaoning, China

Abstract. In order to improve the ability of membership consultants to serve
national fitness and effectively serve potential and actual members of fitness
clubs, 20 professional ability social demand indicators for membership consultants have been determined. After obtaining the survey data, the factor
analysis method has been used to realize the division of the six major factors of
the professional ability of the membership consultants of the fitness clubs.
Furthermore, it provides a scientific basis for exploring ways to improve the
professional ability of membership consultants.

Keywords: Membership consultants · Professional ability · Questionnaire ·
Factor analysis

1 Introduction

The advent of the era of big data will have a huge impact on industry development, unit
decision-making, and personal daily life. In order to realize the "Healthy China 2030"
planning outline and the national policy of Healthy China [1–3], and realize the great
concept of the Chinese people's physical and mental health and a better life, the fitness
clubs put forward more comprehensive and advanced service concepts and Practice
measures. membership consultants play a prominent role in the sales of fitness club, and
their professional abilities directly affect the operating capabilities and economic
benefits of fitness clubs.

2 Establishment of Professional Ability Indicators for Fitness Club Membership Consultants

According to the research of Jin, Tang, Jin, Fang and others on the professional
competence of conference consultants [4–7], a membership consultant professional
competence evaluation index including four categories, three tiers and 20 indicators
was established, which is specifically divided into dedication, enterprising, Memory,
personal image, interpersonal skills, sense of responsibility, willpower, teamwork
ability, integrity, psychological endurance, resilience, affinity, professional knowledge,

sales knowledge, emotional control ability, insight, language expression ability, communication ability, Follow-up service ability, character and ethics are represented by X1, X2,... X20 respectively. Taking membership consultants in fitness clubs as the survey object, a total of 550 questionnaires including 20 evaluation indicators of membership consultants' professional abilities were distributed, 489 questionnaires were collected, and 441 valid questionnaires were sorted out after screening.

3 The Factor Analysis Process of the Professional Ability of Membership Consultants

3.1 Reliability Analysis of the Questionnaire

The survey results were input into SPSS software to obtain the calculation results of KMO and Bartlett test for all the samples. As shown in Table 1, these test indicators meet the significance standard, indicating that the survey results and data are fully suitable for further factor analysis.

Table 1. KMO and Bartlett test

KMO measure of sampling adequacy	.925
Bartlett test for sphericity Approximate chi-square	7631,071
Degree of freedom	190
Significance	000

3.2 Calculate the Principal Factor Solution

SPSS software was used to conduct correlation analysis on 441 sample data of 20 indicators and calculate characteristic values. The contribution rates of the first six factors were 60.148%, 4.673%, 4.174%, 3.766%, 3.574% and 3.024%, respectively. Therefore, when the first 6 factors are extracted as the main factors, the cumulative contribution rate reaches 79.36%. According to the requirements of sociological statistical analysis, these 6 main factors can represent the overall information of 20 original variables. After 25 times of maximal variance rotation, the solution of six major factors after rotation is obtained, as shown in Table 2:

Table 2. The solution of six major factors after rotation

Indicator name (importance)	F1	F2	F3	F4	F5	F6
Dedication	0.744	0.296	0.157	0.226	0.218	0.197
Enterprising	0.652	0.235	0.131	0.375	0.307	0.121
Memory	0.726	0.236	0.297	0.153	0.030	0.290
Personal image	0.353	0.607	0.225	0.252	0.296	0.145
Interpersonal communication	0.229	0.732	0.206	0.221	0.247	0.278
Responsibility	0.260	0.653	0.434	0.195	0.323	0.024
Willpower	0.213	0.502	0.452	0.199	0.091	0.444

(*continued*)

Table 2. (*continued*)

Indicator name (importance)	F1	F2	F3	F4	F5	F6
Teamwork ability	0.412	0.519	0.306	0.199	0.326	0.203
Integrity	0.244	0.286	0.724	0.134	0.159	0.303
Mental capacity	0.238	0.365	0.610	0.391	0.001	0.141
Strain	0.084	0.207	0.609	0.423	0.384	0.117
Affinity	0.548	0.142	0.589	0.153	0.403	0.036
Professional knowledge	0.296	0.368	0.273	0.724	0.220	0.033
Sales knowledge	0.305	−0.001	0.240	0.633	0.284	0.448
Emotion regulation	0.181	0.401	0.229	0.636	0.162	0.302
Insight	0.420	0.176	0.188	0.519	0.498	0.104
Language	0.150	0.319	0.170	0.229	0.783	0.243
Communication	0.274	0.318	0.142	0.253	0.605	0.459
Follow-up services	0.419	0.263	0.457	0.175	0.531	0.074
Morality	0.276	0.259	0.203	0.205	0.283	0.729

3.3 Connotation and Naming of Main Factors

In Table 2, the main factor solution after the maximum variance rotation represents a certain feature of the original data from different sides, and also provides an important reference for further analysis of the usefulness of the research data.

(1) In the first main factor, the variables with larger solution coefficients of the main factor are dedication (X1), enterprising (X2), memory (X3) and other three variables, named "value orientation and skill accumulation factor".

(2) In the second main factor, the variables with larger solution coefficients of the main factor are personal image (X4), interpersonal communication (X5), sense of responsibility (X6), willpower (X7), teamwork ability (X8), etc. Five variables, named "professional image shaping and dedication ability factor".

(3) In the third principal factor, the variables with larger solution coefficients of the principal factor are integrity (X9), mental capacity (X10), strain (X11), affinity (X12) and other four variables, named "professional ethics and job adaptability factor".

(4) In the fourth main factor, the variables with larger solution coefficients of the main factor are professional knowledge (X13), sales knowledge (X14), Emotional regulation (X15), insight (X16) and other four variables, named It is the "professional knowledge reserve and self-management ability factor".

(5) In the fifth principal factor, the variables with the larger solution coefficient of the principal factor are language (X17), communication(X18), follow-up services (X19) and other three variables, named "social communication and customer management ability factor".

(6) In the sixth main factor, the variable of morality (X20) is included, which is named "moral competence factor".

4 Cultivation and Promotion of Professional Ability of Membership Consultants

4.1 Measures to Improve Value Orientation and Skill Accumulation

The contribution rate of the first major factor, "value orientation and skill accumulation factor", reached 60.148%, which was specifically manifested in three aspects: dedication, enterprise and memory. Service member is one of the important work of the membership consultants, must promote membership consultants' professional dedication, work ability and service attitude, through specialized training and promotion methods such as job makes membership consultants to keep pace with The Times learning to master the cutting-edge knowledge and sales skills, attract more members to join, memory, both cognitive skills and marketing knowledge to master and apply, including the basic situation and the demand for members more aspects of memory, in the member to the store again improving customer friendly feeling, achieve market conduction, and customer development.

4.2 Shaping Professional Image and Improving the Ability of Professionalism

The contribution rate of the second major factor, "professional image shaping and dedication ability factor", was 4.673%, which was embodied in five aspects: personal image, interpersonal communication, responsibility, willpower and teamwork ability. A good personal image of membership consultants can guide the fitness vision of members. In the process of communicating with customers, they need to learn to express their professional image and professionalism well, stimulate and mobilize the enthusiasm of customers to participate, and achieve two-way quality improvement. Willpower and team work are also important abilities to expand the market.

4.3 Measures to Improve Professional Ethics and Job Adaptability

The contribution rate of the third major factor, "professional ethics and job adaptability factor", was 4.174%, which was embodied in four aspects: honesty, psychological endurance, strain capacity and affinity. Membership consultants should be responsible and focused on their work, cultivate their market demand-oriented and customer management concepts in the process of education, meet the needs of members for fitness and enjoyment, inspire and influence customers with demonstration effect, gain respect and recognition from members, and improve their performance.

4.4 Measures to Improve Professional Knowledge Reserve and Self-management Ability

The contribution rate of the fourth major factor, "professional knowledge reserve and self-management ability factor", was 3.766%, which was embodied in professional knowledge, sales knowledge, emotional regulation ability and insight. These require membership consultants with strong professional knowledge and sales knowledge reserves and using ability, according to the personality needs of customers targeted promotion of fitness programs, training membership consultants to resolve problems and ease emotional work method, provide high quality, high standard of service for members, promote and guarantee for the club's long-term development.

4.5   Measures to Improve Social Communication and Customer Management Capabilities

The contribution rate of the fifth major factor, "social communication and customer management ability factor", was 3.574%, which was embodied in three aspects: language expression, communication ability and follow-up service. membership consultants can only be recognized by society and members if they communicate information seamlessly with clients. With the increasingly fierce market competition environment, fitness clubs continue to open up new project forms and service contents. The follow-up services that meet customers' personalized requirements will realize the sustainable development of fitness clubs, discover more potential members and open up a broader market on the basis of retaining existing members.

4.6   Measures to Improve Character and Moral Ability Factors

The contribution rate of the sixth major factor, "moral competence factor", was 3.024%, which indicated that this major factor has a certain influence on the self-actualization ability of membership consultants. It is necessary to cultivate the ideal belief, value orientation, dedication and positive psychology and personality of the membership consultants from the student period. Through continuous and long-term training in ordinary jobs, the membership consultants can shape his own conduct, sublimate his own moral quality, and finally achieve the goal of self-realization.

5   Conclusion

According to the professional needs of fitness clubs membership consultants, factor analysis was carried out on 20 professional competence survey data of four categories and three layers, and six main factors of fitness clubs membership consultants' professional competence were obtained. By analyzing the connotation of each main factor for training provide scientific basis for selection and selection of excellent membership consultants for fitness clubs, especially for cultivating applied talents in social sports guidance and management.

Acknowledgements. This work was supported by 2019 Scientific Research Funding Projects of Liaoning Province, "Research on Improving the Innovation and Entrepreneurship Ability of Social Sports Guidance and Management Students from the Perspective of the Integration of Production and Education". (No. LNLGXY2019007); 2019 Teaching Reform Research Project of Liaoning Institute of Science and Engineering, "Research on Improving the Innovation and Entrepreneurship Ability of College Students Majoring in Social Sports Guidance and Management" (No. LG2019008).

References

1. The State Council of the People's Republic of China. National Fitness Program (2016–2020) (2016). (in Chinese)
2. The Central Committee of the Communist Party of China and the State Council. Outline of "Healthy China 2030" Plan (2016). (in Chinese)
3. State Council of the People's Republic of China. Opinions on the Implementation of the Healthy China Action (Guo Fa [2019] No. 13) (2019). (in Chinese)
4. Jin, Y., Zhang, L.: Fitness Club Operation and Management, pp. 166–172. China Labor and Social Security Press, Beijing (2009). (in Chinese)
5. Kun, T.: Investigation on professional ability of employees in fitness club. Sport. **171**, 148–150 (2017). (in Chinese)
6. Liu, J., Liu, Y.: Research on the competency of the membership consultant of fitness club. Bull. Sport Sci. Technol. **5**, 19–20+24 (2019). (in Chinese)
7. Fang, S.: Research on the professional quality of Fitness Club membership consultant. Bull. Sport Sci. Technol. **27**(08), 120–121 (2019). (in Chinese)

Current Status and Industrialization Development of Industrial Robot Technology

Yong Zhao$^{(\boxtimes)}$

Shandong Technician College of Water Conservancy, Zibo, Shandong, China

Abstract. Nowadays, industrial robot has achieved rapid development, and it has been more and more widely used in spraying, assembly, handling and palletizing, grinding and polishing, loading and unloading, welding, machining, automobile manufacturing and other industries. With the rapid development of manufacturing technology, automation technology, intelligent technology and information technology, industrial robots are becoming more intelligent. The application of Industrial robot can greatly reduce the manpower, material and financial costs of the enterprise, and improve the production quality and efficiency of the enterprise. This article mainly expounds the technical status of Industrial robot, proposes the industrialization development strategy of Industrial robot, and analyzes the future development trend of Industrial robot in China.

Keywords: Industrial robot · Technology status · Industrialization development

1 The Concept of Industrial Robot

Industrial robot refers to a multi-degree of freedom or multi-joint manipulator machine device that is widely used in the industrial field. It has a certain degree of autonomy and can complete various industrial processing through its own control function and power energy. It has been widely used in various industries.

2 Current Status of Industrial Robot Technology at Home and Abroad

2.1 Status Quo of Industrial Robot Technology Abroad

Since the 1960s, with the improvement of product processing accuracy requirements, the key process production links have been gradually replaced by Industrial robots, and no longer manual operations. In addition, various countries have stricter requirements on the working environment of workers, and jobs in harsh environments such as toxic and high risk factors are gradually replaced by robots, and the market demand for industrial robots is increasing. In developed countries, complete sets of automated production line equipment and industrial robots have become a key component and development trend of high-end equipment. Industrial robots have been widely used in

© The Author(s), under exclusive license to Springer Nature Switzerland AG 2021
J. Abawajy et al. (Eds.): ATCI 2021, LNDECT 81, pp. 804–808, 2021.
https://doi.org/10.1007/978-3-030-79197-1_117

fields such as manufacturing, logistics, food, rubber and plastics, electrical and electronic industries, machining industries, and automotive industries.

There are many innovations in special robotics technology in the United States. Domestic service robots, medical service robots, and military robots occupy an absolute advantage in the market. Japan and Europe have obvious advantages in the development and production of industrial robots. For example, famous robot companies such as YASKAWA, FANUC, KUKA, ABB, etc. occupy most of the industrial robot market share [1].

2.2   The Status Quo of Industrial Robot Technology in China

Compared with Western developed countries, China's Industrial robot started relatively late. Since the introduction of the Industrial robot in the 1970s, China's Industrial Robot has gone through three stages:

First, the initial stage. After China implemented reform and opening up, it began to pay attention to the development of robots in developed countries abroad, actively introduced industrial robots, and issued a series of related policies to support the development of industrial robots;

Second, the innovation stage. The automotive industry was the first to feel the technical advantages of Industrial robots. The application of Industrial robot in automobile production is already essential. After that, in the 7th Five-Year Plan, the relevant research institutes began to focus on the industrial robot technology, and made great progress in the control system and manipulators. In addition, a number of self-developed technologies can already achieve small-scale production of Industrial robots. Because the technology is not mature enough, industrial robots cannot be produced on a large scale, and industrial robots cannot be put into production on a large scale.

Third, the practical stage. At this stage, industrial robot has been applied in all walks of life, which has improved production quality and production efficiency. At the end of the last century, China invested a lot of financial resources, financial resources, and manpower to research and develop industrial robots, which has achieved a relatively rapid development speed and has basically been able to achieve the actual production needs of all walks of life.

In recent years, China's industrial robot technology has become mature, and industrial robots have been more and more widely used in various industries [2]. For example, welding robots developed in China today are indispensable in industries such as the automobile industry and aircraft manufacturing, and have played a very critical role. In addition, industrial robots have gradually been applied in the commercial field, which has laid a good foundation for the industrial development of Industrial robot technology. China's industrial robot is facing a rare development opportunity in history. Policy dividends and economic and social transformation have made industrial robots have broad development prospects.

3 Development Strategy of Industrial Robot Industrialization in China

Although China's industrial robot has broken through some key technologies, it still lacks in general. Especially in the whole set of equipment and manufacturing process, there is a lack of important components such as high-efficiency, high-speed and high-precision controllers, servo motors and reducers. The author believes that we should vigorously tackle key technologies and grasp the following core technologies: Industrial robot remote fault diagnosis and repair technology, robot dynamics control in complex environments, industrial robot production line integration technology, 3D-based virtual simulation technology, etc.

3.1 Strategies Are Needed

The development of industrial robot is a long-term process, and it cannot be completed overnight. This can only be achieved without a certain industrial foundation. It is necessary to optimize the development of manufacturing materials and their cutting-edge technologies, and accelerate the development of equipment technology and intelligent manufacturing technology.

First, as far as the country is concerned, companies that manufacture industrial robots should be strongly supported, and relevant support and protection policies should be introduced to effectively protect the healthy growth of the companies. It is necessary to build a complete industrial chain, use funds, policies and other methods to continuously improve the industrial robot industrial chain, so as to improve the economic efficiency of industrial robot production; Second, for enterprises, they should have a long-term strategic vision, a strong brand awareness, establish a good corporate image, strengthen technological innovation, and accelerate the development of industrial robots. Through the joint efforts of the state and enterprises, the healthy and orderly development of Industrial robot is guaranteed.

3.2 Standardization and Order

Although China's industrial robot started relatively late, its development speed is relatively fast. Therefore, many companies attach great importance to this and have successively invested in the development and manufacturing of Industrial robots. However, the R&D level and manufacturing level of these companies are uneven, causing malicious competition in the Industrial robot market and disrupting the normal economic order of the market [3]. In addition, the industrial chain of industrial robots produced by some companies is not perfect enough, which increases the production cost of industrial robots. In addition, the industrial robot standard is not uniform enough, which increases its production cost and social cost to a certain extent. If the Industrial robot is to achieve good industrial development, it must be standardized and ordered.

3.3 Reliable Technology and Reasonable Prices Are the Focus of Industrialization

To realize the industrialization development of industrial robot, first, we must have reliable technology to ensure the application efficiency of Industrial robot; second, we must have a reasonable price to ensure that users can obtain economic benefits. In order to realize the reasonableness of the industrial robot price, we should start from the following two aspects: first, establish a brand and expand the industrial chain; second, reduce unnecessary intermediate links, thereby reducing the manufacturing cost of the industrial robot. For the reliable technology of Industrial robot, we must pay attention to the optimization of technology and materials. In addition, components that are prone to problems must be further improved to increase their reliability. The industrial market robot must be recognized by the market through reasonable technology and price, so that its industrialization can achieve good development.

3.4 Scale

In order to achieve industrial development of the industrial robot, it needs to be combined with the actual needs of the market, and the value of the industrial robot must be required to be able to be fully utilized. In terms of specific applications, the input-to-output ratio of industrial robots must be higher than that of automated machines before companies will take the initiative to abandon automated robots to use industrial robots. Only when the industrial robot market is getting larger and larger, can it gradually show its scale effect and produce more economic benefits. Therefore, in the process of industrial robot industrialization, the most critical point is to realize the scale of its application.

4 The Future Development Trend of China's Industrial Robot

With the rapid development of the times, the application of Industrial robot will be more extensive in the future. The following are its future development trends:

First, develop in the direction of bionics. Industrial robots are not flexible enough in specific work processes. However, in the future development process, with the continuous strengthening of dependence on the Industrial robot, the work done by the Industrial robot will be more complicated. Therefore, its development trend must be bionic.

Second, develop in the direction of diversification. Industrial robot will be applied to industries such as automobile industry and machinery manufacturing industry from the very beginning, and gradually applied to industries with stricter requirements and wider demands such as medical treatment [4–7].

Third, develop towards the direction of intelligence. Today, artificial intelligence has entered thousands of households, which not only improves people's quality of life, but also enables industrial robots to achieve further development. The improvement of industrial robot's intelligence will improve its economic benefits [8–10].

5 Conclusion

With the advent of the fourth industrial revolution, today's industrial production is becoming more and more intelligent, along with the stronger and stronger concept of "re-industrialization" of old industrial powers, it will inevitably have a strong impact on China's manufacturing industry. As China's demographic dividend is slowly decreasing, it is very necessary to accelerate China's industrial transformation through technological innovation and promote the development of China's industry. The basis of the individualization and intelligence of China's industrial manufacturing is the industrialization of China's robots, and it is also a key factor in strengthening China's industrial competitiveness. Moreover, China's aging situation is becoming more and more serious, and the number of laborers is slowly decreasing. It is very necessary to transform and upgrade equipment by strengthening China's technological competitiveness. China's Industrial robot has a very large space for development. As it becomes more and more closely integrated with information technology, the functions of robots will inevitably be gradually improved. In the future, they will inevitably enter thousands of households and become popular products. China attaches great importance to industrial robots and intelligent manufacturing. In the future, it will usher in an upsurge in the development of industrial robots, which will form strong support for the industrial transformation and upgrading of China's industries and enable China's industrial robots to achieve better development.

References

1. Sun, X., Guohui, Y., Zhou, W., et al.: Analysis of the development characteristics of industrial robots in the world. Robot. Appl. **3**, 8–9 (2002)
2. Xia, K., Tao, X., Li, J., et al.: Development and application research of industrial robots. Light Ind. Sci. Technol. **024**(008), 63–64 (2008)
3. Cai, Z., Guo, F.: Several problems in the development of industrial robots in China. Robot. Appl. **000**(003), 9–12 (2013)
4. Liu, Z., Li, H., Sun, H.: Technical development and application of industrial robots. Eng. Technol. Res. **17**(01), 127–128 (2018)
5. Zhu, W.: The development trend of industrial robot fault diagnosis technology. South. Agric. Mach. **51**(360(20)), 191–192 (2020)
6. Qin, F., Chen, M.: Research on linear active disturbance rejection control of industrial robot servo system. Ind. Control Comput. **33**(12), 106–107+109 (2020)
7. Li, X., Liu, W., Tao, W., et al.: Analysis of research methods on the mechanism performance of industrial robots. Shandong Ind. Technol. **000**(001), 17–21 (2020)
8. Jing, Z., Xiao, Y., Xie, Q.: Research on automatic control method of industrial robot based on deep learning. Electron. World **605**(23), 61–62 (2020)
9. Liu, Y.: Development and application of intelligent vision guidance technology for industrial robots. Sci. Technol. Econ. Mark. **000**(004), 3–5 (2020)
10. Chen, Q., Ma, B.: Discussion on the rapid calibration technology of industrial robot vision system. Inf. Record. Mater. **21**(11), 188–189 (2020)

Application and Practice of UAV Aerial Survey Technology

Chengxin Liu[✉]

Shandong Technician College of Water Conservancy,
Zibo 255130, Shandong, China

Abstract. China's UAV aerial survey technology is relatively mature and has been well verified in the application and practice fields. Through the analysis of the application and practice characteristics of the technology, the meaning of some mature application fields is deeply explored. As a result, we have more efficient response capabilities in the face of natural disasters that threaten people's lives and property, and can promote the development of some emerging fields, thereby promoting the progress and development of human civilization. In the broad field of application and practice, the continuous discovery of technical problems and corresponding solutions can also promote the mature development of the technology.

Keywords: UAV aerial survey technology · Technical feature · Application and practice

1 Introduction

In the research focus on UAV aerial survey and navigation, from UAV aerial survey photography system applied to UAV aerial survey photography system to the development of UAV aerial survey photogrammetry technology and the practical application of aerial detection, including the research of UAV aerial survey aerial detection technology, we have conducted a lot of theoretical investigations. And we have obtained a lot of research results. The operation of this aerial detection device system has many advantages, such as higher resolution, shorter operation period, lower cost and more accurate data. Therefore, it is widely used in engineering, agriculture, military and other fields. Due to the development of aerial detection technology, UAV aerial survey has greater significance in acquiring and processing space data. At the same time, by continuously improving the resolution of remote sensing data, the production conditions with independent intellectual property rights nationalization have greatly reduced the cost of aerial surveys.

J. Abawajy et al. (Eds.): ATCI 2021, LNDECT 81, pp. 809–813, 2021.
https://doi.org/10.1007/978-3-030-79197-1_118

2 What Is UAV Aerial Survey Technology

The aerial detection of UAV aerial survey is a powerful auxiliary technology for traditional aerial photo survey. It has the outstanding features of flexibility, high efficiency, high precision, low computing cost, wide application range, and short production cycle. This has obvious advantages for quickly obtaining high-resolution images in small areas and complex flight areas. In the development of UAV aerial survey and digital camera technology, digital cameras based on the UAV aerial survey platform are becoming more and more popular, and their unique advantage lies in aerial photography technology [1]. The combination of UAV aerial survey and aerial photo measurement makes UAV aerial survey digital low-altitude remote sensing a new development direction in the field of aerial remote sensing. UAV aerial photographs are widely used in basic surveys such as major civil engineering, disaster emergency treatment, land supervision, resource development, new countryside and small-town construction. It is widely used in surveying and mapping, land resource survey and monitoring, land use dynamic monitoring, digital city construction and emergency rescue survey and map data collection.

3 Features of UAV Aerial Survey Technology

3.1 Break the Traditional Restrictions

UAV aerial survey technology is an important technical support for the development of aircraft. Its birth broke the traditional cognition concept. Since mechanization, our country has made great progress, and the degree of mechanization is relatively good. However, in some special events, mechanization alone cannot meet people's needs. For example, the Wenchuan earthquake in 2008 caused huge loss of life and property to our people. At that time, all roads were destroyed and blocked. Although we mobilized all our strength to clear the way, we still need to understand the situation in the disaster area first. If UAV, the unmanned aerial survey technology at that time, enters the area, and the basic situation of the disaster area is investigated clearly, better preparations can be made. Although the accuracy of the satellite is very high, it is unable to present the situation more completely in terms of angle. The UAV makes good use of its own advantages. First, the UAV is used to quickly understand the terrain, roads and basic conditions of the disaster in different regions, and then provide an important reference for rescue operations, which improves the efficiency of rescue. However, the traditional method needs to push forward little by little from the periphery to carry out rescue, and there is not enough time for us to carry out little by little in the face of natural disasters [2].

3.2 Break the Time Limit

Unmanned aerial survey technology is a technology that is being developed in various developed countries at this stage. Its application range is relatively wide. One of the important reasons is that it can respond quickly in a short period of time and can be put into use quickly. This has provided tremendous help to disaster relief and other needs and also spared valuable time for rescue. In the development of mankind, mankind has always struggled with time, and time is an important criterion for human progress. For example, when drawing a map, manual drawing may take years or even decades to complete the map. The use of UAV aerial survey technology can be completed in a short time. This short concept is a few days, which is surprising compared to the traditional time, but this is true. The cost of UAV is not high, and it can detect terrain, etc. It is very practical in drawing maps. The value of tools is to help humans complete some tasks that humans need more time to complete, and UAV aerial survey technology has played this role well, and its time efficiency is more in line with the actual needs of humans [3].

3.3 Break the Limits of Space

The development of mankind has always been inseparable from the limitation of geographical space. It can be felt from the form and speed of information transmission, especially for a vast country, it is necessary to leave room for space. In a space with relatively complex climate and terrain, UAV aerial survey technology has more advantages. It does not need to use ancient on-walk mapping and satellite methods, because the clarity of satellite methods cannot meet higher requirements. UAV aerial survey technology can achieve this. UAVs can go to spaces that humans cannot go, and UAVs can go to bad links that humans cannot bear. These examples all show that UAV is a means or tool for mankind to break space constraints. This is an important change in the history of human development. For example, the China-India border has recently fallen into a confrontation phase, and the border line is located in the Himalayas. Not only is it high in altitude, cold, and thin, but there is also no clear or relatively safe road, which has caused great trouble for the soldiers. Our country has adopted UAV to break this complicated space restriction, and transport the life displayed by UAV in a hurry. The efficiency is very high, thus effectively ensuring border stability and making outstanding contributions [4].

4   Application and Practice Fields of UAV Aerial Survey Technology

4.1   Application and Practice in the Field of Aerial Photography and Surveying and Mapping

UAV aerial survey technology is not a technology, but a system composed of several technologies, including control flight system, shooting technology, sensor technology and other technologies. These technologies guarantee the application basis of UAV aerial survey technology. Shooting technology is the main technology applied in the field of aerial photography and surveying and mapping. It is different from the camera's photographing function. The shooting technology is more advanced, and it has a special software system that can intelligently recognize and process images. The application field is also widely praised based on this. For example, when people watch the news, they often see introduction videos similar to God's perspective. This is the use of UAV aerial photography technology, which can present text news with videos and images. China has a relatively developed infrastructure industry, in which the surveying and mapping technology is one of the important supports. Therefore, UAV surveying and mapping technology can more efficiently survey and map terrain and landforms [5]. The accuracy is also very high, and it has become the main method of surveying and mapping. The height and viewing angle of surveying and mapping can be freely selected, which can complete surveying and mapping tasks that are difficult for humans to complete and improve the efficiency of surveying and mapping.

4.2   Application and Practice in the Field of Earthquake Relief

No matter how far the development of unmanned aerial survey technology is, it cannot be changed to serve the production and life of human beings. Unmanned aerial survey technology also plays an important role in protecting human life and property. Some natural disasters have caused great harm to humans. In order to protect human life, many measures have been taken to make full use of the limited time. Disasters such as earthquakes, floods, mudslides, landslides, search and rescue have caused great difficulties for direct rescue work. Therefore, it is very important to understand the situation at the first time, formulate rescue plans and focus on giving important reference. The UAV is smaller in size, can adapt to tasks more flexibly, and can adapt to more severe weather conditions. The situation of the disaster site can be transmitted to the control platform in the first time, and then the corresponding response can be made according to the actual situation [6]. The application and practice of this technology in the field of disaster relief, while exerting the advantages of unmanned aerial survey technology, can continuously discover problems in practice and better improve the defects of UAV aerial survey technology.

4.3   Application and Practice in the Field of Plant Protection

Smart agriculture is the future development direction of agriculture, and the industrial chain also includes the application of unmanned aerial survey technology. Traditional plant protection methods are operated manually, but pesticides not only cause little harm to humans, but also cause many poisoning incidents. At the same time, they can liberate labor and improve efficiency. UAV plant protection pesticide spraying is widely used in agriculture. It can change the number and area of pesticide spraying according to the actual conditions of crops and more scientifically control the degree of plant protection. In addition, UAV aerial survey technology can be used to monitor plants, and real-time observation of plant growth [7]. If there are some precursor phenomena of diseases and insects, timely and corresponding measures can be adopted. This may not be necessary for traditional farmers, but it is of great significance to agriculture in the new era. A lot of farmland will be concentrated in the hands of large farmers, which means that farming has also become a batch choice, so the batch of farmland needs the support of unmanned aerial survey technology. First, the cost can be controlled, the price of UAV is not expensive, and the technology is mature.

Secondly, this technology can ensure the healthy growth of plants, which adds a mature application to the application field of UAV aerial survey technology [8].

5 Conclusion

Through the understanding of the characteristics of UAV aerial survey technology, and the three more mature areas of aerial photography and surveying and mapping, earthquake relief, and plant protection, we will continue to improve technical barriers in application and practice to support applications in more fields and practice. [9] In the future, UAV aerial survey technology must cooperate with other fields to play a promoting role, and UAV aerial survey technology as an auxiliary tool is becoming more and more mature. On this path, we still need more investment, including capital investment, talent investment, application and practice investment, etc., continuous innovation is an important foundation for maintaining technological leadership [10].

References

1. Wang, S.: Application and practical analysis of drone aerial survey technology. Urban Constr. Theor. Res. (Electron. Ed.) **264**(18), 109 (2018)
2. Wang, W.: Analysis and application of UAV aerial survey technology. Urban Constr. **000** (003), 275 (2020)
3. Zhao, D.: UAV aerial survey technology and its application analysis. Hous. Real Estate **591** (30), 261–262 (2020)
4. Shao, Q.: The application of drone aerial photography technology in large-scale topographic map surveying and mapping. Geol. Miner. Surv. 3(3) (2020)
5. Yuan, W.: Research on the application of UAV remote sensing surveying and mapping technology in engineering surveying and mapping. China Metal Bull. **1012**(01), 279+281 (2020)
6. Li, X., Zhang, X., Wang, H.: Application and practice of drone technology in agricultural plant protection. Hubei Agric. Mech. **241**(04), 135–136 (2020)
7. Wang, L., Lei, Q.: Application and practice of four-rotor UAV in the field of agricultural plant protection. Ind. Heat. **49**(277(05)), 61–63 (2020)
8. Chaoyi, X., Chu, X.: The application and thinking of drone aerial photography technology in news practice. J. School Electron. Eng. **009**(001), P.121-121 (2020)
9. Jiang, B.: Application analysis of UAV aerial survey system in highway belt topography survey. Smart City **6**(06), 75–76 (2020)
10. Fan, X., Yin, Z., Feng, K., et al.: A forest resource monitoring method based on UAV aerial survey technology (2020)

Pension Fund Financial Balance Model Under the Background of Artificial Intelligence

Fang Yang[1](✉) and Lingni Wan[2]

[1] Department of Business Administration, Wuhan Business University, Wuhan,
Hubei, China
[2] Department of General Education, Wuhan Business University, Wuhan,
Hubei, China

Abstract. The harm of pension fund operation risk is a series of negative effects caused by the operation risk events. The results of risk events are ultimately reflected in the financial imbalance of the fund, which leads to the public endowment insurance system to reduce the security function, increases the financial burden of the relevant subjects, and induces the trust crisis of the pension insurance system. Under the influence of aging and inflation, the negative impact of these crises will be aggravated. In this paper, the financial balance model of pension fund is used to describe the generation and deterioration process of fund operation risk hazards under different fund raising methods, and put forward countermeasures under the background of artificial intelligence.

Keywords: Artificial intelligence · Pension fund · Balance model

1 Introduction

The research on pension risk management focus on the effectiveness of financial information use, improve organizational behavior, establish fraud prevention process, and formulate risk strategies with financial balance as the goal. Pension risk management based on the effectiveness of the use of financial information [1]. There are five key points about the risk management of enterprise annuity [2]. Based on the data of the Australian pension Appeals Board (SCT), and puts forward five risk management points [3]. The pension risk strategy based on financial balance, and considered that the risk mitigation strategy should fully consider the risk factors [4–6]. Artificial intelligence has become the mainstream direction of global science and technology development. Big data resources, computing technology, basic algorithms and AI platform are promoting the rapid development of artificial intelligence [7–10]. Since 2015, China has been continuously promoting the development of artificial intelligence industry at the macro policy level. The government work report of 2019 clearly proposes to upgrade artificial intelligence. The 2020 government work report further proposes to develop industrial Internet, promote intelligent manufacturing and cultivate emerging industrial clusters. China's AI industry extends in the application level, mainly including medical treatment, finance, education, transportation, home

J. Abawajy et al. (Eds.): ATCI 2021, LNDECT 81, pp. 814–819, 2021.
https://doi.org/10.1007/978-3-030-79197-1_119

furnishing, retail, manufacturing, security, government affairs, etc. The planning target scale of artificial intelligence in 12 provinces and cities of China will reach 429 billion by 2020.

This paper builds the pension fund financial balance model, and put forward countermeasures under the background of artificial intelligence.

2 Discussion

2.1 Related Variables and Balance Mechanism

2.1.1 Income Variable

PI, total income of pension fund; PIA, total income of pension fund in period A; Pi, sum of insurance taxes paid by single; PiiA, sum of insurance taxes paid by I in period A; pi, income of pension fund formed by single; piia, insurance taxes paid by single i in period A.

2.1.2 Expenditure Variables

IE: the total expenditure of pension; IEA, the total expenditure of the A-period; ie, the amount of pension benefits enjoyed by single; ieja, the amount of pension received by single j in a period; IeiT the total amount of pension received by single i until issue T.

2.1.3 Insurance Tax Calculation Variables

w, the payment base; wjy the payment base of single j in y period; a, the pension tax rate.

2.1.4 Pension Calculation Variables

$\overline{W}$, the average wage; $\overline{W}_A$, the average wage in period A; Y, the average benefit period of pension treatment in economic society; β, the replacement rate of pension; λ, the proportion included in social pooling account under the partial accumulation system.

2.1.5 Investment Income Variables

R, the total investment income of the current period; r, Return on investment; R_A, the total investment income in period A.

2.1.6 Demographic Variables

n, The number of employees in the economic society; m, the number of beneficiaries in the economic society.

2.1.7 Balance Mechanism

The conditions for realizing the balance of fund revenue and expenditure are as follows:

$$PI = IE \tag{1}$$

2.2   Financial Balance Model

2.2.1   Financial Balance Model of Accounting on the Cash Basis

Accounting on the cash basis is a kind of system which is based on the equal between incomes and pays in the same period.

$$PI_A = \sum_{i=1}^{n} pi_{ia} = \sum_{i=1}^{n} (\alpha \cdot w_{ia}) = \alpha \sum_{i=1}^{n} w_{ia} \tag{2}$$

$$IE_A = \frac{m\beta}{n} \sum_{i=1}^{n} W_{ia} \tag{3}$$

$$\alpha \sum_{i=1}^{n} w_{ia} = \frac{m\beta}{n} \sum_{i=1}^{n} W_{ia} \tag{4}$$

However, this mode can operate normally in the case of young population age structure, narrow scope of security and low payment standard; in the case of economic depression, aging population and continuous improvement of payment scope and standard. There will be problems such as high proportion of payment, heavy contemporary burden and difficulty in fund raising.

2.2.2   Financial Balance Model of Full Accumulation System

The full accumulation model is to balance the total contributions (including contributions, investment income and government subsidies) of each insured person during the working period with the total amount of pension after retirement, that is, to pursue a long-term balance.

$$Pi_{iA} = \sum_{j=1}^{X} pi_{ij}(1+r)^{X-j} \tag{5}$$

$$Ie_{iT} = Tr \cdot Pi_{iA} + Pi_{iA} = (Tr+1) \sum_{j=1}^{X} pi_{ij}(1+r)^{X-j} \tag{6}$$

Due to the small proportion of social co-ordination, the function of income redistribution with full accumulation mode is very weak, and it plays a small role in adjusting the security level. The inconsistency between the prediction and the actual often occurs, which needs to be adjusted in time.

2.2.3   Financial Balance Model of Partial Accumulation System

Partial accumulation system is a kind of old-age insurance system compromising between accounting on the cash basis and full accumulation system. The insurance taxes paid by the insured should be divided into two parts. One part is used to pay the current beneficiary's pension, which is called the pooling account fund, and the other part is used as the fund accumulation, which is called the individual account fund. The equilibrium condition of the public pension fund under the partial accumulation system is that there should be new accumulated funds in each period, so the capital income of each period should be greater than the capital expenditure.

$$PI_A = R_A + \sum_{i=1}^{n} pi_{ia} = R_A + \sum_{i=1}^{n} (\alpha \cdot w_{ia}) = R_A + \alpha \sum_{i=1}^{n} w_{ia} \qquad (7)$$

$$IE_A = \frac{m \cdot \beta}{n} \sum_{i=1}^{n} W_{ia} + \frac{m \cdot Pi_A}{Y} \qquad (8)$$

$$R_A + \alpha \sum_{i=1}^{n} w_{ia} \rangle \frac{m \cdot \beta}{n} \sum_{i=1}^{n} W_{ia} + \frac{m \cdot Pi_A}{Y} \qquad (9)$$

2.2.4   The Main Factors Causing Fund Financial Imbalance

Through the derivation of the above model, we can draw a conclusion that the factors that affect the financial balance of pension fund are different under different financing modes, which are analyzed one by one below (Table 1).

Table 1. The factors under different financing modes

Fund raising mode	Main factors	Mode of action
Accounting on the cash basis	W, α, $\overline{W}$, n, m, β	1. Increasing of W, α, n, is beneficial to maintain the financial balance of the fund, and vice versa 2. Increasing of $\overline{W}$, m, β, is not conducive to maintaining the financial balance of the fund, otherwise it is beneficial
Full accumulation system	r, β,	1. r ≥ 0, the fund keeps in a state of financial balance; r ≤ 0, the fund is in a state of imbalance 2. After adding the expected rate of return factor, r must be greater than or equal to the expected rate of return in order to maintain the financial balance of the fund

According to expression (5), the factors that affect the financial balance of the fund include all the factors that may appear under different financing modes, and their impact on the financial balance of the fund is consistent with the above analysis. In addition, according to China's system, the level of the beneficiaries' pension benefits is also affected by the average benefit period of the society. The higher the value of Y, the lower the amount of expenditure in each period. However, in the long run, Y does not affect the financial balance of the fund.

3 Conclusions

The financial balance of pension fund is the basic premise to ensure the sustainability of pension insurance system. The financial imbalance of the fund refers to the situation that the fund account balance is insufficient to maintain the implementation of pension insurance system. This situation is caused by the fund operation risk and has many negative effects on pension insurance system, including: reducing the solvency of pension fund; At the same time, these negative effects will be intensified under certain conditions and lead to the trust crisis of the system.

From the above analysis, we can see that there are three important mechanisms to maintain the financial balance of the public pension insurance fund: the first is the budget mechanism, the second is the information sharing mechanism, and the third is the institutional operation mechanism.

Budget mechanism is the identification mechanism of fund flow risk. It plays an important role in the prevention of fund market risk and institutional risk by estimating the expected rate of return, financial imbalance gap of fund, accumulated and expenditure.

Information sharing mechanism is the identification mechanism of fund information flow risk. A large amount of information is needed for fund operation, including insurance information, investment information, population information, economic information, policy information and fund risk loss information. These information can help to prevent the internal risk of fund operation and complete the fund budget.

Institutional operation mechanism is the platform of fund information flow and fund flow process. Through member activities, functional departments, governance structure, communication process and internal and external supervision, fund risk can be reduced from the perspective of organizational behavior.

Acknowledgements. This work was supported by Research on the nature and orientation of social insurance agency, a doctoral research fund project of Wuhan Business University (2017KB010).

References

1. Mallet, R.: The pension liability. Financial Management(14719185), pp. 30–31 (2007)
2. Miller, S.: Manage pension plan risk with five expert tips. HR Mag. **57**(3), 12–12 (2012)
3. Wilcocks, J.: Avoiding the risk of fraud. Super Rev. **24**(8), 11–13 (2010)
4. Ruschau, W.J., Williams, D.: Managing risk in a multiemployer plan: new strategies for new times. Benef. Comp. Digest **47**(8), 32–36 (2010)
5. Duijm, P., Bisschop, S.S.: Short-termism of long-term investors? the investment behaviour of Dutch insurance companies and pension funds. Appl. Econ. **50**(31), 3376–3387 (2018). https://doi.org/10.1080/00036846.2017.1420898
6. Tilba, A., Reisberg, A.: Fiduciary duty under the microscope: stewardship and the spectrum of pension fund engagement. Mod. Law Rev. **82**(3), 456–487 (2019)
7. Deloitte. Global Artificial Intelligence Industry Whitepaper (0306750). Deloitte Technology, Media and Telecommunications Industry, 5–6 (2019)

8. Soe, R.-M., Drechsler, W.: Agile local governments: experimentation before implementation. Govern. Inf. Q. **35**(2), 323–335 (2018). https://doi.org/10.1016/j.giq.2017.11.010
9. Iannone, L., Palmisano, I., Fanizzi, N.: An algorithm based on counterfactuals for concept learning in the Semantic Web. Appl. Intell. **26**(2), 139–159 (2007)
10. Valle-Cruz, D., Criado, J.I., Sandoval-Almazán, R., Ruvalcaba-Gomez, E.A.: Assessing the public policy-cycle framework in the age of artificial intelligence: From agenda-setting to policy evaluation. Govern. Inf. Q. **37**(4), N.PAG (2010)

Design and Implementation of Small Commodity Trading System

Xin Guo[1], Tanglong Chen[2], and Wei Huang[3(✉)]

[1] School of Information, Harbin Guangsha University,
Harbin 150030, Heilongjiang, China
[2] Heilongjiang University of Finance and Economics,
Harbin 150030, Heilongjiang, China
[3] Economic and Management School, Jilin Agricultural Science and Technology
University, Jilin 132101, China

Abstract. Nowadays, the way people buy small commodities is not limited to physical stores, and most of them prefer the Internet to buy those items. Therefore, the design and implementation of small commodity system have made the management of small commodities more and more professional and institutionalized. The system not only provides consumers with the opportunity to choose goods, but offers an online sales platform to businesses as well. The system is designed with the framework JSP and SSM, based on the MySQL database, which includes the functions of foreground user and background administrator. Foreground functions contain commodity search, commodity query, user login and registration, shopping cart added, placing orders, modifying personal information, etc. And background functions consist of commodity classification, commodity management, customer list management, order list management and other functions. The system's relatively perfect functions make its shopping environment more humanized, professional and transparent, which will bring better consumption experiences to consumers.

Keywords: Small commodity management · SSM · MySQL

1 Preface

With the growing demands, the design of small commodity system not only provides consumers with the opportunity to choose goods, but also provides an online sales platform for businesses. The system can effectively solve the needs of the people and has firmly occupied a share of the small commodity sales market. Due to its better convenience and price advantages with a wide range of goods, there are more and more people who like to buy small goods online [1]. The system is designed with the framework JSP and SSM, based on the MySQL database, which consists of two parts: foreground and background. The foreground is mainly provided for users, who can register as members of the website. After logging in, they can search for their favorite commodities and view the details. If like, they can also add them to their own shopping cart to complete shopping [2]. Users can also check their own orders, collections and delivery addresses in the personal center. The foreground user flow chart is shown in

Fig. 1. For the background, ordinary users cannot enter it except the accounts with administrators' permissions. It is used to manage the goods and users' information of the website. It can also modify the quantity or classification of goods as well as the users' permissions. The flow chart of background administrator is shown in Fig. 2.

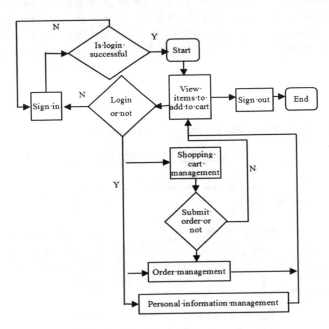

Fig. 1. Flow chart of foreground user

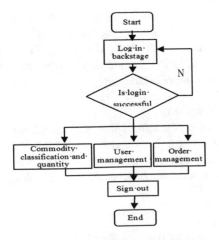

Fig. 2. Flow chart of background administrator

2 System Design

The design of small commodity system is implemented by the framework of JSP and SSM, combined with MySQL database, which is developed in Windows 10 operating system. MyBatis is used to connect with the database so that the system data can be directly stored in the database. Also, the original data of the database can be directly called, and the data can be added, deleted, modified and queried. The system sets up two kinds of role permissions. Users log on to the foreground of the website to shop and browse goods, while the background is only used by the administrator to manage the commodity data and user data of the foreground. The authority of foreground and background is separated. The user authority can only log on to the foreground, but not to background. Meanwhile, the administrator authority is only limited to background login [3]. The structure of small commodity system is shown in Fig. 3.

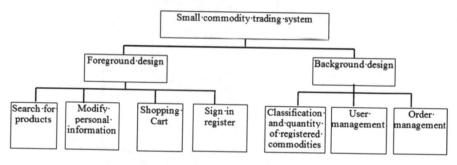

Fig. 3. Small commodity system structure

2.1 Foreground Design

2.1.1 Login Registration

Users can register as users of the small commodity system. Registered users can log in directly, while unregistered users need to register first. After logging in, users can enter the customer center, or click the shopping cart to pay for their preferable small commodities that have been added to the shopping cart before. Also, they can view their own collection of small commodities.

2.1.2 Commodities Searching

Users click their favorite product from the product browsing webpage to enter the product details webpage, where there is the detailed information of the specific commodity, such as unit price, sales volume, users' evaluation and so on. After users have the desire to buy, they can click the related button to add the goods into their shopping cart and purchase there [4].

2.1.3 Shopping Cart

Users can enter the webpage of the shopping cart through the shopping cart function module on the home page. The webpage displays all the products that users have added to the shopping cart. Users can view and check the products that he needs to buy on the webpage and make unified settlement. The system will automatically calculate the overall price of all products.

2.1.4 Modification of Personal Information

In addition to viewing all their orders and collections, users can modify their personal data and passwords as well. Besides, they are able to manage their own delivery addresses, such as adding, modifying or deleting them.

2.2 Background Design

2.2.1 User Management

Administrators can enter the user rights management page, and they can adjust ordinary users to administrators, and vice versa. Ordinary users can only log on to the foreground, while administrators can only log on to the background. This feature makes the system securer. Administrators can manage all registered users, including querying the users' personal information, such as name, password, mailbox and so on. Also, they can modify, add and delete the users' personal information.

2.2.2 Commodity Classification and Management

Administrators can manage the goods through the commodity classification page. The system supports three-level classification, which administrators can delete, add and modify. The system also provides the quick search with the classification name. The commodity classification in the foreground of the system will be updated synchronously when the background changes. The commodity data page covers the detailed information of the name, classification, price, inventory, etc., and administrators can add new commodity data or modify the original commodity data. The webpage supports searching with commodity names and commodity price ranges [5].

2.2.3 Order Management

Through the users' order page, administrators can view all users' orders and purchased goods, and they can also set the status of goods, query and manage the orders placed by users [6]. At the same time, the system also supports quickly searching the corresponding order information with the order number or user's name.

3 Database Design

The database design of this system includes four tables, which are user table, commodity table, commodity classification table and shopping cart table. The user table mainly stores users' personal information, such as ID, user name, password, image address, delivery address and so on. The commodity table is used to store the detailed information of commodities. The commodity ID is set to the primary key auto adding, and the parent class ID is used to indicate the class of the commodity in the three-level classification. The picture URL stores the addresses of mall commodities' pictures on the website. Price, inventory, sales quantity, comments and content of each commodity are stored in the commodity table [7], as shown in Table 1. The commodity

Table 1. Commodity list

Field name	Data type	Primary key	Empty or not	Description
Id	int(11)	Y	N	Commodity ID
Classification id	int(11)	N	Y	Classification ID
Name	varchar(128)	N	N	Commodity name
Age	varchar(32)	N	Y	Classification
Picture URL	varchar(128)	N	N	Picture address
Price	float(8,2)	N	N	Price
Stock	int(8)	N	N	Inventory
Quantity	int(8)	N	N	Sales quantity
Comment	int(8)	N	N	Comments
Content	text	N	Y	Contents

classification table can offer the operations of adding, deleting and modifying. Administrators can modify not only the upper classification ID to reclassify the commodities, but the name of the classification as well. The shopping cart table contains the shopping cart ID, and the user ID, which is the same as that in the user table. Commodity ID, name, picture URL and price are the same as those in commodity table [8]. The quantity field stores the quantity of goods in the users' shopping cart, while the total field stores the total price of goods in the shopping cart after calculation.

4   Conclusion

In order to meet the buyers' purchase demands and increase the sellers' sales significantly, the design and development of this small commodity system has certain practicality. The research and development of the system provides a lot of information resources for more and more consumers and enterprises that like to buy or sell small commodities online. It can meet people's long-term needs, changes people's consumption habits, and promotes the development of all trades and professions, which leads our society into a virtuous circle to a certain extent.

References

1. Zhang, Y.: Computer software java programming characteristics and technical analysis. Comput. Prod. Circ. (01), 23 (2019)
2. Bai, Z.: Design and implementation of campus second-hand commodity trading system based on cloud platform. Guizhou Univ., 31–33 (2018)
3. Ge, M., Huang, S., Ouyang, H.: Java web application based on spring MVC framework. Comput. Moder. (08), 97–101 (2018)
4. Wang, Y.: Application of java programming language in computer software development. Electron. Technol. Softw. Eng. (01), 35 (2019)

5. Zhang, X., Wang, X.: Computer database construction and management. Inf. Technol. Inf. (8), 83–85 (2018)
6. Ma, L.: Design and implementation of commodity sales management system. Jilin Univ., 1–62 (2017)
7. Ye, W.: Application of software development technology in software engineering management. Electron. Technol. Softw. Eng. (18), 60–61 (2017)
8. Gou, W., Yu, Q.: Design and implementation of data management system based on MySQL. Electron. Des. Eng. (06), 62–65 (2017)

Application of Image Recognition Based on Deep Learning Model

Jie Zhao[1], Qiushi Guan[2(⊠)], Hao He[3], Qiuzi He[4], Yuqi Wang[4], and Xinyu Huang[1]

[1] School of Economics and Business Administration,
Central China Normal University, Wuhan, Hubei, China
[2] College of Water Resources and Environment, China Three Gorges University,
Yichang, Hubei, China
CTGUGuan@ctgu.edu.cn
[3] College of Marxism, Central China Normal University, Wuhan, Hubei, China
[4] School of Information Management, Central China Normal University,
Wuhan, Hubei, China

Abstract. Traditional image recognition is mainly based on shallow neural networks, mainly including artificial networks, but these shallow neural networks have limited feature extraction capabilities. This paper can speed up the operation efficiency of the model, eliminate over-fitting, and improve the recognition accuracy by adding batch normalization layers to the neural network. The experiments are based on the MNIST data set, and have fully verified the recognition ability depth model proposed in this article.

Keywords: Deep learning · MNIST · CNN

1 Introduction

Nowadays, image recognition has become an important research field due to its wide application. For image recognition problems such as handwriting classification, the quality of feature extraction is critical to the extraction results [1] extracted the structural features of characters from the strokes and used them.Require manual extraction of features from the image. The predictive ability of the model strongly depends on the prior knowledge of the modeler [2, 3].

Most classification and regression machine learning methods are shallow learning algorithms [4, 5].

Learning is a representation learning method [6]. With enough such conversion combinations, you can learn very complex functions [7–10].

J. Abawajy et al. (Eds.): ATCI 2021, LNDECT 81, pp. 826–831, 2021.
https://doi.org/10.1007/978-3-030-79197-1_121

2 Convolutional Neural Network

2.1 Introduction to Convolutional Neural Network Model

Figure 1 is a simple model. The second layer is the BN layer. It can increase the gradient through the network, increase the learning rate, and greatly increase the training speed of the model. The third layer is the pool layer.

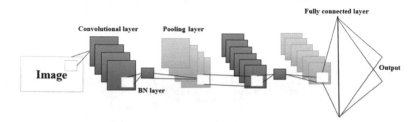

Fig. 1. Convolutional neural network model

2.2 Convolutional Neural Network Theory

Convolution is fixed.

$$y^{l(i,j)} = K_i^l x^{l(r^j)} = \sum_{j'=0}^{W-1} x^{l(j+j')} K^{l(r^j)} \tag{1}$$

$K^{l(r^j)}$ is the layer; $x^{l(r^j)}$ is the j convolution local area of the D layer;.

The batch normalization (BN) layer aims to reduce networks. Conversion process is described as:

$$\hat{y}^{l(i,j)} = \frac{\hat{y}^{l(i,j)} - \mu B}{\sqrt{\delta_B^2 + \varepsilon}} \tag{2}$$

$$z^{l(i,j)} = \gamma^{l(i)} \hat{y}^{l(i,j)} + \beta^{l(i)} \tag{3}$$

In the formula, $\gamma^{l(i)}$ and $\beta^{l(i)}$ are the scale. ε is a value zero.

In CNN architecture, a pooling layer is usually added after the batch normalization layer:

$$P^{l(i,j)} = \max_{(j-i)W+1 \le t \le jW} \left\{ \alpha^{l(i,j)} \right\} \tag{4}$$

CNN can be expressed as:

$$E = \frac{1}{2}\sum_{n=1}^{N}\left(t^n - y^n\right)^2 \tag{5}$$

The loss function and deviation is:

$$V_{b_j}E = \sum \delta_j^l \, V_{K_{ij}^l}E = x_j^{l-1}\delta_j^l \tag{6}$$

In the formula, $\delta_j^l = \left(\sum_{j=1}^{n1} K_{ij}^l \delta_j^{l+1}\right) f'\left(u_i^l\right)$ is the error term, weight according to the following formula:

$$K^l = K^l - \eta\left[\left(\frac{1}{N}\sum_{i,j} V_{K_{ij}^l}E\right) + \lambda K^l\right] \tag{7}$$

$$b^l = b^l - \eta\left(\frac{1}{N}\sum_{i,j} V_{b_{ij}^l}E\right) \tag{8}$$

3   Experimental Verification

3.1   Experimental Data

We choose the deep learning. This shows that data very fragmented. The recognition ability of the model can be sufficiently improved. It has been verified that the image size is 28 × 28. Some actual sample sets in the MNIST handwriting database are shown in Fig. 2.

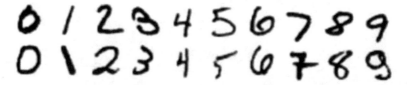

Fig. 2. Some real samples of MNIST database

3.2   Introduction to Convolutional Neural Network Model

First, number of convolution kernels on the recognition accuracy.The recognition accuracy of three different convolutional neural networks is shown in Table 1. All reaching more than 95.

Table 1. Comparison of results of convolutional neural networks with different numbers of convolution kernels

Structure	Learning rate	Iteration steps	Recognition accuracy
784-4-12	0.01	100	97.31%
784-8-24	0.01	100	97.48%
784-16-48	0.01	100	97.71%

From the comparison is Table 1, when number of cores increases from 4, 12 to 8, 24 and then to 16, 48, the accuracy based on the MNIST data set continues to increase, from 97.31% initially to 97.48%, and finally increased To 97.71% (Table 2).

Table 2. Recognition and comparison results of all models

The internet	Recognition accuracy
Artificial neural networks	91.57%
Deep belief network	97.29%
Convolutional neural network	97.71%

Convolutional neural network, this article chooses to use for comparison. Which can fully verify the built convolutional neural network. The image recognition capabilities of the network. The learning rate is 0.05, and the structure of the deep confidence network is 784-150-100-10, the learning rate is 0.1. Figure 3 shows of all networks with the number of iterations.

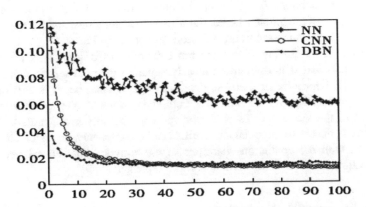

Fig. 3. The error rate of the network changes with the number of iterations

Note: The abscissa is the number of iterations, and the ordinate is the error rate.
From the experimental results, we can see that CNN has unique advantages in image recognition.

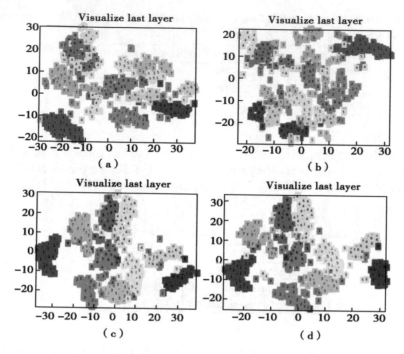

Fig. 4. Visualize the results

3.3 Visualization of Results

In order to intuitively understand the feature learning process of the proposed convolutional neural network, is used deep features learned in different iteration steps. In this paper, a total of the initial state, ten iterations, fifty iterations, and the depth features obtained from the final iteration are selected, as shown in Fig. 4.

It can be seen from Fig. 4 that the ten categories of original data are randomly mixed together, and it is difficult to clearly separate them. After ten iterations, the recognition is slightly better than the original original data, but it is still difficult to separate them directly. As the iteration continues, when it reaches fifty times, the recognition of the ten categories is higher and can basically be separated. When the final iteration effect is reached, all categories can be completely separated clearly. This reflects that with the continuous iteration of the convolutional neural network, the learned features are becoming more and more representative.

4 Conclusion and Discussion

By comparing the experimental results, the feature extraction process of the convolutional neural network is analyzed.

Acknowledgements. This work was supported by YK2020094. We would like to thank Funing people's Hospital for its research support.

References

1. Huang, H.M., Wang, X.J., Yi, Z.J., et al.: A character recognition based on feature extraction. J. Chongqing Univ. (Nat. Sci. Ed.) **23**(1), 66–69 (2001)
2. Rui, T., Shen, C.L., Ding, J., et al.: Handwritten character recognition using principal component analysis. Mini-Microsyst. **26**(2), 289–292 (2005)
3. Walid, R., Lasfar, A.: Handwritten digit recognition using sparse deep architectures. In: International Conference on Intelligent Systems: Theories & Applications, pp. 1–6. IEEE (2014)
4. David, B.: Character recognition using convolutional neural networks. In: Seminar Statistical Learning Theory University of Ulm Germany Institute for Neural Information Processing, pp. 2–5 (2006)
5. Lezoray, O., Cardot, H.: Cooperation of color pixel classification schemes and color watershed: a study for microscopic images. IEEE Trans. Image Process. **11**(7), 783–789 (2002)
6. Wei, L.: Research and Application of Deep Learning in Image Recognition. Wuhan University of Technology, Wuhan (2014)
7. Shafarenko, L., Petrou, M., Kittler, J.: Automatic watershed segmentation of randomly textured color images. IEEE Trans. Image Process. **6**(11), 1530–1544 (1997)
8. Shiji, A., Hamada, N.: Color image segmentation method using watershed algorithm and contour information. In: Proceedings 1999 International Conference on Image Processing (Cat. 99CH36348), Kobe, Japan, 24–28 October 1999, pp. 305–309. IEEE (1999)
9. Yujie, Z.: The application status and advantages of deep learning in the field of image recognition. China Secur. **7**, 75–78 (2016)
10. Dong, L., Li, S., Zhidong, C.: A review of deep learning and its application in image object classification and detection. Comput. Sci. **12**, 13–23 (2016)

On the Transformation and Innovation of Education in the Era of Artificial Intelligence

Wen Wang[1(✉)], Peng Wang[1], Yinping Qi[2], and Xiangrong Shi[3]

[1] Shandong Urban Construction Vocational College,
Jinan 250103, Shandong, China
[2] Shandong Zhengyuan Construction Engineering Co. Ltd,
Jinan 250100, Shandong, China
[3] Shandong Jianzhu University, Jinan 250101, Shandong, China

Abstract. Artificial intelligence (AI), which goal is to make machines think and understand like humans, emphasizing the use of computers to simulate human intelligence. As artificial intelligence has become a national strategy, the development of education in China is facing new opportunities and challenges. In the era of artificial intelligence, human-computer coupling education represents a new form of vocational education in the future. Moving towards the intelligent education environment and education model has become the research of China's education innovation in the new period.

Keywords: Artificial intelligence (AI) · Student management · Embedded system · Computer vision · Auditory information

1 Introduction

Over the past several years, we have seen a revolution of artificial intelligence (AI) in services and products due to the rapid adoption in many industries. AI is a technology that allows machines to perform tasks like humans, which is learned from experience. This is mainly because its AI describes a different kind of technology that provides the ability to learn intelligently on this device [1, 2].

AI see as far as possible to provide an innovative solution, which is been embedded developers. AI sees the development of new things on the Internet that are essential technologies for physical systems such as robots. All aspects of an employee's daily work have been improved by embedded system of AI, which is a broad term that uses deep learning and software platforms. The sustainable development of social economy promotes the progress of science and technology [3, 4].

AI combines multidisciplinary knowledge, including intuitive reasoning (with many social knowledge and physical), natural language perception, computer vision, machine learning and interaction. The cognitive computing of visual and auditory information is Hybrid-augmented's core research content and it intelligence is a typical feature of the next generation of AI. So, it is of great significance for the research and

J. Abawajy et al. (Eds.): ATCI 2021, LNDECT 81, pp. 832–836, 2021.
https://doi.org/10.1007/978-3-030-79197-1_122

development of artificial intelligence to understand the cognitive mechanism of human vision and hearing and establish its computable model [5].

It's not just human beings that are advancing in artificial intelligence, many new ones open up unprecedented experiences for many new future services. This plan can be defined on internally generated information [6]. The information embedded in the processed records acquired the knowledge and helps the organizations provide information in a timely and adequate manner.

2   Article and Citation Trends

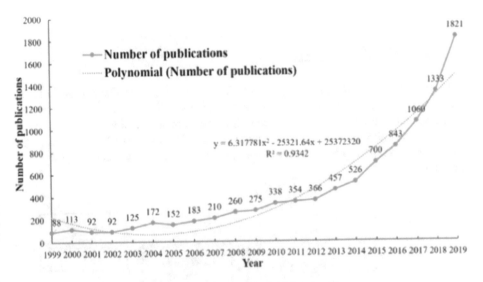

Fig. 1. Publication trend of AI in education.

In general, it can be seen that from 1999 to 2019, the amount of AI in educational publications shows an increasing trend in Fig. 1 [7]. The evolution of the output of published articles has generally gone through three stages. In the first stage, from 1999 to 2002, the growth trend is relatively slow. The second stage shows a steady growth trend from 2003 to 2011. From 2012 to 2019, the third phase is a period of rapid growth. Also, Fig. 1 shows the year of 2019 is the most creative year for citations and publications. And then, research shows that the growing global interest in educational contexts in the application of AI led to a surge in the number of papers on AI in education, with more than 50% published between 2012 and 2019, a rapidly growing area of the research [8].

From major institutions like the National Science Foundation, results of the trend of grants in relation to educational technology indicate a significantly growing interest in the area. In the recent years, governments in various countries have been furiously investing in the development of technological devices which are incorporated with AI technologies in classroom environment [9].

The results of the funding trends for educational technology from major institutions show that there has been a significant increase in interest in the artificial intelligence. Governments of all countries are investing heavily in the development of technological equipment, which is combined with artificial intelligence technology and used in teaching environments, especially in recent years [9].

As a result, the research and education issues related to AI in international education have attracted more and more international interest, not only from researchers and educators, but also from educational institutions and governments. Such as in the year of 2018, the National Natural Science Foundation of China has been added the ``Educational Information Science and Technology Level 2 Code' for the first time to encourage the development of technology-enhanced education research to promote prospects in education in China [10].

3   Method and Materials

In the engineering training process, first of all, the course can be planned accordingly, and the course can also be evaluated and examined. Then the students are been evaluated, and the grades are given according to reasonable standards. From the perspective of teaching methods, the evaluation of this class is part of the learning course also and reflects the teachers' teaching level, in this way, this step can be paid mostly attention to.

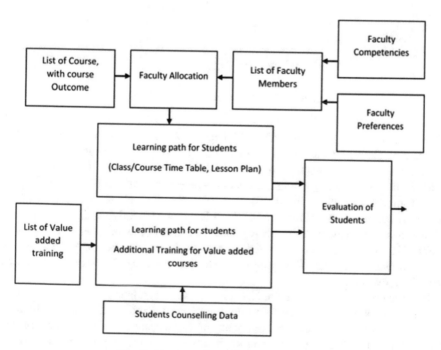

Fig. 2. Process of teaching and learning.

In study planning of a course, the course needs to be developed, which is a very important step. The resulting learning courses must meet the basic requirements of the course, and it must meet the basic requirements of the course at the same time, which possible problems in the classroom should be dealt with. When delivering the course of study, we need to think many times for different students to adopt different teaching methods and their different characteristics [11, 12].

As shown in Fig. 2, the continuous development of AI has brought new changes to school management. First of all, in the changing environment, for training control, the era of AI has brought new possibilities and new requirements. Secondly, intelligent application of AI technology in the field of teaching and training has become even more important. And then, the teaching management system of AI still should be optimized and improved gradually for education administrators, and from many elements new measures must be applied such as control plans, technical approaches and many others.

Secondly, the intelligent application of AI technology in teaching and training has become even more important. Besides, for education managers, the teaching management system of AI should be optimized and improved, and new ways should be taken from multiple aspects such as technical approaches and control plans. In this way, we can achieve efficient control and high-quality of education. Third, the rational application of AI in education will become more and more important. Furthermore, for teaching managers, the teaching control machines of AI need to be gradually optimized and promoted, and new ways should be implemented such as management plans and technical means and other aspects from many aspects. Only in this way can AI be promoted in a civilized and green way, and training and coaching can be fundamentally promoted to achieve high-quality, green control.

4   Conclusion

AI is a hot topic in the rapid development of human technology and science. The continuous progress of AI is not only brought to mankind, a variety of new future services have opened up an unprecedented experience. And then, some new changes should been taken in school management with the continuous development of AI technology. Finally, opportunities and challenges have been brought to the society by the development of AI.

2020 is a testament to the unpredictability of the future. The last few months of the year not only changed the course of life, but also changed the function and purpose of the original plan. What started as a smart speaker with useful university data has become more meaningful by integrating Microsoft's services, including remote connections and new ways to stay informed. The purpose of the speaker goes beyond the original plan, during the COVID-19 period, it transformed into a ' humanoid ' image of students and teachers, while promoting user education and performance.

References

1. Qizhen, W.: Automatic control system regulation scheme of textile air conditioning. Sci. Technol. Inf. **22**, 395–396 (2010). (in Chinese)
2. Geisseler, D., Scow, K.M.: Long-term effects of mineral fertilizers on soil microorganisms-a review. Soil Biol. Biochem. **75**(1), 54–63 (2014)
3. Jianwei, L., Qingchang, R.: Study on supply air temperature forecast and changing machine dew point for variable air volume system. Build. Energy Environ. **27**(4), 29–32 (2008). (in Chinese)
4. Serban, C.: A conceptual framework for object-oriented design assessment. In: UK Sim 4th European Modelling Symposium on Mathematical Modelling and Computer Simulation, Pisa, pp. 17–19, 90–95 (2010)
5. Whitehill, J., Serpell, Z., Lin, Y.C., Foster, A., Movellan, J.R.: The faces of engagement: automatic recognition of student engagement from facial expressions. IEEE Trans. Affect. Comput. **5**(1), 86–98 (2014)
6. Xie, H., Chu, H. C., Hwang, G. J., Wang, C.C.: Trends and development in technology-enhanced adaptive/personalized learning: A systematic review of journal publications from 2007 to 2017. Comput. Educ. **140**, 103599 (2019)
7. Xing, W., Du, D.: Dropout prediction in MOOCs: using deep learning for personalized intervention. J. Educ. Comput. Res. **57**(3), 547–570 (2019)
8. Yang, Q.F., Hwang, G.J., Sung, H.Y.: Trends and research issues of mobile learning studies in physical education: a review of. academic journal publications. Interact. Learn. Environ. **28**(4), 419–437 (2020)
9. Zhang, W., Zhang, Z., Zeadally, S., Chao, H., Leung, V.C.M.: MASM: a multiplealgorithm service model for energy-delay optimization in edge artificial intelligence. IEEE Trans. Ind. Inf. **15**(7), 4216–4224 (2019)
10. Chen, W., Lin, Y., Ng, F., Liu, C., Lin, Y.: RiceTalk: rice blast detection using internet of things and artificial intelligence technologies. IEEE Internet Things J. **7**(2), 1001–1010 (2020)
11. Terzopoulos, G., Satratzemi, M.: Voice assistants and artificial intelligence in education. In: Proceedings of the 9th Balkan Conference on Informatics, BCI 2019, Sofifia, Bulgaria, 26–28, pp. 34:1–34:6. ACM (2019)
12. Trivedi, N.: ProblemPal: generating autonomous practice content in real-time with voice commands and amazon Alexa. In: Proceedings of E-Learn: World Conference on E-Learning in Corporate, Government, Healthcare, and Higher Education, pp. 80–82 (2018)

Application and Development Proposal of Big Data in Targeting Poverty Alleviation

Bailin Pang[✉]

School of Economics, Harbin University of Commerce,
Harbin, Heilongjiang, China

Abstract. The poverty alleviation under extensive concept work uses a regional overall advance method, which gives good to everybody equally like broad irrigation. This mechanism generates a lot of problems. In the strategy of targeting poverty alleviation, the most significant is "accurate". Big data is an essential technological means to realize it. Big data is mainly used to identify problems accurately, make dynamic management in the process, and realizing poverty alleviation through education and medical care by digital technology. In future development, we should solve the security issue, break information isolation, promote information sharing of the whole society, and develop more big data talents.

Keywords: Poverty relief and development · Targeting poverty alleviation · Big data

1 Introduction

For a long time, the Chinese government has accumulated much experience in poverty relief and development. The number of people in poverty and incidence of poverty both slumped, which means an impressive result. However, the poverty alleviation under extensive concept work uses a regional overall advance method, which gives good to everybody equally like broad irrigation. This mechanism generates a lot of problems. For example, the exact number of low residents is not precise, the reasons for poverty are unsharp, the poverty relief and development are not targeted, the directivity of project and fund are not accurate, and so on. Targeted poverty alleviation is based on the experience and defects of traditional poverty relief pattern. To improve the efficiency and effect of poverty relief, targeting poverty alleviation emphasize attention to details. The projects should fit the development of the population in the impoverished area, and the fund should be precise. The targeted poverty alleviation fits the present state of China, which meets the realistic requirement of development work. Under the party and the government's vigorous promotion, the targeted poverty alleviation was developed smoothly and obviously affected. As we know, when a doctor diagnosing an illness, he can target medicine must be prescribed only by understanding the cause. Therefore, the most significant part of targeted poverty alleviation is "target". When the "cause" of poverty is found, the problem can be solved effectively. Nevertheless, the traditional ways of poverty cannot satisfy the need of "target".

© The Author(s), under exclusive license to Springer Nature Switzerland AG 2021
J. Abawajy et al. (Eds.): ATCI 2021, LNDECT 81, pp. 837–840, 2021.
https://doi.org/10.1007/978-3-030-79197-1_123

With the development of information technology (IT), the period of big data has already come. Big data affect the economic society in various ways [1]. Nowadays, big data is a vital resource and technological means of the anti-poverty project.

2 The Use of Big Data in Targeted Poverty Alleviation

2.1 The Accurate Recognition of Big Data

The targeted poverty alleviation is based on accurate recognition. The accurate recognition can make sure to valid identify the poor people. The standard should be considered before recognition. We mainly judge it by the income level. However, this standard is too single. It is hard to screen the really needy people. The phenomenon of omitting and mistake often appears. This is not accurate at all and bad for the target realization of poverty alleviation as well. The accurate identification is challenging because of the multidimensional character of poverty. The poverty of the poor is not limited to the economic sphere, and they also have an obvious deficiency in chances and right of development [2]. It is hard for the traditional method to make comprehensive identification. The poor's development chance and right are always neglected. The check is still basically depending on the source of income. Studies have shown that we have a low accuracy of recognition of the poverty-stricken population. The rough filtrate makes the result fuzzy [3]. While using big data, we can integrate civil administration, medical treatment, education, agriculture, etc. The more comprehensive dimensions of access make the result more accurate.

2.2 Use the Big Data to Make Dynamic Management of Targeted Poverty Alleviation

Whether a person is in poverty or not is always changing over time. Traditional management is static [4]. It cannot be so "targeted" as required. This phenomenon leads to the consequence that some poverty-stricken population's problem cannot be solved while some people who have already got rid of poverty still take up the resources for poverty alleviation, and the new poverty-stricken people hard to be taken into assist. The targeted poverty alleviation needs to take dynamic management objectively. However, the distribution of dynamic management needs to consider many indexes. Besides the incomings and outcomings, housing conditions, degree of education, and some fundamental indexes, it also includes their health condition, balance situation, and so on. It is tough to manage the poverty-stricken population dynamically without the support of big data.

By building a big data management platform, we can update the low-income families' information in season. In this way, the dynamic management of targeted poverty alleviation can be realized [5]. In the platform database, those who should be added will be added as well those who should be deleted will be deleted. This can make the purpose of poverty alleviation come true. At the same time, the database of the platform can reflect the specific flow of the poverty alleviation funds accurately without delay. It can check the funds for another time and avoid the fund used for something else or be occupied by some people or organization. This sort of safeguard increases the utilization efficiency of the poverty alleviation funds.

2.3   Use Digital Technology to Realize Educational Poverty Alleviation

The traditional method of poverty alleviation is always focusing on the outside venture capital of the distressed area. However, it did not mobilize the internal productivity well in poor rural, especially the low-income families. This is mainly because of generally the lack of educational resources and has the backward education level in the distressed area. The poverty-stricken population has less approach and methodology for cultural knowledge, which limit them and make them lake of self-development ability seriously. As a result, improving the population quality is a significant move for changing the current situation of poverty [6]. It is also the fundamental guarantee of solving the problem of poverty. We should consider poverty alleviation through education for the poor as the guarantee of realizing targeted poverty alleviation. In the past, we mainly used the traditional poverty alleviation of education by increasing the investment in the education of the poverty area or afford special skills training, which are made pointed references. But nowadays, with the support of technology of this date, the high-quality educational resources can be shared with the distressed area by remote education. Their requirements of education can be satisfied in this way. The sharing of resources with good quality can be realized by butt joint of schools. They can teach at the same time, do the same exercises, and have the same exam. The copy of good education can improve the problem of educational backwardness [7]. The practice has proved that remote education can afford the distressed area high quality of education with low cost and realize the target of educational poverty alleviation.

2.4   Use Digital Technology to Realize Medical Poverty Alleviation

For a long time, the distribution of healthcare resources is severely uneven. In the distressed area, the shortage of medical resources is severe. It is prevalent to become poverty or regain poverty because of illness. Expensive and difficult for treatment are the biggest problem for prevents poor areas from becoming wealthy. Modern digital technique supplies technical support for telemedicine. It builds a "health line of defense" for poverty-stricken areas to shake off poverty and become rich. Since 2018, Hunan Province began to build remote consultation centers covering counties, townships, and villages. This measure builds a path to health for the poor. This sort of telemedicine provides high-quality medical resources for the poor, while significantly reducing the time and economic costs of time and economic cost of seeking medical care as well [8]. It can effectively reduce the probability of poverty due to illness, which is a strong support for the poverty population to get rid of poverty and get rich.

Besides the aspects mentioned above, there is also conducive to the precise design of poverty alleviation programs, accurate judgments of poverty alleviation performance, advanced prediction of poverty return, etc., by the application of big data.

3 Development Suggestions

First of all, we should solve the problem of data security. Because of the loopholes of big data technically and in management, it is easy to be used by criminals. This will hinder the smooth implementation of poverty alleviation and development. Therefore, it is necessary to accelerate the development of digital technology on which targeted poverty alleviation depends and improve relevant regulations and systems [9]. Besides, we should break the information silos and share information across society. The data and information related to precision poverty alleviation are vast and involve different industries and departments. If this information is operated in isolation and cannot flow effectively across departments and systems, these data will not form valuable information become information islands. Therefore, all data should flow reasonable and be fully shared to improve the efficiency of data resource utilization [10]. Finally, strengthen the training of talents related to big data technology. Precision poverty alleviation in the era of big data requires the support of technical talents such as big data, cloud computing, and the internet of things. The current high scarcity of elite talents in big data has become a severe obstacle to the practice of big data poverty alleviation. The government needs to formulate policies to accelerate the development of relevant talents brought up.

References

1. Baye, M.R., Morgan, J., Scholten, P.: Information, search and price dispersion. Econ. Inf. Syst. **04**, 66–75 (2006)
2. Ahmed, M., Zeng, Y., Ozaki, A., et al.: Poor farmer entrepreneurs and ICT relation in production & marketing of quality vegetables in Bangladesh. J. Fac. Agric. Kyushu Univ. **61**(1), 241–250 (2016)
3. Haythom, C.: Productivity in agriculture and R & D. J. Prod. Anal. **30**(01), 7–12 (2014)
4. McAfee, A., Brynjolfsson, E.: Big data: the management revolution. Harvard Bus. Rev. **33** (9), 30–36 (2015)
5. Temte, M.N.: Blockchain challenges traditional contract law: just how smart are smart contracts. Wyoming Law Rev. **19**(1), 118–187 (2019)
6. Guire, T.M., Manyika, J., Chui, M.: Why big data is the new competitive advantage. Ivey Bus. J. **31**(7), 76–82 (2012)
7. Zetzsche, D.A., et al.: Regulating revolution: from regulatory sandboxes to smart regulation. Fordham J. Corp. Fin. Law **23**(1), 31–104 (2017)
8. Das, B.: ICTs adoption for accessing agricultural information : evidence from Indian agriculture. Agric. Econ. Res. Rev. **27**, 199–208 (2014)
9. Aker, J.C.: Dial, "a" for agriculture: a review of information and communication technologies for agricultural extension in developing countries. Agric. Econ. **42**, 631–647 (2011)
10. Aker, J.C.: information from markets near and far : mobile phones and agricultural markets in Niger. Am. Econ. J. Appl. Econ. **02**, 46–59 (2010)

A Primary Analysis of the Role of Intellectual Media Development in Promoting City Image Communication

Minjing Wang[✉]

Department of Literature and Media, X'ian Fanyi University,
X'ian, Shaanxi, China

Abstract. Artificial intelligence, big data, VR and other technologies have promoted the intelligent development of media. In the context of intellectual media, intelligent algorithm distribution solves the correlation between audience and content; VR/AR and other technologies provide technical support for the "sense of presence" of city image communication. Intelligent software creates the conditions for content production to be convenient and low standard. These new changes will form the three-dimensional and multi-dimensional communication of city image between offline and online, between real and virtual, and open a new paradigm of city image communication.

Keywords: Intellectual Media · Artificial Intelligence · VR · City Image Communication

1 Introduction

As a part of the city's soft power, a good city image can not only promote the city's economic development and open up the city's tourism market, but it can also recruit talents and promote foreign exchanges and cooperation. Therefore, whether the media can be effectively used to construct and spread the image of the city has become a very key indicator to measure the competitiveness of the city. In recent years, driven by various technologies such as artificial intelligence, the Internet of Things, and big data, intelligence has become an important direction for media development. intellectual mediation has the characteristics of everything being a matchmaker, man-machine symbiosis, and self-evolution [1]. Its development has had a profound impact on the link of information dissemination, and also provided a new perspective for the shaping of the city's image.

2 City Image Communication in the Context of Intellectual Media

The advent of the intellectual media era is inseparable from the development of artificial intelligence. The concept of artificial intelligence was first proposed at the Dartmouth Conference in 1956, which means "making a machine react like the intelligence on which a person acts.".". Alan Turing, the father of artificial intelligence, proposed the "Turing test", which is to determine whether a machine has an intelligent method: make the tester talk to the testee. If the tester cannot distinguish whether the testee is a machine or a human, then you can Think that the machine has intelligence [2]. Elaine Ritchie believes that artificial intelligence is "the study of how to make computers do things that humans are better at today" [3]. With the continuous development of technology, artificial intelligence has gradually extended to the field of news production. The initial form of the combination of the two is robot writing. At present, the earliest robot news recognized by the industry and academia is a press release written in 2009 by Stats Monkey developed by the Intelligent Information Laboratory of Northwestern University. In this manuscript, Stats Monkey automatically compiled a game report by studying the game data, summarizing the game process.

Subsequently, artificial intelligence has been increasingly used in the field of communication. AI anchors, robot writing systems, AI cloud editing, and other applications have been launched, which have a profound impact on all aspects of information communication. For example, the Xinhua News Agency's Eagle Eye Discovery system predicts news, and the use of sensors allows news to occur and collection to be synchronized. In the news distribution link, the massive resources in the era of big data provide good fertile ground for artificial intelligence technology. Through a large amount of data resources, media organizations conduct dynamic mining and understanding by analyzing the data retained by users to form user portraits, and then carry out personalized content push and data marketing, and realize the transformation from traditional media to intellectual media. Artificial intelligence is bringing great changes to the field of information dissemination. Among them, the intelligence of information generation, algorithm-based machine writing and accurate information intelligent publishing are the most representative. The combination of VR and the field of communication began in the 1990s. VR technology uses computer-generated three-dimensional virtual environments to allow audiences to immerse and interact with them from a first perspective, creating an "immersive" empathy". From the perspective of communication science, VR has three core characteristics: one is immersion; the other is interaction; the third is imagination [4]. The development of technology has provided unlimited possibilities for the spread of city image In McLuhan's view, the medium is the message, and the medium is the extension of man [5]. In the context of VR technology, what the media disseminates is no longer traditional information and opinions, but an experience implanted in your body. This is the important experience of media as an extension of people. The content of city image dissemination is closely related to people's lives and also closely related to the development of technology. To a certain extent, changes in media technology have changed the symbolic form of city image communication. With the wave of intellectual media, city image communication will show some new communication trends under the development of artificial intelligence, big data, and VR technologies.

3 The Promotion Effect of Intellectual Media Development on City Image Communication

3.1 Algorithm Recommendation Helps the Accurate and Personalized Communication of the City Image

To some extent, city image communication constructs the meaning space of a city. As a cultural symbol and narrative discourse with high recognition degree, the accurate communication of city image is of great significance for shaping the overall brand of a city, transmitting the inner spirit of a city, and constructing the cultural identity of a city [6]. Combining artificial intelligence and algorithm recommendation, it can integrate and analyze the huge content database resources, and make accurate recommendation according to the user's taste, keywords and social relations. In this way, the behavior of the audience to obtain information on the Internet will be informationized, and the search behavior itself has the attribute of information production. The algorithm recommends users according to their social circles and hot topics, which strengthens the personalization, attention and newness. City image promotion, for example, large amount of data and algorithm can recommend according to thumb up and forward quantity, realizes the intelligent distribution and push on the one hand can realize accurate delivery, the image of the city, on the other hand city information in the form of diversification can be embedded in other media form, in the process of upload, browse, sharing and comments, the audience widely and actively involved in the play the function of social interpretation of graphic video content, enhance the value of the public issues, formed the dispersive transmission of the image of the city.

3.2 Intelligent Technology Creates a Sense of Presence for the City Characteristic Landscape

With the increasing complexity of modern society, people's actual scope of action, energy and attention are limited, and it is impossible for them to maintain experiential contact with the whole external environment related to them. With the help of urban landscape constructed by media, people can understand cities, understand urban culture and perceive the differences between cities. Although the information transmitted through the media has a certain understanding of the city, at the same time, there is an insurmountable gap between time and space. In the intelligent era, body sensing technology, interactive projection technology, 3D, AR, VR technology are constantly embedded into the human body, producing more immersive experience. This has become an important driving force to accelerate the communication from "conscious immersion" to "perceptual immersion". "Technology is becoming more and more transparent, deeply embedded in the human body, and thus fully integrated with our physical experience" [7]. Driven by new technology, the boundary between the "real" and the "virtual" has become blurred, and the "presence" of information has been enhanced. Now more and more cities use VR and other technologies to realize the production of city image promotional videos, such as "Hello Chengdu", "360° VR takes you to discover the most beautiful Jiangsu" and other city promotional videos bring strong visual impact and artistic appeal through virtual reality technology, so that the outside world can experience the unique charm of these cities. In the second world constructed by VR technology, the audience can feel the characteristic landscape of the city and realize zero-distance contact with the city, and experience, interaction and immersion become its salient features [8]. Under 5G conditions, holographic technology, VR technology and other video-related technologies will greatly improve the sense of presence experience.

3.3   Intelligent Technology Reduces the Conditions of Image Production and Promotes the Multi-dimensional Communication of City Image

The continuous improvement of intelligent technology has given rise to a variety of live broadcasting and web celebrity, and has also transferred more content originally belonging to professional production to UGC, OGC and MGC of ubiquitous production. With the help of all kinds of intelligent image editing and editing software, the shooting, editing, editing and synthesis of various video materials can be automatically generated, which greatly improves the content quality of the image and converges thousands of individual texts into visually flowing urban texts. Many cities begin to guide the participation of various folk forces, and promote and expand image communication by means of individual perspective and word-of-mouth marketing [9]. "White Paper on Short Video and City Image Research" conducted category statistics on creators of Douyin city image videos, and found that over 80% of the videos with the TOP100 playback volume were created by individual users, and ordinary people became the main creators of city image videos on Douyin platform [10]. Ordinary Internet users use mobile terminal record city image and personal space, the city image by individuals in the propaganda propaganda by government than pay more attention to the emotion, experience, interactive, lower cost and more widespread, the formation and the mainstream media discourse system of multivariate complementary, in the "record" "watch" interactive", "city image successfully realized the audience from the cognitive unity, participants and actors to the viewer. Between offline and online, real and virtual, form the three-dimensional and multi-dimensional communication of city image.

4   Conclusions

With the advent of the era of intellectual media, big data and artificial intelligence technology have brought about the transformation of media ecology and opened a new paradigm of city image communication from "reproduction" to "experience". Intelligent algorithm distribution solves the correlation between people and content; VR/AR and other technologies provide technical support for the "sense of presence" and "sense of immersion" of city image communication. Intelligent software creates the conditions for content production to be convenient and low standard. These are all new directions and new means of city image communication in the era of intellectual media, but at the same time, they also bring some problems. For example, although the algorithm recommendation can help the accurate dissemination of the city image, there are also problems such as "information cocoon room" and personal privacy security. Some scholars argue that algorithms have the power to shape users' experience and even their perception of the world. It is also worth thinking about whether this will deepen the "stereotype" of users on the city image (especially when they are exposed to negative information about the city), and whether it will result in a single and one-sided impression of the city. At the same time, the popularity of VR and other technologies is also limited by users' wearables and the discomfort such as dizziness brought by the technology itself, which is expected to be effectively overcome in the future development.

Acknowledgements. This article is a phased research result of the Shaanxi Provincial Department of Education's 2020 Special Scientific Research Project "Media Communication Strategies in Shaping the Image of Xi'an 'Capital of Hard Technology'" (Project No. 20JK0177).

References

1. Lan, P.: Intellectual media: future media wave – new media development trend report 2016. Int. Press **11**, 7 (2016). (in Chinese)
2. Nick: Brief History of Artificial Intelligence. Posts Telecommun. Press **12**, 11 (2017) (in Chinese)
3. Kurzweil, R.: The Heart of Machines. Tsinghua University Press, Beijing vol. 91 (2014) (in Chinese)
4. Anbin, S.: VR/AR opens the era of human communication media/intelligence media – Key message speech at the first VR/AR industry conference in Jiangsu. Jiangsu Econ. News pp. 6–18 (2017) (in Chinese)
5. McLuhan, M.: Understanding Media – On the Extension of Man He Daokuan, Trans. The Commercial Press, pp. 33–50 (2000) (in Chinese)
6. Hua, M., Qiang, L.: New paradigm of city image communication in the age of intelligence media young journalists **11**, 42 (2020) (in Chinese)
7. Xi, Y., Wenjiao, R.: From consciousness immersion to perceptual immersion: embodied shift of advertising in the intelligent age. Mod. Commun. (Journal of Communication University of China) **1**, 128–132 (2020). (in Chinese)
8. Xin, D.: A new perspective of VR technology for urban brand communication. Young Journalist **1**, 11 (2017). (in Chinese)
9. Lin, H.: Paradigm reform of city image communication in the context of visual communication in 5G Era. J. Sichuan Univ. Light Chem. Technol. (Soc. Sci. Edn.) **35**(12), 90 (2020). (in Chinese)
10. How to use Douyin make your hometown famous. Douy in and Tsinghua University give you advice, pp. 9–11 (2018) (in Chinese)

Generation Mechanism of Internet Altruistic Behavior in Big Data Era

Jing Lin[✉] and Yan Long

School of Management, Wuhan Donghu University,
Wuhan 430000, Hubei, China
elle_lynn@sina.com

Abstract. The advent of big data era had made an important impact on people's lifestyle, mental state and behavior. However, there were few studies on altruistic behavior in social networking under the influence of big data currently. Four mechanisms of the generation and development of internet altruistic behavior were proposed by analyzing the data of internet altruistic behavior during the epidemic period on the two Weibo platforms of CCTV News and Chutian metropolis Daily, which named the motivation mechanism, the restriction mechanism, the promotion mechanism and the guarantee mechanism. It will help expand the advantages of new media in the era of big data and increase positive behavior in the online social environment by using online public opinion data to study the generation mechanism of internet altruistic behavior.

Keywords: Internet altruistic behavior · Generation mechanism · Big data era

1 Introduction

With the rapid development of science and technology in China, the people had entered the era of big data represented by digital media, network media and artificial intelligence. Social media had been currently the main channel for people to exchange information. Through online platforms such as Weibo, QQ and WeChat, people could not only get the latest information, but also use social media to share opinions, insights, experiences and opinions so as to help others anytime and anywhere. During the epidemic period in 2020, netizens had published and reposted more than 200 million epidemic-related blog posts on the Sina Weibo. Massive data had provided a good opportunity and information for exploring the mechanism of internet altruism behavior under the public health emergencies [1]. In view of this, the official microblog of CCTV News and Chutian metropolis Daily were used to sort out the data of public's internet altruistic behavior on the purpose of exploring the generation and changing trend of netizens' internet altruistic behavior, which was of great significance for revealing the nature of altruistic behavior, urging people to show internet altruistic behavior more actively and creating a healthy network environment [2].

2 Internet Altruistic Behavior and Realistic Altruistic Behavior

Internet altruistic behavior was a kind of voluntary behavior such as support, guidance, sharing and reminding, which was shown in the network environment by individuals, it was beneficial to others and society without expecting any return with the characteristics of high frequency, high possibility and wide benefit area [3]. Although internet altruistic behavior occurs in virtual network world, there are many similarities between people's identities in online society and in real life: those who are willing to help others in real life are also likely to show their helpfulness online [4]. As the extension of actual altruistic behavior, it had been connected closely and promote mutually with the internet altruistic behavior.

3 Data Analysis of Internet Altruistic Behavior

As one of the most important sources of public opinion fermentation and hot event gathering places on the Internet in China, Sina Weibo is an indispensable platform for monitoring the dynamics state of Internet public opinion and obtaining opinions. It has more than 220 million daily active users because of its convenience, distribution, and originality. The netizens' comments related to the epidemic from January to June in 2020 were selected for statistical analysis. After the comprehensively consideration on the credibility, influence and specialization of the media, CCTV News and Chutian metropolis Daily were selected as data sources and analysis objects.

In terms of research methods, the event types were divided into the following three categories: digital or information reports (neutral), moving stories (positive) and bad behavior reports (negative) by capturing the content of Weibo comments on typical events during the epidemic. Network behavior could be roughly classified into the following five categories: network support, network guidance, network sharing, network reminder and network abuse. Among which, network support, network guidance, network sharing and network reminder belong to internet altruistic behavior. By screening and categorizing the comments on blog articles related to the epidemic on the platform of CCTV News and Chutian metropolis Daily, the information was integrated and analyzed based on "time–source–title–event type–behavior–quantity".

3.1 Internet Altruistic Behavior on the Platform of Chutian Metropolis Daily

The internet altruistic behavior on the platform of Chutian metropolis Daily from January to June in 2020 showed an overall rise firstly and then a decrease, as shown in Fig. 1. The number and changes of "Network Support" and "Network Sharing" were large and had reached their peaks in February and March respectively. There was smaller overall changes and fluctuations in "Network Guidance", "Network Reminder" and "Network Abuse" which represented internet aggressive behavior. On the whole, internet altruistic behavior showed an overall upward trend from January to March, and then fluctuated rapidly, approaching the axis of abscissa.

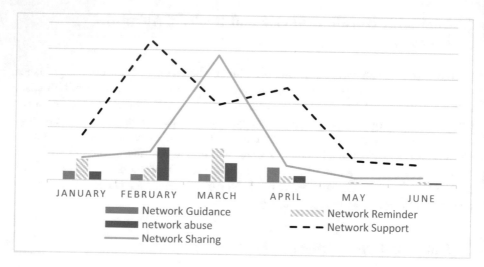

Fig. 1. Internet altruistic behavior in Chutian metropolis daily

In further analysis, the data of internet altruistic behavior on the platform of Chutian metropolis Daily in January and February were selected and divided into 10 stages with an interval of 5 days. The changes were shown in Fig. 2. In addition to "Network Support", the highest value of internet altruistic behavior were all in Stage 4, which was affected by the special factor of unified prevention and control nationwide in this stage. The change curve of "Network Guidance", "Network Sharing" and "Network Reminder" in the 10 stages were an overall increase, but there is fluctuation in the middle.

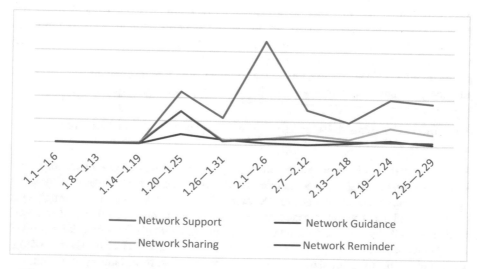

Fig. 2. Internet altruistic behavior in Chutian metropolis daily from January to February

3.2 Internet Altruistic Behavior on the Platform of CCTV News

The data of internet altruistic behavior on the platform of CCTV News from January to June in 2020 was selected, and the analysis results were shown in Fig. 3. It could be seen from the results that the proportion of "Network Support" was the largest during the six months except February and May and the difference between it and the others diminished at a later stage. The participation rate of "Network Reminder" was the highest in February and declined in March, and "Network Sharing" became people's first choice in May. "Network Guidance" was the most stable category, with no significant scale change overall.

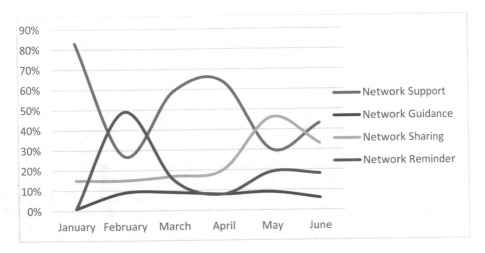

Fig. 3. Internet altruistic behavior in CCTV News

4 Generation Mechanism of Internet Altruistic Behavior in Big Data Era

There is a close relationship between actual altruism behavior and internet altruism behavior, and the occurrence of them can be promoted in a certain way.

4.1 Dynamic Mechanism

The dynamic mechanism of internet altruistic behavior includes the need of reciprocity rules, self-improvement, reducing guilt and alleviating negative emotions through altruistic behavior, obtaining others' praise to achieve self realization, obtaining a broader sense of group belonging, and moral norms, etc. [5]. The epidemic is the common enemy of all mankind and the concept of community with shared future for mankind will drive more countries and people to act mutually beneficial and self-interested behaviors.

4.2   Restriction Mechanism

The restriction mechanism of internet altruistic behavior includes altruistic cognition, moral emotion and moral will. The sense of social morality and professional ethics of special status of people are both favorable restriction mechanisms.

4.3   Promotion Mechanism

The promotion mechanism mainly refers to that some characteristics of the network environment are more conducive to the occurrence of altruistic behavior than the real society. The anonymity of the Internet made people act without caring about the class discrimination, geographical discrimination and group divisions they faced in life, as well as considering the opinions and consequences arising from expressing opinions in the real society. To a certain extent, people can judge the negative phenomena of society from a more objective perspective [6].

4.4   Guarantee Mechanism

The occurrence of internet altruistic behavior needs not only the driving force of dynamic mechanism, the guidance of restriction mechanism and the incentive of promotion mechanism, but also the guarantee mechanism to ensure its normal operation. The guarantee mechanism mainly includes the conscience of the internet altruist and the altruistic skills of them [7, 8]. Conscience can help people make the right moral choice as the core of the internal guarantee system of the network behavior subject, so as to promote the occurrence of altruistic behavior. In addition, altruistic skills and altruistic ability are also the important guarantee.

5   Conclusions

To sum up, exploring the generation mechanism of internet altruistic behavior can not only conducive to giving full play to the positive role of new media, but also guide the network behavior of netizens and prompt more people to participate in the practice of real altruistic behavior and internet altruistic behavior actively. It is good for the whole network environment developing towards a healthy and civilized direction, and promoting the construction of a better network ecosystem.

Acknowledgements. This work was supported by the grants from Youth Foundation Project of Wuhan Donghu University (project number: 2020dhsk003).

References

1. Jie, G., Baolong, Y., Siyu, L.: Internet public opinion analysis of novel coronavirus pneumonia based on micro-blog data. Mod. Sci. **2**, 57–64 (2020). (in Chinese)
2. Yuan, Y.: Public crisis event communication and mainstream media responsibility. China Newspaper Ind. **6**, 10–11 (2020). (in Chinese)

3. Mingwei, Y., Huaiji, T.: Research on the psychological crisis intervention caused by public health emergencies – from the perspective of social work. Mod. Bus. Trade Ind. **41**(14), 54–55 (2020). (in Chinese)
4. Benyuan, X.: The Rational model of altruistic behavior. J. Capital Normal Univ. (Social Sciences Edition) **6**, 138–144 (2019) (in Chinese)
5. Gang, L., Jinghao, C.: A review of online public opinion on public emergencies. Documentation Inf. Knowl. **2**, 111–119 (2014). (in Chinese)
6. Jier, X.: Analysis on ethical anomie in public health events – a case study of covid-19 epidemic. Radio TV J. **4**, 119–120 (2020). (in Chinese)
7. Juncheng, W.: Considerations on mainstream media's ability to enhance public opinion guidance in public crisis. Youth Journalist **3**, 48–49 (2019). (in Chinese)
8. Yifang, W., Ruihong, Z.: Altruistic behavior and intrinsic drive of value in anti-epidemic. Chin. Med. Ethics **34**(01), 1–4 (2021). (in Chinese).

Model of Real-Time Market Segmentation Based on Usage Scenarios Under the Monitoring Platform

Wenbin Wang[✉]

China Automotive Technology and Research Center Co, Ltd, Tianjin, China
wangwenbin@catarc.ac.cn

Abstract. Precise market segmentation is a prerequisite for commercial vehicle companies to effectively carry out product development and marketing. This paper introduces an index-based improved spatial-temporal big data computing platform, uses adjacent continuous storage technology to improve the data reading performance of the monitoring platform, proposes two models of interest point (POI) matching and adaptive interest point clustering, analyzes the actual use of the vehicle, and then provides a real-time segmentation market identification and segmentation model. In response to the characteristics of real-time changes in vehicle usage scenarios, adaptive learning methods are introduced to solve the problems of the generation of new market segments and the refinement of the original market segments. The model is deployed on the actual monitoring platform and verified in actual use scenarios, experimental results and practical applications show that the results obtained are of great significance for studying the actual characteristics of vehicles and guiding product improvements and upgrades.

Keywords: Spatial-temporal big data · Market segmentation · Point of interest

1 Introduction

In July 2021, China will fully implement the GB17691–2018 Phase VI standard, real-time and online supervision of the national VI heavy-duty vehicle emissions. In addition, in accordance with the requirements of the Ministry of Transport on the "Tourist chartered car, passenger shuttle bus, hazardous chemical vehicle" safety management regulations, relevant vehicles must be equipped with on-board terminal equipment [1]. Therefore, most of the existing commercial vehicles have the function of recording and uploading driving trajectory data, which generates a large amount of spatial-temporal data. It has become the consensus of the industry that data mining will generate value and improve work efficiency. However, because spatial-temporal data has attributes such as space and time, which are different from traditional data mining, applying our own mining algorithms to spatial-temporal data requires some major improvements and innovations. Early monitoring platforms were mainly used in vehicle positioning and recognition, behavior trajectory tracking, etc. They did not pay attention to data mining and analysis, let alone matching calculation of spatial location, so there was a problem of generating large amounts of data without effective application.

The research on actual vehicle usage scenarios mainly includes offline question-naire survey and online data analysis. Information can be collected in two ways: manual recording and system automatic collection. Manual collection methods have problems such as high cost, low efficiency, negligence and errors in the filling and entry process, which will affect the accuracy of the survey data. More importantly, the poor real-time performance affects the timeliness of the use scenario research [2]. In order to ensure the accuracy and timeliness of the data, online real-time calculation of usage scenarios based on the real-time location trajectory information uploaded by vehicles on the monitoring platform has become the first choice. The analysis of the trajectory is based on positioning or trajectory navigation, which cannot accurately identify the parking point of the vehicle. Furthermore, it is impossible to calculate the character-istics of the parking point according to the driving trajectory, and it is impossible to distinguish the correlation between the parking point and the market segment, and cannot be based on the market of the vehicle. The scene is used to determine market segments.

The real-time analysis of the fuel quality of the vehicle mainly faces the following problems:

(1) Resolve the computing power of trillion-level spatial-temporal data, and realize high-time analysis and mining;
(2) Identify and analyze parking points based on vehicle sensor data;
(3) Use stop points to perform spatial position matching calculation to obtain a set of points of interest;
(4) Carry out clustering for points of interest and obtain market segmentation results.

In view of the above-mentioned reasons that the existing monitoring platform cannot analyze the current market segmentation based on the vehicle usage scenario, this paper gives an improved spatial temporal big data storage and analysis model based on the continuous storage structure [3, 4], which solves the trillion level Data calculation efficiency issues. Market segmentation is determined through stop-point recognition, spatial location calculation, point-of-interest matching, point-of-interest clustering, etc., and finally the market segmentation analysis and calculation of vehicles based on real-time usage scenarios are obtained.

2 Storage and Calculation Model of Spatial-Temporal Big Data

2.1 Spatial-Temporal Big Data Storage Model

Each packet of data collected by the monitoring platform contains time and location information. The monitoring platform referred to in this article is the national six vehicle model enterprise analysis platform specified by GB17691. The data collection frequency is 1 s, which belongs to high-frequency collection. Taking into account factors such as online rate, number of online vehicles, and duration of use, it is estimated that the daily collected data is tens of billions, and the annual cumulative data scale is trillions, which means that trillions of time and space data need to be analyzed and mined. Since the Morton code can only be used to express a square area, in order to realize the polygon location retrieval, a preliminary screening is first required, and then the screening results are verified and cut again to achieve the purpose of accurate matching. In order to achieve secondary verification and cutting, more random read overhead will be generated, resulting in very poor overall geographic location retrieval performance. In order to solve the above problems, the data of the same geographic location is stored together, and continuous data is constructed to reduce the number of random readings and solve the problem of low data reading efficiency. Through the above improvements, the established model is suitable for the characteristics of time series, spatial relationship, and structure of Internet of Vehicles big data, establishes the relationship between time, space, and objects, and realizes the processing, analysis and mining of Internet of Vehicles big data [5].

2.2   The Calculation Model of Spatial Temporal Big Data

The data flow includes the following steps: ① The vehicle terminal sends the data packet to the receiving gateway through the network; ② After the gateway receives the data, it transmits the data to the message queue; ③ The data packet is parsed according to the agreed protocol and sent to different topics according to the flag bit. To be consumed; ④ Parallel task 1: Alarm status bit data, pushed to the memory database in real time; ⑤ Parallel task 2: Import the parsed trajectory data into the space-time calculation model; ⑥ Read the data through offline calculation for matching calculation, and give each Kind of report output; ⑦ Front-end page output.

3   Real-Time Market Segment Analysis Based on Vehicle Usage Scenarios

The current research on market segments by enterprises is mainly done through offline user research and desk data research. The typical research content mainly includes: description of market segment characteristics; car usage characteristics at all levels of nodes; car usage scale and characteristics; typical user car environment and usage behavior analysis. In traditional research, a large number of car visits and user interviews are required. In the context of the monitoring platform, market segment characteristics, car usage characteristics and usage behavior can be determined by analyzing the spatial-temporal big data of the vehicle's trajectory in the platform, which can solve the problem of inaccurate offline research and can only reflect the research time period [6]. In this study, using the 2020 Baidu map POI data, the material flow rate category under the life service category is selected for matching calculation. Through matching calculation, it will focus on identifying the parking situation of different types of POIs during the use of vehicles, and through clustering, real-time market segmentation analysis results will be given.

The analysis model mainly includes the following steps:

① Obtain real-time vehicle operating data.

The vehicle reads CAN bus data through T-Box, and obtains the latitude, longitude, acquisition time and other information of all online vehicles in the platform per second through analysis;

② Determining the stop position.

The stopping point judgment method mainly includes static matching and dynamic trajectory judgment. For the static judgment algorithm, it is mainly determined based on the POI of gas stations and parking lots; the dynamic judgment algorithm is mainly based on the characteristics of the driving trajectory to extract parking points [7]. This paper divides the monitoring data into two parts: login and logout, and real-time monitoring data. The collected data is analyzed to extract the login and logout data, based on the dwell time interval threshold, and the fuel point determination algorithm is used to remove the fuel and service stations [8]. Wait for the node to get the latitude and position information of the stopping point;

③ Interest Point Matching Model

Pass the latitude and longitude position information of the stopping point to the point of interest matching model, and perform spatial matching calculation with the POI point of interest model library database, and return the calculation result according to the set calculation range;

④ Adaptive Clustering Model of Interest Points

Due to the continuous development and update of usage scenarios, market segmentation will change. After a certain number of market segmentation instances are added, the market segmentation results will undergo adaptive learning such as mergers and splits.

3.1　Interest Point Matching Model

At the parking point, the vehicle needs to perform parking, perform engine shutdown, flameout, etc., and logout data will be uploaded at this time. After the cargo is loaded and unloaded, perform operations such as turning on the engine and ignition, and upload the login data at this time. In response to the above process, the positioning of the latitude and longitude of the stop point W_t is completed. Input the positional parameters into the spatial matching model, in the form of JSON string, input the SYS_JSON_QUERY{query_type:"geo",geo_type:"circle",field:"geodata",list:["lnt, lat"],radius:1000}, The set of all points of interest calculated by the model with the location as the center w_t and the radius r, including the name, category, and location of the point of interest [9].

3.2　Adaptive Clustering Model of Interest Points

Due to the continuous development and update of usage scenarios such as vehicle ownership and operating characteristics, the types of POI points of interest identified in real time will also grow rapidly. Inevitably, a certain amount of new POI matching results will appear, and the new matching results will lead to the original. The clustered market segments will undergo local changes [10]. Mainly include the following changes:

1) The emergence of new market segments: It means that vehicles are used in new business scenarios in actual processes.
2) Refinement of the original market segment: Due to the addition of new POI matching results, the original market segment of a certain category is split into two or more subcategories. It shows that the actual application scenarios of vehicles are gradually clarified and further refined, and new market segments have emerged.
3) The transfer of a certain type of vehicle market segment: The center of the segment market will shift significantly as new matching results increase. It indicates that the direction in which the vehicle is actually used has changed.

4 Experiment and Result Analysis

The analysis model given in this article is encapsulated in the big data analysis module of monitoring platform (CMAP) to support the analysis and debugging of actual online monitoring vehicles. The point of interest data uses a third-party POI database, covering all the data of the third-party map. Select four major categories of car services, catering services, shopping services, and life services, establish a POI point of interest database, and build initial market segments Ontology. This experiment takes Wuhan as an example to perform interest point matching and interest point clustering. The results show that in express, agricultural and sideline, home appliances, supermarkets, building decoration and other market segments with clear POI points of interest, the recognition effect is better; for special operating vehicles such as sweepers, because there is no clear POI information, the market segmentation The recognition cannot be matched based on POI, and can be calculated by driving speed, track similarity, etc.

5 Outlook

This paper uses adjacent continuous storage technology to improve the data reading performance of the monitoring platform, and realizes the calculation and analysis of spatial temporal big data such as vehicle trajectory. By introducing a spatial calculation model, it realizes the matching calculation of the parking point and the point of interest POI. Clustering points of interest to realize the identification of market segments. For vehicle manufacturers, by analyzing the trajectory data, the actual usage scenarios of the models can be judged, and timely updates can be made to help companies upgrade their products according to specific market scenarios. In summary, the models proposed in this paper have good application prospects in monitoring platforms, can be extended through plug-ins, and can also be applied to other spatial data analysis scenarios.

References

1. Jiangtao, L.: Research on Freight OD Information Extraction Method Based on Big Data of Freight Truck Trajectory, Beijing Jiaotong University (2013). (in Chinese)
2. Pluvinet, P., Gonzalez-Feliu, J., Ambrosini, C.: GPS data analysis for understanding urban goods movement. Procedia-Soc. Behav. Sci. **39**, 450–462 (2012)
3. Ren, R., Cheng, J., He, X.-W., et al.: Hybrid tune: spatial-temporal performance data correlation for performance diagnosis of big data systems. J. Comput. Sci. Technol. **34**(6), 1167–1184 (2019)
4. Galić, Z., Mešković, E., Osmanović, D.: Distributed processing of big mobility data as spatial-temporal data streams. GeoInformatica **21**(2), 263–291 (2017)
5. Wang, W.: Research and implementation of big data analysis model for automobile enterprises. China Comput. Commun. **14**, 149–151 (2019). (in Chinese)
6. Yanhon, F., Xiaofa, S.: Research on freight truck operation characteristics based on GPS data. Procedia-Soc. Behav. Sci. **96**, 2320–2331 (2013)
7. Luo, T., Zheng, X., Xu, G., Fu, K., et al.: An improved DBSCAN algorithm to detect stops in individual trajectories. ISPRS Int. J. Geo-Inf. **6**(3), 63 (2017)
8. Wang, W.: Research on Vehicle refueling behavior model based on spatial-temporal big data monitoring platform. J. Adv. Intell. Syst. Comput. **1233**, 708–713 (2020)
9. Quddus, M.A., Ochieng, W.Y., Noland, R.B.: Current map-matching algorithms for transport applications: state-of-the art and future research directions. Transp. Res. Part C Emerg. Technol. **15**(5), 312–328 (2007)
10. Ruiling, Z., Wenbin, W., et al.: Instances drove adaptive ontology learning. Comput. Eng. Appl. **45**(28), 31–34 (2009). (in Chinese)

Intelligent Car Control System Design Based on Single Chip Microcomputer

Shiying Wang[✉]

Department of Infrastructure, Linyi University, Linyi, Shandong, China

Abstract. With the continuous progress of science and technology and the development of social productivity, more and more high-tech products are entering people's lives. The article introduces an intelligent car control system based on ATMEGA32 single-chip microcomputer. The system uses single-chip microcomputer as the core, uses photoelectric sensors for tracking, and ultrasonic sensors for obstacle avoidance detection, so as to control the different operating modes of the motor and realize the different movement modes of the car. The article completed the hardware design and software programming of the system, so that the trolley can complete the functions of automatic tracking and automatic obstacle avoidance, so that it can walk on a predetermined trajectory in a general environment, and avoid obstacles during construction operations. Objects to prevent work errors caused by collisions. The smart car can be used in warehouse transportation, item sorting, sweeping robots and other fields. Through the detection and determination of fixed lines and non-fixed obstacles, it can complete automatic transportation while ensuring its normal and stable operation. This design has more advantages. Wide application value.

Keywords: Single chip microcomputer · Smart car · Tracking · Sensor

1 Introduction

After entering the 21st century, with the rapid development of the world's technological level, various intelligent devices are emerging one after another, and various high-tech are gradually becoming civilians and popularized. This popularization is the development trend of modern society and is also unstoppable. trending. In our country, because the development of artificial intelligence robots has just started, it is still in a very early stage, but with the development of semiconductor technology, it can be seen that more and more intelligent robot products have entered the field of life in the past decade. Smart dishwashers, smart sweeping robots, etc. emerge in endlessly. This not only means the rapid development of China's intelligent robots, but also indicates the tremendous progress of my country's scientific and technological strength [1]. Although today, compared with European and American countries, the performance and technical level of my country's smart products are slightly inferior, with the development of time, it is expected that within five to ten years, the technical level of related products in my country will be able to This trend is becoming more and more obvious in catching up with European and American countries [2].

J. Abawajy et al. (Eds.): ATCI 2021, LNDECT 81, pp. 858–862, 2021.
https://doi.org/10.1007/978-3-030-79197-1_127

2   The Overall Structure of the System

This system uses a single-chip microcomputer as the core and uses the synergy of four photoelectric sensors to detect the current position of the smart car and the black line to determine the current position of the smart car; it uses ultrasonic distance detection to detect obstacles ahead. This method can detect long-distance obstacles and realize the early obstacle avoidance of the smart car; the system has designed a button circuit to realize mode switching [3]. The system has two modes, namely obstacle avoidance mode and tracking mode. To switch between these two modes, you need to use DIP switches or buttons to execute; the main output part of the system is the motor drive part. Since there is no steering gear, the motor differential steering method is designed to realize the steering of the car [4]. The overall structure of the system is shown in Fig. 1.

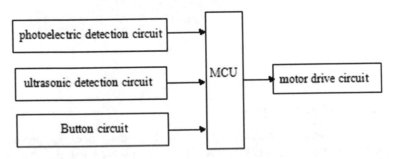

Fig. 1.   Overall structure of the system

3   System Software Design

3.1   System Main Program Module

The main program module includes initialization program and control program, The initialization program refers to the process of configuring the module in advance when calling a program function module, including clock initialization, timer initialization, sensor initialization, servo initial value setting, and motor initial value setting; control program Refers to the main program that controls the operation of the entire single-chip microcomputer system, including the detection program for the front obstacle, the path detection program, the direction calculation program and the motor output program [5].

```
int main(void)
{
    System_Init();
    TIM_SetCompare1(TIM1,75);;
    Duoji_Control(ON);
    Delay_ms(500);
    Duoji_Control  (OFF);
    MOTO_PWM_Out(80, 80);
    Car_Go();
    Delay_ms(10);
    while (1)
    {
            If(S1=0)Wave_BZ2();
            Else Find();
    }
}
```

3.2   Data Collection Procedure

The data collection of the system mainly refers to the obstacle detection procedure and the path detection procedure. For the obstacle detection sensor, the ultrasonic distance measuring module is selected, and the path detection sensor is the photoelectric tube module [6]. For the photoelectric sensor that realizes the tracking function, the data input is the high and low level of the I/O port of the single-chip microcomputer. Therefore, only need to read the high and low level of the photoelectric sensor detection input I/O port to determine the current sensor Is there a black line [7]. The following program is the program for I/O port reading.

```
#define Find_O1 GPIO_ReadInputDataBit(GPIOA, GPIO_Pin_4)
#define Find_O2 GPIO_ReadInputDataBit(GPIOA, GPIO_Pin_5)
#define Find_O3 GPIO_ReadInputDataBit(GPIOA, GPIO_Pin_6)
#define Find_O4 GPIO_ReadInputDataBit(GPIOA, GPIO_Pin_7)
```

For the ultrasonic sensor that realizes ultrasonic distance measurement, the data input needs to use the timer, and the reading of the distance data is completed by setting the timer. For the data output by the ultrasonic sensor, the single-chip microcomputer cannot be used directly as distance data, and algorithm conversion is required to convert time data into distance data [8]. The following procedure is the procedure for data reading and conversion.

```
u16 Wave_Start(void)
{
  u16   Distance;
  Wave_ON();
  Delay_us(20);
  Wave_OFF();
  while(!Wave_State());
  TIM_Cmd(TIM2, ENABLE);
  while(Wave_State());
  TIM_Cmd(TIM2, DISABLE);
  Distance=TIM_GetCounter(TIM2)*5*34/2000;
  TIM_SetCounter(TIM2,0);
  return Distance;
}
```

3.3 Motor Control Program

The motor control program relies on the control of two motors. For the straight movement of the smart car, both motors rotate forward; when the smart car is moving backward, both motors are reversed; when the smart car turns right, the two motors are both reversed. Two motors show that the left motor rotates forward, the right motor reverses, and the smart car turns right on the spot; when the smart car turns left, the two motors appear as the left motor reverses, the right motor rotates forward, and the smart car turns left on the spot; When both motors are stopped, the smart car stops running [9]. The following is the content of this part of the program:

```
void Car_Go(void)
{
  MOTO_Z(GO);        MOTO_Y(GO);
}
void Car_Back(void)
{
  MOTO_Z(BACK);        MOTO_Y(BACK);
}
void Car_Turn_Right(void)
{
  MOTO_Z(GO);        MOTO_Y(BACK);
}
void Car_Turn_Left(void)
{
  MOTO_Z(BACK);        MOTO_Y(GO);
}
void Car_Stop(void)
{
  MOTO_Z(STOP);        MOTO_Y(STOP);
}
```

4 Conclusion

The design of the smart car based on the single-chip microcomputer is a low-cost and functional design. It adopts ultrasonic detection method to realize automatic obstacle avoidance when encountering obstacles, adopts photoelectric black line method to realize automatic driving on a fixed path, and adopts dual-circuit motor differential speed method to realize steering. It has strong scalability and can Directly add related modules to realize sweeping and mowing functions on the basis of the smart car to complete the functions of automatic sweeping and automatic mowing [10]. Carrying out secondary development on the basis of this design can greatly save development time, reduce development cycle, and speed up development. It has strong application value and promotion value.

References

1. Zhang, C.: Design of intelligent vehicle control system based on single chip microcomputer. Materials Science, Energy Technology and Power Engineering II (MEP2018) (2018)
2. Yanquan, et al.: A design of intelligent car based on STC89S52 single chip microcomputer. In: The 2nd International Conference on Computer-Aided Design, Manufacturing, Modeling and Simulation (CDMMS 2012) (2012)
3. Lei, X., Sheng, Y., Guilin, L., Zhen, Z.: Intelligent traffic control system design based on single chip microcomputer. In: Zhu, M. (ed.) ICCIC 2011. CCIS, vol. 236, pp. 232–238. Springer, Heidelberg (2011). https://doi.org/10.1007/978-3-642-24097-3_36
4. Chen, F., Qiu, H., Gao, Y.: Freescale single-chip microcomputer intelligent car voltage control discussed. In: Third International Conference on Digital Manufacturing & Automation, pp. 430–433. IEEE Computer Society (2012)
5. Hao, X.: The design of the four-wheel drive control system of the smart car based on STM32. Jiangsu Sci. Technol. Inf. **36**(9), 51–53 (2019)
6. Xuefei, L., Wanmin, L.: Design and production of intelligent car control system based on 51 single chip microcomputer. Electron. World. **602**(20), 188–189 (2020). (in Chinese)
7. Jun, Z.: Automatic control of smart car. Changzhou Vocat. Coll. Inf. Technol. **4**, 16–18 (2006). (in Chinese)
8. Dekun, Y.: The application of single chip microcomputer technology in sensor design. Electron. Test. **18**, 127–140 (2018). (in Chinese)
9. Li, D., Zhiping, S., Mengmeng, X., et al.: Design of embedded measurement and control system based on STM32. J. Central South Univ. (Natural Science Edition) **2013**(s1), 260–265 (2013)
10. Rivera, D.H., Roldán, G.R., Martínez, R.M.: A capacitive humidity sensor based on an electrospun PVDF/graphene membrane. Sensors. **17**(5), 1009 (2017)

Ship-Borne Antenna Motion Under External Force Based on Data Analysis

Haidong Zou[✉], Jing Wang, and Chunshun Fu

China Satellite Marine Tracking and Control Department,
Jiangyin, Jiangsu, China

Abstract. This paper studies the motion model of ship-borne large reflector antennas under the influence of external force. By deriving the characteristics of the antenna movement under the influence of the synthetic wind, combined with the inertial force brought by the ship's rolling and pitching to the antenna, this paper comprehensively using mathematical analysis methods to study the movement of the ship's antenna under the influence of external force, and a fixed angle antenna motion model is obtained. Through actual data comparison, the effectiveness of the antenna motion model is verified within a certain range of wind speed and ship sway angle.

Keywords: Ship-borne antenna · Wind resistance · Angle

1 Introduction

The main reflecting surface of the shipboard antenna is a parabolic structure, which is axisymmetric when the antenna is upright. The tracking method uses a three-axis tracking system. When the antenna is locked, a special locking mechanism is required to prevent the antenna from being affected by various external forces. The structure is deformed to protect the antenna structure. When locking the antenna, it is necessary to set the azimuth deck angle and the cross deck angle of the antenna to $0°$, the elevation angle to $90°$. The main reflection surface of the antenna is parallel to the deck then. Since there is a time interval of several seconds between the power-off of the drive amplifier and the motor holding device, the relatively stationary antenna is simultaneously effected by the combined wind and the rolling of the ship. The combined action causes the elevation angle and cross angle of the antenna to deviate from the lock angle range at the same time, which affects the locking mechanism. Therefore, it is necessary to power up and power off the antenna drive amplifier repeatedly to adjust the elevation angle and cross angle of the antenna, so that the final locking angle meets the locking requirements for the antenna.

This paper proposes the motion model of the ship-borne parabolic antenna. The antenna motion parameters relate to different wind speeds and wind directions, combines the ship's roll and pitch to the antenna. Through the simulation, the static antenna motion model is obtained. Compared with the actual antenna motion data, the effectiveness of the proposed model is verified within a certain range of wind speed and ship sway angle.

J. Abawajy et al. (Eds.): ATCI 2021, LNDECT 81, pp. 863–868, 2021.
https://doi.org/10.1007/978-3-030-79197-1_128

2 Antenna Motion Model

According to the analysis, the motion of the ship-borne antenna can be basically divided into two categories: one is caused by the ship sway angle, and the other is the synthetic wind, which is brought by the hull navigation and the real wind in the actual environment.

2.1 Antenna Inertial Motion Analysis

Assuming that the ship's sway motion conforms to the sine law, the relationship between these way angle Ω and time t can be expressed by the following formula [1, 2]:

$$\Omega = \Omega_m \cdot \sin\left(2\pi t/T\right) \tag{1}$$

Where Ω_m is the maximum sway amplitude (rad), and T is the sway period (s).
Therefore, the angular acceleration of the ship can be derived as follows:

$$\frac{d\Omega^2}{dt^2} = -\Omega_m \cdot \left(2\pi/T\right)^2 \cdot \sin\left(2\pi t/T\right) \tag{2}$$

Assuming that the distance between the central gravity G of the antenna system and the sway axis of the ship is d, then the tangential acceleration a at the point G is:

$$a = -\Omega_m \cdot \left(2\pi/T\right)^2 \cdot \sin\left(2\pi t/T\right) \cdot d \tag{3}$$

The maximum tangential inertial force f of the antenna appears at the position when $\Omega = \Omega_m$. If the mass of the antenna is m, then the inertial force f of the antenna is:

$$f = m\Omega_m \left(2\pi/T\right)^2 d \tag{4}$$

2.2 Antenna Wind Motion Analysis

According to the numerical analysis and wind tunnel experiment in literature [3, 4], it's known that, the coefficient of wind resistance is rarely changed for a specific type of antenna surface, and the wind load will change nonlinearly with the speed of wind. In a certain direction, the relationship between wind load and wind speed is a second exponential equation. The wind resistance is as follows:

$$w = \frac{1}{2} f \rho v^2 A \tag{5}$$

Where: w is the wind resistance of the antenna, f is the wind resistance coefficient of the antenna, ρ is the air density, v is the wind speed at the height of the antenna sufficiently far away from the antenna, and A is the characteristic area of the antenna.

The above is also a formula for calculating wind load using the structural resistance coefficient obtained from wind tunnel experiments in engineering. It can be seen that for a certain antenna motion with a short duration, the wind speed and direction are fixed, and the wind resistance coefficient is a fixed value, then the wind resistance value of the antenna can be calculated. By decomposing the force, the final torque applied to the antenna structure in the direction of the elevation angle E and the intersection angle C can be obtained.

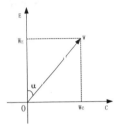

Fig. 1. The decomposition of the force under the synthetic wind direction angle α

Figure 1 shows the force analysis of the antenna in E and C directions under the influence of the wind resistance w and the wind direction (relative to the direction of the ship) α. The wind drag torque in the direction of E and Cis as follows:

$$w_e = w \cos \alpha, w_c = w \sin \alpha \tag{6}$$

2.3 Antenna Motion Analysis

From the previous analysis, it can be seen that the final resultant of the antenna includes the inertia hull sway and the wind resistance. Since the position of antenna is on the fore and aft line of the ship, and when the antenna is locked, the C axis is basically symmetrically to the sway axis. It can be considered that the lateral force will not have a large impact on the C axis of the antenna [5]. Therefore, the total force applied to the E axis of the antenna includes the component of the wind resistance on the E axis (as we) and the force caused by the ship pitching(as Gh).The force on the C axis is the component of the wind resistance W on the C axis (as wc). The two forces are linear in equations [6–8]. Based on this, a linear regression equation can be constructed based on the existing antenna locking data. Then the motion equation of the antenna under the influence of different wind speeds and ship sway angles can be calculated.

According to the foregoing results, multiple sets of actual antenna locking data are taken [9, 10]. The initial antenna angles Eo and Co, and the ultimate antenna angles E and C, the synthetic wind speed Sw (m/s) and the synthetic wind direction Aw (°), the pitch angle Ψm (°), and the antenna mass m (kg) are taken out respectively to construct a linear equation.

$$E = E_0 + \alpha \cdot \left| \cos\left(\pi \cdot A_w / 180 \right) \right| \cdot (S_w)^2 + \beta \cdot \left(m / T^2 \right) \cdot \cos\left(\pi \cdot \Psi_m / 180 \right)$$

$$C = C_0 + \gamma \cdot \sin\left(\pi \cdot A_w / 180 \right) \cdot (S_w)^2, (A_w > 180°) \tag{7}$$

$$C = C_0 - \gamma \cdot \sin\left(\pi \cdot A_w / 180 \right) \cdot (S_w)^2, (0° \leq A_w < 180°)$$

In the formulas above, α, β, and γ are all coefficients. In the real environment, the initial angle of the antenna, the synthetic wind speed and the synthetic wind direction, the pitch angle of the ship, and the quality of the antenna are all known. Therefore, for the above two equations, the values of α, β, and γ can be calculated within two sets of testing data, respectively.

3 Result Analysis

3.1 Determination of the Parameters

Taking multiple sets of testing data in the real environment, recording them in Table 1. Substituting them into the Eq. (7). We can get $\alpha = 0.01$, $\beta = -0.04$, and $\gamma = 0.01$.

Table 1. Antenna motion testing records in different environments

No	Eo	Co	E	C	Sw	Aw	Ψ_m	T
1		−0.23		0.56	11	300	−0.02	8
2		−0.73		−0.02	11	300	−0.30	8
3	89.17		89.87		10	40	−0.29	7
4	89.13		89.56		10	40	0.44	7
5	89.21		89.74		10	40	−0.21	7
6		0.38		−0.30	10	40	0.13	11
7		0.62		−0.09	10	40	−0.30	11
8	89.60		89.74		8	20	0.19	7

3.2 Verification and Analysis

According to the values of α, β, and γ, as well as the parameter values of each group in Table 1 (Eo, Co, Sw, Aw, Ψ_m, T), they are substituted into formula 7 for verification respectively. The results are shown in Table 2.

It can be seen from Table 2 that the maximum difference between the theoretical calculation value and the actual value is not more than 0.20°, and the minimum is 0.02°. This can basically confirm the validity of the ship antenna motion equation above. However, due to the limitation of the accuracy and timeliness of the wind speed and wind direction values recorded, the coefficients of the motion equations have a certain deviation, which in turn leads to certain errors in the theoretical values of the

Table 2. Antenna motion testing records in different environments

No	Theoretical value	Actual value	Difference
1	0.38	0.56	−0.18
2	−0.13	−0.02	−0.11
3	89.85	89.87	−0.02
4	89.71	89.56	0.15
5	89.88	89.74	0.14
6	−0.39	−0.30	−0.09
7	−0.15	−0.09	−0.06
8	89.79	89.74	0.05

subsequent calculations. In addition, the movement caused by the structural gap of the antenna itself will also introduce some errors to the result. The superposition of the two errors causes the error to be slightly larger in some cases. In engineering, an anemometer can be added to the antenna to perform real-time measurement, as to eliminate errors caused by the time-varying nature of synthetic wind. The structural gap error is a fixed deviation, which can be eliminated by specifying an offset in the calculation software.

4   Conclusions

This paper studies the motion feature of the ship-borne antenna under the influence of synthetic wind and hull sway. Then the antenna motion equations are constructed. Based on the existing experiments, the equations are solved and verified. The actual data tests show that, this method can effectively calculate the motion of the antenna, within synthetic wind speed range of 8 m/s–11 m/s, and the hull sway angle is within ± 4°.

Follow-up work should be carried out to study the feature of antenna motion with larger wind speeds and larger sway ranges to improve the scope of application. In addition, intelligent algorithms should be considered to recognize different wind speeds and hull sway angles. All this can be used to improve the reliability of system applications.

References

1. Wanxuan, X.: Determination of antenna load for ship-borne radar. Mod. Radar. **20**(5), 69–77 (1998). (in Chinese)
2. Qiang, D., Pingan, D.: Numerical simulation analysis of antenna wind load. Mod. Radar. **31**(3), 77–80 (2009). (in Chinese)
3. Jiao, H., Zou, H.: Study of real-time monitoring for ship-borne antenna locking mechanism. In: 2019 9th International Conference on Information and Social Science (ICISS 2019), pp. 68–71 (2019)

4. Jung, H., Jung, J., Lim, Y.: Low side-lobe beamforming antenna for earth stations in motion. J. Korean Inst. Electromagn. Eng. Sci. **31**(8), 693–700 (2020)
5. Zhu, L., Yang, X.-B., Xu, C., Xu, T.-T., Jiang, L.: Analysis on the motion characteristics of dynamic aircraft by dual-line-array TDI CCD optical camera. Optoelectron. Lett. **16**(88(1)), 1–6 (2020). https://doi.org/10.1007/s11801-020-9015-3
6. Long, T., et al.: Effect analysis of antenna vibration on GEO SAR Image. IEEE Trans. Aerosp. Electron. Syst. **56**(3), 1708–1721 (2020)
7. Liu, Y.: A Progressive motion-planning algorithm and traffic flow analysis for high-density 2D traffic. Transp. Sci. **53**(6), 1501–1525 (2019)
8. Chen, Z., Tse, K.T., Kwok, K.C.S., et al.: Modelling unsteady self-excited wind force on slender prisms in a turbulent flow. Eng. Struct. **202**(1), 109855.1–109855.11 (2020)
9. Zulli, D., Piccardo, G., Luongo, A.: On the nonlinear effects of the mean wind force on the galloping onset in shallow cables. Nonlinear Dyn. **103**(4), 3127–3148 (2020). https://doi.org/10.1007/s11071-020-05886-y
10. Zhao, B., Fang, R., Shi, W.: Modeling of motion characteristics and performance analysis of an ultra-precision piezoelectric inchworm motor. Materials **13**(18), 3976 (2020)

Realization of Remote Monitoring System for Antenna Locking System Based on C#

Haidong Zou[✉], Jing Wang, and Chunshun Fu

Satellite Marine Tracking and Control Department, Jiangyin, Jiangsu, China

Abstract. Remote real-time monitoring system of antenna locking mechanism is studied by using C# programming in this paper. Remote data of the locking pin is read and displayed from the serial port. The current of the working motor is used to judge the working state of the locking mechanism in real time. Through the application of this system, real-time monitoring of the working state of the antenna locking system is realized. This system helps to improve the operation and maintenance efficiency of the antenna system.

Keywords: Antenna motion · Remote monitoring · Serial communication

1 Introduction

Three axis tracking system is used in ship borne antenna, and special mechanical locking mechanism is needed to prevent antenna structure damage caused by external force, so as to protect antenna structure. The existing locking system uses an AC motor to drive the reduction gear, then the locking pin is driven to advance or retreat [1]. The locking system is located at the antenna, so it is inconvenient for the operator to operate on site, which leads to the operator hard to know the working status of the locking mechanism in real time. In order to solve this problem, it is necessary to design a remote monitoring system of ship-borne antenna locking system.

In this paper, the remote real-time monitoring system of the working state of the antenna locking system is programmed by using C# [2–4]. The data collected from the remote is read from the serial port [5]. After unpacking and processing, the distance of the locking pin and the current of the working motor are displayed in real-time graphics. Combined with its working logic, the working state of the locking mechanism is comprehensively determined, so as to remote real-time monitoring of state. Through this system, it can effectively reflect the real-time working state of the antenna locking system, and helps to improve the operation and maintenance efficiency of the antenna system.

© The Author(s), under exclusive license to Springer Nature Switzerland AG 2021
J. Abawajy et al. (Eds.): ATCI 2021, LNDECT 81, pp. 869–874, 2021.
https://doi.org/10.1007/978-3-030-79197-1_129

2 System Design

According to the functional requirements of the system, combined with the idea of modular design, the overall design of the system software, including the serial communication module, motor current display module, lock pin travel distance display module, lock mechanism working state judgment module, etc.

As a human-computer interface, the remote server processing software needs to complete the collection and processing of the whole field data, analyze and display it [6]. After the software starts, initialization is carried out, which mainly completes the initial setting of the control and window resources involved in the software, so as to facilitate the call of subsequent program modules. After that, the system sets some variables that need to be used globally and the properties of specific controls. The serial communication module sets the properties of the serial port that need to be communicated and calls the serial communication program [7, 8]. When it is judged that the serial communication is normal, it receives the serial data stream reported by the remote actively, and analyzes and stores the variables according to the communication protocol for the use of the calculation and graphic display module. When the program needs to be closed after completing the above work, the system releases the open serial port resources and exits the program module [9]. According to these processes, the basic processing flow chart of the system software can be drawn, as shown in Fig. 1.

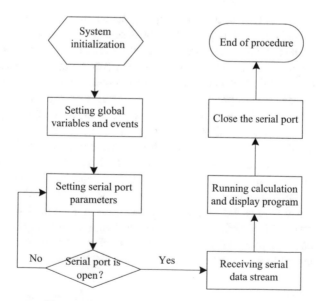

Fig. 1. System software processing flow chart

3   System Implementation

3.1   Serial Data Reading

Using serial communication in C #, you can directly use the serialport control provided by the system to achieve the required functions. The common properties of serial port control include judging whether the serial port is open or not, obtaining or setting the communication port name, etc. The common methods of serial port control include serial port open (), serial port close (), serial port dispose (), getPortnames (), readLine (), etc. [9, 10].

The initialization, data reading and release methods of serial port are as follows.

```
//Initialize the serial port
private void Form1 Load(object sender, EventArgs e)
{
serialPort1.Open();
// When the serial port receives the data, the graphic event will be triggered
private        void        serialPort1_DataReceived(object        sender,
System.lO.Ports.SerialDataReceivedEventArgs e)
{
data = serialPort1.ReadLine();
this.Invoke(new EventHandler(DisplayTex));
}
// When closing the program, destroy the opened serial port at the same time
private void Form1_FormClosing(object sender, FormClosingEventArgs e)
{
serialPort1.Dispose();
}
```

3.2   Realization of Real Time Data Graphic Display

The data graphic display of the system mainly includes the measurement data of the travel distance of the locking pin and the real-time display of the voltage and current data of the working motor.

```
//Graphical display of data read by serial port
private woid Displaytext(object sender, EventArgs e)
{
float temp= float.Parse(data);
int dd=(int)temp;
short q= (short)(dd * 20);
Graphics pic = this.pictureBox1.CreateGraphics():
pictureBox1.BackColor = Color.White;
if(x>1)                        // When more than 2 values are read, start drawing
{
pic.DrawLine(Pens.Red, x, pictureBox1.Height-qtemp, x+1, pictureBox1.Height-q);
}
qtemp=q;
x+;
it(x>500)// When the value exceeds 500, refresh the page and start drawing again
{
x=0;
pic.Clear(Color.White);
}
}
```

3.3   Realization of State Judgment of Locking Mechanism

The state judgment of locking mechanism mainly depends on three indexes: distance of locking pin, motor current and duration. Because in normal operation, the total stroke of the locking pin is 1 (CM) and the total operation time is t (s), a basic constant delta = L / T can be set as the basic reference for the operation judgment of the locking pin. The average value of the working current of the motor is set to avgcurrent as the operating parameter of the motor. Take the difference deltadis between the current travel value and the previous travel value. The basic judgment logic is as follows.

```
if (motorCurrent==0 && deltaDis <delta)    state= Not running;
if (motorCurrent>0 && motorCurrent<= avgCurrent *1.2)
{
if (deltaDis>=delta 0.9) state= Normal operation;
else state= The locking pin needs maintenance;
duration++;
}
else state= Motor Locked Rotor;
```

4 System Applications

According to the design principle of the antenna locking mechanism condition monitoring system, a test system is built, and the actual test effect of the system is obtained,, and the actual display effect is shown in Fig. 2.

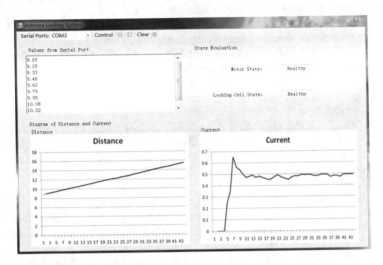

Fig. 2. System application test

After running the software, select the "COM3" serial port in the ports drop-down list, and then click the "start" button in the control bar to complete the communication between the software and the serial port. At the same time, the "start" button becomes gray and unavailable. After that, the software starts to receive the data sent by the serial port of the single chip microcomputer, and triggers the data received event. After a series of calculation and processing, the corresponding graph of "distance" and "current" will be displayed.

The "distance" curve shows the actual running distance of the locking pin, while the "current" graph shows the working current curve of the locking motor with time.

5 Conclusions

In this paper, the remote real-time monitoring of antenna locking system is realized by using C # software. This paper focuses on the detailed analysis of the realization mechanism of the key modules, such as the serial port reading remote data acquisition, the real-time graphical display and the performance judgment of the antenna locking mechanism, so as to complete the basic functions of the software. The application results show that the system can effectively reflect the real-time travel state of the antenna locking pin, timely reflect the working state of the driving motor and the lubrication of the mechanical parts of the system, and realize the purpose of improving the operation and maintenance efficiency of the antenna system.

References

1. Jiao, H., Zou, H.: Study of real-time monitoring for ship-borne antenna locking mechanism. In: 2019 9th International Conference on Information and Social Science (ICISS 2019), pp. 68–71 (2019)
2. Zhou, J.: Artificial intelligence driven wireless network remote monitoring based on Diffie-Hellman parameter method. Comput. Commun. **160**, 132–138 (2020)
3. Catini, A., Papale, L., Capuano, R., et al.: Development of a sensor node for remote monitoring of plants. Sensors (Basel Switzerland) **19**(22), 4865 (2019)
4. Gordillo, G.C., Ruiz, G.R., Stauffer, Y., et al.: EplusLauncher: an API to perform complex EnergyPlus simulations in MATLAB and C#. Sustainability **12**, 59–68 (2020)
5. Li, K.J., Xie, Y.Z., Zhang, F., et al.: Statistical inference of serial communication errors caused by repetitive electromagnetic disturbances. IEEE Trans. Electromagn. Compat. **62**, 1160–1168 (2019)
6. Han, M., Zhu, Y., Yang, D., et al.: Soil moisture monitoring using GNSS interference signal: proposing a signal reconstruction method. Remote Sens. Lett. **11**, 373–382 (2020)
7. Tesfamariam, B.G., Melgani, F., Gessesse, B.: Rainfall retrieval and drought monitoring skill of satellite rainfall estimates in the Ethiopian Rift Valley Lakes Basin. J. Appl. Remote Sens. **13**, 014522 (2019)
8. Kwon, S.W., Lee, S.W., Kim, Y.R., et al.: A Filament supply system capable of remote monitoring and automatic humidity control for 3D printer. J. Sens. **2020**(2), 1–10 (2020)
9. Mili, S.D., Babi, B.M.: Toward the future—upgrading existing remote monitoring concepts to IIoT concepts. IEEE Internet Things J. **7**, 11693–11700 (2020)
10. Ray, P.P., Dash, D., Moustafa, N.: Streaming service provisioning in IoT-based healthcare: an integrated edge, loud perspective. Trans. Emerg. Telecommun. Technol. **5**(6), 34–42 (2020)

Innovative Thinking of Natural Resources Management Model Based on Big Data of Land and Resources

Hongpin Zhang[✉]

Yunnan Vocational College of Land and Resource,
Kunming 650000, Yunnan, China

Abstract. Big data plays an important role in promoting the development of natural resource management technology and natural resource management. The implementation of the protection of arable land's quantity, quality, and ecology should be empowered by land resource data, relevant economic and social development data, and network data; fueled by data sharing, opening, and analysis, together with the mining and application of big data, the development of natural resource management technology is advanced and the quality of natural resource management has stepped up. The application of big data still needs to further strengthen scientific and technological innovation and demonstration application. What's also in need is to raise awareness, deepen research, innovate technology and strengthen demonstration in data-driven decision-making of natural resource management, practical application of natural resource management and construction of the disciplines of natural resource management.

Keywords: Big data technology · Natural resource management · Innovation model

Preface

Against the backdrop that China continues to advance the construction of big data technology, in order to use land resources more rationally and improve the effectiveness of land resources, we proposed the 12th Five-Year Plan of Big Data Technology for Land Resources in 2012: China is striving to establish a system of national big data-based natural resource management.With the continuous development of information technology, new technologies such as big data, Internet of Things, and cloud computing have become innovative drivers of urban development and have led to intelligent upgrading of land and resource management information. In the era of "Internet+", it is required to ensure multi-channel data collection, comprehensive analysis and multi-category application, and big data technology to actively carry out mutual cooperation with other departments on the basis of big data technology of land and resources.

J. Abawajy et al. (Eds.): ATCI 2021, LNDECT 81, pp. 875–879, 2021.
https://doi.org/10.1007/978-3-030-79197-1_130

1 Explanation of the Concept of Land Resources Management in the Context of Big Data

Large-scale database construction is a fundamental project. In the context of big data, land resources and natural resource management requires extensive and complete data bank to monitor and manage various land resources, minerals and environment in cities. In the past, most of the cities in China have gone through the process of building big data technology for land, collected extensive land information, and created specialized databases, so as to support urban construction by classifying the soil industry and digital land phases. In the context of big data, the creation of smart cities in the natural resource management phase further accelerates the creation and exchange of data and emphasizes the continuous cyclical processing of data.

Today's large land and resource databases should not only be limited to historical and thematic data, but should also be integrated to form information resources for use by government decision-making departments, businesses, individuals, and other social entities. Once data collection is complete, the service delivery process is processed by providing data for real-time traffic and disaster warning functions. In the context of big data, the focus of building big data on land resources and land resources management is to establish professional management and supervision databases based on big data and basic data on land resources, and to perform data mining and statistical analysis to provide scientific data support for the work and life of urban entities through screening and integration [1].

2 The Current Situation of Land Resources Management Based on Big Data of Land Resources

2.1 Insufficient Attention to Information Sharing by Staff

During the operation of big data technology in natural resource management, the work of big data technology in land and resource departments at all levels has been handled by the bureau information center, especially the information sharing led by the center to manage and maintain the implementation of urban big data technology construction projects. In accordance with the integrated planning of the provincial government land and resource bureaus, it coordinates the management of the information exchange of the third land department as required. However, as the business of land and resource management grows and the content of the information-sharing platform is enhanced, information sharing will expand not only to independent information centers but also to more business departments. However, in the actual operation process, some employees have a limited understanding of big data technology due to the integration of traditional work methods. There are some employees who have some difficulties in using the new technology, which leads to a resistance to the new working model. Such a working attitude makes some obstacles to the implementation of information sharing of big data information platform.

2.2   Difficulty in Collecting Scattered Information

The construction of natural resource management for big data on land resources is still in its initial stage, and the implementation of various standards is still in progress. Under the government's unified plan, ministries are increasingly aware of the importance of information sharing. However, there is still no clear concept of the extent and mechanism of information sharing. In terms of individual departments to establish information exchange mechanisms, in the land sector, the required data includes more urban planning information with certain confidential characteristics and profit value. The Ministry of Natural Resources has indicated that the information disclosed on public information platforms is incomplete. For example, it is difficult for the official website of the Ministry of Natural Resources to collect and classify certain detailed data about home security and environmental governance, and the coverage of information on certain topics is insufficient in the process of sharing queries by unit [2].

2.3   Unclear Division of Functions

Due to the natural resources department has not adjusted and improved the organizational structure for many years, the new function of land and resources information management and control has not had a clear main body responsible agency, and the old organizational structure has been unable to meet the current work requirements of land and resources information and data management and control. The specific performance is as follows: Each office entrusts technical units to carry out work when necessary, but when there is a problem with the data, no department is willing to take responsibility. Because the original division of labor was not clear and there was no coordination before the work started, work efficiency was not improved. The most worrying problem of unclear division of labor is the security of information and data Information data, as the core resource of the natural resources department, cannot be guaranteed for safety, which directly affects all aspects of the land and resources management business [3].

3   Optimization Strategy of Land Resources Management by Big Data of Land Resources

3.1   Innovative Ways of Popular Participation

With the increasing use of new Internet media in urban management and people's life, the new information distribution system of land department will help to expand the scope of land information services and enlarge the service channels so that people can easily receive applied information. On the other hand, in the process of advancing big data technology for land management in life, it is also helpful to allow land management departments to be monitored by people [4]. We keep urging the Ministry of Land and Transportation to improve the quality of its services and to build a solid foundation for the construction of land resources management with big data. The Ministry of Natural Resources is accelerating the use of new media to innovate ways for people to monitor and participate in order to prevent the possibility of individuals applying in the name of the government, falsifying or imitating official Weibo and WeChat accounts. Since the standardization of official information disclosure channels such as Internet portals and spokespersons of the Ministry of Land, Infrastructure, Transport and Tourism and local government organizations may adversely affect the image of the Ministry of Land, Infrastructure, Transport and Tourism, innovative ways of people's participation to achieve openness and transparency of big data is also one of the key areas for reform and innovation.

3.2   Expanding Data Development and Promoting Information Sharing

The development and utilization of land resources data is not merely a part of value-added knowledge, but also the key to the success of land resources management. The effective information contained in the data has been fully explored, while the data and the use of the value-added information should have the vitality. Today is the era of information and knowledge economy, sharing of open data owned by government functions is a basic requirement, and the role of open data in promoting social economy is immeasurable. Data resources should be easier to find, access and use; public sharing of government data should be expanded for greater societal benefit. The openness of public data resources is promoted in an orderly manner by listing national land resources data under the premise of ensuring security and protecting privacy [5].

3.3   Further Improve the System and Process

Establish and complete a new system and process, taking the supervision and use of land and resources information data as an important part of it. Improve the ideological awareness of the majority of cadres and employees, and realize the importance and necessity of information and data in land and resources management under the new situation. Encourage them to learn new advanced management methods and possess new skills in land and resources management under the new situation. Third-party institutions can be introduced to test the systems and processes, and ultimately promote the realization of each position in the natural resources department to perform their duties, and jointly improve work quality and efficiency.

3.4   Strengthening Talent Training

In the work related to big data management, the use of talents must be considered at multiple levels and in various aspects. From the perspective of information science, computer and large database management talents are essential. From the perspective of industrial management, talents in geographic information systems, natural resource management, and public administration are also beneficial for big data management of land resources. In short, data management includes not only database management knowledge in the information field, but also expertise in land and resource management, as well as the application of 3S technology along with many other methods. Therefore, the construction of a talent pool is a top priority [6].

4 Conclusion

The Ministry of Natural Resources has already established data informatization plat-forms for aerial images, satellite images, land use status, basic farmland, and dynamic remote sensing monitoring of land use by implementing big data technology projects. Land resource data management is a fundamental, systematic and long-term task, and it is necessary to establish a more ambitious framework to improve data sharing and information public service mechanisms, while meeting the tasks of land and resource management. We hope that future research will combine professional perspectives on the use of information technology with more detailed data available through multiple sources to conduct in-depth studies that examine and test the scientific nature of the above strategies.

References

1. He, L.: Analysis of innovation in natural resource management model with the application of land resources big data. Constr. Eng. Technol. Des. **9**, 2640 (2020)
2. Li, N.: Analysis of innovation in natural resource management model with the application of land resources big data. New Technol. New Prod. China **21**, 119–120 (2019)
3. Du, N.: Research on innovation in natural resource management model with the application of land resources big data. Constr. Eng. Technol. Des. **4**, 3631 (2020)
4. Wei, H.: Research on the innovation of natural resource management model based on the application of big data of land resources. Constr. Eng. Technol. Des. **6**, 50 (2018)
5. Ma, B.: Research on the innovation of natural resource management model based on the application of big data of land resources. Legal Expo **14**, 210 (2019)
6. Jiangxi, L.: Analysis of natural resource management model innovation with the application of big data of land resources. Constr. Eng. Technol. Des. **6**, 3683 (2020)

Tea Marketing Strategy Based on Big Data

Shuyu Hu and Ming Huang[✉]

School of Economics and Management, Hunan Open University, Changsha, Hunan, China

Abstract. With the continuous changes in the economic environment and consumer demand, the traditional tea market is also undergoing transformation and upgrading. This research mainly discusses tea marketing strategies based on big data. It can grasp the needs of consumers more accurately, and monitor the situation of competitors and monitor the market. The future development is predicted to a certain extent. Competitor products are clearly differentiated. 37% of the respondents like to buy tea in tea specialty stores, and 33% choose supermarkets or specialty stores to buy tea. This research is helpful to provide reference and reference for other enterprises dedicated to the marketing.

Keywords: Big data · Tea marketing · Big data mining · Market positioning

1 Introduction

Since ancient times, our country has been an important producing area of tea, but also a big consumer country. Through thousands of years of development, the tea industry has been prosperous and has become one of the first products to participate in international trade. With the strengthening of the frequency of international trade, people's demand for high-quality life continues to increase.

However, if the company lacks a set of reasonable and scientific marketing strategies, the development of the company has entered a bottleneck stage. For example, the sales channels of the company's products are not stable enough, the market positioning is not clear enough, and the company's competitiveness is low, etc. [1, 2]. Therefore, if the tea company wants to get out of the bottleneck, it cannot only focus on tea production and planting, but also needs to focus on the tea marketing level [3, 4]. Therefore, when designing products, Tea Co., Ltd. should pay attention to many aspects such as packaging design, new product development, and price improvement to ensure that the brand can maintain close contact with the core consumer groups, otherwise the product brand will gradually lose its vitality and eventually die [5, 6]. Finally, ensure product quality [7, 8]. Local sample surveys are meaningless [9, 10]. Accuracy is more effective than accuracy [11].

On the one hand, companies need to improve their brand, connotation and image; on the other hand, they need to explore scientific and effective marketing strategies, have a rational positioning of tea products and the market, expand marketing channels, fully meet the multi-level needs of consumers, and cultivate consumption brand loyalty. Therefore, for tea companies, use scientific marketing methods and comprehensive use of various marketing strategies to ensure that more consumers can understand the

J. Abawajy et al. (Eds.): ATCI 2021, LNDECT 81, pp. 880–885, 2021.
https://doi.org/10.1007/978-3-030-79197-1_131

products of tea companies, thereby enhancing the company's market competitiveness. This is a need for tea companies, and research important topics.

2   Tea Marketing Strategy

2.1   Tea Marketing

From the perspective of the brand of enterprise products, relevance refers to the relevance to emerging consumer groups and grades. With the development of the times, the income structure of consumer groups is also changing. Consumer taste usually determines the consumer's preference for a certain product. The amount of consumption is usually called a secondary consideration. Regardless of whether it is from a technical or quality point of view, it must be improved, so as to improve consumer satisfaction with the product, and more satisfy the actual needs of consumers. First of all, in terms of product quality, it is necessary to accurately control the quality of the supply, and effectively monitor to ensure the qualified rate of the product. Secondly, according to the development and changes of the market and the changes in customer demand, adjust the packaging style and price system of the company's products. Finally, in customer service, we should pay attention to the gradual transformation from traditional satisfying tea customers' tea tasting needs to the reflection of tea customers' tea value.

If $D(X_i, C_j)$ meets the following conditions:

$$D(X_i, C_j) = \min\left\{D(X_i, C_j)'\right\} \tag{1}$$

The data object X_j is placed in the C_j class.

Online shopping transaction data belongs to a large data set with a data volume of one million, which requires extremely strong scalability and extremely high processing efficiency.

$$Guan[i] = \max\{w_i | w_i \in d_i, i = 1, 2, 3, \ldots, n\} \tag{2}$$

Most of the online shopping transaction data are numerical characteristic variables, and non-numerical characteristic variables can also be converted into numerical data.

$$MI(d_i, C) = \log\frac{TD(Guan[i], C) * |D|}{TD(Guan[i]) * TD(C)} \tag{3}$$

Among them, $i = 1, 2, 3 \ldots n$.

2.2   Big Data

For a long time, many limited factors rely on conducting market research and analysis. However, they did not consider that the matter of data sampling itself has many inherent limitations, such as not enough data samples, not representative, time lag, etc.

3   Tea Marketing Experiment

3.1   Selection of Target Market

Enables it to be quantitatively measured based on more accurately, but also the situation of competitors.

3.2   Market Positioning Based on Big Data

Shape the distinctive feeling of the product, paving the way for subsequent precision marketing strategies. Table 1 shows the uses of consumers buying tea pages online.

Table 1.   Purposes of consumers buying tea pages online

Use	Features	Consumer groups
Self-drinking	Small amount	Personal
health	lose weight	Obesity, health
Gift	Refreshing	Entertain
Grade	Entertainment	Upper class

4   Tea Marketing Analysis

4.1   Age Analysis of Tea Consumers

According to the difference of consumer age, it can be divided into young (below 20 years old), middle-aged and young (20–35 years old), middle-aged (35–65 years old) and old (over 65 years old) consumer markets. Young people under the age of 20 are full of interest in novel things, so they have a strong interest in novel product packaging and tastes; the young and middle-aged groups of 20–35 years old are not as strong as young people in pursuit of new things, so they are not pay too much attention to the novelty of the taste and pay more attention to the function and quality of the tea itself; the middle-aged and elderly people aged 35–65 do not like novel things and tend to use traditional things all the year round. They have relatively high requirements for product quality; 65 years old the above-mentioned elderly people not only require products with complete functions, but also require low prices and relatively conservative consumption tendencies. The age difference of tea consumers is shown in Table 2.

Table 2. Age differences among tea consumers

Age	Group	Product demand
<20	Young group	1 Product novelty
		2 Good taste
20–35	Young and middle-aged	1 Product function
		2 Product quality
35–65	Middle-aged and elderly people	1 Product quality
		2 Products that are used to drinking
>65	The elderly	1 Product quality
		2 Low price
		3 Product features

4.2 Channels for Consumers to Buy Tea

The channels for consumers to buy tea are shown in Fig. 1. It can be seen from Fig. 1 that 37% of the respondents prefer to buy tea in tea specialty stores, 33% choose supermarkets or specialty stores to buy tea, while only 10% of consumers choose to buy tea online. The respondents believe that they buy tea online. There are too few channels, and 15% of consumers choose other consumers to buy directly from middlemen's acquaintances. It can be seen that tea is mainly concentrated in more traditional marketing channels, and new marketing channels are still relatively few developed. Therefore, consumers believe that the current channels of tea are not enough to satisfy consumers. According to Fig. 1, it can be seen that 5% of the customers are buying from other marketing channels. Improve users' shopping experience, and ultimately increase users' loyalty to the system.

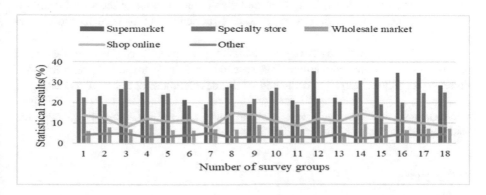

Fig. 1. Channels for consumers to buy tea

Figure 2 shows the regional ratio of the unit price of tea purchased by consumers online. Consumers buy products online in the price range of 50–100 yuan, 100–200 yuan, and the proportion is as high as 61%; then, below 50 yuan, 200–400 yuan, and 400–500 yuan account for 14% and 10% respectively, 10%; Finally, there are not many consumers over 500 yuan, only 5%. It can be seen that consumers prefer the price range of 50–200 yuan for products purchased online. Each end of business means that the next new business relationship is taking shape. In fact, the precision of precision marketing is also based on the constantly occurring business relationships, that is, it is not achieved all at once, but is gradually accurate business dealings with customers.

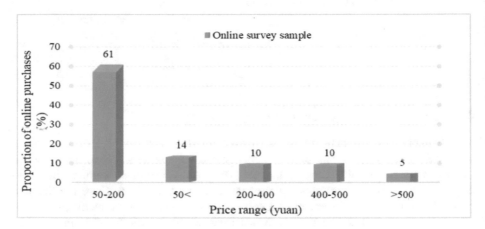

Fig. 2. Regional ratio of the unit price of tea purchased by consumers online

5   Conclusion

Strengthen the functional design of the website and provide timely feedback to customers' doubts or problems. Long-term and stable maintenance of the website can not only provide customers with a good sense of experience, but also collect more information from customers.

Therefore, the company's regular website maintenance can ensure the better development of online marketing. This research is helpful to provide reference and reference for other enterprises dedicated to the development of network marketing.

Acknowledgements. This work was financially supported by Hunan Vocational Education Teaching Reform Research Project (ZJGB2020410).

References

1. Xu, W., Zhou, H., Cheng, N., et al.: Internet of vehicles in big data era. IEEE/CAA J. Automatica Sinica **5**(1), 19–35 (2018)
2. Wang, X., Zhang, Y., Leung, V.C.M., et al.: D2D Big data: content deliveries over wireless device-to-device sharing in large scale mobile networks. IEEE Wirel. Commun. **25**(1), 32–38 (2018)

3. Yudong, C., Yuejie, C.: Harnessing structures in big data via guaranteed low-rank matrix estimation. IEEE Signal Process. Mag. **35**(4), 14–31 (2018)
4. Gu, K., Tao, D., Qiao, J.F., et al.: Learning a no-reference quality assessment model of enhanced images with big data. IEEE Trans. Neural Netw. Learn. Syst. **29**(4), 1301–1313 (2018)
5. Al-Ali, A.R., Zualkernan, I.A., Rashid, M., et al.: A smart home energy management system using IoT and big data analytics approach. IEEE Trans. Consum. Electron. **63**(4), 426–434 (2018)
6. Zhang, N., Yang, P., Ren, J., et al.: Synergy of big data and 5G wireless networks: opportunities, approaches, and challenges. IEEE Wirel. Commun. **25**(1), 12–18 (2018)
7. Mohammadi, M., Al-Fuqaha, A.: Enabling cognitive smart cities using big data and machine learning: approaches and challenges. IEEE Commun. Mag. **56**(2), 94–101 (2018)
8. Duan, M., Li, K., Liao, X., et al.: A parallel multiclassification algorithm for big data using an extreme learning machine. IEEE Trans. Neural Networks Learn. Syst. **29**(6), 2337–2351 (2018)
9. Kasradze, T.: Challenges Facing Financial Inclusion Due to the COVID-19 Pandemic. Europ. J. Mark. Econ. **3**(2), 63–74 (2020)
10. Eon, W.J., Yeon, C.J., Caroline, C.H., et al.: Safety and effectiveness of once-daily tadalafil (5 mg) therapy in Korean men with benign prostatic hyperplasia/lower urinary tract symptoms in a real-world clinical setting: results from a post-marketing surveillance study. World J. Mens Health **36**(2), 161–170 (2018)
11. Bitama, P.C., Lebailly, P., Ndimanya, P., et al.: Global value chain governance and relation between local actors in the burundian tea sector. Int. Rev. Manag. Mark. **9**(6), 105–111 (2019)

Applying Big Data in the Evaluation and Management of Research Performance in Universities

Lihong Li[1(✉)] and Wanbing Shi[2]

[1] School of Humanities and Law, Northeastern University, Shenyang, Liaoning, China
lilihong1127@sina.com
[2] Special Education Research Center, Nanjing Normal University of Special Education, Nanjing, Jiangsu, China

Abstract. Research has been a prominent function of universities, so the evaluation and management of research performance in universities has become a key issue to be solved, but there are still some problems like the efficiency and reliability. The utilization of big data has made it possible for the evaluation and management of research performance in technologies, ideas and critical changes to deal with the existing problems. On the base of big data, identifying which scientific knowledge and technological needs are the most critical to plan scientific research. Through the establishment of university personnel databases, results databases, funding databases, facts laboratory database, teaching and scientific research database to a rational allocation of sources provides scientific and effective management means. So it is necessary to improve big data infrastructure and establish big data standards in sharing, teachers and researchers in universities have to be aware of the development and application of big data technology.

Keywords: Big data · Research · Research performance · Universities

1 Introduction

Higher education institutions and universities have developed the function of research since the founding of the University of Berlin in 1810. Research means to uncover and generate new knowledge, or to solve theoretical and applied problems. Research is of central importance in modern society. With dramatic expansion in student enrollment in higher education systems, the lack of funding and operating costs, public accountability, the effectiveness and efficiency of research in universities couldn't be more emphasized. Researches need systematic evaluations for optimizing their research resource allocations, reorienting their research support, rationalizing research organizations, restructuring research in particular fields, or augmenting research productivity. The application of big data brings a great convenience and opportunity to evaluation and manage the research performance.

© The Author(s), under exclusive license to Springer Nature Switzerland AG 2021
J. Abawajy et al. (Eds.): ATCI 2021, LNDECT 81, pp. 886–890, 2021.
https://doi.org/10.1007/978-3-030-79197-1_132

1.1   The Development of Big Data

Big data is used in all walks of life, ranging from health management, policy-making, marketing and promoting, to education management. Standard conceptions of big data focus on various "V's"–the volume, velocity, value, veracity, and variety of what are petabytes of data. [1] "Big Data is less about data that is big than it is about a capacity to search, aggregate, and cross-reference large data sets...We define Big Data as a cultural, technological, and scholarly phenomenon that rests on the interplay of: Technology: maximizing computation power and algorithmic accuracy to gather, analyze, link, and compare large data sets; Analysis: drawing on large data sets to identify patterns in order to make economic, social, technical, and legal claims; Mythology: the widespread belief that large data sets offer a higher form of intelligence and knowledge that can generate insights that were previously impossible, with the aura of truth, objectivity, and accuracy" [2].

The utilization of big data has made it possible in operating and managing higher education. Large quantities of information about their students' enrollment, research programmes and experimental facilities has been collected in universities to facilitate the management and strategy. With the emergence of new sources of digital 'big data' and systems to manage it, universities are increasingly being enmeshed in networks of digital data technologies and expert technical practices, and reimagined as 'smarter universities' [3] Quantitative big data have become significant for market-led HE (higher education) policy-making too. [4] Over the past twenty years universities has been transformed by the tide of digital data. The era of big data has changed the research and knowledge generating and production totally. The datafication of higher education has previously materialized in the evolution of methods and applications of learning analytics, educational data mining, intelligent tutoring systems and even artificial intelligence [5].

1.2   The Importance of Research performance in Universities

Research has long been the priority and the core of strategic development in universities, therefore a large amount of money from government and private donation has been invested into it. Good research can create new knowledge, and cultivate young researchers as well to ensure the sustainable development of our society. So it requires effective evaluation and proper criticism to secure good research in the long run. Whether the assessment methods and procedure is effective and economical should be considered carefully.

1.3   The Evaluation and Management Practices in Research Performance in Universities

Scientific research evaluation is a kind of cognitive activity that uses scientific methods to judge scientific research activities and their input and output on the basis of certain scientific research objectives, to manage, supervise, predict and control scientific research activities, and to provide basis for decision-making. In universities evaluation of research performance is a part of quality management and an instrument of quality assurance. The decision was made on the base of evaluation results. Most countries allocate budget and money on the base of research performance.

2   The Application of Big Data in Management of Research Performance

2.1   The Application of Big Data in Research Planning

The topic selection of scientific research project is the starting point of scientific work. Scientificity directly affects the feasibility and innovation of project implementation. On the base of big data, identifying which scientific knowledge and technological needs are the most critical to improve scientific research. It is urgent and the social and economic benefits are extremely prominent. The pertinence of the topic selection can realize the scientific research topic selection. The application of big data technology guide researchers to hot issues to be solved. Consequently, many institutions have made known that analytics can help meaningfully advance their universities in strategic fields as proper means of allocation & utilization, student success, and finance is taken into consideration [6]. Identifying which scientific knowledge and technological needs are the most critical to improve scientific research. It is urgent and the social and economic benefits are extremely prominent. The pertinence of the topic selection can realize the scientific research topic selection. The application of big data technology guide researchers to hot issues to be solved.

2.2   The Application in Allocating Research Resources

Many universities have the problem of unbalanced allocation of research resources, Some departments have not made good use of research resources. The lack of research resources seriously restricts its development. How to make rational distribution of scientific research resources among different departments and disciplines is a problem for decision makers. Development and application of big data technology provides an effective means of big data through the establishment of university personnel databases, results databases, funding databases, facts laboratory database, teaching and scientific research database, a rational allocation of sources.

2.3   The Supervision of Research Quality

The national and/or regional research performance system is usually measured by publication output and citation scores as indicators. There are three different systems to evaluate research quality: Dutch system, Britain system and New Zealand system. In the three systems, there is one common indicator: research output. Research quality illustrates originality and scientific relevance of research performance, as well as adequacy of methods. The scientific research evaluation index system usually focuses on the evaluation of scientific research output quality and scientific research publishing activities.

Big data technology can control scientific research and whole process. Various types of process information will follow the development of research project continuously collected to the relevant information systems, these information covers the implementation of project plan, and various research funds use, performance and workload of various scholars, Use of various experimental instruments and equipment, various literature use, papers published by various categories of personnel and publicly published textbooks, the awards of participants in various competitions, etc. These information are objective and true, often the quality of research can be good or bad effectively reflected through data, through the analysis of scientific research data. And mining can effectively control the quality of and research.

3 Conclusions

In the era of knowledge economy, research in universities is indispensable with the development of economy and comprehensive national power. So it is essential to build a research performance system to ensure the high quality of research and scholarship in responding to stakeholders, government, media and public. Policy-makers regard higher education institutions as the global competition center, research is the core of the center. Utilizing big data is to ensure the research in high performance state. So big data has to be developed further.

3.1 Establishing Big Data Infrastructures for Research Evaluation and Development

At present, digital campus is now built massively in many universities, but it is just a beginning, it is still a long way to fully utilize big data to serve the development of higher education and economy. And the two or more information systems cannot be linked to share information, and the function design is not completed and perfect. Matthew Goldstein said: "I think technology is a theme that is going to course through our planning over the next several years at all of our institutions" [7].

It is essential to develop a science-driven procedure for the evaluation of data research infrastructures. First, the procedure point out the success of data research infrastructures depends not only on scientific potential, but the competitive environment. Second, projects of all areas of science are evaluated in the same way and the results are expressed in a standardized mode. For example, Federal Ministry of Education and Research of Germany has decided to invest research data infrastructure in humanities and science: exceptional fundamental and application-oriented research on social problems requires efficient research data infrastructures as well as appropriately processed data. The fundamental principles for optimal dissemination and follow-up use of research data and findings are the FAIR principles (find-able, accessible, interoperable, reusable).

3.2   Establish Data Standards in Sharing

The quick integration and analysis of data can be achieved duo to cloud computing technology. It is common customs to construct personal database for teachers on the base of big data for collecting data and analyzing research progresses. Data sharing agreements need to be linked into the mechanisms for data protection and privacy, including anonymization for open data, access control, rights management, and data usage control [8]. So it is urgent to establish data standards. Data standards are an essential prerequisite to any large scale information infrastructure because they provide benchmarks for data quality and define how information are formatted and categorized in order to be stored, managed, searched for and used across many different software applications [9].

3.3   Improve Researcher's Big Data Quality and Awareness

Researchers and teachers should learn to utilize the big data technologies and information technologies for selection of research plan, synchronization information, and control research the speed of process and budget use. Though the quality of big data need to be standardized, completed and improved, big data cannot be avoided for any researcher in universities. So Technology sage Mark Weiser has said: "The most profound technologies are those that disappear. They weave themselves into the fabric of everyday life until they are indistinguishable from it." In the case of big data, it is straightforward to make the case that although the hype cycle is over, the technologies themselves are no longer big news because they are widely deployed. It is not a stretch to say that to be successful today, every researcher must become a data scientist! [10].

References

1. Landon-Murary, M.: Big data intelligence: applications, human capital and education. J. Strateg. Secur. **9**(2), 94–123 (2016)
2. Boyd, D., Crawford, K.: Critical questions for big data. Inf. Commun. Soc. **15**(5), 662–679 (2012)
3. Williamson, B.: The hidden architecture of higher education: building a big data infrastructure for the 'smarter universities.' Int. J. Educ. Technol. High. Educ. **15**(12), 1–26 (2018)
4. Komljenovic, J., Robertson, S.L.: The dynamics of 'market-making' in higher education. J. Educ. Policy. **31**(5), 622–636 (2016)
5. Dillenbourg, P.: The evolution of research on digital education. Int. J. Artif. Intell. Educ. **26**(2), 544–560 (2016)
6. Dede, C., Ho, A., Mitros, P.: Big data analysis in higher education: promises and pitfalls. Educause Rev. **51**(9), 22–34 (2016)
7. Bharucha, J., Grabois, N., Zimmer, R., Van Zandt, D.: What ought universities look like in 20 to 30 years? Soc. Res. **79**(3), 551–572 (2012)
8. From the editors: Big data and management. Acad. Manage. J. **47**(2), 321–326 (2014)
9. Easterling, K.: Extrastatecraft: the power of infrastructure space. Verso, London (2016)
10. Ramakrishnan, N., Kumar, R.: Big data. IEEE **49**(4), 20–22 (2016)

Application Off Line Programming of ABB Industrial Robot

Yongcheng Huang[✉]

Guang Dong Polytechnic College, Zhaoqing 526100, Guangdong, China

Abstract. With the development of science and technology, the utilization of robot is more and more extensive, and industrial robot is a machine device for industrial control. This paper on account of the robot Studio software, which uses the method of manual off-line programming to draw the six side graphics. After the simulation software verifies the accuracy of the program, it enters the site for debugging.

Keywords: Industrial robot · Robot studio · Off line programming

1 Introduction

Robot is "the Pearl at the top of the crown of manufacturing industry". Its R&D, manufacturing and application are important symbols to measure a country's scientific and technological innovation and high-end manufacturing level [1]. Industrial robot integrates automation, computer and machinery science and technology. The talents needed for the design and development of industrial robot not only need systematic scientific knowledge, but also need industrial professional experience and knowledge [2]. In order to meet the flexible needs of robot learning and evaluation, this paper adopts the flexible design of industrial robot related skills learning and evaluation module, which has a good guiding significance for beginners.

2 Model Building of Industrial Robot

This paper uses the latest generation of ABB Robot 6 axis industrial robot. The Fig. 1 is the calibration of 6 axis. The smallest multi-purpose industrial robot launched by ABB so far has the following characteristics: maximum working stroke 580 mm, compact, light weight, load 3 kg, weight only 25 kg, flexible, and picking distance 112 mm under the base. Flexible automation industry based on robot is specially designed for 3C [3].

J. Abawajy et al. (Eds.): ATCI 2021, LNDECT 81, pp. 891–894, 2021.
https://doi.org/10.1007/978-3-030-79197-1_133

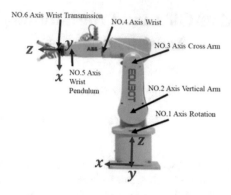

Fig. 1. IRB120 six axis calibration

3 System Construction

In order to facilitate beginners to learn industrial robots, this example establishes a virtual module according to the 1:1 basic module of the field module in the three-dimensional software. HD1XKB is chosen as the training platform of industrial robot skills. Import the HD1XKB training platform into robotstudio, and walk the regular hexagon in Fig. 2 [4] on the module platform. The IRB120 in ABB model library is introduced into the system and placed on the corresponding position of HD1XKB training platform; the tool is installed on the flange plate of IRB 120, and the system is established, as shown in Fig. 2 [5].

Fig. 2. System construction

4 Parameter Setting

For industrial robots, it's necessary to have a system, and determine position of seven points P10, P20, P30, P40, P50, P60 and P70. In industrial robots, the position is determined by the teaching point of the hand handle. The regular hexagon of starting point is P10, P20, P30, P40, P50, P60 and P70 are regular hexagon points, and other parameters take the default Value [6].

5 Programming

The off line programming of industrial robot is to reconstruct the three-dimensional virtual environment of the working scene in the computer through the robotstudio software, and then according to the regular hexagon outline, through the teaching device to teach each point to establish routine program, determine the teaching point, track, program and test process. The established routine is called liubianxing, and its routine is such as:

```
PROC liubianxing( )
    MoveAbsJ  jps10\NoEOffs,  v500,  z50;  [6]
    MoveJ  P10,  V500,  z50;
    MoveL  P20,  V200,  fine;
    MoveL  P30,  V200,  fine;
    MoveL  P40,  V200,  fine;
    MoveL  P50,  V200,  fine;
    MoveL  P60,  V200,  fine;
    MoveL  P70,  V200,  fine;
    MoveL  P20,  V200,  fine;
    MoveJ  P10,  V500,  z50;
ENDPROC
```

The starting point of the movement is jps10. When the regular hexagon is executed, it finally returns to P10. The trajectory line diagram is shown in Fig. 3. Using offline programming, it can run according to the predetermined graph and import it into the field to run the program. The final verification and off line programming trajectory are completely consistent, and the field practice is shown in Fig. 4 [7].

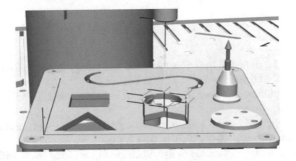

Fig. 3. Trajectory diagram

Fig. 4. Practice chart

6 Conclusion

Industrial robot is the need of the development of the times. Modular learning is used to verify the correctness of the program through off-line programming simulation, and then imported into the field for verification, which greatly improves the learning interest of beginners. In a word, it has a certain guiding role for the learning of industrial robot programming and the training of ABB industrial robot programming application related talents.

Acknowledgements. Young innovative talents project of colleges and universities in Guangdong Province in 2018 (2018KQNCX314), offline programming experiment of industrial robot in Guangdong Institute of technology in 2020 (XNFZ20201), teaching reform of industrial robot application technology in Guangdong Institute of technology in 2020 (JXGG202028), multi index optimization analysis of injection molding process parameters in Guangdong Institute of technology in 2020 (2020GKJZK004).

References

1. Yanxia, L., Bin, L.: Industrial robot application technology based on ABB Robot irb120. Harbin Institute of Technology Press, Harbin (2018)
2. Chuhong, Y.: Exploration on teaching reform of industrial robot technology. Sci. Technol. Wind **8**, 54 (2017). (in Chinese)
3. Bin, Y., Yongcheng, H.: ABB industrial robot programming application research. Mech. Electr. Eng. Technol. (10), 20–21 (2019). (in Chinese)
4. Di, Q.: Construction and application of hybrid course of industrial robot offline programming. Sci. Technol. Wind (33), 56 (2019). (in Chinese)
5. Huixiang, Z.: Analysis of mechanical structure design methods in current industrial robot applications. South. Agric. Mach. **16**, 43 (2018). (in Chinese)
6. Hui, G., Shikai, D.: Research on Application of SolidWorks modeling software in industrial robot offline programming software. Sci. Tech. Inform. (36), 25–26 (2018). (in Chinese).
7. Shuangming, L.: Application of information instructional design in offline programming of industrial robot. Electron. World **05**, 26–27 (2019). (in Chinese)

The Logistics Distribution Models of Chain Management in the Network Era

Huilin Zhang[✉]

Fujian Polytechnic of Information Technology, Fuzhou, Fujian Province, China
huilin_zhang@yeah.net

Abstract. Chain management logistics distribution in the network era is a way for chain management logistics distribution centers to centralize and unify procurement and distribution based on the market's low-cost competitive advantage, so as to realize the distribution of goods. This paper is to propose countermeasures to optimize the logistics distribution mode of chain management enterprises in the network era to realize of the goal of improving distribution efficiency.

Keywords: Network era · Chain management · Logistic distribution mode · Countermeasure research

1 Introduction

The network era provides basic technical support for the rapid development of modern logistics. This is an inevitable requirement for chain management companies to expand their own scale. It is also the need to better meet the people's growing yearning for a better life by choosing a distribution model that suits the company's own characteristics [1]. Chain operation enterprises should choose a scientific and reasonable network distribution model based on the characteristics of the network era, the reality of the primary stage of China's socialism and the needs of financial service information development, so as to improve the core competitiveness of chain operation enterprises. Chain management logistics distribution in the network age is a distribution method in which distribution centers centrally purchase and distribute commodities with low-cost competitive advantages. Aiming at the problems of chain logistics distribution centers, the countermeasures for optimizing the locations of distribution centers and optimizing the vehicle dispatching and distribution system are proposed to improve the realization of distribution efficiency.

2 The Meaning of Chain Management Logistics Distribution in the Network Era

2.1 The Meaning of Chain Management Logistics Distribution

The logistics distribution of chain operation in the network age is a logistics activity process, according to customer requirements, of realizing the activities of sorting,

J. Abawajy et al. (Eds.): ATCI 2021, LNDECT 81, pp. 895–899, 2021.
https://doi.org/10.1007/978-3-030-79197-1_134

processing, packaging, dividing, and arranging items, and delivering them to the destination on time. It is based on network technology and within a reasonable area [2]. The particularity and comprehensive activity characteristics of chain management logistics distribution in the network age have realized the combination of business flow and logistics activities as well as several functional elements.

Chain management logistics distribution in the network era covers all the elements of the logistics function, which is the epitome of logistics in the chain management activities [3]. Logistics distribution can achieve the completion of the series of activities of loading and unloading, packaging, storage, and transportation, and the target requirements for the delivery of goods to the destination. The special distribution of logistics supported by processing activities includes a wider range of aspects. General logistics is the process of realizing transportation and storage. Logistics distribution is the process of realizing transportation, sorting and distribution. The special distribution is the process of realizing the sorting of goods. It is also a characteristic activity which is the transportation process for the purpose of simplifying distribution.

2.2   Logistics Distribution Form of Chain Operation in the Network Era

2.2.1   Self-operated Distribution Model

The self-operated distribution model of chain operation refers to the links of sorting, processing, packaging, segmentation, and assembly in logistics distribution of chain operation enterprises are prepared and operated by themselves, so as to realize the internal and external goods distribution model.

The advantage of this model is that it can achieve a high systematic activity level that integrates supply, production and sales of an enterprise, internally meet the distribution needs of raw materials, semi-finished products and finished products, and externally meet the needs of expanding markets. The disadvantage is that since the scope of investment in the distribution system is expanded. it costs higher when it is small-scale distribution.

2.2.2   Supplier Direct Distribution Model

The chain supplier direct distribution model is a logistics activity in which the goods purchased by the chain retail enterprises are directly delivered to chain stores by the production enterprises at a specified time.

2.2.3   Third-Party Logistics Distribution Model

The third-party logistics of chain operation refers to the operation model in which the logistics distribution of the buyers and the sellers entrust a third party to complete the distribution. The advantages are: By using a third-party logistics company, the risk of chain operation enterprises can be reduced and logistics and transportation costs can be saved. The disadvantage is that it is difficult for chain operation enterprises to choose a suitable third-party logistics enterprise, which hinders mutual cooperation.

2.2.4   Joint Distribution Mode

The joint distribution model of chain operation refers to the distribution model in which many chain operation enterprises realize the rationalization of logistics distribution based on the principle of mutual benefit, jointly fund the establishment of distribution centers, and jointly formulate plans, distribute user products in the distribution area, and share distribution vehicles.

3 Research on the Countermeasures of Optimizing the Logistics Distribution Models of Chain Management Enterprises in the Network Age

3.1 Construct a Perfect Distribution Model Centered on Third-Party Logistics in the Network Era

In actual operation, chain operation enterprises must enhance their core competitiveness to fully guarantee their economic benefits. For this reason, chain operation enterprises must fully improve the third-party logistics distribution model. They should improve consumer loyalty on the basis of meeting consumer needs, reduce commodity inventory costs, and strictly control distribution cost management [4]. when Introducing the third-party logistics, on the basis of the integration of the supply chain, enterprises should fully grasp the resources and plan design. The best plan is to realize the combination of outsourcing and self-operated distribution, integrate the strength of mixed resources, and achieve the best service effect of the third-party logistics distribution through cooperation.

3.2 Construct a Model Innovation Based on the Logistics Alliance of Chain Management Enterprises in the Network Era

In order to meet the development needs of small and medium-sized chain enterprises, the cooperation of many logistics alliances has been carried out in the new era in order to achieve a comprehensive grasp of the logistics investment system. In this process, through the formation of vertical logistics alliances and horizontal logistics alliances, the system integration of self-operated distribution and third-party distribution models can be realized [5]. This diversified logistics distribution service reduces the number of service transactions and improves the service level of the logistics alliance, especially the long-term investment to optimize the operation of the logistics system, fully realize the cooperation of market entities and complete the innovation of the logistics delivery work models.

3.3 Research on Logistics Distribution Optimization Strategy of Chain Management Enterprises in the Network Era

3.3.1 Strategic Research Based on Qualitative Analysis of the Distribution Center Location

The choose of the location should consider the risk factors of natural conditions for the construction of a logistics distribution center [6]. First, it is necessary to consider the geological conditions of the construction site of the distribution center to meet the standards for building a logistics distribution center; the second is to consider the impact of natural disasters such as water supply and drainage, traffic and storms at the distribution center. Then, the factor of life and transport facilities should be considered. First of all, the power and communication infrastructure of the center must be complete. It must also be capable of handling sewage and solid waste. The transportation of the logistics distribution center is related to the issue of distribution cost and efficiency. The transportation of the distribution center should meet the needs of warehousing, delivery and employee transportation, ensure that the delivery vehicles deliver goods in time, and realize the normal operation and profitability of the chain operation enterprise. The impact of land cost factors is also very important. Land is the key to the logistics and distribution of chain business enterprises. If the center wants to handle the goods anytime and anywhere, sufficient operating space is required. However, the scarcity of land resources increases the cost of land. The location of the logistics distribution center of a chain operation enterprise should comprehensively consider the location of the center and the cost of land.

3.3.2 The Strategy of Optimizing the Strategy of Logistics Distribution Routes for Chain Enterprises

First, the basic idea of optimizing routes is to use the shortest time, shortest distance, and lowest cost to achieve the best route with the highest efficiency of delivery [7]. Second, the conditions required for the optimal route are: First, to meet all consumer requirements. Second, to meet the requirements that any vehicle must not be overloaded. Third, to meet the operating time or mileage of each vehicle must be within the specified range. Fourth, to meet all requirements Consumer arrival time requirements. Third, optimizing vehicle scheduling requirements. Logistics distribution is a key link connecting producers and consumers. Vehicle transportation is indispensable in logistics distribution [8]. Therefore, it is very important to optimize vehicle transportation scheduling links to improve the economic benefits of chain logistics distribution, reduce costs, and realize scientific logistics management.

4 Conclusion

In short, chain operation enterprises in the network era must choose appropriate logistics distribution models, integrate social resources and model innovation, comprehensively improve and design diversified distribution services, strengthen the cooperation with the third-party distribution companies, realize social resource sharing, and risk sharing, enhance the overall coordinated development of various chain operation enterprises and logistics enterprises, and enhance the modernization level.

References

1. Tang, W., Zhou, B., Hui, X.: Product market competition, R&D investment and financial performance: A comparative perspective based on property rights heterogeneity. East China Econ. Manage. **32**(07), 110–119 (2018)

2. Jia, J., Wei, Y.: Research on product market competition, customer relationship governance and corporate innovation——Based on the competition degree of industry and the market position of enterprise. Soft Sci. **33**(12), 66–71 (2019)
3. Xu, C., Wang, Z., Wang, D.: Research on the combination of elements in disruptive innovation and development of smart logistics. J. Beijing Jiaotong University (Soc. Sci. Edn.) **20**(01), 105–115 (2021)
4. Zhou, C., Zhao, X., Liu, X.: Supply chain relational resources and corporate debt financing capacity: empirical evidence based on the perspective of earnings management. Stat. Dec. **06** (07), 167–170 (2020)
5. Shen, Z., Peng, Y.: Research on distribution route optimization of chain supermarket based on improved saving algorithms. J. Liuzhou Vocat. Techn. Coll. **19**(02), 27–31 (2019)
6. Zhao, H., Jiang, X.: Research on the application of Internet of Things technology in supermarket chain distribution operation. Market Weekly **10**, 01–02 (2018)
7. Wenjie, W., Zhang, R.: Optimization of urban logistics distribution route based on multi-attribute decision model. Logistics Eng. Manage. **41**(07), 73–76 (2019)
8. Yan, B.: Research on optimal scheduling problems of modern logistics transportation vehicles. Internal Combust. Engine Parts **14**, 194–198 (2018)

The Teaching Reform of Information Security Specialty Driven by Innovative Practice Project Under the Background of Big Data

Weiqi Shi[1,2], Qing Lu[1,2(✉)], Weijiang Wen[4], Junxiang Wan[4],
Dayan Hu[3], and Jian Liu[4]

[1] Hunan Police Academy, Changsha, Hunan, China
[2] Hunan Provincial Key Laboratory of Network Investigational Technology,
Changsha, Hunan, China
[3] Public Security Department of Hunan Province, Changsha, Hunan, China
[4] Public Security Bureau of Yueyang City, Yueyang, Hunan, China

Abstract. In the era of big data, the rapid development of information society and the application of Internet of Things, cloud computing and artificial intelligence have brought about changes in people's lives. As we all know, at the present stage ominated by big data, China attaches great importance to the cultivation of scientific and technological talents. The government and universities support students' innovation and entrepreneurship greatly, and integrate off-campus teachers to participate in the cultivation of innovative and entrepreneurial talents. In order to cultivate a group of high-end professional talents with innovative thinking, entrepreneurial passion and leading the development of big data, many domestic universities have put forward innovative practice projects, aiming to enable students to gain certain experience on campus through innovative practice projects, thus cultivating a group of innovative talents in science and technology in the field of big data. In addition, from the perspective of information security teaching, the big data innovation practice project can provide the driving force for the reform of information security teaching, make this major more practical, and closely meet the development of the Internet era and the country's demand for future talents. Therefore, this paper will elaborate on the teaching reform of information security specialty driven by innovative practice projects under the background of big data.

Keywords: Internet plus · Big data · Innovative practice project · Information security major · Teaching reform

1 Preface

Like a raging fire in Internet plus, AI, big data analysis and deep mining algorithms, it has provided an exceptional external environment for information, Internet, big data analysis, deep mining algorithm and information security. In recent years, universities in China have also built up students' active participation in innovative practice projects. Good atmosphere. Information security specialty is relatively late in China, but it is also one of the more rapidly developing specialties. In line with the original intention, it

J. Abawajy et al. (Eds.): ATCI 2021, LNDECT 81, pp. 900–905, 2021.
https://doi.org/10.1007/978-3-030-79197-1_135

has become a trend for students to accept the teaching process of safety education in combination with the good practice of professional education.

2 Analysis on the Teaching Reform of Information Security Specialty Driven by Innovation Practice Project

2.1 Analysis of Teaching and Training Objectives of Information Security Major

At present, the training goal of information security specialty of full-time undergraduate students in our country is to cultivate senior information security professionals who can engage in computer, communication, e-commerce, e-government, e-finance and other fields [1]. The details are as follows:

1) The establishment and setting of firewall can provide users with a good and safe Internet environment and avoid the worries of users;
2) The setting of intrusion detection system can make users monitor the security risk of computer network system in real time and find out the foreign invaders in time;
3) The security assessment analysis tool can evaluate and analyze the current network environment of users and put forward feasible suggestions;
4) The anti-virus system can eliminate the harm of virus and prevent the virus events, so as to realize the complete intelligence of anti-virus;
5) The server protection system can protect the user's core data;
6) Deploy and maintain user information management (OA, exchange) system, UNIX system, etc.;
7) Professional data backup and restore system to protect the most critical data and resources of enterprise users;
8) Operating systems (Linux, windows), database products (SQL server, Oracle) and other secure domain environment design, and formulate strict security policies and personnel security requirements according to different business requirements.

2.2 Analysis on the Current Situation of Teaching Market of Information Security Specialty

On June 18, 2015, "information security major" officially became the first-class discipline of engineering. Therefore, the information security major has been stationed in China's science and engineering universities. At present, there are about 10000 information security professionals trained by domestic colleges and universities every year, but no more than half of them have really entered the information security industry. Through the survey, it is found that many graduates have turned to jobs related to software development, system operation and maintenance, and sales, which shows that the current employment market in the security industry is not big enough. From this, the current security industry market is less than 30 billion yuan. According to the average per capita output value of the safety industry of 300000–500000 yuan, only 100000 people can be supported. Among them, the number of professional safety technicians will not exceed 50000, and the number will increase to 100000 in the next five years. This is the real situation of the industry demand. Although there is an urgent shortage of safety professionals and colleges and universities are trying to cultivate high-end security professionals, especially offensive and defensive talents, from the perspective of educators, the more urgent market is to let the software R & D personnel, operation and maintenance personnel, product technical personnel and management personnel of various enterprises understand the safety related professional knowledge and apply them in their work [2]. This market capacity expanded dozens of times at a time, which reminds educators that the teaching of information security major is not only staying in the stage of learning textbook knowledge, but also must adapt to the new market through new teaching reform, because the essence of education is to enable the majority of students to build our country through knowledge learning, and make the society become more and more OK. With the rapid development of Internet information in our country at this stage and the national call for innovation and entrepreneurship, the enthusiasm and enthusiasm of middle school students for innovation projects are getting higher and higher. Therefore, taking innovation projects as the driving force to promote the teaching reform of information security specialty will be a wise move to make the information security specialty integrate into the information security industry in China more quickly.

The following table shows the statistics of projects applied/participated by students in a university in recent years (Table 1):

Table 1. A statistical table of the application/participation of students in a certain university in recent years

Year	Total number of projects	National level projects	Provincial level projects	Hospital level
2011	28	2	10	16
2012	37	3	15	20
2013	40	4	17	19
2014	70	9	21	40
2015	77	9	30	38
2016	83	11	28	44
2017	108	13	30	65
2018	120	10	29	81

2.3 Analysis on the Specific Ideas of Teaching Reform of Information Security Specialty Driven by Innovation Practice Project

In fact, at this stage of school education, whether it is college, undergraduate or vocational education, in terms of Internet information, all provide students with a lot of innovation and entrepreneurship practice projects, giving students sufficient space for self-development. From the perspective of practice, students' growth through innovation practice projects is very fast, so based on this, schools can create As an internal driving force, the new project aims to provide students with more high-quality professional learning resources and help, and carries out the teaching reform of information security specialty on the basis of traditional teaching. The specific measures are as follows:

2.3.1 Innovation of Online Education Form of Information Security Specialty

First of all, returning to the essence of talent training and student learning, the purpose of all people receiving education and training is to hope that their abilities can be increased, the society can recognize them, they can find good jobs, get promoted and pay rise, and through the training of innovative practice projects, they can be closer to the practical application of professional knowledge and more calmly deal with the future [3]. In addition, in the interviews with students doing innovative practice projects, they said that there is a greater resistance, that is, at this stage, the degree of knowledge and experience is limited, and they can't be comfortable in the innovation and practice projects. Then, in the teaching arrangement, we should pass on more experience to the students, let them obtain the certificate recognized by the society, and help them do a good job in career planning. Online and offline education has its own advantages. There is more interaction between offline and lecturers, focusing on the transfer of experience, and online focusing on the transfer of knowledge. Therefore, we should combine them and take into account various forms to create excellent courses. For example, for CISA and CISSP, the main certification courses of information security major, students need to participate in face-to-face courses for several days to interact with the lecturer, and then watch the video course review in the online school. At the same time, there will be a live broadcast once a week to explain the key points and difficulties of the course to help students understand, and cooperate with the mobile app question library to do exercises, so that all-round interaction can help students quickly grasp Holding a knowledge system, the passing rate of certification examination is also very high. Through the combination of online and offline course teaching mode, redesign the course teaching content and form, help students learn quickly. With professional knowledge as the backing, students can be more confident in doing innovative practice projects, at the same time; it can also improve the quality of innovative practice projects, obtain a sense of achievement, and make more efforts and in-depth professional learning.

2.3.2 In Depth Cooperation Between Colleges and Enterprises to Cultivate Practical Talents of Information Security Specialty

The teaching mode of college enterprise joint training is to emphasize the deep cooperation between colleges and enterprises, participate in the whole process of personnel training, and instill engineering practice consciousness into students imperceptibly; the second is to realize the combination of learning and application and repeatedly train students' practical engineering skills through the teaching mode of "theory practice theory re practice". Most of the practice methods are innovative practice projects, the main purpose of which is more advanced cognition and richer curriculum. Systematic education runs through the four years of undergraduate education. At the beginning of professional learning, students will lead them to visit the information center of Huawei, Huasan and other enterprises and colleges, so that students can have a real understanding and understanding of the major at an early stage, and establish perceptual cognition. The teaching of some professional courses has also been advanced to the freshman stage, which not only helps students lay a solid theoretical foundation, but also reserves more sufficient time for the subsequent corresponding practice. Innovation practice projects are generally completed through group cooperation from the enterprise real case design; students need to start from the needs analysis, complete the network architecture design, and then carry out the specific implementation [4]. Students need to independently design engineering solutions including actual measurement, wiring, equipment placement, equipment configuration, etc., and teachers and professionals of the college information (Network) center jointly guide and evaluate the rationality of the solutions proposed by students. In this process, students' participation in innovation practice project greatly cultivates students' exploration consciousness, improves their team cooperation awareness and engineering practice ability.

3 Conclusions

To sum up, the teaching reform of information security specialty driven by innovation practice project is based on the social demand for talents of information security specialty, and has a clear professional teaching goal. Nowadays, information is an important strategic resource for social development. With the rapid development of network information technology, information security is the most important subject. Information security graduates can be engaged in the research, design, development and management of various information security systems and computer security systems in government agencies, national security departments, banks, finance, securities and communication fields, as well as computer applications in it fields. Through the internal driving force of innovation practice project, combined with the teaching reform of online and offline teaching mode and college enterprise joint teaching mode, students of this major can be more selective in knowledge system learning, and better teaching effect can be obtained by combining theory with practice through the experience of enterprise real innovation practice project.

Acknowledgments. This work was supported in part by Educational science planning project of Hunan Education Department under Grant XJK20CGD053, in part by in part by Innovation Platform Open Fund Project of Hunan Education Department under Grant 20K048, in part by the Open Research Fund of Key Laboratory of Network Crime Investigation of Hunan Provincial Colleges under Grant 2020WLFZZC002, in part by Innovative Project of Science and Technology Planning Application of Ministry of Public Security under Grant 2018YYCXHNST048, in part by the Science and Technology Project of Hunan Province of China under Grant 2017SK1040, in part by Teaching Reform Project Fund of Hunan Police College, in part by the Scientific Research Excellent Youth Project of Hunan Education Department under Grant 18B549 and Grant 16B048.

References

1. Xiuxia, T., Yuan, P., Chaochao, S., et al.: Teaching reform of information security specialty driven by innovation practice project. Comput. Educ. **25**(23), 30–33 (2015)
2. Bin, L., Xia, Z.: Application of project driven innovation in practice teaching. Times Educ. **10** (19), 162 (2017)
3. Nan, J., Jing, L., Dejun, Y., et al.: Exploration of project driven practical teaching mode. China Mod. Educ. Equipment **37**(15), 88–90 (2015)
4. Yanbin, L., Yugang, L.: Research on the cultivation of project driven practical innovation ability. Contemp. Educ. Pract. Teach. Res. **15**(5), 156 (2017)

Possible Future: Trend of Industrial Design Based on Artificial Intelligence

Jingsong Gui[1,2] and Yueyun Shao[3(✉)]

[1] School of Design, Ningbo Tech University, Ningbo 315100, Zhejiang, China
[2] College of Humanities, Tongji University, Shanghai 200092, China
[3] Zhejiang Fashion Institute of Technology, Ningbo 315211, Zhejiang, China

Abstract. Artificial Intelligence (AI) is the latest achievement in the development of human technology. Intelligent design driven by artificial intelligence technology is the inevitable trend of future design development. Intelligent design has the characteristics of user experience as the center, universality to the industry and inevitable risk. It plays an important role in industrial practice and may bring out of control risk [1,2].

Keywords: AI · Design · Risk

1 Introduction

2016 From March 9 to 15, 2009, Google AI "Alphago" held a competition with the world's top chess player Li Shishi at four seasons hotel in Seoul, and "Alphago" finally defeated the top players in the world with an absolute advantage of 4–1. Alphago's victory over Li Shishi has made AI technology the focus of public opinion and science and technology circles, and also indicates the development direction of technology in the future [3, 4].

Starting with Alphago, AI has spread to agriculture, medical care, finance, transportation, big data processing, social governance and even artistic creation, becoming a powerful tool for human beings to extend their intelligence. Industrial design with technology as the core driving factor has always been the creative field of the traditional human brain [5]. Due to the intervention of artificial intelligence, important changes have taken place. Judging from the current trend, intelligent design based on artificial intelligence will become the most important technology development direction of industrial design in the future [6].

2 Concept and Technical Background of Intelligent Design

In the modern sense, the idea of computer-based artificial intelligence originated from the theory of "let machines do the same behavior as human beings" put forward by American computer scientist John McCarthy and others in the 1950s, and thus produced the concept of "artificial intelligence (AI)" [7]. Since then, scientists have been trying to incorporate human thinking into the computer and "try to let the computer

J. Abawajy et al. (Eds.): ATCI 2021, LNDECT 81, pp. 906–910, 2021.
https://doi.org/10.1007/978-3-030-79197-1_136

automatically adjust the data". Through machine learning technology, the computer can realize self-training and automatic data adjustment, especially deep learning, so that the computer can have a certain ability of human beings, and even surpass human beings in some fields.

At present, many technology giants have started to set up artificial intelligence laboratories, investing huge resources in artificial intelligence research and planning for the future of artificial intelligence. From the government to science and technology companies, AI has been regarded as the leading strategy for the future development in China. From the national level to the enterprise development, AI has been promoted as a whole to meet the coming intelligent society. At the same time, the design industry is also facing the trend of intelligence.

The artificial intelligence is divided into "strong intelligence" and "weak intelligence". Weak intelligence refers to the ability to respond to the environment under certain conditions, such as the ability to avoid obstacles and walk in a specific environment. "Strong intelligence" refers to the ability to make judgments and decisions like human beings. We call the former "narrow AI" or "weak AI", and the latter "general AI" or "strong AI" or "artificial general AI (AGI)". The artificial intelligence we use and study today, such as driverless cars, intelligent voice systems (such as Siri, Apple's intelligent voice assistant) and urban brain, all belong to narrow or weak artificial intelligence.

From the concept of artificial intelligence put forward to today's application in practice of artificial intelligence technology, this is a great progress of human beings, but also the highest achievement of human thousands of years of technological evolution. For industrial design, artificial intelligence also has an immeasurable impact.

3 Characteristics of Intelligent Design

The development of science and technology in today's world is fast changing. In the future, industrial design should actively embrace the intelligent direction of human technology development; fully exploit the possibility of using technology, so as to seek the well-being of human development.

Specifically, what are the characteristics of intelligent design in the future? I think it has the following three characteristics:

First, user experience. That is, the core is user value and experience. Intellectualization is based on the needs of users, using technology to understand and simulate people's thinking and needs, and to serve human beings. The product is actually the extension of human's physical ability and wisdom. However, the traditional products are only the low-level form of human physical intelligence. However, intellectualization does aim at the highest form of human intelligence, and its final form dispels the estrangement between things and people, making things and people in one. For traditional products, the change of user experience is a great revolution. The function should be consumer-centered, emphasize the humanization of function, make the beauty of function closer to people, and make people happy. This is the problem that should be paid attention to in the future product function transformation, and also the starting point of intelligent design. Second, universality. Intelligent + is an important direction of product design in the future. Intellectualization refers to the process of making the machine (object) have the characteristics of human intelligence and ability. Specifically, it is the process of making the object have the sensitive perception function, the correct thinking and judgment function, and the accurate and effective execution function. It is the Huawei watch mentioned above. In fact, it is called intelligent, but it is more humanized, far from reaching the level of intelligence. In the future, intelligent design and product + is an important trend, which has become a trend in the automobile and daily household appliances industry. For example, Tesla, Xiaomi TV and intelligent speaker have been realized on a large scale in the market.

With the in-depth development of Internet and artificial intelligence technology, human society has entered the era of "artificial intelligence 2.0" from the information age. Artificial intelligence from the emphasis and pursuit of individual intelligence, to focus on network-based swarm intelligence, the formation of swarm intelligence internet services. With the combination of Internet, Internet of things, big data, multimedia and virtual reality, artificial intelligence will shift from the original machine intelligence to swarm intelligence and human-computer integration intelligence to solve various problems faced by human society. In this era, the way of human innovation has also undergone profound changes. More and more attention has been paid to the use of public wisdom to solve some problems through the Internet. The way of industrial design innovation has also changed from individual innovation to group intelligent innovation.

Third, risk. That is to say, intellectualization may bring unpredictable risks such as materialization of human nature and ethical conflicts between people and things. The most basic assumption of human culture is that man is the highest value and goal on earth. Although this kind of anthropocentrism is impacted by modern natural rights and ecologism, its basic thought and value still hold. So, intelligent things may have the characteristics of intelligent creatures, so how to deal with conflicts between people and their relationships? This is one and two sides of things, which is inevitable. The possible problems and consequences of this kind of intellectualization are worthy of our consideration and vigilance.

4 The Future of Intelligent Design

Intelligent design has had a huge impact in the industry. Luban is an intelligent design product bred and developed in Alibaba intelligent design laboratory. The core function of Luban is based on image intelligent generation technology, which changes the traditional design mode and efficiency. It can complete a large number of banner drawings, poster drawings and venue drawings in a very short time, and greatly improves the design efficiency. Users can input the desired style and size according to their own needs, and the system can replace manual to automatically complete the most time-consuming design projects in traditional design, such as material analysis, matting, color matching, and generate massive design solutions that meet the requirements and standards in real time. In the "double 11" in 2017, Luban generated 8000 posters per second, and completed a total of 410 million advertising pictures, refreshing people's cognition of AI creative ability. In addition to Alibaba's Luban intelligent design system, Jingdong also has its own intelligent design platform, Linglong system.

There is no doubt that the future trend of intelligent design is to improve the efficiency of industrial design. So there is an optimism that the future of intelligence (Artificial Intelligence) will replace human work. Warren g. Bennis, a master of leadership theory, once predicted the future of society: "the future factory will have only two employees a man and a dog." [8] The person's job is to feed the dog, and the dog's role is to keep the person away from the equipment. In other words, there will be nothing that robots (Artificial Intelligence) can't do in the future society. The premise of this optimism is to believe that human beings will eventually develop a strong artificial intelligence, so that it can replace all human work. However, in the foreseeable future, the ability boundary of artificial intelligence is only as the assistant of human beings, and cannot really replace human beings.

Of course, some scientists warn that artificial intelligence will do harm to human beings. In an interview with the BBC in December 2014, the late famous British physicist Hawking said, "making machines that can think is undoubtedly a great threat to the existence of human beings. When artificial intelligence develops completely, it will be the end of mankind." In an article of the same year, Hawking said, "when artificial intelligence technology develops to the extreme, we will face the best or worst things in human history." In a conference held by Competence in April 2017, Hawking expressed his concern about the future of artificial intelligence in the form of a video speech: "the rise of artificial intelligence is likely to lead to the end of human civilization." Similarly suffering from "AI anxiety disorder" are the famous CEO and CTO of Space, CEO of Tesla and chairman of the board of directors of Solar City, Elon Musk, who is also worried that the development of AI out of control will destroy mankind [9, 10].

From the perspective of design, artificial intelligence provides new possibilities and ways for design, which is the trend of future design development. Artificial intelligence based on computational rationality pursues certainty and accuracy, while design and art require uncertainty and fuzziness. How to coordinate the relationship between artificial intelligence and design has become the key of artificial intelligence design and the growth point of the future development of artificial intelligence design.

References

1. Artificial intelligence, Competence Research Institute. China Renmin University Press, Beijing (2017)
2. Reich, B.: Philosophy of Artificial Intelligence. Wenhui Publishing House, Shanghai (2020)
3. Shijian, L.: Swarm intelligence Innovation: a new innovation paradigm in AI 2.0 era. Packaging Eng. 41(06) (2020)
4. Herali, Y.: A Brief History of the Future. CITIC press, Beijing (2017)
5. Xiang, F.: My Name is Alpha: On Law and Artificial Intelligence. China University of Political Science and Law Press, Beijing (2018)
6. Yamamoto, S.: A Brief History of Artificial Intelligence. Beijing Daily Press, Beijing (2019)

7. Gang, Q.: A Brief History of Silicon Valley: the Road to Artificial Intelligence. China Machine Press, Beijing (2018)

8. Quan, L.: Blockchain and Artificial Intelligence: Building an Intelligent Digital Economic World. People's Posts and Telecommunications Publishing House, Beijing (2019)

9. Zhou, S., Yong, Z.: Introduction to Artificial Intelligence. China Machine Press, Beijing (2020)

10. Fenghui, Y.: Financial Technology: The Application and Future of Big Data, Blockchain and Artificial Intelligence. Zhejiang University Press, Hangzhou (2018)

Image Recognition Technology Based on Artificial Intelligence

Yu Fu[✉]

Xijing University, Xi'an 710123, Shaanxi, China
20180141@xijing.edu.cn

Abstract. In the new era, the continuous development of artificial intelligence technology has laid a good foundation for the improvement of people's life quality. In the process of effective research of AI technology, we must start from many aspects. In order to improve the level of image recognition, we should pay attention to the effective application of artificial intelligence technology, in order to further ensure the level of image recognition.

Keywords: Artificial intelligence · Image recognition technology · Application analysis

1 Introduction

With the continuous development of computer technology, people's research on image recognition technology is more and more in-depth. Through the effective introduction of artificial intelligence means, the efficiency of image recognition can be greatly improved [1]. Therefore, in the process of practical exploration, we should actively improve the efficiency of the research on artificial intelligence technology, so as to promote its more effective application to the image recognition process. We hope that through the Effective research can further improve the level of artificial intelligence image recognition technology.

2 Analysis of Image Recognition Technology Based on Artificial Intelligence

2.1 Overview of Image Recognition and Artificial Intelligence

Image recognition mainly includes human image recognition and computer image recognition [2]. There is no essential difference between the two forms in the application process. The main difference is that computer image recognition technology will not be affected by human sensory differences. In the application process of human image recognition technology, it will not only rely on the image stored in the brain for recognition, but also rely on the feature classification of image recognition, and apply according to different categories to complete the final recognition. Artificial intelligence is a new concept gradually explored by human beings in the process of development. It has gradually infiltrated into various fields. Based on big data, computer and

© The Author(s), under exclusive license to Springer Nature Switzerland AG 2021
J. Abawajy et al. (Eds.): ATCI 2021, LNDECT 81, pp. 911–915, 2021.
https://doi.org/10.1007/978-3-030-79197-1_137

information technology, it provides human beings with high-quality services and adapts to the development of the times [3].

2.2   Basic Principles of Image Recognition Technology

The visual effect produced by the human eye is a very magical phenomenon. The recognition ability of the human eye to the image is very strong. When the distance, position and angle of an image change, people's senses will also change accordingly, and the size and shape of the image on the retina of the human eye will also change correspondingly, but this change is not enough to affect people's judgment of the image, and people can judge a certain image through various senses, for example, when writing on the back of the hand, they can also judge this through the senses Fonts [4]. Image recognition technology is an important technology in the field of artificial intelligence, so its recognition principle is similar to that of human eye recognition, which is based on the prominent features of image [5]. The focus of image recognition is to find out the features of the image [6]. For example, in the English capital letters, a has a prominent sharp angle, O has a circle, and y can be regarded as basically composed of lines and sharp angles and obtuse angles. For special information, that is, the capture and recognition of prominent features, the image can be recognized by the effective information in the image, and the content and nature of the image can be judged, And analyze the meaning it represents. In order to imitate the principle of human eye recognition and achieve the effect of human eye recognition, scientists have compiled computer programs to simulate human image recognition activities, so as to obtain many recognition models about image recognition. When the computer captures an image, if the image features match the sensory stimulation of human brain in memory, the image is considered to have been recognized.

3   Analysis of Image Recognition Technology of Artificial Intelligence

3.1   Pattern Recognition

The so-called pattern recognition is to do image recognition from the perspective of a large number of information and data, which is also an effective model in image recognition technology [7]. The recognition model of this technology is based on the premise of image recognition cognition, using computer calculation, using mathematical principles to reasoning, automatically completing the recognition of image, curve, number, character and other features, and then evaluating the integrity of the recognition.

For recognition pattern, there are two main stages: learning stage and realization stage [8]. First of all, we can understand the learning stage as a process of storage. First, we use computer to store image information, sample collection and other contents, and integrate the relevant information for specific classification and effective recognition. Finally, we can form image recognition program.

The second is the realization stage of recognition pattern. At this stage, we pay more attention to the consistency between the image and the template in the brain, which is conducive to the further completion of the recognition program. From the practical application, computer recognition is different from human brain recognition. He just matched the newly captured image information with the previous content according to the relevant data, information and characteristics of the memory stage when the computer was identifying. If it can be completed as time required, it means that the image has been identified. However, it should be noted that sometimes the content of recognition is limited. For similar content, it may cause the problem of false recognition.

3.2 Neural Network Form

At present, image recognition technology in the form of neural network is more popular and widely used [9]. On the basis of image recognition, such image recognition technology perfectly integrates modern neural network algorithm, and then realizes a new recognition form. Image recognition is a technology in the field of artificial intelligence, so neural network is an artificial neural network, which simulates the distribution characteristics of human neural network. Compared with the previous image recognition technology, the neural network algorithm in the image recognition program, although the cost is high, the program is relatively complex, but it can better play a good effect.

In the neural network program, the extracted and captured image features can be reflected in the practical application, and they can be classified after the accurate recognition of the image. In the traffic management system, the intelligent vehicle monitoring and shooting recognition uses the neural network program technology, which can quickly distinguish the license plate and identify the relevant information content at the moment of shooting, so that the traffic management can be implemented smoothly.

3.3 Nonlinear Dimension Reduction Form

Nonlinear recognition technology has its own advantages [10]. It is not only the main recognition technology of high-dimensional form, but also can effectively identify the low resolution graphics. For the image recognition of nonlinear dimensionality reduction, the computer should not only do a lot of calculation in a short time, but also do a good job in linear and nonlinear processing, and the nonlinear dimensionality reduction is simple, but the effect is quite obvious. For example, in a high-dimensional space, the face image distribution is not very uniform, and it is difficult to extract the information content of the representative character features, which makes the recognition more difficult. At this time, we can give full play to the good role of nonlinear dimensionality reduction and improve the face recognition.

4 Application and Prospect of Image Recognition Technology

With the continuous development of image recognition technology in intelligent network, it will be applied in many fields such as public security, biology, industry, agriculture, transportation, medical treatment and so on, which will bring a series of more positive impacts on our people's life. For example, in the field of public security, the application of face recognition system can better improve the safety and convenience of our society; in the field of medicine, the recognition of ECG and B-ultrasound will greatly promote the development of our medical cause; in the field of agriculture, the application of seed recognition technology and food quality detection technology will greatly improve the production quality of our agricultural products The people will benefit directly from it; and in the people's life of our country, the application of image recognition technology in the refrigerator will greatly improve the convenience of people's life in our country. This application can realize the functions of automatic refrigerator food list generation, food preservation status display, food optimal storage temperature judgment, etc., which will greatly improve the quality of life of people in our country. With the continuous development of science and technology in the future, the image recognition technology of artificial intelligence will achieve more rapid development, and this development will also enable our people to better accept the services brought by image recognition technology, and ultimately greatly improve their quality of life.

5 Conclusion

In short, in the future development, image recognition technology will be further popularized, and with the increase of users, the technology will be further updated to meet the needs of people's production and life. At present, image recognition technology has become an important technology that can serve the society, promote economic development and ensure property security. In the future, it will have a broader development space and be more and more deeply understood and mastered by people. It is hoped that the above practical research can further improve the application efficiency of artificial intelligence technology in the process of image recognition, so as to continuously improve the level of image recognition technology.

References

1. Zhao, R.: Research on Application of Artificial Intelligence Technology in Electrical Automation Control (2019)
2. Wenpeng, C.: Research on computer intelligent image recognition algorithm. Wireless Internet Technology (2019)
3. Xiao, G.: Research on computer network information security system based on big data. In: 2020 IEEE International Conference on Artificial Intelligence and Computer Applications (ICAICA). IEEE (2020)

4. Labhishetty, V., Cholewiak, S.A., Banks, M.S.: Contributions of foveal and non-foveal retina to the human eye's focusing response. J. Vis. **19**(12), 18 (2019)
5. Hu, Y., Yang, D., Zhang, Z., et al.: Application of artificial intelligence in dynamic image recognition. J. Phys. Conf. Ser. **1533**, 032093 (2020)
6. Sandler, W.: Deutsche Heimat in Afrika: Colonial revisionism and the construction of Germanness through photography. J. Women's Hist. **25**(1), 37–61 (2013)
7. Sinha, A., Namdev, N.: Feature selection and pattern recognition for different types of skin disease in human body using the rough set method. Netw. Model. Anal. Health Inform. Bioinform. **9**(1), 1–11 (2020). https://doi.org/10.1007/s13721-020-00232-z
8. Guigues, V.: Inexact stochastic mirror descent for two-stage nonlinear stochastic programs. Math. Program. **187**(1–2), 533–577 (2020). https://doi.org/10.1007/s10107-020-01490-5
9. Chen, J.: Image recognition technology based on neural network. IEEE Access **8**, 157161–157167 (2020)
10. Khouni, S., Hemsas, K.E.: Nonlinear system identification using uncoupled state multi-model approach: application to the PCB soldering system. Eng. Technol. Appl. Sci. Res. **10**(1), 5221–5227 (2020)

Application of LIM Technology in Landscape Design

Ziru Zhang[(⊠)]

Xi'an FanYi University, Xi'an, Shaanxi, China
zhangziru@xafy.edu.cn

Abstract. Under the background of rapid urbanization in China, urban environmental deterioration and ecological restoration are the important research directions of landscape, the research on digitalization of landscape information makes it possible to exchange project information without damage, reduce construction cost and shorten construction period. Based on the related research on the concept and development status of LIM, this paper explores its application in the landscape industry, so as to provide reference and basis for the future research and practice of LIM.

Keywords: Landscape Information Modeling · Landscape · Applied technology

1 Related Concepts of Landscape Information Modeling

In 2009, scholar Stephen M Ervin put forward the concept of Landscape Information Modeling (LIM), which is an extension of BIM in Landscape industry and a digital information technology and project cycle process with landscape characteristics. Building Information Modeling (BIM) is a combination of digital technology and engineering design [1].

He discussed in detail how BIM-related technologies can be applied to landscape design, and how these technologies can enable seamless connections between design, planning, construction, and operations management. With BIM, the entire life cycle of a building can be demonstrated before construction, resulting in significant cost savings and increased building efficiency [2]. LIM is an application of BIM to landscape objects, facing the requirements and conditions of landscape. Its carrier is a digital three-dimensional model. Its objects are mainly engineering projects, and the scope of its functions includes planning, design, construction and operation, that is, the "life cycle" of the project. The focus of its role is to rely on data and information for management and optimization. Its connotation is not only a technology to builda digital model, but also a method system for applying technology and a process of implementing a method system [3]. In addition, LIM is being defined by a growing number of foreign organizations: The Landscape Information Modeling (LIM), developed by the British Landscape Architecture Association, demonstrates how collaboration with other professionals and clients can streamline information flow, improve decision-making, and deliver landscapes on schedule and within budget. In 2012, the BFL (BIM For Landskaps Arkitektur) group at the Norwegian Institute of Landscape Architects

J. Abawajy et al. (Eds.): ATCI 2021, LNDECT 81, pp. 916–921, 2021.
https://doi.org/10.1007/978-3-030-79197-1_138

came up with the term Landskaps Informasjon. Modell (LIM), a term used to fill a gap in the landscape industry's vocabulary. LIM establishes various standards in public projects, IFCCity GML, Lands XML, etc. (Table 1).

Table 1. Definitions of LIM

Individual or organization	Time/year	Definition	Content
Cypress	2008	Land Information Models	LIM will be used to build information models at a community or regional scale, while SIM(Site Information Models) will add site-oriented applications
Owen	2009	Landscape Information Modeling	LIM makes it possible to exchange information without damage between design, planning, construction, and operation management
British Association of Landscape Architecture	–	Landscape Information Modeling	Use BIM for landscape design and architecture, its concept is synonymous with BIM, but for a specific subset
Norwegian Association of Landscape Architects	2012	Landskaps Informasjon Modell	Appropriate vocabulary when applying BIM to landscape architecture

However, the development of LIM is lagging behind due to the lack of interest in LIM by owners, the lack of demand for construction, the limitation of industry scale, the lack of quantitative analysis and evaluation system in planning and design, and the technical difficulty of LIM [4] (Table 2).

Table 2. Activities of organizations adopting LIM

Country	Organization	Related events
USA	American Landscape Architecture Association	Introducing BIM in landscape architecture Quoting the definition of LIM and SIM
UK	British Association of Landscape Architecture	Ensure that members of the Landscape Architecture Association meet Level 2 BIM by 2016 Identify key technical requirements and supply gaps that need to be met Understand the BIM language Ensure that the space and construction elements of the construction project are properly resolved Appropriately enable registration practices that comply with BIM standards Ensure that landscape objects are included in the national BIM database and COBie data Setting up BIM related courses in university courses Provide a simple spreadsheet to evaluate BIM maturity
Norway	Norwegian Association of Landscape Architects	Enable relevant agencies to use BIM in public projects Establish various standards for LIM in public projects, such as IFC, CityGML, LandsXML and other standards Develop landscape objects to make LIM modeling more efficient Feedback landscape architects' opinions on BIM tools to BIM software companies

2 Application of LIM in Landscape Design

LIM is a new concept and technology, its essence is to build a platform for multiple participants to manipulate the same data model.

So as to solve the problems of information non-sharing and exchanging and the disconnection of information, process and application in the traditional project. The application of LIM in landscape design is mainly embodied in three aspects: smart landscaping, information management and parametric design.

2.1 Smart Landscape

"Smart landscaping" is an extension of the concept of "smart construction" pioneered by BIM.

The essence of smart landscaping is that it is part of the infrastructure BIM, and an efficient measure to create and apply electronic data model in the field of landscape design. Peter Packek, professor, La Pazville University of Applied Sciences, Switzerland, developed the Autodesk 123 Catch-3ds Max Design-Civil 3D technical framework. Autodesk 123D Catch, a three-dimensional scanning software, analyzes the acquired data and looks for characteristic elements such as object boundaries, light and shadow shapes, and other characteristic points [5]. Based on this, it establishes the coordinates of the object and creates a shape based on the coordinates. Then, it imports the generated model into 3ds Max Design, reduces the file to a workable size, and then imports it into Civil 3D for related engineering calculations, and then imports the optimized model into 3ds Max Design for cloud rendering, thereby generating STL files and accurate DTM models that can be used for print output. Using LIM in the pre-construction phase allows scheduling and workflow coordination in the preliminary stage before construction, cost estimation of site cranes and materials, and construction of virtual logistics [6]. Since LIM is used in the pre-construction stage to establish and evaluate various construction plans, it is necessary to add planning data to the 3D landscape design model to obtain a 4D model with time as the fourth dimension. The 4D landscape design model is a technology that integrates the 3D model with the construction schedule. It can show the construction sequence clearly and detect the conflict of the construction plan automatically, finally, the survey technicians of the construction team copy the data into the 3D mechanical control system, and the bulldozers and excavators carry out on-site construction (Table 3).

2.2 Information Management

LIM can provide information about project quality, construction progress and construction cost simultaneously to ensure the full realization of the objectives in the implementation phase of the project. Through LIM [7].

Table 3. Landscaping SMART workflow.

	Data output: 3DMachine control	
Data output: Paper plane		No data: Project start
Data output: 3D printer	Data Model	Data input
Data output: 3Dmodel. Google Earth		DataModel: Electronic terrain model
	Data Model CloudService	

Project managers is able to evaluate the construction status of the project, synchronize the climate information, hydrological information, tree conditions, earthwork information, facility performance, and financial information in the landscape project, evaluate the operation and maintenance of the project, and modify the maintenance plan based on this, so as to improve the level of project revenue and cost management.

2.3 Parametric Design

With the development of computer technology and the maturity and perfection of programming language, the application value of programming technology in the landscape industry has been gradually reflected, and the use of programming language to aid planning and design has been gradually applied to practical projects [8].

Parametric design has also been effectively applied in landscape, for example, the Rizhao Pedestrian Landscape Bridge of HH Design (Beijing) adopts parametric design technology in the process of spatial organization, structure optimization, facade treatment, project management, and modeling design, which greatly creates rich landscape effects. In the design stage, LIM can be used to make schematic diagrams and related design details, so that the design can be quickly "visualized", so that it is more convenient to introduce the plan to the owner and increase the communication of the project team, so that it can make better decisions [8].When modifying the conceptual model, LIM can ensure that the data is accurate and automatically updated, thereby reducing the man-hours of the space plan and effectively connecting the work of the parties. Parametric design has been widely used in landscape sketches, exhibition halls, curtain walls, stage backgrounds, etc., such as the glacier pavilion designed by Didzis

Jaunzems Architecture, the Motril pedestrian bridge designed by Guallart Architects, the Pauhu performance pavilion designed by Tampere Architecture Week, etc., indicating that the application of parametric design abroad has involved all aspects of landscape design [9].

3 Conclusion and Outlook

In recent years, due to the earth-shaking development of "digital landscape" technical methods, the application of landscape information model has played a certain role in improving the efficiency of landscape industry, and it has gradually gained the attention of landscape architects [10].

With the improvement of LIM technology, scholars at home and abroad have a deeper and more comprehensive understanding of LIM. That is, although LIM is an extension and extension of BIM technology, the core elements of LIM are characteristic of landscape[10]. In the context of "VR technology" and "Internet of Things", LIM provides a relatively objective solution to the problem of inefficient information exchange in landscape projects, and provides the unification of landscape information standards and data compatibility, achieving coordination among various disciplines and full life cycle operation management, which exerts the greatest effectiveness of LIM. LIM is an extension of BIM in the landscape industry. It is a digital information technology and project cycle process with landscape characteristics. By analogy with the impact of BIM on the construction industry, it can be foreseen that LIM is a new trend in the future sustainable development of the landscape industry. But at present, the application of landscape information models is mainly concentrated in the three stages of parametric design, smart landscaping, and information management. Due to the lack of coherence of model information and data, garden practice projects basically apply LIM in stages. It is worth noting that the development of LIM is still in its infancy, requiring in-depth research and practical application by landscape architects [11].

References

1. Sipes, J.L.: Integrating BIM Technology into Landscape Architecture. American Society of Landscape Architects, Berlin/Offenbach: Wichmann, pp. 65–72
2. Nessel, A.: The place for information models in landscape architecture, or a place for landscape architects in information models. In: Digital Landscape Conference, Bernburg (2013)
3. Song, L., Wen, S.: Digital landscape technology research progress: international digital landscape conerence development overview. Chin. Landscape Archit. **2**, 45–50 (2015)
4. Pechek, P., Yong, G.: Landscaping smart. Landscape Archit. **1**, 33–37 (2013)
5. Research Progress: International digital landscape conference development overview. Chin. Landscape Archit. **2**, 45–50 (2015)
6. Kim, B.Y., Son, Y.H.: The current status of BIM in the field of landscape architecture and the issues on the adoption of LIM. J. Korean Inst. Landscape Archit. **42**(3), 50–63 (2014)
7. Guo, Y.: Study on the prospect of landscape information model for sustainable site design. Dynamic (eco-city and green building) **2014**(4), 62–65 (2014)

8. Ruiqing, B., Yong, L.: Programming technology to assist the planning and design methods to explore. Landscape Archit. **2**, 26–32 (2016)
9. Wei, K.: Landscape parametric planning and design of the status quo and thinking. Landscape Archit. **1**, 58–64 (2013)
10. Mohammad, A., Adamu, A.: The Need for Landscape Information Modelling (LIM) in landscape architecture. In: Dessau & Bernburg, Digital Landscape Conference (2012)
11. Jitao, Z., Lei, L.: China's garden industry building information model development prospects. Landscape Archit. **1**, 91–94 (2012)

Author Index

J. Abawajy et al. (Eds.): ATCI 2021, LNDECT 81, pp. 923–926, 2021.
https://doi.org/10.1007/978-3-030-79197-1

Printed in the United States
by Baker & Taylor Publisher Services